# Methods in Enzymology

Volume 228

AQUEOUS TWO-PHASE SYSTEMS

# METHODS IN ENZYMOLOGY

*Methods in Enzymology*

*Volume 228*

# *Aqueous Two-Phase Systems*

EDITED BY

*Harry Walter*

LABORATORY OF CHEMICAL BIOLOGY
VETERANS AFFAIRS MEDICAL CENTER
LONG BEACH, CALIFORNIA

*Göte Johansson*

BIOCHEMISTRY, CHEMICAL CENTER
UNIVERSITY OF LUND
LUND, SWEDEN

ACADEMIC PRESS
A Division of Harcourt Brace & Company
San Diego New York Boston London Sydney Tokyo Toronto

This book is printed on acid-free paper. ∞



Academic Press, Inc.
1250 Sixth Avenue, San Diego, California 92101-4311

*United Kingdom Edition published by*
Academic Press Limited
24–28 Oval Road, London NW1 7DX

International Standard Serial Number: 0076-6879

International Standard Book Number: 0-12-182129-3

PRINTED IN THE UNITED STATES OF AMERICA
94 95 96 97 98 99 EB 9 8 7 6 5 4 3 2 1

# Table of Contents



## Section I. General Methodology and Apparatus



## Section II. Partitioning and Affinity Partitioning of Macromolecules

### A. Proteins

## B. Nucleic Acids



## Section III. Studies on Protein Interactions, Molecular Structure, Charge, Hydrophobicity, and Conformational Changes



## Section IV. Partitioning and Affinity Partitioning of Particulates

## A. Cells

### Separation and Fractionation of Cell Populations and Obtaining Information on Their Surface Properties

## Affinity Partitioning



## B. Organelles: Separation and Subfractionation

## C. Membranes
### Separation and Subfractionation



### Affinity Partitioning



### Membrane Domain Analysis



## Section V. Aqueous Phase Separations in Biological Systems



## Section VI. Aqueous Two-Phase Systems in Large-Scale Processes: Biotechnology

### A. Extractive Bioconversions

## B. Proteins: Downstream Processing



## C. Design of Proteins for Enhanced Extraction



## D. Other Applications of Aqueous Phases in Biotechnology

# Contributors to Volume 228

Article numbers are in parentheses following the names of contributors. Affiliations listed are current.


HANS-ERIK ÅKERLUND (8), *Department of Plant Biochemistry, University of Lund, S-220 07 Lund, Sweden*

PER-ÅKE ALBERTSSON (1, 7, 8, 45, 49), *Department of Biochemistry, Chemical Center, University of Lund, S-221 00 Lund, Sweden*

BERTIL ANDERSSON (39), *Department of Biochemistry, Arrhenius Laboratories for Natural Sciences, University of Stockholm, S-106 91 Stockholm, Sweden*

EVA ANDREASSON (45), *Department of Biochemisry, Chemical Center, University of Lund, S-221 00 Lund, Sweden*

FRANCES H. ARNOLD (15), *Division of Chemistry and Chemical Engineering, California Institute of Technology, Pasadena, California 91125*

LARS BACKMAN (22), *Department of Biochemistry, University of Umeå, S-901 87 Umeå, Sweden*

DARWIN BAILEY (15), *Division of Chemistry and Chemical Engineering, California Institute of Technology, Pasadena, California 91125*

GERD BIRKENMEIER (14, 24, 34), *Institute of Biochemistry, University of Leipzig, D-04103 Leipzig, Germany*

ANDREAS BLENNOW (16), *Department of Biochemistry, Chemical Center, S-221 00 Lund, Sweden*

GÖRAN BLOMQUIST (32, 66), *National Institute of Occupational Health, Analytical Chemistry Division, S-907 13 Umeå, Sweden*

MICHAEL E. BRADLEY (42), *Department of Physiology and Biophysics, University of Southern California, School of Medicine, Los Angeles, California 90033*

DONALD E. BROOKS (2, 36), *Departments of Pathology and Chemistry, University of British Columbia, Vancouver, British Columbia, Canada V6T 2B5*

JOHN S. BRUSCA (17), *Texas/American Veterinary Labs, San Angelo, Texas*

BONG H. CHUNG (15), *Genetic Engineering Research Institute, KIST, DaeDuk Science Town, DaeJeon, South Korea*

JOHN I. CLARK (51), *Departments of Biological Structure and Ophthalmology, University of Washington, School of Medicine, Seattle, Washington 98185*

ARNO CORDES (59), *ASA Spezialenzyme GmbH, D-38329 Salzgitter, Germany*

CRISTINA DELGADO (37), *Molecular Cell Pathology, Royal Free Hospital School of Medicine, London NW3 2PF, England*

DONNA L. DONNELLY (35), *Space Science Laboratory, NASA/Marshall Space Flight Center, Huntsville, Alabama 35812*

SVEN-OLOF ENFORS (64), *Department of Biochemistry and Biotechnology, Royal Institute of Technology, S-100 44 Stockholm, Sweden*

DEREK FISHER (19, 37), *Department of Biochemistry and Molecular Cell Pathology, Royal Free Hospital School of Medicine, London NW3 2PF, England*

DANIEL FORCINITI (20), *Department of Chemical Engineering, University of Missouri, Rolla, Missouri 65401*

GILLIAN E. FRANCIS (19, 37), *Molecular Cell Pathology, Royal Free Hospital School of Medicine, London NW3 2PF, England*

PER GARDESTRÖM (41), *Department of Plant Physiology, University of Umeå, S-901 87 Umeå, Sweden*

J. PETER GIEROW (50), *Department of Physiology and Biophysics, University of Southern California, School of Medicine, Los Angeles, California 90033*

LENA HAMMAR (63), *Department of Veterinary Virology, Center for HIV Research in Uppsala, Biomedical Center, S-751 23 Uppsala, Sweden*

JEFFREY C. HANSEN (25), *Department of Biochemistry, The University of Texas Health Science Center at San Antonio, San Antonio, Texas 78284*

J. MILTON HARRIS (35), *Department of Chemistry, University of Alabama, Huntsville, Alabama 35899*

PATRICK HUBERT (26), *Laboratoire de Chimie-Physique Macromoléculaire, CNRS URA 494, ENSIC, F-54001 Nancy, France*

HELMUT HUSTEDT (56, 58, 60), *Gesellschaft für Biotechnologische Forschung, D-38124 Braunschweig, Germany*

BENGT JERGIL (47), *Biochemistry, Chemical Center, University of Lund, S-221 00 Lund, Sweden*

MONICA JOELSSON (6, 12), *Department of Biochemistry, University of Lund, S-221 00 Lund, Sweden*

GÖTE JOHANSSON (3, 5, 6, 21, 48, 55), *Department of Biochemistry, Chemical Center, University of Lund, S-221 00 Lund, Sweden*

LAUREL J. KARR (35), *Space Science Laboratory, NASA/Marshall Space Flight Center, Huntsville, Alabama 35812*

RAJNI KAUL (54), *Department of Biotechnology, Chemical Center, University of Lund, S-221 00 Lund, Sweden*

KRISTINA KÖHLER (62), *Kabi Pharmacia BioScience Center, S-112 87 Stockholm, Sweden*

GERHARD KOPPERSCHLÄGER (11, 13, 24), *Institute of Biochemistry, University of Leipzig, D-04103 Leipzig, Germany*

ANTONI KOZLOWSKI (35), *Department of Chemistry, University of Alabama, Huntsville, Alabama 35899*

MARIA-REGINA KULA (57, 59), *Institute of Enzyme Technology, University of Düsseldorf, D-(W)5170 Jülich, Germany*

MANUEL J. LÓPEZ-PÉREZ (38), *Departamento de Bioquíimica, Biología Molecular y Celular, Facultad de Veterinaria, Universidad de Zaragoza, 50013 Zaragoza, Spain*

ROSS W. LAMBERT (42), *Department of Physiology and Biophysics, University of Southern California, School of Medicine, Los Angeles, California 90033*

CHRISTER LARSSON (4, 40, 44), *Department of Plant Biochemistry, University of Lund, S-220 07 Lund, Sweden*

PER-OLOF LARSSON (10), *Department of Pure and Applied Biochemistry, Chemical Center, University of Lund, S-221 00 Lund, Sweden*

SUSANNE LUNDBERG (22), *Department of Biochemistry, University of Umeå, S-901 87 Umeå, Sweden*

BOEL MÅRTENSSON (31), *Department of Plant Physiology, University of Lund, S-220 07 Lund, Sweden*

IAN M. MØLLER (41), *Department of Plant Physiology, University of Lund, S-220 07 Lund, Sweden*

KARL-ERIC I. MAGNUSSON (29), *Department of Medical Microbiology, Faculty of Health Sciences, University of Linköping, S-581 85 Linköping, Sweden*

BJÖRN MÄLARSTIG (66), *Pegasus Separation AB, S-751 03 Uppsala, Sweden*

BO MATTIASSON (54), *Department of Biotechnology, Chemical Center, University of Lund, S-221 00 Lund, Sweden*

JESUS MENDIETA (33), *C.E.C. Joint Research Center, Central Bureau for Nuclear Measurements, B-2440 Geel, Belgium*

ULRICH MENGE (61), *Gesellschaft für Biotechologische Forschung, D-38124 Braunschweig, Germany*

AUSTIN K. MIRCHEFF (42), *Department of Physiology and Biophysics, University of Southern California, School of Medicine, Los Angeles, California 90033*

D. JAMES MORRÉ (39, 43), *Department of Medicinal Chemistry and Pharmacognosy, Life Sciences Research Building, Purdue University, West Lafayette, Indiana 47907*

DOROTHY M. MORRÉ (43), *Department of Foods and Nutrition, Purdue University, West Lafayette, Indiana 47907*

WERNER MÜLLER (9, 18), *E. MERCK, D-(W)6100 Darmstadt, Germany*

RAYMOND NORRIS-JONES (2), *Department of Pathology, Faculty of Medicine, University of British Columbia, Vancouver, British Columbia, Canada V6T 2B5*

NEOPHYTOS PAPAMICHAEL (56), *Gesellschaft für Biotechnologische Forschung, D-38124 Braunschweig, Germany*

JENS PAULSEN (58), *Department of Biotechnology, Gesellschaft für Biotechnologische Forschung, ZWE-Biotechnik, D-38124 Braunschweig, Germany*

ANDERS PERSSON (47), *Biochemistry, Chemical Center, University of Lund, S-221 00 Lund, Sweden*

INGRID PERSSON (52), *Department of Applied Microbiology, Chemical Center, University of Lund, S-221 00 Lund, Sweden*

PATRICE X. PETIT (41), *BFMV Biosystèmes Membrànaires, CNRS, UPR 39, F-91198 Gif-sur-Yvette, France*

JUSTIN D. RADOLF (17), *Departments of Internal Medicine and Microbiology, University of Texas, Southwestern Medical Center at Dallas, Dallas, Texas 75235*

TIMOTHY REUST (43), *Department of Medicinal Chemistry, Life Sciences Research Building, Purdue University, West Lafayette, Indiana 47907*

DIETER SCHUSTOLLA (60), *Schering AG, D-W 3340 Wolfenbüttel, Germany*

HORST SCHÜTTE (60), *Technische Fachhochschule Berlin, FB 3, Biotechnikum, D-W 1000 Berlin, Germany*

VITHALDAS P. SHANBHAG (23), *Department of Biochemistry, University of Umeå, S-901 87 Umeå, Sweden*

PAUL T. SHARPE (30), *Molecular Embryology Laboratory, School of Biological Sciences, University of Manchester, Manchester M13 9PT, England*

ANNA-LISA SMEDS (64), *Kabi-Pharmacia, 31-6, S-112 87 Stockholm, Sweden*

MARIANNE SOMMARIN (44), *Department of Plant Biochemistry, University of Lund, S-220 07 Lund, Sweden*

HREINN STEFANSSON (45), *Department of Biochemistry, Chemical Center, University of Lund, S-221 00 Lund, Sweden*

S. JILL STOCKS (36), *Departments of Pathology and Chemistry, University of British Columbia, Vancouver, British Columbia, Canada V6T 2B5*

GUNNAR B. STRÖM (32, 66), *Pegasus Separation AB, Department of Environmental Medicine, University of Umeå, S-901 87 Umeå, Sweden*

FOLKE TJERNELD (1, 12, 52, 53), *Department of Biochemistry, Chemical Center, University of Lund, S-221 00 Lund, Sweden*

ANDRES VEIDE (62), *Department of Biochemistry and Biotechnology, Royal Institute of Technology, S-100 44 Stockholm, Sweden*

JÖRG VERNAU (57), *Institute of Enzyme Technology, University of Düsseldorf, D-(W)5170 Jülich, Germany*

HANS VILTER (65), *Gesellschaft für Biotechnologische Forschung D-38124 Braunschweig, Germany*

HARRY WALTER (4, 20, 27, 28, 34), *Laboratory of Chemical Biology, Veterans Affairs Medical Center, Long Beach, California 90822*

SUSANNE WIDELL (31, 44), *Department of Plant Physiology, University of Lund, S-220 07 Lund, Sweden*

KIM E. WIDEN (34), *Laboratory of Chemical Biology, Veterans Affairs Medical Center, Long Beach, California 90822*

RAINER WOKER (57), *Institute of Enzyme Technology, University of Düsseldorf, D-(W)5170 Jülich, Germany*

LOUIE WOLLENBERGER (45), *Department of Biochemistry, Chemical Center, University of Lund, S-221 00 Lund, Sweden*

SHIZUO YOSHIDA (46), *Institute of Low Temperature Science, Hokkaido University, Sapporo 060, Japan*

# Preface

Partitioning of biomaterials in aqueous two-phase systems is a general and sensitive method for purification of and analytical studies on cell constituents of various sizes including proteins, nucleic acids, membranes, and cell organelles. Whole cells can also be fractionated and aspects of their surface properties gauged from their partitioning behavior. The two-phase systems often provide an excellent milieu in which biochemical activities of biomaterials are not only retained but are often stabilized.

The use of two-phase systems is simple, and can be carried out without the use of sophisticated equipment. Section I in this volume deals with fundamental properties of the phases and techniques of partitioning, including a description of devices used in multistep extraction processes. Section II presents an assortment of purification procedures for proteins and nucleic acids primarily by use of phase-constricted affinity ligands.

In addition to their use in the purification of biomaterials, aqueous two-phase systems can also be used to study physical properties of macromolecules and their behavior in solution. Such methods are described in Section III with examples given for determining protein hydrophobicity or charge, protein–protein interaction and protein–ligand binding, isoelectric point, and conformational changes.

The absence of solid supports or adsorptive surfaces, as exist in chromatography, makes aqueous two-phase systems ideal for the separation and study of cell particles by partitioning. Section IV gives procedures for partitioning of particulates both for purification, including the isolation of highly purified plasma membranes (even of vesicles of desired sidedness), and for analytical studies such as the tracing of cell surface changes that accompany normal or abnormal biological processes.

The ease with which procedures using the two-phase systems can be scaled up makes them valuable tools in biotechnology. Section V presents selected downstream processes for the extraction of specific proteins of commercial interest and of other large-scale applications such as the concentration of viruses and the removal of microorganisms from technically employed liquids.

We hope that this volume, which represents the sole, currently available, collection of procedures devoted to partitioning in aqueous two-phase systems, will prove useful to those who wish to test this versatile methodology on their own specific problems by guiding them along a shorter and smoother path to success.

HARRY WALTER
GÖTE JOHANSSON

# METHODS IN ENZYMOLOGY

VOLUME I. Preparation and Assay of Enzymes
*Edited by* SIDNEY P. COLOWICK AND NATHAN O. KAPLAN

VOLUME II. Preparation and Assay of Enzymes
*Edited by* SIDNEY P. COLOWICK AND NATHAN O. KAPLAN

VOLUME III. Preparation and Assay of Substrates
*Edited by* SIDNEY P. COLOWICK AND NATHAN O. KAPLAN

VOLUME IV. Special Techniques for the Enzymologist
*Edited by* SIDNEY P. COLOWICK AND NATHAN O. KAPLAN

VOLUME V. Preparation and Assay of Enzymes
*Edited by* SIDNEY P. COLOWICK AND NATHAN O. KAPLAN

VOLUME VI. Preparation and Assay of Enzymes (*Continued*)
Preparation and Assay of Substrates
Special Techniques
*Edited by* SIDNEY P. COLOWICK AND NATHAN O. KAPLAN

VOLUME VII. Cumulative Subject Index
*Edited by* SIDNEY P. COLOWICK AND NATHAN O. KAPLAN

VOLUME VIII. Complex Carbohydrates
*Edited by* ELIZABETH F. NEUFELD AND VICTOR GINSBURG

VOLUME IX. Carbohydrate Metabolism
*Edited by* WILLIS A. WOOD

VOLUME X. Oxidation and Phosphorylation
*Edited by* RONALD W. ESTABROOK AND MAYNARD E. PULLMAN

VOLUME XI. Enzyme Structure
*Edited by* C. H. W. HIRS

VOLUME XII. Nucleic Acids (Parts A and B)
*Edited by* LAWRENCE GROSSMAN AND KIVIE MOLDAVE

VOLUME XIII. Citric Acid Cycle
*Edited by* J. M. LOWENSTEIN

VOLUME XIV. Lipids
*Edited by* J. M. LOWENSTEIN

VOLUME XV. Steroids and Terpenoids
*Edited by* RAYMOND B. CLAYTON

VOLUME XVI. Fast Reactions
*Edited by* KENNETH KUSTIN

VOLUME XVII. Metabolism of Amino Acids and Amines (Parts A and B)
*Edited by* HERBERT TABOR AND CELIA WHITE TABOR

VOLUME XVIII. Vitamins and Coenzymes (Parts A, B, and C)
*Edited by* DONALD B. MCCORMICK AND LEMUEL D. WRIGHT

VOLUME XIX. Proteolytic Enzymes
*Edited by* GERTRUDE E. PERLMANN AND LASZLO LORAND

VOLUME XX. Nucleic Acids and Protein Synthesis (Part C)
*Edited by* KIVIE MOLDAVE AND LAWRENCE GROSSMAN

VOLUME XXI. Nucleic Acids (Part D)
*Edited by* LAWRENCE GROSSMAN AND KIVIE MOLDAVE

VOLUME XXII. Enzyme Purification and Related Techniques
*Edited by* WILLIAM B. JAKOBY

VOLUME XXIII. Photosynthesis (Part A)
*Edited by* ANTHONY SAN PIETRO

VOLUME XXIV. Photosynthesis and Nitrogen Fixation (Part B)
*Edited by* ANTHONY SAN PIETRO

VOLUME XXV. Enzyme Structure (Part B)
*Edited by* C. H. W. HIRS AND SERGE N. TIMASHEFF

VOLUME XXVI. Enzyme Structure (Part C)
*Edited by* C. H. W. HIRS AND SERGE N. TIMASHEFF

VOLUME XXVII. Enzyme Structure (Part D)
*Edited by* C. H. W. HIRS AND SERGE N. TIMASHEFF

VOLUME XXVIII. Complex Carbohydrates (Part B)
*Edited by* VICTOR GINSBURG

VOLUME XXIX. Nucleic Acids and Protein Synthesis (Part E)
*Edited by* LAWRENCE GROSSMAN AND KIVIE MOLDAVE

VOLUME XXX. Nucleic Acids and Protein Synthesis (Part F)
*Edited by* KIVIE MOLDAVE AND LAWRENCE GROSSMAN

VOLUME XXXI. Biomembranes (Part A)
*Edited by* SIDNEY FLEISCHER AND LESTER PACKER

VOLUME XXXII. Biomembranes (Part B)
*Edited by* SIDNEY FLEISCHER AND LESTER PACKER

VOLUME XXXIII. Cumulative Subject Index Volumes I–XXX
*Edited by* MARTHA G. DENNIS AND EDWARD A. DENNIS

VOLUME XXXIV. Affinity Techniques (Enzyme Purification: Part B)
*Edited by* WILLIAM B. JAKOBY AND MEIR WILCHEK

VOLUME XXXV. Lipids (Part B)
*Edited by* JOHN M. LOWENSTEIN

VOLUME XXXVI. Hormone Action (Part A: Steroid Hormones)
*Edited by* BERT W. O'MALLEY AND JOEL G. HARDMAN

VOLUME XXXVII. Hormone Action (Part B: Peptide Hormones)
*Edited by* BERT W. O'MALLEY AND JOEL G. HARDMAN

VOLUME XXXVIII. Hormone Action (Part C: Cyclic Nucleotides)
*Edited by* JOEL G. HARDMAN AND BERT W. O'MALLEY

VOLUME XXXIX. Hormone Action (Part D: Isolated Cells, Tissues, and Organ Systems)
*Edited by* JOEL G. HARDMAN AND BERT W. O'MALLEY

VOLUME XL. Hormone Action (Part E: Nuclear Structure and Function)
*Edited by* BERT W. O'MALLEY AND JOEL G. HARDMAN

VOLUME XLI. Carbohydrate Metabolism (Part B)
*Edited by* W. A. WOOD

VOLUME XLII. Carbohydrate Metabolism (Part C)
*Edited by* W. A. WOOD

VOLUME XLIII. Antibiotics
*Edited by* JOHN H. HASH

VOLUME XLIV. Immobilized Enzymes
*Edited by* KLAUS MOSBACH

VOLUME XLV. Proteolytic Enzymes (Part B)
*Edited by* LASZLO LORAND

VOLUME XLVI. Affinity Labeling
*Edited by* WILLIAM B. JAKOBY AND MEIR WILCHEK

VOLUME XLVII. Enzyme Structure (Part E)
*Edited by* C. H. W. HIRS AND SERGE N. TIMASHEFF

VOLUME XLVIII. Enzyme Structure (Part F)
*Edited by* C. H. W. HIRS AND SERGE N. TIMASHEFF

VOLUME XLIX. Enzyme Structure (Part G)
*Edited by* C. H. W. HIRS AND SERGE N. TIMASHEFF

VOLUME L. Complex Carbohydrates (Part C)
*Edited by* VICTOR GINSBURG

VOLUME LI. Purine and Pyrimidine Nucleotide Metabolism
*Edited by* PATRICIA A. HOFFEE AND MARY ELLEN JONES

VOLUME LII. Biomembranes (Part C: Biological Oxidations)
*Edited by* SIDNEY FLEISCHER AND LESTER PACKER

VOLUME LIII. Biomembranes (Part D: Biological Oxidations)
*Edited by* SIDNEY FLEISCHER AND LESTER PACKER

VOLUME LIV. Biomembranes (Part E: Biological Oxidations)
*Edited by* SIDNEY FLEISCHER AND LESTER PACKER

VOLUME LV. Biomembranes (Part F: Bioenergetics)
*Edited by* SIDNEY FLEISCHER AND LESTER PACKER

VOLUME LVI. Biomembranes (Part G: Bioenergetics)
*Edited by* SIDNEY FLEISCHER AND LESTER PACKER

VOLUME LVII. Bioluminescence and Chemiluminescence
*Edited by* MARLENE A. DELUCA

VOLUME LVIII. Cell Culture
*Edited by* WILLIAM B. JAKOBY AND IRA PASTAN

VOLUME LIX. Nucleic Acids and Protein Synthesis (Part G)
*Edited by* KIVIE MOLDAVE AND LAWRENCE GROSSMAN

VOLUME LX. Nucleic Acids and Protein Synthesis (Part H)
*Edited by* KIVIE MOLDAVE AND LAWRENCE GROSSMAN

VOLUME 61. Enzyme Structure (Part H)
*Edited by* C. H. W. HIRS AND SERGE N. TIMASHEFF

VOLUME 62. Vitamins and Coenzymes (Part D)
*Edited by* DONALD B. MCCORMICK AND LEMUEL D. WRIGHT

VOLUME 63. Enzyme Kinetics and Mechanism (Part A: Initial Rate and Inhibitor Methods)
*Edited by* DANIEL L. PURICH

VOLUME 64. Enzyme Kinetics and Mechanism (Part B: Isotopic Probes and Complex Enzyme Systems)
*Edited by* DANIEL L. PURICH

VOLUME 65. Nucleic Acids (Part I)
*Edited by* LAWRENCE GROSSMAN AND KIVIE MOLDAVE

VOLUME 66. Vitamins and Coenzymes (Part E)
*Edited by* DONALD B. MCCORMICK AND LEMUEL D. WRIGHT

VOLUME 67. Vitamins and Coenzymes (Part F)
*Edited by* DONALD B. MCCORMICK AND LEMUEL D. WRIGHT

VOLUME 68. Recombinant DNA
*Edited by* RAY WU

VOLUME 69. Photosynthesis and Nitrogen Fixation (Part C)
*Edited by* ANTHONY SAN PIETRO

VOLUME 70. Immunochemical Techniques (Part A)
*Edited by* HELEN VAN VUNAKIS AND JOHN J. LANGONE

VOLUME 71. Lipids (Part C)
*Edited by* JOHN M. LOWENSTEIN

VOLUME 72. Lipids (Part D)
*Edited by* JOHN M. LOWENSTEIN

VOLUME 73. Immunochemical Techniques (Part B)
*Edited by* JOHN J. LANGONE AND HELEN VAN VUNAKIS

VOLUME 74. Immunochemical Techniques (Part C)
*Edited by* JOHN J. LANGONE AND HELEN VAN VUNAKIS

VOLUME 75. Cumulative Subject Index Volumes XXXI, XXXII, XXXIV–LX
*Edited by* EDWARD A. DENNIS AND MARTHA G. DENNIS

VOLUME 76. Hemoglobins
*Edited by* ERALDO ANTONINI, LUIGI ROSSI-BERNARDI, AND EMILIA CHIANCONE

VOLUME 77. Detoxication and Drug Metabolism
*Edited by* WILLIAM B. JAKOBY

VOLUME 78. Interferons (Part A)
*Edited by* SIDNEY PESTKA

VOLUME 79. Interferons (Part B)
*Edited by* SIDNEY PESTKA

VOLUME 80. Proteolytic Enzymes (Part C)
*Edited by* LASZLO LORAND

VOLUME 81. Biomembranes (Part H: Visual Pigments and Purple Membranes, I)
*Edited by* LESTER PACKER

VOLUME 82. Structural and Contractile Proteins (Part A: Extracellular Matrix)
*Edited by* LEON W. CUNNINGHAM AND DIXIE W. FREDERIKSEN

VOLUME 83. Complex Carbohydrates (Part D)
*Edited by* VICTOR GINSBURG

VOLUME 84. Immunochemical Techniques (Part D: Selected Immunoassays)
*Edited by* JOHN J. LANGONE AND HELEN VAN VUNAKIS

VOLUME 85. Structural and Contractile Proteins (Part B: The Contractile Apparatus and the Cytoskeleton)
*Edited by* DIXIE W. FREDERIKSEN AND LEON W. CUNNINGHAM

VOLUME 86. Prostaglandins and Arachidonate Metabolites
*Edited by* WILLIAM E. M. LANDS AND WILLIAM L. SMITH

VOLUME 87. Enzyme Kinetics and Mechanism (Part C: Intermediates, Stereochemistry, and Rate Studies)
*Edited by* DANIEL L. PURICH

VOLUME 88. Biomembranes (Part I: Visual Pigments and Purple Membranes, II)
*Edited by* LESTER PACKER

VOLUME 89. Carbohydrate Metabolism (Part D)
*Edited by* WILLIS A. WOOD

VOLUME 90. Carbohydrate Metabolism (Part E)
*Edited by* WILLIS A. WOOD

VOLUME 91. Enzyme Structure (Part I)
*Edited by* C. H. W. HIRS AND SERGE N. TIMASHEFF

VOLUME 92. Immunochemical Techniques (Part E: Monoclonal Antibodies and General Immunoassay Methods)
*Edited by* JOHN J. LANGONE AND HELEN VAN VUNAKIS

VOLUME 93. Immunochemical Techniques (Part F: Conventional Antibodies, Fc Receptors, and Cytotoxicity)
*Edited by* JOHN J. LANGONE AND HELEN VAN VUNAKIS

VOLUME 94. Polyamines
*Edited by* HERBERT TABOR AND CELIA WHITE TABOR

VOLUME 95. Cumulative Subject Index Volumes 61–74, 76–80
*Edited by* EDWARD A. DENNIS AND MARTHA G. DENNIS

VOLUME 96. Biomembranes [Part J: Membrane Biogenesis: Assembly and Targeting (General Methods; Eukaryotes)]
*Edited by* SIDNEY FLEISCHER AND BECCA FLEISCHER

VOLUME 97. Biomembranes [Part K: Membrane Biogenesis: Assembly and Targeting (Prokaryotes, Mitochondria, and Chloroplasts)]
*Edited by* SIDNEY FLEISCHER AND BECCA FLEISCHER

VOLUME 98. Biomembranes (Part L: Membrane Biogenesis: Processing and Recycling)
*Edited by* SIDNEY FLEISCHER AND BECCA FLEISCHER

VOLUME 99. Hormone Action (Part F: Protein Kinases)
*Edited by* JACKIE D. CORBIN AND JOEL G. HARDMAN

VOLUME 100. Recombinant DNA (Part B)
*Edited by* RAY WU, LAWRENCE GROSSMAN, AND KIVIE MOLDAVE

VOLUME 101. Recombinant DNA (Part C)
*Edited by* RAY WU, LAWRENCE GROSSMAN, AND KIVIE MOLDAVE

VOLUME 102. Hormone Action (Part G: Calmodulin and Calcium-Binding Proteins)
*Edited by* ANTHONY R. MEANS AND BERT W. O'MALLEY

VOLUME 103. Hormone Action (Part H: Neuroendocrine Peptides)
*Edited by* P. MICHAEL CONN

VOLUME 104. Enzyme Purification and Related Techniques (Part C)
*Edited by* WILLIAM B. JAKOBY

VOLUME 105. Oxygen Radicals in Biological Systems
*Edited by* LESTER PACKER

VOLUME 106. Posttranslational Modifications (Part A)
*Edited by* FINN WOLD AND KIVIE MOLDAVE

VOLUME 107. Posttranslational Modifications (Part B)
*Edited by* FINN WOLD AND KIVIE MOLDAVE

VOLUME 108. Immunochemical Techniques (Part G: Separation and Characterization of Lymphoid Cells)
*Edited by* GIOVANNI DI SABATO, JOHN J. LANGONE, AND HELEN VAN VUNAKIS

VOLUME 109. Hormone Action (Part I: Peptide Hormones)
*Edited by* LUTZ BIRNBAUMER AND BERT W. O'MALLEY

VOLUME 128. Plasma Lipoproteins (Part A: Preparation, Structure, and Molecular Biology)
*Edited by* JERE P. SEGREST AND JOHN J. ALBERS

VOLUME 129. Plasma Lipoproteins (Part B: Characterization, Cell Biology, and Metabolism)
*Edited by* JOHN J. ALBERS AND JERE P. SEGREST

VOLUME 130. Enzyme Structure (Part K)
*Edited by* C. H. W. HIRS AND SERGE N. TIMASHEFF

VOLUME 131. Enzyme Structure (Part L)
*Edited by* C. H. W. HIRS AND SERGE N. TIMASHEFF

VOLUME 132. Immunochemical Techniques (Part J: Phagocytosis and Cell-Mediated Cytotoxicity)
*Edited by* GIOVANNI DI SABATO AND JOHANNES EVERSE

VOLUME 133. Bioluminescence and Chemiluminescence (Part B)
*Edited by* MARLENE DELUCA AND WILLIAM D. MCELROY

VOLUME 134. Structural and Contractile Proteins (Part C: The Contractile Apparatus and the Cytoskeleton)
*Edited by* RICHARD B. VALLEE

VOLUME 135. Immobilized Enzymes and Cells (Part B)
*Edited by* KLAUS MOSBACH

VOLUME 136. Immobilized Enzymes and Cells (Part C)
*Edited by* KLAUS MOSBACH

VOLUME 137. Immobilized Enzymes and Cells (Part D)
*Edited by* KLAUS MOSBACH

VOLUME 138. Complex Carbohydrates (Part E)
*Edited by* VICTOR GINSBURG

VOLUME 139. Cellular Regulators (Part A: Calcium- and Calmodulin-Binding Proteins)
*Edited by* ANTHONY R. MEANS AND P. MICHAEL CONN

VOLUME 140. Cumulative Subject Index Volumes 102–119, 121–134

VOLUME 141. Cellular Regulators (Part B: Calcium and Lipids)
*Edited by* P. MICHAEL CONN AND ANTHONY R. MEANS

VOLUME 142. Metabolism of Aromatic Amino Acids and Amines
*Edited by* SEYMOUR KAUFMAN

VOLUME 143. Sulfur and Sulfur Amino Acids
*Edited by* WILLIAM B. JAKOBY AND OWEN GRIFFITH

VOLUME 144. Structural and Contractile Proteins (Part D: Extracellular Matrix)
*Edited by* LEON W. CUNNINGHAM

VOLUME 145. Structural and Contractile Proteins (Part E: Extracellular Matrix)
*Edited by* LEON W. CUNNINGHAM

VOLUME 182. Guide to Protein Purification
*Edited by* MURRAY P. DEUTSCHER

VOLUME 183. Molecular Evolution: Computer Analysis of Protein and Nucleic Acid Sequences
*Edited by* RUSSELL F. DOOLITTLE

VOLUME 184. Avidin–Biotin Technology
*Edited by* MEIR WILCHEK AND EDWARD A. BAYER

VOLUME 185. Gene Expression Technology
*Edited by* DAVID V. GOEDDEL

VOLUME 186. Oxygen Radicals in Biological Systems (Part B: Oxygen Radicals and Antioxidants)
*Edited by* LESTER PACKER AND ALEXANDER N. GLAZER

VOLUME 187. Arachidonate Related Lipid Mediators
*Edited by* ROBERT C. MURPHY AND FRANK A. FITZPATRICK

VOLUME 188. Hydrocarbons and Methylotrophy
*Edited by* MARY E. LIDSTROM

VOLUME 189. Retinoids (Part A: Molecular and Metabolic Aspects)
*Edited by* LESTER PACKER

VOLUME 190. Retinoids (Part B: Cell Differentiation and Clinical Applications)
*Edited by* LESTER PACKER

VOLUME 191. Biomembranes (Part V: Cellular and Subcellular Transport: Epithelial Cells)
*Edited by* SIDNEY FLEISCHER AND BECCA FLEISCHER

VOLUME 192. Biomembranes (Part W: Cellular and Subcellular Transport: Epithelial Cells)
*Edited by* SIDNEY FLEISCHER AND BECCA FLEISCHER

VOLUME 193. Mass Spectrometry
*Edited by* JAMES A. MCCLOSKEY

VOLUME 194. Guide to Yeast Genetics and Molecular Biology
*Edited by* CHRISTINE GUTHRIE AND GERALD R. FINK

VOLUME 195. Adenylyl Cyclase, G Proteins, and Guanylyl Cyclase
*Edited by* ROGER A. JOHNSON AND JACKIE D. CORBIN

VOLUME 196. Molecular Motors and the Cytoskeleton
*Edited by* RICHARD B. VALLEE

VOLUME 197. Phospholipases
*Edited by* EDWARD A. DENNIS

VOLUME 198. Peptide Growth Factors (Part C)
*Edited by* DAVID BARNES, J. P. MATHER, AND GORDON H. SATO

Volume 199. Cumulative Subject Index Volumes 168–174, 176–194

VOLUME 200. Protein Phosphorylation (Part A: Protein Kinases: Assays, Purification, Antibodies, Functional Analysis, Cloning, and Expression)
*Edited by* TONY HUNTER AND BARTHOLOMEW M. SEFTON

VOLUME 201. Protein Phosphorylation (Part B: Analysis of Protein Phosphorylation, Protein Kinase Inhibitors, and Protein Phosphatases)
*Edited by* TONY HUNTER AND BARTHOLOMEW M. SEFTON

VOLUME 202. Molecular Design and Modeling: Concepts and Applications (Part A: Proteins, Peptides, and Enzymes)
*Edited by* JOHN J. LANGONE

VOLUME 203. Molecular Design and Modeling: Concepts and Applications (Part B: Antibodies and Antigens, Nucleic Acids, Polysaccharides, and Drugs)
*Edited by* JOHN J. LANGONE

VOLUME 204. Bacterial Genetic Systems
*Edited by* JEFFREY H. MILLER

VOLUME 205. Metallobiochemistry (Part B: Metallothionein and Related Molecules)
*Edited by* JAMES F. RIORDAN AND BERT L. VALLEE

VOLUME 206. Cytochrome P450
*Edited by* MICHAEL R. WATERMAN AND ERIC F. JOHNSON

VOLUME 207. Ion Channels
*Edited by* BERNARDO RUDY AND LINDA E. IVERSON

VOLUME 208. Protein–DNA Interactions
*Edited by* ROBERT T. SAUER

VOLUME 209. Phospholipid Biosynthesis
*Edited by* EDWARD A. DENNIS AND DENNIS E. VANCE

VOLUME 210. Numerical Computer Methods
*Edited by* LUDWIG BRAND AND MICHAEL L. JOHNSON

VOLUME 211. DNA Structures (Part A: Synthesis and Physical Analysis of DNA)
*Edited by* DAVID M. J. LILLEY AND JAMES E. DAHLBERG

VOLUME 212. DNA Structures (Part B: Chemical and Electrophoretic Analysis of DNA)
*Edited by* DAVID M. J. LILLEY AND JAMES E. DAHLBERG

VOLUME 213. Carotenoids (Part A: Chemistry, Separation, Quantitation, and Antioxidation)
*Edited by* LESTER PACKER

VOLUME 214. Carotenoids (Part B: Metabolism, Genetics, and Biosynthesis)
*Edited by* LESTER PACKER

VOLUME 215. Platelets: Receptors, Adhesion, Secretion (Part B)
*Edited by* JACEK J. HAWIGER

VOLUME 216. Recombinant DNA (Part G)
*Edited by* RAY WU

VOLUME 217. Recombinant DNA (Part H)
*Edited by* RAY WU

VOLUME 218. Recombinant DNA (Part I)
*Edited by* RAY WU

VOLUME 219. Reconstitution of Intracellular Transport
*Edited by* JAMES E. ROTHMAN

VOLUME 220. Membrane Fusion Techniques (Part A)
*Edited by* NEJAT DÜZGÜNEŞ

VOLUME 221. Membrane Fusion Techniques (Part B)
*Edited by* NEJAT DÜZGÜNEŞ

VOLUME 222. Proteolytic Enzymes in Coagulation, Fibrinolysis, and Complement Activation (Part A: Mammalian Blood Coagulation Factors and Inhibitors)
*Edited by* LASZLO LORAND AND KENNETH G. MANN

VOLUME 223. Proteolytic Enzymes in Coagulation, Fibrinolysis, and Complement Activation (Part B: Complement Activation, Fibrinolysis, and Nonmammalian Blood Coagulation Factors)
*Edited by* LASZLO LORAND AND KENNETH G. MANN

VOLUME 224. Molecular Evolution: Producing the Biochemical Data
*Edited by* ELIZABETH ANNE ZIMMER, THOMAS J. WHITE, REBECCA L. CANN, AND ALLAN C. WILSON

VOLUME 225. Guide to Techniques in Mouse Development
*Edited by* PAUL M. WASSARMAN AND MELVIN L. DEPAMPHILIS

VOLUME 226. Metallobiochemistry (Part C: Spectroscopic and Physical Methods for Probing Metal Ion Environments in Metalloenzymes and Metalloproteins)
*Edited by* JAMES F. RIORDAN AND BERT L. VALLEE

VOLUME 227. Metallobiochemistry (Part D: Physical and Spectroscopic Methods for Probing Metal Ion Environments in Metalloproteins)
*Edited by* JAMES F. RIORDAN AND BERT L. VALLEE

VOLUME 228. Aqueous Two-Phase Systems
*Edited by* HARRY WALTER AND GÖTE JOHANSSON

VOLUME 229. Cumulative Subject Index Volumes 195–198, 200–227

VOLUME 230. Guide to Techniques in Glycobiology
*Edited by* WILLIAM J. LENNARZ AND GERALD W. HART

VOLUME 231. Hemoglobins (Part B: Biochemical and Analytical Methods)
*Edited by* JOHANNES EVERSE, ROBERT M. WINSLOW, AND KIM D. VANDEGRIFF

VOLUME 232. Hemoglobins (Part C: Biophysical Methods) (in preparation)
*Edited by* JOHANNES EVERSE, ROBERT M. WINSLOW, AND KIM D. VANDEGRIFF

VOLUME 233. Oxygen Radicals in Biological Systems (Part C) (in preparation)
*Edited by* LESTER PACKER

VOLUME 234. Oxygen Radicals in Biological Systems (Part D) (in preparation)
*Edited by* LESTER PACKER

VOLUME 235. Bacterial Pathogenesis (Part A: Identification and Regulation of Virulence Factors) (in preparation)
*Edited by* VIRGINIA L. CLARK AND PATRIK M. BAVOIL

VOLUME 236. Bacterial Pathogenesis (Part B: Integration of Pathogenic Bacteria with Host Cells) (in preparation)
*Edited by* VIRGINIA L. CLARK AND PATRIK M. BAVOIL

VOLUME 237. Heterotrimeric G Proteins (in preparation)
*Edited by* RAVI IYENGAR

VOLUME 238. Heterotrimeric G Protein Effectors (in preparation)
*Edited by* RAVI IYENGAR

VOLUME 239. Nuclear Magnetic Resonance (Part C) (in preparation)
*Edited by* THOMAS L. JAMES AND NORMAN J. OPPENHEIMER

VOLUME 240. Numerical Computer Methods (Part B) (in preparation)
*Edited by* MICHAEL L. JOHNSON AND LUDWIG BRAND

VOLUME 241. Retroviral Proteases (in preparation)
*Edited by* LAWRENCE C. KUO AND JULES A. SHAFER

# Section I

# General Methodology and Apparatus

# [1] Phase Diagrams

*By* PER-ÅKE ALBERTSSON and FOLKE TJERNELD

## Introduction

Aqueous two-phase systems are generated by mixing aqueous solutions of two structurally different polymers or by mixing one polymer with certain salts at high concentrations. Such mixtures settle in 5–30 min and give rise to two liquid layers, both of which have a high water content (80–95%). The two phases are in equilibrium. In two-polymer systems one phase is enriched with respect to one polymer while the other is enriched with respect to the second polymer. In polymer–salt systems one phase is enriched with polymer and the other with salt. Aqueous two-phase systems have been widely used in the separation of biological materials by partitioning.[1,2] A large number of applications can be found in other chapters of this volume. For nearly all applications it is of considerable value to consult the phase diagram of a phase system. The diagram indicates whether a certain mixture will give rise to one or two phases, and also, in the latter case, gives the composition of the phases. Here we describe how a phase diagram can be constructed. References are also given to phase diagrams available from the literature (Table I).

## Phase Diagrams

Phase diagrams are used to characterize phase systems. Each phase in a two-phase system has a distinct composition, with respect to the phase constituents, which can be compared with the composition of the entire system. Thus, in a two-phase system there are three compositions to consider, namely, the compositions of the top phase, the bottom phase, and the total system. If these are determined experimentally for a number of mixtures, differing in the concentrations of constituents, a phase diagram can be constructed which can predict the composition of the phases for any mixture within the two-phase region. Usually the phase diagram for a system with three components is shown in a triangular form. Aqueous

[1] P.-Å. Albertsson, "Partition of Cell Particles and Macromolecules," 3rd Ed. Wiley (Interscience), New York, 1986.

[2] H. Walter, D. E. Brooks, and D. Fisher (eds.), "Partitioning in Aqueous Two-Phase Systems: Theory, Methods, Uses, and Applications to Biotechnology." Academic Press, Orlando, Florida, 1985.

TABLE I
AQUEOUS TWO-PHASE SYSTEMS

| System | Refs. |
|---|---|
| Polymer–polymer | |
| Nonionic | |
| Dextran–poly(ethylene glycol) (PEG) | *a* |
| Dextran–poly(ethylene glycol propylene glycol) copolymer (Ucon) | *b* |
| Dextran–poly(vinyl alcohol) (PVA) | *a* |
| Dextran–ethylhydroxyethyl cellulose | *c* |
| Dextran–benzoyldextran | *d* |
| Dextran–hydroxypropyl dextran | *a* |
| Dextran–Ficoll | *a* |
| Dextran–polyvinylpyrrolidone (PVP) | *a* |
| Benzoyldextran–PEG | *d* |
| Hydroxypropyl starch–PEG | *e* |
| PVA–PEG | *c, f* |
| Maltodextrin–PEG | *g* |
| Pullulan–PEG | *h* |
| Poly(vinyl methyl ether)–PEG | *i* |
| Ionic | |
| Dextran sulfate–PEG | *a* |
| Carboxymethyl dextran–PEG | *a* |
| Dextran sulfate–poly(styrene sulfonate) | *j* |
| Dextran sulfate–DEAE-dextran | *a* |
| PVA–acrylic copolymers | *k* |
| Poly(acrylic acid)–PEG | *l* |
| Poly(acrylamide)–PVP | *l* |
| Salt–polymer | |
| Potassium phosphate–PEG | *a* |
| Ammonium sulfate–PEG | *a* |
| Potassium citrate–PEG | *m* |
| Magnesium sulfate–PEG | *a* |

[a] P.-Å. Albertsson, "Partition of Cell Particles and Macromolecules," 3rd Ed. Wiley (Interscience), New York, 1986.

[b] P. A. Harris, G. Karlström, and F. Tjerneld, *Bioseparation* **2,** 237 (1991).

[c] F. Tjerneld, *in* "Separations Using Aqueous Phase Systems" (D. Fisher and I. A. Sutherland, eds.), p. 429. Plenum, New York, 1989.

[d] M. Lu, F. Tjerneld, G. Johansson, and P.-Å. Albertsson, *Bioseparation* **2,** 247 (1991).

[e] S. Sturesson, F. Tjerneld, and G. Johansson, *Appl. Biochem. Biotechnol.* **26,** 281 (1990).

[f] A. Kokkoris, J. B. Blair, and J. A. Shaeiwitz, *Biochim. Biophys. Acta* **966,** 176 (1988).

[g] D. C. Szlag and K. A. Giuliano, *Biotechnol. Tech.* **2,** 277 (1988).

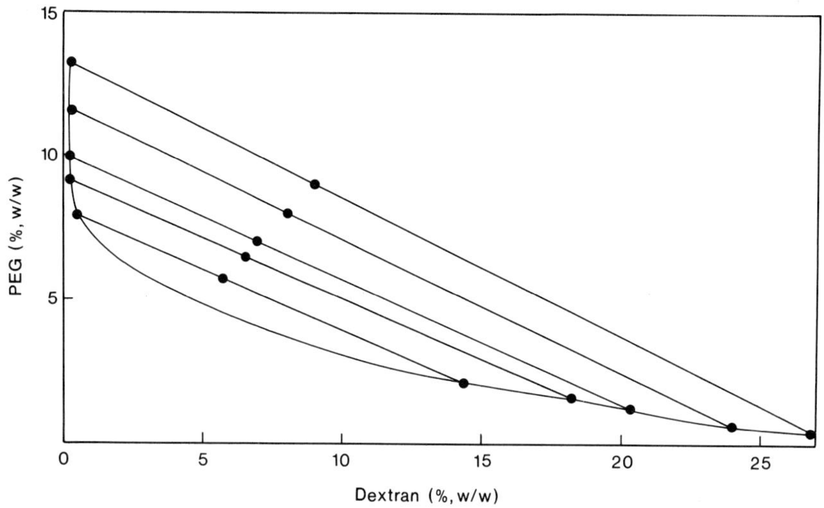


FIG. 1. Phase diagram for the dextran T500 (MW 500,000)–PEG 3350 (MW 3350)–water system at 0°. Data from Table II. The curved line is the binodal curve; the lines connecting top and bottom phases are tie lines. [From P.-Å. Albertsson, "Partition of Cell Particles and Macromolecules," 3rd Ed. Wiley (Interscience), New York, 1986, with permission.]

polymer phase systems are often represented in diagrams with two axes in order to simplify use.[1]

A typical phase diagram for a polymer–polymer–water phase system is shown in Fig. 1. It is based on the experimental data given in Table II which show the compositions of the top and bottom phases of different mixtures of dextran T500 (Pharmacia Biotechnology AB, Uppsala, Sweden) and poly(ethylene glycol) (PEG) 3350 (Union Carbide, New York, NY) in water. The curved line (the binodal curve) separates two regions of compositions. All compositions represented by points at or below the curved line give a one-phase system. All compositions above the binodal curve give a two-phase system. For example, the phase system 5.70% dextran and 5.70% PEG consists of an upper phase with 0.40% dextran and 7.92% PEG and a lower phase with 14.38% dextran and 2.57% PEG.

[h] A. L. Nguyen, S. Grothe, and J. Luong, *Appl. Microbiol. Biotechnol.* **27,** 341 (1988).
[i] J. N. Baskir, T. A. Hatton, and U. W. Suter, *J. Phys. Chem.* **93,** 2111 (1989).
[j] L. Piculell, L. Falck, S. Nilsson, and F. Tjerneld, *Polym. Commun.* **32,** 158 (1991).
[k] P. Hughes and C. R. Lowe, *Enzyme Microb. Technol.* **10,** 115 (1988).
[l] M. B. Perrau, I. Iliopoulos, and R. Audebert, *Polymer* **30,** 2112 (1989).
[m] J. Vernau and M.-R. Kula, *Biotechnol. Appl. Biochem.* **12,** 397 (1990).

TABLE II

COMPOSITION OF PHASES FOR DIFFERENT PHASE SYSTEMS OF DEXTRAN–POLY(ETHYLENE GLYCOL)–WATER SYSTEM[a,b]

| System | Total system Dextran (%, w/w) | Total system PEG (%, w/w) | Total system Water (%, w/w) | Bottom phase Dextran (%, w/w) | Bottom phase PEG (%, w/w) | Bottom phase Water (%, w/w) | Top phase Dextran (%, w/w) | Top phase PEG (%, w/w) | Top phase Water (%, w/w) |
|---|---|---|---|---|---|---|---|---|---|
| A | 5.70 | 5.70 | 88.00 | 14.38 | 2.57 | 83.05 | 0.40 | 7.92 | 91.68 |
| B | 6.50 | 6.50 | 87.00 | 18.21 | 2.05 | 79.74 | 0.18 | 9.13 | 90.69 |
| C | 7.00 | 7.00 | 86.00 | 20.31 | 1.81 | 77.88 | 0.20 | 9.96 | 89.84 |
| D | 8.00 | 8.00 | 84.00 | 23.96 | 1.49 | 74.55 | 0.35 | 11.53 | 88.12 |
| E | 9.00 | 9.00 | 82.00 | 26.80 | 1.26 | 71.94 | 0.26 | 13.20 | 86.54 |

[a] At 0°. The phase diagram is shown in Fig. 1. Dextran is dextran T500 (Pharmacia Biotechnology AB), and poly(ethylene glycol) is Carbowax PEG 3350 (Union Carbide).

[b] From P.-Å. Albertsson, "Partition of Cell Particles and Macromolecules," 3rd Ed. Wiley(Interscience), New York, 1986, with permission.

Points representing the top or bottom phase all lie on the binodal curve. The line connecting the composition of two phases in equilibrium is called a tie line. The line becomes shorter as the polymer concentrations of the two-phase system are reduced. At a certain point, the critical point, the compositions of the two phases become the same; the differences between them vanish at the critical point.

The tie line has points which show the compositions of the total system and of the top and bottom phases. If the polymer concentrations are expressed in percent weight/weight (w/w), the weight ratio top/bottom phase is determined by calculating the ratio between lengths on the tie line: (bottom − total)/(top − total). The densities of the polymer phases are not very different from that of water (in the range 1.0–1.1), and, therefore, the approximate volume ratio top/bottom phase can be obtained.[1]

## Determination of Phase Diagrams

### *Binodal Curve*

The binodal curve can be determined by turbidometric titration. Phase systems with compositions in the two-phase region of the phase diagram are weighed into test tubes. Water is gradually added with shaking until the turbid mixture just clears, and the final weight of the system is then determined. The polymer concentrations at the point of phase transition are calculated. That point lies on the binodal curve.

### *Tie Lines*

The tie lines are determined by removing samples from the top and bottom phases after phase separation has occurred and measuring the polymer concentrations in each phase. Typically, 5 g of phase is diluted with water to 25 ml in a volumetric flask, and the solution is used for the analysis (see below).

## Analytical Procedures

### *Dextran Concentration*

In phase systems where dextran is the only polymer with optical rotation (e.g., in systems with PEG), the dextran concentration is determined by using a polarimeter. The specific optical rotation of dextran at the sodium D line (589 nm) is +199° ml dm$^{-1}$ g$^{-1}$ (at 25°) for dextran with molecular weights above 10,000.[1]

*Poly(ethylene Glycol) Concentration*

In phase systems containing PEG and dextran, it is advantageous first to determine the dextran concentration in samples from both phases by polarimetry. The concentration of PEG is then determined with a refractometer. The increment in refractive index relative to water is $1.33 \times 10^{-4}$ liter/g for PEG 8000 and $1.47 \times 10^{-4}$ liter/g for dextran T500 at 22°. It is convenient to make standard curves for the refractive index of pure PEG and dextran solutions. The PEG concentration in samples from the phases is obtained by subtracting the dextran contribution to the refractive index readings of the samples. The refractive index is sensitive to the presence of salts. When salts are present, the standard curves must be made at the salt concentrations used in the phase system.

Another method for determination of PEG concentration is to freeze-dry aliquots from both phases. The total polymer concentration in each phase is obtained in this way, and subsequently the dextran concentration, previously determined by polarimetry, is subtracted.[1] The two methods described can be used equally well for phase systems composed of PEG and other carbohydrate polymers with optical rotation (e.g., PEG–Ficoll, PEG–hydroxypropyl starch).

*Hydroxypropyl Starch Concentration*

In phase systems with PEG the hydroxypropyl starch concentration is determined by polarimetry in the same way as described for dextran. The specific optical rotation of the hydroxypropyl starch polymers Reppal PES 100 and Reppal PES 200 (Reppe AB, Växjö, Sweden) at the sodium D line (589 nm) is +190° ml dm$^{-1}$ g$^{-1}$ (at 25°).

*Ficoll Concentration*

In phase systems with PEG the Ficoll (Pharmacia Biotechnology AB) concentration is determined by polarimetry in the same way as is described for dextran. The specific optical rotation of Ficoll of 400,000 molecular weight at the sodium D line (589 nm) is +54.9° ml dm$^{-1}$ g$^{-1}$ (at 25°).[1]

*Ucon Concentration*

Ucon (Union Carbide) is a random copolymer of ethylene oxide and propylene oxide which phase separates in water solution on an increase in temperature. Ucon has been used, in phase systems with dextran, for

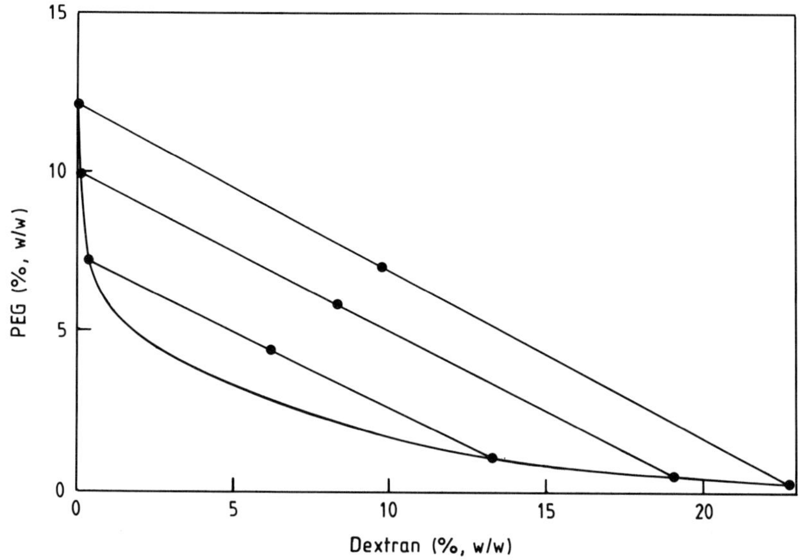


FIG. 2. Phase diagram for the dextran T500 (MW 500,000)–PEG 8000 (MW 8000)–water system at 20°. [From P.-Å. Albertsson, "Partition of Cell Particles and Macromolecules," 3rd Ed. Wiley (Interscience), New York, 1986, with permission.]

protein purification using temperature-induced phase separation.[3,4] The concentration of Ucon in phase systems with carbohydrate polymers (i.e., dextran or hydroxypropyl starch) is determined by refractive index measurements or by freeze-drying in the same way as described for PEG. The concentration of the carbohydrate polymer is determined separately by polarimetry.[3]

### *Salt Concentration*

PEG–salt systems are analyzed by freeze-drying samples from both phases and measuring the dry weight. The salt is determined separately by measuring the conductivity. Phosphate can be determined by titration of the acid form.[1] The PEG concentration is obtained by subtraction.

Examples of aqueous two-phase systems are listed in Table I, and phase diagrams of a few selected systems are presented in Figs. 1–5.

[3] P. A. Harris, G. Karlström, and F. Tjerneld, *Bioseparation* **2,** 237 (1991).

[4] P. A. Alred, F. Tjerneld, A. Kozlowski, and J. M. Harris, *Bioseparation* **2,** 363 (1992).

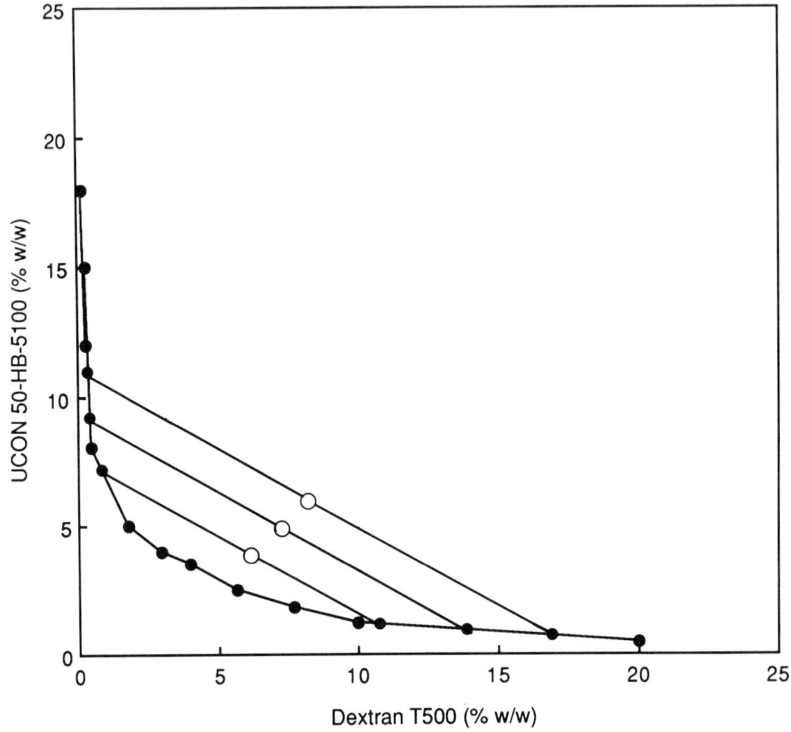


FIG. 3. Phase diagram for the dextran T500 (MW 500,000)–Ucon 50-HB-5100 (MW 4000)–water system at 22°. [From P. A. Harris, G. Karlström, and F. Tjerneld, *Bioseparation* **2,** 363 (1992), with permission.]

## Properties of Aqueous Two-Phase Systems

The properties of a phase system depend on the type and concentration of the polymer(s) and the salt, the temperature, and also the presence of extra added salts or other low molecular weight substances such as sucrose.[1]

### *Molecular Weight of Polymers*

The higher the molecular weight of the polymers, the lower is the concentration required for phase separation. The binodal curve becomes more asymmetric when the difference in molecular weight between the two polymers is increased.[1]

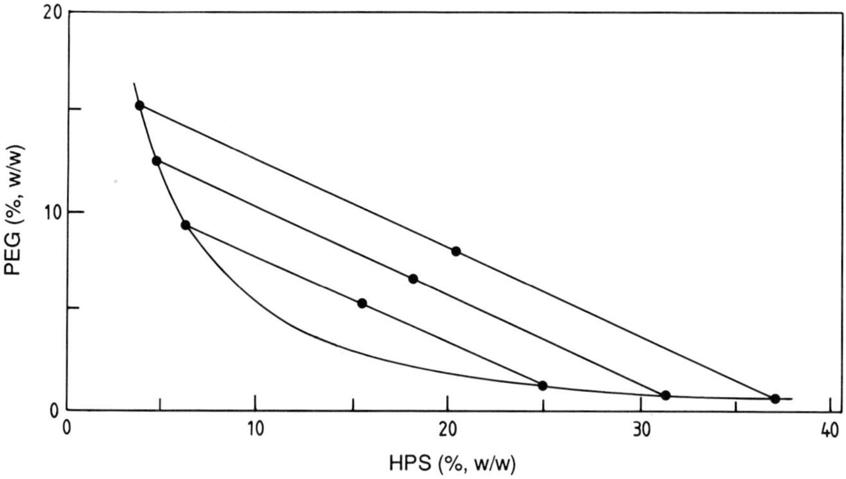


FIG. 4. Phase diagram of the hydroxypropyl starch (MW 100,000)–PEG 8000 (MW 8000)–water system at 20°. The hydroxypropyl starch is Reppal PES 100 (Reppe AB). [From S. Sturesson, F. Tjerneld, and G. Johansson, *Appl. Biochem. Biotechnol.* **26,** 281 (1990), with permission.]

### *Polymer Concentration*

The higher the polymer concentration, the larger is the difference in properties between the phases (also reflected in a longer tie line). Thus, differences in density, refractive index, and viscosity will increase with an increase in polymer concentration.[1]

### *Time of Phase Separation*

The time of phase separation depends on the type of polymer, polymer concentration, the volume ratio, the rate of coalescence of the droplets, and also the presence of particulate material in the system. Typical times of phase separation for the dextran–PEG system are 5–30 min at 1 *g*. Low-speed centrifugation reduces this time to under 1 min.[1] In large-scale use of PEG–salt systems centrifugal separators are often used to reduce the time of phase separation.[2]

### *Effect of Temperature*

Temperature affects the composition of the two phases of a given system and, hence, the phase diagram.[1] Stringent control of temperature is therefore recommended in all partitioning experiments. For PEG–dextran systems phase separation is obtained at lower polymer concentrations

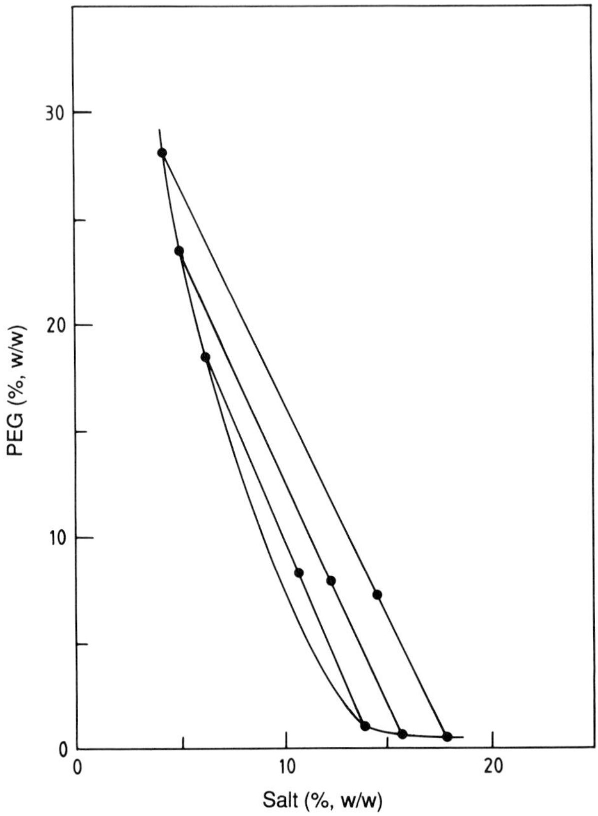


FIG. 5. Phase diagram of the potassium phosphate–PEG 8000 (MW 8000)–water system at 20°. [From P.-Å. Albertsson, "Partition of Cell Particles and Macromolecules," 3rd Ed. Wiley (Interscience), New York, 1986, with permission.]

when the temperature is lowered.[1] For PEG–salt systems an opposite effect is obtained. By increasing the temperature the solubility of PEG in water is reduced, and lower concentrations of PEG and salt can be used to obtain phase separation.[3]

### *Interfacial Tension*

The interfacial tension between the two phases of an aqueous polymer two-phase system is extremely low compared to the interfacial tension between an aqeous phase and an organic solvent phase. Thus, the interfacial tension of the dextran–PEG system varies between 0.1 and 0.0001 mN $m^{-1}$ depending on the polymer composition. The higher the polymer concentration, the larger is the interfacial tension.[1,5]

### *Ionic Composition*

Addition of salts at moderate concentrations affects the phase diagram of nonionic polymer–polymer systems only marginally.[1,5,6] In contrast the behavior of phase systems containing polyelectrolytes, for example, DEAE-dextran, is strongly dependent on the ionic composition. Generally, in such systems a much lower polymer concentration is needed for phase separation to occur when the salt concentration is increased.[1,7]

## Conclusions

It is necessary to know the phase diagram for almost all work with aqueous two-phase systems. The phase diagram shows the distance from the critical point for the system used. This information indicates whether the system is sensitive to small changes in, for example, temperature or polymer concentrations. This is the case for systems close to the critical point. In systems farther from the critical point the difference in properties between the phases is large. The selection of phase systems at different distances from the critical point permits the manipulation of the partitioning behavior of a biomaterial. The phase diagram provides the composition of the two phases in equilibrium. Such data can be used to prepare top or bottom phases of the appropriate composition separately to use in extraction or multistep partitioning applications.

[5] S. Bamberger, D. E. Brooks, K. A. Sharp, J. M. Van Alstine, and T. J. Webber, *in* "Partitioning in Aqueous Two-Phase Systems: Theory, Methods, Uses, and Applications to Biotechnology" (H. Walter, D. E. Brooks, and D. Fisher, eds.), p. 11. Academic Press, Orlando, Florida, 1985.

[6] B. Y. Zaslavsky, L. M. Miheeva, Y. P. Aleschko-Ozhevski, A. U. Mahmudov, T. O. Bagirov, and E. S. Garaev, *J. Chromatogr.* **439,** 267 (1988).

[7] L. Piculell, L. Falck, S. Nilsson, and F. Tjerneld, *Polym. Commun.* **32,** 158 (1991).

# [2] Preparation and Analysis of Two-Phase Systems

*By* DONALD E. BROOKS and RAYMOND NORRIS-JONES

## Introduction

Partitioning in aqueous two-phase systems provides a powerful approach to a variety of analytical and preparative problems. The partition coefficient, $K$, is highly sensitive to the surface properties of the material being examined and is a reproducible characteristic of the material, provided the two-phase system in which the distribution is performed is itself faithfully reproduced from experiment to experiment. The $K$ value can be extremely dependent on the composition of the system, however, as illustrated below.

Consider, for instance, the partition ratio of whole cells and other particulates, which typically distribute between the interface and one of the bulk phases rather than between the two bulk phases. It has been shown[1] that the partition ratio (equal to the number of cells in the bulk phase/number of cells in the interface) varies exponentially with the interfacial tension of the interface, $\gamma_{tb}$. The interfacial tension is low in two-polymer systems, but it is strongly dependent on the difference in polymer concentrations between the phases. A measure of the mean concentration difference is provided by the tie-line length, TLL, as described in [1] in this volume. We have shown[2] that $\gamma_{tb}$ varies approximately as (TLL $\times$ $10^4$). This means that the partition ratio varies roughly as exp(TLL $\times$ $10^4$) for particulates, a strong dependence. The $K$ value for proteins and other macromolecules (defined for soluble material as concentration in top phase/concentration in bottom phase) also varies exponentially with the difference in polymer concentration between phases,[3] again underlining the sensitivity of partitioning behavior to system composition. Hence, considerable care is required in composing the systems to obtain reproducibility.

[1] D. E. Brooks, K. A. Sharp, and D. Fisher, *in* "Partitioning in Aqueous Two-Phase Systems: Theory, Methods, Uses, and Applications to Biotechnology" (H. Walter, D. E. Brooks, and D. Fisher, eds.), p. 11. Academic Press, Orlando, Florida, 1985.

[2] S. Bamberger, G. V. F. Seaman, K. A. Sharp, and D. E. Brooks, *J. Colloid Interface Sci.* **99,** 194 (1984).

[3] A. D. Diamond and J. T. Hsu, *Biotechnol. Bioeng.* **34,** 1000 (1989).

## Preparation of Phase Systems

### *Polymers*

By far the most common components of aqueous two-phase systems are dextran [Dx; predominantly poly($\alpha$-1,6-glucose)], poly(ethylene glycol) (PEG), and salts such as sodium phosphates, sodium citrate, and sodium or magnesium sulfate which are incompatible with PEG and form two-phase systems with this polymer in the absence of a second macromolecular species. Numerous other water-soluble polymers will form two-phase systems, however, as it is rare to find two polymers that will codissolve in water without phase separating at moderate concentrations.[1] There are generally no difficulties in making up stock solutions of any such material and combining them to form two-phase solutions. The techniques and handling procedures described below can be applied to virtually all polymers currently in use in partitioning applications. Difficulties often are encountered in measuring the concentrations of individual components in the phases, but a phase diagram can always be constructed from determinations of cloud points, as described in [1] in this volume.

The greatest single problem in dealing with most aqueous polymer solutions is the viscosity of both the stock solutions and the phases themselves. High viscosity inhibits polymer dissolution, increases the length of time required for phase separation and partition equilibrium to occur, and renders the solutions impractical to quantitate volumetrically. Polymer solutions of the concentrations and types typically used in partitioning cannot be pipetted with any accuracy owing to the length of time thin films of the viscous solutions take to drain from pipette tips. Hence, for all manipulations in which the amount of polymer stock or phase system is required, distribution or dilutions should be carried out by weight. Volumes of the polymer solutions may then be obtained from the solution density, which is measured or calculated as described below.

### *Preparation and Storage of Stock Solutions*

Stocks of Dx and PEG are discussed as examples, but the methods apply to most polymers of practical interest. Not all polysaccharides are amenable to quantitation by polarimetry, some (e.g., hydroxypropyl cellulose) having insufficient specific optical rotation to be of practical value. In such cases refractive index measurements may be used provided a specific refractive index increment is known. If not available, the refractive index increment readily may be measured by thoroughly drying a lyophilized sample of the material over phosphorus pentoxide under vacuum and then measuring the refractive index of a series of dilutions of a stock solution carefully made up by weight from the dried material. The slope of the least-squares fit to a linear plot of refractive index as a function of concentration (%, w/v) provides the desired increment.

*Dextran Stock Solutions*. Dextran is available in a wide range of molecular weights and grades from Pharmacia LKB (Uppsala, Sweden), Pfeifer and Langen (Dormagen, Germany), and secondary suppliers. The fraction with a molecular weight of approximately 500,000 is a popular choice since it phase separates with PEG of low molecular weight (and therefore low viscosity) at moderately low concentrations but is not too viscous to handle, owing to the compact nature of the slightly branched polysaccharide in solution. Although the reasonably narrow molecular weight distribution fractions available from Pharmacia are somewhat expensive, for laboratory work the ability to vary molecular weight and manipulate $K$ values is of significant advantage. The relatively narrow molecular weight distribution is useful since some fractionation on the basis of molecular weight occurs between phases, which is minimized by using tightly distributed polymer fractions. Dextran also has the advantage that there are a number of different ways in which it may be assayed (see [1] in this volume) when its concentration in a phase or complex solution is needed. In particular, it is straightforward to measure its concentration in the presence of PEG and salts by polarimetry.

For most applications, Dx is supplied in a sufficiently pure form that the powder may be used directly to form stock solutions. However, in cases where the material being partitioned is extremely sensitive to impurities, such as multivalent cations, it may be necessary to dialyze the Dx before use or to treat the solution with a mixed-ion exchanger. Care should be taken to dialyze at moderate concentrations to avoid rupture of the dialysis sacs owing to the osmotic pressure of the polymer solution.[4]

Most often Dx is supplied as a powder that contains 5–10% water. The stock solution is made up in distilled water by weight, typically at an apparent concentration of slightly more than four times the desired final concentration in order to allow easy addition of stocks of a second polymer and buffer [e.g., 22% (w/w) for a phase system containing 5% (w/w) Dx]. It helps to make a paste of the powder before adding the final amount of water; gentle heating with continuous stirring will take it into

[4] S. Bamberger, D. E. Brooks, K. A. Sharp, J. M. Van Alstine, and T. J. Webber, *in* "Partitioning in Aqueous Two-Phase Systems: Theory, Methods, Uses, and Applications to Biotechnology" (H. Walter, D. E. Brooks, and D. Fisher, eds.), p. 85. Academic Press, Orlando, Florida, 1985.

solution. Addition of 0.03% (w/w) sodium azide will increase the useful lifetime of the stock solution. The Dx concentration must be measured, most easily by polarimetry on a suitably diluted sample (see [1] in this volume), because of the uncertainty introduced by variable hydration of the solid as supplied.

Dextran stock solutions should be stored at 4°. Stock solutions containing azide usually last 1 month under such conditions. Heating the solution to approximately 90° for 15 min and cooling (before measuring Dx concentration) also will prolong the shelf-life at 4°. However, if Dx solutions are always stored in a particular refrigerator, microorganisms selected for dextranases may predominate and necessitate shorter storage periods. Stock solutions may be frozen and stored indefinitely if tightly sealed and thoroughly remixed before use after thawing.

*Poly(ethylene Glycol) Stock Solutions.* Poly(ethylene glycol) is supplied as reasonably narrow molecular weight distribution fractions by a number of manufacturers, including Union Carbide (New York, NY) and Fluka AG (Buchs, Switzerland). Typically low molecular weight fractions of PEG are used (approximately 4000 to 20,000) owing to the very high viscosities associated with solutions of higher molecular weight fractions at concentrations sufficient to phase separate with polysaccharides or salts. Fractions of the above molecular weights are supplied as powders or flakes that hydrate much less than Dx (typically in the range 0.1 to 2%). In most cases PEG as supplied contains an antioxidant that retards the oxidation of the terminal hydroxyl groups and the main chain of the native polymer. Oxidation can be detected by the presence of acidic groups on the PEG, which reduce its pH in solution, and, in some cases, by a color change of the powder toward gold.

Stock solutions of PEG are made up by weight in distilled water, typically at approximately 30% (w/w). The lower molecular weight fractions dissolve readily at room temperature with continuous stirring. The concentration of the stock solution should be measured via the change in refractive index (see [1] in this volume) to give the stock concentration in percent weight/volume. The concentration in weight/weight is then obtained by dividing by the density of the stock solution, calculated as described below.

For consistency, PEG stock solutions we make also contain 0.03% (w/w) sodium azide, although the polymer is not attacked by microorganisms. Inclusion of the bacteriostatic agent at the same concentration in all stock solutions and in the water used to finish the composition ensures that all phase systems made from them contain a constant azide concentration.

The density, $\rho_{PEG}$, of a solution of PEG of a concentration $C_{PEG}$ (g/100 ml) at 25° is given by[4]

$$\rho_{PEG} = [0.997 + (0.169C_{PEG}/100)] \quad \text{g/ml} \tag{1}$$

where the partial specific volume of PEG of 0.833 ml/g has been used. Similarly, the density of a Dx solution, $\rho_{Dx}$, of concentration $C_{Dx}$ (g/100 ml) at 25° can be calculated from

$$\rho_{Dx} = [0.997 + (0.391C_{Dx}/100)] \quad \text{g/ml} \tag{2}$$

which utilizes the specific volume of Dx of 0.611 ml/g. The density given by Eq. (2) may be used to calculate the concentration in % (w/w) of Dx solutions whose concentration is known in % (w/v).

Stock solutions of PEG and other polyethers are much less prone to degradation than are those of polysaccharides. Storage in sealed containers at 4° is effective for many months.

*Salt Stock Solutions.* It is useful to make up concentrated buffer solutions for use in composing two-polymer phase systems. The stock solutions should be made up at least four times more concentrated than that required in the final system. If a system consisting of PEG and salt only is desired, however, it is easiest to simply weigh dry salt into the system as it is made up. Buffer stock solutions should be made up by weight and the concentrations described in moles per kilogram to make calculations of final concentrations simple following dilution by weight. For consistency we add 0.03% (w/w) azide to salt stock solutions, although they may be safely stored without bacteriostatic agents for long periods of time at 4°.

### *Preparation of Two-Phase Systems*

Phase systems are made up by weight from stock solutions of, for example, 20% (w/w) Dx, 30% (w/w) PEG, and concentrated salt solutions. Stock solutions should be mixed well before weighing out to incorporate any condensate uniformly. The stock solutions are added in decreasing order of their densities: Dx, then PEG, salt, and water. Careful layering of one on the other allows slight overweighing to be easily corrected. It is not necessary to have all the stock solutions at the temperature of the partitioning experiment when making up the systems since no volumetric measurements are made. Most commonly used phase systems can be sterilized by filtration through 0.22- or 0.45-$\mu$m filters at this point. Composed phase systems may be stored frozen indefinitely if tightly sealed. They can be thawed by overnight storage at 4° (or another temperature at which they are to be used), followed by thorough mixing and equilibration as described in the following passage.

Completed systems should be shaken thoroughly and left to equilibrate at the temperature at which partitioning is to be performed. Particulate impurities generally accumulate at the interface, from where they can be collected and discarded. Phase separation may be allowed to take place by standing, most conveniently in a separatory funnel, overnight at the temperature of the experiment, or it may be speeded by centrifugation. Care should be taken that the temperature is not altered while samples are in the centrifuge. It is preferable to use swinging bucket heads to avoid mixing during deceleration as the interface moves from vertical to horizontal orientation.

If it is required to make up a number of systems in which the polymer concentrations are held constant, but the buffer or other component is systematically altered, it is efficient to make up a phase system at double the concentration of both polymers required. Provided this system is kept thoroughly mixed, it may be weighed out in aliquots to which are added by weight the desired salt stock solutions and the systems completed with water to double the original weight of the two-phase solution.

## Measurement of Physicochemical Properties of Two-Phase Systems

### *Phase Composition*

Techniques for measuring polymer concentrations in isolated phases in order to construct a phase diagram have been described in [1] in this volume. It may be of interest in some cases to determine the distribution of other components in the systems such as buffers, proteins, or affinity ligands. Specific assays for particular systems are given in other chapters in this volume. However, it is appropriate here to outline some general methods.

Usually the most trouble-free method for measuring low concentrations or concentration ratios of any ion, protein, or affinity ligand is to radiolabel an aliquot of the material. Including an appropriate number of labeled molecules in a system allows aliquots of each phase to be assayed with the appropriate type of nuclear counter. Some problems may be encountered in $\beta$ counting Dx-rich phases owing to quenching by insoluble polysaccharide precipitated by the fluor, but use of newer solubilizers containing surfactants usually alleviates such difficulties. The great advantages of radiolabeling are sensitivity and the ability to assay material independent of any denaturation, complex formation, or degradation which may have taken place.

Protein concentrations can also be measured in the presence of phase polymers (which interfere in many chemical assays) utilizing fluorescamine[5] (Aldrich, Milwaukee, WI), a sensitive fluorescence assay based on detection of N-terminal amino groups. Salt concentration ratios in two-polymer systems can be estimated by diluting aliquots of each phase at least 10-fold with deioinized water and measuring the ratio of electrical conductivities. The viscosity difference contributed by phase polymers has very little effect at this level, and the conductivity ratio provides a reasonable approximation to the salt distribution, certainly one which will indicate whether significant partitioning is occurring.

### *Osmotic Pressure*

The contribution of the polymers to the osmotic pressure of a phase in a two-polymer system is typically small, 10–20 mOs*m*, owing to the high molecular weights and low molar concentrations of the polymers. Most of the osmotic pressure is contributed by the buffer salts. Hence, in systems containing cells, organelles, or liposomes which respond to the absolute osmotic pressure of the phases, it usually suffices to ensure that the low molecular weight components of the system are isotonic with the contents of the membrane-bound species being partitioned. If it is necessary to measure the osmotic pressure, either a vapor pressure or membrane osmometer should be used. Freezing point depression, on which most clinical osmometers are based, is not a reliable method owing to the unusual effects of polymer, particularly PEG, on freezing kinetics.

### *Viscosity*

Viscosity of phase systems has a strong effect on the rate at which the phases separate and on the energy required to mix, centrifuge, or pump the systems, which can be important in large-scale applications. The viscosity of single-polymer solutions increases strongly with concentration and molecular weight. When a polymer is incorporated in two-phase systems with a second polymer, however, the dependence of phase viscosity on these parameters is affected by the location of the system on the phase diagram since the concentrations of each polymer in each phase are determined by the phase equilibrium. Higher molecular weight systems phase separate at lower concentrations, tending to balance out the increase associated with this parameter. The viscosity of the PEG-rich top phase of a Dx/PEG system is almost independent of tie-line length for a wide range of phase compositions and molecular weights, for instance.[4]

Because there does not seem to be any indication of non-Newtonian

[5] S. J. Stocks, A. J. M. Jones, C. W. Ramey, and D. E. Brooks, *Anal. Biochem.* **154,** 232 (1986).

TABLE I
VISCOSITIES OF DEXTRAN–POLY(ETHYLENE GLYCOL) TWO-PHASE SYSTEMS[a,b]

| System | Phase viscosity (mPa sec) | |
|---|---|---|
| | Top (PEG-rich) | Bottom (Dx-rich) |
| No additive | 3.25 | 26.51 |
| 0.15 *M* NaCl | 3.27 | 28.89 |
| 0.11 *M* Sodium phosphate buffer, pH 7.5 | 3.24 | 28.55 |
| 0.3 *M* sucrose | 4.40 | 37.25 |
| 0.3 *M* Sorbitol | 3.75 | 31.12 |

[a] All systems contain 5% (w/w) dextran T500 (Pharmacia) and 4% (w/w) PEG 8000 (Union Carbide).

[b] From S. Bamberger, D. E. Brooks, K. A. Sharp, J. M. Van Alstine, and T. J. Webber, *in* "Partitioning in Aqueous Two-Phase Systems: Theory, Methods, Uses, and Applications to Biotechnology" (H. Walter, D. E. Brooks, and D. Fisher, eds.), p. 85. Academic Press, Orlando, Florida, 1985.

behavior of isolated phases (i.e., viscosity is independent of shear rate), virtually any standard method may be used to measure viscosities of the polymer solutions or phases. For the more viscous solutions a variable shear rate rotational viscometer is appropriate,[6] whereas for less viscous phases and low molecular weight PEG stocks simple capillary viscometry[6] is technically straightforward, inexpensive, and sensitive. Representative data obtained by both techniques have been published.[4,7] Data illustrating the effects on phase viscosity of low molecular additives are given in Table I.

### *Interfacial Tension and Contact Angles*

As discussed above, the partitioning behavior of cells and particles is controlled to a large extent by the interfacial tension of the system, $\gamma_{tb}$, since this parameter determines adsorption at the interface. Hence, if it is required to reproduce the properties of a particular system subsequent,

[6] J. R. Van Wazer, J. W. Lyons, K. Y. Kim, and R. E. Colwell, "Viscosity and Flow Measurement: A Laboratory Handbook of Rheology." Wiley (Interscience), New York, 1963.

[7] H. Hustedt, K. H. Kroner, and M.-R. Kula, *in* "Partitioning in Aqueous Two-Phase Systems: Theory, Methods, Uses and Applications to Biotechnology" (H. Walter, D. E. Brooks, and D. Fisher, eds.), p. 529. Academic Press, Orlando, Florida, 1985.

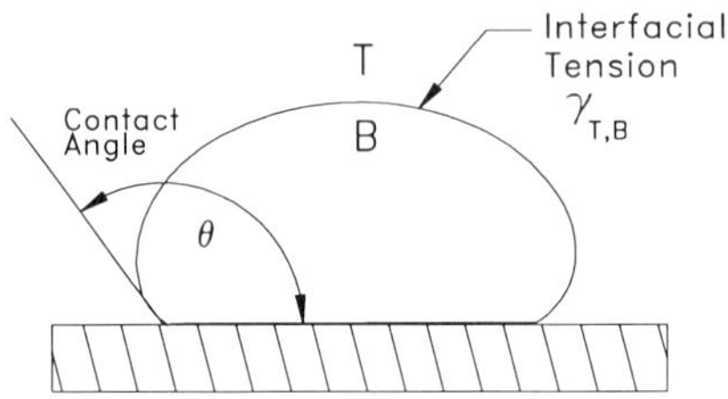


FIG. 1. A sessile drop of bottom phase (B) in top phase (T) at equilibrium with a solid surface, illustrating the parameters associated with the drop shape. The contact angle, $\theta$, is determined by a balance between the difference in surface affinity felt by each phase and the magnitude of the interfacial tension.

for instance, to a change in the source or type of polymer, it is useful to be able to measure $\gamma_{tb}$. Because $\gamma_{tb}$ depends so strongly on phase composition through the tie-line length, its measurement can provide a useful measure of the state of a system if the method is available on a routine basis. Two of the three common methods for measuring $\gamma_{tb}$ also can provide a measure of the contact angle formed by the interface between the phases and the solid surface with which they are in mutual contact (Fig. 1). Measurement of the contact angle provides fundamental information on the relative affinities of the phases for the surface because it obeys the following relationship[1]:

$$\cos\theta = \frac{\gamma_{ts} - \gamma_{bs}}{\gamma_{tb}} \tag{3}$$

where $\theta$ is the interior angle between interface and solid surface and $\gamma_{ts}$, $\gamma_{bs}$, and $\gamma_{tb}$ represent the free energy per unit area of interface between top phase and surface, bottom phase and surface, and top and bottom phases (the interfacial tension), respectively.

For materials that accumulate significantly at the interface during a partitioning experiment, such as cells or other particulates, $\theta$ is the major determinant of partitioning behavior. Because it is a true equilibrium parameter, its value is a more reliable indicator of the difference in chemical affinity for the particle exhibited by the phases than is the partition ratio.[1] The partition ratio of particles (as opposed to the $K$ of macromolecules) is not a true thermodynamic parameter independent of the conditions of measurement, but is known to depend to some degree on the time of sampling, the geometry of the container, and the history of the sample.[1,8]

[8] H. Walter, E. J. Krob, and L. Wollenberger, *J. Chromatogr.* **542,** 397 (1991).

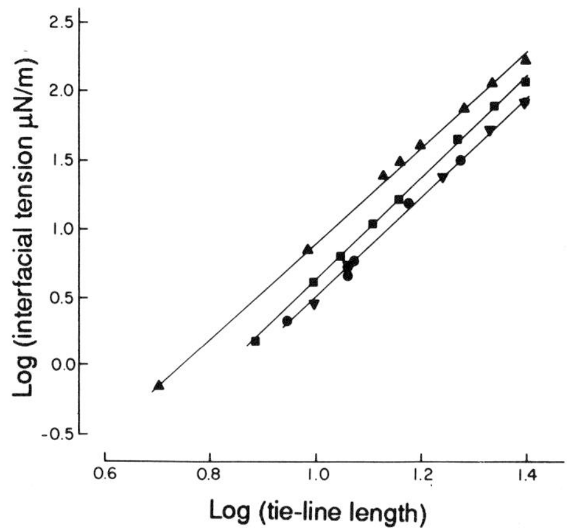


FIG. 2. Dependence of the interfacial tension on the tie-line length at 22° in systems containing dextran T40 (Pharmacia) and PEG 8000 (Union Carbide) with no salt added (●), with 0.15 *M* NaCl (▼), with 0.11 *M* sodium phosphate buffer, pH 7.5 (■), and with 0.22 *M* sodium phosphate buffer, pH 7.5 (▲). [From D. E. Brooks, K. A. Sharp, S. Bamberger, C. H. Tamblyn, G. V. F. Seaman, and H. Walter, *J. Colloid Interface Sci.* **102,** 1 (1984).]

Moreover, although the partition ratio is clearly dependent on the chemical properties of the phases, its dependence on the contact angle does not obey thermodynamic predictions.[1] Hence, if a completely objective measure of surface–phase interaction is required, measurement of $\theta$ is the approach of choice.

The interfacial tensions of two-polymer aqueous systems are extremely low, some of the lowest ever measured, and are typically in the 1–50 $\mu$N/m range (Fig. 2). These may be compared with values of the order of 50 mN/m for an oil/water interface. Hence, traditional techniques developed to work on oil/water or air/water interfaces, which depend on measurement of relatively large forces, may not be appropriate for two-polymer systems. Single-polymer/salt/water systems typically have tensions near 1 mN/m, however, so more options are available for study of such mixtures.

Three techniques have been used successfully to measure $\gamma_{tb}$ on aqueous systems, namely, the rotating drop,[2,9] pendant drop,[2,10] and sessile drop techniques.[11] In all cases the tension is calculated from the deformation of a drop of one phase in the other as the result of centrifugal or gravitational forces. In the rotational technique the drop is suspended in

[9] J. Rydén and P.-Å. Albertsson, *J. Colloid Interface Sci.* **37,** 219 (1971).

[10] S. Schürch, D. F. Gerson, and D. J. L. McIver, *Biochim. Biophys. Acta* **640,** 557 (1981).

[11] Y. Rotenberg, L. Boruvka, and A. W. Neumann, *J. Colloid Interface Sci.* **93,** 169 (1983).

the second phase in a rotating tube, whereas in the other two cases the drop is in contact with a solid surface. Simple geometric measurements of axial ratios are generally sufficient for the determination of $\gamma_{tb}$; standard texts[12] should be referred to for the necessary formulas and tables.

All these techniques rely on a difference in density between the phases to allow a deforming force to be applied to the interface and distort it from the spherical shape which surface tension alone would produce, as illustrated in Fig. 1. Hence, if phase compositions close to the critical point are used, which produce small density differences, relatively small deformations will be produced by gravity, and considerable inaccuracy in measuring the small deviations from sphericity will result. The spinning drop method holds an advantage for such systems in that the deforming force can be controlled by varying the rotation rate. However, there are other difficulties associated with secondary flows in the spinning tube and lengthy equilibration times that make ultralow tension measurements difficult by this method as well.[10]

Other considerations when comparing the techniques include the possibility of nonuniformities or time dependence in the properties of the surfaces with which the drops are in contact in the sessile and pendant drop approaches, either of which will reduce accuracy. Surfaces should be cleaned rigorously: Chromate washing followed by exhaustive rinsing in distilled water is appropriate for glass. For all techniques evaporation of solvent must be prevented since any change in composition is reflected dramatically in the interfacial tension. Evaporation is generally not a problem in the rotating drop case because the tube is filled and sealed. The sessile drop method is susceptible to difficulties from this source, however, because drops are generally placed on the surface via a pipette through the free surface of the top phase. Floating a film of hexadecane on the surface has been found to be successful at inhibiting evaporation; however, it must be recognized that hexadecane has a finite water solubility, and if hydrophobic solids are being used there is the possibility of time-dependent properties being introduced by alkane adsorption.

For accurate results temperature control is necessary, although the interfacial tensions of aqueous polymer systems are not as temperature sensitive as air/water or oil/water interfaces since the temperature dependence of phase compositions is weak.[4] The rotating drop apparatus is somewhat more difficult to keep at constant temperature because of the nature of the motion involved and because the heat produced at the bearings holding the rotating tube can conduct into the sample during long runs.

[12] A. W. Adamson, "Physical Chemistry of Surfaces," 3rd Ed. Wiley (Interscience), New York, 1976.

An additional useful feature of the sessile drop technique is that the contact angle formed with the solid surface may be obtained simultaneously using a goniometer or by geometric analysis of photographs of the drop taken horizontally. Alternatively, in a more sophisticated and accurate approach, the profile of the drop is digitized and both $\gamma_{tb}$ and $\cos \theta$ obtained by application of the axisymmetric drop analysis approach of Rotenberg *et al.*[11] The last method has been used to determine the molecular weight dependence of the contact angle against glass surfaces derivatized with various sized dextrans in phase systems composed of different molecular weight Dx fractions.[13]

### *Electrostatic Potential Differences between Phases*

Although most two-polymer phase systems utilized routinely in separation work consist of polymers that bear no detectable numbers of ionogenic groups, in the presence of certain salts they behave as if the phases have an electrostatic potential difference between them. The presence of such a potential is supported by the partitioning behavior of charged macromolecules in such systems,[14] the unequal partitioning of the salts that cause the apparent potentials in these systems,[15,16] and measurements of electrostatic potential differences between reversible electrodes connected by salt bridges to the two phases of systems that show the above effects.[16,17] The potential difference measurement technique using salt bridges is described below. It provides accurate, reproducible data if care is taken in the process. Interpretation of the potential values is not straightforward, however, and has been somewhat controversial.[1] Efforts to write fundamental theories for the partition coefficients of proteins[18] have focused more attention on the issue.[19]

The problem in interpreting the potential measurements is that it is well known that such measurements do not provide the thermodynamic electrostatic potential difference between the phases, owing to the possible presence of liquid junction potentials at the tips of the salt bridges. This

[13] J. F. Boyce, B. A. Hovanes, J. M. Harris, J. M. Van Alstine, and D. E. Brooks, *J. Colloid Interface Sci.* **149,** 153 (1992).

[14] H. Walter, R. Garza, and R. P. Coyle, *Biochim. Biophys. Acta* **156,** 409 (1968).

[15] S. Bamberger, G. V. F. Seaman, J. A. Brown, and D. E. Brooks, *J. Colloid Interface Sci.* **99,** 187 (1984).

[16] D. E. Brooks, K. A. Sharp, S. Bamberger, C. H. Tamblyn, G. V. F. Seaman, and H. Walter, *J. Colloid Interface Sci.* **102,** 1 (1984).

[17] R. Reitherman, S. D. Flanagan, and S. H. Barondes, *Biochim. Biophys. Acta* **297,** 193 (1973).

[18] H. Walter, G. Johansson, and D. E. Brooks, *Anal. Biochem.* **197,** 1 (1991).

[19] C. A. Haynes, H. W. Blanch, and J. M. Prausnitz, *Fluid Phase Equilib.* **53,** 463 (1989).

TABLE II
ELECTROSTATIC POTENTIAL DIFFERENCES FOR DEXTRAN–POLY(ETHYLENE GLYCOL) TWO-PHASE SYSTEMS[a,b]

| Polymer composition (%, w/w) | | | | | |
|---|---|---|---|---|---|
| Dx | | PEG | | | |
| 40[d] | 500[d] | 8[d] | 20[d] | TLL[c] (%, w/w) | Potential (mV) |
| 7.5 | — | 4.5 | — | 12.7 | 0.26 |
| — | 6.0 | 4.0 | — | 13.1 | 0.19 |
| — | 7.0 | 4.4 | — | 15.7 | 3.53 |
| 8.0 | — | — | 4.0 | 13.4 | 3.70 |
| 10.0 | — | — | 5.0 | 18.8 | 5.6 |

[a] From S. Bamberger, D. E. Brooks, K. A. Sharp, J. M. Van Alstine, and T. J. Webber, *in* "Partitioning in Aqueous Two-Phase Systems: Theory, Methods, Uses, and Applications to Biotechnology" (H. Walter, D. E. Brooks, and D. Fisher, eds.), p. 85. Academic Press, Orlando, Florida, 1985.
[b] All systems contain 0.11 *M* sodium phosphate buffer, pH 7.5.
[c] Tie-line length (TLL).
[d] Molecular weight given by $10^3$ times number indicated; Dx is from Pharmacia, PEG from Union Carbide.

limitation enters the interpretation of all such electrode measurements, including the widely applied membrane potential measurements that have been so successful in explaining fundamental electrophysiological phenomena.[20] The problem is that the potentials in aqueous two-phase systems are predicted, and found, to be small, a few millivolts at most, yet, because all interaction energies in these systems are small, electrostatic interactions of the magnitude implied can dominate partitioning behavior. Hence, a detailed understanding of the processes responsible for the electrical properties is required to settle the outstanding issues.

Questions of interpretation notwithstanding, it is relatively straightforward to obtain reliable potential measurements. The technique we have successfully used is described in detail by Brooks et al.[16] Glass microelectrodes are drawn with internal tip diameter of the order of 30 $\mu$m and filled with 0.1 *M* KCl to produce electrodes with a resistance of about 1.5 MΩ. Ag–AgCl electrode wires or cylinders[4] are inserted in the micropipettes and the electrodes conditioned by passing 100 mV ac through them

[20] B. Katz, "Nerve, Muscle and Synapse." McGraw-Hill, New York, 1966.

for 1 hr or more with the tips immersed in 1 *M* KCl. The electrodes are then connected to a millivoltmeter with a high input resistance ($>10^7\ \Omega$) and immersed in the top phase of at least 20 ml of an equivolume phase system equilibrated in a thermostatted water bath. One of the pair should be mounted on a manipulator that will allow smooth vertical motion of the electrode. The entire apparatus must be placed inside a grounded metal Faraday cage to reduce stray voltages. The voltage showing between the electrodes when they are in the top phase should not be more than 5 mV; higher values indicate nonequilibrated phase systems, blocked electrodes, or poorly plated Ag–AgCl.

To make a set of measurements, one electrode is slowly moved through the interface into the bottom phase and the change in voltage noted with time. There is a transient period during which the top phase drains away from the tip and the potential difference increases to a steady value, which is taken as the reading. The electrode is then returned to the top phase and the potential difference again taken at steady state. At least 10 such readings should be averaged to provide the potential difference for that particular system. Some typical potential difference readings obtained by this technique are listed in Table II for various Dx/PEG systems.

## Conclusion

To optimize the strength of the partitioning approach in solving separation or analytical problems, it is necessary to pay close attention to the properties of the phase systems and, in particular, to reproduce carefully these properties if comparisons between runs are to be made. Appropriate use of the techniques outlined will allow comparisons to be made with confidence and will aid in the interpretation of experimental data.

# [3] Partitioning Procedures and Techniques: Small Molecules and Macromolecules

*By* GÖTE JOHANSSON

## Introduction

### *Usefulness*

Partitioning in aqueous two-phase systems can be used to separate all kinds of water-soluble molecules present in the living cell.[1,2] Such compounds are usually adequately soluble in both aqueous phases to permit their partitioning at relatively high concentrations. It is often observed that low molecular weight substances partition fairly evenly between the phases. However, weak interactions between one of the phase-forming polymers and the partitioned material may cause an uneven partitioning. Polymer–salt systems have greater differences between the physical properties of the two phases, and thus the partitioning of small molecules is often unequal and favors one of the phases. With molecules of increasing molecular weight (peptides, proteins, nucleic acids) the partitioning will be more uneven and also more sensitive to phase system composition. Proteins can therefore be fractionated by partitioning by use of either a few partitioning steps or a multistep procedure (e.g., countercurrent distribution).[3,4] To manipulate the partition coefficient, $K$ (see below), of proteins and to gain greater selectivity in the extraction, the interplay between electrostatic and hydrophobic forces is commonly utilized. The $K$ of nucleic acids is easier to adjust than the $K$ of proteins because of the high net charge of the nucleic acids; extreme partitioning can be obtained by appropriate selection of salts.[5]

Partitioning of low molecular weight substances and macromolecules can be used to study their interactions,[6] charge,[7] isoelectric point,[8] hydro-

[1] P.-Å. Albertsson, "Partition of Cell Particles and Macromolecules," 3rd Ed. Wiley(Interscience), New York, 1986.

[2] H. Walter and G. Johansson, *Anal. Biochem.* **155,** 215 (1986).

[3] G. Johansson, *in* "Partitioning in Aqueous Two-Phase Systems: Theory, Methods, Uses, and Applications to Biotechnology" (H. Walter, D. E. Brooks and D. Fisher, eds.), p. 161. Academic Press, Orlando, Florida, 1985.

[4] M.-R. Kula, *Bioseparation* **1,** 181 (1990).

[5] W. Müller, this volume [18].

[6] S. Lundberg and L. Backman, this volume [22].

[7] G. Johansson, this volume [21].

[8] H. Walter and D. Forciniti, this volume [20].

phobicity,[9] purity,[10] and the presence of multiple forms of enzymes,[11] and it provides a means for their preparation as well. It is important that studies be carried out in the absence of gel matrices that would influence the results because of surface adsorption.

*Expressions Describing Partitioning Behavior*

The partitioning of soluble materials, unlike that of particulates, takes place between the two bulk phases. The most common expressions[12] describing partitioning behavior include (1) the partition coefficient, $K$, defined as the ratio of the concentrations of partitioned material in the upper phase and in the lower phase; (2) the partition ratio, $G$, defined as the ratio of the mass of partitioned material in the two phases (top to bottom) and whose value is dependent on the volume ratio of the phases and related to the $K$ value via Eq. (1):

$$G = KV_T/V_B \qquad (1)$$

where $V_T$ and $V_B$ are the volumes of top and bottom phase, respectively; and (3) the fractions, $p$ and $q$, of partitioned material in the top phase and bottom phase, respectively, which are useful for calculating purification and recovery of material when repeated extractions are carried out. The fractions are related to the $G$ value via Eqs. (2) and (3):

$$p = G/(1 + G) \qquad (2)$$
$$q = 1/(1 + G) \qquad (3)$$

By definition $p + q = 1$. Also, the percentage partition fractions, $p(100)$ and $q(100)$, are frequently used to describe partitioning.

*Factors Determining Partitioning of Proteins*

The protein partition coefficient, $K$, can be written as the product of two factors, $K_0$ and $K_Z$. $K_0$ is related to the relative solvation of the protein in the phases. $K_Z$ depends on the net charge, $Z$, of the protein molecule and its interaction with the phases, which depends on the electrolytes

[9] V. P. Shanbhag, this volume [23].

[10] G. Johansson, M. Joelsson, and H.-E. Åkerlund, *J. Chromatogr.* **298,** 483 (1984).

[11] G. Blomquist, A. Hartman, V. Shanbhag, and G. Johansson, *Eur. J. Biochem.* **48,** 63 (1974).

[12] C. J. O. R. Morris and P. Morris, *in* "Separation Methods in Biochemistry," 2nd Ed., p. 639. Pitman, London, 1976.

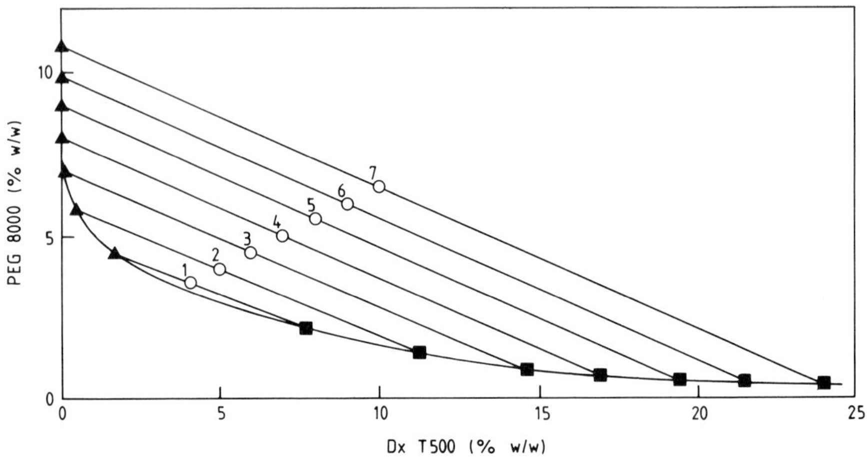


FIG. 1. Phase diagram for dextran (Dx) T500 and poly(ethylene glycol) (PEG) 8000 at 4°. Systems 1–7 have the total composition of polymers (○) as given in Table I, with the corresponding tie lines connecting the compositions of top phases (▲) and bottom phases (■). [Based on data from P.-Å. Albertsson, "Partition of Cell Particles and Macromolecules," 3rd Ed. Wiley(Interscience), New York, 1986. Copyright © 1986 by John Wiley & Sons, Inc., reprinted by permission of John Wiley & Sons, Inc.]

present.[1,13,14] This relation is expressed in Eq. (4):

$$K = K_0 K_Z \tag{4}$$

but is often used in its logarithmic form [Eq. (5)]:

$$\log K = \log K_0 + \log K_Z \tag{5}$$

Furthermore,

$$\log K_Z = \gamma Z \tag{6}$$

where $\gamma$ is a factor determined by the types and concentrations of polymers used, the types of electrolytes and their concentrations, and the temperature.[15] It should be noted that the $K_0$ value may vary with the pH of the system.[13]

*Polymer Concentration.* Increasing polymer concentrations result in greater differences between the compositions of the two phases. This can be seen in the phase diagram[16] of a two-polymer system (Fig. 1). A way

[13] G. Johansson, *Mol. Cell. Biochem.* **4,** 169 (1974).

[14] D. E. Brooks and R. Norris-Jones, this volume [2].

[15] G. Johansson, *Acta Chem. Scand. Ser. B* **B28,** 873 (1974).

[16] P.-Å. Albertsson and F. Tjerneld, this volume [1].

TABLE I
RELATION BETWEEN POLYMER CONCENTRATION OF DEXTRAN T500–POLY(ETHYLENE GLYCOL) 8000 SYSTEMS[a] AND LENGTH[b] OF CORRESPONDING TIE LINE[c]

| System | DX T500 (%, w/w) | PEG 8000 (%, w/w) | Tie-line length[d] |
|---|---|---|---|
| 1 | 4.00 | 3.50 | 6.4 |
| 2 | 5.00 | 4.00 | 11.8 |
| 3 | 6.00 | 4.50 | 15.7 |
| 4 | 7.00 | 5.00 | 18.4 |
| 5 | 8.00 | 5.50 | 21.2 |
| 6 | 9.00 | 6.00 | 23.4 |
| 7 | 10.00 | 6.50 | 26.0 |

[a] At 4°.
[b] Measured in units of polymer concentration (%, w/w).
[c] In the phase diagram depicted in Fig. 1.
[d] The tie-line lengths have been estimated from the phase diagram of the systems in P.-Å. Albertsson, "Partition of Cell Particles and Macromolecules," 3rd Ed. Wiley(Interscience), New York, 1986. Copyright © 1986 by John Wiley & Sons, Inc., reprinted by permission of John Wiley & Sons, Inc.

to quantitate the difference between the phases is to use the tie-line length ($s$) of the system (Table I). Figure 2 shows the change of $K_0$ of a protein with $s$ in a dextran (Dx)–poly(ethylene glycol) (PEG) system.[17] In this kind of system $K_0$ usually decreases with increasing $s$ owing to the reduced solubility of the protein in the upper phase with increasing PEG concentration. The polymer concentration can thus be used to adjust $K_0$ values.

*Kinds of Salts.* Different salts in the phase system cause $K$ values to be either increased or decreased.[15] The order in which different ions affect partitioning in a Dx–PEG system is indicated in Table II. Of all possible combinations of ions in Table II, the one that gives the greatest increase in $K$ for a negatively charged protein is tetrabutylammonium phosphate. The greatest decrease in $K$ for the same protein is obtained with potassium perchlorate. Increasing polymer concentrations result in a larger $\gamma$ value.[17] As an approximation, $\gamma$ can be assumed to be proportional to $s$. With certain salts, and depending on the polymer type and temperature of the phase system with $\gamma = 0$, the charge dependence of partitioning is

[17] G. Johansson, *J. Chromatogr.* **150,** 63 (1978).

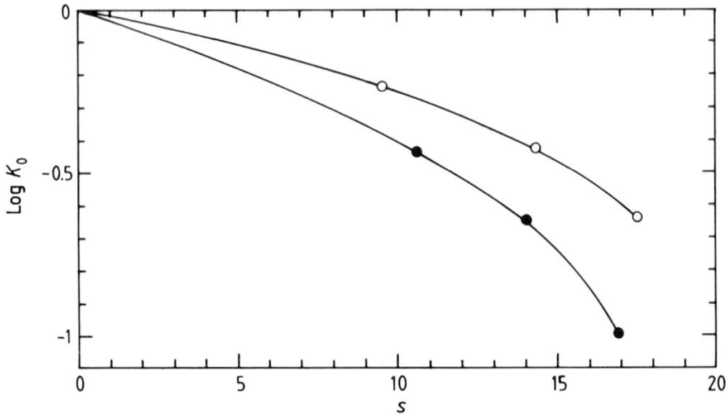


FIG. 2. $K_0$ value for bovine serum albumin, partitioned at its isoelectric point, as a function of tie-line length, *s*, in systems containing dextran T500 and PEG 8000 (○) or PEG 35,000 (●). Temperature, 20°. [Reproduced from G. Johansson, *J. Chromatogr.* **150,** 63 (1978), with permission.]

TABLE II
SPECIFIC VALUES, $\gamma_+$ AND $\gamma_-$, FOR CATIONS AND ANIONS FOR CALCULATION OF RELATIVE $\gamma$ VALUES[a,b]

| Cation | $\gamma_+$ | Anion | $\gamma_-$ |
|---|---|---|---|
| $K^+$ | 58 | $H_2PO_4^-/HPO_4^{2-}$ (1 : 1) | 79 |
| $Na^+$ | 56 | $SO_4^{2-}$ | 77 |
| $NH_4^+$ | 49 | $F^-$ | 69 |
| $Li^+$ | 43 | Acetate | 56 |
| $(C_2H_5)_4N^+$ | 32 | $Cl^-$ | 50 |
| $(C_4H_9)_4N^+$ | 6 | $Br^-$ | 44 |
| | | $I^-$ | 34 |
| | | $SCN^-$ | 30 |
| | | $ClO_4^-$ | 28 |

[a] Obtained with different salts in Dextran T500–poly(ethylene glycol) 8000 phase systems at 0–4°. A relative $\gamma$ value, $\gamma_{rel}$, for a salt is obtained as the difference of the $\gamma_+$ and $\gamma_-$ values for the cation and anion, respectively, of the salt: $\gamma_{rel} = \gamma_+ - \gamma_-$.

[b] The $\gamma_{rel}$ values obtained give approximate values that can be used to estimate the relative steering effects of different salts. The values are proportionate to the true values with a factor that depends on the concentrations of the polymers.

eliminated, and the $K$ observed for the protein is equal to its $K_0$ value.[13,15,17,18]

*pH Dependence.* Because the net charge of proteins and other biopolymers changes with the acidity of solution, pH can be used to adjust the $K$ value, especially in combination with salts having large $\gamma$ values or with charged polymers.[13]

*Use of Charged Polymers.* Charged polymers, for example, trimethylamino-PEG, PEG sulfonate, diethylaminoethyl-Dx, and Dx sulfate, can be used to obtain systems in which the partitioning of polyelectrolytes is strongly dependent on their net charge.[19–21]

*Hydrophobic Partitioning.* The binding of hydrophobic groups (e.g., palmitate) to one of the phase-forming polymers allows the extraction of biopolymers with hydrophobic sites.[9,21,22] This hydrophobic partitioning effect can be described by a factor $K_{hp}$. This factor is a part of $K_0$, that is, $K_0 = K_0'K_{hp}$. Equation (4) is then modified to reflect this, as shown in Eq. (7):

$$K = K_0'K_ZK_{hp} \tag{7}$$

$K_{hp}$ increases with polymer concentration, for example, with the tie-line length (Fig. 3A).

*Affinity Partitioning.* "Immobilizing" affinity ligands for biomaterials in one phase by attaching them to the phase-forming polymer, rich in that phase, is the basis for affinity partitioning. Specific extractions of enzymes and other proteins can thus be performed.[23–25] The specific effect due to the affinity group can be described by a factor $K_{aff}$ which is part of $K_0$, namely, $K_0 = K_0''K_{aff}$. The $K$ of the partitioned substance is then

$$K = K_0''K_ZK_{aff} \tag{8}$$

$K_{aff}$ increases with length of the tie line (Fig. 3B).

*Temperature.* Temperature has a complex effect on partitioning. Lowering the temperature generally favors phase separation of two-polymer systems, whereas the opposite applies to salt–polymer systems.[1] Two-polymer systems which form at high temperatures are, however,

[18] H. Walter and D. Forciniti, this volume [20].

[19] G. Johansson, A. Hartman, and P.-Å. Albertsson, *Eur. J. Biochem.* **33,** 379 (1973).

[20] L. Cheng, M. Joelsson, and G. Johansson, *J. Chromatogr.* **523,** 119 (1990).

[21] G. Johansson, this volume [21].

[22] G. Johansson, *J. Chromatogr.* **451,** 517 (1976).

[23] G. Kopperschläger, this volume [11].

[24] S. D. Flanagan and S. H. Barondes, *J. Biol. Chem.* **250,** 1484 (1975).

[25] G. Johansson and F. Tjerneld, *J. Biotechnol.* **11,** 135 (1989).

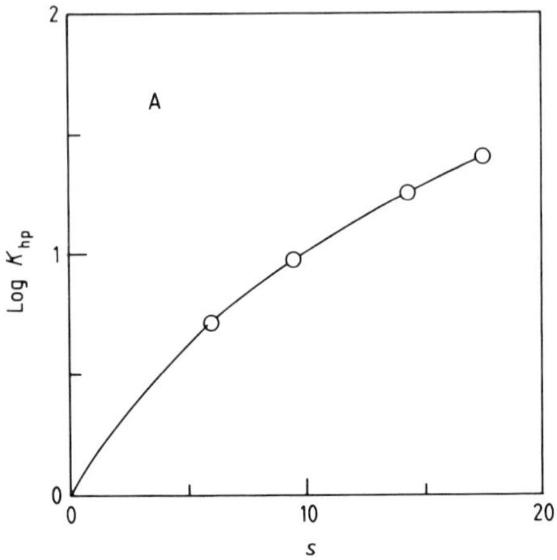


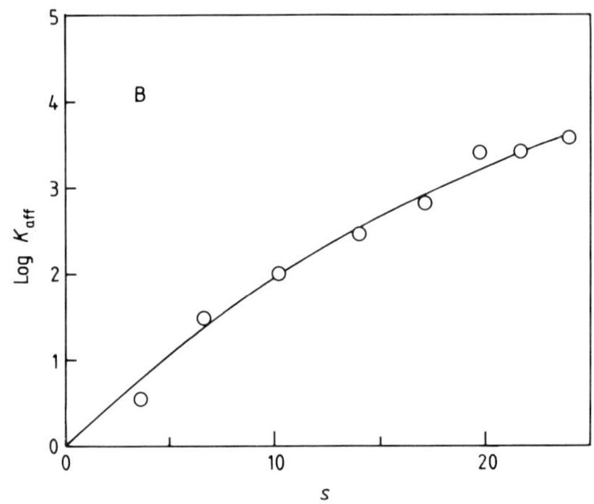


FIG. 3. (A) Variation of $K_{hp}$, for bovine serum albumin using PEG–palmitate, with increasing tie-line length, $s$. Phase system: dextran T500, PEG 8000, and 50 m$M$ $K_2SO_4$. Temperature, 20°. [Based on data from G. Johansson, *J. Chromatogr.* **150,** 63 (1978), used with permission.] (B) Variation of $K_{aff}$, for phosphofructokinase from yeast using Cibacron Blue F3G-A–PEG, with increasing $s$ values. Phase system: dextran T500, PEG 8000, and 50 m$M$ sodium phosphate buffer, pH 7.0. Temperature, 0°. [Based on data from G. Johansson, G. Kopperschläger, and P.-Å. Albertsson, *Eur. J. Biochem.* **131,** 589 (1983), and published with permission.]

also known.[26,27] The $\gamma$ values are affected by temperature, as well as the binding strength of hydrophobic and affinity groups to biomaterials. Often, low temperatures (0–4°) are used to maintain biological activities.

[26] P. A. Harris, G. Karlström, and F. Tjerneld, *Bioseparation* **2,** 237 (1991).

[27] P. A. Alred, F. Tjerneld, A. Kozlowski, and J. M. Harris, *Bioseparation* **2,** 363 (1992).

Log $K_{aff}$

| (Molecular weight of dextran) × $10^{-3}$ | 1000 | 3400 | 8000 | 35,000 |
|---|---|---|---|---|
| 2000 | 0.2 | 0.9 | 1.4 | 2.0 |
| 500 | 0.1 | 0.8 | 2.9 | 2.2 |
| 70 | Not determined | 2.7 | 3.1 | 3.9 |
| 40 | Not determined | Not determined | 3.5 | 4.0 |

Molecular weight of PEG

FIG. 4. Effect of molecular weights of PEG and dextran on $K_{aff}$ when phosphofructokinase (from yeast) is extracted with Cibacron Blue F3G-A–PEG 8000. Systems contained 25 m*M* sodium phosphate buffer, pH 7.0, and excess ligand–PEG. The systems had tie lines with comparable lengths. [Based on data from G. Johansson, G. Kopperschläger, and P.-Å. Albertsson, *Eur. J. Biochem.* **131,** 589 (1983), and published with permission.]

*Molecular Weights of Polymers.* Polymers of higher molecular weight give rise to phase systems at lower concentrations. In some cases, the molecular weights of the polymers influence partitioning. An example of such an effect in affinity partitioning is shown in Fig. 4.[28]

*Different Kinds of Polymers.* A large variety of polymer combinations useful in generating two-phase systems has been studied.[16,29] Important properties to consider when choosing polymers (other than those most commonly used, i.e., Dx and PEG) are the amount necessary to achieve phase separation, viscosity of the phases, solubility of biopolymers in their presence, price, etc.

### Determination of Partitioning

In principle, almost any analytical method can be used to determine the concentrations of partitioned material in the phases. However, great

[28] G. Johansson, G. Kopperschläger, and P.-Å. Albertsson, *Eur. J. Biochem.* **131,** 589 (1983).
[29] F. Tjerneld and G. Johansson, *Bioseparation* **1,** 255 (1990).

TABLE III
EFFECT OF PROCION YELLOW HE-3G POLY(ETHYLENE GLYCOL) 8000 (PrY–PEG) ON PROTEIN DETERMINATION[a]

| PrY–PEG (%) | Relative $A_{595}$ |
|---|---|
| 0 | 100 |
| 0.05 | 40 |
| 0.10 | 32 |
| 0.20 | 25 |
| 0.30 | 16 |

[a] According to Bradford.[31]

care must be taken to check for possible influences of the polymers on an assay. These effects include changes in absorption spectrum, light scattering, and inability or alteration in the formation of colored complexes. An example of the latter is the failure to detect aromatic amino acids with the ninhydrin reaction in the upper (PEG-rich) phase (G. Johansson, 1968, unpublished observation). Polymer-bound triazine dyes at higher concentrations[30] influence the reaction of proteins with Coomassie Brilliant Blue according to Bradford[31] (Table III).

Analytical problems can in many cases be circumvented by using calibration curves in which upper and lower phases are present. The excess in absorbance due to light scattering must be subtracted, either by performing a separate measurement of the diluted phase or by using such a solution as a blank. The light scattering increases strongly when the wavelength is decreased and influences absorbance measurements in the UV region. This point is illustrated for diluted phases of a Dx–PEG two-phase system in Fig. 5. The apparent absorbance due to light scattering is not linearly dependent on phase concentration. It is also affected by the spectrophotometer used, as it depends on the distance between cuvette and photomultiplier.

## Methods

### *Sampling and Determination of the K Value*

For many applications the determination of $K$ with precision is fundamental. *Equal* volumes of the two phases are collected and diluted with

[30] G. Johansson and M. Joelsson, *Appl. Biochem. Biotechnol.* **13,** 15 (1986).
[31] M. M. Bradford, *Anal. Biochem.* **72,** 248 (1976).

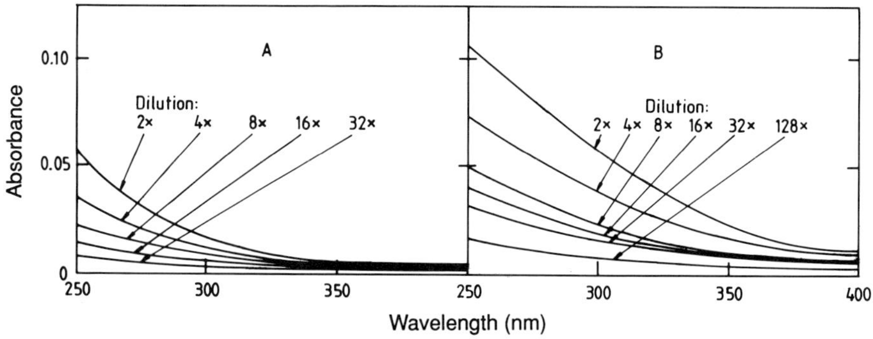


FIG. 5. Light scattering, measured as absorbance, of diluted top phase (A) and bottom phase (B) from a two-phase system as a function of wavelength. The system contained 7% dextran T500, 5% PEG 8000, and 25 m*M* sodium phosphate buffer at 25°. A Hitachi spectrophotometer (Model 100-60) was used.

known volumes of water, buffer, or assay solution. Because of the rather high viscosities of the phases, the method recommended for sampling is to collect a given volume of each phase using the same pipette (automatic pipette or constriction pipette). The phase volume is slowly drawn into the pipette, and the outside of the pipette is wiped dry. The contents are then emptied into the diluting liquid (of known volume). The phase usually sediments through the liquid and forms a distinct layer at the bottom. The pipette is rinsed by washing it several times with the dilution liquid to recover any portion of phase adhering to its inner surface. If traces of phase are washed out this can be seen, owing to differences in refractive index, as wormlike structures leaving the pipette. For PEG phases normally 2–3 washings are enough, whereas for Dx phases 5–10 washings may be necessary. In the latter case the washing should not be done too quickly because of the slow flow along the pipette wall. Plastic tips or glass pipettes in which the liquid tends to form drops (instead of an even surface film) should be cleaned or discarded.

A homogeneous solution is then obtained by careful mixing. In some cases this can be done directly in a measuring cuvette. Again, differences in refraction indicate inhomogeneous regions which show that mixing was not adequate. Sampling from the lower Dx-rich phase requires special care: It is possible to go through the upper phase with the tip of the pipette, but in doing so one should observe the process by holding the phase system up to the light. No upper phase should enter the pipette, which can occur as a result of hydrostatic pressure. This can be avoided by maintaining a positive air pressure in the tip while it is moved down through the upper phase. The interface behaves much like a membrane and has to be "punctured." Thereafter, the end of the pipette tip is moved slowly into the middle of the lower phase. Otherwise the tip may be surrounded by a liquid film of the upper phase which may form a channel around the pipette and, owing to its much lower viscosity, enter the pipette. By strictly following the indicated procedures, accurate determinations of the relative concentrations of partitioned material can be obtained. By calculating the ratio of the values for upper and lower phase, corrected for difference in dilution, the $K$ value is obtained.

*Preparation of Systems*

To get a general idea of how a particular, soluble compound, either pure or in a mixture (if a specific method for its analysis exists), partitions, its $K$ value is determined in a number of arbitrarily selected phase systems.[14] These are generally used for partitioning of proteins, but the systems can be used equally well for partitioning nucleic acids and other biological substances.

*Stock Solutions*

The two-phase systems are made up from the following stock solutions:

20% (w/w) Dx T500 (Pharmacia LKB, Uppsala, Sweden), determined by polarimeter[14]

40% (w/w) PEG 8000 (BP Chemicals, Hythe, UK)

0.5 *M* Sodium phosphate buffer, pH 6.0, 6.5, 7.0, 7.5, 8.0

0.5 *M* Tris-HCl buffer, pH 8.0, 8.5, 9.0

0.5 or 1 *M* Salt solutions (LiCl, NaCl, KCl, KSCN, $NaClO_4$, sodium acetate, $Na_2SO_4$, $Li_2SO_4$, tetrabutylammonium phosphate buffer, pH 6, 7, and 8)

Solutions of polymer derivatives [2% or 5% (w/w) PEG–palmitate (Aqueous Affinity, Arlöv, Sweden), DEAE-Dx (Pharmacia LKB), Dx sulfate (Pharmacia LKB), and PEG carrying an affinity ligand, e.g., Cibacron Blue F3G-A (ASA Spezialenzyme, Salzgitter, Germany), 40% (w/w) trimethylamino-PEG (TMA-PEG, Aqueous Affinity), and PEG sulfonate (Aqueous Affinity)].

*Estimation of $K_0$ Value*

Systems of 4 g final weights are prepared in 10-ml centrifuge tubes by weighing in 1.40 g of 20% Dx and 0.55 g of 40% PEG; then 0.40 ml of 1 *M* sodium acetate is added and, finally, water to give a total weight of 3.00 g. The solution is mixed and cooled in an ice bath (0°) or refrigerated water bath (2–4°). One-gram portions of similarly cooled samples of protein, 0.1–5 g/liter, in buffer, not more concentrated than 15 m$M$, are added to give the final 4 g systems containing 7% Dx T500, 5% PEG 8000, and 100 m$M$ sodium acetate. The amount of protein required depends on the sensitivity of the subsequent analytical method used. The systems are mixed by 10 inversions, and the phases are permitted to settle either by standing for 10–20 min or by low-speed centrifugation for 2–5 min, always maintaining the equilibrium temperature. If a pure protein is partitioned, its concentration may be determined via absorbance (280 nm) or by the Bradford method.[31] The $K$ value determined by sampling and analyzing the phases is close to the $K_0$ of the protein due to the presence of sodium acetate.[15]

### *Variation of K as Function of Salt Type at Given pH*

To study the effects of different salts on the partitioning behavior of a protein a number of systems are prepared as described above but using salts other than acetate. The salts can be chosen from the following series: tetrabutylammonium phosphate, sodium phosphate, $Li_2SO_4$, $Na_2SO_4$, NaCl, KSCN, and $NaClO_4$. The first salts in this series (phosphates) decrease $K$ values for positively charged proteins and increase $K$ values for negatively charged proteins.[32] The last salts in the series (thiocyanate and perchlorate) have the opposite effect on $K$. By using systems with different pH values the protein net charge can be varied, resulting in changes in $K$.

### *Effects of Polymer Concentration*

The effects of salts can be accentuated by increasing the polymer concentrations. The amounts of Dx and PEG stock solutions may be increased (see above) to 1.6 g + 0.55 g, 1.8 g + 0.6 g, or 2 g + 0.65 g. This yields systems with the percentages of Dx/PEG equal to 8/5.5, 9/6, and 10/6.5, respectively. The increase in polymer may reduce the space available for the protein sample. To use even higher polymer concentrations more concentrated stock solutions must be prepared or more concentrated protein solutions must be used. When less extreme partitioning is preferred, systems with a lower content of polymers (see Table I) are chosen.

[32] D. E. Brooks, K. A. Sharp, and D. Fisher, *in* "Partitioning in Aqueous Two-Phase Systems: Theory, Methods, Uses, and Applications to Biotechnology" (H. Walter, D. E. Brooks, and D. Fisher, eds.), p. 11. Academic Press, Orlando, Florida, 1985.

*Effects of Polymer Derivatives*

By using systems such as those discussed above but with low buffer concentrations (1–5 m*M*) and replacing part of the PEG (10–50%) with a charged PEG (TMA-PEG or PEG sulfonate), greater changes in the *K* value due to protein charge can be obtained.[3,19] Small quantities of charged Dx derivatives (constituting 0.01–0.5% of the system) can also be tried.[20] PEG with hydrophobic groups (e.g., PEG–palmitate) can be used at any salt concentration, and 1–10% of the PEG is usually replaced by such a derivative.[9] The same quantity of PEG derivative can be used when specific affinity ligands are bound to PEG, but in this case the kind of salt and its concentration may prove to be important.[28]

## Countercurrent Distribution

For fractionation and studies of complex mixtures of macromolecules, the two-phase systems can be used in a countercurrent distribution (CCD) process. The sample is included in one phase system, which is equilibrated by mixing, and the phases are allowed to settle. A number ($n$) of bottom phases of the same volume as in system 0 are prepared and numbered 1 to $n$. The upper phase is then transferred to bottom phase No. 1, and the same volume of fresh top phase is added to the original bottom phase No. 0. After mixing and settling, the upper phases (of systems 0 and 1) are moved to the next bottom phases (1 and 2, respectively) and fresh upper phase is added to bottom phase No. 0. The process is continued until the original top phase has reached bottom phase No. $n$. After mixing and settling of all the $n + 1$ (No. 0 to No. $n$) systems, the CCD is completed. The contents of partitioned substances are then analyzed along the train of phase systems.

The CCD process can be carried out by manual mixing and transferring, which is feasible for $n$ values up to 10. For CCD studies with a greater number of transfers (or number of experiments), special CCD apparatuses are available.[33] Their practical use has been detailed by Walter and Larsson.[34] The best separation of a mixture is obtained when, in the first partition step (system No. 0), approximately equal amounts of partitioned materials are present in the transferred part (the top phase) and in the stationary part (the bottom phase), that is, the $G$ value is equal to 1. This can be achieved by applying some of the above means to adjust the overall $K$ value of the mixture as well as by choosing the phase volume ratio according to Eq. (1).

[33] H.-E. Åkerlund and P.-Å. Albertsson, this volume [8].

[34] H. Walter and C. Larsson, this volume [4].

*Experimental Use of Countercurrent Distribution*

One of the thin-layer apparatuses[33,34] (Workshop, Chemical Center, University of Lund, Sweden) used for CCD studies with aqueous systems contains 60 chambers with a bottom (stationary) phase capacity of 0.9 ml. The two-phase system is added to all chambers except those which are to contain the samples. The system is well mixed and delivered in volumes such that the bottom phase is 90–95% of the capacity of the stationary part of each chamber (i.e., 0.81–0.85 ml).

For the sample cavities two-phase systems of the same composition are prepared (i.e., polymers, buffer, salts, other additives) which also include the sample (i.e., 3–30 mg protein per ml). When preparing the sample systems it is best to mix all components first except for the sample which is added last. Thus, contact of sample with high concentrations of PEG, which might cause protein precipitation, is avoided. Alternatively, the sample can be included in a bottom phase (i.e., by evaporating water from the phase and then adding sample to restore its original weight) which is added, together with an appropriate amount of upper phase, in the cavities for the sample systems.

The settings of time for mixing and settling may differ depending on the viscosity of the phases (a function of phase polymer concentrations), but typical times are 30 sec mixing and 10 min settling. After the CCD run has been completed the two-phase systems are converted to homogeneous solutions by addition of an equal volume (normally 1–2 ml) of water to each chamber. The CCD plates are shaken (1 min), and the contents from each chamber are collected either by simple withdrawal by pipette or by use of a collection ring with 60 plastic test tubes into which the mixtures are transferred in a single operation.[33] The chamber into which the sample was loaded (or the first chamber if sample was loaded into several chambers) is given the number zero. The solutions are analyzed for components or properties of interest, for example, proteins (absorbance at 280 nm or according to Bradford[31]), nucleic acids (absorbance at 260 nm), or specific reactions (e.g., enzyme activities). For these measurements enzyme-linked immunosorbent assay (ELISA) photometers are, because of the large number of samples involved, both useful and time-saving. The concentrations and/or biological activities obtained are plotted versus the numbers of the cavities (tube numbers), resulting in a CCD diagram.

## Discussion

The guidelines provided for partitioning can easily be expanded to include other polymers, other molecular weights,[30] and other temperatures. The technique can also be used to study other biomaterials (nucleic acids, polysaccharides, peptides, toxins, steroids, etc.). Careful studies on a small scale on the partitioning behavior of materials of interest are of the utmost importance as a basis for the design of large-scale extraction procedures and for optimizing extractions. They are also the initial step in adjusting the $K$ of a complex biological extract which is to be subjected to countercurrent distribution.[33]

Acknowledgments

This work has been supported by grants from the Swedish Natural Science Research Council.

# [4] Partitioning Procedures and Techniques: Cells, Organelles, and Membranes

*By* HARRY WALTER and CHRISTER LARSSON

## Introduction

### *Usefulness*

Partitioning in two-polymer aqueous phase systems is a highly sensitive method for the separation and fractionation of biological particulates (cells, organelles, membranes) on the basis of surface properties.[1,2] The sensitivity stems from the fact that parameters involved in determining the partition ratio of the particulate (see definition below) are *exponentially* related to it.[3] In most other separatory procedures the analogous relationship is *linear* (e.g., cell surface charge density and electrophoretic mobility).

Appropriate selection of polymers [dextran (Dx) and poly(ethylene glycol) (PEG) represent the most widely used combination for partitioning particulates] and of conditions suited to the biomaterial of interest (e.g.,

[1] H. Walter, *in* "Partitioning in Aqueous Two-Phase Systems: Theory, Methods, Uses, and Applications to Biotechnology" (H. Walter, D. E. Brooks, and D. Fisher, eds.), p. 327. Academic Press, Orlando, Florida, 1985.

[2] H. Walter, this volume [27].

[3] H. Walter, D. E. Brooks, and D. Fisher (eds.), "Partitioning in Aqueous Two-Phase Systems: Theory, Methods, Uses, and Applications to Biotechnology." Academic Press, Orlando, Florida, 1985.

tonicity, pH) yields aqueous two-phase systems that are not only mild but also actually often protective of the partitioned particulate. Thus, for example, erythrocytes can withstand hypotonic solutions in the presence of the polymers in which they would otherwise lyse immediately. Cells that have been subjected to partitioning [or multiple extraction procedures such as countercurrent distribution (CCD); see below and Ref. 4] in Dx–PEG systems are viable (i.e., can be cultured, retain metastatic potential, produce antibody, phagocytose), whereas organelles and membranes maintain their biological functions (e.g., oxidative and photosynthetic phosphorylation, carbon dioxide fixation by chloroplasts, transport capacities).

*Basic Phenomena*

Unlike the partitioning of soluble materials which distribute according to their solubilities between the two bulk phases, particulates partition, as their size increases, among the two bulk phases and the interface and, finally, between one bulk phase and the interface. The normal or resting position of particulates is the interface because interface formation requires energy, and the presence of particulates reduces its size. Thus particulates partition only if some interaction with the chemical or physical properties of the phases serves to pull them out of the interface and into a bulk phase.[1,2] In most of the cell partitions to be related here, partitioning takes place between the top phase and interface (Fig. 1).

The partitioning behavior of particulates can be expressed in terms of a partition ratio, which is the quantity of cells in the top phase divided by that at the interface plus bottom phase. Particulate partitioning can also be described in terms of a *P* value, defined as the quantity of particulate in the top phase, at time of sampling, as a percentage of the total particulate added.[1] Separation of one particulate from a mixture, or of two particulates from one another, can be effected in a single (or bulk) extraction step when their *P* values differ significantly. One material (high *P*) will be recovered in the top phase and the other (low *P*) at the interface or bottom phase (see also Ref. 5). When smaller differences in *P* values exist, separation may still be feasible, but one must resort to a multiple extraction procedure (e.g., CCD). Efficient separation of two populations of particulates is effected when the *P* value of one is below and that of the other is above 50% in a bulk (top) phase. Subfractionation by CCD of a single particulate population (e.g., cultured cells in different parts of their cell cycle) is best when about 50% of the material is in the top phase.

[4] H.-E. Åkerlund and P.-Å. Albertsson, this volume [8].

[5] C. Larsson, M. Sommarin, and S. Widell, this volume [44].

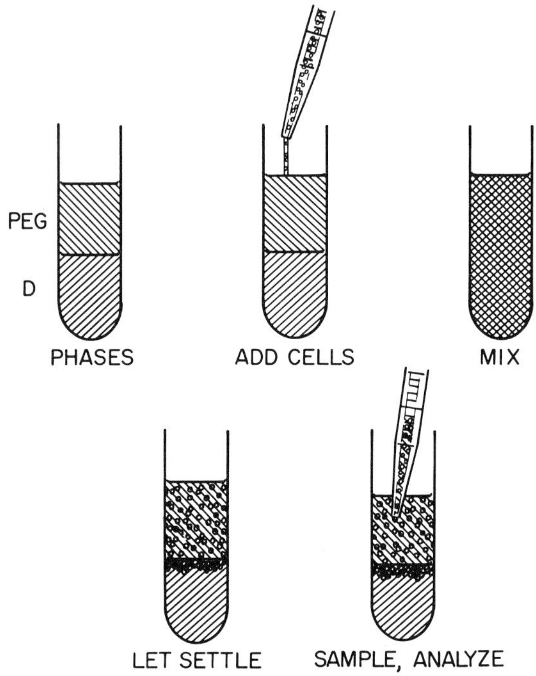


FIG. 1. Diagrammatic presentation of partitioning procedure with cells. PEG, PEG-rich phase; D, dextran-rich phase. A known quantity of cells is added to the phases, which are then mixed and permitted to settle by the clock. At the end of the settling time an aliquot is withdrawn from the top phase, and the cell quantity is determined. The partition is expressed as the quantity of cells in the top phase (percentage of total cells added). [From H. Walter, *in* "Red Cell Membrane: Structure and Function" (G. A. Jamieson and T. J. Greenwalt, eds.), p. 368. Lippincott, Philadelphia, Pennsylvania, 1969.]

Because the partitioning of particulates depends on the interaction of their surface properties with the physical properties of the phase system selected, use of an appropriate phase system can serve two ends: (1) to obtain a desirable $P$ value (see above) for the biomaterial(s) to be partitioned and (2) to yield information on the surface properties of a particulate (see below and Refs. 2 and 5).

The physical properties of the phase systems are determined by the polymer composition and concentration and by the ionic composition and concentration.[3,6,7] Dextran–PEG aqueous phase systems are comprised

[6] D. E. Brooks, K. A. Sharp, and D. Fisher, *in* "Partitioning in Aqueous Two-Phase Systems: Theory, Methods, Uses, and Applications to Biotechnology" (H. Walter, D. E. Brooks, and D. Fisher, eds.), p. 11. Academic Press, Orlando, Florida, 1985.

[7] D. E. Brooks and R. Norris-Jones, this volume [2].

of three types: charge-sensitive, non-charge-sensitive, and affinity systems [the latter containing (more or less) specific affinity ligands], and also mixtures thereof. Although charge-associated or non-charge-related differences in cell surface properties may predominate when charge- or non-charge-sensitive phase systems are used, respectively, the interaction of cells with the physical properties of the phases is complex, and the nature of the surface properties reflected by partitioning cannot be unequivocally stated. In the cases of organelles and membranes (see below) it is even more difficult to know the extent to which hydrophobic/hydrophilic properties and net surface charge density are determinant in the partitioning behavior of these particulates.

*Charge-Sensitive Phase Systems.*[3,6,7] Even though Dx and PEG are nonionic polymers, ions of some salts (e.g., $Li_2HPO_4$) have different affinities for the two phases. A Donnan potential between the two phases results (top phase positive for the salt mentioned).[7a] Other salts (e.g., NaCl) have ions that have essentially equal affinities for the two phases, and there is thus virtually no potential difference between them.[8] When erythrocytes (representative of cells since all are negatively charged) are partitioned in a series of Dx–PEG phase systems at some distance from the critical point[1] (i.e., having higher polymer concentrations), in which the ratio of phosphate/chloride is systematically reduced (while keeping the tonicity, pH, and temperature of the systems relatively constant), the highest $P$ value is found in the system containing phosphate and the lowest in NaCl (Fig. 2). This change correlates with the experimentally determined diminution of Donnan potential between the phases as the indicated salt ratio is changed.[1,6,7]

Figure 3A shows a correlation for red blood cells from different species between their partition ratios in a charge-sensitive phase system and their

[7a] The electrostatic potential difference measured between the phases of Dx–PEG systems containing alkali monohydrogen phosphate (or sulfate), alkali halides, and their mixtures correlate well with the observed partitioning behavior of cells.[1,6] The partitioning behavior of some other materials, however, for example, that of a positively charged polymer (DEAE-Dx) in phases containing NaCl,[1] of proteins in phases containing $K_2SO_4$ (V. P. Shanbhag, this volume [23]), and the phenomenon of cross-partitioning (H. Walter and D. Forciniti, this volume [20]), differs from that expected from electrode measurements of the potential difference between phases. The latter materials behave as if Dx–PEG systems containing alkali halides have a potential difference (bottom phase positive) and systems containing $K_2SO_4$ have little or no potential difference. (See Ref. 3 for a detailed discussion of these points.) The partitioning of organelles and membranes is complex and can respond to both alkali phosphate buffers and chlorides as if the phases were charge-sensitive (in opposite directions). In addition, the hydrophobic/hydrophilic properties may outweigh those related to charge, giving rise to partitioning which is, as mentioned later, largely empirical.

[8] H. Walter, E. J. Krob, and D. E. Brooks, *Biochemistry* **15,** 2959 (1976).

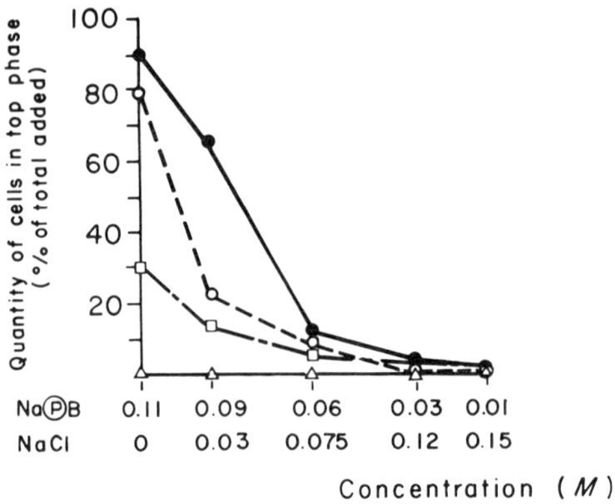


FIG. 2. Quantity of erythrocytes from four different species found, at room temperature, in the top phase of a system containing 5% (w/w) dextran T500 and 4% (w/w) PEG 8000 at different concentrations of NaCl and sodium phosphate, pH 6.8. ●, Dog; ○, human; □, sheep; △, rabbit. [From P.-Å. Albertsson and G. D. Baird, *Exp. Cell Res.* **28,** 296 (1962), with permission.]

relative electrophoretic mobilities. The exceptions to the correlation indicate that the surface charge measured by electrophoresis and partitioning can differ and that surface properties other than charge can also affect the partition ratio obtained (see Ref. 1 for detailed discussion).

*Non-Charge-Sensitive Phase Systems.* When the polymer concentrations are reduced (i.e., in systems closer to the critical point), the interfacial tension diminishes and the partition ratios increase. Even in phases containing NaCl measurable partition ratios often result. Because phases containing NaCl as the predominant salt have almost no potential difference between the phases,[7a] it would appear that the partitioning is due to parameters other than surface charge.[8] In the case of erythrocytes from different species, this is evidenced by correlations between their partition ratios in such a non-charge-sensitive phase system and the cell membrane ratio of polyunsaturated/monounsaturated fatty acids (Fig. 3B) as well as other membrane lipid components.[8]

*Phase Systems Containing Affinity Ligands.* Dextran–PEG phase systems can be made into affinity systems by attaching to one of the phase-forming polymers (usually to PEG) a ligand so selected that it will bind to a receptor or region on the particulate of interest. The ligand (if not too large or with a number of polymer molecules attached) will partition into the phase in which the polymer to which it is bound predominates. In a phase system containing, for example, NaCl and the lowest polymer concentrations at which virtually all of the particulates are at the interface, addition of a PEG–ligand to the phase system or its substitution for unreacted PEG can result in the extraction of the desired material into the top, PEG-rich, phase.[9–13]

*Mixed Phase Systems.* Clearly Dx–PEG systems that combine physical properties of the above-delineated types can also be used to advantage. Systems with polymer concentrations such as those used in the non-charge-sensitive phases described above into which phosphate, for example, is incorporated will have both charge-sensitive and non-charge-sensitive characteristics. Charge-sensitive systems into which a polymer–ligand is incorporated may also reflect those properties.

## Methods

### *Preparation of Particulate Material for Partitioning*

*Cells.* As indicated above the partitioning behavior of cells depends on their surface properties. If one is interested primarily in effecting a cell separation, and not in the original surface properties of the cells, the preparation of single cells by any means (enzymatic, mechanical) that does not alter the biological property of interest can be used. The separation obtained will be empirical and may reflect possible differential effects on different cell populations and subpopulations by the preparatory treatment employed. Thus, surface and partitioning differences among cell populations may, as a consequence of the treatment, be diminished, enhanced, or left intact.

When the interest centers on studying surface properties of cells (or their alterations as a function of normal or abnormal *in vivo* processes or *in vitro* treatments[1,2]), care must be taken not to alter these in the course of cell preparation for partitioning. In the cases of cells, for example, from blood, bone marrow, spleen, adenoids, and intestinal epithelium, as well as those that can be grown in suspension culture, no particular problem in single cell preparation exists. Bacterial, yeast, and many algal cells can also be readily prepared. In the cases of other, primarily tissue (e.g., liver) cells or cells cultured in monolayers, no totally satisfactory method for maintaining original surface properties during preparation is available.

*Organelles and Membranes.* For organelles and membranes, standard methods are used to obtain crude pellets of chloroplasts, mitochondria,

[9] G. Birkenmeier, H. Walter, and K. E. Widen, this volume [34].
[10] L. J. Karr, D. L. Donnelly, A. Kozlowski, and J. M. Harris, this volume [35].
[11] D. E. Brooks and S. J. Stocks, this volume [36].
[12] C. Delgado, G. E. Francis, and D. Fisher, this volume [37].
[13] G. Johansson, this volume [48].

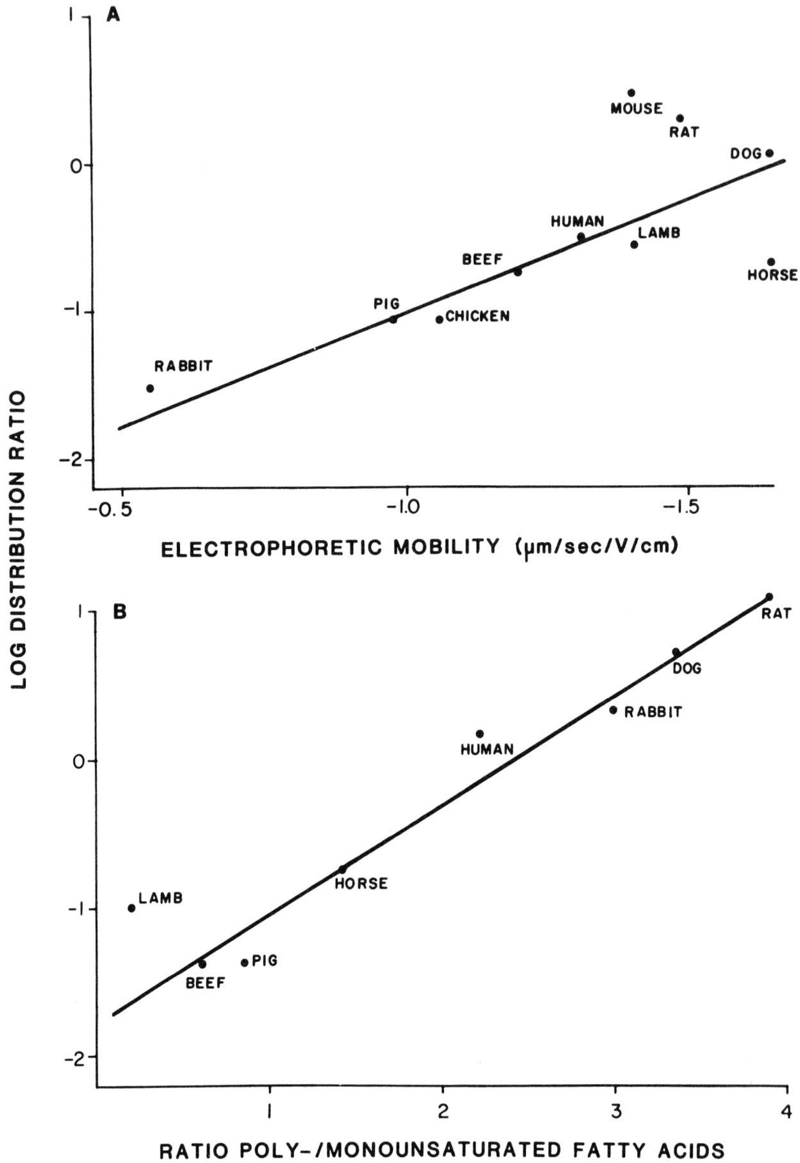


FIG. 3. (A) Logarithmic distribution ratio (i.e., the ratio of the number of cells in the top phase divided by the number of cells at the interface plus bottom phase) in a charge-sensitive phase system [containing 5% (w/w) dextran T500, 4% (w/w) PEG 8000, 90 m*M* sodium phosphate, pH 6.8, and 30 m*M* NaCl], at 21–24°, plotted against the relative electrophoretic mobility of erythrocytes from different species. [Redrawn with permission from H. Walter, F. W. Selby, and R. Garza, *Biochim. Biophys. Acta* **136,** 148 (1967).] (B) Logarith-

microsomal fractions, etc. These procedures generally involve homogenization of tissue (or cultured cells) in a suitable medium (with regard to tonicity, pH, and protective agents) and relatively rapid subfractionation of the homogenate using differential centrifugation. Mechanical methods for homogenization are preferred so as to minimize the effect on the surface properties of membranes that are to be separated by phase partitioning. Thus, use of enzyme mixtures to digest cell walls or soften tissue may introduce artifacts and give results that are difficult to reproduce.

### *Preparation of Phase Systems*

Preparation and analysis of phase systems is detailed by Brooks and Norris-Jones in this volume.[7] In short, systems are compounded by weight from stock solutions which are severalfold the final concentrations desired. For example, systems described below (and in Fig. 1) for cells are prepared from stocks of 20% (w/w) Dx T500 (Pharmacia-LKB, Piscataway, NJ), with the concentration determined by polarimetry; 40% (w/w) PEG 8000 (Union Carbide, New York, NY); 0.44 *M* sodium phosphate, pH 6.8; and 0.60 *M* NaCl. Thus, if one wishes to prepare 10 g of a phase system containing 5% Dx, 4% PEG, 90 m*M* sodium phosphate, pH 6.8, and 30 m*M* NaCl, the following quantities of stock and water are weighed into a tube:

| | |
|---|---|
| | 2.50 g of 20% (w/w) Dx stock solution |
| | 1.00 g of 40% (w/w) PEG stock solution |
| | 2.05 g of 0.44 *M* Sodium phosphate, pH 6.8 |
| | 0.50 g of 0.60 *M* NaCl solution |
| | 3.95 g of Water |
| Total: | 10.00 g of phase system of above-indicated composition |

The tube is capped, and the contents are mixed thoroughly by repeated inversion. The phases are then permitted to reach the temperature at which partitioning is to be carried out [in most experiments either room (21–24°) or cold-room (4–5°) temperature], are mixed again, and are then allowed to settle.

mic distribution ratio in a non-charge-sensitive phase system nearer the critical point, that is, with lower polymer concentration [containing 5% (w/w) dextran T500, 3.5% (w/w) PEG 8000, 0.15 *M* NaCl, and 10 m*M* sodium phosphate buffer, pH 6.8], at 21–24°, plotted against the ratio of membrane polyunsaturated to monounsaturated fatty acids of erythrocytes from different species. [Redrawn with permission from H. Walter, E. J. Krob, and D. E. Brooks, *Biochemistry* **15,** 2959 (1976). Copyright 1976 American Chemical Society.]

*Single-Tube Partitioning Procedure and Determination of P Value*

*Cells.* The partitioning procedure with cells is depicted, in schematic form, in Fig. 1. In most cases described here the partitioning of cells takes place between top phase and interface. We have found it convenient to partition cells in 10 g of phase system (selected as described in the next section). The phase system is permitted to come to the temperature at which partitioning is to be carried out. It is then well mixed and poured to the top mark in a 10-ml calibrated tube. The tubes are centrifuged to speed phase settling, the top and bottom phase volumes are adjusted to be equal, and the top phase volume is recorded.

When working with cells that can be pipetted even when packed (e.g., erythrocytes), 0.1 ml of washed, packed cells are pipetted into the partition tube and an equal volume into a known volume of suspending medium (e.g., saline, water) in a second tube. (The capacity of the phases for cells is great, and the *P* value obtained is essentially the same over the range of 0.01 to 0.5 ml packed cells.) The partition tube is capped, then inverted at least 10 times to mix the contents, and the phases are permitted to settle by the clock until a sharp, clear interface between the bulk phases appears. For phases somewhat removed from the critical point [e.g., 5% (w/w) Dx T500, 4% (w/w) PEG 8000], a 20-min settling time is suitable; at higher polymer concentrations 15 min is adequate. For phases closer to the critical point, where phase separation takes a long time, partition tubes can be capped and the phases permitted to settle with the tubes in horizontal position [e.g., 7 min horizontal plus 1 min vertical for a system composed of 5% (w/w) Dx T500, 3.5% (w/w) PEG 8000]. Reducing the phase column height speeds phase separation (although somewhat reducing efficiency of cell separation[14]).

At the end of the settling time a known aliquot of cell suspension (e.g., 1 ml) is removed from the middle of the top phase layer (for greater reproducibility) and analyzed for cell number or some other parameter related to the number of cells present (e.g., hemoglobin absorbance in the case of erythrocytes, turbidity, etc.). The quantity of total cells added to the partition tube is obtained by analysis of the second tube (see above). The *P* value is calculated as the quantity of cells present in the top phase as a percentage of total cells added.

Cells that cannot readily be pipetted when packed can, instead, be incorporated directly into the phase system. They can be suspended in a salt solution with a composition identical to that used in the phase system. Such a cell suspension is then weighed directly into the phase system

[14] H. Walter, E. J. Krob, and L. Wollenberger, *J. Chromatogr.* **542,** 397 (1991).

when the latter is prepared and the quantity of salt stock solution and water adjusted in a manner to yield the desired final composition (see outline of preparation above and in Ref. 7). Alternatively, the phase system may be prepared, equilibrated in a separatory funnel, and top and bottom phases separated. Washed, packed cells can then be suspended (e.g., by use of a Pasteur pipette) directly in an aliquot of top phase. A known volume of the latter can be added to an equal volume of bottom phase and the mixture used for cell partitioning.

If one wishes to conserve polymers, polymer–ligands (in the case of affinity partitioning), and/or cells, the partitioning procedure can be scaled down. Thus phase systems of 2, 3, and 4 ml have been used. In these cases it may be best to prepare the phases, equilibrate them, separate top and bottom, and pipette equal top and bottom phases into a tube. Once the cells have been added (or incorporated as indicated above) the tubes can be mixed and a small aliquot of the mixture removed that serves to determine the "total" quantity of cells added. Partitioning is then carried out as usual.

*Organelles and Membranes*. As with cells, the partitioning of organelles and membranes takes place mainly between top phase and interface (see Fig. 1), but material may also partition to the bottom phase. For convenience, the interface and the bottom phase are usually collected together, and the partitioning of material is calculated as being between the top phase and the rest of the phase system. In initial experiments, relatively small phase systems (e.g., 4 g) may be used to save material. Many biological activities are better preserved at low temperature, so experiments should be performed in a cold room or in an ice bucket. Once a temperature is chosen (e.g., 4°) it should be maintained throughout the experiment since partitioning varies with temperature.[7,15]

The material is usually added as a relatively dense suspension in a medium as used in the phase system. Thus, for a phase system containing 0.3 *M* sucrose, 5 m*M* potassium phosphate, pH 7.8, 3 m*M* KCl, 0.1 m*M* EDTA, and 1 m*M* dithiothreitol (DTT), a solution is made of identical composition but without polymers. To prepare the 4 g phase system one calculates the quantity of polymer stock solutions needed on the basis of 4 g (see above) but all other components (i.e., salt, buffer, etc.) on the basis of 3.5 g, to which weight the system is finally brought by addition of water. When 0.5 g of membrane suspension in the indicated medium is added to this mixture, the desired two-phase system composition is obtained. In practice, for convenience, 0.5 ml of the membrane suspension is pipetted rather than weighing 0.5 g. The tube is then capped (Parafilm)

[15] P.-Å. Albertsson and F. Tjerneld, this volume [1].

and its contents mixed by 20 inversions of the tube. Phase settling may be facilitated by low-speed centrifugation (1500 *g* for 3–5 min), or the tubes may just be left standing for 10–20 min. For large and heavy organelles that sediment rather rapidly (such as intact chloroplasts), it may be wise to compare results obtained at unit gravity with those obtained after centrifugation.

Once a distinct interface has formed, the top phase is carefully removed with a Pasteur pipette and added to a new tube. Usually it is only possible to remove about 90% of the top phase without disturbing the material at the interface, and, therefore, up to 10% of the top phase material will be included with that in the bottom phase plus interface. This is of little practical importance but can, if one wishes, be corrected by calculation. The top phase and the rest of the phase system are now diluted with a suitable medium to the same final volume (or, more conveniently, to the same final height in identical tubes); samples are withdrawn, and markers are assayed to determine the partitioning of different organelles and membrane vesicles. If the material is too dilute to obtain reliable readings, or if some component in the phase system interferes with an assay, the material may be pelleted and resuspended in a small volume. In the latter case the bottom phase must be diluted about 10-fold before pelleting due to its high viscosity. (Note that the polymers usually do not interfere with enzyme assays.) The partitioning of a particular marker is calculated as the percentage in the top phase of the total amount recovered after phase separation.[5]

### *Selection of Phase Systems*

*Special Requirements.* Some particulates have special requirements for the suspension medium such as tonicity, pH, ionic composition, protective agents, and temperature. Usually conditions deemed most important for the material of interest can be incorporated into or employed with the phase system to be used.

*Manipulation of Phase System Variables to Obtain Desired P Value.* To find a system suitable for the partitioning of the material of interest, the material is first partitioned in a series of arbitrarily selected phases that differ in polymer concentration and ionic composition and concentration. A series of simple rules then permits manipulation of the phase system components in a manner that increases or decreases the *P* value as desired so that a "usable" phase system (as defined earlier) results.

CELLS. A good place to start, for cells, is with four different systems having the following compositions: 5% (w/w) Dx T500, 4% (w/w) PEG 8000, and 0.11 *M* sodium phosphate, pH 6.8 (system 1); the same polymer concentrations as system 1 and 0.15 *M* NaCl plus 10 m*M* sodium phosphate, pH 6.8 (system 2); 5% Dx, 3.5% PEG, and 0.11 *M* sodium phosphate, pH 6.8 (system 3); and the same polymer concentrations as system 3 and 0.15 *M* NaCl plus 10 m*M* sodium phosphate, pH 6.8. These phases are essentially isotonic (for erythrocytes), are in a pH range reasonable for mammalian cells, and differ in interfacial tension (different polymer concentrations) and electrostatic potential difference between the phases (see above). If usable *P* values are not obtained in any of these systems, the magnitude of the *P* value can be adjusted by following rules set out below.

Lower polymer concentrations give higher *P* values; increasing the phosphate/chloride ratio will increase the *P* value of negatively charged biomaterials (e.g., cells); cations and anions decrease or increase the *P* value of negatively charged particulates in a sequence similar to the Hofmeister series (thus the *P* values of particulates decrease with lithium > sodium > potassium and with sulfate > citrate > chloride); lowering the phosphate concentration (which also reduces interfacial tension[6]), and, if necessary, maintaining tonicity by incorporation of sucrose, increases the *P* value; increasing pH with phosphate buffer increases the *P* value; lowering the molecular weight of one of the phase-forming polymers (e.g., substituting Dx T40 for Dx T500 or PEG 3350 for PEG 8000) causes cells to partition more into the phase rich in that polymer; loading cells initially into bottom phase yields lower *P* values than loading into top phase[16] and permitting phases to settle during partitioning in horizontal rather than vertical position (permitting settling to a time where a clear bulk interface first appears) yields higher *P* values. If a usable *P* value cannot be obtained by the indicated manipulations, a charged or hydrophobic PEG–ligand can be incorporated into the system. Small quantities [e.g., <0.001% (w/w)] of PEG–fatty acid esters (primarily PEG–palmitate) have proved highly effective in extracting cells based on their hydrophobic properties,[1] whereas trimethylamino-PEG, a positively charged polymer, can be used in the charge-based extraction of particulates.

The phase system selection process just outlined makes use of parameters irrespective of whether the resultant phase system is charge- or non-charge-sensitive. If one wishes to restrict partitioning to a charge- *or* non-charge-sensitive phase system, the parameters manipulated are restricted to those that maintain the desired physical properties. Thus, for charge-sensitive phases sodium phosphate (or another salt containing ions differing in their affinity for the two phases) must be the predominant salt, and the *P* value can be adjusted by changing polymer concentrations (with the lower limit being the concentrations at which virtually all cells are at

[16] H. Walter, T. J. Webber, and E. J. Krob, *Biochim. Biophys. Acta* **1105**, 221 (1992).

the interface in a non-charge-sensitive system), salt concentration, pH with phosphate buffer, molecular weight of phase-forming polymers, horizontal or vertical settling position, cell loading, or incorporation of a charged polymer (all as described above).

For non-charge-sensitive phases a salt with ions that have the same relative affinity for the two phases must be predominant. The $P$ value can be adjusted by changing polymer concentrations, altering the molecular weight of a phase-forming polymer, permitting cells to partition in the horizontal or vertical position, and loading cells in the top or bottom phase.

Phase systems for affinity partitioning, usually of the non-charge-sensitive type, are first selected to have the lowest polymer concentrations at which virtually but not quite all cells (e.g., about 95%) are at the interface. Under such conditions cells are only weakly held at the interface and can be extracted by use of an appropriate ligand. A polymer–ligand (e.g., PEG–ligand) is introduced into the phase system either by supplementing the PEG already in the phase or, in part, substituting for it. When the PEG–ligand is effective in extracting a particulate of interest at a concentration so low [e.g., PEG–palmitate is often effective at 0.001% (w/w)] that the original polymer concentration is virtually unaffected by its addition, supplementation is used; if larger quantities of the ligand are necessary to effect extraction (e.g., PEG–metal chelate[9]), substitution must be employed. [*Note:* When substituting PEG–ligand for PEG it is important that the PEG with the ligand has essentially the same molecular weight as the PEG initially used in the phase system; otherwise, the physical properties of the phases will be altered with the possible result that either the particulate will not be extracted (with higher molecular weight) or the extraction obtained will not depend solely on the ligand (with lower molecular weight).[9]] Although the capacity of the phase systems for large quantities of particulate is high (see section on partitioning procedures above), in affinity partitioning cell quantity is clearly one of the variables. Thus in a phase system selected as indicated above the cell quantity should, initially, be small and constant while the quantity of polymer–ligand is systematically increased to test its effectiveness in extracting cells of interest. Increased cell quantities generally require a corresponding increase in ligand concentration to effect adequate extraction.

ORGANELLES AND MEMBRANES. With organelles and membranes the usual goal is to separate one particular kind from all the others in a complex mixture rather than to compare surface properties. Relatively rapid preparative procedures are desired so as to better preserve biological activities. The best scenario is when the material of interest partitions into either the top phase or to the interface plus bottom phase and the rest of the material to the opposite phase. This is thus the condition for which to strive.

Phase systems containing 5.5–6.5% (w/w) Dx T500 and the same concentration of PEG 3350 (Union Carbide) are commonly used. These are prepared in a medium suitable for the material of interest, for example, 0.3 *M* sorbitol, 5 m*M* potassium phosphate, pH 7.8, for intact chloroplasts.

Close to the critical point of the above-indicated phase system (below 5.5% of each polymer at 4°), all membrane particles partition almost completely to the top phase. As the polymer concentrations are increased, material starts to collect at the interface. This response to increasing polymer concentrations is most pronounced for mitochondrial membranes, endoplasmic reticulum, and the chloroplast envelope (intact chloroplasts), whereas chloroplast thylakoids, Golgi vesicles, and particularly plasma membranes (right-side-out vesicles) are less sensitive.[5] Interestingly, the behavior is similar irrespective of the source of material (plant, fungal, animal). These differences in response to increasing polymer concentrations may, at least in part, reflect differences in hydrophobic/hydrophilic properties of the membrane surfaces.[17] Varying the polymer concentrations in a series of phase systems is, thus, one way to find optimal conditions for separation.

When membrane particles are partitioned in a phase system without salt present, they usually collect at the interface. On addition of increasing amounts of a suitable salt (e.g., potassium phosphate, pH 7.8), the material will partition more and more to the top phase (Fig. 4, left). The marked salt effect on partitioning is due to the electrostatic potential difference between the bulk phases.[7] This Donnan potential, as mentioned earlier, is a consequence of the different affinities of different ions for the two bulk phases. For example, $HPO_4^{2-}$ has a stronger affinity for the bottom phase than $K^+$. Therefore, with $K_2HPO_4$ a "positively charged" top phase is obtained. Because all membranes have a net negative charge at neutral pH, they will partition to the top phase in the presence of $K_2HPO_4$. The interfacial potential created by one salt can be balanced by addition of another salt. This is shown in Fig. 4 (middle), where addition of KCl reduces the potential created by the phosphate and makes the material collect at the interface. By addition of more $K_2HPO_4$, the positive potential is restored (Fig. 4, right). Examples of potassium salt ions which give a positive potential are sulfate and citrate, whereas acetate and dihydrogen

[17] P.-Å. Albertsson, "Partition of Cell Particles and Macromolecules." Wiley(Interscience), New York, 1986.

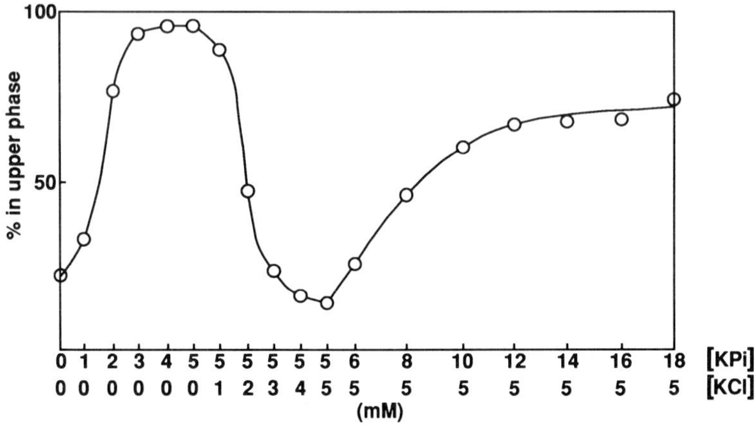


FIG. 4. Effects of potassium phosphate (KPi), pH 7.8, and KCl on the partitioning of spinach chloroplast thylakoids in a polymer two-phase system composed of 5.7% (w/w) dextran T500, 5.7% (w/w) PEG 3350, and 0.3 *M* sorbitol. Temperature, 4°. The percentage of material recovered in the upper phase is plotted versus salt composition. [From C. Larsson, *in* "Isolation of Membranes and Organelles from Plant Cells" (J. L. Hall and A. L. Moore, eds.), p. 277. Academic Press, London, 1983.]

phosphate, like chloride, reduce the potential.[18] Note that $KH_2PO_4$ has an effect opposite to that of $K_2HPO_4$. This means that the partitioning in a phase system containing phosphate buffer is strongly affected by the ratio of $HPO_4^{2-}$ to $H_2PO_4^-$ (i.e., the pH).

The experiment in Fig. 4 shows how partitioning of a membrane can be controlled by manipulating the salt composition of a phase system. A suitable salt composition is usually found by starting with a phase system containing 5 m*M* potassium phosphate buffer, pH 7.8, as the only salts and then making a series of phase systems with increasing KCl concentrations (1–5 m*M*, see Fig. 4, middle). With increasing $Cl^-$ concentration, material starts to collect at the interface (Fig. 4, middle). Because different membrane surfaces are differently charged, they will respond differently to the interfacial potential. The effect of increasing $Cl^-$ concentration is similar to that of increasing polymer concentration (see above); that is, mitochondrial membranes and endoplasmic reticulum are strongly affected and plasma membranes (right-side-out vesicles) are least affected, and remain in the top phase even at 5 m*M* $Cl^-$ (also see Larsson *et al.*[5]).

[18] E. Eriksson and G. Johansson, *in* "Cell Populations, Methodological Surveys (B): Biochemistry" (E. Reid, ed.), Vol. 8, p. 81. Ellis-Horwood, Chichester, 1979.

The partitioning of each particular membrane is a function of its hydrophobic/hydrophilic surface properties and net surface charge density, and it is usually difficult to know which factor predominates. Furthermore, any change in salt composition of the phase system may affect not only the electrostatic potential difference but also the hydrophobic/hydrophilic properties of the phases. Analogously, any change in polymer concentration changes not only the hydrophobic/hydrophilic properties of the phases but also the interfacial potential since the ion distribution between the phases is also affected. Thus, phase partitioning is empirical, and we generally try to optimize separations with respect to both polymer and salt composition. By making series of phase systems with increasing polymer concentrations and changing salt composition (e.g., increasing $Cl^-$ concentration), suitable conditions for separation are usually found.

*Countercurrent Distribution*

*Introduction.* When the differences in $P$ values are inadequate to resolve particulate populations by a single extraction step, one must resort to a multiple extraction procedure.[4,19] For CCD studies the most widely used equipment with aqueous two-phase systems is the apparatus designed by Albertsson.[4] The advantage of this unit is that it is a thin-layer apparatus and that phase settling thus proceeds more rapidly than in units utilizing higher phase columns (as in the Craig CCD apparatus). The thin-layer CCD apparatus (Workshop, Chemical Center, University of Lund, Sweden) consists of two circular Plexiglas plates, one stator and one rotor, having 120 cavities with a bottom phase capacity of 0.7 ml. Depending on requirements one can carry out 120 transfers with one sample or fewer transfers with a commensurately larger number of samples which one wishes to compare. In many of our experiments, 30 transfers were carried out permitting examination of four different samples without overlap. Procedurally, preliminary phase system selection for CCD studies of particulates is made by obtaining a usable $P$ value in single-tube partitioning as described above. An adequate quantity of phase system is then prepared (usually 225 to 275 g), permitted to come to the temperature at which the CCD study is to be carried out, placed into a separatory funnel, and mixed thoroughly. The phase system is then allowed to settle (usually overnight), and top and bottom phases are separated.

*Procedure.* Because particulates partition between a bulk (usually top) phase and the interface, the loading of the CCD unit must be in a manner such that the top phase is separated from the interface plus

[19] W. Müller, this volume [9].

bottom phase at each extraction step. This is done by loading less bottom phase than the capacity of each bottom chamber to leave space for the interface to be retained in the latter. For example, 0.25 ml of packed cells are suspended in 2.75 ml of top phase ("load mix"). All cavities receive 0.5 ml of bottom phase by use of a repeat pipettor. Cavities 0–2 receive 0.9 ml of one load mix, cavities 30–32 receive 0.9 ml of the second load mix, cavities 60–62 receive 0.9 ml of the third load mix, and cavities 90–92 the fourth load mix. All other cavities receive 0.9 ml of top phase. [Alternatively, load mixes can, as in the case of single-tube partitioning, be prepared either by suspending material, which cannot be pipetted, directly in top phase or by incorporating it in a small quantity of phase system, the top and bottom phase volumes of which are then adjusted to be the same as the top/bottom phase volume ratio (e.g., 0.9/0.5) loaded onto the rest of the CCD unit, and loaded as a mixture into one or a few adjacent cavities as indicated above.]

The CCD experiment consists of repeated cycles of *shaking* (to mix the phases and, in the load cavities, the phases and material to be partitioned), *settling* (a time period during which phases are permitted to separate, carrying material that prefers the top phase to the top and material that prefers the interface to the latter), and a *transfer* (in which the top plate moves clockwise 3°, thus separating material in the top phases from material at the interfaces by moving every top phase to be over the bottom phase of the next adjacent cavity). The CCD study is carried out at a selected temperature (e.g., 4–5° or 21–24°) using a settling time of 5–8 min (with the longer settling times needed for systems closer to the critical point) and a shaking time of 15–30 sec (with the longer shaking times needed for systems with higher polymer concentrations).

As a rule of thumb one can load material into up to one-tenth the number of cavities that one does transfers (e.g., one can load 12 cavities for a 120-transfer run) with only minor loss of resolution. In the case of cells, not more than 0.1 ml of packed cells should be loaded per load cavity (i.e., about $10^9$ erythrocytes, $10^8$ lymphocytes, etc.). Unlike the behavior of cells in single-tube partitioning where the same $P$ value is obtained over a range of cell quantities (see above), smaller quantities of cells (e.g., $<10^8$ erythrocytes) give, on CCD, lower apparent partition ratios or $G$ values [$=r_{max}/(n - r_{max})$, where $r_{max}$ is the cavity number of the peak of a CCD curve and $n$ is the number of transfers] than larger quantities.[1] This may be due to adsorption of such small quantities of cells to the plates and/or adsorption of small quantities of some contaminant present in the phases on cell surfaces. It has been found that incorporation of protein [e.g., 2% (w/w) fetal bovine serum] into the phases resolves this problem and gives rise to $G$ values for small cell quantities that are indistinguishable from those obtained with larger amounts.

Results from an initial CCD experiment are used to adjust the composition of the phase system, by manipulating phase system parameters as outlined above, to yield desirable $G$ values. In addition, the $G$ value can be adjusted by changing the top/bottom phase volume ratio loaded on the CCD apparatus (e.g., to 0.8 ml top/0.6 ml bottom), thereby altering the percentage of cells transferred. One of the reasons why an adjustment is necessary is that the phase system selected, based on the $P$ values obtained on single-tube partitioning, tends to give higher $G$ values on CCD because the difference in geometry of the phase columns in the partition tube and in the thin-layer cavity affects phase settling.[14]

For tissue culture and other experiments that require sterile conditions, the unseparated two-phase systems [at least up to polymer concentrations corresponding to 5% (w/w) Dx and 4% (w/w) PEG] can be filtered, while stirring periodically to prevent phase settling, through a 0.45-$\mu$m Millipore filter (Millipore, Bedford, MA). Phase systems can also be autoclaved, but care must be taken to replace any water lost from the system during this process. The Plexiglas CCD plates can be wiped with alcohol (but should *not* be left standing in it). Alternatively, the plates can be rendered sterile by gassing with ethylene oxide. We have found that use of such procedures, together with use of sterile glassware and CCD collection tubes, generally suffices in giving cells that are free of contamination (as evidenced by prolonged cell culture).

Because of the length of the described CCD procedure, up to 9 hr for 60 transfers, losses of biological activities associated with organelles and membranes may be so large as to be unacceptable. A centrifugal CCD apparatus has therefore been developed that reduces the time for a run considerably (2 hr for 60 transfers).[4] Phase settling, which is the most time-consuming step in a separation cycle, is accelerated by spinning the entire unit at 3000 $g$ to separate the phases rapidly.

*Analysis of Particulate Distribution Curves and Presentation of Data.* Upon completion of a CCD run, the contents of each cavity of the apparatus are transferred separately but simultaneously directly into centrifuge tubes by use of a fraction collector placed over the plates, which are then inverted. A few tenths milliliter (e.g., 0.7 ml) of a suitable solution is added to each tube to convert, by dilution, the two-phase system into a single homogeneous suspending medium. [For better recovery of the partioned material it can be of advantage to dilute the phases directly in each chamber prior to collection. By doing this one sacrifices, however, the possibility of visual evaluation of whether fractionation of cells under a distribution curve has taken place (see next section).] Depending on the

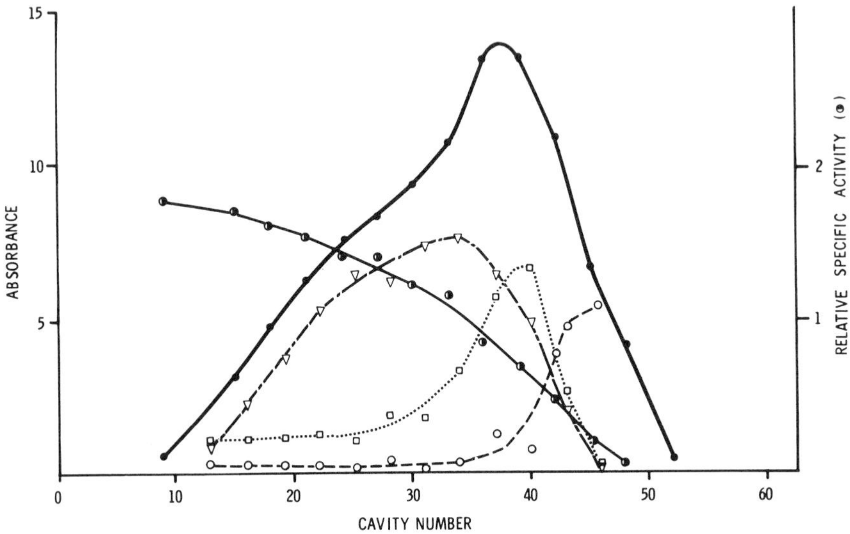


FIG. 5. Countercurrent distribution of a red blood cell population (from a phenylhydrazine-injected rat) containing 83% reticulocytes (●) in a phase system containing 5% (w/w) dextran T500, 4% (w/w) PEG 8000, 90 m*M* sodium phosphate, pH 6.8, and 30 m*M* NaCl. Sixty transfers were carried out at 4–5°. Distributions (obtained by microscopic counts) of heavily reticulated reticulocytes (▽), lightly reticulated reticulocytes (□), and erythrocytes (○) are indicated. The cells were also incubated with [$^{14}$C]valine. High relative specific activities (◑) show that the location of cells most active in protein synthesis is to the left. [From H. Walter, A. Miller, E. J. Krob, and G. S. Ascher, *Exp. Cell. Res.* **73,** 145 (1972).]

particulate being studied and the nature of the experiment, determination is made of some parameter reflecting the quantity of the material in cavities along the extraction train (e.g., absorbance at 680 nm for chlorophyll; electronic cell count of aliquots of every or every *n*th tube; centrifugation of cells in selected individual tubes or in pools of two or three adjacent tubes followed by suspension in a suitable medium and analysis by turbidimetry or, in the case of erythrocytes, by lysis and measurement of hemoglobin absorbance at 540 nm). Properties of interest are then studied on, for example, centrifuged, washed, and resuspended cells from different parts of the CCD either directly (e.g., isotope distribution of labeled cells, electrophoretic mobility) or after cell culture (e.g., antibody production, metastatic potential). See Ref. 2 for numerous examples.

The data are presented as a distribution curve in which the quantity of the material (or a parameter relating thereto) as well as other investigated properties are plotted against cavity number (see, e.g., Fig. 5).

*Tests for Subfractionation by Countercurrent Distribution*

Separation by CCD of two or more particulate populations having different surface properties (i.e., $G$ values) is apparent by the presence of two or more distribution curves. Fractionation by CCD of a population of particulates into subpopulations, of great analytical interest, is not necessarily obvious from inspection of the distribution curve. Relatively symmetrical distribution curves obtained with a particulate population can be statistical or due to differences in surface properties of the constituent subpopulations (e.g., cultured cells in different parts of their cell cycle, rat erythrocytes of different ages).

There are a number of simple tests that establish whether fractionation of a population has taken place. For populations that are visible with the naked eye (e.g., erythrocytes, chloroplasts) visual inspection of the CCD collection tubes is useful. If the distribution is statistical, the partitions of the material visible in the individual tubes shortly after collection, when the phases have just settled, should all look the same; a distribution based on surface properties, on the other hand, is revealed by increases in $P$ values with increasing tube number (i.e., tubes to the left will have the material more "down" and tubes to the right will have the material more "up"). When visual inspection is not feasible, the $P$ value of the particulate material in tubes from different parts of the distribution curve can be measured to determine whether it is constant or increases through the distribution. The particulate material can also be pooled separately from the left and right ends of a CCD run and each subjected to a second CCD experiment. If the two distributions now obtained overlap, the particulate is statistically distributed and there was no fractionation of the particulate population; if the material from the left has a lower mean $G$ value than the material from the right, the distribution is on the basis of surface properties and a fractionation is obtained.

Fractionation of a population can be established by determining any suitable physical/biological property of the material through the distribution curve (e.g., cell electrophoretic mobility, microscopic examination of cell type, cell size, metastatic potential, susceptibility to antibiotics, activity of marker enzymes). If a difference is found fractionation is established. The reverse does not hold.

*Detection of Surface Differences between Closely Related (Cell) Populations*

To determine whether closely related cell populations (e.g., cultured cell sublines, red blood cells from different individuals) differ in surface properties, a simple method based on a combination of isotopic labeling and CCD can be used, which is described in detail elsewhere in this volume.[2]

## Some Cautionary and Concluding Remarks

### *Cells*

Cell partitioning is a dynamic process in which some cells attach to droplets of one phase suspended in the other after mixing, in a cell- and phase-specific manner, and are delivered on the droplets to the bulk interface.[20] For reproducible results one should select a sampling time that is not so short that the kinetic aspects of phase and cell settling are still appreciable, nor so long that cell sedimentation effects become superimposed on partitioning.[14] The time used is one at which a sharp bulk interface first appears. The speed of phase settling also affects the *P* value of cells. If partitioning is carried out in phases with low column height (e.g., the thin-layer CCD unit), the phases separate more rapidly than in phases with high columns (as in the Craig CCD unit, which has tubes arranged in vertical position). The rapidity of phase separation in the former case does not permit cells to attach to phase droplets to the same extent as in high phase columns and thereby interferes with the cell partitioning mechanism. Higher *P* values are observed in thin layers, and cell separation is less efficient.[14] One will have to balance the benefits of increased efficiency of separation against possible detrimental effects of the longer times required to attain it.

When initiating a study on cells by partitioning, an investigator should determine the partition ratios he/she obtains in a given series of phase systems rather than rely on the numerical values published by others. One important reason for this is that differences exist in molecular weight distributions, irrespective of nominal designation, of polymers purchased from different sources.

Particular attention should be paid to obtain suitable *P* values for cell populations to be studied and not to compare the relative partitioning behavior of two cell populations under conditions in which both have extreme *P* values (either high or low). Although the surface differences of the populations may be significant, they cannot be observed if both cell populations are predominantly either in the same bulk phase or at the interface.

Although the size (molecular weight, surface area) of macromolecules or small particulates affects their partition coefficient or ratio, the partition ratio of particulates with surface areas greater than about 0.2 $\mu m^2$ is determined by surface properties per unit area rather than by surface area per se.[21]

[20] H. Walter, F. D. Raymond, and D. Fisher, *J. Chromatogr.* **609,** 219 (1992).

[21] H. Walter, D. Fisher, and C. Tilcock, *FEBS Lett.* **270,** 1 (1990).

Occasions arise in which the physical properties of phases and the surface properties of cells are "balanced" in a manner such that cell separation or resolution does not occur. Such phases have been called "nondiscriminatory."[1] It is thus important that, if no segregation of cells is obtained with a given phase system, other phase compositions be tested before the conclusion is reached that separation by partitioning is not feasible.

To reiterate, a major advantage of partitioning is its great sensitivity to particulate surface properties, permitting subtle differences and changes in them to be detected. Although charge-associated or non-charge-related differences in surface properties may predominate when charge- or non-charge-sensitive phase systems are used, respectively, the interaction of cells with the physical properties of the phases is complex, and, hence, the properties reflected by partitioning can neither be generally defined nor unequivocally stated.

### *Organelles and Membranes*

A number of organelles and membrane fractions have now been isolated by phase partitioning, often using batch procedures of just a few steps. Large-scale preparations of plant plasma membrane vesicles of opposite sidedness[5] and right-side-out and inside-out thylakoid vesicles[22] are now possible. A particular advantage of phase partitioning is that it permits separation of vesicles of opposite sidedness by exploiting the fact that opposite surfaces of a membrane differ in some properties. Such preparations have facilitated the study of transverse asymmetry of membrane components as well as of vectorial properties, such as transport processes.

In analytical work different domains of the same membrane have been separated by CCD experiments and characterized.[23,24] This constitutes another area in which phase partitioning is particularly useful because the method is able to discriminate between relatively small differences in membrane surface properties.

## Acknowledgments

Work in the laboratory of Harry Walter was supported by the Medical Research Service of the U.S. Department of Veterans Affairs, and work in the laboratory of Christer Larsson was supported by the Swedish Natural Science Research Council and the Swedish Council for Forestry and Agricultural Research.

[22] P.-Å. Albertsson, E. Andreasson, H. Stefansson, and L. Wollenberger, this volume [45].

[23] J. P. Gierow, this volume [50].

[24] P.-Å. Albertsson, this volume [49].

# [5] Uses of Poly(ethylene Glycol) with Charged or Hydrophobic Groups

*By* GÖTE JOHANSSON

## Introduction

Two polymers that give an aqueous biphasic system are each concentrated in opposite phases. This property has been used for the "immobilization" of certain groups in primarily one phase, by attaching them to one of the phase-forming polymers. This chapter deals with charged or hydrophobic groups bound to a polymer. Other, more specific ligands are treated in other parts of this volume.

The most commonly used polymer for carrying groups is poly(ethylene glycol) (PEG), which normally is concentrated in the upper phase. There are several reasons for choosing PEG as carrier. One is that properties of a polymer change with increasing number of bound groups. Therefore, it is advantageous that not more than one group be attached per polymer molecule. PEG has only two reactive sites, namely, the hydroxyl groups at the ends of the linear polymer. There are also PEG derivatives in which the chains have only one end group (e.g., methoxy-PEG). The solubility of PEG in water as well as in many organic solvents is another reason for its use because this facilitates synthetic procedures for attaching a group and for the purification of the product by crystallization. A third reason is that proteins usually favor the non-PEG phase, but when a hydrophobic group liganded to PEG is used the proteins with hydrophobic pockets can be extracted into the PEG phase and separated from nonhydrophobic proteins.

## Synthesis

A number of PEG derivatives are easily prepared by first introducing a reactive group in the terminal position of the polymer chain (Fig. 1).

*Monobromo-PEG.* A general way to activate PEG is to first introduce a good leaving group, for example, the bromo group.[1] Other useful leaving groups are tosylate[2] and tresylate.[3] Monobromo-PEG or dibromo-PEG can be obtained by reacting PEG with thionyl bromide (Fluka, Buchs,

[1] G. Johansson, *Biochim. Biophys. Acta* **222,** 381 (1970).
[2] J. M. Harris and M. G. Case, *J. Org. Chem.* **48,** 5390 (1983).
[3] K. Nilsson and K. Mosbach, this series, Vol. 104, p. 56.

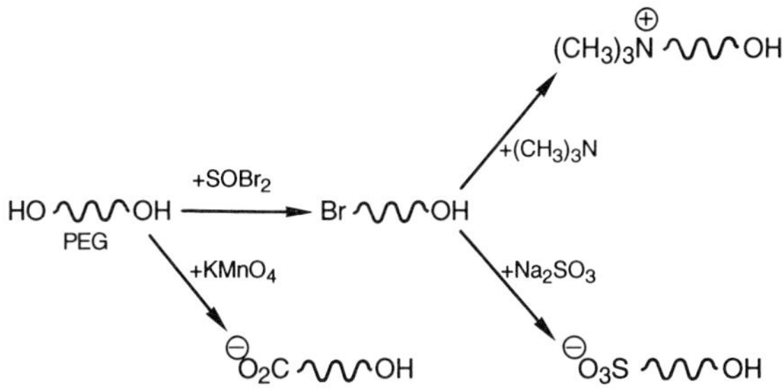


FIG. 1. Scheme for synthesis of charged PEG derivatives.

Switzerland). Two hundred fifty grams of PEG 3400 or PEG 8000 (Breox, BP Chemicals, Hythe, UK) is dissolved in 1.5 liters of toluene, and then 100 ml solvent is distilled off to remove water. The solution is allowed to cool to 30–35° and mixed with 12 ml (for PEG 3400) or 6 ml (for PEG 8000) of dry triethylamine. Newly distilled $SOBr_2$, 7 g (for PEG 3400) or 3.5 g (for PEG 8000), dissolved in 50 ml toluene is added dropwise with stirring. The reaction mixture is then brought to gentle boiling and boiled for 10 min. The amount of colored impurities produced during the reaction can be reduced by flushing dry nitrogen through the mixture during the addition of the thionyl bromide. The salt formed is removed by filtration and the polymer crystallized at 3° overnight. After collection of polymer by suction filtration and washing with cold toluene the bromo-PEG is recrystallized from 1 liter of toluene, followed by treatment with active carbon to remove dark-colored by-products. Traces of toluene are removed by evaporation under vacuum at 100°. In some batches of PEG the presence of small amounts of hydroxyl compounds, which are not removed by distillation, may make it necessary to increase the quantity of amine and thionyl bromide used.

*Dibromo-PEG.* The disubstituted PEG is obtained by using 2–3 times as much triethylamine and thionyl bromide as described for the monobromo-PEG.

*Trimethylamino-PEG.* Bromo-PEG is reacted in ethanol solution, with trimethylamine to yield tri-methylamino-PEG (TMA-PEG, a quaternary amine), which is positively charged[1] (Fig. 1). One hundred grams of monobromo-PEG (3400 or 8000) is dissolved in 500 ml of absolute ethanol in a 1-liter round-bottomed flask. Trimethylamine in excess, 14 g (for PEG 3400) or 7 g (for PEG 8000), is added either by introducing the gas directly into the polymer solution at room temperature or adding the amine dissolved in 100 ml cold absolute ethanol. After 1 hr at room temperature the solution is gently refluxed for 24 hr. Excess amine is removed by passing a strong stream of nitrogen through the solution for 20–60 min. Absolute ethanol is added to a final volume of 1–1.2 liters, and the polymer is crystallized overnight at 3°, collected on a Büchner funnel, and washed with ice-cold ethanol. The product is recrystallized twice from 1-liter portions of absolute ethanol, and remaining solvent is removed by use of a rotational vacuum evaporator at 100°.

*PEG Sulfonate*. Reacting bromo-PEG in water–ethanol solution with sodium sulfite ($Na_2SO_3$) yields PEG sulfonate (S-PEG), which is negatively charged[1] (Fig. 1). One hundred grams of monobromo-PEG (3400 or 8000) is dissolved in 500 ml water plus 100 ml ethanol together with 14 g (for PEG 3400) or 7 g (for PEG 8000) of analytical grade $Na_2SO_3$. The solution is kept in a 60° water bath for 24 hr with gentle stirring. The pH of the solution is then, if necessary, adjusted with phosphoric acid to 6.5–7. To remove salts present, all solvent is evaporated under vacuum at 100°, the remaining molten polymer is dissolved in 1 liter of toluene, and traces of water are removed by azeotropic distillation. The crystallized salts are removed by filtering the hot solution, and the S-PEG is allowed to crystallize at 3° overnight. The product collected by filtration is freed from traces of toluene by evaporation under vacuum and recrystallized twice from 1-liter portions of absolute ethanol.

*Carboxymethyl-PEG*. Another method of introducing negative charges on PEG is to oxidize carefully the —$CH_2OH$ groups at the ends of the polymer chain to carboxylic groups,[4] which yields carbomethoxy-PEG (CM-PEG). Thirteen grams of $KMnO_4$ dissolved in 200 ml water is added to a solution of 250 g PEG 8000 and 0.8 g NaOH in 600 ml of water. After 15 min on a boiling water bath, the brown precipitate that forms ($MnO_2$) is removed by suction filtration. The clear filtrate is adjusted to pH 2.6 with 1 *M* HCl. The water is removed by evaporation under vacuum, and the remaining polymer is dissolved in 1.2 liters of toluene. One hundred milliliters of toluene is distilled from the solution to remove traces of water. The precipitated salts are removed by filtration and the polymer is allowed to crystallize overnight at 3°. The polymer collected by filtration is recrystallized from 1.5 liters of absolute ethanol. The solid polymer is recovered by filtration and allowed to dry at room temperature. Before use, this acid form of CM-PEG is dissolved in water and brought into salt form by addition of NaOH or KOH solution to pH 6–7. To remove possible traces of $Mn^{2+}$, the solution is passed through an ion exchanger (Dowex 50WX-100) loaded with $Na^+$ or $K^+$, respectively. During the synthetic

[4] G. Johansson and A. Hartman, *in* "Proceedings International Solvent Extraction Conference 1974" (T. C. J. Gribnau, J. Visser, and R. J. F. Nivard, eds.), Vol. 1, p. 927. Society of Chemical Industry, London, 1974.

process the mean molecular weight of PEG may be slightly reduced (less than 10%) owing to minor oxidative splitting of polymer chains.

Carboxylic groups can also be introduced by forming the alcoholate by reacting PEG under absolutely dry and oxygen-free conditions (i.e., under pure nitrogen) with sodium naphthalene. The alcoholate is then reacted with ethyl chloroacetate, and the resulting ester is hydrolyzed in alkaline water solution to yield carboxymethyl-PEG. This polymer has a $pK_a$ of 3.6 and can therefore be regarded as being completely ionized at pH values above 5.

*Esters of PEG with Fatty Acids.* Fatty acid can be esterified with one or both of the terminal hydroxyl end groups of PEG by reacting it in an inert solvent (e.g., toluene) with the chloride[5] or anhydride of the fatty acid. Alternatively, the free acid can be condensed with PEG by using an ester-forming agent (e.g., dicyclohexyl-carbodiimide).[6] Because the disubstituted fatty acid–PEG gives aqueous solutions of extremely high viscosity even at low concentrations, it is best to keep the degree of substitution at no more than one.

The following procedure gives a PEG–fatty acid ester with approximately one acid per PEG molecule. One hundred grams of PEG (3400 or 8000) is dissolved in 600 ml toluene, and traces of water are removed by azeotropic distillation (about 100 ml of toluene are collected). Triethylamine, 5 ml (for PEG 3400) or 2.5 ml (for PEG 8000), is added followed by a solution of acyl chloride, for example, palmitoyl chloride, 30 mmol (for PEG 3400) or 15 mmol (for PEG 8000), which is added dropwise while mixing, The solution is refluxed for 30 min, and the salt formed is removed by filtration. Five hundred milliliters toluene is added to the filtrate, and the polymer is allowed to crystallize at 3°. It is then recrystallized 2–3 times from 1-liter portions of absolute ethanol as described above for TMA-PEG. The diester can be obtained by using 2–3 times the amounts of amine and acyl chloride.

## Phase Formation

Introduction of charged groups on PEG strongly affects the phase-forming properties in the absence of salt.[1,7] Table I shows the effects of charge on the phase transition point. Addition of salt restores the binodal curve to its original position. The effect of polymer-bound charged groups on the partitioning of charged biopolymers (e.g., proteins) is reduced by

[5] V. P. Shanbhag and G. Johansson, *Biochem. Biophys. Res. Commun.* **61,** 1141 (1974).

[6] G. Johansson, *Biochim. Biophys. Acta* **451,** 517 (1976).

[7] P.-Å. Albertsson and F. Tjerneld, this volume [1].

TABLE I
PHASE TRANSITION POINT FOR DEXTRAN T500 AND POLY(ETHYLENE GLYCOL) 3400[a]

| PEG type or mixture | Salt | Transition point % (w/w) PEG = % (w/w) dextran |
|---|---|---|
| PEG | None | 6.2 |
| | 100 m*M* Sodium phosphate buffer, pH 6.8 | 6.2 |
| TMA-PEG | None | 10.0 |
| | 100 m*M* Sodium phosphate buffer, pH 6.8 | 6.2 |
| S-PEG | None | 10.7 |
| | 100 m*M* Sodium phosphate buffer, pH 6.8 | 6.2 |
| 75% PEG–25% TMA-PEG | None | 6.9 |
| | 30 m*M* Sodium acetate buffer, pH 4.7 | 6.4 |
| 75% PEG–25% S-PEG | None | 7.1 |
| | 30 m*M* Sodium acetate buffer, pH 4.7 | 6.5 |
| 50% TMA-PEG–50% S-PEG | None | 6.2 |

[a] Plus disubstituted charged PEG, at 23°. Adapted from G. Johansson, *Biochim. Biophys. Acta* **222,** 381 (1970), by permission of Elsevier.

increasing salt concentrations,[8] as shown in Fig. 2. This effect depends on the type of salt present. Phosphate enhances the effect of TMA-PEG, whereas chloride (e.g., Tris-HCl) enhances that of S- or CM-PEG.[4] Salt concentrations of 2–7 m*M* are recommended. Zwitterions (e.g., glycine) can be used to increase the solubility of proteins at extremely low salt concentrations without reducing the effect of charged PEG.[8]

## Uses of Charged Poly(ethylene Glycol) Derivatives

Charged PEG can be used to obtain extreme partitioning of proteins, which may allow the purification of certain proteins in a mixture by batch extraction. To analyze the partitioning behavior, the protein mixture is included in a large system containing charged PEG. The pH is successively adjusted and measured, and sample systems are withdrawn. The partitioning is determined after phase separation of the sampled system by analyzing the phases for content of enzymes or other proteins. The procedure is illustrated by the following example.

*Extraction Curves.* A two-phase system of 40 g containing 6.4% (w/w) dextran (Dx) T500 (Pharmacia, Uppsala, Sweden), 6.6% (w/w) charged PEG (TMA-PEG, CM-PEG, or S-PEG), protein mixture (e.g., centrifuged cell homogenate), and 5 m*M* buffer is weighed into a 50-ml beaker. The

[8] G. Johansson, A. Hartman, and P.-Å. Albertsson, *Eur. J. Biochem.* **33,** 379 (1973).

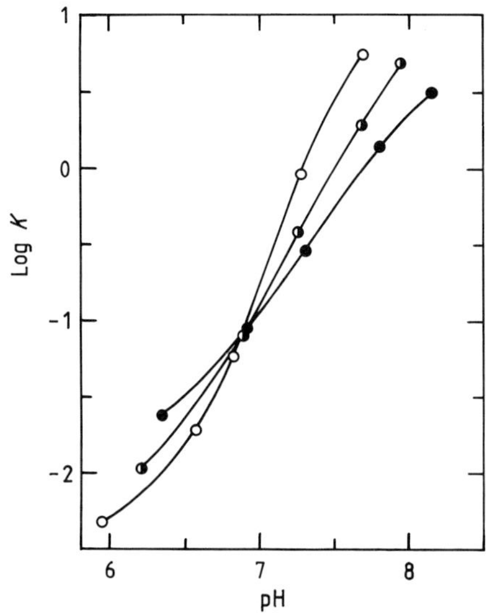


FIG. 2. Extraction profile of human CO–hemoglobin by using TMA-PEG at different concentrations of phosphate buffer. System composition: 8% (w/w) dextran T500, 8% (w/w) di-TMA-PEG 8000, 6.2 g hemoglobin per liter, and potassium phosphate buffer, 2 m*M* (○), 5 m*M* (◑), and 10 m*M* (●), at different pH values. Temperature, 20°. [From G. Johansson, A. Hartman, and P.-Å. Albertsson, *Eur. J. Biochem.* **33,** 379 (1973), by permission.]

buffer can be introduced by first dialyzing the protein mixture against a more concentrated buffer (10–25 m*M*). The beaker is placed in a water bath at the desired temperature. The glass electrode of a pH meter is introduced in the phase system. With careful but efficient mixing (e.g., magnetic stirrer) the pH value of the system is adjusted by slow addition of 2 *M* solutions of NaOH, $H_3PO_4$, or HCl using a plastic syringe or an automatic pipette. Steps of 0.3–0.6 pH units are desirable. After each pH adjustment a small sample of mixed system, 0.5–2.5 ml, is withdrawn, and the phases are permitted to settle by standing, at the water bath temperature, for 10 min or by centrifugation for 1–2 min. The concentrations of partitioned substances, for example, enzymes and bulk protein, are determined in upper and lower phases.

The partition coefficients, *K*, taken as the ratio of the concentrations of analyzed substances in the upper and lower phase, are calculated. The logarithm of the partition coefficients, log *K*, is plotted against the measured pH, yielding the extraction curve of the substance. Figure 3 shows extraction curves of some enzymes present in the soluble protein mixture from swine heart. It is also useful to determine the *K* values of

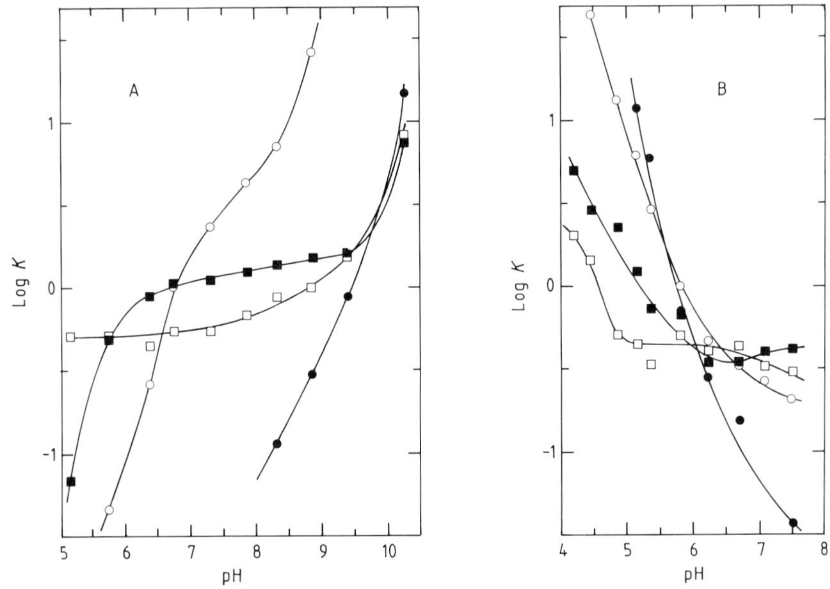


FIG. 3. Extraction curves of enzymes at 22° when soluble proteins from swine heart homogenate were partitioned in a system with (A) TMA-PEG and (B) CM-PEG. System composition: 6.4% (w/w) dextran T500, 6.6% (w/w) charged PEG, and 12.5% (w/w) of protein sample obtained by homogenizing the heart in twice its volume of 10 m*M* potassium phosphate buffer, pH 6.8, and centrifuging at 73,500 *g* for 2 hr at 3° to remove particles. Enzymes: ○, fumarase (fumarate hydratase); ●, enolase (phosphopyruvate hydratase); □, malate dehydrogenase; and ■, aspartate aminotransferase. [From G. Johansson and A. Hartman, *in* "Proceedings International Solvent Extraction Conference 1974" (T. C. J. Gribnau, J. Visser, and R. J. F. Nivard, eds.), Vol. 1, p. 927. Society of Chemical Industry, London, 1974, with permission.]

total proteins by use of the Bradford method.[9] From the extraction curves the optimal pH for the purification of a certain enzyme can be obtained as well as its extent.

*Partitioning of Particles*. Biological particles, which may be sensitive to pH changes, may be extracted at constant pH by increasing the concentration of charged PEG. When the particles are in the upper phase the negatively charged S- or CM-PEG is used, as shown for *Chlorella pyrenoidosa*,[10] and when they are concentrated in the lower phase or at the interface, as for *Saccharomyces cerevisiae*, the positively charged TMA-

[9] M. M. Bradford, *Anal. Biochem.* **72**, 248 (1976).

[10] G. Johansson, *in* "Methodological Developments in Biochemistry" (E. Reid, ed.), Part 2, Preparative Techniques, p. 155. Longman, London, 1973.

TABLE II
SERIES OF PHASE SYSTEMS USEFUL FOR ADJUSTING INITIAL PARTITIONING OF PARTICLES[a]

| Temperature (°) | Molecular weight of polymers: Dextran | PEG | System composition [% (w/w) dextran/% (w/w) PEG]: I | II | III | IV | V |
|---|---|---|---|---|---|---|---|
| 4 | T40 | 3400 | 9.6/5.6 | 10.0/5.8 | 10.5/6.1 | 11.1/6.4 | 12.0/6.6 |
| | T500 | 3400 | 6.0/4.6 | 6.5/4.7 | 7.0/4.8 | 7.7/5.1 | 8.6/5.3 |
| | T500 | 8000 | 4.5/3.2 | 5.0/3.3 | 5.5/3.4 | 6.2/3.6 | 7.2/4.0 |
| 20 | T500 | 3400 | 5.6/6.3 | 6.1/6.4 | 6.6/6.6 | 7.5/6.6 | 8.5/6.7 |
| | T500 | 8000 | 4.4/3.6 | 4.8/3.7 | 5.2/3.8 | 6.0/4.0 | 7.0/4.3 |

[a] When cell organelles or cells without cell walls are partitioned, the systems have to be made isotonic by including 0.35 *M* sorbitol, mannitol, or sucrose or by using isotonic salt concentrations at the appropriate pH. The systems give roughly equal volumes of the two phases.

PEG is used.[11] Charged PEG has also been used to estimate the relative charge of microorganisms[12,13] and for determining the isopartition point (which may be close to the isoelectric point) of biological particles.[14,15]

## Uses of Hydrophobic Poly(ethylene Glycol) Derivatives

Two variables that are useful for adjusting partitioning changes arising from hydrophobic interaction are (1) the length of the hydrophobic tail attached to PEG and (2) the concentration of the hydrophobic element in the system. The effects of fatty acid PEG esters on the partitioning of proteins are discussed by Shanbhag.[16]

*Hydrophobic Extraction of Membranes and Organelles.* When a sample of particles is to be investigated it is useful to first examine its partitioning behavior in the absence of hydrophobic ligands in a series of systems (e.g., the ones given in Table II) incorporating buffer (buffer type and concentration can be chosen with great freedom). By increasing the polymer concentrations the particles normally are decreased in the upper

[11] G. Johansson, *Mol. Cell. Biochem.* **4,** 169 (1974).

[12] K.-E. Magnusson and G. Johansson, *FEMS Microbiol. Lett.* **2,** 225 (1977).

[13] O. Stendahl, L. Edebo, K.-E. Magnusson, C. Tagesson, and S. Hjertén, *Pathol. Microbiol. Scand. Sect. B* **85B,** 334 (1977).

[14] H. Miörner, P.-Å. Albertsson, and G. Kronvall, *Infect. Immun.* **36,** 227 (1982).

[15] H. Walter and D. Forciniti, this volume [20].

[16] V. P. Shanbhag, this volume [23].

TABLE III

PARTITIONING OF SPINACH CHLOROPLASTS[a] IN PHASE SYSTEMS[b] CONTAINING POLY(ETHYLENE GLYCOL)-BOUND FATTY ACID[c]

| Fatty acid group | Number of C atoms | Chloroplasts (%) in/at: Upper phase | Interface | Lower phase | $A_{550}/A_{680}$ of incubated chloroplasts |
|---|---|---|---|---|---|
| None | 0 | 0 | 21 | 79 | 0.71 |
| Acetate | 2 | 2 | 24 | 74 | 0.74 |
| Butyrate | 4 | 8 | 52 | 40 | 0.73 |
| Caproate | 6 | 14 | 72 | 14 | 0.72 |
| Caprylate | 8 | 9 | 68 | 23 | 0.71 |
| Caprate | 10 | 21 | 65 | 14 | 0.71 |
| Laurate | 12 | 67 | 20 | 13 | 0.71 |
| Myristate | 14 | 82 | 11 | 7 | 0.72 |
| Palmitate | 16 | 80 | 16 | 4 | 0.63 |
| Stearate | 18 | 83 | 11 | 4 | 0.65 |

[a] Chloroplasts consisted of 40% whole and 60% broken chloroplasts.

[b] System composition: 5% (w/w) dextran T500, 4% (w/w) PEG 8000, 50 mmol/kg NaCl, 5 mmol/kg sodium phosphate buffer, pH 7.0, and 250 m*M* sucrose. The concentration of monosubstituted PEG–fatty acid was (depending on the derivative) 3.5–6% of the total PEG. Temperature, 3°.

[c] From G. Johansson and H. Westrin, *Plant Sci. Lett.* **13,** 201 (1978), by permission of Elsevier.

phase. A system in which not more than 10% of the particles are in the top phase offers an excellent starting system for incorporation PEG–hydrophobic ligand for hydrophobic extraction. Another possibility to decrease the partitioning of particles (in the absence of PEG–ligand) is to increase the concentration of chloride in the system. In this case the concentration of other salts (e.g., buffer) should not be too high (e.g., below 10 m*M*).

To study the effect of fatty acid modified PEGs, they are substituted, in different amounts, for part of the unreacted PEG. In a preliminary trial 0.1, 0.33, 1, 3.3, and 10% of total PEG is replaced by modified PEG. This will normally give a reasonable extraction curve and indicate the amount of PEG ester necessary to obtain saturation. The lengths of fatty acids bound to PEG (Table III) is another parameter that affects specificity in the extraction of certain particles and allows their extraction without affecting membrane structure. The fatty acid must have a critical length to interact with the biological particles.[17] If, on

[17] G. Johansson and H. Westrin, *Plant Sci. Lett.* **13,** 201 (1978).

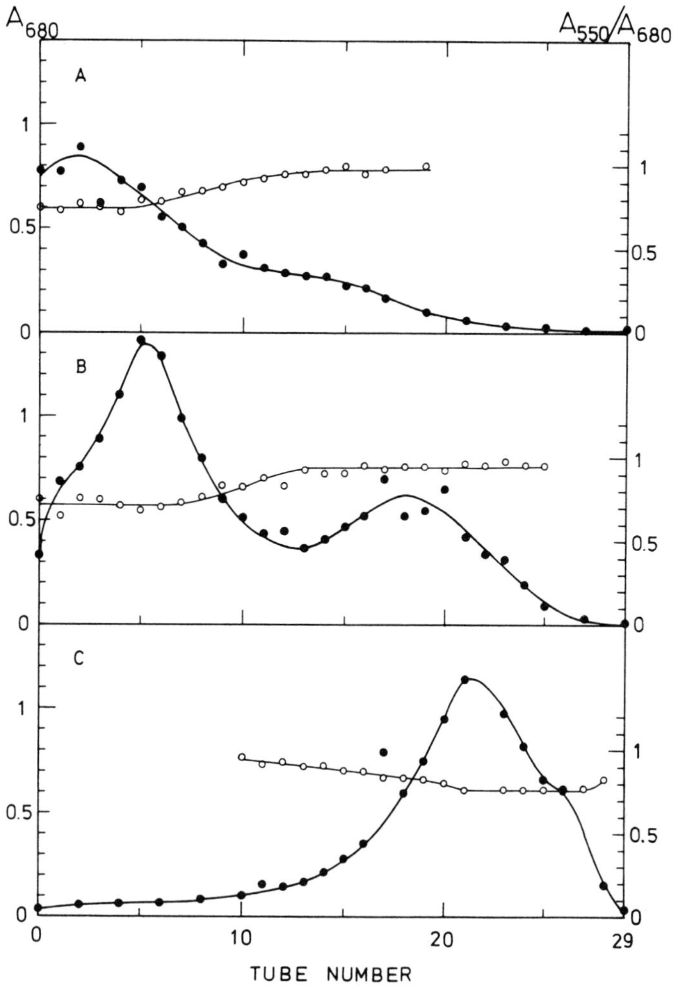


FIG. 4. Countercurrent distribution of a chloroplast mixture containing whole chloroplasts (class I) and broken chloroplasts (class II) using dextran–PEG aqueous systems containing PEG esters of (A) caprylic (octanoic) acid, (B) capric (decanoic) acid, and (C) lauric (dodecanoic) acid. ●, $A_{680}$; ○, $A_{550}/A_{680}$. Systems as in Table III. [From G. Johansson and H. Westrin, *Plant Sci. Lett.* **13,** 201 (1978), by permission of Elsevier.]

the other hand, the hydrophobic chain is too long it may act as a detergent and partially dissolve membranes. This effect can be seen in Table III from the ratio of the apparent absorbances of chloroplasts at 550 and 680 nm. A decrease in this ratio is a consequence of the increase of broken chloroplasts.

That the differences in extraction depend on the fatty acid bound has been demonstrated by using these systems for countercurrent distribution studies of a mixture of intact and broken chloroplasts[17] (Fig. 4). In addition, a system containing the $C_{10}$ acid bound to PEG permits the extraction of chloroplasts from crude extract of spinach. Using single-step partitioning followed by two washings of the upper phase with fresh lower phase, the intact chloroplasts are obtained within 45 min.[17]

Hydrophobic partitioning has been used to probe the hydrophobic properties of the surface of microorganisms in relation to their infectiousness.[18]

## Conclusions

PEG to which either charged or hydrophic groups have been attached can be used to obtain extreme partitioning of proteins as well as of biological particles when incorporated in two-polymer aqueous two-phase systems. In addition to their use in separations, modified PEGs also provide valuable information on the physicochemical nature of the partitioned species.

## Acknowledgments

This work was supported by grants from the Swedish Natural Science Research Council.

[18] H. Miörner, G. Johansson, and G. Kronvall, *Infect. Immun.* **39,** 336 (1983).

# [6] Systems Containing Organic Solvents

*By* GÖTE JOHANSSON and MONICA JOELSSON

## Introduction

The formation of two liquid phases when two polymers are dissolved together in a solvent is not restricted to water. Phase separations have been observed with organic solvents. It is therefore possible to use two-polymer two-phase systems containing an organic solvent for separation of hydrophobic substances by partitioning.[1] The solvation in aqueous two-phase systems can also be modified by replacing part of the water with a

[1] G. Johansson and M. Joelsson, *J. Chromatogr.* **464,** 49 (1989).

TABLE I
SOLUBILITY OF POLYMERS IN SERIES OF SOLVENTS[a,b]

| Solvent | Dextran T500 | Ficoll 400 | Hydroxypropyl-starch | Hydroxypropyl-dextran | PEG 8000 |
|---|---|---|---|---|---|
| Water | *sol* | *sol* | *sol* | *sol* | *sol* |
| Glycerol | *sol* | *sol* | *sol* | *sol* | *ns*(84) |
| Ethylene glycol | *sol* | *sol* | *sol* | *sol* | *ns*(>95) |
| Dimethyl sulfoxide | *sol* | *sol* | *sol* | *sol* | *sol* |
| Ethanol | *ns*(30) | *ns*(61) | *ns*(50) | *ns*(91) | *sol* |
| Tetrahydrofuran | *ns*(30) | *ns*(47) | *ns*(42) | *ns*(59) | *sol* |
| Dioxane | *ns*(46) | *ns*(69) | *ns*(64) | *ns*(80) | *sol* |
| Acetonitrile | *ns*(22) | *ns*(36) | *ns*(31) | *ns*(44) | *sol* |
| Dimethylformamide | *ns*(62) | *sol* | *sol* | *sol* | *sol* |
| Acetone | *ns*(24) | *ns*(46) | *ns*(33) | *ns*(61) | *sol* |

[a] At 22°. From G. Johansson and M. Joelsson, *J. Chromatogr.* **464,** 49 (1989), by permission of Elsevier Science Publishers.

[b] *sol* indicates solubility of polymer (>5%, w/w) in solvent. *ns* indicates that the polymer is practically insoluble in pure solvent. The solubility limits for polymer in organic solvent–water mixtures are shown by the numbers in parentheses. These values give the cloud points (as a percentage of organic solvent in total solvent) when 10% (w/w) aqueous solutions of the polymer were titrated with the solvent to permanent turbidity.

water-soluble organic solvent.[1–3] The hydrophobicity of both phases is thus increased and results in a higher solubility of hydrophobic substances.[1] The freezing points of such two-phase systems are reduced, allowing partitioning at sub-zero temperatures.[3]

## Formulation of Two-Phase Systems

### *Choice of Polymers*

Before selecting polymers for preparation of two-phase systems based on an organic solvent, or an organic/aqueous solvent mixture, one must know the solubilities of polymers in the solvent. The solubilities of some polymers in a number of solvents are given in Table I. The next step is to find polymer pairs which lead to phase separation in the solvent. Poly(ethylene glycol) (PEG) is soluble in several solvents (at room temperature) and can be used as one of the phase-forming polymers. Dextran (Dx) is less useful because of its extremely low solubility in most organic solvents (with the exceptions of ethylene glycol, glycerol, and dimethyl

[2] G. Johansson and G. Kopperschläger, *J. Chromatogr.* **388,** 295 (1987).

[3] G. Johansson, M. Joelsson, and B. Olde, *Biochim. Biophys. Acta* **1029,** 295 (1990).

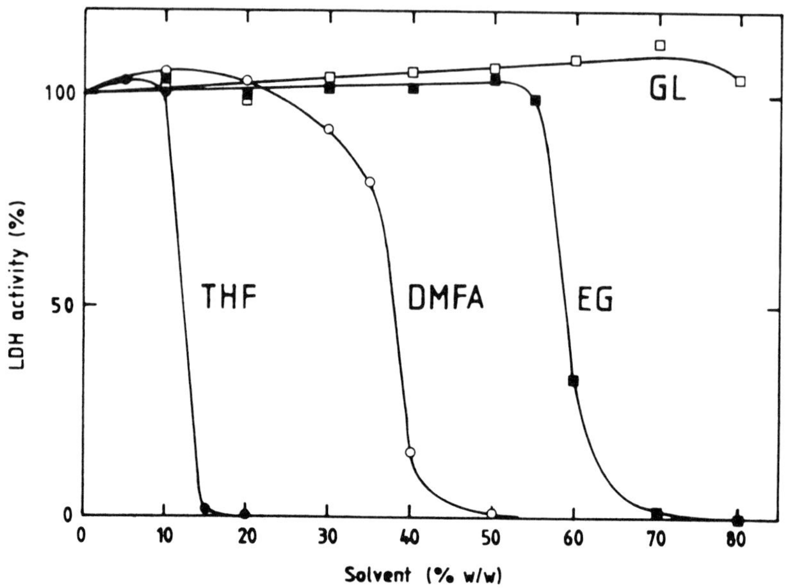


FIG. 1. Effect of organic solvents in water on enzyme activity of lactate dehydrogenase (LDH). Solvents: tetrahydrofuran (THF), *N,N*-dimethylformamide (DMFA), ethylene glycol (EG), and glycerol (GL). The water–solvent mixtures contained 25 m*M* sodium phosphate buffer, pH 7.0. The enzyme activity was measured after 1 min of incubation at 22°. The LDH concentration was 90 U/ml, and the enzyme was added in the form of a crude extract from muscle. [From G. Johansson, M. Joelsson, and B. Olde, *Biochim. Biophys. Acta* **1029,** 295 (1990), by permission of Elsevier Science Publishers.]

sulfoxide). The solubility of Dx is, however, markedly increased by modifying the polymer with moderately hydrophobic groups (e.g., hydroxypropyl groups).

### *Choice of Solvent*

The solvent used must be nonreacting with and nondenaturing for the material being partitioned. For water-soluble solvents, dilution with water can be used to avoid denaturation. The denaturing effects of some water–solvent mixtures on the enzyme lactate dehydrogenase are illustrated in Fig. 1. The solubilities of hydrophobic substances increase significantly when only a small part of the water is replaced by an organic solvent, as shown with benzoic acid (Fig. 2).

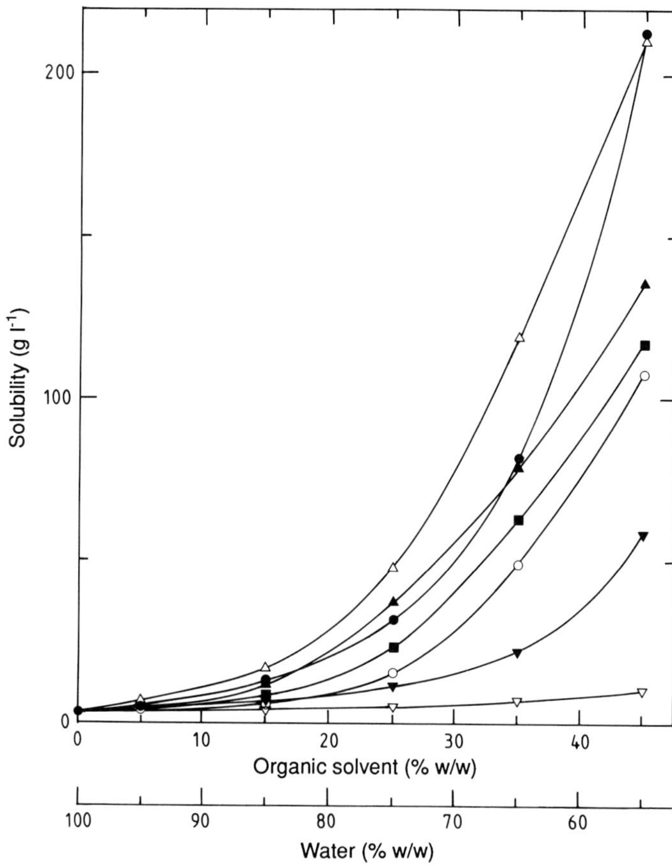


FIG. 2. Solubility of benzoic acid in mixtures of organic solvent and water at 22°. Solvents: ▽, ethylene glycol; ▼, dimethyl sulfoxide; ○, ethanol; ■, dioxane; ▲, acetonitrile; ●, *N,N*-dimethylformamide; △, acetone. [From G. Johansson and M. Joelsson, *J. Chromatogr.* **464,** 49 (1989), by permission of Elsevier Science Publishers.]

## *Phase Diagram of Two Polymer–Two Solvent Systems*

Replacing some of the water with an organic solvent may drastically change the phase diagram[4] for the polymer pair. A time-saving way to estimate the position of the binodal curve at constant organic solvent concentration is as follows. Known quantities of systems giving two phases and which contain a given concentration of organic solvent are,

[4] P.-Å. Albertsson and F. Tjerneld, this volume [1].

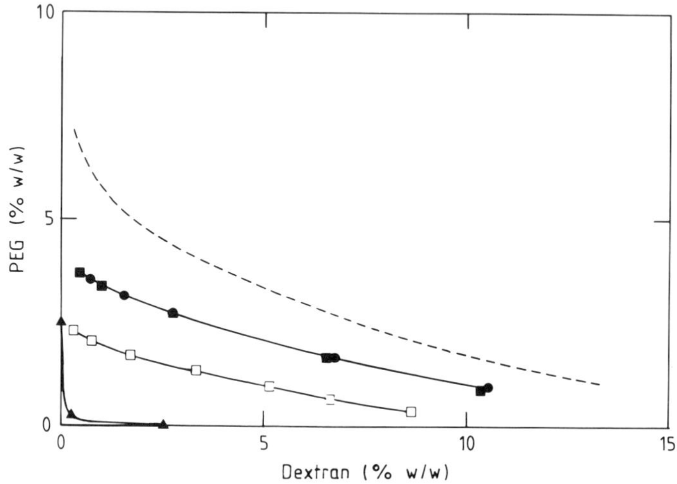


FIG. 3. Transition curves (between one and two phases) at 22° for systems composed of dextran T500, PEG 8000, water, and organic solvent (25%, w/w, of total system). Solvents: ■, dioxane; ●, *N,N*-dimethylformamide; □, tetrahydrofuran; ▲, acetonitrile. The dashed line indicates the corresponding curve with water as the solvent. [From G. Johansson and M. Joelsson, *J. Chromatogr.* **464,** 49 (1989), by permission of Elsevier Science Publishers.]

while mixing, diluted dropwise with the organic solvent in aqueous solution at the same concentration. When the transition into one phases occurs (i.e., turbidity disappears), the final weights are recorded and the polymer concentrations are calculated. Five to ten determinations (at different positions in the phase diagram) are adequate to locate the "binodal curve"[4] (Fig. 3). The curves represent cross sections (at constant solvent concentration) of the binodal *surface,* which is the border between one- and two-phase spaces of a three-dimensional phase diagram made up by the concentrations of polymer 1, polymer 2, and organic solvent. The composition of the phases of a given system are represented as points on the surface, but the points are usually not found on the cross-sectional curve. Instead, they can be found on two other curves parallel to the first curve but on opposite sides (Fig. 4).

Complete phase diagrams are obtained by analyzing the compositions of the two phases of a large number of systems. Concentrations of the phase components can be determined by measuring the refractive index[5] after their separation by chromatography (e.g., gel chromatography). This can conveniently be done by use of a refractive index monitor. The phase

[5] D. E. Brooks and R. Norris-Jones, this volume [2].

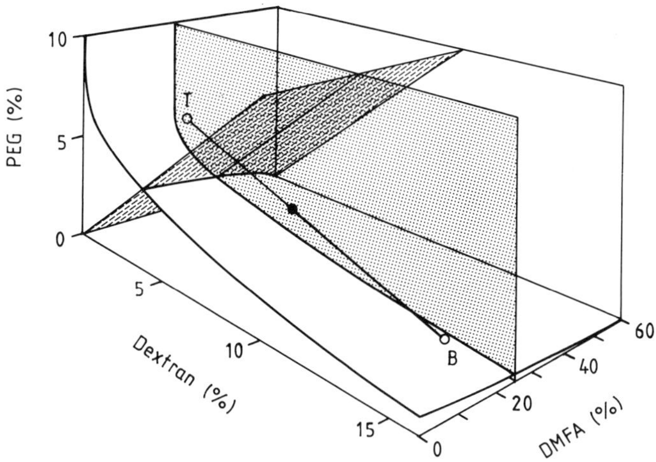


FIG. 4. Phase diagram for the system dextran T500, PEG 8000, water, and *N,N*-dimethylformamide (DMFA) at 22°. The binodal surface is cut by two planes: one where the concentrations of PEG and Dx are equal, the other with constant concentration of DMFA (25%, w/w). The total composition of a system (6.9% dextran, 2.7% PEG, and 25% DMFA) in the latter plane is marked with ●. The composition of the top phase (T) lies behind the plane. The composition of the lower phase (B) is located in front of the same plane. The three points fall on a straight line. [From G. Johansson and M. Joelsson, *J. Chromatogr.* **464,** 49 (1989), by permission of Elsevier Science Publishers.]

compositions generate the binodal surface and give the orientation of the tie lines. Alternatively, the location of the binodal surface can be found by constructing sets of curves, as in Fig. 3, using different contents of organic solvent. An example of a phase diagram is shown in Fig. 4. For the system indicated in Fig. 4 the concentration of organic solvent (dimethylformamide, DMFA) is higher in the top phase than in the bottom phase, as is the concentration of water. The DMFA/water ratios are practically the same in both phases (0.38 and 0.39 for top and bottom phase, respectively).

Polymers soluble in both water and organic solvent may have a greater solubility when a mixed solvent is used (Table II).

## Partitioning of Low Molecular Weight Substances

Introducing organic solvents into two-phase systems increases the solubility of hydrophobic substances, which can thus be partitioned at reasonable concentrations. This is of importance in the purification by partitioning, as well as in binding studies involving hydrophobic moieties.

TABLE II

TRANSITION POINTS (BETWEEN ONE AND TWO PHASES) OF FICOLL 400, POLY(ETHYLENE GLYCOL) 8000, WATER, AND DIMETHYLFORMAMIDE SYSTEM USING EQUAL CONCENTRATIONS OF POLYMERS[a,b]

| Solvent mixture | | System composition at phase transition point | | | |
|---|---|---|---|---|---|
| Water (%)[a] | DMFA (%) | Ficoll (%) | PEG (%) | Water (%) | DMFA (%) |
| 100 | 0 | 7.7 | 7.7 | 84.6 | 0 |
| 90 | 10 | 8.5 | 8.5 | 74.7 | 8.3 |
| 80 | 20 | 9.3 | 9.3 | 65.1 | 16.3 |
| 70 | 30 | 11.4 | 11.4 | 54.0 | 23.2 |
| 60 | 40 | 16.0 | 16.0 | 40.8 | 27.2 |
| 50 | 50 | >20 | >20 | <30 | <30 |
| 30 | 70 | 16 | 16 | 20 | 28 |
| 20 | 80 | 10.6 | 10.6 | 15.8 | 63.0 |
| 10 | 90 | 8.0 | 8.0 | 8.4 | 75.6 |
| 0 | 100 | 7.5 | 7.5 | 0 | 85.8 |

[a] At 22°.

[b] Adapted from G. Johansson and M. Joelsson, *J. Chromatogr.* **464,** 49 (1989), by permission of Elsevier Science Publishers.

Addition of an organic solvent may also be useful in systems used as enzyme reactors[6] (with the enzyme predominantly in one phase) when the substrate and/or product are hydrophobic. The partition coefficients, $K$ (=concentration ratio top to bottom), of low molecular weight substances are in the range 1.2–2.5 (Table III). Aromatic, heteroaromatic, aliphatic (saturated or with conjugated double bonds), and alicyclic compounds all show a higher affinity for the PEG-rich phase. By using systems with DMFA, strongly hydrophobic substances, such as borneol, stearic acid, and cholesterol, can be partitioned at concentrations of at least 10 g/liter.[1]

## Partitioning of Non-Water-Soluble Proteins

A number of plant storage proteins are not soluble in water, but can be extracted by alcohol. One of these, zein (Sigma, St. Louis, MO), has been partitioned in systems containing DMFA (Table III).[7] In a multiple extraction procedure (countercurrent distribution) using such a system, zein was found to be heterogeneous.[1]

[6] F. Tjerneld, this volume [53].

[7] H.-E. Åkerlund and P.-Å. Albertsson, this volume [8].

TABLE III
PARTITIONING OF HYDROPHOBIC COMPOUNDS IN SYSTEMS BASED ON DIMETHYLFORMAMIDE[a,b]

| Partitioned compound | Concentration (g/liter) | Partition coefficient | | |
|---|---|---|---|---|
| | | A | B | C |
| Benzoic acid | 2 | 1.28 | 1.18 | 1.35 |
| Borneol | 10 | — | 1.3 | 1.5 |
| β-Carotene | 2 | 2.09 | 1.69 | 2.14 |
| Chlorophyll *a* | 0.05 | 2.08 | 1.68 | 2.43 |
| Cholic acid | 10 | — | 1.9 | 2.3 |
| β-Estradiol | 2 | 1.45 | 1.26 | 1.36 |
| β-Pinene | 10 | — | 1.5 | 1.9 |
| Stearic acid | 30 | — | 1.4 | 2.4 |
| Zein (corn protein) | 10 | 1.20 | 0.76 | 0.54 |

[a] Reproduced from G. Johansson and M. Joelsson, *J. Chromatogr.* **464,** 49 (1989), by permission of Elsevier Science Publishers.
[b] A: 8.5% Ficoll 400 and 8.5% PEG 8000 in DMFA. B: 8.1% hydroxypropyl-dextran 500 and 8.1% PEG 8000 in DMFA. C: 10% hydroxypropyl-dextran 500 and 8.5% PEG 8000 in DMFA. At 22°.

### *Preparation of System*

A system consisting only of Ficoll, PEG, and an organic solvent can be obtained. Preparation and use of such systems using DMFA as the solvent is described here. Ficoll 400 (Pharmacia, Uppsala, Sweden) is dissolved in DMFA (Fluka, Buchs, Switzerland) to a 15% (w/w) concentration by careful heating and mixing. It is essential to use a high grade of DMFA (one which does not give high pH values when dissolved in water). PEG 8000 (BP Chemicals, Hythe, UK) is dissolved to 40% (w/w) in DMFA. Then 2.27 g of Ficoll and 0.85 g of PEG stock solutions are mixed in 10-ml glass centrifuge tubes. A 0.88-g sample, dissolved in DMFA, is added. The tubes are closed with inert stoppers and equilibrated at room temperature while mixing. The systems are centrifuged for 15 min at 3000 *g* and room temperature, and the phases are sampled and analyzed. Instead of DMFA, dimethyl sulfoxide (DMSO, Fluka), can be used with the same polymer concentrations.

## Lowering of Freezing Point

An organic solvent can be used to depress the freezing point of a phase system. Glycerol and ethylene glycol have been used for this purpose.[2,3] Most of the water can be replaced with these solvents without causing a

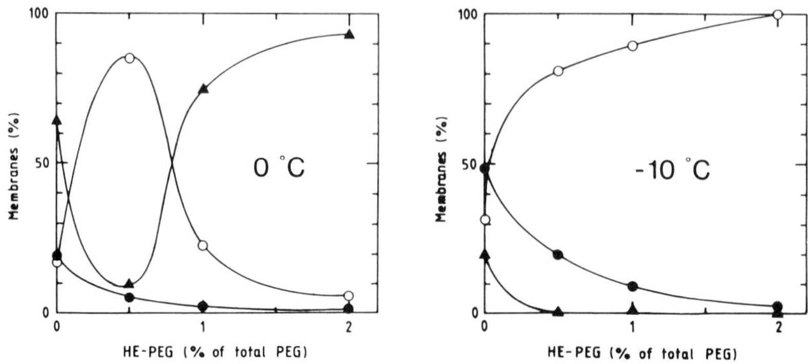


FIG. 5. Partitioning of synaptic membranes with a PEG-bound affinity ligand, hexaethonium (HE–PEG), at 0 and $-10°$. The membrane concentration was measured by light scattering at 400 nm. ○, Upper phase; ●, lower phase; and ▲, interface. System composition: 10.1% (w/w) dextran T40, 6.9% (w/w) PEG 3400 at 0°; 9.0% (w/w) dextran T40, 6.14% (w/w) PEG 3400 at $-10°$. The systems also contained different amounts of cross-linked HE–PEG 8000, 20% (w/w) glycerol, and 2 m*M* Tris–phosphate buffer (pH 7.8). The total concentration of membranes corresponded to 1 g protein per liter of the system. [From G. Johansson, M. Joelsson, and B. Olde, *Biochim. Biophys. Acta* **1029,** 295 (1990), by permission of Elsevier Science Publishers.]

noticeable change in the phase composition or the volume ratio. Enzymes are active at high concentrations of these solvents (Fig. 1). Freezing point depressions to $-30°$ have been reported.[2] Affinity partitioning of enzymes may be affected both by the presence of the solvent and by the temperature. Yeast phosphofructokinase, for example, is not extracted by Cibacron Blue F3G-A–PEG at $-20°$, but is instead excluded from the upper phase[2]; the affinity extraction of alkaline phosphatase does not depend on the temperature, but the effect is reduced by increasing concentrations of ethylene glycol.[3] In the case of lactate dehydrogenase, the presence of organic solvent shows only marginal effects on affinity partitioning.[3]

## Cryopartitioning of Synaptic Membranes

The decrease in temperature from 0 to $-10°$ has a stabilizing effect on synaptic membrane structures. In the cold (3°) synaptic membranes are gradually affected by the two-phase system and change their partitioning properties, manifested by accumulation at the interface.[8] This phenomenon has been observed after countercurrent distribution of the membranes, and it is exacerbated by certain PEG–ligands with affinity for cholinergic

[8] G. Johansson, H.-E. Åkerlund, and B. Olde, *J. Chromatogr.* **311,** 277 (1984).

receptors, for example, hexaethonium–PEG of the cross-linked type.[9] The effect is shown in Fig. 5 at a temperature of 0°. It is assumed that the PEG–ligand, when bound to receptor proteins, causes a rearrangement within the mosaic structure of the fluid membrane. Exposed large hydrophobic portions of lipid bilayer may then cause membrane aggregation. However, at −10° (Fig. 5) no changes occur. This might indicate a "freezing" of the membrane structure in a native state.

### *Preparation of System*

Four-gram systems containing 9% (w/w) Dx T40 (Pharmacia), 6.5% (w/w) PEG 3400 (BP Chemicals), and 20% glycerol (Fluka) are prepared from glycerol (with the concentration of trace water known) and aqueous stock solutions of polymers (30% Dx T40 and 40% PEG 3400). To prepare the two-phase systems, 1.2 g Dx solution, 0.65 g PEG solution, and 0.6 g glycerol (if 100%, otherwise corrected for water) are weighed into 10-ml glass tubes, and water is added to give a final weight of 3.00 g. The systems are mixed in closed tubes and brought to −10°. Samples (e.g., membranes), suspended in 20% glycerol containing a 4-fold concentration of buffer, are cooled to −10°. One-gram aliquots of the sample suspension are added to the 3-g systems. These are equilibrated in such a manner that the temperature does not rise. Preferably the systems should be kept in a cold bath of constant temperature in which they then can also be left for settling. Use of a centrifuge to speed up phase settling is convenient as well as time-saving, but it requires close temperature control.

## Discussion

The introduction of an organic solvent can be used as a means for modifying solvating properties as well as freezing points of aqueous two-phase systems. The handling of the systems requires special care because of the potential for evaporation of solvent. In cryopartitioning, the condensation of water vapor forming ice above the surface of the system may cause a progressive crystallization of the phases. Low temperature and the presence of glycerol also markedly increase the viscosity of the phases.

## Acknowledgments

This work was carried out with financial support from Carl Trygger's Foundation for Scientific Research and from NFR, the Swedish Natural Science Research Council.

[9] G. Johansson, R. Gysin, and S. D. Flanagan, *J. Biol. Chem.* **256,** 9126 (1981).

# [7] Chiral Resolution

*By* PER-ÅKE ALBERTSSON

## Introduction

There is a growing need for optically pure substances, particularly in the pharmaceutical industry, and this has led to the development of several techniques for analytical and preparative chiral separation. In chromatography, proteins included in the stationary phase have allowed the resolution of several racemic compounds.

Under suitable conditions, a protein can be confined almost completely to one of the phases of an aqueous polymer two-phase system, whereas low molecular weight compounds partition more equally between the phases. Hence, if the protein binds one of the enantiomeric forms of a racemic mixture selectively, the phase partitioning yields a chiral separation.[1,2] This separation can be magnified by a multiple extraction process such as countercurrent distribution (CCD).[1] Liquid–liquid partitioning has the advantage that the separation can be easily scaled up.[2]

Chiral ligands bound to one of the polymers can be used for selective interaction with a protein. This has been demonstrated with chiral copper–chelate complexes, which alter the selectivity in metal affinity protein partitioning.[3] This chapter describes a preparative procedure for the enantiomeric separation of kynurenine.[2]

## Reagents

Dextran T40 is obtained from Pharmacia (Uppsala, Sweden), poly(ethylene glycol), (PEG) 8000 from Union Carbide (New York, NY), and bovine serum albumin (BSA) and DL-kynurenine from Sigma (St. Louis, MO).

## Phase System

The phase system is composed of 8% (w/w) dextran T40, 7% (w/w) PEG 8000, 6% (w/w) bovine serum albumin, 50 m*M* sodium phosphate buffer, pH 8.5, and 0.1 *M* sodium chloride.

[1] B. Ekberg, B. Sellergren, and P.-Å. Albertsson, *J. Chromatogr.* **333,** 211 (1985).

[2] B. Sellergren, B. Ekberg, P.-Å. Albertsson, and K. Mosbach, *J. Chromatogr.* **450,** 277 (1988).

[3] G. E. Wuenschell, E. Wen, R. Todd, and F. H. Arnold, *J. Chromatogr.* **543,** 345 (1991).

## Procedure

### *Countercurrent Extractions*

1. Eight separatory funnels (500 ml) are each filled with 80 ml of upper phase (PEG-rich). The first funnel contains in addition 70 ml of lower, dextran-rich, phase and 100 mg of DL-kynurenine.
2. After shaking the phases and permitting them to settle, the lower phase is transferred to the second funnel, and 70 ml of fresh lower phase is added to the first funnel. This completes the first transfer.
3. The first and second funnels are shaken, and, after phase settling, the lower phase of the second funnel is transferred to the third and the lower phase of the first funnel transferred to the second. Thereafter fresh lower phase is added to the first funnel. This completes the second transfer.
4. In this manner eight transfers are completed.

Systems from each funnel are taken to analyze the optical purity by high-performance liquid chromatography (HPLC)[2,4] and for determination of the amount of substance in each fraction.

### *Removal of Phase Polymers and Salt*

Pooled fractions (funnels 2–4 and funnels 7–8) are diluted with an equal volume of water, and the pH is adjusted to about 4 with acetic acid.[2] The solution is recycled through an ultrafiltration unit (Pelicon laboratory cassette system; Millipore, Bedford, MA) with a molecular weight cutoff of 10,000.[5] The filtrate is freeze-dried, and a small amount of methanol is added to extract the kynurenine from the salt.[2]

## Results and Discussion

The partitioning of serum albumin, like that of other proteins, depends among other factors on the molecular weight of the phase-forming polymers and on the ionic composition of the phase system.[6] For the purpose of confining the serum albumin to the lower phase, a dextran of low molecular weight was selected because it promotes a lower partition coef-

[4] S. Allenmark, B. Bomgren, and H. Borén, *J. Chromatogr.* **316,** 617 (1984).

[5] F. Tjerneld, I. Persson, P.-Å. Albertsson, and B. Hahn-Hägerdal, *Biotechnol. Bioeng. Symp.* **15,** 419 (1985).

[6] P.-Å. Albertsson, "Partition of Cell Particles and Macromolecules," 3rd Ed. Wiley (Interscience), New York, 1986.

TABLE I
OPTICAL PURITY AND AMOUNT OF KYNURENINE IN POOLED FRACTIONS[a]

| Fractions | Kynurenine (mg) | Optical purity of D or L form |
|---|---|---|
| 2–4 | 22 | 90 (D) |
| 5 | 13 | 83 (D) |
| 6 | 35 | 55 (D) |
| 7–8 | 30 | 90 (L) |

[a] From B. Sellergren, B. Ekberg, P.-Å. Albertsson, and K. Mosbach, *J. Chromatogr.* **450,** 277 (1988), with permission.

ficient, $K$, for serum albumin.[6–8] In addition, the selection of 0.1 $M$ sodium chloride as the dominant salt also leads to a lower $K$ for serum albumin. In the system used more than 95% of the serum albumin partitions into the lower phase, whereas kynurenine partitions almost equally between the phases.

Table I shows the total amount of kynurenine and the optical purity of the different fractions. In the countercurrent distribution procedure the lower phase, containing serum albumin, is the moving phase, and because the L form is found in the fractions with higher numbers it follows that L-kynurenine is bound preferentially by serum albumin. From the concentration of the serum albumin and the L and D forms in the different fractions, one can calculate the binding constant, as demonstrated by a similar countercurrent distribution experiment with tryptophan.[1]

The difference in binding constants for the D and L forms determines the number of transfers required in the countercurrent distribution process. In principle the selectivity of chiral partitioning could be increased by employing two different binding proteins, one of which preferentially binds the L form and partitions into the bottom phase while the other protein preferentially binds the D form and partitions into the top phase.

As an alternative to countercurrent distribution one might also use liquid–liquid partition chromatography if, for example, the stationary phase contains a protein that selectively binds one of the enantiomeric forms. Aqueous two-phase partitioning has a great potential for chiral separation on a preparative basis because, quite generally, it is relatively easy to scale up liquid–liquid extraction methods.

[7] V. P. Shanbhag, R. Södergård, H. Carstensen, and P.-Å. Albertsson, *J. Steroid Biochem.* **4,** 537 (1973).

[8] P.-Å. Albertsson, A. Cajarville, D. E. Brooks, and F. Tjerneld, *Biochim. Biophys. Acta* **926,** 87 (1987).

# [8] Thin-Layer Countercurrent Distribution and Centrifugal Countercurrent Distribution Apparatus

*By* HANS-ERIK ÅKERLUND and PER-ÅKE ALBERTSSON

## Introduction

A crucial step in many biochemical studies is the separation and purification of the individual structure of interest. For cells, cell organelles, and membrane fragments the use of aqueous two-phase partitioning has become an important tool. The basis for separation in these phase systems is the surface properties of the material, and the method therefore complements the more traditional centrifugation techniques. Excellent separations have sometimes been obtained in just one or a few partitioning steps. However, in most instances a single partitioning step only gives a partial or incomplete separation of components. By use of a multiple extraction procedure, the resolving power can be drastically increased, and components with only small differences in surface properties can be separated.

Countercurrent distribution (CCD) is, in theory, very similar to chromatography. However, an outstanding feature of CCD is that it is based on several clearly defined steps, each involving an equilibrium partitioning between two liquid phases. This is particularly advantageous when macromolecules are involved, because for these it is thought to be easier to achieve equilibrium between two liquid phases than between a liquid and a solid phase. There are also no steric problems that pertain when using a solid phase. In CCD each substance gives rise to a partition curve that may be compared with a theoretical curve. Such a comparison gives information on the purity of the distributed substance. CCD is therefore an important analytical tool. Because different substances are separated from each other, CCD may also be used for preparative purposes. It is of particular interest that particles can be subjected to CCD for separation and for the analysis of complex mixtures such as cells or cell organelles. The power of CCD in combination with aqueous polymer two-phase partitioning can be illustrated by a few examples. The technique has been used to demonstrate changes in surface properties of red blood cells with age[1]; to show weak interactions between glycolytic enzymes and filamentous

[1] H. Walter and E. J. Krob, *Br. J. Haematol.* **38,** 43 (1978).

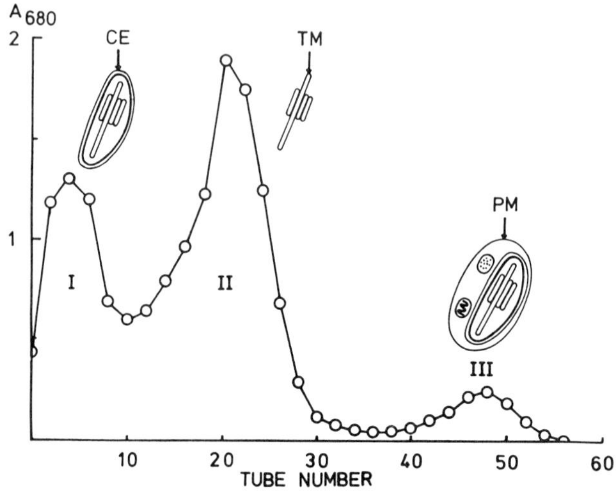


FIG. 1. CCD of a chloroplast sample from spinach cotelydons. The membrane surface exposed to the phase system is indicated. CE, Chloroplast envelope; TM, thylakoid membrane; PM, plasma membrane. [Reproduced and adapted from C. Larsson, B. Andersson, and H.-E. Åkerlund, *in* "Partitioning in Aqueous Two-Phase Systems" (H. Walter, D. E. Brooks, and D. Fisher, eds.), p. 507. Academic Press, Orlando, Florida, and London, 1985, by permission.]

actin[2]; and to obtain separation of isoenzymes differing slightly in charge,[3] three types of chloroplasts (Fig. 1; see also Ref. 4), thylakoid vesicles with opposite sidedness,[5,6] and blood proteins by an affinity technique.[7]

Several different types of CCD apparatuses have been constructed for use with organic solvents. The most widely known design is the all-glass machine of Craig and Post.[8] This can be used for polymer aqueous two-phase systems for both liquid–liquid and liquid–interface CCD.[9,10] A drawback with polymer-phase systems when using the Craig glass machine is the long time needed for phase separation to occur, which is due to the small density difference between the phases and also their high viscosities.

[2] H. Westrin and L. Backman, *Eur. J. Biochem.* **136,** 407 (1983).

[3] G. Blomquist, A. Hartman, V. P. Shanbhag, and G. Johansson, *Eur. J. Biochem.* **48,** 63 (1974).

[4] C. Larsson, this volume [40].

[5] H.-E. Åkerlund and B. Andersson, *Biochim. Biophys. Acta* **725,** 34 (1983).

[6] P.-Å. Albertsson, E. Andreasson, H. Stefansson, and L. Wollenberger, this volume [45].

[7] G. Birkenmeier, this volume [14].

[8] L. C. Craig and O. Post, *Anal. Chem.* **21,** 500 (1949).

[9] P.-Å. Albertsson and E. J. Nyns, *Nature (London)* **184,** 1465 (1959).

[10] P.-Å. Albertsson and G. D. Baird, *Exp. Cell. Res.* **28,** 296 (1962).

This is an obvious disadvantage for the CCD separation of unstable biological material, particularly when many transfers are required for a good separation. Different ways to shorten the separation time have been attempted. Albertsson constructed an apparatus in which the phases form thin layers, about 2 mm, so reducing the settling time by a factor of about 10.[11,12] Attempts have also been made to reduce the settling time by centrifugation, either for small amounts of sample in a continuous system[13] or in a manual system employing a few steps.[14] Åkerlund developed a fully automated apparatus utilizing the increased speed of phase separation obtained in a centrifugal acceleration field.[15] This has allowed the use of phase systems with high polymer concentrations, which are important in affinity partitioning.[16]

## Theoretical Treatment

The principle of the CCD method is well known and has been treated in books and review articles.[17–20] Therefore, only a short description and the relevant formulas will be given here. In CCD a number of extraction steps are carried out in order to separate substances having different partition coefficients, $K$, where $K$ is defined as the concentration ratio, for the partitioned substance, between the two phases. In the following example two substances are distributed in a two-phase system, having equal top and bottom phase volumes, in which their $K$ values are 0.33 and 3, respectively (Fig. 2A). After reaching equilibrium, the amount of substance A in the top phase is one-third of that in the bottom phase. For substance B the amount in the top phase is three times that in the bottom phase. If the total amount of each substance is 100, distributions of the

[11] P.-Å. Albertsson, *Anal. Biochem.* **11,** 121 (1965).

[12] P.-Å. Albertsson, "Partition of Cell Particles and Macromolecules," 3rd Ed. Wiley(Interscience), New York, 1986.

[13] I. A. Sutherland, D. Heywood-Waddington, and T. J. Peters, *J. Liq. Chromatogr.* **8,** 2315 (1985).

[14] B. W. Morris and T. J. Morris, *Eur. J. Biochem.* **121,** 421 (1982).

[15] H.-E. Åkerlund, *J. Biochem. Biophys. Methods* **9,** 133 (1984).

[16] G. Johansson, M. Andersson, and H.-E. Åkerlund, *J. Chromatogr.* **298,** 483 (1984).

[17] L. C. Craig and D. Craig, *in* "Techniques of Organic Chemistry" (A. Weissberger, ed.), 2nd Ed., Vol. 3, Part 1, Interscience, New York, 1956.

[18] L. C. Craig, *in* "Laboratory Manual of Analytical Methods of Protein Chemistry" (P. Alexander and R. J. Block, eds.), Vol. 1, p. 121. Pergamon, Oxford, 1960.

[19] E. Hecker, "Verteilungsverfahren im Laboratorium," Monographien zu "Angewandte Chemie" und "Chemie-Ingenieur-Technik," No. 67. Verlag-Chemie, GmbH, Weinheim, Germany, 1955.

[20] T. P. King and L. C. Craig, *in* "Methods of Biochemical Analysis" (D. Glick, ed.), Vol. 10, p. 20. Wiley(Interscience), New York and London, 1962.

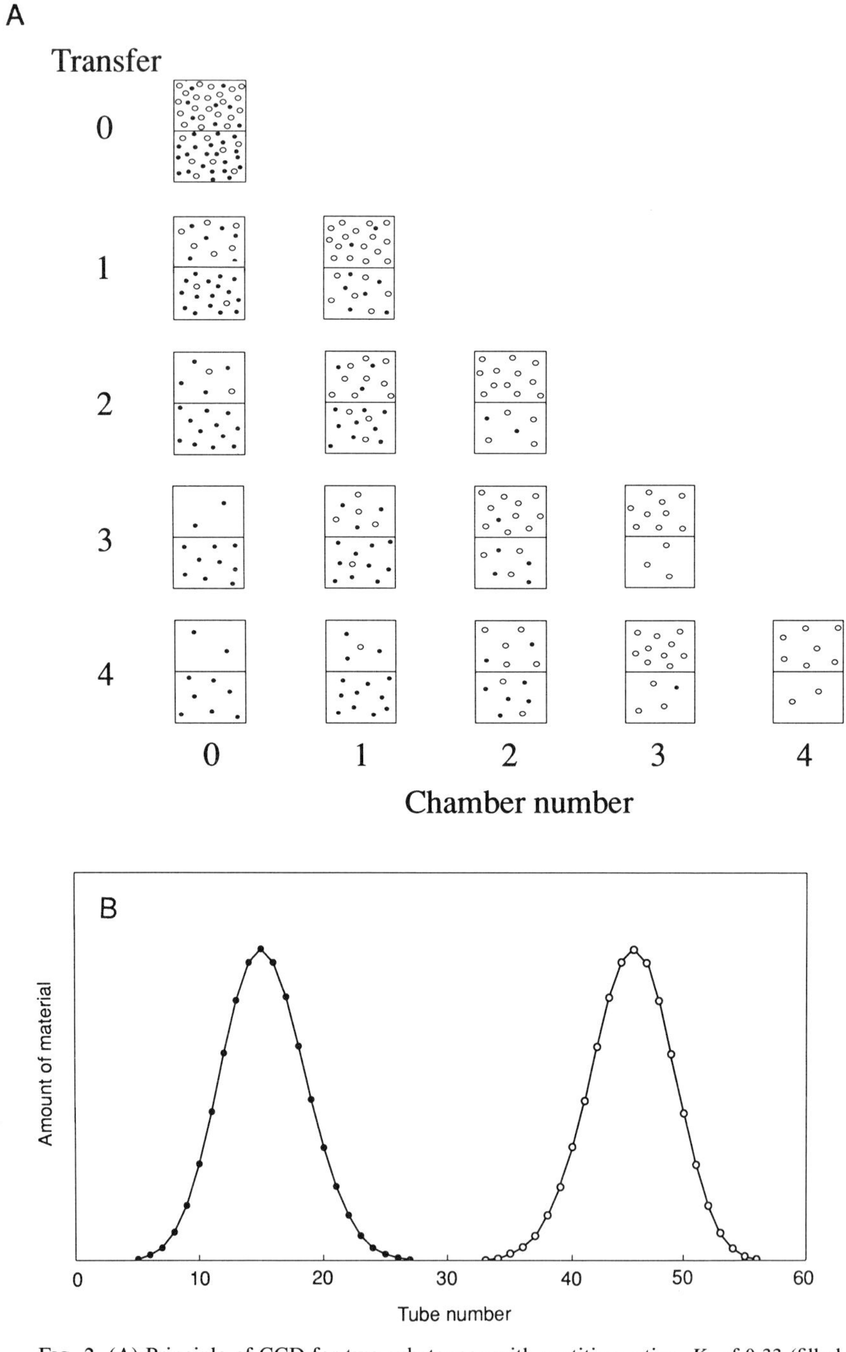


FIG. 2. (A) Principle of CCD for two substances with partition ratios, $K$, of 0.33 (filled symbols) and 3 (open symbols). (B) Theoretical particle distribution obtained after 60 transfer steps.

two substances will be 25/75 and 75/25, respectively (chamber No. 0 in Fig. 2A). The top phase is now transferred to a new chamber, No. 1, containing fresh bottom phase of equal volume as the first one. To the bottom phase of chamber No. 0 fresh top phase of equal volume as that removed is added. The two chambers are shaken, and, after reaching equilibrium, the substances will be distributed as shown in the second row of Fig. 2A. One transfer has now been completed. The top phase of chamber No. 1 is then transferred to chamber No. 2 containing fresh bottom phase; the top phase of chamber No. 0 is transferred to chamber No. 1; and fresh top phase is added to chamber No. 0. After equilibration the material will distribute in accord with the third row of Fig. 2A. The second transfer is thereby completed. The procedure may be repeated several times. With each transfer the number of chambers is increased by 1. The top phases are each transported to the adjacent chamber with higher number. In Fig. 2B the theoretical distribution curves for substances A and B are shown after 60 transfers. The two substances can clearly be separated.

For a large number of transfers, each solute, behaving ideally, gives rise to an approximately symmetrical peak. Its position is determined by the number of transfers and by the partition ratio, $G$ (which is the product of $K$ and the volume ratio). Figure 3 shows examples of some theoretical distribution curves. The following relation holds between $G$, the ratio between the amount of substance in the top phase and the bottom phase of the solute in the chambers, the number of transfers, $n$, and the number of the chamber, $r_{max}$, at peak maximum.

$$r_{max} = nG/(1 + G) \tag{1}$$

The width of the distribution curve increases with the square root of $n$, whereas $r_{max}$ increases linearly with $n$. Thus, substances with only small differences in $G$ can be separated by increasing the number of transfers. For a given number of transfers maximal separation of two substances is obtained if their peaks are symmetrically located on opposite sides of the center of the diagram, that is, $G_1G_2 = 1$.

The distribution scheme so far discussed is based on partitioning between the two bulk phases and is referred to as liquid–liquid CCD. This is the normal procedure for soluble substances like proteins and nucleic acids.[21] In the case of cells, cell organelles, and membranes, however, partitioning usually takes place between one of the bulk phases and the interface.[22] This type of distribution is called liquid–interface CCD. Usu-

[21] G. Johansson, this volume [3].

[22] H. Walter and C. Larsson, this volume [4].

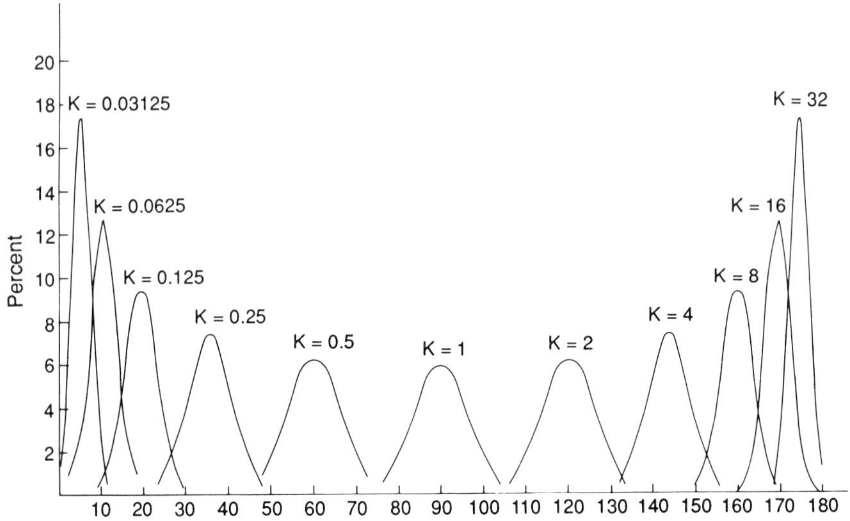


FIG. 3. Theoretical distribution curves as a function of the $K$ value for a substance. The volume ratio is 1, and the number of transfers is 180. [From P.-Å. Albertsson, "Partition of Cell Particles and Macromolecules," 3rd Ed., p. 123. Wiley(Interscience), New York, 1986, by permission.]

ally the distribution takes place between the top phase and the interface. In this case the countercurrent principle can also be applied. The bottom phase, together with the interface and a small layer of the top phase just above the interface, is kept stationary in the apparatus while the top phase is the mobile phase. If we let the volume of the top phase be $V_t$, the volume of the layer of the top phase that is kept stationary in the distribution train be $v$, the concentration of the solute in the top phase be $C_t$, and the amount of the solute adsorbed at the interface be $a$, then the ratio of interest for the CCD is

$$G_i = (V_t - v)C_t/(vC_t + a) \tag{2}$$

When a substance behaves ideally, that is, its adsorption at the interface is reversible and proportional to $C_t$, it will be transported along the distribution train according to $G_i$, and, for calculation of theoretical curves, this value may be used in the same manner as $G$ is used in liquid–liquid CCD. In an analogous manner a distribution between the bottom phase and the interface may be used for CCD. The phase volumes then chosen are such that the top phase, together with the interface and a small layer below it, becomes the mobile layer in the countercurrent experiment while the remaining bottom phase is the stationary phase.

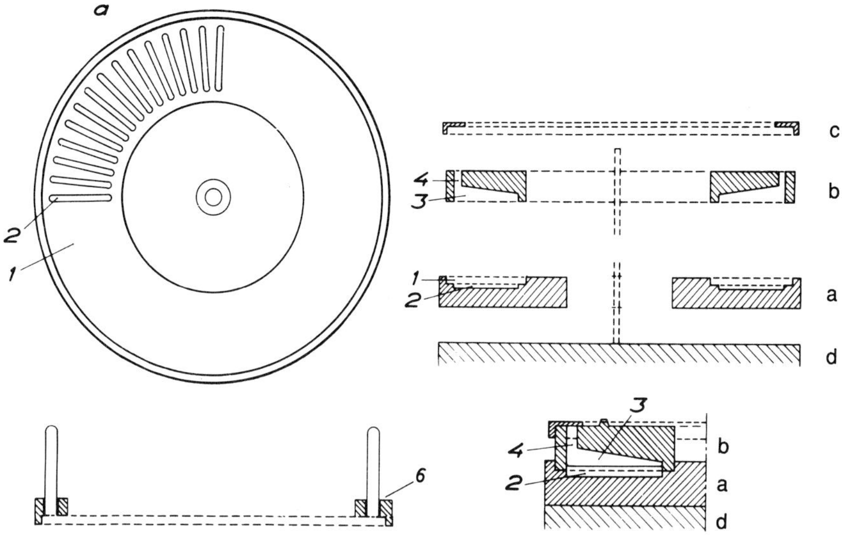


FIG. 4. Details of the partition cell block of the thin-layer CCD apparatus. See text for explanations. [From P.-Å. Albertsson, "Partition of Cell Particles and Macromolecules," 3rd Ed., p. 126. Wiley(Interscience), New York, 1986, with permission.]

## Elution Countercurrent Distribution

The procedure described above is called the fundamental process. It is limited to the same number of transfers as there are partition chambers. An alternative scheme is the elution CCD (or single withdrawal mode) in which the top phase of the last chamber, on each additional transfer, is collected in a fraction collector and new fresh top phase is added to chamber 0. This procedure is reminiscent of chromatography and has been used with thylakoid fragments.[23]

## Thin-Layer Countercurrent Distribution

### *Technical Description*

The thin-layer countercurrent procedure takes place in a partition cell block consisting of two cylindrical plates and constructed of Plexiglas (Fig. 4). The lower plate (Fig. 4, a) has a shallow annular groove (1) that is concentric with respect to the vertical axis. A number of shallow cavities (2) in this groove form the lower parts of the partition chambers and will

[23] P.-Å. Albertsson and E. Hultin, *Anal. Biochem.* **166,** 218 (1987).

contain the bottom phase. The depth of the cavities is normally 2 mm. The upper plate (Fig. 4, b) rests in the groove of the lower plate. The upper plate can rotate about its axis and is guided by the inner and outer edges of the groove. The lower surface of the upper plate, which is in contact with the groove, also has cavities (3) of the same horizontal cross section as the lower plate cavities. The cavities in the upper plate will contain the top phase.

By turning the upper (rotor) plate relative to the lower plate, each of its cavities can be brought into coincidence with a cavity of the lower (stator) plate. A circular sequence of partition chambers thus results. The circular arrangement of the chambers allows recycling procedures so that the number of transfers can be increased to obtain maximum separation. The upper cavities are deeper than the lower cavities to provide air space which permits efficient mixing in each partition chamber during shaking. Each upper cavity is provided with a hole (4, Fig. 4) in the top for filling and for emptying. The upper part of the inner surface of each upper cavity slopes to facilitate the latter process. A circular top (c, Fig. 4) is used to cover the holes.

The partition block rests on a horizontal shaking table (d, Fig. 4) with a driving motor and a device to permit variation in shaking intensity. Mixing of the phase systems is achieved by a planetary movement of the shaking table. In the center of the table there is a unit which drives the rotation of the upper plate while the lower plate remains fixed. All procedures including shaking, settling, and phase transfer are guided automatically by a control unit. One typical setup is shown in Fig. 5.

### *Operation*

The lower and upper plates (Fig. 4, a and b) are put together on the shaking table and adjusted so that the cavities of the two plates coincide, thus forming chambers (Fig. 4, lower right). The sample, in a two-phase mixture (containing the desired top to bottom phase volume ratio), is added to one or a few of the chambers. A suitable volume of top and bottom phase is added to each of the remaining chambers. The cover plate is placed on top and then kept in place by spring-enforced wheels. The apparatus is turned on.

Each cycle in the CCD includes the following. The phases are mixed thoroughly by shaking with a frequency of 180 turns per minute for a suitable period of time. Typically 15–30 sec is sufficient, but, for very viscous phase systems, several minutes may be required. The phases are then allowed to settle. This typically takes 5–15 min, but the time may have to be increased considerably if very viscous phase systems are chosen

FIG. 5. Bench model of a thin-layer CCD apparatus.

or if the density difference between the top and bottom phases is small. After the phases have separated, the lighter phase is transferred to the adjacent chamber by a turn of the upper plate relative to the lower. All this is automatic.

After completion of the CCD, the cover plate is removed and the fractions are collected by use of a collector ring. This unit consists of a circular holder for test tubes with tube positions corresponding to the cavities in the partition block (Fig. 4, lower left). The collector unit is aligned on top of the partition block, and both are jointly and rapidly turned upside down. To improve the recovery in the collecting stage the phases in the partition block can be diluted by adding a sufficient amount of a suitable diluent to each chamber so as to obtain a single homogeneous suspending medium before collection.

## Centrifugal Countercurrent Distribution

### *Technical Description*

The centrifugal CCD appparatus consists of units for centrifugation (Fig. 6, a–c), separation (Fig. 6, d–f), and shaking (Fig. 6, n, o). The separation unit is made up of three Plexiglas parts: an annular housing

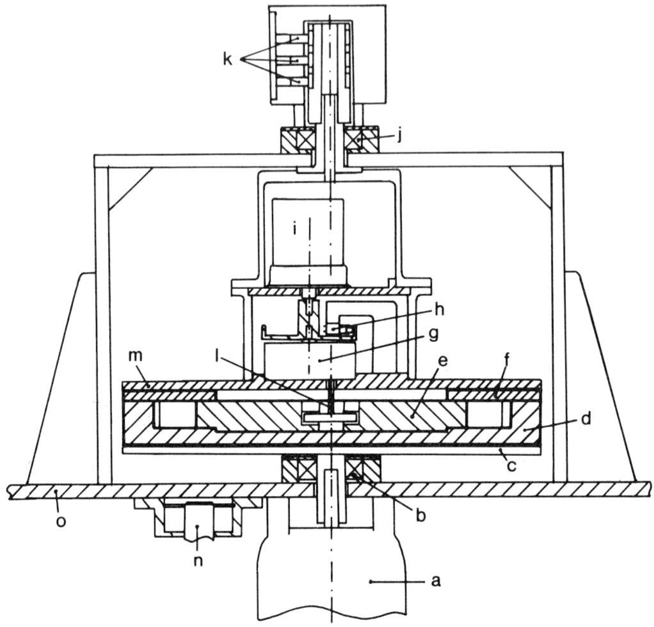


FIG. 6. Cross-sectional view of the central part of the centrifugal CCD apparatus. See text for explanations. [Reproduced from H.-E. Åkerlund, *J. Biochem. Biophys. Methods* **9,** 133 (1984), by permission.]

(Fig. 7, a), a cylindrical body (Fig. 7, d), and a cover plate (Fig. 7, e). Both the cylindrical body and the annular housing have the same number of cavities. Each cavity has the form depicted (Fig. 7, d). The cavities in the cylindrical body have a much larger volume to allow changes in the amount of top phase used and to allow movement of liquid during mixing. The cylindrical body can be inserted in the annular housing. The cavities will coincide and form chambers with a close fit to avoid leakage between chambers. A cover plate is fixed to the annular housing to cover the chambers. To avoid leakage from the chambers during centrifugation an O ring (Fig. 7, f) is located between the annular housing and the cover plate. When mounted, the cylindrical body can be rotated with respect to the annular housing.

The separation unit is connected to the centrifugal unit through the annular housing, and it can be rotated with a speed of up to 3000 rpm, corresponding to an acceleration field of 3000 *g*. The annular housing and the cylindrical body rotate with the same speed. In the original construction a transfer motor (Fig. 6, i) is mounted on top of the separation unit

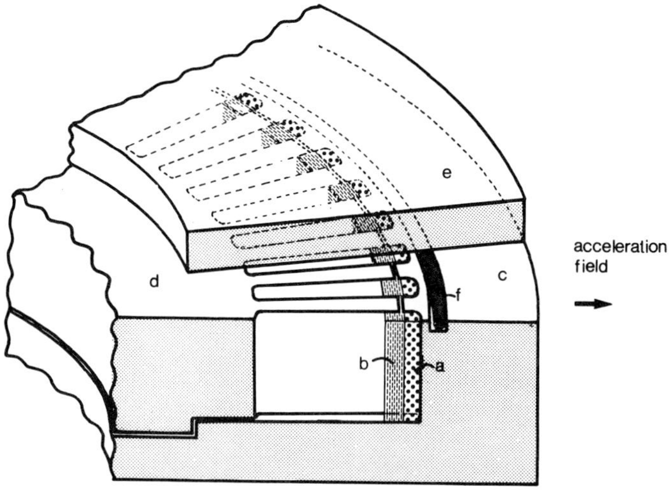


FIG. 7. Section of the separation unit of the centrifugal CCD apparatus. See text for explanations. [Reproduced from G. Johansson, H.-E. Åkerlund, and B. Olde, *J. Chromatogr.* **311,** 277 (1984), by permission.]

and fixed to the annular housing. The axis of this motor is connected to the cylindrical body of the separation unit. Power to the transfer motor is obtained through sliding contacts (Fig. 6, k). The transfer motor is arranged with a gear (Fig. 6, g) and a microswitch (Fig. 6, h) in such a way that each time the relay gets a pulse from the steering unit the cylindrical body turns one step with respect to the annular housing. The units for separation, centrifugation, and transfer are all mounted on a horizontal shaking table (Fig. 6, o). Mixing is achieved by a planetary movement of the shaking table. This setup is shown in Fig. 8. In more recent constructions the transfer motor has been placed below the separation unit. The transfer motor is still fixed to the annular housing. The motor axis is connected to the cylindrical body through a hole in the bottom of the annular housing. All procedures, including shaking, settling, and phase transfer, are guided automatically by a control unit.

### *Operation*

The annular housing and the cylindrical body are put together and adjusted so that the cavities of the annular housing and cylindrical body coincide, thus forming chambers. The sample, in a two-phase mixture (containing the desired top to bottom phase volume ratio), is added to one or a few of the chambers. A suitable volume of top and bottom phase

FIG. 8. Centrifugal CCD apparatus, with protective cover removed.

is added to each of the remaining chambers. The cover plate is placed on top and attached with screws. The whole apparatus is then assembled and started.

Each cycle in the CCD experiment includes the following. The phases are mixed thoroughly by shaking at unit gravity for a suitable period of time. Typically 30 sec is sufficient, but for very viscous systems several minutes may be required. The phases are then separated by centrifugation. Usually an acceleration field of around 100 *g* is sufficient. The separation time required is typically 1 min, including approximately 20 sec for acceleration, but the time may need to be increased for very viscous systems. After the phases have separated and while the separation unit is still rotating at full speed, the lighter phase is transferred to the adjacent chamber by a turn of the cylindrical body relative to the annular housing. The centrifuge motor is then decelerated, and a new cycle can begin. All this is automatic. After completion of the CCD, the apparatus is disassembled and the fractions collected in a manner similar to that described for thin-layer CCD.

## Maintenance and Performance Tests

To achieve good results with the equipment described it is important to handle the Plexiglas components with particular care. These parts have been manufactured with a high degree of precision to prevent leakage between chambers. They are, however, prone to deformation and should only be washed in lukewarm water and, when not in use, stored horizontally on a planar surface. To reduce the possibility of leakage, caused by hydrostatic pressure differences, the separation unit should be adjusted horizontally with the aid of a level. Another way to reduce leakage is to use a very thin layer of silicone grease on relevant surfaces.

To evaluate the performance of the equipment the following experiments may be helpful. For testing the efficiency of mixing or the possibility of leakage between chambers, a CCD experiment is run using a pure substance with a $K$ value of approximately 1. Insufficient mixing or leakage leads to a peak broader than the theoretical one. To check whether the settling time for a particular phase system is adequate, experiments may be run with two different settling times. Insufficient settling will lead to a movement of peaks toward the center of the CCD diagram when the settling time is decreased. A more sensitive test for the settling time is to run a CCD experiment with the particular phase system chosen using a pure substance with an extremely small or large $K$ value.

## Availability

CCD apparatuses can be made by the Mechanical Workshop at the Chemical Center, Box 124, S-221 00 Lund, Sweden, as well as at the Central Mechanical Workshop, Hicks, University of Sheffield, Hounsfield Road, S10 2UN, Sheffield, UK.

# [9] Columns Using Aqueous Two-Phase Systems

*By* WERNER MÜLLER

## Liquid–Liquid Partition Chromatography

Liquid–liquid partition chromatography (LLPC) involves the partitioning of material between two (or more) equilibrated liquid phases while the material migrates through a chromatographic bed. The partitioning of the sample components is maintained, according to Nernst's law, by diffusion of the analytes through both the phases and the phase borders. The presence of a liquid stationary phase in LLPC is the fundamental difference between it and chromatography based on adsorption–desorption phenomena.[1]

The LLPC technique was first described by Martin and Synge.[2] They prepared a two-phase column using water-coated silica as the support material for the stationary phase. The mobile phase consisted of water-saturated chloroform/ethanol mixtures. The analyte to be separated was a silk hydrolyzate. Besides describing the successful separation of the amino acids in the digest, they ascribed the basis of the separatory process to the theory of countercurrent distribution (CCD), which had been developed.

Until the 1970s the LLPC technique was one of the most important separation techniques available for the isolation of low molecular weight natural compounds, especially antibiotics. The support materials were improved because unmodified silica was too adsorptive for many compounds. Cellulose or cross-linked dextran was found to be an excellent substitute that provided sufficient binding capacity for the stationary aqueous phase with good accessibility for small molecules. Furthermore, partitioning was combined with complexation for modifying the partition coefficient, $K$, of the target molecule. For example, crude mixtures of actinomycin antibiotics (molecular weight about 1300), in which the main constituents differed by only one methyl group, could easily be separated in quantities of 15 g on a column ($60 \times 4$ cm) in a few hours. In this case the aqueous phase contained sodium cresotinate (as complexing agent), and the organic phase consisted of butanol and di-*n*-butyl ether.[3]

[1] J. F. K. Huber, M. Pralowska, and P. Markl, *Chromatographia* **19,** 19 (1984).
[2] A. J. P. Martin and R. L. M. Synge, *Biochem. J.* **35,** 1358 (1941).
[3] H. Brockmann, *in* "Fortschritte der Chemie Organischer Naturstoffe" (H. Zechmeister, ed.), Vol. 18, p. 1. Springer-Verlag, Wien, 1960.

For biopolymers like proteins and nucleic acids LLPC could not be used because these materials are either insoluble in organic solvents or denature in such media. When Albertsson introduced aqueous polymer two-phase systems[4] in 1958 their potential for LLPC was obvious. However, the classic support materials exhibited essentially no preference for binding one of the two aqueous phases so as to render it stationary in the column bed. Only cellulose bound reasonable quantities of the lower, dextran (Dx)-rich, phase of the two-phase system composed of poly(ethylene glycol) (PEG) and Dx. However, most of the phase bound turned out not to be accessible for large molecules like proteins or nucleic acids. The phase support problem was finally solved by making use of the principles underlying phase formation in aqueous polymer two-phase systems.[5]

## Liquid–Liquid Partition Chromatography with Aqueous Two-Phase Systems: Basic Principles

### *Principal Setup of Columns Operating with Aqueous Polymer Two-Phase Systems*

An LLPC column consists of an ordinary chromatographic tube filled with a support material coated with a stationary phase which is suspended in a mobile phase. A schematic description of such a column is shown in Fig. 1. The only unusual feature of such a column is the support material. It consists of either cross-linked cellulose or a hydrophilic polymer or silica–diol onto which non-cross-linked polyacrylamide chains have been grafted or Dx chains have been fixed. Such supports "bind" the Dx-rich lower phase of Albertsson's PEG–Dx system so tightly as to prevent its loss during perfusion of the mobile PEG-rich phase. The volume ratio of the stationary phase to the mobile phase often reaches 1 : 1 when measured using low molecular weight test compounds. With larger test compounds, which are unable to penetrate the smaller pores of the support, values of 1 : 4 are not uncommon.

The sample to be fractionated, dissolved in mobile phase, is applied to the column. The column outlet should be connected to a UV monitor and, for preparative purposes, to a fraction collector. When labile compounds are to be separated, the column can be run in a cold room. To obtain reproducible elution volumes the column should be thermostatted. When very large molecules, like nucleic acids, that have very small diffusion coefficients are loaded, it is best to run the column at 40° or even 50°.

[4] P.-Å. Albertsson, *Biochim. Biophys. Acta* **27,** 378 (1958).

[5] W. Müller, *Eur. J. Biochem.* **155,** 213 (1986).

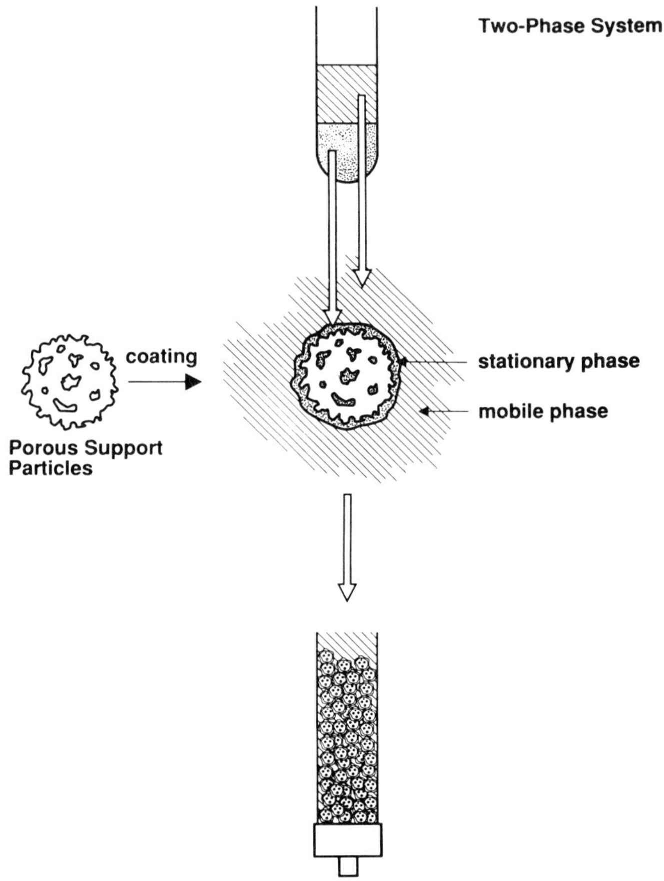


FIG. 1. Scheme visualizing the principle of liquid–liquid partition chromatography columns. [From W. Müller, *Bioseparation* **1,** 265 (1990), with permission.]

*Fundamental Rules of Chromatography*

The basic relations governing LLPC were established by Martin and Synge in 1941[2] as follows:

$$V_{el} = V_m + V_s K_A \tag{1}$$

where $V_{el}$ is the elution volume of a compound A applied to a LLPC column, $V_m$ is the volume of mobile phase, $V_s$ is the volume of stationary phase, and $K_A$ is the partition coefficient defined as the concentration of A in the stationary phase over the concentration of A in the mobile phase.

*Performance.* The width of a peak, that is, the zone within which A migrates through the column, is a function of the performance of the column. This parameter consists of the packing quality (determined by the particle size, size distribution, and arrangement in the tube), the deviation of the analyte from the equilibrium state when moving through the column, and the longitudinal diffusion of the analyte (which plays only a minor role with macromolecules). The only analyte-independent parameter of a column is the quality of the packing. The performance is expressed in terms of the number of theoretical plates per unit length of the column; the greater this number is, the smaller will be the peak width.

The performance is determined from the elution profile of a noninteracting analyte by measuring the width (at the base) of the peak, $W$, and its elution volume, $V_{el}$. The number of theoretical plates ($N$) of the column is calculated as follows:

$$N = 16(V_{el}/W)^2 \tag{2}$$

*Selectivity.* Selectivity is another parameter used for describing the separation potential of a column. The term selectivity indicates the distance between the peaks of two compounds, A and B, eluting from the column. In LLPC the selectivity depends[5] on the volume of the stationary phase in the column and the relative affinities of A and B for the stationary phase as manifested by their partition coefficients, $K_A$ and $K_B$:

$$V_{el,B} - V_{el,A} = V_s(K_B - K_A) \tag{3}$$

*Resolution.* The last parameter used to characterize a chromatographic separation is resolution. This most important parameter for practical purposes indicates the degree of separation of two compounds, A and B, and is a function of the performance of the column and the selectivity:

$$R_s = N^{1/2}V_s(K_B - K_A)/(V_m + V_sK_B) \tag{4}$$

From the $R_s$ values the cross-contamination of two compounds after separation can be calculated. For practical purposes it may be sufficient to know that $R_s = 0.5$ is equivalent to a 20% cross-contamination and a value of 1 corresponds to 4% cross-contamination. Good separations with cross-contaminations less than 0.3% are characterized by $R_s$ values greater than 1.5.

## Liquid–Liquid Partition Chromatography Columns Using Aqueous Two-Phase Systems: Conditions for Stability

In a classic LLPC column using aqueous/organic or organic/organic two-phase systems, a continuous release of stationary phase (“bleeding”) can often be observed that results in a continuous change in the ratio of the mobile phase to stationary phase. This effect can make it difficult to use such columns for analytical purposes (i.e., to identify a compound according to its elution position), because of changes in $V_m$ and $V_s$. In contrast, when aqueous two-phase polymer systems were used in columns, with cellulose as the support material, it was observed that (after a relatively short equilibration time) a constant value (of about 0.4 bed volumes of stationary phase) was established, which did not change with further use of the column. On the other hand, equilibrating untreated cellulose by percolation with mobile phase caused a continuous formation of stationary phase in the column. These effects pointed toward a strong affinity of the cellulose for Dx, the main component of the stationary phase. All attempts to demonstrate such an affinity directly (e.g., by passing a diluted Dx solution through packed cellulose) failed; no retardation of the polysaccharide could ever be observed. Silica-diol, on the other hand, shows a pronounced binding capacity for the PEG-rich phase, which is maintained even when the Dx component is replaced by a salt, such as potassium phosphate or ammonium sulfate. This binding of phase displays all the properties of an adsorption process (i.e., a rather unstable coating).

When silica–diol is coated with polyacrylamide by grafting,[5,6] the Dx- or salt-rich phase is bound to the support instead of the PEG-rich phase. This behavior can be explained by the fact that non-cross-linked polyacrylamide (2–3%, w/v) and PEG 8000 (10%, w/v) in an aqueous solution form a two-phase system. Thus, PEG and polyacrylamide are incompatible.[7] Dextran and polyacrylamide, however, appear to be fully compatible in the concentration ranges given above. These findings strongly suggest that the "phase-binding" effects observed actually result from incompatibility effects rather than from affinities. A support material coated with polyacrylamide, by repelling the PEG-rich phase, promotes the highly preferential wetting by the other, either Dx- or salt-rich, phase.

From the above it is clear that the coating by the Dx-rich phase in a LLPC column run with PEG–Dx systems is much more stable than is the case with stationary phases in the classical LLPC columns. The incompatibility between PEG and polyacrylamide is so strong that the amount of Dx present in the PEG-rich phase on equilibrium is extracted by the support as it builds up an insulating Dx-rich phase against the PEG-rich phase. This results in an equilibrium situation that promotes the high, long-term stability of the Dx-phase coatings.

[6] G. Mino and S. Kaizerman, *J. Polym. Sci.* **31,** 242 (1958).

[7] W. Müller, *in* "Macromolecular Symposia" (H. Höcker, ed.), Vol. 17, p. 257. Hüthig & Wepf Verlag, Basel, Heidelberg, and New York, 1988.

## Aqueous Two-Phase Systems Suitable for Liquid–Liquid Partition Chromatography of Biopolymers

### *Poly(ethylene Glycol)–Dextran Systems*

The most widely used aqueous polymer two-phase systems, namely, PEG–Dx systems, have been applied in LLPC. Besides Dx and PEG they contain a buffer and various salts. All the components should be reasonably UV transparent. PEG, which is a polyether, is susceptible to peroxide formation that may cause chain breaks and result in aldehyde and carboxyl groups at the breaks. It is recommended that fresh material be purchased directly from the manufacturer and that it be stored in the dark in a cold, dry place so as to avoid decomposition. For long storage, fast melting under reduced pressure or under argon and then keeping the resolidified polymer in small blocks is advisable. PEGs with a molecular weight of 1500 to 20,000 melt between 62° and 65°.

Dextran is offered by a number of manufacturers. The product quality is adequately high so as not to require further purification. The molecular weight distribution is not narrow, in general, which is documented by ratios of weight average molecular weight ($M_w$) over number average molecular weight ($M_n$) of about 2. The chemical stability of Dx is higher than that of PEG; nevertheless, carboxyl functions may also form by end-group oxidation, especially when Dx is stored as aqueous solutions. Saturating the solution with nitrogen or, better still, with argon provides excellent protection.

Selection of the quantity of polymers to be used in a two-phase system for LLPC requires a compromise to be made between the concentration and the phase viscosity. Higher polymer concentrations (corresponding to longer tie lines in the phase diagrams[8]) generally increase the selectivity of the system. Increased phase viscosity, on the other hand, reduces the mass transfer rate in the chromatographic process, which manifests itself in a broadening of the peak. This effect becomes more pronounced as the diffusion coefficient of the analytes becomes smaller.

Most of the applications described here have been performed with the following two systems (also supplied with various salts):

System I: 9.8% (w/w) Dx T40 (molecular weight 40,000, from Pharmacia-LKB, Uppsala, Sweden), 5.4% PEG 6000 (molecular weight 6000, from Merck, Darmstadt, Germany)

System II: 6.2% (w/w) Dx T500 (molecular weight 500,000, from Pharmacia-LKB), 4.4% PEG 8000 (molecular weight 8000, from Union Carbide, New York, NY)

[8] P.-Å. Albertsson and F. Tjerneld, this volume [1].

The polymer content in the individual (i.e., mobile or stationary) phases is as follows:

System I: Mobile phase: 9.3% (w/w) PEG, 1.2% (w/w) Dx; stationary phase: 1.6% (w/w) PEG, 18.3% (w/w) Dx
System II: Mobile phase: 7.3% (w/w) PEG, 0.35% (w/w) Dx; stationary phase: 1.1% (w/w) PEG, 13.6% (w/w) Dx

The PEG–Dx systems provide a gentle and nondenaturing environment for proteins and permit, in most cases, LLPC separations at room temperature that would normally require cold room conditions. A disadvantage of these systems is the incompatibility between some of the larger proteins and PEG. This causes protein precipitation (e.g., of immunoglobulin or aldolase). Because the solubility of the immunoglobulins G (IgGs) in the PEG-rich phase is only 0.2 mg/ml, LLPC with such compounds may only be used for analytical purposes.

The selectivity of polymer two-phase systems is partly a result of the fact that one of the polymers is slightly more hydrophobic than the other. This implies that the surface hydrophobicity of the analyte forms one of the selection criteria of the system. Another important property of the system is the electrostatic potential difference between the phases. This property, related to the asymmetric affinity of ions for the two phases, provides a basis for separating analytes according to their net charges.

*Poly(vinylpyrrolidone)–Dextran Systems*

Two-phase systems based on poly(vinylpyrrolidone) (PVP) were studied so as to overcome the incompatibility of PEG with large proteins and thus provide higher capacities for preparative chromatography.[9] The PVP reagent is easily available in molecular weight ranges from several thousand to several hundred thousand. A PVP with $M_w$ of 45,000 and $M_n$ of 10,000 (KollidonR 30, BASF, Ludwigshafen/Rhein, Germany has been used in the separations described. The commercial material is, in general, not very clean and produces additional yellow, UV-absorbing products when exposed to light. This effect and the noticeable decrease in selectivity of PVP–Dx systems in comparison to PEG–Dx systems justifies its use only in cases in which a strong incompatibility between analyte and PEG must be overcome. The phase diagram of the most widely used PVP–Dx systems) can be obviated by exchanging 25% of the PVP present in the system for PEG 6000.[9] The phase diagram of the PVP–Dx system is not appreciably affected by this polymer substitution.

[9] F. Engelhard, Diplomarbeit, Fachhochschule Darmstadt, Germany, 1987.

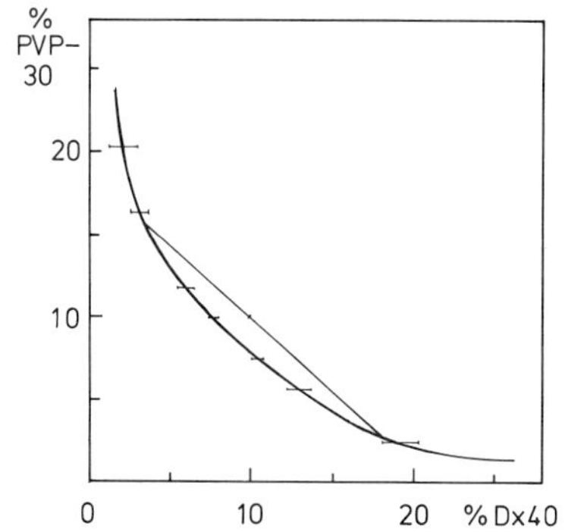


FIG. 2. Phase diagram of poly(vinylpyrrolidone) (PVP 30)–dextran (Dx T40) in water at 23°. [From F. Engelhard, Diplomarbeit, Fachhochschule Darmstadt, 1987.]

*Poly(vinyl alcohol)–Dextran Systems*

Exchanging poly(vinyl alcohol) (PVA) for PEG results in systems in which protein–polymer incompatibilities have not been observed but which have a rather high viscosity. A PVA having a molecular weight of 10,000 and containing 11.5% *O*-acetyl groups (Aldrich, Steinheim, Germany) has been used for LLPC[10] in combination with Dx having a molecular weight of 40,000. The phase diagram of such a system is shown in Fig. 3. The PVA–Dx systems do not give rise to electrostatic potential differences between the phases in response to incorporation of salts which induce such potential differences in PEG–Dx systems. Thus, high molecular weight DNA cannot be shifted into the PVA-rich phase by incorporation of salt.

*Polymer–Salt Systems*

Three PEG–salt systems have been selected for use in LLPC[11]:

System I: PEG 1500–potassium phosphate, pH 7.1
System II: PEG 1500–trisodium citrate, pH 8.65
System III: PEG 1500–potassium phosphate, pH 9.0

The compositions for the systems are given in Table I.

All three systems show a limited protein capacity owing to protein incompatibility with PEG, resulting in $K$ values below 0.1 for most proteins. Raising the pH from 7 to 9 (system I versus system III) raises the

[10] M. Ruppel, Diplomarbeit, Fachhochschule Darmstadt, Germany, 1989.
[11] S. Böse, Diplomarbeit, Fachhochschule Darmstadt, Germany, 1991.

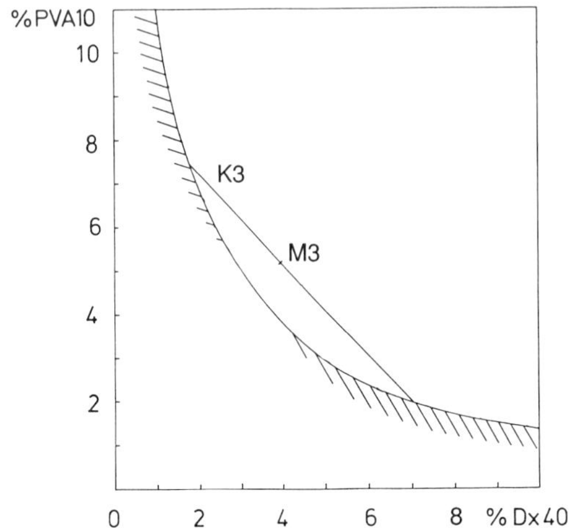


FIG. 3. Phase diagram of poly(vinyl alcohol) (PVA)–dextran (Dx T40) in water at 23°. [From M. Ruppel, Diplomarbeit, Fachhochschule Darmstadt, 1989, with permission.]

*K* values for a number of proteins without improving the situation. The large amount of salt present in all these systems alters the chromatographic mode from one of LLPC to that of hydrophobic interaction chromatography (HIC) in which the hydrophobicity of the support plays an important role. Consequently, the close relation between *K* and the elution position of a compound, as given by Eq. (1), is no longer valid. This means that one of the most interesting features of LLPC, namely, the prediction of a separation of two compounds based on their *K* values, is lost.

## Liquid–Liquid Partition Chromatography with Aqueous Two-Phase Systems: A Practical Guide

### *Setting up and Testing Columns*

The polymer-based support materials for LLPC were marketed as LiParGel by E. Merck until the end of 1990. Two types were available

TABLE I
COMPOSITION OF POLY(ETHYLENE GLYCOL)–SALT SYSTEMS USED IN LIQUID–LIQUID PARTITION CHROMATOGRAPHY[a]

| System I | System II | System III |
|---|---|---|
| 11.25% PEG 1500 | 16.75% PEG 1500 | 14.80% PEG 1500 |
| 12.75% Potassium phosphate | 16.0% Sodium citrate | 10.8% Potassium phosphate |
| pH 7.10 | pH 8.65 | pH 9.0 |

[a] From S. Böse, Diplomarbeit, Fachhochschule Darmstadt, 1991, with permission.

(LiParGel 650 and 750) differing only in pore size. The second, having wider pores, was recommended for separating larger proteins (>80,000 Da) or nucleic acids and, especially, when two-phase systems containing large polymers like Dx T500 were used. These materials will now have to be prepared like the silica-based supports by a grafting polymerization process as described in detail in the literature.[5] Superdex (Pharmacia) may be used instead[12] if pronounced nonspecific adsorption effects can be tolerated. For coating the support material, an appropriate amount of wet LiParGel is rinsed on a Büchner funnel with three to four bed volumes of stationary phase using a water pump. For effective coating the flow rate obtained through a packing of equal height and diameter is essentially correct. The saturation of the support may be tested by adding 1–2 drops of mobile phase to about 20 drops of flow-through phase. An immediately resulting turbidity indicates saturation. It is important to prevent air from entering the gel bed because this effectively blocks the coating process.

The excess of stationary phase present in the gel bed is removed by rinsing the latter with 2–3 volumes of mobile phase; the process may be stopped as soon as the flow-through phase becomes clear. For packing the chromatographic column the gel is suspended in about three volumes of mobile phase and packed in a glass column fitted with at least one adjustable plunger. The column must be able to withstand a pressure of at least 50 bar at an internal diameter of 1 cm. Columns of the Superformance series (E. Merck) have been specially designed for this purpose. LiParGels may be packed at flow rates producing a back-pressure of up to 15 bar; silica-based supports may be packed at back-pressures of up to 70 bar; working at 50 bar is generally adequate for 10-$\mu$m particles.

The column is equilibrated with mobile phase at a maximum linear flow of 1 cm/min until the UV noise of the effluent has dropped below 0.005 OD units at 280 nm. The sample dissolved in the mobile phase may then be applied to the column with a valve common in high-performance liquid chromatography (HPLC) equipped with an adequately large sample loop. The sample volume should not exceed 2–3% of the bed volume in analytical runs; in preparative runs the sample volume may be twice as large, depending on the separation.

To test the general functioning of the column, two test compounds in the same molecular weight range as the analytes, but differing by at least a factor of 2 in their $K$ values, should be separated in a test run. Well-characterized and easily available standard proteins in the weight range

[12] K. Andersson, C. Wingren, and U.-B. Hansson, *J. Chromatogr.* **603,** 73 (1992).

of 15,000 to 70,000 are suitable. Horse heart myoglobin and horseradish peroxidase have been used for this purpose.[13]

Data from such a test run provide values for the volume of stationary and mobile phase per unit column volume and for the number of theoretical plates. The two specific phase volumes are calculated from the following two expressions[5] derived from Eqs. (1) and (3):

$$V_s = (V_{el,B} - V_{el,A})/(K_B - K_A) \tag{5}$$

$$V_m = V_{el,B} - V_s K_B \tag{6}$$

The number of theoretical plates of the column, standardized for a 1-m column length, may be calculated from Eq. (2) or, if the test compound contains some minor impurities broadening the peak base, from the half-width of the peak using the following well-known expression:

$$N = 5.5(V_{el}/\beta)^2 \tag{7}$$

in which $\beta$ is the half-width of the peak.

Typical values for $V_s$ (in percent of total column volume) for LiParGel columns prepared with the PEG–Dx system I range from 18 to 25; the corresponding values for $V_m$ range from 25 to 35, using the two proteins mentioned above. Values for the performance of LLPC columns are typically between 1500 and 2300 plates/meter for LiParGel columns set up in the Superformance series (Merck) with the PEG–Dx system I at a linear flow rate of 0.2 cm/min.

### *Sample Preparation and General Run Parameters*

In LLPC the sample is not concentrated at the top of the column as in ion-exchange or hydrophobic chromatography. This means that the minimum zone width of an analyte is given by the volume of the sample loaded as in gel permeation chromatography. Therefore, the ideal preparation of the sample solution consists of dissolving the salt-free, dry analytes in as little mobile phase as possible. Unfortunately there is an upper limit for the concentration of high molecular weight compounds in the sample which is raised by the viscosity of the sample. If sample viscosity exceeds solvent viscosity by more than a factor of 3, the equilibration time of the sample constituents between the phases becomes large compared to the flow rate. This results in a major loss in column performance, a phenomenon especially severe when large nucleic acids are to be separated. Reducing the flow rate and raising the temperature often help to resolve this problem.

[13] W. Müller, "Liquid–Liquid Partition Chromatography of Biopolymers." GIT-Verlag GmbH, Darmstadt, Germany, 1988.

In practice the analyte mixture to be separated hardly ever exists in a dry, salt-free state. In the worst case the mixture will be an aqueous solution which contains various salts of unknown concentrations. A gel-permeation step or microdialysis against a medium containing the same salts as present in the mobile phase may be used to adjust the electrolytes and their concentrations, provided the dilutions normally accompanying these operations are tolerated. If not, ultrafiltration can be applied if the nonspecific adsorption to the filter is adequately small so as not to cause too much loss of analyte. Such manipulations will finally provide the analyte(s) in an aqueous solution containing the appropriate salts in correct amounts. The solution is then adjusted to the concentration of the main polymer present in the mobile phase by dissolving the appropriate amount of dry polymer in the solution. The second (minor) polymer present in the mobile phase may be omitted.

The sample thus prepared may be loaded on the column with the usual application device common in HPLC (e.g., an inert Rheodyne valve) equipped with an adequately large sample loop. The flow rate used depends to some extent on the size of the analytes that determine the bulk viscosity of the sample. In general, a linear flow rate of 0.4 cm/min for proteins up to 200,000 Da at a concentration of 2 mg/ml can be used at room temperature. For DNA samples above $3 \times 10^6$ Da the flow rate should be reduced to about two-thirds or one-half of the above value and the temperature should be raised to 35–40°. It is clear that running the column at a temperature which differs from the system equilibration temperature requires a short preequilibration of the column with the mobile phase from a phase pair equilibrated at the newly selected temperature.

Scaling up a LLPC column is easy if one considers a thick column to be like a bundle of thin columns fed simultaneously with the same eluant and by the same pump. At constant length the resolution of the thick column, in an isocratic run, will be the same as that of the thin one.

### *Recovering Biopolymers from Mobile Phases*

Recovery of analytes separated by the LLPC process from the eluates means separating them from the phase-forming polymer. This can be done by passing the eluate, after addition of water (to dilute the two-phase system and render it homogeneous and to reduce the salt concentration), through a small bed of a good cation or anion exchanger. Because the phase-forming polymer is, in general, nonionic, it passes through the column and may be washed off by rinsing the bed at low ionic strength. The analyte may then be recovered by a salt elution step from the column at high concentration and may be dialyzed against the final medium. Eluting the analyte in a small volume is crucial to obtain the final product in an adequate concentration. Very fast ion exchangers free of nonspecific side activities are required for this step; the "tentacle"-type ion exchangers (E. Merck) are especially suited for this purpose.

## Conclusions

The most attractive feature of LLPC is its unusual selectivity, which is based on a balance between electrostatic and hydrophobic effects. One extraordinary property, in contrast to the "phases" usually applied in biochromatography, is that the analyte is never fixed on a solid surface in LLPC, but always remains in solution. This reduces the probability of nonspecific side reactions caused by surface interaction and aggregation. For low molecular weight compounds with high diffusion coefficients the transition from one phase into the other is fast compared to the usual flow rates, and the advantage of having the analytes in a dissolved state can be fully exploited. With macromolecules the small diffusion coefficients may sharply reduce the mass transfer and thus limit the applicable flow rates. In addition, incompatibility effects often limit capacities to values below the limits required for economic reasons. Thus, applying LLPC for protein purification should only be considered when unusual selectivity is required or if the method presents an appreciable advantage. The problems may be ameliorated, however, as new polymers more suited for LLPC become available.

# [10] Magnetically Enhanced Phase Separation

*By* PER-OLOF LARSSON

## Introduction

One prerequisite for achieving fast separation/isolation of a desired protein using aqueous two-phase systems is that the phase separation itself is rapid. Thoroughly mixed, standard aqueous two-phase systems, such as dextran (Dx)poly(ethylene glycol) (PEG), will fairly rapidly separate into two distinct phases when the mixture is allowed to stand. However, this description is valid only for "clean" systems. When large amounts of proteins and other cellular components are present, stabilized emulsions can form that will slow phase separation considerably (hours instead of minutes). The standard way for dealing with such a situation

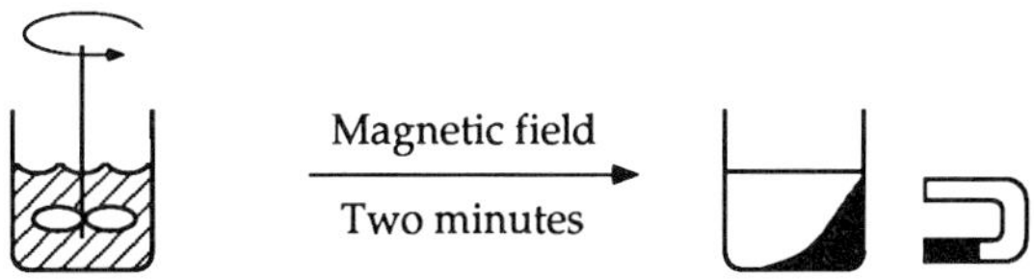


FIG. 1. Magnetically enhanced phase separation. The magnetic phase contains paramagnetic material, either micron-sized iron oxide particles or nanometer-sized, coated iron oxide particles (ferrofluid).

is to use some kind of centrifugal separator. Figure 1 shows an alternative approach: magnetically enhanced phase separation. This method seems to be especially useful when (1) two phases have very similar densities, (2) the volumetric ratio between the phases is very high or low, or (3) the systems are viscous. Magnetically enhanced phase separation usually increases the speed of phase separation by a factor of about 10 in "well-behaved" systems, but it may increase by a factor of many thousands in "difficult" systems.[1]

## Materials

### *Micron-Sized Iron Oxide Particles*

Micron-sized iron oxide particles are available from various sources (e.g., as precipitated $Fe_3O_4$ from BDH, Poole, UK). The main advantages of this material are its availability and the fact that it is strongly influenced even in moderate magnetic fields. The particles have been used with about 25 aqueous two-phase systems, and in all except 1 the particles partitioned into one of the phases exclusively.[1] One distinguishing property of the micron-sized particles is that they may be collected from the phase into which they partition. After the completion of a magnetically enhanced phase separation the system consists of two layers, for example, a PEG layer and a Dx layer with particles as shown in Fig. 1. If the magnetic field continues to be applied, particles aggregate close to the magnet and can be isolated as a third phase. This may be of advantage in many cases. The phenomenon is more pronounced the stronger the magnetic field.

[1] S. Flygare, P. Wikström, G. Johansson, and P.-O. Larsson, *Enzyme Microb. Technol.* **12,** 95 (1990).

*Dextran-Based Ferrofluid*

The iron oxide particles in a ferrofluid are of molecular dimensions (typically a few hundred angstroms) and cannot be removed from the ferrofluid. This is of advantage because there is no danger that a strong magnetic field could accidentally aggregate the particles. Major aggregation of particles would cause the procedure to become inoperative, since the magnetic field would force the particles out of the two-phase system, a phenomenon observed with micron-sized particles as mentioned above. Ferrofluids have to be prepared using special techniques, and comparatively strong magnetic fields are required to obtain good results.

Ferrofluid stabilized with Dx[2] may be prepared by the following procedure, which is a modification of the one given by Molday and Mackenzie[3] for starch. $FeCl_3 \cdot 6H_2O$ (7.55 g) and $FeCl_2 \cdot 4H_2O$ (3.20 g) are dissolved in 50 ml distilled water and mixed with 50 ml of a 20% (w/v) solution of Dx T500 (Pharmacia-LKB Biotechnology, Stockholm, Sweden). The temperature is raised and maintained at 60°. Aqueous ammonia (100 ml, 7.5%) is added in small portions during a 5-min period with vigorous stirring. After the addition, the stirring is continued for 15 min at 60°. The suspension is cooled to room temperature and dialyzed against distilled water to neutral pH. Solid material is removed by centrifugation (12,000 *g* for 15 min at room temperature). The iron content of the Dx–ferrofluid preparation is typically 7 mg/ml (determined by atomic absorption), and the dry weight content is about 50 mg/ml.

It has been noted[2] that some preparations of ferrofluid affect the activity of enzymes when the enzymes are present at low concentrations and in the absence of other proteins, owing to poorly coated iron oxide particles. Because such particles are prone to aggregation to larger entities, they are usually removed by an additional centrifugation step (12,000 *g* for 15 min at room temperature).

## Separation Experiments

*Separation of Enzymes in Multistage Magnetic Separators*

A test mixture of lactate dehydrogenase and β-galactosidase has been separated by partitioning in a DxPEG two-phase system using a multistage (15 stages) magnetic separator (Fig. 2), operated in semiautomatic fashion.[2] The peristaltic pump and stirrer (Fig. 2) are controlled by an electronic switch, allowing the times for pumping, stirring, and settling to be indepen-

[2] P. Wikström, S. Flygare, A. Gröndalen, and P.-O. Larsson, *Anal. Biochem.* **167,** 331 (1987).
[3] R. S. Molday and D. MacKenzie, *J. Immunol. Methods* **52,** 353 (1982).

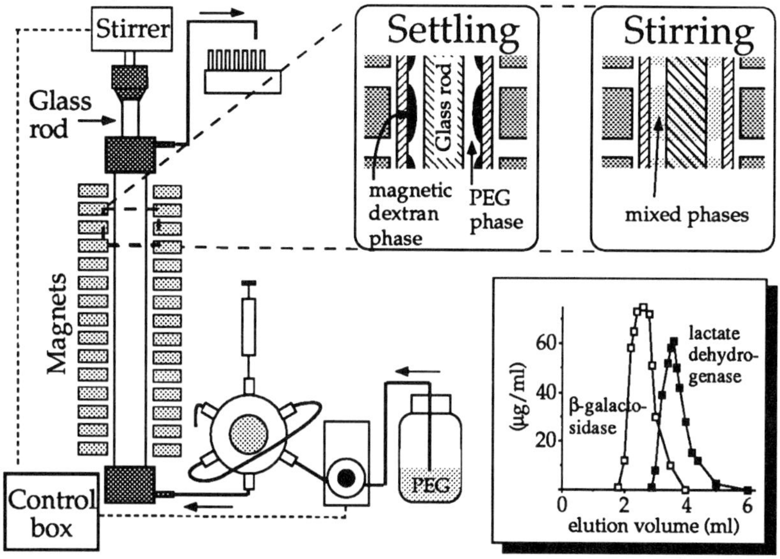


FIG. 2. Separation of a test mixture of bovine heart lactate dehydrogenase and *Escherichia coli* $\beta$-galactosidase with a multistage magnetic separator. A precision-bore glass tube (0.60 × 25 cm) was provided with thirty permanent magnets (1.5 × 5.0 × 0.6 cm) giving a magnetic field of about 0.2 tesla in the middle of the tube. Teflon end pieces kept a stirring rod (0.53 × 30 cm) centered in the tube. The internal volume was 1.5 ml. Operating conditions: Stationary magnetic Dx, 1.5 ml; separation cycle time, 30 sec (stirring 10 sec, settling 5 sec, pumping of PEG phase 15 sec); stirring rate, 400 rpm; pumping rate, 1.5 ml/min. The enzyme test mixture (90 + 90 $\mu$g) was dissolved in 0.1 ml PEG phase (7%, w/w, PEG 6000 equilibrated with Dx phase) and introduced in the separator via a high-performance liquid chromatography (HPLC) injector. Fractions (0.1 ml) were collected and analyzed.

dently set. During the stirring period the phases are thoroughly mixed and equilibrated (see inset in Fig. 2). For the subsequent settling period the stirrer is stopped and the Dx phase containing ferrofluid coalesces and is drawn toward the magnets (see inset in Fig. 2). Finally, the pump is started, and the continuous PEG phase is pumped while the Dx phase remains close to the magnets. The pump is stopped and a new cycle begun. The time for a cycle is 30 sec.

The test mixture is resolved into two fairly sharp peaks (Fig. 2). The number of plates is calculated[4] and found to be in the range of 50–100, clearly above 15. This indicates that each magnetic stage corresponds to more than one separation step.

[4] L. R. Snyder and J. J. Kirkland, "Introduction to Modern Liquid Chromatography." Wiley, New York, 1979.

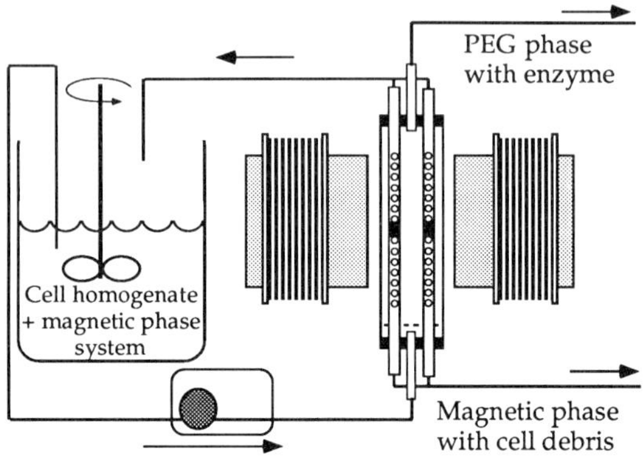


FIG. 3. Preparative-scale continuous separator. A Perspex tube (3.3 × 30 cm) was placed in the air gap of an electromagnet. The pole surfaces facing the air gap had the dimensions 1.5 × 2.5 cm. In typical runs the field strength in the middle of the separator was about 0.3 tesla. The electromagnet was driven by a power supply for electric welding capable of a continuous output of 1200 W (direct current). Magnetic Dx phase was withdrawn via two perforated tubes placed inside the separator tube. Each tube was divided into two sections, and the magnetic phase was withdrawn from the separator in two directions. The magnetic phase removed via the top part of the tubes was contaminated with PEG phase and was recirculated.

### *Preparative, Continuous Separation of Two-Phase Systems Containing Cell Debris*

Baker's yeast (750 g) is homogenized in a Waring blendor together with crushed dry ice.[1] The carbon dioxide is allowed to dissipate, and the homogenate is diluted with sodium phosphate buffer (50 m*M*, pH 7.0, containing 5 m*M* 2-mercaptoethanol and 2 m*M* EDTA). Crude Dx (final concentration 7%, w/w; obtained from Sorigona AB, Staffanstorp, Sweden), PEG 6000 (final concentration 5%, w/w; obtained from E. Merck, Darmstadt, Germany), and Cibacron Blue F3G-A–PEG[5,6] (2% of the total amount of PEG) is added and thoroughly mixed. The Dx contains 100 mg/ml of 1-$\mu$m iron oxide particles. The total volume of the mixture is 3.0 liters, and the volume ratio between the phases equals 1. The mixture is phase separated in a preparative-scale magnetic separator (Fig. 3).

The phase mixture is pumped into the separator and gives rise to two exiting streams: one PEG phase stream (33 ml/min) containing enzymes with affinity for Cibacron Blue and one Dx phase stream (33 ml/min) containing cell debris and iron oxide particles. The separation is satisfac-

[5] G. Johansson and M. Joelsson, *Biotechnol. Bioeng.* **27,** 621 (1985).

[6] G. Kopperschläger, this volume [11].

tory with the first 800 ml. Subsequently, both phases are slightly cross-contaminated. It is noted that the cross-contaminating droplets are large and could easily be removed in a gravity settler. This suggests an alternative use of the continuous magnetic separator, namely, to run it very fast as a droplet coalescer in combination with a gravity settler. The final phase separation will be achieved at a much improved rate in the gravity settler owing to the large phase droplets generated in the coalescer. Initial experiments indicate that flow rates of 500 ml/min will be possible.

The separated phases are analyzed with respect to several enzymes known to have affinity for the Cibacron Blue ligand: phosphofructokinase, alcohol dehydrogenase, glucose-6-phosphate dehydrogenase, and hexokinase. The total recovery of enzymes (Dx phase plus PEG phase) is satisfactorily high (87–93%), indicating that the procedure is mild, whereas the recovery of enzymes in the Cibacron Blue-containing PEG phase is lower than expected. Thus the observed partition coefficient, $K$, for phosphofructokinase is 0.44 in the presence of Cibacron Blue–PEG. Apparently the distribution of the enzymes is affected by the presence of the other cell constituents, as has been reported for other systems.[7] In clean systems a $K$ value of about 10 would be expected.

## Conclusions

Magnetically enhanced phase separation seems especially attractive in "difficult" systems in which the phase settling may be very slow with standard conditions of phase separation (gravity, centrifugation). Multistage separations are also conveniently handled using the magnetic approach.

[7] F. Tjerneld, G. Johansson, and M. Joelsson, *Biotechnol. Bioeng.* **30,** 809 (1987).

# Section II

## Partitioning and Affinity Partitioning of Macromolecules

A. Proteins
*Articles 11 through 17*

B. Nucleic Acids
*Articles 18 and 19*

# [11] Affinity Extraction with Dye Ligands

*By* GERHARD KOPPERSCHLÄGER

## Principle of Affinity Partitioning

Affinity partitioning combines the property of biological macromolecules to partition in aqueous two-phase systems with that of biorecognition. The latter results from the interaction of a specific ligand with its complementary binding site on the biomolecule. As a rule, the ligand is bound to one of the phase-forming polymers, and therefore its partitioning occurs in a rather extreme manner, being either in favor of the upper or the lower phase. If a protein has affinity for the ligand, the complex formation, after the ligand is incorporated into the system, leads to a change in the partitioning of the protein.

Equation (1) indicates that the partition coefficient of a protein, $K_p$, in a two-phase system can be described, on a logarithmic scale, as the sum of different terms[1]:

$$\ln K_p = \ln K_{el} + \ln K_{conf} + \ln K_{hphob} + \ln K_{hphil} + \ln K_{lig} + \dots \quad (1)$$

where the subscripts refer to effects on the partitioning owing to electrostatic effects (el), conformational state of the protein (conf), hydrophobic interactions (hphob), hydrophilic interactions (hphil), and specific ligand interactions (lig), respectively. When the strength of the ligand–protein interaction is high the term $\ln K_{lig}$ becomes dominant and the partitioning of the protein in a two-phase system is governed by the formation of the ligand–protein complex. At saturating ligand concentration and optimal conditions the target protein is transferred completely from one phase to the other.

Based on this behavior, affinity partitioning has been exploited as a new branch of affinity separation technology. The method also turns out to be a sensitive tool in the study of protein–ligand interaction.[2] The effect of affinity partitioning is quantified by measuring the difference of the partition coefficients, $K$, of a biomolecule obtained in the presence and in the absence of the ligand given on a logarithmic scale (i.e., $\Delta \log K$).

When $\Delta \log K$ is plotted against increasing concentrations of the ligand–polymer, saturation functions result in most cases from which $\Delta$ log

[1] P.-Å. Albertsson, "Partition of Cell Particles and Macromolecules," 3rd Ed. Wiley(Interscience), New York, 1986.

[2] G. Kopperschläger and G. Birkenmeier, *Bioseparation* **1,** 235 (1990).

$K_{max}$ (maximum extraction power) and the half-saturation point $0.5 \times \Delta \log K_{max}$ can be calculated. The parameter $\Delta \log K_{max}$ was found to be related to the number of binding sites for the ligand.[3–5]

In Eq. (2) a general expression is given describing the partitioning of a protein in the presence of a ligand[4]:

$$K_c = K_p[1 + (C_L/k_d)_1]^N/[1 + (C_L/k_d)_2]^N \qquad (2)$$

in which $K_c$ is the partition coefficient of the ligand–polymer–protein complex, $K_p$ is the partition coefficient of the free protein, $(k_d)_1$ and $(k_d)_2$ are the dissociation constants of the ligand–polymer–protein complex in the upper (1) and in the lower (2) phase, $C_L$ is the concentration of the ligand in the upper (1) and the lower (2) phase, respectively, and $N$ is the number of binding sites of the ligand. Equation (2) reduces to Eq. (3) at high ligand concentration and assuming $(k_d)_1 = (k_d)_2$:

$$K_c = (K_L)^N K_p \qquad (3)$$

in which $K_L$ is the partition coefficient of the free ligand. Equation (3) should permit the calculation of the number of binding sites of a protein under saturating conditions of the ligand–polymer. However, much of the experimental data[5,6] published on affinity partitioning do not fit Eq. (3).

## Group-Specific Ligands: Definition and Properties

Group-specific ligands bind to complementary domains of proteins. In contrast to ligands which interact selectively with proteins because of specific binding sites, group-specific ligands bind to a certain class of proteins which share a common motif on their surface. For example, dinucleotide- or nucleotide-dependent enzymes exhibit a common structural element known as the nucleotide-binding domain. Other group-specific ligands are hydrophobic and thiophilic ligands,[7] metal chelates,[8] quaternary ammonium compounds,[9] or dyes.[10] Among the classes of

[3] S. D. Flanagan and S. Barondes, *J. Biol. Chem.* **125,** 1484 (1975).

[4] D. E. Brooks, K. A. Sharp, and D. Fisher, *in* "Partitioning in Aqueous Two-Phase Systems" (H. Walter, D. E. Brooks, and D. Fisher, eds.), p. 11. Academic Press, Orlando, Florida, 1985.

[5] A. Cordes, J. Flossdorf, and M.-R. Kula, *Biotechnol. Bioeng.* **30,** 514 (1987).

[6] G. Johansson, G. Kopperschläger, and P.-Å. Albertsson, *Eur. J. Biochem.* **131,** 589 (1983).

[7] J. Porath, J. Carlsson, I. Olsson, and G. Belfrage, *Nature* (*London*) **258,** 598 (1975).

[8] G. E. Wuenschell, E. Naranjo, and F. H. Arnold, *Bioprocess Eng.* **5,** 199 (1990).

[9] G. Johansson, R. Gysin, and S. D. Flanagan, *J. Biol. Chem.* **256,** 9126 (1981).

[10] G. Kopperschläger, H.-J. Böhme, and E. Hofmann, *in* "Advances in Biochemical Engineering" (A. Fiechter, ed.), p. 101. Springer-Verlag, Berlin, Heidelberg, and New York, 1982.

dye ligands many mimic, more or less precisely, structural motifs of nucleotides because of the presence of hydrophobic parts, charged groups, and high flexibility. Such compounds have been used as pseudo-biospecific ligands in affinity separation techniques including partitioning.

Although cofactors of enzymes and other proteins are potentially more biospecific than dye ligands, their practical use is not widespread. One reason might be the laborious chemical procedures required to maintain a functional active state while coupling to one of the phase-forming polymers. Moreover, natural biological ligands are generally expensive because the ligands themselves often require purification and are chemically and biologically labile.

## Dye Ligands: Chemical Nature and Mode of Coupling to Polymers

Biomimetic reactive dyes have several advantages over their natural counterparts. These include a moderate to high selectivity for enzymes and other proteins, availability as low-cost chemicals having a large diversity of structures, high reactivity to permit nucleophilic substitution in many polymers, and chemical and biological stability. Basically, reactive dyes contain a specific chromophore and a reactive group which is capable of effecting the nucleophilic substitution. Both parts of the reactive dyes are linked by auxochromic groups in a variety of chemical combinations.[11] There are anthraquinone dyes including the well-known Cibacron Blue F3G-A (Ciba-Geigy, Basel, Switzerland), azo dyes like Procion Red HE-3B, phthalocyanine dyes like Procion Green H-4G, and dye–metal complexes like Procion Brown MX-5BR (all from I.C.I., Organics Division, Blackley, UK). Triazine dyes contain monochloro- or dichloro-substituted triazine rings as reactive groups, whereas the class of Remazol dyes (Höchst, Frankfurt, Germany) react via the vinylsulfone group with the polymer of choice. In Fig. 1 some typical structures of reactive dyes are depicted.

To achieve asymmetric partitioning of the dyes in a two-phase system it is advantageous to immobilize the ligand by covalent coupling to one of the phase-forming polymers, preferably poly(ethylene glycol) (PEG) (Serva, Heidelberg, Germany), under moderate chemical conditions to yield primarily the monosubstituted polymer.[12] The replacement of the terminal hydroxyl group of PEG by primary amines results in higher yields of substitution.

[11] C. V. Stead, *J. Chem. Biotechnol.* **37,** 55 (1987).

[12] G. Johansson and M. Joelsson, *Biotechnol. Bioeng.* **27,** 621 (1985).

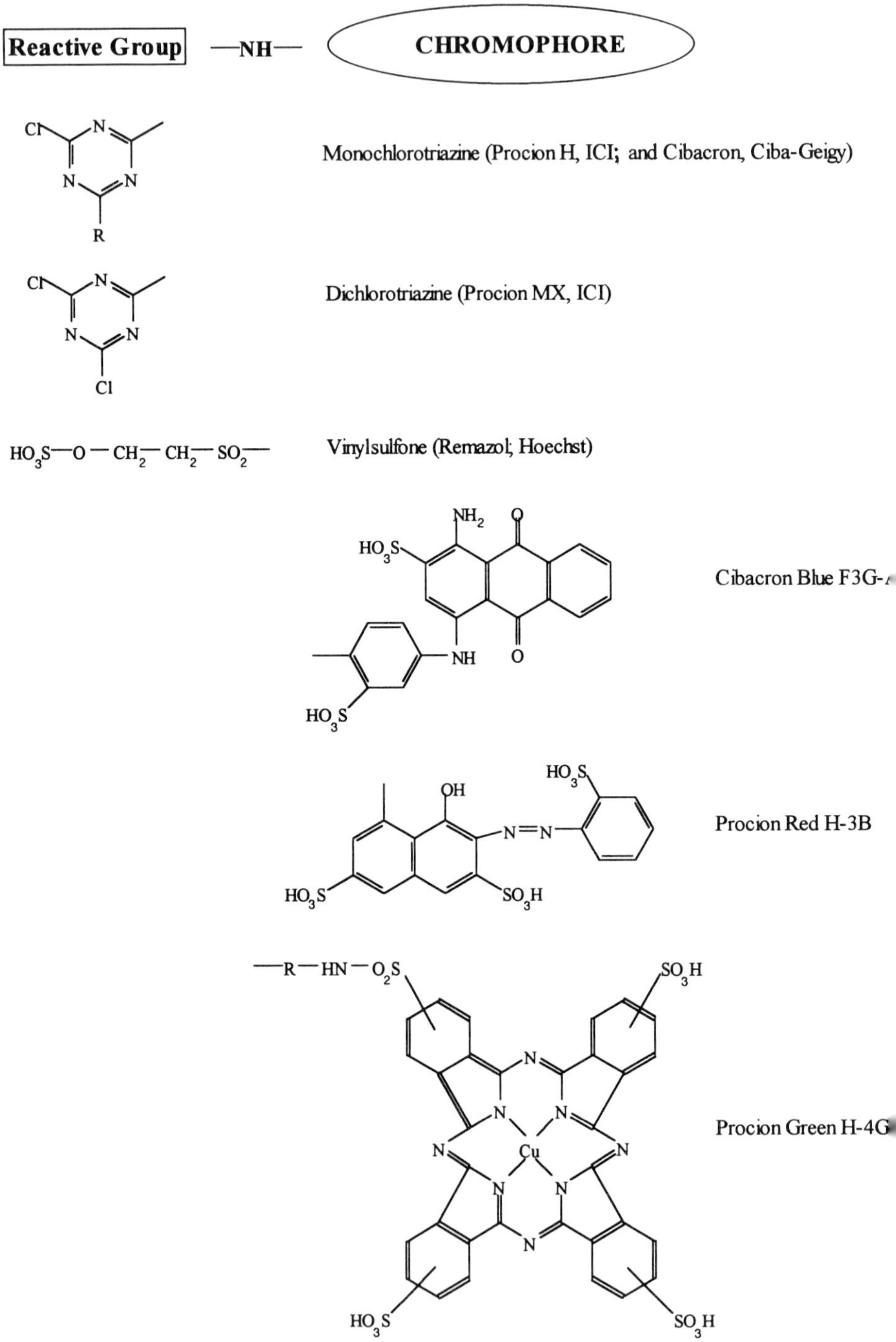


FIG. 1. Structures of reactive dyes.

When using poly(vinylpyrrolidone) (PVP) 360 (Sigma Chemical Co., St. Louis, MO) and maltodextrin (Grain Processing Corp., Muscatine, IA) as phase-forming polymers it is not necessary to immobilize the dye ligand because of its extreme partitioning into the PVP-rich top phase, both in the absence and presence of the protein.[13] In an analogous manner free DNA, when partitioned in favor of the upper phase, can be used for the extraction of restriction endonucleases from the bottom phase.[14]

## Screening of Dye–Protein Interactions by Affinity Partitioning

To obtain the maximum effect of affinity partitioning the two-phase systems must be optimized. In systems composed of PEG and dextran (Dx) (Pharmacia, Uppsala, Sweden) most proteins partition in favor of the Dx-rich phase to yield a $K$ value of 0.1 to 0.01. By stepwise replacement of PEG by dye–PEG the $K$ of the protein increases until saturation is achieved. Figure 2 demonstrates the affinity partitioning of different enzymes with Cibacron Blue F3G-A as the dye ligand. The enzymes are clearly distinguished both in the extent to which they are extracted and in their affinity to the dye, characterized by the steepness of the curves. A double-reciprocal plot of such curves permits the calculation of $\Delta \log K_{max}$ and the half-saturation point $0.5 \times \Delta \log K_{max}$. This approach is adequately sensitive to distinguish among high, moderate, low, or no dye–protein interactions.

In Table I the $\Delta \log K_{max}$ values are listed for a number of enzymes and different dye ligands. In general, the higher the $\Delta \log K_{max}$, the greater is the efficiency for separation of the bound protein from the nonbound bulk protein in a single extraction step. The $\Delta \log K_{max}$ value can be regarded as a measure of a dye to act as a biomimetic ligand also in other affinity techniques. This has been found with phosphofructokinase,[15] alkaline phosphatase,[16] lactate dehydrogenase,[17] glucose-6-phosphate dehydrogenase,[18,19] albumin,[20] and prealbumin.[21] However, $\Delta \log K_{max}$ cannot be used to predict the efficiency of this ligand in other affinity separation

[13] K. A. Giuliano, *Anal. Biochem.* **197,** 333 (1991).

[14] G. Vlatakis and V. Bouriotis, *J. Chromatogr.* **538,** 311 (1991).

[15] G. Kopperschläger and G. Johansson, *Anal. Biochem.* **124,** 117 (1982).

[16] J. Kirchberger and G. Kopperschläger, *Bioseparation* **1,** 33 (1990).

[17] M. Naumann, R. Reuter, P. Metz, and G. Kopperschläger, *J. Chromatogr.* **466,** 319 (1989).

[18] R. Reuter, P. Metz, G. Lorenz, and G. Kopperschläger, *Biomed. Biochim. Acta* **49,** 151 (1990).

[19] R. Reuter, M. Naumann, P. Metz, and G. Kopperschläger, *Biomed. Biochim. Acta* **49,** 539 (1990).

[20] G. Birkenmeier and G. Kopperschläger, *Mol. Cell. Biochem.* **73,** 99 (1987).

[21] G. Birkenmeier, B. Tschechonien, and G. Kopperschläger, *FEBS Lett.* **174,** 162 (1984).

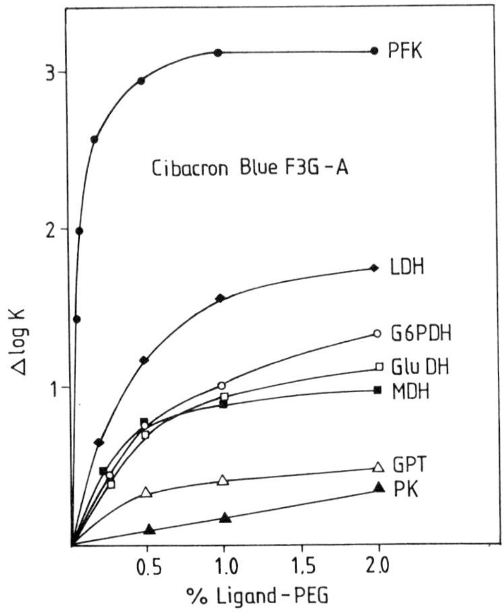


FIG. 2. Affinity of enzymes in an aqueous two-phase system containing increasing amounts of Cibacron Blue F3G-A–PEG. The system is composed of 5% (w/w) PEG 6000, 7% (w/w) dextran T500, and 50 m*M* sodium phosphate buffer, pH 7.0, supplemented with 5 m*M* 2-mercaptoethanol. The system was equilibrated at 0°. PFK, phosphofructokinase; LDH, lactate dehydrogenase; G6PDH, glucose-6-phosphate dehydrogenase; GluDH, glutamate dehydrogenase; GPT, glutamic-pyruvic transaminase; MDH, malate dehydrogenase; PK, pyruvate kinase. [From G. Kopperschläger and G. Birkenmeier, *Bioseparation* **1,** 235 (1990), with permission.]

techniques. Although a low value of $\Delta \log K_{max}$ reflects a low interaction, a high value does not, per se, assure its usefulness in practice.

Another important parameter which results from the saturation function in Fig. 2 is the ligand concentration, $c_{0.5}$, giving half-saturation, $0.5 \times \Delta \log K_{max}$. This $c_{0.5}$ can be regarded as a measure for the relative affinity of dyes toward enzymes under identical experimental conditions. As an example (see Table II), two isoenzymes of lactate dehydrogenase (Sigma, France), namely LDH1 (from rabbit heart) and LDH5 (from rabbit muscle) revealed similar $\Delta \log K_{max}$ values within the row when they were partitioned in systems with different dye ligands. They show, however, significant differences in their relative affinity for the dyes. A high difference in $c_{0.5}$ was found between LDH1 and LDH5 with either Procion Blue H-5R or Procion Red 3BA.[22]

[22] J. Kirchberger, G. Kopperschläger, and M. A. Vijayalakshmi, *J. Chromatogr.* **557,** 325 (1991).

A similar behavior has been found when multiple forms of human alkaline phosphatase were partitioned in systems containing dye ligand. Desialylation of the isoenzymes changes only the relative affinity of the dye for the enzyme but not the maximum extraction power.[22,23]

## Factors Influencing Affinity Partitioning

### *pH*

The partitioning of several enzymes having affinity for the dye–PEG is not markedly influenced by changing pH in the range of 6 to 9.[6] This behavior is due to the fact that the $pK_a$ values of the dyes are apart from the pH applied. Because, however, the partitioning of enzymes in the absence of dye–PEG is more dependent on the pH of the buffer, the calculated $\Delta \log K$ is often changed by varying the pH. An exception is the significant increase of the $K$ of phosphofructokinase from rat erythrocytes in PEG–Dx systems containing Cibacron Blue F3G-A–PEG.[24]

### *Temperature*

The affinity partitioning effect decreases for a number of enzymes with an increase in temperature. Table III summarizes data for bulk protein, three purified enzymes from baker's yeast, and alkaline phosphatase from calf intestine. The reduction in $\Delta \log K$ is caused mainly by the increase of $K$ in systems free of ligand–PEG. On the other hand, the partitioning of alkaline phosphatase was essentially unchanged with an increase in temperature.

### *Competitive Effectors*

Affinity partitioning permits the detection of any interference with the dye–protein interaction. The technique is well suited to discern even small effects of substrates or cofactors of enzymes, metal ions, changes in the ionic strength or the dielectric constant of the buffer, etc.[24–26] Competitive ligands result, as a rule, in a reduction of the effect of affinity partitioning; under certain circumstances $\Delta \log K$ is completely abolished.

[23] J. Kirchberger, U. Domar, G. Kopperschläger, and T. Stigbrand, *J. Chromatogr.* **574,** 237 (1992).

[24] M. C. Tejedor, C. Delgado, M. Grupeli, and J. Luque, *J. Chromatogr.* **589,** 127 (1992).

[25] G. Kopperschläger, G. Lorenz, and E. Usbeck, *J. Chromatogr.* **259,** 97 (1983).

[26] G. Kopperschläger, J. Kirchberger, and T. Kriegel, *Makromol. Chem., Makromol. Symp.* **17,** 373 (1988).

TABLE I

MAXIMUM AFFINITY PARTITIONING EXPRESSED AS $\Delta \log K_{max}$ FOR ENZYMES AND POLY(ETHYLENE GLYCOL) DYES[a,b]

| | $\Delta \log K_{max}$ | | | | | | |
|---|---|---|---|---|---|---|---|
| Dye | >3.0 | 2.5–3.0 | 2.0–2.4 | 1.5–1.9 | 1.0–1.4 | 0.5–0.9 | <0.5 |
| Cibacron Blue F3G-A | PFK; NR | LDH5 | Y-G6PDH; GLDH | LDH; b-G6PDH | c-IAP; MDH; GPT; Y-ADH; LDH1 | GOT | PK; PGK; HK; GK |
| Cibacron Brilliant Blue BRP | | PFK | | LDH; GLDH | c-IAP | PK; GPT; GOT | HK; GK |
| Procion Blue MX-R | | | PFK | GLDH | LDH | MDH; GPT; Y-ADH; GOT | PK; GK |
| Procion Blue SP-3R | | | | | Y-G6PDH; Y-ADH | c-IAP, L-ADH | |
| Procion Navy H-ER | PFK; NR | | Y-G6PDH; PK | c-IAP; LDH; GLDH | GPT; LDH1; b-G6PDH | MDH; HK; GOT | GK |
| Procion Red HE-3B | PFK | PK | Y-G6PDH; LDH5 | c-IAP, GLDH | MDH; LDH1; b-G6PDH | GPT; GOT | Y-ADH; HK; GK |
| Procion Red HE-7B | PFK | GLDH | | Y-G6PDH | MDH | c-IAP; GPT; GOT | PK; HK; Y-ADH; GK |
| Procion Red MX-G | | | | | | c-IAP; GLDH; Y-G6PDH | |
| Procion Scarlet MX-G | | | | | Y-G6PDH | c-IAP | |
| Procion Blue H-5R | LDH5 | h-LAP | | | LDH1; c-IAP | | h-IAP |
| Procion Orange MX-G | | | | | | LDH; PGK | MDH; GLDH; PK Y-G6PDH; c-IAP |
| Procion Yellow H-3R | | | | | Y-G6PDH | Y-ADH | |
| Procion Yellow MX-GR | | | | | Y-G6PDH | c-IAP; Y-ADH | |

| | | | | | | | |
|---|---|---|---|---|---|---|---|
| Procion Yellow MX-4G | NR | | | | | | c-IAP; Y-G6PDH |
| Procion Yellow MX-4R | | | | PFK; LDH; PK | c-IAP; PGK; Y-G6PDH | MDH | |
| Procion Yellow HE-3G | | LDH5 | Y-G6PDH; LDH | PGK | c-IAP; h-IAP; b-G6PDH | h-LAP | |
| Procion Navy MX-RB | | h-LAP | | | LDH5; c-IAP | LDH1 | h-IAP |
| Procion Blue MX-2G | | | | LDH5 | c-IAP, b-G6PDH | LDH1 | h-IAP; h-LAP |
| Procion Brown HE-G | NR | PFK | | LDH; Y-G6PDH; PK | MDH; GOT; GPT; GLDH | | HK; GK |
| Procion Olive MX-3G | | | | LDH; Y-G6PDH | | PGK | |
| Procion Green H-4G | PFK; NR | | Y-G6PDH | LDH; GLDH | PK; MDH; b-G6PDH | | LDH |
| $Cu^{2+}$-Red Violet 2KT | | | | Y-ADH; L-ADH | | | |
| Bright Red 6C | | | | L-ADH | | Y-ADH | |
| $Cu^{2+}$-Light-resistant Yellow 2KT | | Y-ADH | | L-ADH | | | |

[a] Y, Yeast; h, human; c, calf; b, bacterial. ADH, Alcohol dehydrogenase; GK, glycerol kinase; GPT, glutamic-pyruvic transaminase; GOT, glutamic-oxaloacetic transaminase; GLDH, glutamate dehydrogenase; G6PDH, glucose-6-phosphate dehydrogenase; IAP, intestinal alkaline phosphatase; HK, hexokinase; LDH; lactate dehydrogenase; LDH1, LDH5, isoenzymes of lactate dehydrogenase; LAP, liver alkaline phosphatase; MDH, malate dehydrogenase; NR, nitrate reductase; PK, pyruvate kinase; PFK, phosphofructokinase; PGK, phosphoglycerate kinase.

[b] Reference citations: For h-IAP, h-LAP: J. Kirchberger, U. Domar, G. Kopperschläger, and T. Stigbrand, *J. Chromatogr.* **574,** 237 (1992). For c-IAP: J. Kirchberger and G. Kopperschläger, *Bioseparation.* **1,** 33 (1990). For Y-ADH, L-ADH: J.-H. J. Pesliakas, V. D. Zutautas, and A. A. Glemža, *Chromatographia* **26,** 85 (1988). For Y-G6PDH: G. Kopperschläger and G. Lorenz, *Biomed. Biochim. Acta* **44,** 517 (1985), and G. Johansson and M. Joelsson, *Enzyme Microb. Technol.* **7,** 629 (1985). For b-G6PDH: R. Reuter, P. Metz, G. Lorenz, and G. Kopperschläger, *Biomed. Biochim. Acta* **49,** 151 (1990). For NR: J. Schiemann and G. Kopperschläger, *Plant Sci. Lett.* **36,** 205 (1984). For PGK: G. Johansson and M. Andersson, *J. Chromatogr.* **303,** 39 (1984). For PFK, PK, MDH, HK, GK, GOT, GPT, GLDH, LDH: G. Kopperschläger, G. Lorenz, and E. Usbeck, *J. Chromatogr.* **259,** 97 (1983), and G. Johansson and M. Joelsson, *Appl. Biochem. Biotechnol.* **13,** 15 (1986). For LDH1, LDH5: J. Kirchberger, G. Kopperschläger, and M. A. Vijayalakshmi, *J. Chromatogr.* **557,** 325 (1991).

TABLE II
MAXIMAL AFFINITY PARTITIONING EFFECT ($\Delta \log K_{max}$) OF LACTATE DEHYDROGENASE ISOENZYMES AND CONCENTRATION ($c_{0.5}$) OF DYE–POLY(ETHYLENE GLYCOL) IN SYSTEM[a] GIVING HALF-SATURATION[b]

| DyePEG | LDH1 | | LDH5 | |
|---|---|---|---|---|
| | $\Delta \log K_{max}$ | $c_{0.5}$[c] | $\Delta \log K_{max}$ | $c_{0.5}$[c] |
| Procion Red HE-3B | 1.47 | 0.15 | 2.34 | 0.10 |
| Procion Blue H-5R | 1.33 | 1.02 | 3.33 | 0.10 |
| Cibacron Blue F3G-A | 1.61 | 0.65 | 2.45 | 0.12 |
| Cibacron Red 3BA | 1.43 | 2.38 | 2.86 | 0.44 |
| Vilmax Dye 1 | 1.25 | 0.92 | 2.86 | 0.31 |

[a] Systems (2 g) contained 9% (w/w) dextran 500, 6% (w/w) PEG 6000, 50 m*M* sodium phosphate buffer, pH 7.5, and 5 units of enzyme, equilibrated at 25°.
[b] From J. Kirchberger, G. Kopperschläger, and M. A. Vijayalakshmi, *J. Chromatogr.* **557,** 325 (1991), with permission.
[c] The term relative affinity $c_{0.5}$ is defined as that dye–PEG concentration (expressed as percentage of the total PEG) which yields affinity partitioning of $0.5 \times \Delta \log K_{max}$.

TABLE III
EFFECT OF TEMPERATURE ON AFFINITY PARTITIONING OF SELECTED ENZYMES EXPRESSED AS $\Delta \log K$

| Enzyme | Ligand | $\Delta \log K$ | | | | | Ref. |
|---|---|---|---|---|---|---|---|
| | | 0° | 10° | 20° | 30° | 40° | |
| Phosphofructokinase (yeast) | Cibacron Blue F3G-A | 3.42 | 3.22 | 3.14 | 2.75 | 2.15 | *a* |
| Glucose-6-phosphate dehydrogenase (yeast) | Procion Yellow HE-3G | 2.40 | 2.0 | 1.56 | 0.98 | 0.61 | *b* |
| 3-Phosphoglycerate kinase (muscle) | Procion Yellow HE-3G | 1.62 | 1.48 | 1.18 | 0.50 | 0.17 | *b* |
| Alkaline phosphatase (calf intestine) | Procion Red HE-3B | 0.98 | n.d. | 0.86 | n.d. | 0.80 | *c* |
| Protein (yeast) | Procion Yellow HE-3G | 0.93 | 0.73 | 0.48 | 0.33 | 0.21 | *b* |

[a] G. Johansson, G. Kopperschläger, and P.-Å. Albertsson, *Eur. J. Biochem.* **131,** 589 (1983).
[b] G. Johansson and M. Andersson, *J. Chromatogr.* **303,** 39 (1984).
[c] G. Johansson and G. Kopperschläger, *J. Chromatogr.* **388,** 295 (1987). n.d., Not determined.

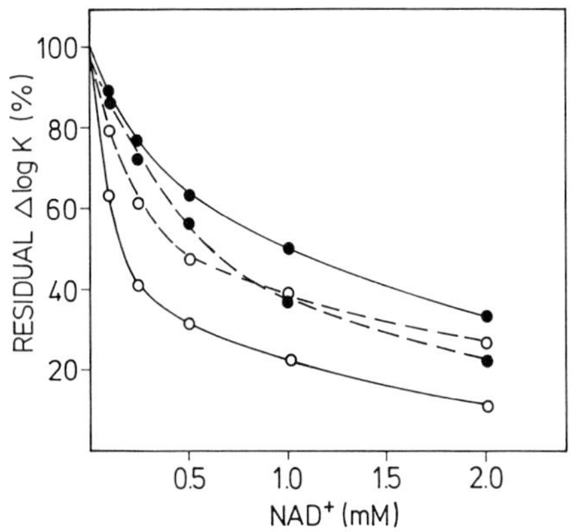


FIG. 3. Affinity partitioning of isoenzymes of lactate dehydrogenase as a function of the amounts of $NAD^+$ added to the system. The systems were composed of 6% (w/w) PEG 6000, 9% (w/w) dextran T500, and 50 m*M* sodium phosphate buffer, pH 7.0, including 0.3% Procion Red HE-3B–PEG (----) and 2% Procion Blue H-5R–PEG (——) for LDH1 (○) and 0.2% Procion Red HE-3B–PEG and 0.2% Procion Blue H-5R–PEG for LDH5 (●). Five units of the respective isoenzyme were incorporated in the systems and then equilibrated at 25°. [From J. Kirchberger, G. Kopperschläger, and M. A. Vijayalakshmi, *J. Chromatogr.* **557,** 325 (1991), with permission.]

As an example, Fig. 3 demonstrates the competitive action of $NAD^+$ on the partitioning of LDH1 and LDH5 applying two dye–polymers, Procion Blue H-5R–PEG and Procion Red HE-3B–PEG. In the case of Procion Red HE-3B the competitive effect of $NAD^+$ on both isoenzyme–dye complexes is similar. However, a significant difference is found when LDH1 and LDH5 are compared with Procion Blue H-5R–PEG. This leads to the conclusion that $NAD^+$ acts less competitively with the LDH5–dye complex than with the LDH1–dye complex.[22]

The $NAD^+$–LDH interaction is strengthened by sulfite ions and gives rise to a highly stable ternary $NAD^+$–sulfite–LDH complex. This behavior has been confirmed by a comparison of the competitive action of free $NAD^+$ and $NAD^+$ with sulfite on the Cibacron Blue F3G-A–LDH interaction.[25]

## *Chemical Modification of Dye Ligand*

To study the chemistry of dye–protein interactions, chemical modification of the dye can help in evaluating the requirements of dye structure for complex formation. Modifications of the dyes have included splitting off functional residues, exchanging peripheral residues without changing the basic dye structure, and reducing azo groups to obtain charged hydraz-

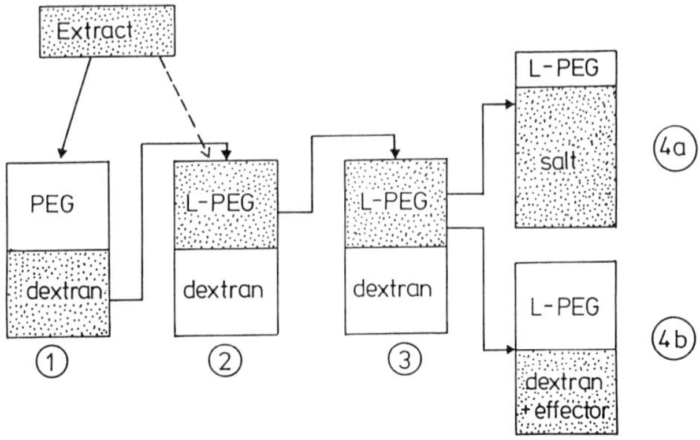


FIG. 4. Scheme for affinity partitioning for enzyme purification. Step 1: Partitioning of an extract in a system without ligand–PEG. Step 2: Partitioning of the enzyme from the dextran phase of step 1 or, alternatively, of the extract in the presence of ligand–PEG. Step 3: Extraction step of the top phase of step 2 with a fresh bottom phase. Step 4a: Formation of a PEG–salt two-phase system by adding solid salt to the top phase of step 3. Step 4b: Partitioning in a system composed of the top phase of step 3 and a fresh bottom phase containing a competitive effector. [From G. Kopperschläger and G. Birkenmeier, *Bioseparation* **1,** 235 (1990), with permission.]

ino groups and to change flexibility. All these modifications can be performed after covalent attachment of the dye to the PEG, thus making available sets of specifically designed ligand–PEG derivatives.[27]

## Purification of Enzymes by Single-Step Affinity Partitioning

Affinity partitioning has the potential for enzyme or protein enrichment from crude material by single- or multistep extraction. In contrast to other affinity separation techniques, which are intrinsically batch procedures, liquid–liquid phase systems can operate continuously and can easily be scaled up. To achieve sufficient yields and a high degree of purification a number of parameters must be optimized: Affinity ligands must yield an adequate specificity and high power of interaction ($\Delta \log K_{max}$), the volume ratio of the phases and the $K$ of proteins must be optimal, and competitive effectors must be able to dissociate the enzyme–ligand complex.

The theoretical yield ($Y$) of a single-step extraction for the top (t) and bottom (b) phase is given by Eqs. (4) and (5):

[27] J. Kirchberger, F. Cadelis, G. Kopperschläger, and M. A. Vijayalakshmi, *J. Chromatogr.* **483,** 289 (1989).

TABLE IV
PURIFICATION OF ENZYMES BY MEANS OF DYE LIGAND AFFINITY PARTITIONING

| Enzyme | Ligand | Purification ($n$-fold) | Ref. |
|---|---|---|---|
| Phosphofructokinase (yeast) | Cibacron Blue F3G-A | 10 | *a* |
| Phosphofructokinase (rat erythrocytes) | Cibacron Blue F3G-A | 10 | *b* |
| Glucose-6-phosphate dehydrogenase (yeast) | Procion Yellow HE-3G | 22 | *c* |
| | Procion Olive MX-3G | 12 | *c* |
| Lactate dehydrogenase (muscle) | Procion Yellow HE-3G | 30 | *d* |
| | Cibacron Blue F3G-A | 25 | *d* |
| Formate dehydrogenase (*Candida boidinii*) | Procion Red HE-3B | 6 | *e* |
| Alkaline phosphatase (calf intestine) | Procion Red HE-3B | 13 | *f* |
| Alcohol dehydrogenase (yeast) | Light-resistant Yellow 2KT-Cu(III) | 6 | *g* |
| Restriction endonuclease | | | |
| *Eco*RV | Procion Blue MX-4GD | 3 | *h* |
| *Eco*RI | Procion Blue MX-4GD | 10 | *h* |

[a] G. Kopperschläger and G. Johansson, *Anal. Biochem.* **124,** 117 (1982).
[b] M. C. Tejedor, C. Delgado, M. Grupeli, and J. Luque, *J. Chromatogr.* **589,** 127 (1992).
[c] G. Johansson and M. Joelsson, *Enzyme Microb. Technol.* **7,** 629 (1985).
[d] G. Johansson and M. Joelsson, *Appl. Biochem. Biotechnol.* **13,** 15 (1986).
[e] A. Cordes and M.-R. Kula, *J. Chromatogr.* **376,** 375 (1986).
[f] J. Kirchberger, unpublished.
[g] J.-H. J. Pesliakas, V. D. Zutautas, and A. A. Glemža, *Chromatographia* **26,** 85 (1988).
[h] G. Vlatakis and V. Bouriotis, *J. Chromatogr.* **538,** 311 (1991).

$$Y_t\ (\%) = 100/(1 + V_b/V_tK) \quad (4)$$
$$Y_b\ (\%) = 100/(1 + V_tK/V_b) \quad (5)$$

in which $K$ gives the partition of the protein and $V_t$ and $V_b$ are the volumes of the top and bottom phases, respectively. Both $K$ and the volume ratio ($V_b/V_t$) determine the net mass transfer between the phases. Figure 4 shows a general scheme for employing affinity partitioning in protein purification.

It is of advantage to precipitate the bulk protein as a first step. This removes low molecular weight material from the cell extract, which often interferes with the protein–dye interaction, and concentrates the protein. Alternatively, the extract is first partitioned in a system free of affinity ligand so as to concentrate the bulk protein in one phase. After removal of the opposite phase a fresh phase containing the respective dye–polymer or the free ligand is added.

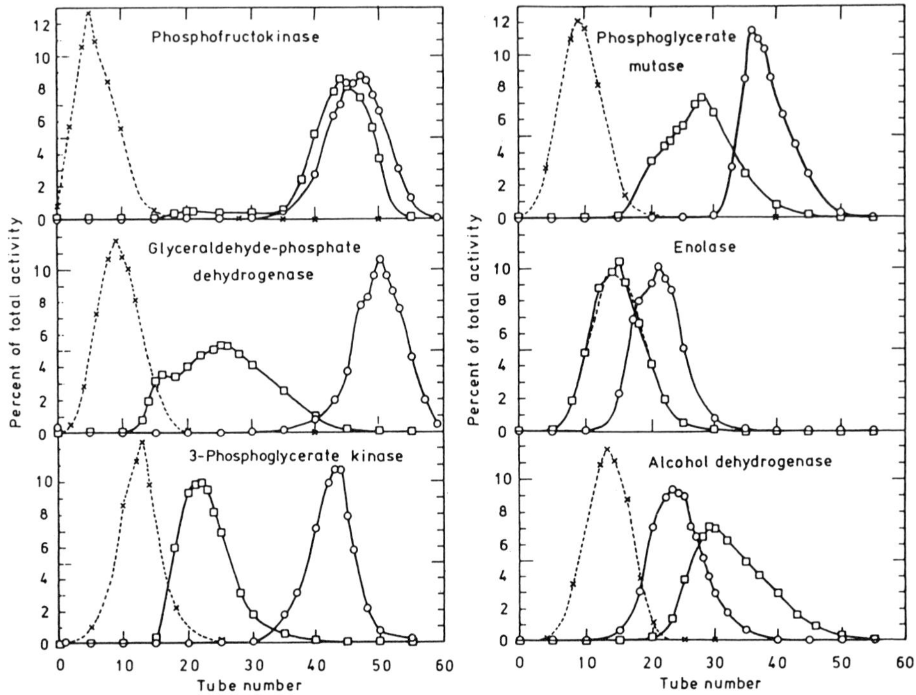


FIG. 5. Countercurrent distribution of yeast extract. Systems were composed of 5% PEG 6000, 7% dextran T500, 50 m*M* sodium phosphate buffer, pH 7.0, 0.2 m*M* EDTA, and 5 m*M* 2-mercaptoethanol (x) or with Procion Yellow HE-3G–PEG (○) or Procion Olive MX-3G–PEG (□) replacing 1% of the total PEG. The systems were equilibrated at 3°. [From G. Johansson, M. Andersson, and H.-E. Åkerlund, *J. Chromatogr.* **298,** 483 (1984), with permission.]

The phase containing the ligand is washed with a fresh bottom phase, and after permitting the phases to settle the enzyme is separated from the dye ligand by dissociation. This can be effected by addition of solid salt (e.g., ammonium sulfate or potassium phosphate) to the separated top phase so as to form a new biphasic PEG–salt system in which the enzyme partitions to the salt phase. Alternatively, the replacement of the bottom phase by a fresh phase which contains a competitive effector leads to the transfer of the target enzyme into the bottom phase.

As a rule, affinity partitioning of crude extracts cannot provide highly purified enzymes in a single-step extraction. The procedure has to be integrated into a series of steps also involving ion-exchange chromatography to remove the phase-forming polymers, because these often interfere in the separation of proteins by precipitation with ammonium sulfate. In

TABLE V
RELATIVE $G$ VALUES[a]

| Enzyme | $G_y/G_n$ | $G_0/G_n$ | $G_y/G_0$ |
|---|---|---|---|
| Hexokinase 2 | 2.9 | 1.0 | 2.9 |
| Glucose-6-phosphate dehydrogenase | 41 | 14 | 2.8 |
| Phosphofructokinase | 64 | 50 | 1.3 |
| Glyceraldehyde-phosphate dehydrogenase | 48 | 4.5 | 10.6 |
| 3-Phosphoglycerate kinase | 12 | 2.1 | 5.5 |
| Phosphoglycerate mutase | 48 | 4.5 | 2.0 |
| Enolase | 1.8 | 1.1 | 1.7 |
| Alcohol dehydrogenase | 12 | 2.1 | 0.64 |

[a] For enzyme peaks in Fig. 5. $G_n$, without ligand; $G_y$, with Procion Yellow HE-3G; and $G_0$, with Procion Olive MX-3G. The $G$ value is defined as the percentage of the enzyme (at equilibrium) in the mobile phase (upper phase) divided by the percentage in the stationary phase (lower phase). From G. Johansson, M. Andersson, and H.-E. Åkerlund, *J. Chromatogr.* **298,** 483 (1984), with permission.

most cases only one of the polymers carries the ligand. However, use of two ligands which are bound to the polymers of the top and bottom phases and which have different affinities for the target protein can increase the efficiency of purification.[28,29] In Table IV a number of enzymes that have been purified by dye ligand affinity partitioning are listed.

## Multistep Affinity Partitioning

Enzymes that are not separated by a single extraction in aqueous two-phase systems can often be isolated by a multistep extraction procedure. The rationale for this approach is the difference in the affinity of enzymes for a certain ligand effecting the extraction under nonsaturating conditions of the ligand.

In Fig. 5 countercurrent distribution experiments with diverse enzymes present in a yeast extract and with different dyes are depicted.[30] In general, Procion Yellow HE-3G showed a greater separation effect than Procion Olive MX-3G except for phosphofructokinase. As shown, there is a significant difference in the distribution of individual enzymes. This can be quantified by calculation of $G$ values, that is, the partition ratio of the amount of enzymes in the top and the bottom phases.[31]

[28] G. Johansson, M. Joelsson, and H. E. Åkerlund, *J. Biotechnol.* **2,** 225 (1985).
[29] J.-H. J. Pesliakas, V. D. Zutautas, and A. A. Glemža, *Chromatographia* **26,** 85 (1988).
[30] G. Johansson, M. Andersson, and H.-E. Åkerlund, *J. Chromatogr.* **298,** 483 (1984).
[31] H.-E. Åkerlund and P.-Å. Albertsson, this volume [8].

Table V shows the relative $G$ values of the experiments in Fig. 5, calculated from the $G$ values of the enzymes without and with the dyes. These ratios give a measure of the effectiveness of enzyme separation using countercurrent distribution.

Phosphofructokinase from yeast and muscle, which both bind to Cibacron Blue F3G-A but with different affinities, are completely separated from each other by countercurrent distribution. The dye–ligand polymer of the top phase shifts the yeast enzyme into the mobile phase, reaching a $G$ value of 1.6, but the ligand–polymer concentration is not adequate to transfer substantial amounts of the muscle enzyme into the mobile phase ($G$ value 0.017).[32]

## Conclusions

Affinity partitioning of proteins is a simple and adequate approach to study protein–ligand interaction qualitatively and quantitatively. The influence of competing effectors on protein–ligand interaction can easily by followed by determination of the actual $K$ values in the system. Moreover, the method offers advantages for protein purification, particularly when carried out on a large scale.

[32] G. Kopperschläger, unpublished results, 1985.

# [12] Purification of Lactate Dehydrogenase from Pig Muscle by Affinity Partitioning

*By* MONICA JOELSSON and FOLKE TJERNELD

## Introduction

The use of aqueous two-phase systems is an efficient method for protein purification. Liquid–liquid extraction with aqueous phases is a rapid process giving high yields.[1,2] The use of affinity ligands coupled to one of the phase-forming polymers gives a selective extraction of proteins. Affinity ligands in widespread use are triazine dyes.[3] These dyes can be obtained in a number of molecular structures with varying affinities for the nucleotide

[1] H. Hustedt, K. H. Kroner, U. Menge, and M.-R. Kula, *Trends Biotechnol.* **3,** 139 (1985).
[2] P.-Å. Albertsson, G. Johansson, and F. Tjerneld, *in* "Separation Processes in Biotechnology" (J. A. Asenjo, ed.), p. 287. Dekker, New York and Basel, 1990.
[3] G. Kopperschläger, this volume [11].

binding site of dehydrogenases and kinases. By screening a number of dyes the most effective can be selected for extraction of a target protein.[4–7] A method for purification of lactate dehydrogenase (LDH) from pig muscle has been developed using affinity partitioning. The partition of LDH in a dextran (Dx)–poly(ethylene glycol) (PEG) system can be strongly influenced by using the triazine dye Procion Yellow HE-3G covalently bound to PEG.[8]

The optimal conditions for the purification of LDH from pig muscle are here described. Important parameters are the concentrations of polymers, affinity ligand, and buffer as well as the pH.

## Methods

### *Preparation of Muscle Extract*

Pig muscle, 200 g, is cut into pieces and homogenized with 360 ml ice-cold sodium phosphate buffer (40 m*M*, pH 7.9), in a standard household mixer with rotating knives, for 5 min. The mixture is diluted with 300 ml of the same buffer and further treated with an Ultra-Turrax homogenizer (Staufen, Germany) for four 5-sec periods. After the homogenization steps the mixture is centrifuged at 16,300 *g* for 20 min at 4°. The supernatant solution is recovered (550 ml) and filtered to remove traces of fat. This extract (505 ml) contains 11–13 g protein/liter and 190–266 kU lactate dehydrogenase/liter, corresponding to 3.15–4.45 $\mu$kat (microkatal/liter).

### *Two-Phase Systems*

For the determination of optimal extraction conditions 4-g phase systems of various compositions are used. Systems are prepared from stock solutions of Dx T500 (20%, w/w) and PEG 8000 (40%, w/w). The systems also contain Procion Yellow He-3G–PEG 8000 (stock solution 4%, w/w), muscle extract, water, and, in appropriate cases, sodium phosphate buffer. The polymer concentrations are given in percent weight/weight of the total system. The partitioning is carried out at 0°. The systems are equilibrated by careful mixing and centrifuged at 1500 *g* for 5 min at 4° to speed

[4] G. Kopperschläger, G. Lorenz, and E. Usbeck, *J. Chromatogr.* **259,** 97 (1983).
[5] G. Johansson and M. Andersson, *J. Chromatogr.* **291,** 175 (1984).
[6] M. Joelsson and G. Johansson, *Enzyme Microb. Technol.* **9,** 233 (1987).
[7] G. Kopperschläger and G. Birkenmeier, *Bioseparation* **1,** 235 (1990).
[8] G. Johansson and M. Joelsson, *Appl. Biochem. Biotechnol.* **13,** 15 (1986).

phase settling. Samples of the phases are analyzed for protein,[9] LDH,[10] and other enzymes. The partition coefficients, $K$ (concentration ratio, top to bottom), are calculated.

For preparative extraction of LDH a 1-kg system is used containing 10% Dx and 7.1% PEG including dye–PEG. The system is equilibrated on ice, and the centrifugations (5000 $g$, 3–5 min) are carried out at 4°. LDH is concentrated in the upper yellow phase. Lower phases for the washing steps are prepared from a phase system without dye–PEG.

### *Affinity Ligand*

The triazine dye, Procion Yellow HE-3G (ICI, Macclesfield, UK), is bound to PEG 8000 in alkaline water solution just as previously described for Cibacron Blue F3G-A.[11] The PEG derivative contains 140 $\mu$mol Procion Yellow HE-3G per gram polymer. The dye concentration is determined photometrically at 400 nm ($\varepsilon = 35.6\ \text{m}M^{-1}\ \text{cm}^{-1}$)[12] and the PEG concentration from the refractive index.

### *Effect of Dye–Polymer Concentration*

To get the optimal conditions for extraction, an important parameter is the concentration of ligand–PEG. Proteins in general show $K$ values lower than 1, that is, they are concentrated in the lower phase. By introducing increasing amounts of Procion Yellow–PEG, the $K$ of LDH increases gradually up to 740 times, whereas the $K$ of total protein only changes moderately (Fig. 1). Thus the enzyme can be extracted with a high degree of purification. It should be noted that high concentrations of the dye–PEG affect the Bradford[9] protein determination and that suitable corrections must therefore be made.[13,14]

### *Effects of Polymer Concentration*

The degree of purification of LDH is strongly enhanced by an increase in polymer concentration (Fig. 2). This effect depends mainly on a more effective exclusion of bulk protein from the upper phase, and LDH is, to a large extent, recovered from this phase. At polymer concentrations

[9] M. M. Bradford, *Anal. Biochem.* **72,** 248 (1976).

[10] H. U. Bergmeyer, K. Gawehn, and M. Grassl, *in* "Methoden der Enzymatischen Analyse" (H. U. Bergmeyer, ed.), Vol. 1, 2nd Ed., p. 441. Verlag Chemie, Weinheim, 1970.

[11] G. Johansson and M. Joelsson, *Biotechnol. Bioeng.* **27,** 621 (1985).

[12] G. Johansson and M. Joelsson, *J. Chromatogr.* **393,** 195 (1987).

[13] F. Tjerneld, G. Johansson, and M. Joelsson, *Biotechnol. Bioeng.* **30,** 809 (1987).

[14] G. Johansson, this volume [3].

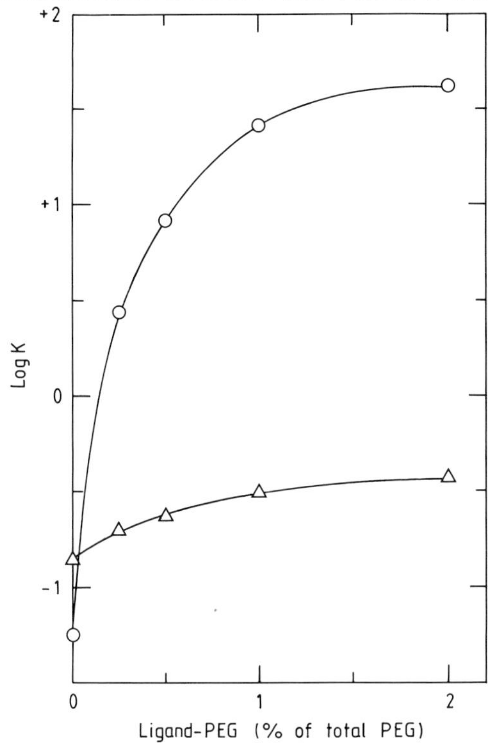


FIG. 1. Effect of the concentration of PEG-bound Procion Yellow HE-3G on the partition of muscle protein extract: protein (△) and lactate dehydrogenase (○). The two-phase system contained 7% (w/w) dextran, 5% (w/w) PEG, 25 m*M* sodium phosphate buffer, pH 7.9, including 50% (w/w) muscle extract. Temperature, 0°. [From G. Johansson and M. Joelsson, *Appl. Biochem. Biotechnol.* **13,** 15 (1986), with permission.]

higher than 10% Dx and 7.1% PEG, the recovery of the target enzyme is reduced. This is due to the lowering of the *K* of LDH (in the presence of ligand–PEG) with increasing polymer concentrations.

*Effects of pH and Buffer Concentration*

In affinity extraction with PEG–triazine dyes into the top phase, the degree of purification is usually increased at higher pH.[15,16] With LDH an opposite effect is observed. The optimal pH for LDH extraction is 7.5, and the degree of purification is negatively influenced when the pH is

[15] G. Johansson and M. Andersson, *J. Chromatogr.* **303,** 39 (1984).
[16] G. Johansson, this series, Vol. 104, p. 356.

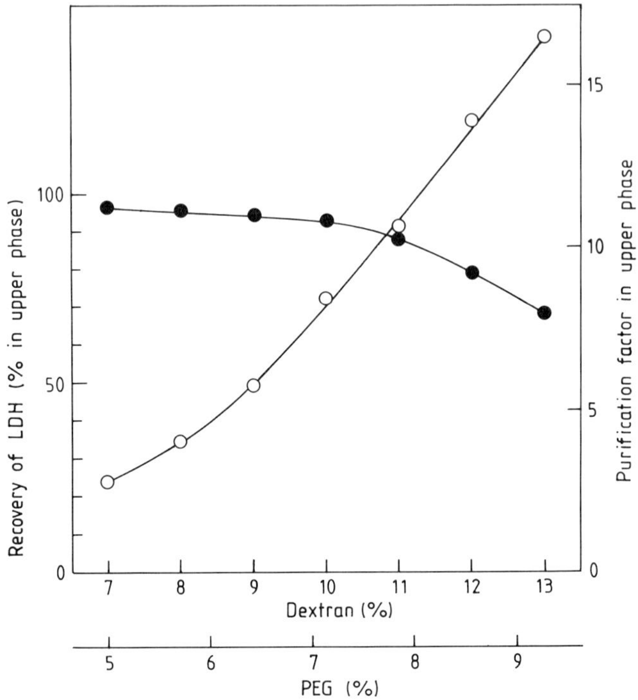


FIG. 2. Recovery (●) and purification factor (○) of LDH in the upper phase, using systems with different concentrations of dextran and PEG. Systems contained, besides the polymers, 25% (w/w) protein extract and 12.5 m*M* sodium phosphate buffer, pH 7.9. Approximately 1% of the PEG was carrying Procion Yellow HE-3G. Temperature, 0°. [From G. Johansson and M. Joelsson, *Appl. Biochem. Biotechnol.* **13,** 15 (1986), with permission.]

raised from 7.5 to 9.5. The specific activity is reduced from 64.2 to 45.6 U/mg protein and the recovery from 100 to 87%.[8] The concentration of phosphate buffer has only a moderate effect on the partitioning of both LDH and bulk protein. When the concentration of sodium phosphate buffer, pH 7.9, is increased from 10 to 100 m*M*, the log *K* of LDH decreases from 1.4 to 1.2. The log *K* of total protein changes from −0.6 to −0.8.[8]

### *Affinity Extraction of Enzyme*

The isolation of LDH is typically made from 100 g of muscle tissue. The extract, prepared as above, is included in a 1.0-kg system (10% dextran, 7.03% PEG, 0.07% Procion Yellow–PEG, 50 m*M* sodium phosphate buffer, pH 7.9, and 25% muscle extract). Affinity extraction of LDH

TABLE I
EXTRACTION OF LACTATE DEHYDROGENASE FROM MUSCLE EXTRACT

| Partition step[a] | Phase | Specific LDH activity[b] (U/mg protein) | Purification factor (-fold) | Recovery of LDH (%) |
|---|---|---|---|---|
| Extract | — | 23.6 | 1 | 100 |
| Affinity partition | Upper | 246 | 10.4 | 92 |
| Wash step 1 | Upper | 428 | 18.1 | 86 |
| Wash step 2 | Upper | 485 | 20.6 | 79 |

[a] System composition: 10% dextran, 7.03% PEG, 0.07% Procion Yellow–PEG, 50 m*M* sodium phosphate buffer, pH 7.9, and 25% muscle extract. Temperature, 0°. After the affinity partitioning step, the upper phase of the system was washed twice with the same volume of pure lower phase.

[b] LDH activity is expressed as micromoles substrate converted per minute (= 1 U) at 25°.

is shown in Table I. The mixed two-phase system is divided into two or four centrifugal beakers (e.g., for the Beckman GSA rotor), cooled on ice, and equilibrated. A short centrifugation (3–5 min at 5000 *g*) at 4° gives sufficient separation of bulk phases. Alternatively, the phases may be permitted to separate by standing 30–60 min.

The upper phase of the initial system is extracted twice with fresh lower phase. These washing steps remove proteins with low or no interaction with the ligand. In the initial affinity extraction step a 10-fold purification of the enzyme is obtained. The purification factor is increased to 21-

TABLE II
CONTAMINATING ENZYMES IN LACTATE DEHYDROGENASE PREPARATION COMPARED TO ORIGINAL HOMOGENATE[a]

| Enzyme | Activities relative to LDH[b] | |
|---|---|---|
| | Homogenate | LDH product |
| Pyruvate kinase | 50 | 0.006 |
| 3-Phosphoglycerate kinase | 19 | 0.06 |
| Myokinase | 14 | 4.7 |
| Malate dehydrogenase | 58 | 19 |
| Glutamate-aspartate transaminase | 5.6 | 1.2 |

[a] From G. Johansson and M. Joelsson, *Appl. Biochem. Biotechnol.* **13,** 15 (1986), with permission.

[b] Activities, originally calculated in micromoles per minute, are given as percentages of LDH activity.

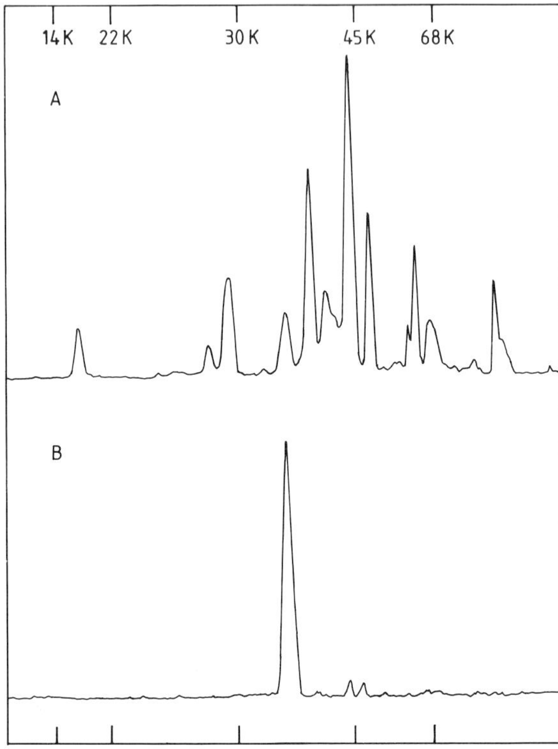


FIG. 3. Polypeptide pattern obtained by sodium dodecyl sulfate gel electrophoresis and analyzed by photometric scanning. (A) Original protein extract from porcine muscle. (B) LDH preparation obtained by affinity extraction. Molecular weights are indicated at top. [From G. Johansson and M. Joelsson, *Appl. Biochem. Biotechnol.* **13,** 15 (1986), with permission.]

fold following the two washing steps. The extraction is completed in 80 min, including three 20-sec periods of mixing and three 10-min periods of centrifugation (including acceleration and deceleration). The remaining time is used for separation of phases as well as loading the centrifuge.

Enzyme and polymers (PEG and Procion Yellow–PEG) in the upper phase can easily be separated. This is done by addition of 213 g of a 50% (w/w) solution of sodium phosphates ($Na_2HPO_4 \cdot H_2O$ and $NaH_2PO_4$, with a molar ratio of 1 : 1) to the recovered yellow upper phase and equilibrating this PEG–salt system at room temperature. The phases separate within a few minutes without centrifugation. LDH is recovered in the lower, salt-rich phase in good recovery (95%). Approximately 95% of the PEG together with Procion Yellow–PGE, present in the upper phase, is recovered. The amounts of contaminating enzymes in the final lower phase (after desalting by dialysis) are given in Table II. The electrophoresis patterns of the final product and the original protein extracts are shown in Fig. 3.

## Discussion

Lactate dehydrogenase can effectively be extracted into the upper, PEG-rich phase when muscle extract is included in an aqueous two-phase system. The extraction is performed by binding a ligand for the enzyme, in this case the triazine dye Procion Yellow HE-3G, to PEG. The choice of dye is of importance since the triazine dyes have varying selectivities.[3] For LDH extraction Procion Yellow HE-3G was found to have favorable properties.

The aqueous two-phase systems offer some important features for the preparation of LDH: (1) extremely rapid enzyme extraction (10–30 min in any scale with adequate equipment); (2) high capacity (i.e., 100 g tissue/liter system with only 1% of total PEG in liganded form); (3) very good recovery of the target enzyme; and (4) easy scale up because of the simplicity of the process. The gel electrophoresis pattern (Fig. 3) shows that LDH can be obtained in good purity by using affinity partitioning exclusively. The yield is also high (79%). Other enzymes activities in muscle are in some cases effectively eliminated (e.g., pyruvate kinase and 3-phosphoglycerate kinase), whereas other activities are partly coextracted (e.g., malate dehydrogenase and myokinase) (Table II).

A low-cost starch polymer can be substituted for Dx in the extraction of lactate dehydrogenase.[13] The polymer is a hydroxypropyl starch derivative which is commercially available as Reppal PES 100 (earlier trade name Aquaphase PPT) from Reppe AB (Växjö, Sweden). This polymer can be used with PEG to generate two-polymer phase systems with properties similar to those of Dx–PEG systems.[17,18] LDH has been extracted in large scale (30 liters) using the hydroxypropyl starch polymer and the PEG-bound dye ligand Procion Yellow HE-3G.[13]

[17] F. Tjerneld, S. Berner, A. Cajarville, and G. Johansson, *Enzyme Microb. Technol.* **8,** 417 (1986).

[18] S. Sturesson, F. Tjerneld, and G. Johansson, *Appl. Biochem. Biotechnol.* **26,** 281 (1990).

# [13] Phosphofructokinase from Baker's Yeast

*By* GERHARD KOPPERSCHLÄGER

## Introduction

In studying the interaction of biomimetic dyes with nucleotide-dependent enzymes, phosphofructokinase from baker's yeast was found to be one of the most interesting enzymes. The complex formation of Cibacron Blue F3G-A with this kinase was discovered in the late 1960s, and affinity separation techniques employing the immobilized dye have been developed over a short time because of its potential for nucleophilic substitution on diverse matrices and under mild chemical conditions.[1-3]

The binding of Cibacron Blue F3G-A and structurally related dyes to phosphofructokinase probably occurs on the ATP-binding sites of the enzyme. This is of particular interest because ATP functions not only as a substrate but also as an allosteric inhibitor. There are kinetic and binding experiments which indicate that in addition to the substrate binding site an ATP regulatory binding site exists.[4]

Analysis of the interaction of Cibacron Blue F3G-A with phosphofructokinase revealed that both ATP and MgATP counteract the dye–enzyme complex formation. The binding of the enzyme to the dye can also be abolished nonspecifically by increasing ionic strength (e.g., addition of ammonium sulfate up to 1 *M*). Based on these results an effective purification procedure, which includes affinity chromatography on immobilized Cibacron Blue F3G-A, has been worked out[5,6] to yield essentially undegraded enzyme, as most of the endogenous proteinases in the yeast extract are rapidly removed by the affinity chromatographic step. Here I present a rapid and efficient purification method for phosphofructokinase using aqueous two-phase systems containing biomimetic dyes as affinity ligands.

[1] H.-J. Böhme, G. Kopperschläger, J. Schulz, and E. Hofmann, *J. Chromatogr.* **69,** 209 (1972).

[2] G. Kopperschläger, H.-J. Böhme, and E. Hofmann, *in* "Advances in Biochemical Engineering" (A. Fiechter, ed.), p. 101. Springer-Verlag, Berlin, Heidelberg, and New York, 1982.

[3] C. R. Lowe and J. C. Pearson, this series, Vol. 104, p. 97.

[4] K. Nissler, W. Schellenberger, and E. Hofmann, *Acta Biol. Med. Ger.* **36,** 1027 (1977).

[5] W. Diezel, H.-J. Böhme, K. Nissler, R. Freyer, W. Heilmann, G. Kopperschläger, and E. Hofmann, *Eur. J. Biochem.* **38,** 479 (1973).

[6] E. Hofmann and G. Kopperschläger, this series, Vol. 90, p. 49.

## Methods

### *Assay Methods*

*Phosphofructokinase.* The enzyme phosphofructokinase phosphorylates fructose 6-phosphate to fructose 1,6-bisphosphate according to Eq. (1):

$$\text{Fructose 6-phosphate} + \text{ATP} \xrightarrow{\text{Mg}^{2+}} \text{fructose 1,6-bisphosphate} + \text{ADP} \quad (1)$$

Enzyme activity is measured in the optical test coupled with aldolase, $\alpha$-glycerophosphate dehydrogenase, and triose-phosphate isomerase by following the oxidation of NADH at 340 nm. To achieve optimum conditions the allosteric activator AMP is added to the assay mixture.

The assay solution contains 100 m*M* imidazole hydrochloride, pH 7.2, 3 m*M* fructose 6-phosphate, 0.6 m*M* ATP, 1 m*M* AMP, 5 m*M* $MgSO_4$, 5 m*M* $(NH_4)_2SO_4$, 0.2 m*M* NADH, 1 unit/ml aldolase, 1.5 units/ml $\alpha$-glycerophosphate dehydrogenase, and 5 units/ml triose-phosphate isomerase. The assay is started by addition of 20 to 50 $\mu$l of adequately diluted enzyme into 1.5 ml of the assay mixture.

*Protein assay.* Protein is assayed according to Bradford[7] or spectrophotometrically at 279 nm by using the absorption coefficient of 0.881 ml $mg^{-1}$ $cm^{-1}$.[8]

### *Aqueous Two-Phase Systems*

Two-phase systems are prepared from aqueous solutions of poly(ethylene glycol) (PEG) 6000 ($M_r$ 6000–8000; Serva, Heidelberg, Germany), 40% (w/w), and dextran (Dx) T500 ($M_r$ 500,000; Pharmacia, Uppsala, Sweden), 20% (w/w). The concentration of the Dx solutions is determined polarimetrically.[9] Unless otherwise noted the standard buffer contains 50 m*M* sodium phosphate, pH 7.1, 0.5 m*M* EDTA, and 5 m*M* 2-mercaptoethanol. The partitioning of the enzyme is determined by withdrawing aliquots of the phases, which are added either directly or after dilution to the assay mixture. The partition coefficient, $K$, is defined as the ratio of the activities in the upper and lower phases when equal volumes are compared. The term $\Delta \log K$ is calculated as the difference between the $\log K$ values of systems with and without Cibacron Blue F3G-A–PEG.

[7] M. M. Bradford, *Anal. Biochem.* **72,** 248 (1976).
[8] G. Kopperschläger, J. Bär, K. Nissler, and E. Hofmann, *Eur. J. Biochem.* **81,** 317 (1977).
[9] D. E. Brooks and R. Norris-Jones, this volume [2].

*Preparation of Cibacron Blue F3G-A–Poly(ethylene Glycol)*

The synthesis of Cibacron Blue F3G-A–PEG has been described elsewhere.[10-12] Briefly, 50 g PEG, 2.5 g anhydrous sodium sulfate, and 5.5 g sodium hydroxide are dissolved in 50 ml water. Two grams Cibacron Blue F3G-A (Ciba-Geigy, Basel, Switzerland) are added incrementally, and the mixture is stirred for about 7 hr at room temperature. The pH is then adjusted to 6–7 by addition of concentrated phosphoric acid, and the reaction mixture is diluted to 1.5 liters or exhaustively dialyzed against water. Activated DEAE-cellulose (50 g, wet weight) is added and the slurry mixed until the colored material is adsorbed. The ion exchanger is then washed with an excess of water by suction filtration. The Cibacron Blue F3G-A–PEG is eluted with 2 *M* KCl (0.5 liter) and the liganded polymer extracted from the eluate with chloroform. After drying the organic phase with anhydrous $Na_2SO_4$, the chloroform is evaporated; the remaining dark blue mass routinely weighs about 2.5 g. It contains 0.155 mmol Cibacron Blue F3G-A per gram polymer when measured photometrically at 612 nm.[13]

## Results

*Partitioning of Phosphofructokinase in Poly(ethylene Glycol)–Dextran System*

Phosphofructokinase partitions in favor of the Dx-rich phase when the concentration of the two polymers constituting the phase system are increased in a constant proportion (Fig. 1).[11] At 5% PEG and 7.5% Dx more than 90% of the total activity is found in the lower, Dx-rich phase. The partitioning behavior of the enzyme is nearly independent of the state of purity. As expected from theory,[14] a lower molecular weight of PEG favors the partitioning of the enzyme to the top phase.

The partitioning of phosphofructokinase strongly depends on pH in the range from 6 to 8. The $K$ shifts from 0.01 at pH 6.0 to about 1.0 at pH 8.0.[11] An increase in $K$ is also observed with an increase in temperature. Increasing ionic strength has little or no effect on the partitioning of the enzyme in the absence of the dye ligand.

[10] G. Kopperschläger and G. Johansson, *Anal. Biochem.* **124,** 117 (1982).

[11] G. Johansson, G. Kopperschläger, and P.-Å. Albertsson, *Eur. J. Biochem.* **131,** 589 (1983).

[12] G. Johansson and M. Joelsson, *Biotechnol. Bioeng.* **27,** 621 (1985).

[13] V. Bouriotis and P. D. G. Dean, *J. Chromatogr.* **206,** 521 (1981).

[14] P.-Å. Albertsson, A. Cajarville, D. E. Brooks, and F. Tjerneld, *Biochim. Biophys. Acta* **926,** 87 (1987).

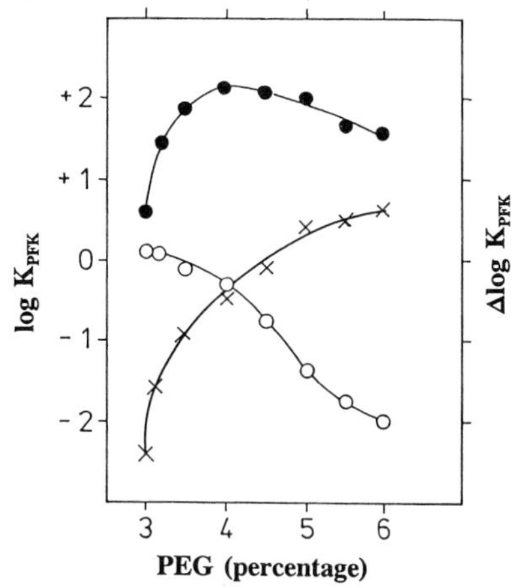


FIG. 1. Dependence of the partitioning of phosphofructokinase on the concentration of the polymers PEG and dextran. The system was formed with PEG 6000 and dextran T500 at a constant ratio of 1 : 1.5 but different amounts in 25 m*M* sodium phosphate buffer, pH 7.0, 0.5 m*M* EDTA, and 5 m*M* 2-mercaptoethanol. Plotted are values of log *K* in the absence (○) and presence (●) of Cibacron Blue–PEG (3% of the total amount of PEG) along with the calculated Δ log *K* values (x). [From G. Johansson, G. Kopperschläger, and P.-Å. Albertsson, *Eur. J. Biochem.* **131,** 589 (1983), with permission.]

## *Affinity Partitioning of Phosphofructokinase Using Poly(ethylene Glycol)-Bound Dye*

Affinity partitioning of phosphofructokinase is carried out in aqueous two-phase systems in which a small portion of PEG is replaced by PEG–dye ligand. In a system that contains 5% (w/w) PEG 6000 and 7% (w/w) Dx T500 (the resulting volume ratio $V_t/V_b$ is 1.5), 92.2% of the total enzyme activity is found in the lower phase ($K = 0.056$). By replacing increasing amounts of PEG with Cibacron Blue F3G-A–PEG the *K* value increases incrementally toward a limiting value of 50, which corresponds to a maximum increase in Δ log *K* of 2.9 (Fig. 2). Using an inverse plot of 1/Δ log *K* versus 1/concentration of Cibacron Blue F3G-A–PEG, a straight line is obtained (Fig. 2, inset) from which the maximum extraction effect at infinite concentration of Cibacron Blue F3G-A–PEG (intercept with the ordinate, i.e., 0.345) and the relative binding strength (intercept with the abscissa, i.e., 17.4 inverse percentage units) can be obtained. The latter value gives the concentration of Cibacron Blue F3G-A–PEG necessary to obtain the half-maximum of Δ log *K* (i.e., 0.06% of total PEG or 4 μ*M* of Cibacron Blue–PEG). Close to the critical point of the two-phase system the effect of affinity partitioning is less pronounced.

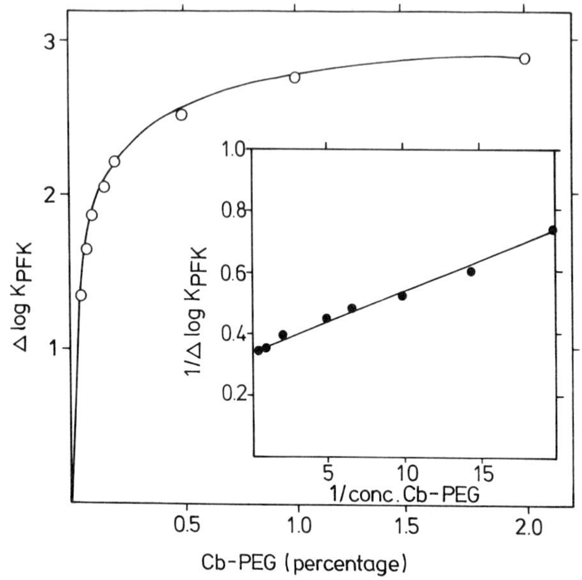


FIG. 2. Partition coefficient of phosphofructokinase as a function of the Cibacron Blue F3G-A–PEG concentration. The system contained 5% PEG 6000, 7% dextran T500, 50 m*M* sodium phosphate buffer, pH 7.0, 0.5 m*M* EDTA, 5 m*M* 2-mercaptoethanol, and various amounts of Cibacron Blue F3G-A–PEG (Cb–PEG). *Inset:* Inverse plot of the data. [From G. Johansson, G. Kopperschläger, and P.-Å. Albertsson, *Eur. J. Biochem.* **131,** 589 (1983), with permission.]

Among the classes of reactive dyes there is a large number of compounds which interact with phosphofructokinase under conditions similar to those with Cibacron Blue F3G-A (Table I). Most of these biomimetic dyes gave rise to a saturation function when $\Delta \log K$ was plotted against increasing concentrations of the ligand. Under saturated conditions a maximum $\Delta \log K$ of approximately 3 was found for several dyes, meaning a 1000-fold change in the $K$ value of phosphofructokinase.

The effect of affinity partitioning can be reduced by addition of competitive effectors that counteract the dye–enzyme binding. As seen in Fig. 3 MgATP is most effective in this regard, whereas the effect of AMP is negligibly small. The nonlinearity of the double-reciprocal plot with free ATP points to the existence of two different classes of binding sites for ATP.[4,15]

In addition to the specific interference of the dye–phosphofructokinase

[15] A. F. Chaffotte, M. Laurent, A. Tardieu, C. Roucous, F. Seydoux, and J. M. Yon, *Biochimie* **66,** 49 (1984).

TABLE I
MAXIMUM AFFINITY PARTITIONING EFFECT OF PHOSPHOFRUCTOKINASE USING DYE–POLY(ETHYLENE GLYCOL) COMBINATIONS[a,b]

| Dye ligand | $\Delta \log K_{max}$ |
|---|---|
| Procion Red HE-7B | 3.4 |
| Procion Red HE-3B | 3.4 |
| Procion Navy HE-R | 3.3 |
| Procion Green H-4G | 3.3 |
| Cibacron Blue F3G-A | 3.2 |
| Procion Brown HE-G | 3.0 |
| Procion Olive MX-3G | 2.8 |
| Cibacron Brilliant Blue BRP | 2.4 |
| Procion Yellow HE-4R | 2.3 |
| Procion Blue MX-R | 2.0 |
| Procion Red H-3B | 0.9 |

[a] From G. Kopperschläger, unpublished, 1985.
[b] Experimental conditions are as given in Fig. 2. The dye–poly(ethylene glycol) represented 1.5% of the total PEG concentration.

interaction by competitive natural ligands, higher salt concentrations, particularly of those with chaotropic ions, also tend to dissociate the dye–enzyme complex (Table II).

*Purification of Phosphofructokinase*

*Extraction of Baker's Yeast.* One kilogram of fresh baker's yeast is disrupted in a blender in 100-g portions with a 5-fold excess (w/w) of crushed dry ice for 4 min. The resulting powder is spread in 1 cm thick layers on trays to let the carbon dioxide dissipate. As soon as the material is liquefied, 1 liter of ice-cold 50 m*M* sodium phosphate buffer, pH 7.1, containing 5 m*M* 2-mercaptoethanol, 0.5 m*M* EDTA, and 0.5 m*M* phenylmethylsulfonyl fluoride (buffer A) is added. Alternatively, the fresh yeast can be suspended in ice-cold buffer A to give a cell concentration of about 60% (v/v), and cell disruption is achieved by sonication at 4°. To do this the yeast suspension is pumped through the jacketed flow-through vessel of an ultrasonic disintegrator at a frequency of 20 kHz, a sonication density of 50 W/cm$^2$ (ultrasonic disintegrator, type 250, Schöller & Co., Frankfurt/Main, Germany), and a flow rate of about 2 liters/hr.

*Fractional Precipitation with Poly(ethylene Glycol) 6000.* Solid PEG 6000 (4%, w/w) is added, over a period of 15 min with vigorous stirring, directly to the cell homogenate at 4° so as to obtain the first precipitation.

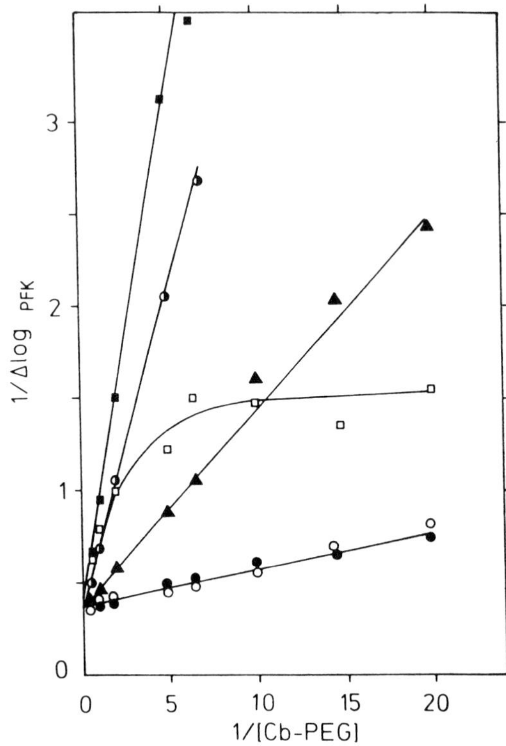


FIG. 3. Double-reciprocal plot of Δ log $K$ of phosphofructokinase and the concentration of Cibacron Blue F3G-A–PEG as a function of various substrates and effectors. Systems as described in Fig. 2 were used. (●) No effector; (○) 1 m$M$ AMP; (▲) 1 m$M$ ITP and 2.5 m$M$ $MgCl_2$; (□) 1 m$M$ ATP; (◑) 1 m$M$ ADP and 2.5 m$M$ $MgCl_2$; (■) 1 m$M$ ATP and 2.5 m$M$ $MgCl_2$. [From G. Johansson, G. Kopperschläger, and P.-Å. Albertsson, *Eur. J. Biochem.* **131,** 589 (1983), with permission.]

The homogenate is stirred for an additional 15 min and then centrifuged at 5000 $g$ for 30 min at 4°. The supernatant solution is retained, and solid PEG is added as before to give a final concentration of 8–10% (w/w). The turbid solution is allowed to stand for another 15 min before centrifugation at 4° (5000 $g$, 30 min). The pellet is dissolved in 120 ml of 25 m$M$ sodium phosphate buffer, pH 7.1, containing 0.5 m$M$ EDTA, 5 m$M$ 2-mercaptoethanol, and 0.5 m$M$ phenylmethylsulfonyl fluoride.

*Affinity Partitioning.* The dissolved pellet (120 ml) is added to a mixture of 104.9 g of 20% (w/w) Dx T500, 33.4 g of 40% (w/w) PEG 6000, 6.67 ml of 0.5 $M$ sodium phosphate buffer, pH 7.1, 65 $\mu$l of 0.5 $M$ phenylmethylsulfonyl fluoride in dioxane, and 100 $\mu$l of 2-mercaptoethanol. All solutions

TABLE II
EFFECT OF NEUTRAL SALTS ON AFFINITY PARTITIONING OF PHOSPHOFRUCTOKINASE[a,b]

| Salt | Δ log *K* (relative) at different salt concentrations | | | |
|---|---|---|---|---|
| | 25 m*M* | 63 m*M* | 125 m*M* | 250 m*M* |
| Sodium phosphate, pH 7.0 | 100 | 103 | 92 | — |
| Sodium formate, pH 7.0 | — | — | — | 88 |
| Sodium acetate, pH 7.0 | 105 | 97 | — | 88 |
| Sodium propionate, pH 7.0 | — | — | — | 89 |
| Sodium sulfate | 98 | 85 | — | 33 |
| Potassium chloride | 104 | 80 | — | 21 |
| Lithium sulfate | — | 77 | — | — |
| Triethanolammonium chloride, pH 7.0 | — | 76 | — | — |
| Potassium iodide | 97 | 69 | — | 1 |
| Sodium rhodanide | — | 67 | — | — |
| Potassium phthalate, pH 7.0 | — | 48 | — | — |
| Sodium perchlorate | — | 42 | — | — |

[a] From G. Johansson, G. Kopperschläger, and P.-Å. Albertsson, *Eur. J. Biochem.* **131,** 589 (1983), with permission.

[b] Phosphofructokinase was partitioned in the same two-phase system as in Fig. 2 with 10 m*M* sodium phosphate buffer, pH 7.0, and addition of various salts. The dye–poly(ethylene glycol) was 2% of the total PEG concentration.

[c] The affinity partitioning effect is given as the log *K* of phosphofructokinase, in the presence of salt (in addition to 10 m*M* sodium phosphate buffer), relative to the log *K* value of the enzyme in the presence of only the 10 m*M* buffer.

are kept at 4°, and the system is mixed carefully for 30 sec. Phase separation is speeded by centrifugation at 4° for 5 min at 3000 *g*. The upper phase is removed and the lower phase containing the enzyme is washed by addition of fresh upper phase from a system containing 7.5% (w/w) Dx, 5% (w/w) PEG, 25 m*M* sodium phosphate buffer, pH 7.1, 0.5 m*M* EDTA, and 0.5 m*M* phenylmethylsulfonyl fluoride. The system is mixed and allowed to settle at 4° as above.

Phosphofructokinase is then extracted from the lower phase by treating it with 135 ml of an upper phase containing 8.5% (w/w) Cibacron Blue F3G-A–PEG 6000, 25 m*M* sodium phosphate buffer, pH 7.1, 0.5 m*M* EDTA, 5 m*M* 2-mercaptoethanol, and 0.5 m*M* phenylmethylsulfonyl fluoride. The blue upper phase is washed with 100 ml of a fresh lower phase (not containing Cibacron Blue F3G-A–PEG) to remove proteins that have little or no affinity for the dye. The upper phase is collected and adjusted to room temperature before 15 g of solid $KH_2PO_4$ and $K_2HPO_4$ (molar ratio 1 : 1.2) is dissolved in it. Within 5 min a colorless salt phase (95

TABLE III
PURIFICATION OF PHOSPHOFRUCTOKINASE FROM BAKER'S YEAST[a]

| Purification step | Volume (ml) | Total activity (U) | Total protein (mg) | Specific activity (U/mg) | Yield (%) | Purification factor (-fold) | Total proteolytic activity (U) |
|---|---|---|---|---|---|---|---|
| Homogenate | 1370 | 5400 | 13,170 | 0.41 | 100 | 1 | 13,370[b] 3690 |
| Fractional precipitation with PEG | 120 | 4810 | 1836 | 2.62 | 89 | 6.4 | 670 |
| Affinity partitioning | 120 | 3610 | 153 | 23.6 | 67 | 58 | 34 |
| DEAE-cellulose batch treatment | 40 | 2520 | 63 | 40 | 47 | 98 | 15 |
| Gel filtration | 4 | 1625 | 28 | 58 | 30 | 142 | 1.7 |

[a] Wet weight: 1 kg. From G. Kopperschläger and G. Johansson, *Anal. Biochem.* **124,** 117 (1982), with permission.
[b] Proteolytic activity in absence of phenylmethylsulfonyl fluoride.

ml) separates from the mixture which contains the phosphofructokinase activity. The Cibacron Blue F3G-A–PEG present in the upper phase is recovered by extraction of the polymer with chloroform and rotary evaporation. The dye–polymer can be used repeatedly for additional extractions.

*Ion-Exchange Chromatography with DEAE-Cellulose.* The salt phase of the foregoing step is desalted at 4° by dialysis against buffer A containing 1 m*M* fructose 6-phosphate or by passing it through a Sephadex G-25 column (2.5 × 60 cm) equilibrated with dialysis buffer. To the solution

TABLE IV
PROPORTION OF ENZYMES IN YEAST EXTRACTS AND IN THE FINAL PREPARATION[a,b]

| Enzyme | In homogenate[c] | In purified phosphofructokinase[c] |
|---|---|---|
| Glucose-6-phosphate dehydrogenase | 51 | 0.2 |
| Hexokinase | 235 | 0.03 |
| Adenylate kinase | 30 | 0.015 |
| Enolase | 668 | 0.01 |
| Phosphoglycerate phosphomutase | 795 | 0.004 |
| Glyceraldehyde-3-phosphate dehydrogenase | 226 | 0.01 |
| 3-Phosphoglycerate kinase | 2640 | 0.05 |

[a] Expressed as percentage of the total activity of phosphofructokinase.
[b] From G. Kopperschläger and G. Johansson, *Anal. Biochem.* **124,** 117 (1982), with permission.
[c] All activities have been calculated in micromoles substrate converted per minute.

containing the enzyme, 7–10 g of activated DEAE-cellulose (wet weight) is added, and the slurry is stirred for 15 min. After suction filtration the ion exchanger is washed four times with 100 ml of the dialysis buffer, and the washings are discarded. The enzyme is then eluted by treating the DEAE-cellulose, packed in a Büchner funnel, with five 5-ml portions of buffer A containing 1 m*M* fructose 6-phosphate and 200 m*M* potassium chloride. The filtrates are collected and concentrated either by ultrafiltration using the membrane XM300 (Amicon Ltd., Stonehouse, UK) or by precipitation with ammonium sulfate.

*Gel Filtration.* The enzyme is concentrated to a volume of approximately 5 ml and subjected to gel filtration on a Sepharose 6B-CL (Pharmacia) column (2.5 × 100 cm) which has previously been equilibrated with buffer A containing 1 m*M* fructose 6-phosphate and 1.27 *M* ammonium sulfate. The fractions containing the enzyme activity are collected and concentrated again by ammonium sulfate precipitation. Table III summarizes the purification protocol with yield, specific activity, and purification factor. In addition, the separation of phosphofructokinase from the endogenous proteolytic activities is indicated. In Table IV the effectiveness of the procedure with regard to the removal of other glycolytic enzymes is shown.

## Discussion

The quality and yield of phosphofructokinase obtained by affinity partitioning method are similar to those of other methods.[6,8,15] This is true for the specific activity, the recovery, and the subunit pattern after sodium dodecyl sulfate–polyacrylamide gel electrophoresis. The quantity of partially degraded phosphofructokinase caused by endogenous yeast proteinases is less than 10% of the native form.

The application of affinity partitioning using Cibacron Blue F3G-A–PEG turned out to be of particular advantage for the rapid removal of contaminating proteolytic activities and for producing a high degree of purification of the enzyme. The effectiveness of affinity partitioning permits the extraction of the enzyme from most other cellular components in one step. Moreover, the time required to carry out affinity partitioning is less than 1 hr. Starting from 1 kg of fresh yeast, extraction in the two-phase system requires a total volume of only 250 ml and can easily be scaled up 10-fold or more. PEG has the additional feature of stabilizing the enzyme during purification and can be used for the precipitation of phosphofructokinase.

# [14] Partitioning of Blood Proteins Using Immobilized Dyes

*By* GERD BIRKENMEIER

## Introduction

The increasing demand for blood products for use as therapeutic agents necessitates the development of effective protein separation techniques. Purification of individual plasma proteins is often associated with problems because of similarities in their physicochemical properties and because many of them are present at low concentrations. Affinity chromatography techniques have greatly improved protein purification.[1] Thus, a large number of proteins have been purified from human plasma by use of reactive dyes as affinity ligands.[2,3] Immobilized Cibacron Blue F3G-A was applied to the fractionation and purification of albumin,[4,5] blood clotting factors,[6] $\alpha$-fetoprotein,[7] $\alpha_1$-proteinase inhibitor and $\alpha_1$-acid glycoprotein,[8] and other proteins.[2]

Since their introduction by Albertsson,[9] aqueous two-phase systems have been applied for the extractive separation of biological materials, such as proteins, membranes, organelles, and cells.[10] The partitioning of a macromolecule is influenced by many factors, including the molecular weight and charge of the biomolecule, concentration and molecular weight of the polymer, type and concentration of added salt, pH, and temperature.[11] Addition of affinity ligands covalently coupled to one of the phase-forming polymers can often be used to alter the partitioning of a protein

[1] C. R. Lowe and P. D. G. Dean, "Affinity Chromatography." Wiley, New York, 1974.

[2] G. Kopperschläger, H.-J. Böhme, and E. Hofmann, *in* "Advances in Biochemical Engineering" (A. Fiechter, ed.), p. 101. Springer-Verlag, Berlin, 1982.

[3] G. Kopperschläger, this volume [11].

[4] J. Travis and R. Pannell, *Clin. Chim. Acta* **49,** 49 (1973).

[5] G. M. Ghiggeri, G. Candiano, G. Delfino, and C. Queirolo, *Clin. Chim. Acta* **145,** 205 (1985).

[6] A. C. W. Swart, B. H. M. Kop-Klassen, and H. C. Hemker, *Haemostasis* **1,** 237 (1972/1973).

[7] K. Huse, M. Himmel, G. Birkenmeier, M. Bohla, and G. Kopperschläger, *Clin. Chim. Acta* **133,** 335 (1983).

[8] G. Birkenmeier and G. Kopperschläger, *J. Chromatogr.* **235,** 237 (1982).

[9] P.-Å. Albertsson, "Partition of Cell Particles and Macromolecules." Wiley, New York, 1960.

[10] H. Walter, D. E. Brooks, and D. Fisher (eds.), "Partitioning in Aqueous Two-Phase Systems." Academic Press, Orlando, Florida, 1985.

[11] G. Johansson, this volume [3].

in a selective manner.[12] As in affinity chromatography, textile dyes are also useful ligands in affinity partitioning. These dyes can be obtained in a great variety of molecular structures, and some have a striking affinity for certain proteins. Affinity partitioning has also proved to be a sensitive tool in the analytical study of the mechanisms of dye–protein interactions. For example, the topography of dye-binding sites on the protein surface,[12] the dynamics of protein structure,[13] and the influence of ligand immobilization on the binding mechanism have all been studied in detail by use of this method.[14] Here I focus on the applications of aqueous two-phase systems in the study of interactions of dyes with individual serum proteins and on serum protein fractionation by use of affinity ligands and with multistep extractions [i.e., countercurrent distribution (CCD)] or partitioning in three-phase systems.

## Materials and Methods

Human serum albumin (HSA) and prealbumin are prepared from fresh human serum by dye affinity chromatography.[15] Human serum is obtained from presumably healthy blood donors. Prior to use, the serum is dialyzed against the respective buffer used in the partitioning experiments. Small precipitates are removed by centrifugation. Antisera to human serum proteins are obtained from Behringwerke AG (Marburg, Germany). Poly(ethylene glycol) (PEG) is purchased from Serva (Heidelberg, Germany) (PEG 6000) or from Union Carbide (New York, NY) (PEG 8000). Dextran (Dx) T500, Dx T70, Dx T40, and Ficoll 400 are obtained from Pharmacia-LKB (Uppsala, Sweden).

### *Single-Tube Partitioning*

Phase systems are prepared essentially as outlined by Brooks and Norris-Jones.[16] To prepare a 10-g two-phase system of the composition given in Fig. 1, and using 20% (w/w) Dx and 40% (w/w) PEG stock solutions, the following solutions are weighed out in a centrifuge tube: 5 g of 20% Dx T500, 1.875 g of 40% PEG 6000, 1 g buffer (10-fold concentrated), and 0–2.125 g of the protein sample. If less than 2.125 g of protein solution is added the difference is made up with water. The systems are carefully mixed by 20–40 inversions of the sealed tube, and the phases

[12] G. Johansson, G. Kopperschläger, and P.-Å. Albertsson, *Eur. J. Biochem.* **131,** 589 (1983).

[13] G. Kopperschläger and G. Johansson, *Biomed. Biochim. Acta* **44,** 1047 (1985).

[14] G. Johansson and M. Joelsson, *J. Chromatogr.* **537,** 219 (1991).

[15] G. Birkenmeier, B. Tschechonien, and G. Kopperschläger, *FEBS Lett.* **174,** 162 (1984).

[16] D. E. Brooks and R. Norris-Jones, this volume [2].

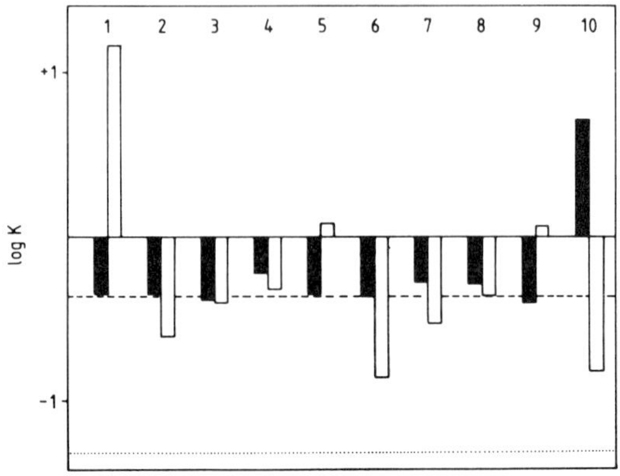


FIG. 1. Effect of different dye–PEG derivatives on partitioning of prealbumin and albumin. Prealbumin (1.6 $\mu M$, black bars) and albumin (13.1 $\mu M$, white bars) were partitioned together at 0° in 2-g systems composed of 10% (w/w) dextran T500, 7.5% (w/w) PEG 8000, and 10 m$M$ sodium phosphate buffer, pH 7.0. The respective dye–PEG derivatives constituted 1.6% of total PEG in the system. Partitioning of prealbumin (---) and albumin (·····) in the absence of dye–PEG is indicated. Dye–PEG derivatives: (1) Cibacron Blue F3G-A; (2) Procion Red H-3B; (3) Procion Orange MX-G; (4) Procion Yellow MX-R; (5) Procion Scarlet HR-N; (6) Procion Yellow MX-4G; (7) Procion Scarlet MX-G; (8) Procion Yellow MX-GR; (9) Procion Brown MX-5BR; (10) Remazol Yellow GGL. [From G. Birkenmeier and G. Kopperschläger, *Mol. Cell. Biochem.* **73,** 99 (1987). Reprinted by permission of Kluwer Academic Publishers.]

are then allowed to separate. The settling time may be reduced to less than 1 min by brief centrifugation at low speed.

In case of affinity partitioning one of the polymers is substituted by a desired percentage of a liganded polymer. For example, if the system above was to contain 1.6% liganded PEG, 1.6% (w/w) of the total PEG would be replaced by liganded PEG, and thus the system would contain 1.845 g PEG and 0.03 g liganded PEG.

### *Single-Tube Partitioning in Three-Phase Systems*

Three-phase systems of 5 g are prepared by mixing aqueous stock solutions of Dx T40 (20%), Ficoll 400 (40%), and PEG 6000 (40%) including liganded polymer solution and sample to yield the desired total composition of 10% (w/w) Dx T40, 7% (w/w) Ficoll 400, and 3.5% (w/w) PEG 6000. The systems are equilibrated at 22°. Centrifugation for 10 min at 1000 *g* is used to speed phase separation. Samples for analysis are withdrawn from the phases, appropriately diluted with barbital buffer, and subjected to electroimmunodiffusion.

*Countercurrent Distribution of Serum Proteins*

A two-phase system composed of 5% (w/w) PEG 8000, 7.5% (w/w) Dx T70, 20 m*M* sodium phosphate buffer, pH 7.0, and 20 m*M* NaCl is used for countercurrent distribution. A stock system of 150 g is prepared and equilibrated at 3° overnight. A thin-layer CCD apparatus with 60 chambers is used.[9] The volume of each cavity in the lower phase is 0.79 ml. The chambers are loaded with 0.95 ml of the upper phase and 0.7 ml of the lower phase of the separated phase system. A sample system of the same composition weighing 10 g is prepared by including 0.45 ml of serum per gram of the two-phase system. Protein precipitation visible at the interface after mixing the phase system is removed by centrifugation (5000 *g* for 2 min). The two phases of the sample system are separated, and volumes as indicated above are added to chambers 0 and 1 of the plates. The number of transfers is chosen to be 58. A shaking time of 30 sec and a separation time of 10 min per transfer cycle are applied. All steps are carried out at 3°. After the run, 1 ml of 50 m*M* sodium phosphate buffer, pH 7.0, is added to each chamber for two-phase dilution to a single phase. The fractions are analyzed for total protein content and for individual serum proteins (see below).

The partitioning of serum proteins in CCD is expressed as the partition ratio, $G$, defined as the percentage of protein in the mobile part of the system divided by the percentage of protein in the stationary part. An approximate calculation of $G$ can be made from the peak position in the CCD run according to the relationship $G = \hat{\imath}/(n - \hat{\imath})$, where $\hat{\imath}$ is the distance in the number of the tubes that the protein (peak position) has moved and $n$ is the number of transfers.

*Calculation of Partitioning in Single-Tube Experiments*

The partition coefficient, $K$, is calculated as the ratio of the concentration ($c$) in the top (t) and the bottom (b) phases, $K_{t/b} = c_t/c_b$. The effect of a polymer-bound ligand on the partitioning of a protein is expressed in terms of $\Delta \log K$, given by $\Delta \log K = \log K_{aff} - \log K_0$, where $K_0$ and $K_{aff}$ are the partition coefficients of the protein in the absence and presence of the polymer-bound ligand in the system, respectively, other conditions being identical. The maximum $\Delta \log K$ ($\Delta \log K_{max}$) is obtained by plotting $1/\Delta \log K$ versus 1/ligand concentration.

*Preparation of Dye–Polymer Conjugates*

*Dye–Poly(ethylene glycol) Conjugates.* The coupling of reactive dyes to PEG proceeds at 90° in aqueous solution under alkaline conditions by an nucleophilic reaction between reactive groups of the dyes and the OH group of the polymer.[12] For the coupling, 20 g of PEG 6000 and 0.2 g LiOH are dissolved in 46 ml distilled water and heated to 90°. With continuous stirring 4.5 g of dye and 0.56 g LiOH are added stepwise over a period of 1 hr. After continued incubation for 1 hr the solution is stirred for an additional 16 hr at room temperature. Then 200 ml of distilled water and 200 ml of 3 *M* KCl are added, and the polymer is extracted three times with chloroform (200 ml each time). Chloroform is evaporated, and the resulting polymer (18 g) is dissolved in 400 ml water, mixed with DEAE-cellulose (Reanal, Budapest, Hungary) (100 g wet weight), and stirred for 1 hr at room temperature. Unsubstituted PEG is removed by washing the material with excess water on a sintered glass filter. Elution of dye-substituted PEG is accomplished by adding 3 *M* KCl solution. Again, the dye–PEG is extracted with chloroform (3 times 200 ml each time). The organic solutions (600 ml) are combined, and 100 g anhydrous $Na_2SO_4$ (Merck, Darmstadt, Germany) is added to remove excess water. The solution is filtered, and the chloroform is removed by evaporation. The yield is between 200 mg and 2 g of dye–PEG depending on the reactivity of the particular dye used. The degree of substitution ranges between 0.8 and 1.1 mol dye/mol PEG, indicating monosubstitution.

*Dye Conjugates of Dextran and Ficoll.* Dx T40 ($M_r$ 40,000) and Ficoll 400 ($M_r$ 400,000) can be liganded to Remazol Yellow GGL (Hoechst AG, Frankfurt, Germany) according to the method of Johansson and Andersson.[17] Ten grams Dx T40 or Ficoll 400 is dissolved in 50 ml distilled water containing 4 g $Na_3PO_4 \cdot 12H_2O$, 1 g NaCl, and 0.3 g Remazol Yellow GGL. The temperature of the mixture is slowly raised to 70° in a water bath, and the mixture is stirred for 1 hr. After cooling to room temperature the solution is dialyzed against excess distilled water to remove free dye. The polymer content in the final solution is 5.9% Ficoll 400 of 6.5% Dx T40 (based on polarimetric determination using $[\alpha]_D^{25}$ values of +199° and +54.9° ml g$^{-1}$ dm$^{-1}$ for Dx and Ficoll, respectively). The degree of dye substitution as analyzed photometrically using the extinction coefficient of 13.4 m*M* cm$^{-1}$ (396 nm) for Remazol Yellow GGL is 37 μmol dye/g Ficoll and 11 μmol dye/g Dx.

Coupling of Cibacron Blue F3G-A (Serva) to polymers is accomplished in an analogous manner. After the reaction is completed the polymers are precipitated by the slow addition of ethanol (500 ml). The precipitate is dissolved in water and reprecipitated. This cycle is repeated three times so as to remove the uncoupled dye. The last precipitate is dialyzed against excess distilled water. The resulting solutions contain 4.32% Ficoll 400

[17] G. Johansson and M. Andersson, *J. Chromatogr.* **303,** 39 (1984).

and 5.6% Dx T40, respectively. The degree of substitution is 12 μmol dye/g Ficoll and 21 μmol dye/g Dx, respectively, using the extinction coefficient of 13.6 m*M* $cm^{-1}$ (610 nm) for Cibacron Blue F3G-A.

### *Analysis of Serum Proteins in Phase Systems*

*Immunological Analysis.* Protein concentration in the top and bottom phases can be obtained by rocket electroimmunoassay according to Laurell.[18] Glass plates (8 × 9 cm) are coated with 15 ml of 1.5% agarose (Agarose Serva standard EEO) in sodium barbital buffer (0.1 *M*, pH 8.6). The concentration of the antiserum in the agarose ranges from 0.1 to 1% (v/v) depending on titer. Samples withdrawn from the phase system are diluted appropriately with barbital buffer and placed, in 5-μl aliquots, into holes punched in the agarose. The appropriate dilutions of the standard serum and the samples are determined by the height of the precipitation peaks (1 to 5 cm) in the agarose. Electrophoresis can be performed in a Multiphor apparatus (Pharmacia-LKB) at 10 V/cm for 3 hr at 12°. After the run the plates are washed with saline solution overnight and finally air-dried. Staining of the precipitation lines is by Coomassie Blue R-250. Protein concentrations of the samples are calculated from a calibration curve using a double logarithmic plot of concentration versus peak height.

*Photometric Analysis.* Quantitation of protein is based on the Bradford method.[19] Aliquots from sample tubes obtained after CCD runs are mixed with 300 μl of Bradford reagent in titer plates (96 wells) for 5 min. The absorbance is read at 610 nm in the ELISA-Reader Titertek Multiskan (Flow Laboratories, Meckenheim, Germany), and the protein concentration is calculated from a standard curve using bovine serum albumin (Boehringer, Mannheim, Germany) as standard.

## Results and Discussion

The partitioning of proteins in aqueous two-phase systems depends on a number of factors, which include the composition of the two-phase system used and the surface properties of the protein. Resolution of serum proteins by partitioning in two-phase systems, which can be achieved by changing parameters such as pH, salt and ion composition, polymer concentration, and/or molecular weight of the polymers, is based on differences in surface properties of proteins such as charge or size.

More selective partitioning of a particular protein can be effected by the use of specific ligands covalently bound to one of the phase-forming

[18] C.-B. Laurell, *Anal. Biochem.* **15,** 45 (1966).

[19] M. M. Bradford, *Anal. Biochem.* **72,** 248 (1976).

polymers. Thus, ligands with high specificity should extract a target protein from the bulk protein phase. Figure 1 depicts the result when 10 different dyes were screened for their ability to bind human serum albumin and prealbumin. Of the dyes tested only Remazol Yellow GGL–PEG affects the partitioning of prealbumin, thus indicating its specificity for binding to this protein. In contrast, the $K$ of albumin undergoes significant changes when PEG derivatives with different dyes are used, with Cibacron Blue F3G-A-PEG found to have the greatest effect. The relatively high affinity of albumin for different dyes[20] with great differences in chemical structure is not surprising since albumin is known to bind many different compounds including aromatics and fatty acids.[21]

The selectivity of protein binding to Remazol Yellow GGL led us to study its quantitative aspects. This can be done by constructing partitioning curves as shown in Fig. 2. Proteins are partitioned in phases with increasing substitution of dye–PEG for bulk PEG. By use of an inverse plot of $1/\Delta \log K$ versus 1/dye concentration (Fig. 2B) the following binding data were obtained: $\Delta \log K_{max}$ of 0.6 and 0.54 and a relative affinity ($0.5 \times \Delta \log K_{max}$) of 110 and 45 $\mu M$ for binding of albumin and prealbumin to Remazol Yellow GGL, respectively. When prealbumin is partitioned in the presence of excess albumin the binding of prealbumin to the dye is significantly increased, yielding a $\Delta \log K_{max}$ of 1.12, whereas the relative affinity remains constant. This increase has been explained by complex formation between albumin and prealbumin in the presence of the dye.[15,22] Such analytical data obtained by partitioning are valuable parameters to compare and evaluate the affinity and extraction power of ligands for proteins.

Competitive inhibition studies in two-phase systems were carried out so as to elucidate the Remazol Yellow GGL–protein interaction in more detail. As shown in Fig. 3, the binding of prealbumin to the dye can be reduced by the thyroid hormones T3 (L-3,3′,5-triiodothyronine) and T4 (L-thyroxine), which are naturally bound ligand. These hormones were less effective in reducing the affinity partitioning of albumin. In agreement with data obtained by other methods,[22] the results indicate that Remazol Yellow GGL binds directly or close to the hormone-binding sites in prealbumin. Similarities in the structural and stereochemical features of thyroid hormones and dye may account for this assumption.

By use of more than two polymers, systems consisting of several phases can be obtained.[9] Three-phase systems composed of Dx, Ficoll,

[20] G. Birkenmeier, E. Usbeck, and G. Kopperschläger, *Anal. Biochem.* **136,** 264 (1984).

[21] F. W. Putnam, "The Plasma Proteins." Academic Press, New York, 1975.

[22] G. Birkenmeier and G. Kopperschläger, *Mol. Cell. Biochem.* **73,** 99 (1987).

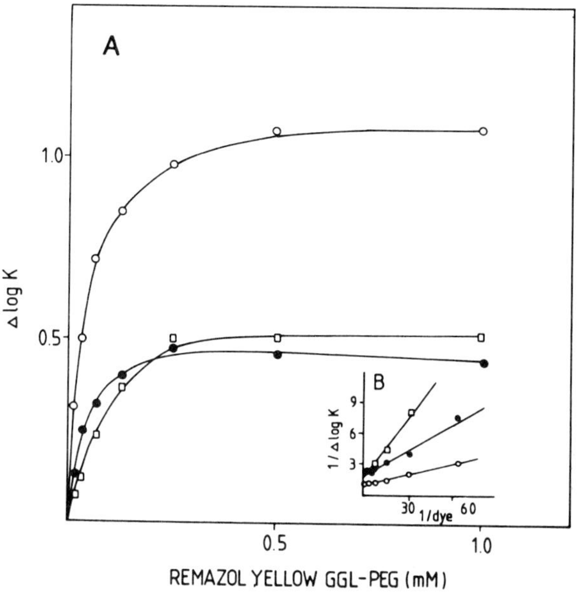


FIG. 2. Effect on the partition coefficients of prealbumin and albumin of increasing concentrations of Remazol Yellow GGL–PEG. (A) Systems of 2 g were composed of 10% (w/w) dextran T500, 7.5% (w/w) PEG 6000 containing different quantities of Remazol Yellow GGL–PEG, 10 m*M* sodium phosphate buffer, pH 7.0, and protein. The log *K* of prealbumin and albumin in the absence of dye–PEG was −0.39 and −1.37, respectively. The concentration of 1 m*M* immobilized dye corresponds to a total replacement of 3.2% of the PEG by dye–PEG. Partitioning was carried out at 0°. Protein added: (□) albumin (4.3 μ*M*); (●) prealbumin (1.6 μ*M*); (○) prealbumin (1.6 μ*M*) plus albumin (13.1 μ*M*). (B) Reciprocal plot of the data in (A) (same symbols). [From G. Birkenmeier and G. Kopperschläger, *Mol. Cell. Biochem.* **73,** 99 (1987). Reprinted by permission of Kluwer Academic Publishers.]

and PEG allow proteins to be resolved among three phases. The possibility of binding specific ligands to each of the polymers could lead to increases in the selectivity of protein partitioning. In this context we have studied the distribution of albumin and prealbumin in a three-phase system composed of 10% (w/w) Dx T40, 7% (w/w) Ficoll 400, and 3.5% (w/w) PEG 6000.[23] The volumes of the PEG-rich top phase, the Ficoll-rich middle phase, and the Dx-rich bottom phase were 24, 25, and 51%, respectively, of the total volume at 22°. Remazol Yellow GGL and Cibacron Blue F3G-A attached to different polymers were used as ligands because of their differential interaction with prealbumin and albumin.

[23] P.-Å. Albertsson and G. Birkenmeier, *Anal. Biochem.* **175,** 154 (1988).

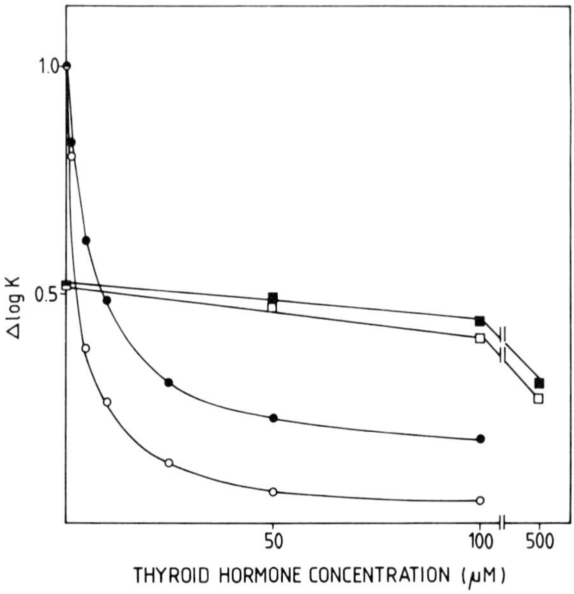


FIG. 3. Effect of thyroid hormones on affinity partitioning of prealbumin and albumin. The proteins were partitioned at 0° in 2-g systems composed of 10% (w/w) dextran T500, 7.5% PEG (w/w) 6000, 10 m*M* sodium phosphate buffer, pH 7.25, and increasing hormone concentrations. The Δ log *K* refers to the difference in partitioning of proteins in the absence and presence of 0.5 m*M* Remazol Yellow GGL–PEG. T4 (L-thyroxine) (○, □) and T3 (L-3,3′,5-triiodothyronine) (●, ■) were dissolved in 10 m*M* NaOH, incubated with the proteins for 15 min at 22°, and added to the phase system. (○, ●) Prealbumin (1.6 μ*M*) in the presence of 30 μ*M* albumin; (□, ■) albumin (30 μ*M*). [From G. Birkenmeier and G. Kopperschläger, *Mol. Cell. Biochem.* **73,** 99 (1987). Reprinted by permission of Kluwer Academic Publishers.]

Table I shows the relative distributions of the two proteins among the three phases in the absence and presence of the dyes. The rationale for testing two different ligands in one phase system was to achieve separation of the two proteins in a single extraction on the basis of specific dye interactions. This offers the advantage of steering the distribution of two proteins of a mixture into different phases simultaneously. As seen, in the absence of the ligands both proteins are found in the Dx-rich bottom phase. A very effective separation of albumin and prealbumin is achieved in the phase system containing the ligands Cibacron Blue F3G-A–Ficoll and Remazol Yellow GGL–Dx. More than 90% of the prealbumin is found in the bottom phase, whereas albumin partitions in favor of the Ficoll-rich middle phase (81%). However, the yield of a protein in each phase depends on the volume ratio. Thus, even in the experiment with Cibacron

TABLE I
PARTITIONING OF ALBUMIN AND PREALBUMIN IN THREE-PHASE SYSTEMS[a] IN PRESENCE OF TWO DIFFERENT IMMOBILIZED DYES[b]

| Dye–polymer combination[c] | Relative amount of protein in the phases | | | | | |
|---|---|---|---|---|---|---|
| | Albumin | | | Prealbumin | | |
| | Top | Middle | Bottom | Top | Middle | Bottom |
| Without dye | 1 | 16 | 83 | 6 | 19 | 75 |
| Cb-Fic/RY-Dx | 2 | 81 | 17 | 1 | 5 | 94 |
| Cb-Dx/RY-Fic | 0 | 2 | 8 | 4 | 27 | 69 |
| Cb-Dx/RY-PEG | 0 | 2 | 98 | 16 | 24 | 60 |
| Cb-PEG/RY-Dx | 32 | 34 | 34 | 1 | 7 | 92 |
| Cb-PEG/RY-Fic | 35 | 33 | 32 | 6 | 30 | 64 |
| Cb-Fic/RY-PEG | 2 | 76 | 22 | 17 | 26 | 57 |

[a] Composition of the system: 10% (w/w) dextran T40, 7% (w/w) Ficoll 400, 3.5% (w/w) PEG 6000, 20 m*M* sodium phosphate buffer, pH 7.0, and 20 m*M* NaCl. The concentrations of albumin and prealbumin in the system were 5 and 1.1 $\mu M$, respectively. Temperature, 22°.

[b] From P.-Å. Albertsson and G. Birkenmeier, *Anal. Biochem.* **175,** 154 (1988), by permission.

[c] The concentrations of the dye–polymers in the phases were as follows: Cibacron Blue F3G-A–PEG (Cb-PEG) (1.6%), Cibacron Blue F3G-A–Ficoll (Cb-Fic) (6%), Cibacron Blue F3G-A–Dx (Cb-Dx) (1%), Remazol Yellow GGL–PEG (RY-PEG) (1.6%), Remazol Yellow GGL–Ficoll (RY-Fic) (6%), and Remazol Yellow GGL–Dx (RY-Dx) (6%). The percentage of the respective dye–polymer derivatives represents the fraction of the total amount of the polymer in the system that is replaced by the dye-liganded polymer.

Blue F3G-A–PEG/Remazol Yellow GGL–Dx a 5-fold increase in the volume of the top phase would cause 71% albumin and 90% prealbumin to be extracted into the top and bottom phases, respectively. In this context the advantage of a three-phase system compared to a two-phase system is that at least one phase remains that can extract the impurities.

Compared to single-step partitioning, further improvements in separation of serum protein mixtures can be achieved by use of a multistep extraction procedure such as CCD.[9] Figure 4 demonstrates the separation of total serum after 58 transfers in a two-phase system composed of 5% (w/w) PEG 6000, 7.5% (w/w) Dx T70, 20 m*M* sodium phosphate buffer, pH 7.0. The protein peak with a maximum at tube 15 consists mainly of the albumin fraction. Immunological analysis of other proteins revealed significant differences in their partitioning behavior. In the two-phase system used, most of the proteins partition in favor of the lower phase (i.e., are to the left in the CCD train). However, almost complete separa-

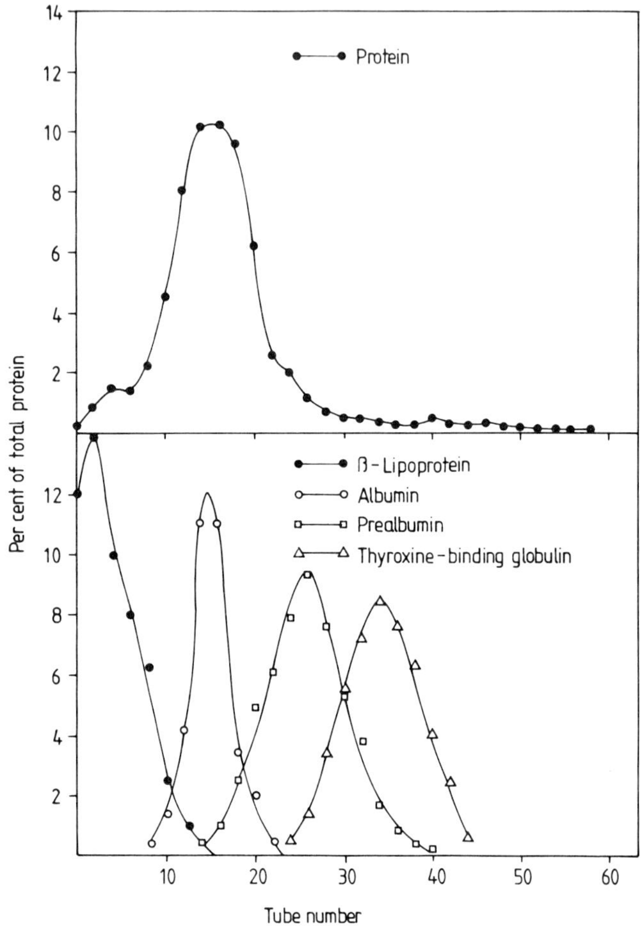


FIG. 4. Thin-layer countercurrent distribution of human serum in a two-phase system composed of 5% (w/w) PEG 6000, 7.5% (w/w) dextran T70, and 20 m*M* sodium phosphate buffer, pH 7.0. The ordinate shows the relative amounts of different proteins recovered in each tube. [From G. Birkenmeier, G. Kopperschläger, P.-Å. Albertsson, G. Johansson, F. Tjerneld, H.-E. Åkerlund, S. Berner, and H. Wickstroem, *J. Biotechnol.* **5,** 115 (1987), by permission.]

tion from albumin was achieved in the cases of $\alpha_1$-acid glycoprotein, thyroxine-binding globulin, $\alpha_1$-antichymotrypsin, and $\beta$-lipoprotein. The extent of protein separation for individual serum components is given in Table II by the separation factor log $\beta_{\mathrm{Alb}}$, defined as the logarithm of the ratio between the $G$ values of a protein and of albumin.

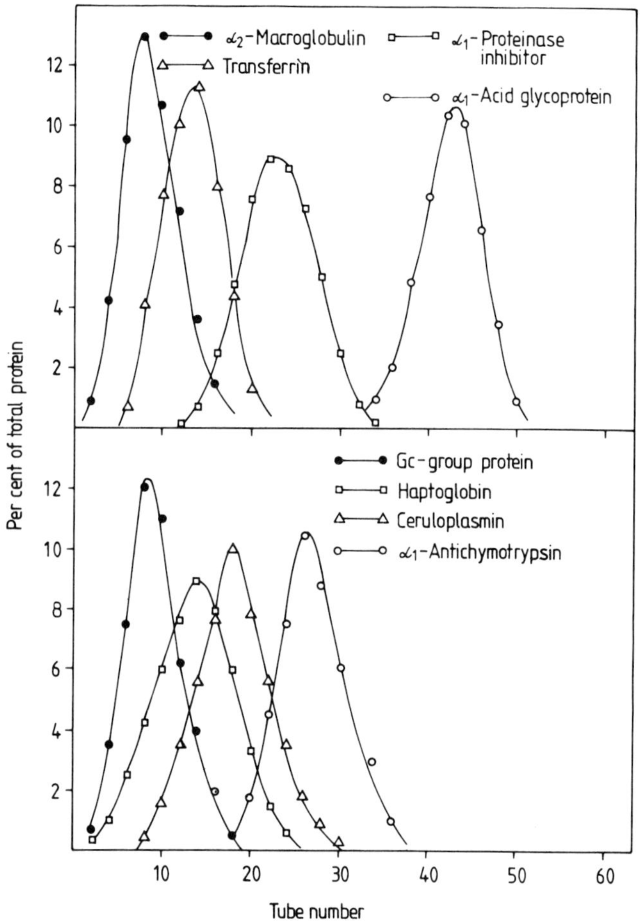


FIG. 4. (*Continued*)

Proteins with similar partitioning behavior such as albumin, transferrin, and haptoglobin can be resolved successfully by introducing dye–PEG into a two-phase system. Replacement of 1/20 of the total PEG by Cibacron Blue F3G-A–PEG, for instance, changes the partitioning of albumin and $\beta$-lipoprotein considerably, whereas relatively small partitioning changes result with transferrin, $\alpha_1$-proteinase inhibitor, prealbumin, and ceruloplasmin. Even a small retardation of movement of $\alpha_1$-acid glycoprotein could be detected, which is probably due to both the result of repulsive ionic forces between the negatively charged dye and the protein (p*I* 2.7) and the lack of any affinity to the dye.

TABLE II

*G* VALUES OF SERUM PROTEINS OBTAINED BY COUNTERCURRENT DISTRIBUTION IN POLY(ETHYLENE GLYCOL)–DEXTRAN SYSTEM[a] IN ABSENCE AND PRESENCE OF POLYMER-BOUND DYES[b]

| | Without dye | | Cibacron Blue F3G-A | | Procion Yellow HE-3G | | Procion Red HE-3B | |
|---|---|---|---|---|---|---|---|---|
| Protein | *G* value | log $\beta_{Alb}$[c] | *G* value | log $\beta_{Alb}$[c] | *G* value | log $\beta_{Alb}$[c] | *G* value | log $\beta_{Alb}$[c] |
| Albumin | 0.33 | 0 | 4.52 | 0 | 3.29 | 0 | 1.18 | 0 |
| Prealbumin | 0.78 | 0.37 | 0.84 | −0.73 | 1.03 | −0.50 | 0.78 | −1.18 |
| $\alpha_1$Acid glycoprotein | 2.74 | 0.92 | 1.83 | −0.39 | 2.74 | −0.08 | 2.51 | 0.33 |
| $\alpha_1$-Proteinase inhibitor | 0.63 | 0.28 | 0.59 | −0.88 | 0.59 | −0.74 | 0.59 | −0.30 |
| Transferrin | 0.27 | −0.09 | 0.36 | −1.09 | 0.54 | −0.78 | 0.30 | −0.59 |
| Haptoglobin | 0.30 | −0.04 | 0.55 | −0.91 | 3 | −0.04 | 0.84 | −0.15 |
| Ceruloplasmin | 0.43 | 0.11 | 0.55 | −0.91 | 1.36 | −0.38 | 0.59 | −0.30 |
| $\beta$-Lipoprotein | 0.02 | −1.22 | 11.2 | 0.39 | 0.68 | −0.68 | *d* | *d* |
| $\alpha_2$-Macroglobulin | 0.17 | −0.29 | I 0.33 | −1.13 | 0.63 | −0.72 | 0.30 | −0.59 |
| | | | II 0.50 | −0.95 | 2.74 | −0.08 | | |
| $\alpha_1$-Antichymotrypsin | 0.78 | 0.37 | 2.74 | −0.22 | 4.52 | 0.14 | 4.52 | 0.58 |
| Gc-Group protein | 0.15 | −0.34 | 0.59 | −0.88 | 0.59 | −0.74 | 0.43 | −0.43 |

[a] Composition of the two-phase systems: 5% (w/w) PEG 6000, 7.5% (w/w) dextran T70, and 20 m*M* sodium phosphate buffer, pH 7.0. All other conditions are described in the materials and methods section. In the CCD runs containing dye–PEG, 5% of the total PEG was replaced by dye-liganded PEG.

[b] From G. Birkenmeier, G. Kopperschläger, P.-Å. Albertsson, G. Johansson, F. Tjerneld, H.-E. Åkerlund, S Berner, and H. Wickstroem, *J. Biotechnol.* **5,** 115 (1987), by permission.

[c] The logarithmic separation factor, log $\beta_{Alb}$, relative to albumin, is defined by log $\beta_{Alb}$ = log $G_{Protein}$ − log $G_{Albumin}$

[d] No distinct peak position.

Table II also delineates the effect of a number of dyes on the resolution of serum proteins by CCD. These dyes show remarkable differences in fractionation properties. For example, Cibacron Blue F3G-A and, even more, Procion Yellow HE-3G separate serum $\alpha_2$-macroglobulin into two distinct fractions, each having a different affinity for the dyes, whereas Procion Red HE-3 is less effective in binding to this protein. Dyes in conjunction with CCD are, therefore, sensitive tools for the detection of protein heterogeneities.[24]

The capacity of the phase systems for protein is high. Sample systems prepared for CCD contained serum in a concentration of 20–30 g/liter. Two to five times more material may be included in the phase system provided that it does not precipitate.[8] The precipitates that occurred were mainly immunoglobulins and high molecular weight plasma proteins. However, the solubility of these proteins increased in systems containing

[24] G. Birkenmeier, G. Kopperschläger, P.-Å. Albertsson, G. Johansson, F. Tjerneld, H.-E. Åkerlund, S. Berner, and H. Wickstroem, *J. Biotechnol.* **5,** 115 (1987).

dye–PEG. This is in keeping with results showing that precipitation of proteins by PEG can be prevented by adding dye–PEG.[25,26]

## Conclusion

Aqueous phase systems without or with affinity ligands are highly useful tools for the purification and study of serum proteins. A large number of dyes are commercially available and can be coupled easily to different polymers without loss of protein-binding properties. The simplicity and rapidity of partitioning make this method especially attractive for prepurification of whole plasma as well as for the isolation of high value–low volume therapeutic products from human blood.

[25] G. Johansson, *in* "Protein–Dye Interaction: Developments and Applications" (M. A. Vijayalakshmi and O. Bertrand. eds.), p. 165. Elsevier, London, 1989.
[26] G. Birkenmeier and G. Kopperschläger, *J. Biotechnol.* **12,** 93 (1991).

# [15] Metal Affinity Partitioning

*By* BONG H. CHUNG, DARWIN BAILEY, and FRANCES H. ARNOLD

## Introduction

The partitioning of biological materials in aqueous two-phase systems, for example, poly(ethylene glycol) (PEG)–dextran (Dx) systems, can be altered in a reproducible and predictable fashion using affinity ligands attached to one of the phase-forming polymers. Metal ion complexes such as copper(II) iminodiacetate [Cu(II)IDA] have several very attractive features as affinity ligands: they are stable under a wide range of solvent conditions and temperatures and are therefore easy to recycle; they can be incorporated into chromatographic supports with very high loading capacities; their interactions with target molecules are reasonably specific and are reversible under mild conditions; and they are inexpensive.[1]

Metal ion complexes have been used as affinity ligands in aqueous

[1] F. H. Arnold, *Bio/Technology* **9,** 151 (1991).

two-phase systems for partitioning of proteins[2–4] and cells.[5] Partition coefficients ($K$) for proteins can be dramatically enhanced with PEG–IDA–metal ion complexes in PEG–Dx and PEG–salt two-phase systems. The increased partitioning to the PEG-rich phase is selective for proteins that contain surface-accessible metal-coordinating ligands, notably histidine residues. For example, hemoglobin (>20 surface histidines) partitions very strongly to the PEG-rich phase of a PEG–sodium sulfate two-phase system containing Cu(II)IDA–PEG, whereas human serum albumin remains in the bottom, salt phase,[3] allowing efficient, single-step separation.

## Mechanisms of Metal Affinity Partitioning

The affinity ligands most commonly used in metal affinity separations are the IDA complexes of first row transition metals. As illustrated in Fig. 1, IDA is a tridentate chelator, occupying three coordination sites on the metal ion. By using the metal ion complex covalently attached to PEG, molecules with coordinating ligands accessible to the IDA-bound metal ion can be preferentially extracted to the PEG-rich phase. The histidine imidazole moiety binds Cu(II) with high affinity through its electron-rich nitrogen(s); therefore, the $K$ values for proteins with surface-accessible histidines will be increased by Cu(II)IDA–PEG.

To avoid protein precipitation during extraction, it is necessary to use metal affinity ligands that are essentially monofunctional, for example, by selecting PEG monomethyl ether (M-PEG) in the synthesis of the IDA–PEG or by derivatizing bifunctional PEG with a very small amount of IDA. Interactions between the metal ion complex and histidine-containing proteins can be reversed relatively easily by Lewis acids ($H^+$, $Zn^{2+}$), which compete for binding to the protein, or the protein can be displaced by a Lewis base (imidazole, *N*-acetylhistidine), which competes for coordination to the metal. In practice, however, the enhanced partitioning is not always completely reversible when these competitors are added, and other measures may be required. A strong chelating agent such as EDTA will break up the ternary protein–metal–IDA complex. In fact, a preliminary fractionation (e.g., ultrafiltration) is often necessary to remove from very crude protein mixtures small compounds that are capable of seques-

[2] G. E. Wuenschell, E. Naranjo, and F. H. Arnold, *Bioprocess Eng.* **5,** 199 (1990).

[3] S. D. Plunkett and F. H. Arnold, *Biotechnol. Tech.* **4,** 45 (1990).

[4] G. Birkenmeier, M. A. Vijayalakshmi, T. Stigbrand, and G. Kopperschläger, *J. Chromatogr.* **539,** 267 (1991).

[5] H. Goubran-Botros, G. Birkenmeier, A. Otto, G. Kopperschläger, and M. A. Vijayalakshmi, *Biochim. Biophys. Acta* **1074,** 69 (1991).

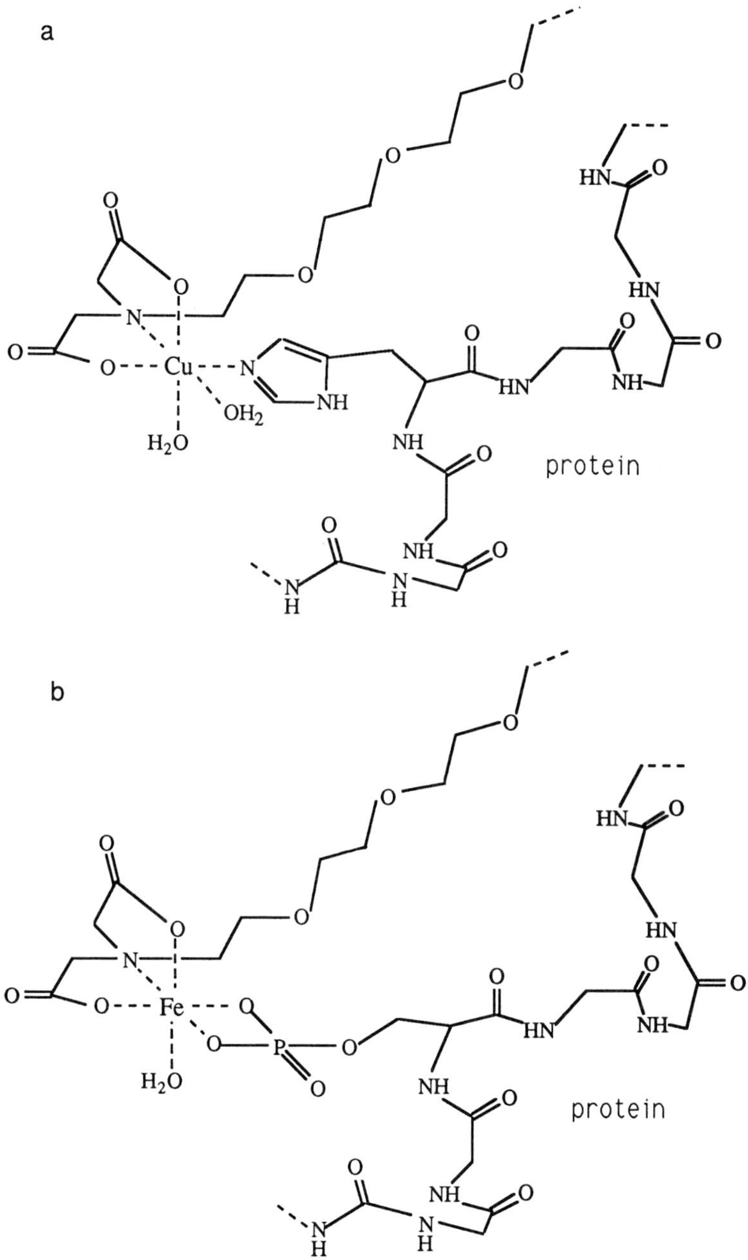


FIG. 1. (a) Coordination of histidine-containing proteins by Cu(II)IDA–PEG. (b) Coordination of proteins containing phosphoryl groups by Fe(III)IDA–PEG.

tering the metal ion before a metal affinity extraction can be performed. A change in the oxidation state of the metal ion can also disrupt the protein–metal interaction.

Fe(III) has a high affinity for phosphoryl groups (Fig. 1b) and can be used to target phosphorylated proteins in partitioning experiments.[6] Proteins rich in carboxyl groups and tyrosines also show marked affinity for Fe(III)IDA in chromatographic experiments[7]; we have observed that some nonphosphorylated proteins are also strongly partitioned by Fe(III)-IDA–PEG in aqueous two-phase systems.[6] Other first-row transition metal ions commonly used in metal affinity chromatography [Zn(II), Ni(II)] are not well-suited for partitioning in aqueous two-phase systems, however, because the stability constants are an order of magnitude smaller than for Cu(II).[8]

Metal ion coordination is promoted by phase-forming salts, provided the anions are not good ligands for the metal ion or the cation does not effectively compete for the targeted ligand. Thus, partitioning of histidine-rich proteins with Cu(II)IDA–PEG can be greatly enhanced in PEG–salt compared to PEG–Dx systems. In a PEG–$Na_2SO_4$ two-phase system, the $K$ value for human hemoglobin increases by more than a factor of 1000 when only 1% of the PEG is replaced by Cu(II)IDA–PEG.[3]

## Experimental Methods

### *Materials*

PEG 5000 monomethyl ether (M-PEG) and bifunctional PEG 5000 are purchased from Fluka (Ron Bonkoma, NY). PEG 10,000 monomethyl ether is the generous gift of Dow Chemical Company (Midland, MI). Dextran T500 is purchased from Pharmacia (Uppsala, Sweden), and cellulose acetate filters are from Schleicher & Schuell (Keene, NH). Proteins are obtained from Sigma (St. Louis, MO).

### *Synthesis of Iminodiacetate–Poly(ethylene Glycol)*

The PEGs are derivatized with iminodiacetate and metallated according to Scheme I. The synthesis is conducted as follows.

[6] B. H. Chung and F. H. Arnold, *Biotechnol. Lett.* **13,** 615 (1991).

[7] G. Muszynska, G. Dobrowolska, A. Medin, P. Ekman, and J. Porath, *J. Chromatogr.* **604,** 19 (1992).

[8] A. E. Martell and R. M. Smith, "Critical Stability Constants," Vol. 1. Plenum, New York, 1974.

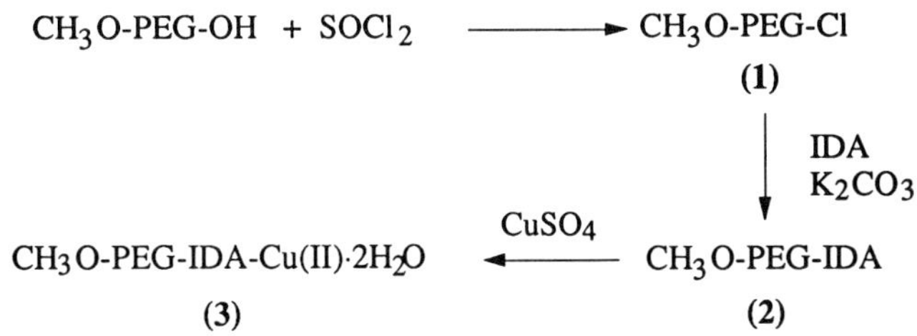


SCHEME I

*PEG 10,000 Chloride*. PEG chloride **(1)** is synthesized following the procedure of Töke *et al.*[9] and Bayer *et al.*[10] Solid PEG 10,000 monomethyl ether (100 g) is dried under vacuum at 50° for 5 hr. Then 3.6 ml thionyl chloride is added, and the melt is reheated at 65° for 5 hr under an argon atmosphere. Excess thionyl chloride is removed under vacuum, and the melt is poured into a polyethylene weigh boat. Yield: 99.3 g (99%). Analysis calculated for C, H: C 54.37, H 9.14; found: C 54.14, H 8.95.

*PEG 10,000–IDA*. To 5 g (0.5 mmol) of the PEG 10,000 chloride in 50 ml of water are added 2.1 g (15.8 mmol) iminodiacetic acid (IDA) and 2 g potassium carbonate. The solution is refluxed for 48 hr. Sodium sulfate (5 g) is added to the hot reaction mixture, which is allowed to cool and separate into two phases. The PEG phase is retained and diluted to a volume of 60 ml. Solid impurities are removed by filtration through an 0.4-$\mu$m filter. The solution is dialyzed against 1% sodium bicarbonate, 0.5% sodium bicarbonate, and finally extensively against water over a period of 48 hr. The solution is lyophilized to give 3.7 g (73%) of the product PEG 10,000–IDA **(2)**. Anal. calcd. for C, H, N: C 54.33, N 9.11, H 0.14; found: C 53.72, N 8.83, H 0.34.

Gel-permeation chromatography of **2** indicated that this derivatization procedure does not result in significant polymer cleavage or degradation.

*Cu(II)IDA–PEG*. Next, 1.09 g of **2** (0.1 mmol) and 0.069 g (0.28 mmol) of copper sulfate are dissolved in 5 ml water and allowed to equilibrate for 24 hr. The solution is diluted to 20 ml, filtered through a 0.4-$\mu$m filter, and then dialyzed against water for 48 hr. The contents are then lyophilized to give 1 g (92% yield) of Cu(II)IDA–PEG **(3)**. Anal. Calcd. for C, H, N, Cu: C 53.71, N 9.05, H 0.14, Cu 0.62; found: C 53.74, N 9.38, H 0.05, Cu 0.59.

[9] L. Töke, G. T. Szabo, and K. Aranyosi, *Acta Chim. Acad. Sci. Hung.* **100,** 257 (1979).
[10] E. Bayer, H. Zheng, and K. Geckeler, *Polym. Bull.* **8,** 585 (1982).

Fe(III)IDA–PEG is prepared by combining equal molar quantities of **2** and $FeCl_3$ in the PEG stock solutions immediately prior to the partitioning experiments.

### *Partitioning Experiments*

Aqueous two-phase systems are prepared in 15-ml tubes from stock solutions of PEG 5000 (40%, w/w) and Dx T500 (13.5%, w/w). For the experiments with copper, a portion of the PEG is replaced by Cu(II)IDA–PEG in the stock solution. Two-phase systems (4.0 g total weight) are made by combining 0.70 g of the PEG stock solution, 1.30 g of the Dx solution, and 2.00 g of a protein solution in a buffer (typically 1.0 mg/g protein in 45 m*M* sodium phosphate, 0.45 *M* NaCl at various pH values). The final polymer compositions are 7.0% PEG and 4.4% Dx.

For the partitioning experiments using Fe(III)IDA–PEG, metal-containing stock solutions are prepared immediately before each experiment by combining equal molar quantities of $FeCl_3$ and PEG 10,000–IDA to a PEG stock solution. For experiments containing the metal ion, 4% of the PEG 5000 is replaced by Fe(III)IDA–PEG 10,000, unless otherwise specified.

The tubes are shaken at room temperature for 10–30 min, taking care to thoroughly mix the polymer phases. The phases are then separated by centrifugation for 15–25 min. Aliquots from the two phases are collected and diluted with distilled water for analysis of the protein concentrations photometrically at 410 nm (for heme proteins) or by the Bradford[11] protein assay. The corresponding phase from a similar two-phase system containing no protein is used as a reference. Copper concentrations in the two phases are determined either by atomic absorption or spectrophotometrically, following the procedure of Woelfel.[12]

## Results of Partitioning Experiments

*K* values for various histidine-containing proteins in the absence ($K_0$) and presence ($K$) of Cu(II)IDA–PEG in a PEG–Dx two-phase system are presented in Table I. The enhanced partitioning in the presence of the metal ion, as indicated by $K/K_0$, increases steadily with the number of accessible histidines. Higher concentrations of the metal ion affinity ligand can give proportionally larger enhancements in partitioning.[13]

[11] M. M. Bradford, *Anal. Biochem.* **72,** 248 (1976).

[12] W. Woelfel, *Anal. Chem.* **20,** 722 (1948).

[13] S.-S. Suh and F. H. Arnold, *Biotechnol. Bioeng.* **35,** 682 (1990).

TABLE I
METAL AFFINITY PARTITIONING OF PROTEINS USING Cu(II)IDA–PEG IN PEG–DEXTRAN TWO-PHASE SYSTEMS[a]

| Protein | $K_0$ | $K$ | $K/K_0$ | Surface histidines |
|---|---|---|---|---|
| *Candida krusei* cytochrome *c* | 0.56 | 0.75 | 1.3 | 2 |
| Horse myoglobin | 0.36 | 0.64 | 1.8 | 4 |
| Whale myoglobin | 0.42 | 0.85 | 2.0 | 5 |
| Bovine hemoglobin | 0.25 | 6.0 | 24 | 20 |
| Human hemoglobin | 0.38 | 14 | 37 | 24 |

[a] In the absence ($K_0$) and presence ($K$) of Cu(II)IDA–PEG 5000 ($1.6 \times 10^{-4}$ *M*), pH 8.0. Composition of two-phase systems: 4.4% dextran T500, 7% PEG 5000 containing 1% Cu(II)IDA–PEG, 0.1 *M* NaCl, and 10 m*M* sodium phosphate, pH 8.0. From G. E. Wuenschell, E. Naranjo, and F. H. Arnold, *Bioprocess Eng*. **5,** 199 (1990), with permission.

Partitioning of various phosphorylated and nonphosphorylated proteins using Fe(III)IDA–PEG is shown in Table II. The small, nonphosphorylated proteins cytochrome *c*, myoglobin, and ribonuclease A have little interaction with the metal ion ($K/K_0 < 2$), even when four or five accessible histidines are present (myoglobins). In contrast, large increases in $K$ were observed for hemoglobin and serum albumin in the presence

TABLE II
METAL AFFINITY PARTITIONING OF PROTEINS USING Fe(III)IDA–PEG IN PEG–DEXTRAN TWO-PHASE SYSTEMS[a]

| Protein | $K_0$ | $K$ | $K/K_0$ |
|---|---|---|---|
| Ovalbumin | 0.40 | 1.02 | 2.6 |
| Phosvitin | 0.013 | $\gg 100$[b] | $\gg 1000$ |
| Human albumin | 0.13 | 0.64 | 4.9 |
| Bovine hemoglobin | 0.17 | 2.47 | 14.5 |
| *Saccharomyces cerevisiae* cytochrome *c* | 0.67 | 0.84 | 1.25 |
| Horse heart cytochrome *c* | 0.59 | 0.78 | 1.32 |
| Horse myoglobin | 0.44 | 0.53 | 1.20 |
| Sheep myoglobin | 0.43 | 0.48 | 1.12 |
| Ribonuclease A | 0.88 | 0.87 | 0.99 |

[a] In the absence ($K_0$) and presence ($K$) of Fe(III)IDA–PEG 10,000 ($2.76 \times 10^{-4}$ *M*). Composition of systems: 4.4% dextran T500, 7% PEG 5000, 0.5 mg/ml protein, 25 m*M* acetate buffer, pH 5.0, and 0.25 *M* NaCl. Approximately 4% M-PEG 5000 was replaced by Fe(III)IDA–PEG 10,000.

[b] FE(III)IDA–PEG concentration of $2.76 \times 10^{-3}$ *M* [40% Fe(III)IDA–PEG content in total PEG].

of Fe(III)IDA–PEG, despite the fact that these proteins are also not phosphorylated.

Phosphorylated proteins are clearly preferentially extracted in the presence of Fe(III)IDA–PEG. Phosvitin, which is 40% phosphate, is quantitatively partitioned to the PEG-rich phase of a PEG–Dx system in which 40% of the PEG is derivatized with Fe(III)IDA. This overwhelming preference for the PEG-rich phase is not sensitive to pH in the range 3 to 6.[6] Ovalbumin, with 0–2 phosphoryl groups, is also extracted by Fe(III)-IDA–PEG, but this preference is strongly pH sensitive (see below).

## Modeling Metal Affinity Partitioning

### *Copper(II)–Histidine*

The partitioning behavior of most proteins in Cu(II) metal affinity partitioning experiments follows very closely that predicted by a simple mathematical model [Eq. (1)].[13]

$$\ln\left(\frac{K}{K_0}\right) = n \ln\left[\frac{1 + K_{\mathrm{H}}[\mathrm{H}] + K'_{\mathrm{a}}[M_{\mathrm{tot}}]\dfrac{(R+1)}{(R+1/K_{\mathrm{M}})}}{1 + K_{\mathrm{H}}[\mathrm{H}] + K''_{\mathrm{a}}[M_{\mathrm{tot}}]\dfrac{(R+1)}{(RK_{\mathrm{M}}+1)}}\right] \tag{1}$$

In Eq. (1) [H] is the proton concentration, $K_{\mathrm{M}}$ is the partition coefficient of the affinity ligand [Cu(II)IDA–PEG], $R$ is the phase ratio, and $M_{\mathrm{tot}}$ is the total (average) metal ion concentration in the two-phase system. The assumptions made in deriving Eq. (1) are (1) the protein has $n$ independent binding sites, (2) all the sites have identical binding constants in the top ($K_{\mathrm{a}}'$) and bottom ($K_{\mathrm{a}}''$) phases, (3) protons compete with the metal ion for binding to a histidine, and the proton affinities ($K_{\mathrm{H}}$) are the same for all the sites.

Equation (1) predicts that $\ln(K/K_0)$ is proportional to the number of surface histidines, whereas the slope reflects the binding constants. The linear dependence has been demonstrated in several experiments[2,13,14] and is also shown in Fig. 2. Although this seeming lack of interaction between the PEG–ligands at the protein surface appears surprising, it should be remembered that binding is weak relative to most other affinity ligands ($10^{3.6}$ $M^{-1}$ for histidyl-Cu(II)IDA), such that only a fraction of the sites are occupied by PEG at any given time. The pH dependence of the parti-

[14] R. Todd, M. Van Dam, D. Casimiro, B. L. Haymore, and F. H. Arnold, *Proteins Struct. Funct. Genet.* **10,** 156 (1991).

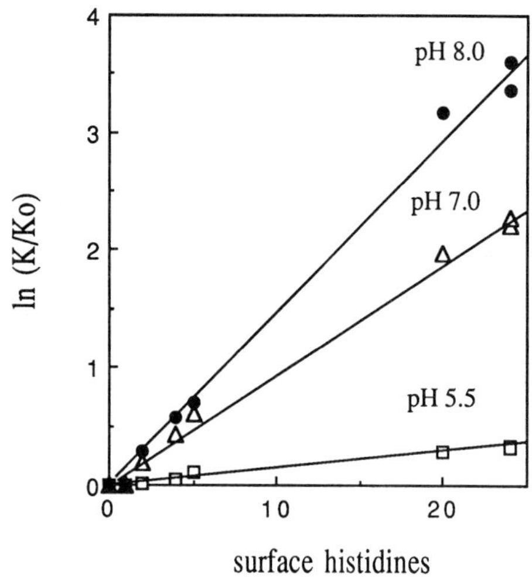


FIG. 2. Linear dependence of $\ln(K/K_0)$ on number of surface histidines. [From Wuenschell *et al.*[2]]

tioning of histidine-containing proteins seen in Fig. 2 reflects protonation of surface histidines ($pK_a$ ~6.4–6.6).

From an analysis of the experimental partition coefficients for a variety of proteins at different values of pH and metal concentration, values of the association constants for the complexation of Cu(II)IDA with an average histidine on the protein surface have been obtained.[13] Slightly different values are obtained for interactions in the PEG-rich and Dx-rich phases. In the Dx-rich phase, the stability constant ($10^{3.6}$ $M^{-1}$) is very similar to those measured for analogous small molecule complexes with histidine derivatives by conventional potentiometry techniques and by observing the effects of paramagnetic ions on proton nuclear magnetic resonance spectra. The mathematical model has been extended to describe interactions between Cu(II)IDA and dihistidine metal-chelating sites incorporated by site-directed mutagenesis into protein surfaces.[14,15]

### *Iron(III)–Phosphate*

The partitioning models presented above can be modified in order to describe the Fe(III)–phosphate interaction and partitioning of phosphorylated proteins. The presumed structure of the phosphoryl-Fe(III)IDA com-

[15] S. S. Suh, B. L. Haymore, and F. H. Arnold, *Protein Eng.* **4,** 301 (1991).

plex shown in Fig. 1b indicates bivalent chelation of two oxygens to the Fe(III) ion, in analogy to Ca(II) and Mg(II) complexes with phosphoserine.[16] Assigning proton affinities $K_{H1}$ and $K_{H2}$ to the two phosphoryl oxygens, one obtains Eq. (2).

$$\ln\left(\frac{K}{K_0}\right) = n \ln \left[ \frac{(1 + K_{H_1}[H])(1 + K_{H_2}[H]) + K_a'[M_{tot}] \dfrac{(R+1)}{(R + 1/K_M)}}{(1 + K_{H_1}[H])(1 + K_{H_2}[H]) + K_a''[M_{tot}] \dfrac{(R+1)}{(RK_M + 1)}} \right] \quad (2)$$

The strong affinity of hydroxide ion for Fe(III) must be included in the mathematical description in order to describe the rapid decrease in partitioning at high pH. In addition, two assumptions were made. First, we assumed that monodentate protein–metal complexes (binding by a single oxygen) are much less favored than the bidentate complex depicted in Fig. 1b and can be neglected. Second, because the first ionization of the phosphoryl group is characterized by p$K$ values below 1.0,[17] the product $K_{H1}$[H] is essentially equal to 0 in the pH range of interest. The resulting expression is given by Eq. (3).

$$\ln\left(\frac{K}{K_0}\right) = n \ln \left[ \frac{1 + K_{H_2}[H]) + K_a'\left(\dfrac{[M_{tot}]}{1 + K_{OH}[OH]}\right) \dfrac{(R+1)}{(R + 1/K_M)}}{1 + K_{H_2}[H]) + K_a''\left(\dfrac{[M_{tot}]}{1 + K_{OH}[H]}\right) \dfrac{(R+1)}{(RK_M + 1)}} \right] \quad (3)$$

$K$ values for the two phosphorylated ovalbumins in a PEG–Dx system containing Fe(III)IDA–PEG are shown in Fig. 3. Ovalbumin A1, with two phosphoryl groups, is more strongly partitioned into the PEG phase at all pH values than is ovalbumin A2, which contains a single phosphoryl group. As expected, the pH dependence of partitioning of both phosphorylated components is bell shaped: partitioning decreases at very low pH as protons compete with the metal ion for binding to the phosphate groups, whereas at high pH hydroxide ion competes very favorably for binding to the metal ion.

Equation (3) was fit to the experimental data in Fig. 3 to obtain estimates for the Fe(III)–phosphate association constants. The solid curves in Fig. 3 represent a fit of Eq. (3) obtained for log $K_a' = 3.8$, log $K_a'' = 4.4$, log $K_{H2} = 4.0$, and log $K_{OH} = 8.8$. The association constants for the protein–Fe(III)IDA–PEG interaction, $K_a'$ and $K_a''$, are 3–5 times greater than those of the Cu(II)IDA–PEG–histidine interaction. The p$K_a$ for the

[16] C. W. Childs, *Can. J. Chem.* **49,** 2359 (1971).

[17] G. Fölsch and R. Osterberg, *J. Biol. Chem.* **234,** 2298 (1959).

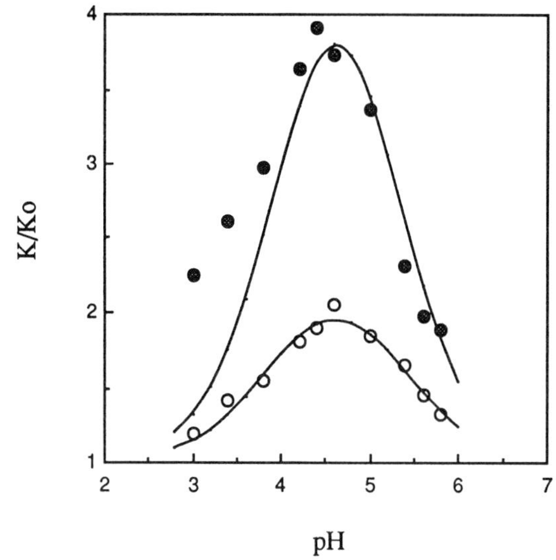


FIG. 3. pH dependence of $K$ values for ovalbumin A1 (●) and ovalbumin A2 (○).

second ionization of the phosphoryl group ($K_{H_2}$), however, is significantly lower than the p$K_a$ values of phosphorylated compounds measured by titration (~5.5),[17] whereas the value of p$K_{OH}$ (=8.8) for the complexation of hydroxide ion by Fe(III)IDA–PEG is much smaller than that for complexation to free ferric ion (p$K_{OH}$ = 11.1).[8]

## Improving Metal Affinity Partitioning

### *Enhancing Partition Coefficient of Affinity Ligand ($K_M$)*

The models presented above can be used to identify ways of enhancing the selective partitioning of proteins in metal affinity two-phase systems. Increasing the strength of metal–protein interactions is only partially effective since the stability constants appear in both the numerator and denominator of Eq. (1) [and Eq. (3)]. Clearly, one would prefer to enhance the stability constant for the complexation in the top phase ($K_a'$) relative to that in the bottom phase ($K_a''$).

$K_M$ can be increased without altering other system parameters by attaching the metal ion complex to a molecule that partitions more strongly to the top phase, (e.g., a higher molecular weight PEG). When Cu(II)-IDA–PEG 10,000 is used in place of an equal (molar) concentration ($1.6 \times 10^{-4}$ $M$) of Cu(II)IDA–PEG 5000, $\ln(K/K_0)$ values for histidine-containing

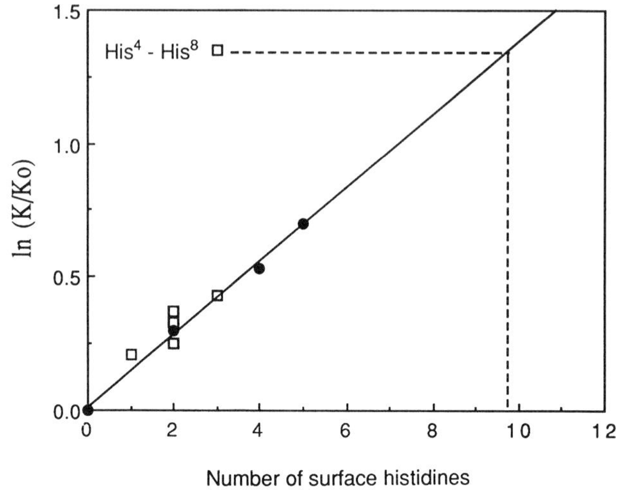


FIG. 4. Plot of $\ln(K/K_0)$ values for a series of natural proteins (●) and several engineered cytochromes *c* containing different numbers of surface histidines (□).

proteins are enhanced by a factor of 2. $K_M$ for Cu(II)IDA–PEG 5000 in these PEG–Dx two-phase systems is only 3.6, whereas for the affinity ligand based on PEG 10,000 $K_M$ is 6.7.

## *Engineered Metal-Binding Sites*

We have been interested in the design of dihistidine metal-chelating sites on protein surfaces for applications in protein purification and stabilization.[18] A strong metal-binding site engineered into the surface of a recombinant protein provides a unique affinity for the metal ion complex that can be exploited in metal affinity extraction. This is illustrated in Fig. 4. As histidines are added or removed from the surface, there is a corresponding increase or decrease in $\ln(K/K_0)$, as predicted by Eq. (1). A chelating site consisting of two histidines separated by three amino acid residues in an $\alpha$ helix (His-$X_3$-His) shows very different behavior: the $His^4$-$His^8$ cytochrome *c* variant partitions as if it had 9–10 surface histidines, when in reality it contains only three. As it is rare for natural proteins to have more than a few surface histidines, this two-histidine site confers unique metal-binding properties and therefore unique behavior during metal affinity chromatography or partitioning.[14] Similar results have been observed for recombinant proteins containing poly(histidine)

[18] F. H. Arnold and B. L. Haymore, *Science* **252,** 1796 (1991).

"tails"[19] or chelating peptides[20] at their N or C termini in metal affinity chromatography. Such constructs would be expected to partition very favorably in metal affinity extraction. Metal affinity partitioning also offers a useful technique for measuring the strength of the metal-binding interaction in natural or engineered surface metal-binding sites. His-$X_3$-His sites in bovine somatotropin and cytochrome *c* exhibit binding constants of $10^4$ to $10^6$ $M^{-1}$ for Cu(II)IDA–PEG.[15]

Acknowledgments

This research is supported by the National Science Foundation, Grant No. EET-8807351, and a Presidential Young Investigator Award (F.H.A.). F.H.A. gratefully acknowledges a fellowship from the David and Lucile Packard Foundation. D.B. is the recipient of a fellowship from the Irvine Foundation.

[19] E. Hochuli, W. Bannwarth, H. Dobeli, R. Gentz, and D. Stuber, *Bio/Technology* **6,** 1321 (1988).
[20] M. C. Smith, T. C. Furman, T. D. Ingolia, and C. Pidgeon, *J. Biol. Chem.* **263,** 7211 (1988).

# [16] Starch Synthase from Potato Tubers

*By* ANDREAS BLENNOW

## Introduction

Soluble starch synthases in plants catalyze the formation of unbranched α-glucan by condensing the glucose portion of adenosine diphosphoglucose (ADP-glucose) to the C-4 position of an α-glucan primer, giving an α-glucosidic bond. The soluble starch synthase when acting together with branching enzyme (Q-enzyme) produces amylopectin, a branched product that contains both α-1,4- and α-1,6-glucosidic linkages. Approximately 70 to 80% of normal, nonmutant starch is amylopectin.[1]

Soluble starch synthases from several plants have been successfully purified by affinity chromatography using aminoalkyl groups or analogs to ADP-glucose (e.g., ADP-hexanolamine) as ligands.[2] However, these affinity methods do not work well for the soluble starch synthase from potato.[3] For purification of the latter, α-glucans (cyclodextrin[3] and malto-

[1] J. Preiss, *in* "The Biochemistry of Plants" (P. K. Stumpf and E. E. Conn, eds.), p. 181. Academic Press, New York, 1988.
[2] A. M. Smith, *in* "Methods in Plant Biochemistry" (P. J. Lea, ed.), Vol. 3, p. 93. Academic Press, San Diego, 1990.
[3] A. Ponstein, J. G. de Wit, and B. Witholt, *Food Hydrocolloids* **5,** 229 (1991).

TABLE I
PURIFICATION OF POTATO SOLUBLE STARCH SYNTHASE BY USE OF α-GLUCANS AS AFFINITY LIGANDS

| α-Glucan used | Method | Purification factor (-fold) | Recovery (%) |
|---|---|---|---|
| Maltotriose[a] | Chromatography | 4 | 6 |
| Cyclodextrin[b] | Chromatography | >20 | Not determined |
| Hydroxypropyl starch[c] | Partitioning | 16 | 28 |

[a] T. Baba, M. Noro, M. Hiroto, and Y. Arai, *Phytochemistry* **29,** 719 (1990).
[b] A. Ponstein, J. G. de Wit, and B. Witholt, *Food Hydrocolloids* **5,** 229 (1991).
[c] A. Blennow, Doctoral Thesis, Lund University, Lund, Sweden (1992).

triose[4]) have been used as affinity ligands (Table I). Potato soluble starch synthase is, however, often recovered in low quantities when using conventional or high-performance column chromatography (HPLC), probably because of adsorption to the column matrices.[4–6] Aqueous two-phase partitioning overcomes this problem because separation proceeds exclusively between liquid phases. Potato tuber soluble starch synthase has affinity for starch,[5,7] dextrins,[4,5] and cyclodextrin,[3] and, as is shown here by partitioning, in a poly(ethylene glycol) (PEG)–hydroxypropyl starch (Aquaphase PPT) aqueous two-phase system, starch synthase also has a strong affinity for derivatized starch.

## Methods

The chemicals used are of analytical grade if not otherwise mentioned. All steps are performed at 0–5°. Potato tubers (1.2 kg) are washed and homogenized in a fruit juicer, filtered through four layers of nylon cloth (0.3 mm pore diameter), and immediately mixed with 50 ml of 0.5 *M* dipotassium hydrogen phosphate, to which has been added 2.5 ml nicotine (Merck, Darmstadt, Germany), technical grade), 2.5 g sodium dithionite, 0.4 g EDTA, and 50 g sucrose. This solution (500 ml) is first centrifuged for 5 min at 15,000 *g* and then the supernatant liquid is brought to 36% saturation with ammonium sulfate, incubated for 15 min, and centrifuged for 30 min at 15,000 *g*.

[4] T. Baba, M. Noro, M. Hiroto, and Y. Arai, *Phytochemistry* **29,** 719 (1990).
[5] A. Blennow, Doctoral Thesis, Lund University, Lund, Sweden (1992).
[6] J. S. Hawker, J. L. Ozbun, and J. Preiss, *Phytochemistry* **11,** 1287 (1972).
[7] T. Akazawa and T. Murata, *Biochem. Biophys. Res. Commun.* **19,** 21 (1965).

The precipitate contains starch synthase, typically enriched 2-fold. The pellet is dissolved in 75 ml of an aqueous two-phase system prepared as follows: A total weight of 300 g is prepared containing 9% (w/w) PEG 8000 (BP Chemicals, Hythe, UK), 7% (w/w) Aquaphase PPT (available as Reppal PES 100, Reppe AB, Växjö, Sweden), 50 mmol/kg potassium phosphate buffer, pH 7.6, 1 mmol/kg dithiothreitol (DTT), and 10% (w/w) sucrose in water. After phase settling, 20 ml of bottom phase (Aquaphase enriched) and 55 ml of top phase (PEG enriched) are withdrawn and mixed to give 75 ml of the two-phase system with a volume ratio of approximately 3 (top/bottom), in which the pellet is dissolved. The system is mixed and then centrifuged for 5 min at 10,000 *g*. The top phase is discarded, new top phase (55 ml) is added, and the system is mixed and centrifuged again. In this way the bottom phase is "washed" a total of 3 times. The bottom phase can then be "desalted" by washing once more with buffer-free top phase, a necessary step if subsequent adsorption of the starch synthase to hydroxylapatite is to take place.

## Results

Using aqueous two-phase partitioning for purification gives a 28% recovery of starch synthase and a 16-fold purification (Table I). The contents of contaminating enzymes involved in starch synthesis, for example, Q-enzyme ($\alpha$-1,4-glucan branching enzyme), debranching enzyme (amylopectin 6-glucano-hydrolase), ADP (and UDP)glucose pyrophosphorylase, phosphorylase, and phosphoglucomutase, can be greatly reduced by combining aqueous two-phase partitioning with ammonium sulfate precipitation and hydroxylapatite chromatography.[5] Further purification by chromatography with hydroxylapatite (Bio-Rad, Richmond, CA) and $\omega$-aminopentyl-agarose (Sigma, St. Louis, MO) enriches starch synthase activity 200-fold but with poor recovery (0.4%).[5] The major advantages and drawbacks of the two-phase method for purification of starch synthase are summarized in Table II.

TABLE II
PURIFICATION OF SOLUBLE STARCH SYNTHASE BY PARTITIONING IN AQUEOUS TWO-PHASE SYSTEMS

| Advantages | Disadvantages |
|---|---|
| Enzyme recovery is high | Other starch-metabolizing enzymes are copurified (though to a small degree) |
| Conditions are mild | Phase-forming polymers must be removed |
| Procedure can easily be scaled up | |
| Method is inexpensive | |
| Standard laboratory devices can be used | |

## Discussion

Because the partitioning method can be used early in the purification process, a high degree of enzyme purification is rapidly obtained, thereby reducing the potentially harmful effects of, for example, proteases that are present in crude extracts from plants. The preparation time can be further reduced by dissolving the polymers directly in the raw homogenate and performing the partitioning in a single step. In this case a two-phase system with a volume ratio of about 10 is used.

The affinity of the starch synthase for the starch derivative in the lower phase is so strong that potassium phosphate concentrations between 5 and 50 mmol/kg influence the partitioning of starch synthase only moderately. Bulk proteins, however, partition more to the PEG phase at high (50 mmol/kg) potassium phosphate concentrations, which enhances the purification of starch synthase in the bottom phase.

# [17] Isolation of Integral Membrane Proteins by Phase Partitioning with Triton X-114

*By* JOHN S. BRUSCA and JUSTIN D. RADOLF

## Introduction

Integral membrane proteins are anchored to membranes by hydrophobic stretches of amino acids or by amphiphilic groups covalently linked to the polypeptide chains. Such proteins can be identified by a number of different strategies, including inspection of DNA sequences for the encoding of transmembrane domains (i.e., hydrophobic stretches), copurification with membranes on cell fractionation, and localization to membranes by immunoelectron microscopy. These procedures can be difficult, time-consuming, and subject to inaccuracies or artifacts. This chapter describes Triton X-114 phase partitioning, a technique for the identification and isolation of membrane proteins that exploits their amphiphilic properties. Extensive use of Triton X-114 phase partitioning has proved that this technique accurately predicts membrane localization of proteins and provides a convenient means for obtaining enriched preparations of membrane proteins from complex biological systems (Table I).

TABLE I
SELECTED EXAMPLES OF TRITON X-114 PHASE PARTITIONING

| Protein | Comments | Ref. |
|---|---|---|
| *Pseudomonas* exotoxin | Partitions to aqueous phase at pH 7 and detergent phase at pH 4 | *a, b* |
| *Mycoplasma hyopneumoniae* surface proteins | Detergent-phase proteins identified as immunogenic lipoproteins | *c* |
| *Escherichia coli* penicillin-binding protein 4 | Identified as hydrophilic by phase partitioning (penicillin-binding proteins are expected to be membrane proteins) | *d* |
| Human platelet glycoproteins and phosphoproteins | All but two glycoproteins partition to detergent phase; phosphoproteins partition to aqueous phase | *e* |
| Carboxypeptidase E | Soluble form partitions to aqueous phase when pH is less than 6 | *f* |
| GPI-linked proteins of *Leishmania* and *Trypanosoma* | Proteins shift from detergent to aqueous phase following removal of GPI group | *g, h* |
| *Entamoeba histolytica* 29-kDa surface antigen | Identified as a hydrophilic peripheral membrane protein by sequence analysis and partitioning to aqueous phase | *i* |
| Interleukin 2 | Native cytokine partitions to aqueous phase; some genetically engineered forms partition to detergent phase | *j* |

[a] K. Sandvig and J. O. Moskaug, *Biochem. J.* **245,** 899 (1987).
[b] T. Idziorek, D. FitzGerald, and I. Pastan, *Infect. Immun.* **58,** 1415 (1990).
[c] K. S. Wise and M. F. Kim, *J. Bacteriol.* **169,** 5546 (1987).
[d] H. Mottl and W. Keck, *Eur. J. Biochem.* **200,** 767 (1991).
[e] K. J. Clemetson, D. Bienz, M.-L. Zahno, and E. F. Lüscher, *Biochim. Biophys. Acta* **778,** 463 (1984).
[f] L. D. Fricker, B. Das, and R. H. Angeletti, *J. Biol. Chem.* **265,** 2476 (1990).
[g] C. Bordier, R. J. Etges, J. Ward, M. J. Turner, and M. L. Cardoso de Almeida, *Proc. Natl. Acad. Sci. U.S.A.* **83,** 5988 (1986).
[h] C. Hernandez-Munain, M. A. Fernandez, A. Alcina, and M. Fresno, *Infect. Immun.* **59,** 1409 (1991).
[i] S. L. Reed, B. M. Flores, M. A. Batzer, M. A. Stein, V. L. Stroeher, J. E. Carlton, D. L. Diedrich, and B. E. Torian, *Infect. Immun.* **60,** 542 (1992).
[j] C. A. Bergmann, B. J. Landmeier, and D. R. Kaplan, *Mol. Immunol.* **28,** 99 (1991).

## Properties of Triton X-114

The Triton X series of nonionic detergents consists of a hydrophobic *p-tert*-octyl phenol group attached to a hydrophilic polyoxyethylene head group; members of this series are considered to be nondenaturing.[1] The

[1] A. Helenius and K. Simons, *Biochim. Biophys. Acta* **415,** 29 (1975).

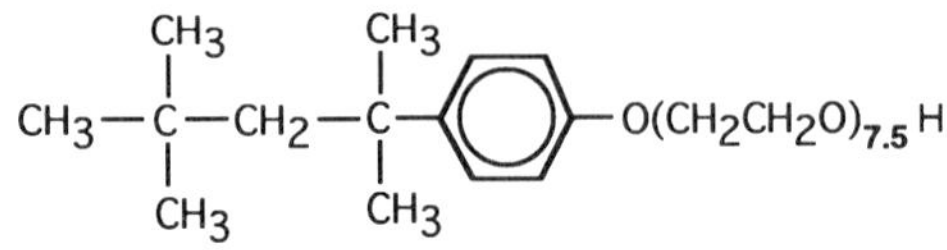

FIG. 1. Structure of Triton X-114. The more hydrophilic Triton X-100 has an average of 9.5 head groups, whereas the more hydrophobic Triton X-45 has an average of 5 head groups.

average number of hydrophilic oxyethylene head groups in Triton X-114 is 7.5, as opposed to an average of 9.5 for the more commonly used Triton X-100 (Fig. 1). Thus, Triton X-114 has a hydrophile–lipophile balance (HLB)[2] of 12.4, whereas the more water-soluble Triton X-100 has an HLB value of 13.5. The critical micelle concentration (CMC) of Triton X-114 is 210 $\mu M$ (0.0113%, w/v), slightly lower than that (250 $\mu M$ or 0.0156%, w/v) of Triton X-100.[1] Triton X-114 thus exists primarily as stable micelles rather than monomers at normal working concentrations. To our knowledge, the micellar weight of Triton X-114 has not been determined. Extrapolating from Triton X-100, which forms micelles containing at least 140 molecules,[3] the micellar weight of Triton X-114 should be at least 75,000. Because of the high micellar weight and low monomer concentration, Triton X detergents are difficult to remove by dialysis; for this reason, other methods are often employed to facilitate their removal (see Detergent Exchange and Removal).

Triton X detergents form a homogeneous solution of micelles within an optimum temperature range limited at the low end by the freezing point of the solvent and at the high end by the cloud point.[1] In 150 m$M$ NaCl, the cloud point of Triton X-114 is 22°,[4] whereas that of Triton X-100 is 64°.[5] At temperatures above the cloud point, most nonionic detergents form a cloudy suspension, presumably as a result of increased micellar size or micellar aggregation owing to decreased hydration of the polar head groups.[6] The aggregated micelles can be pelleted by centrifugation to yield a denser detergent-rich phase and a lighter detergent-poor phase containing about 12 and 0.04% Triton X-114, respectively.[4] Because Triton X detergents selectively bind to amphiphilic proteins, when a membrane system is solubilized in Triton X-114, the detergent-rich phase will contain 90% or more of the soluble integral membrane proteins. If a 1% initial concentration of Triton X-114 is used, the final volume of the detergent

[2] J. M. Neugebauer, this series, Vol. 182, p. 239.

[3] L. M. Kushner and W. D. Hubbard, *J. Phys. Chem.* **58,** 1163 (1954).

[4] C. Bordier, *J. Biol. Chem.* **256,** 1604 (1981).

[5] W. Maclay, *J. Colloid Sci.* **11,** 272 (1956).

[6] E. J. Staples and G. J. T. Tiddy, *Faraday Soc. Trans.* **74,** 2530 (1978).

phase will be about 5% of the starting volume.[7] We customarily refer to the lighter detergent-poor phase as the aqueous phase and to the denser detergent-rich phase as the detergent phase.

## Triton X-114 Phase Separation of Integral Membrane Proteins

Bordier was the first to recognize that the low cloud point of Triton X-114 could be exploited as a means of separating hydrophilic and integral membrane proteins.[4] In a landmark paper he demonstrated this initially with purified hydrophilic and integral membrane proteins and then extended the procedure to resolve the membrane proteins from a complex mixture extracted from whole cells. He also deserves credit for precisely defining the physical properties of Triton X-114 with respect to phase separation (see the miniprint addendum to the paper[4]). Hooper and Bashir showed that a single membrane-spanning domain or hydrophobic group is sufficient to cause a protein to partition into the detergent phase.[8] Almost 400 publications utilizing Triton X-114 for protein and membrane analysis have appeared since the original paper by Bordier (see Table I).[9] This technique has been used to characterize the membrane proteins of the spirochetal bacterial pathogens *Treponema pallidum*[7,10–13] and *Borrelia burgdorferi*,[14,15] the etiologic agents of syphilis and Lyme disease, respectively (Fig. 2).

## Methods

### *Precondensation of Triton X-114*

The number of oxyethylene head groups in commercial preparations of Triton detergents usually ranges plus or minus one group from the average. In the original procedure by Bordier, a precondensation step was included to remove the more hydrophilic components from commercial

[7] J. D. Radolf, N. R. Chamberlain, A. Clausell, and M. V. Norgard, *Infect. Immun.* **56,** 490 (1988).

[8] N. M. Hooper and A. Bashir, *Biochem. J.* **280,** 745 (1991).

[9] J. G. Pryde, *Trends Biochem. Sci.* **11,** 160 (1986).

[10] T. M. Cunningham, E. M. Walker, J. N. Miller, and M. A. Lovett, *J. Bacteriol.* **170,** 5789 (1988).

[11] B. K. Purcell, M. A. Swancutt, and J. D. Radolf, *Mol. Microbiol.* **4,** 1371 (1990).

[12] M. A. Swancutt, J. D. Radolf, and M. V. Norgard, *Infect. Immun.* **58,** 384 (1990).

[13] L. Weigel, M. Brandt, and M. Norgard, *Infect. Immun.* **60,** 1568 (1992).

[14] M. E. Brandt, B. S. Riley, J. D. Radolf, and M. V. Norgard, *Infect. Immun.* **58,** 983 (1990).

[15] J. S. Brusca, A. W. McDowall, M. V. Norgard, and J. D. Radolf, *J. Bacteriol.* **173,** 8004 (1991).

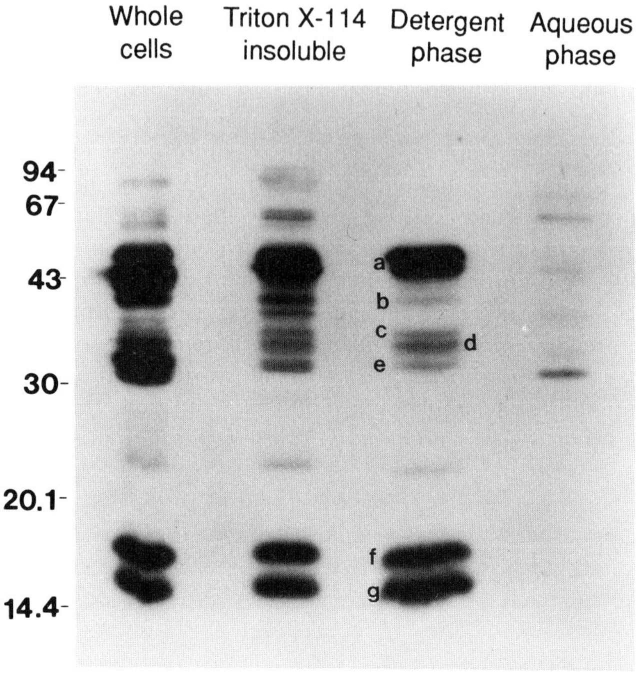


FIG. 2. Immunoblot analysis of *Treponema pallidum* polypeptides after extraction with 2% Triton X-114 and phase partitioning. The immunoblots were analyzed using human secondary syphilitic serum. Major polypeptides detected in the detergent phases are labeled a through g. The results indicate that immunodominant proteins selectively partition to the detergent phase. [Reprinted from J. D. Radolf, N. R. Chamberlain, A. Clausell, and M. V. Norgard, *Infect. Immun.* **56,** 490 (1988), with permission.]

preparations.[4] The precondensation step lowered the cloud point by 2° and reduced the detergent concentration in the aqueous phase by 75%. The precondensation procedure is described below.

1. Twenty milliliters of commercially purchased Triton X-114 is added to 980 ml of 10 m*M* Tris-HCl, pH 7.4, 150 m*M* NaCl. The solution is mixed at 0° until clear and then heated to 30° until cloudy.
2. The mixture is allowed to separate into aqueous and detergent phases by incubation at 30° overnight. A centrifugation step at 30° may be substituted for the overnight incubation.
3. The aqueous phase is discarded and replaced by an equal volume of 10 m*M* Tris-HCl, pH 7.4, 150 m*M* NaCl. The condensation is repeated two more times.

The final detergent phase is used as a stock for extractions. Bordier measured the final concentration of precondensed Triton X-114 to be

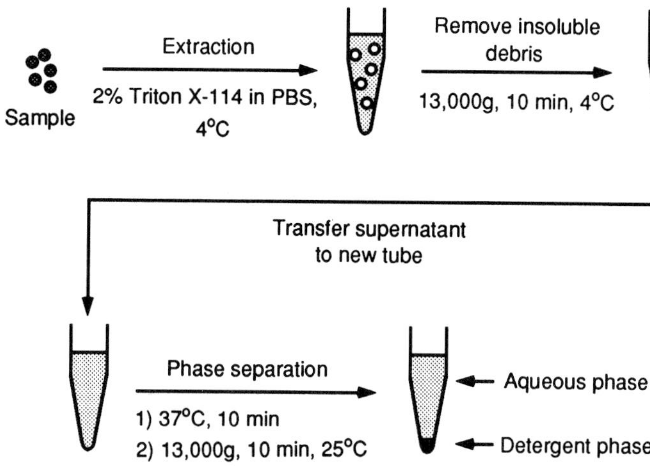


FIG. 3. Triton X-114 phase-partitioning procedure. The resulting aqueous and detergent phases are supplemented with Triton X-114 or PBS, respectively, and washed three to five times by repeating the phase-separation step.

11.4%.[4] We have found the precondensation procedure to be unnecessary for efficient phase partitioning of proteins. In addition, the procedure produces a stock with a potentially variable detergent concentration. For these two reasons, we prefer to omit the precondensation procedure and, instead, directly dilute a commercial preparation of Triton X-114 in phosphate-buffered saline (PBS; 10 m*M* $NaPO_4$, 145 m*M* NaCl) to a 10% (v/v) working solution.

To prevent oxidation of protein sulfhydryl groups by contaminants present in Triton X-114,[16–18] Bordier added 0.0016% (w/v) butylated hydroxytoluene to the 2% solution of detergent prior to precondensation. We have found that freshly diluted membrane grade Triton X-114 (available from Boehringer-Mannheim Biochemicals, Indianapolis, IN) does not contain contaminants that interfere with protein purification.

*Detergent Extraction of Proteins*

Our standard extraction conditions employ 2% (v/v) Triton X-114 in PBS and a detergent-to-protein ratio of 10 mg detergent per milligram of protein (Fig. 3).[7] However, we have used levels of Triton X-114 as low as 0.1% to extract selectively the outer membranes of *T. pallidum* and *B. burgdorferi*.[7,15] For complete extraction of whole cells or tissue, the

[16] H. W. Chang and E. Bock, *Anal. Biochem.* **104,** 112 (1980).
[17] Y. Ashani and G. N. Catravas, *Anal. Biochem.* **109,** 55 (1980).
[18] D. Schubert, K. Boss, H.-J. Dorst, J. Flossdorf, and G. Pappert, *FEBS Lett.* **163,** 81 (1983).

extraction should be conducted for a minimum of 1 hr, more commonly overnight. We do not routinely include protease inhibitors, but their use should be considered if unexpectedly low yields or degradation of proteins are observed.

### *Phase Separation of Aqueous and Detergent Phases*

Detergent-insoluble material should be removed by centrifugation at 4° prior to phase separation (Fig. 3). When extracting with low concentrations of Triton X-114, the detergent should be increased to 2% prior to phase separation to increase the volume of the resulting detergent phase.

*Sucrose Cushion Technique.* In Bordier's original protocol, the phase separation is performed in the presence of a sucrose cushion. Because of its intermediate density the sucrose cushion forms a layer between the phases that facilitates removal of the aqueous phase without detergent contamination. The Bordier procedure for sucrose cushion phase separation is as follows:

1. A 300-$\mu$l cushion of 6% (w/v) sucrose, 10 m*M* Tris-HCl, pH 7.4, 150 m*M* NaCl, and 0.06% Triton X-114 is added to a 1.5-ml conical Eppendorf tube. The cold Triton X-114 extract is placed on top of the cushion, and the tube is incubated at 30° for 3 min.
2. The cloudy solution is centrifuged at 300–13,000 *g* for 3 min at room temperature or 30°. The detergent phase appears as a viscous clear layer at the bottom of the tube.
3. The upper aqueous phase is removed and supplemented with Triton X-114 stock to a final concentration of 0.5%.
4. The solution is mixed at 0°, overlaid on the sucrose solution used previously, incubated for 3 min at 30°, and centrifuged onto the previous detergent phase.
5. The aqueous phase is then removed, and Triton X-114 stock is added to a final concentration of 2%. The solution is then mixed at 0°, incubated at 30° for 3 min, and centrifuged in the absence of a sucrose cushion. The final aqueous phase is transferred to a new tube, and the detergent phase is discarded.
6. The sucrose cushion is removed from the original detergent phase. The washed aqueous and detergent phases are diluted in 10 m*M* Tris-HCl, pH 7.4, 150 m*M* NaCl to the original sample volume.

*Multiple Washing Technique.* We have found the sucrose cushion technique to be somewhat tedious and insufficient to produce clean resolution of aqueous and detergent phases. Instead we perform multiple (three to five) washings of the aqueous and detergent phases as follows:

1. After removal of any insoluble material (Fig. 3), the sample is incubated at 37° for 10 min followed by centrifugation for 10 min at room temperature.
2. The aqueous phase is transferred to a new tube, Triton X-114 is added to a final concentration of 2%, the phases are separated, and the aqueous phase is collected.
3. The detergent phase (~50 $\mu$l) is thoroughly mixed with 1 ml of PBS at 0°, warmed to 37° for 10 min, and centrifuged as before. The aqueous phase is discarded.

### *Sample Processing after Phase Separation*

*Gel Electrophoresis, Chromatography, and Tissue Culture Bioassays.* The versatility of Triton X-114 phase partitioning is exemplified by the number of analytical and preparative procedures that can be performed with the proteins still in the detergent phase. These include isoelectric focusing (when performed alone or as the first part of two-dimensional gel electrophoresis), chromatofocusing, ion-exchange chromatography, immunoprecipitation, and immunoaffinity chromatography.[7,19,20] However, sodium dodecyl sulfate–polyacrylamide gel electrophoresis (SDS–PAGE) analysis of proteins while still in the detergent phase is complicated by the fact that Triton X-114 causes a smearing artifact. Prior to SDS–PAGE it is necessary to precipitate the proteins from the detergent phase with 10 volumes of acetone at −20° for 1 hr, followed by pelleting of the precipitate at 13,000 *g* for 10 min. Acetone-precipitated proteins may contain residual detergent that may cause toxic effects in assays utilizing tissue culture cells. This problem may be avoided by washing the precipitated proteins several times with PBS. Membrane lipids also remain noncovalently associated with the detergent-phase proteins following acetone precipitation. These may be removed by extraction with chloroform–methanol.[21]

*Protein Assay.* Triton detergents interfere with the Folin–Ciocalteu phenol (Lowry) and Coomassie Brilliant Blue (Bradford) protein assays. Although the BCA protein assay (Pierce Chemical Co., Rockford, IL) is reported to be compatible with 1% Triton X-114, the 37° incubation temperature of the standard assay is above the cloud point of Triton X-114.

[19] B. K. Purcell, N. R. Chamberlain, M. S. Goldberg, L. P. Andrews, E. J. Robinson, M. V. Norgard, and J. D. Radolf, *Infect. Immun.* **57,** 3708 (1989).

[20] P.-L. Hsu, N. R. Chamberlain, K. Orth, C. Moomaw, L.-Q. Zhang, C. A. Slaughter, J. D. Radolf, S. Sell, and M. V. Norgard, *Infect. Immun.* **57,** 196 (1989).

[21] J. D. Radolf, M. V. Norgard, M. E. Brandt, R. D. Isaacs, P. A. Thompson, and B. Beutler, *J. Immunol.* **147,** 1968 (1991).

For this reason, we assay detergent-phase proteins following acetone precipitation and resolubilization in PBS containing 1% SDS.

*Detergent Exchange and Removal.* Under some circumstances, it may be advantageous to exchange Triton X-114 for a detergent from a class of surfactant with different properties. Although the Triton detergents may be more difficult to exchange than other nonionic detergents, procedures for accomplishing this have been described.[22,23] Alternatively, it may be necessary to remove the detergent in order to reconstitute the membrane proteins into liposomes for assay of biological activity or topographical analysis. Triton X-114 is easily removed from protein solutions by batch absorption with hydrophobic polystyrene resin beads, (Bio-Beads SM-2, Bio-Rad Laboratories, Richmond, CA) in conjunction with dialysis.[24,25] Reconstitution of proteins into liposomes is often facilitated by the addition of phospholipids solubilized in a dialyzable detergent prior to removal of detergent.[26,27]

### *Modifications of Procedure*

*Cloud Point Alteration.* Triton X-114 solutions must be handled below the cloud point to avoid inadvertent phase separation. When desired, cloud points above room temperature can be achieved by mixing Triton X-114 and Triton X-100.[4]

Manipulation of thermolabile proteins at temperatures above the cloud point of Triton X-114 may be detrimental to biological activities. For this reason, two modifications of the Triton X-114 phase separation procedure that lower the cloud point of the detergent solution have been developed.

Ganong and Delmore[28] used Triton X-45, a detergent with only 5 oxyethylene head groups and a HLB of 10.4, to lower the cloud point of Triton X-114 solutions. The relationship between the relative proportions of Triton X-45 and Triton X-114 and the resultant cloud point is described by the equation

$$C = 21.4 - 1.23W$$

where $C$ is the cloud point (in °C) and $W$ is the weight percentage of total detergent that is Triton X-45. This procedure allows phase separations to

[22] L. M. Hjelmeland, this series, Vol. 182, p. 277.

[23] A. Furth, *Anal. Biochem.* **109,** 207 (1980).

[24] P. W. Holloway, *Anal. Biochem.* **53,** 304 (1973).

[25] D. Levy, A. Bluzat, M. Seigneuret, and J.-L. Rigaud, *Biochim. Biophys. Acta* **1025,** 179 (1990).

[26] T. D. Madden, *Chem. Phys. Lipids* **40,** 207 (1986).

[27] H. Scheer, G. Prestipino, and J. Meldolesi, *EMBO J.* **5,** 2643 (1986).

[28] B. R. Ganong and J. P. Delmore, *Anal. Biochem.* **193,** 35 (1991).

be performed at any temperature between 0 and 21°. Werch-Reichhart *et al.* reported that the addition of glycerol lowers the cloud point of Triton X-114 solutions.[29] In the presence of 20% glycerol, the cloud point was lowered about 10°. At glycerol concentrations between 20 and 30%, the similar densities of the detergent and aqueous phases preclude satisfactory phase separation. At 30% glycerol, the detergent and aqueous phases are inverted, and phase separation occurs spontaneously at 4°. The phases do not form when 30% glycerol is added if the detergent-to-protein ratio is less than 3.5.

*Use of Reduced Triton X-114.* A major disadvantage of the use of Triton X detergents is the high extinction coefficient (owing to the phenyl ring) at the UV wavelength (280 nm) used to monitor proteins during column chromatography. One solution to this problem is to exchange the Triton X detergent prior to column chromatography for a detergent that does not absorb at that wavelength. A second approach is to employ a form of Triton X-114 in which the UV-absorbing aromatic ring has been reduced (available from Aldrich Chemical Co., Milwaukee, WI).[30] We have found that reduced Triton X-114 extracts and partitions proteins effectively.

## Discussion

As pointed out by Maher and Singer,[31] Triton X-114 phase partitioning is deceptively powerful because cell fractionation and detergent extraction are combined with an appropriate 20-fold enrichment of the proteins that enter the detergent phase. Although this chapter has emphasized the preparation of membrane proteins in the detergent phase, the aqueous phase and insoluble material are also potentially valuable. Material not solubilized by extensive extraction was used as a source of peptidoglycan from *Treponema pallidum*.[32] Kalmokoff *et al.* isolated flagella from the aqueous phase of Triton X-114-extracted *Methanococcus* spp.[33] Triton X-114 phase partitioning has also been used as a highly effective method to remove endotoxin from solutions containing water-soluble proteins.[34]

Triton X-114 phase partitioning can provide crucial insights regarding

[29] D. Werck-Reichhart, I. Benveniste, H. Teutsch, F. Durst, and B. Gabriac, *Anal. Biochem.* **197,** 125 (1991).

[30] G. E. Tiller, T. J. Mueller, M. E. Dockter, and W. G. Struve, *Anal. Biochem.* **141,** 262 (1984).

[31] P. A. Maher and S. J. Singer, *Proc. Natl. Acad. Sci. U.S.A.* **82,** 958 (1985).

[32] J. D. Radolf, C. Moomaw, C. A. Slaughter, and M. V. Norgard, *Infect. Immun.* **57,** 1248 (1989).

[33] M. L. Kalmokoff, K. F. Jarrell, and S. F. Koval, *J. Bacteriol.* **170,** 1752 (1988).

[34] Y. Aida and M. J. Pabst, *J. Immunol. Methods* **132,** 191 (1990).

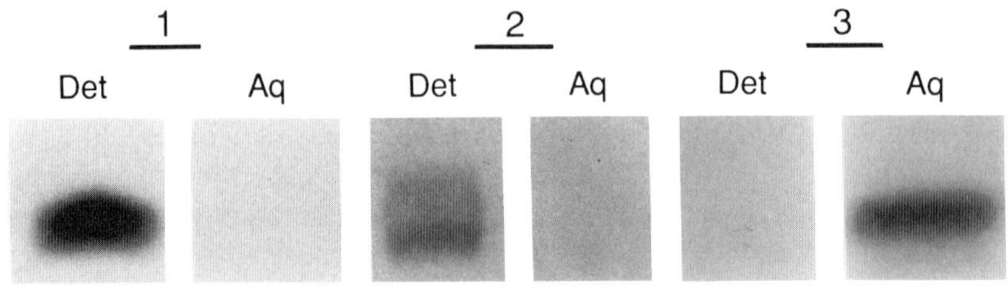


FIG. 4. Triton X-114 phase partitioning of the 15-kDa antigen of *Treponema pallidum* expressed in *T. pallidum, Escherichia coli,* and a cell-free transcription–translation system. Detergent (Det) and aqueous (Aq) phases of *T. pallidum* (**1**) and *E. coli* RR1 harboring a plasmid expressing the 15-kDa antigen (**2**) were separated by 15% SDS–PAGE and analyzed by immunoblotting with pooled 15-kDa immunogen-specific monoclonal antibodies; following Triton X-114 phase partitioning $^{35}$S-labeled amino-acid products in the cell-free system (**3**) were immunoprecipitated with the same pool of monoclonal antibodies and separated by 10% SDS–PAGE. Only the relevant portion of each gel is shown. The results show that the mature lipidated form partitions to the detergent phase, whereas the nonlipidated precursor partitions to the aqueous phase. These results also suggest that the 15-kDa antigen is anchored to the membrane by the covalently bound lipid.[11] Identical results have been obtained for the 34-kDa,[12] 47-kDa,[13,35] and 17-kDa[35a] antigens of *T. pallidum*. [Reprinted from B. K. Purcell, M. A. Swancutt, and J. D. Radolf, *Mol. Microbiol.* **4,** 1371 (1990), with permission.]

the means by which proteins are tethered to membranes. We have used this procedure extensively to investigate the posttranslational modification of spirochetal membrane lipoproteins. In fact, it was the discrepancy between the observed hydrophobicity of the major membrane immunogens of *T. pallidum,* as determined by Triton X-114 phase partitioning, and their hydrophilicity, as predicted by DNA sequence analyses, which led to the demonstration that these molecules contain covalently bound lipids.[11–13,35] Our data also indicate that the lipid groups are responsible for the membrane localization of these *T. pallidum* proteins (Fig. 4). The finding that glycophosphatidylinositol (GPI)-modified proteins shift from the detergent to the aqueous phase on digestion by phosphatidylinositol-specific phospholipase C led to the similar conclusion that the GPI group provides the membrane anchor for this class of integral membrane proteins.[36–44]

[35] N. R. Chamberlain, L. DeOgny, C. Slaughter, J. D. Radolf, and M. V. Norgard, *Infect. Immun.* **57,** 2878 (1989).

[35a] D. R. Akins, B. K. Purcell, M. M. Mitra, M. V. Norgard, and J. D. Radolf, *Infect. Immun.* **61,** 1207 (1993).

[36] C. Bordier, R. J. Etges, J. Ward, M. J. Turner, and M. L. Cardoso de Almeida, *Proc. Natl. Acad. Sci. U.S.A.* **83,** 5988 (1986).

[37] C. Hernandez-Munain, M. A. Fernandez, A. Alcina, and M. Fresno, *Infect. Immun.* **59,** 1409 (1991).

Investigators employing this technique should be aware that there are rare examples of membrane proteins that are either insoluble in Triton X-114 or that partition anomalously into the aqueous phase. The two best examples of anomalous partitioning of membrane proteins to the aqueous phase are the acetylcholine receptor of *Torpedo californica*[31] and the nonlipidated membrane bound form of the immunoglobulin heavy chain.[45] Hooper and Bashir also reported that proteins modified by GPI are not solubilized by Triton X-114 and can be pelleted from a cold detergent solution.[8] The significance of that report is unclear since numerous other examples have been reported of GPI-linked proteins which are soluble in the detergent phase.[36–44]

[38] E. Paul, F. A. Leblond, and D. LeBel, *Biochem. J.* **277,** 879 (1991).

[39] D. LeBel and M. Beattie, *Biochem. Biophys. Res. Commun.* **154,** 818 (1988).

[40] M. J. Rindler, S. S. Naik, N. Li, T. C. Hoops, and M. N. Peraldi, *J. Biol. Chem.* **265,** 20784 (1990).

[41] S. Das, A. Traynor-Kaplan, D. S. Reiner, T. C. Meng, and F. D. Gillin, *J. Biol. Chem.* **266,** 21318 (1991).

[42] S. Y. Sauma and M. Strand, *Mol. Biochem. Parasitol.* **38,** 199 (1990).

[43] E. J. Pearce, A. I. Magee, S. R. Smithers, and A. J. Simpson, *EMBO J.* **10,** 2741 (1991).

[44] K. Kodukula, D. Cines, R. Amthauer, L. Gerber, and S. Udenfriend, *Proc. Natl. Acad. Sci. U.S.A.* **89,** 1350 (1992).

[45] S. Pillai and D. Baltimore, *Proc. Natl. Acad. Sci. U.S.A.* **84,** 7654 (1987).

# [18] Separation of Proteins and Nucleic Acids

*By* WERNER MÜLLER

## Fractionation of Protein Mixtures

### *Introduction*

Proteins can be separated by use of aqueous two-phase systems for liquid–liquid partition chromatography (LLPC) on columns.[1] The proteins behave essentially in accord with their partition coefficients, $K$, in the two-phase systems. The poly(ethylene glycol) (PEG)–dextran (Dx) systems have been used most frequently for LLPC with polymer aqueous two-phase systems, but their deficiencies have also led to experiments with other systems.

[1] W. Müller, this volume [9].

TABLE I
COMPARISON OF MEASURED PARTITION COEFFICIENTS ($K$) WITH THOSE CALCULATED FROM ELUTION VOLUMES FOR SELECTED PROTEINS[a]

| | $K$ | |
|---|---|---|
| Protein | Measured | Calculated from $V_{el}$ |
| Horseradish peroxidase | 1.236 (±0.02) | 1.255 (±0.06) |
| Myoglobin | 2.532 (±0.03) | 2.500 (±0.1) |
| Ovalbumin | 3.584 (±0.05) | 3.597 (±0.2) |
| Bovine serum albumin | 9.43 (±0.3) | 8.33 (±1.0) |

[a] From W. Müller, "LLPC of Biopolymers," p. 22. GIT-Verlag, Darmstadt, Germany, 1988, with permission.

*Separations of Standard Proteins*

Most of the standard proteins examined here are available in a pure and salt-free dry state and thus readily provide data for correlating $K$ values and elution volumes of analytes for establishing the LLPC mode of the column. Table I shows a correlation between the $K$ values of four proteins and chromatographic data obtained from a polyacrylamide (PAA)-coated silica–diol column (1000 Å pore size and 10 $\mu$m particle size, Merck, Darmstadt, Germany) for which the specific phase volumes $V_s$ and $V_m$ were determined with tRNA and 5s RNA.[2] The deviations of measured from calculated values are caused by the limited accessibility of protein to the stationary phase, when LiParGel (Merck) is used, and by nonspecific side affinities (mostly electrostatic) for PAA-coated silica–diol[3] (Fig. 1). To minimize the concentration effects of the proteins used, the measurements of $K$ in a test tube should mimic the situation in the column, that is, the concentration of the test compounds in the partitioning experiment should be within the concentration range it will have during the column run. The separations of some standard proteins on LiparGel and and PAA-coated silica–diol of 10 $\mu$m particle size (Fig. 2) show that a much higher resolution is obtained for the finer and structurally better defined silica support. Figure 2a also shows that basic proteins may behave quite well on silica supports. The high resolution observed for the separation of ovalbumin from human serum albumin, which is in full agreement with their $K$ values, is unique for LLPC.

[2] W. Müller, A. Heubner, and J. Kaniecki, *Kontakte (Darmstadt)* **1988,** 55 (1988).

[3] A. Walsdorf, Doctoral Thesis, University of Düsseldorf, Germany (1990); A. Walsdorf, D. Forciniti, and M. R. Kula, *J. Chromatogr.* **523,** 103 (1990).

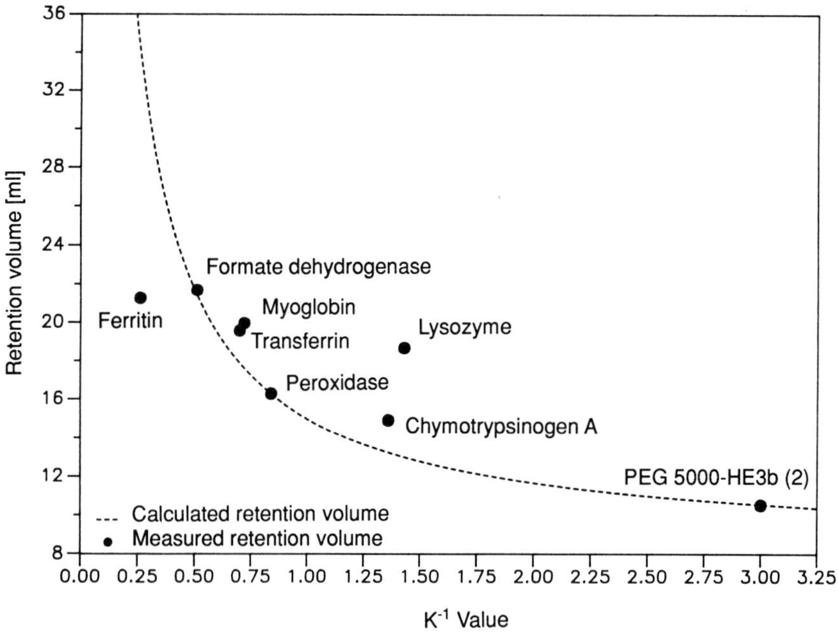


FIG. 1. Plot of the retention volumes of some standard proteins against the inverse partition coefficients, $K^{-1}$, for a LiParGel column containing 6.71 ml of stationary and 8.26 ml of mobile phase. The system contained dextran T500 and PEG 20,000 (Serva, Heidelberg, Germany) as phase-forming polymers. The dotted line represents the position at which all the retention volumes could be expected to be based on the partition coefficients of peroxidase and PEG–red, a PEG-bound red dye. The deviation of lysozyme and ferritin from expectation may be explained by the slight acidity of the support and a size-exclusion effect, respectively. [From A. Walsdorf, D. Forciniti, and M.-R. Kula, *J. Chromatogr.* **523,** 103 (1990), with permission.]

Proteins with larger molecular weights may also be separated if the ionic strength of the system is kept low. Figure 3 shows the separation of three dehydrogenases using a phosphate concentration gradient. The breadth of the last peak may be due either to the well-known isoenzymes of lactate dehydrogenase or to nonspecific interactions between this basic protein and the LiParGel matrix enhanced by the low ionic strength. Use of the poly(vinylpyrrolidone) (PVP)–Dx systems instead of the PEG–Dx systems does not change the elution order, but it reduces the resolution.[4] The loss of selectivity may be due, in part, to the fact that the available salt concentration range in these systems is restricted because they tend to become monophasic. Thus, a two-phase system containing 4% Dx T40

[4] F. Engelhard, Diplomarbeit, Fachhochschule Darmstadt, Germany, 1987.

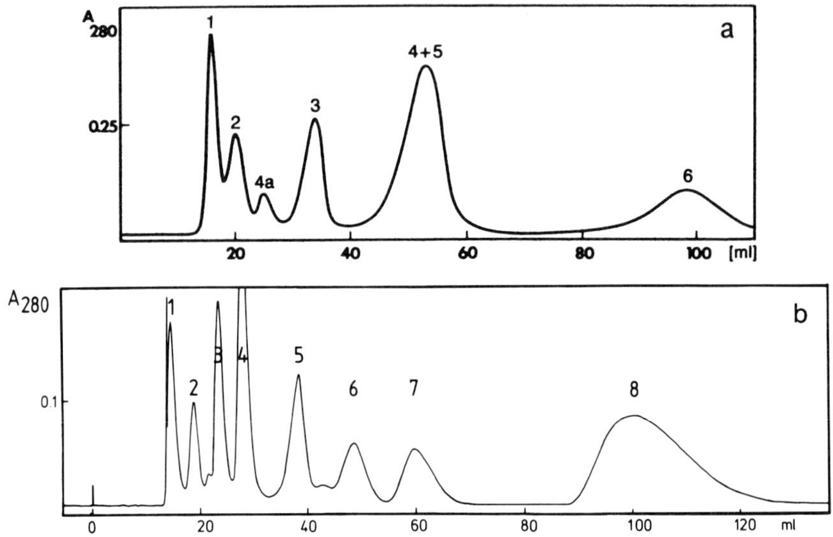


FIG. 2. (a) Separation of some standard proteins on LiParGel 750. Column size, 32 × 1 cm. The system contained 9.8% (w/w) dextran T40, 5.4% (w/w) PEG 6000, 75 m*M* KBr, 10 m*M* sodium phosphate, pH 7.2, and 1 m*M* $NaN_3$. Temperature, 23°. 1, Chymotrypsinogen A; 2, peroxidase; 3, carbonate dehydratase; 4 and 4a, β-lactoglobulin; 5, ovotransferrin; 6, ovalbumin. (b) Separation of eight standard proteins on PAA-coated silica (1000 Å/10 μm). The system contained 9.8% (w/w) dextran T40, 5.4% (w/w) PEG 6000, 75 m*M* $NaClO_4$, 25 m*M* lithium acetate, 10 m*M* sodium cacodylate, pH 7.2. Temperature, 23°; flow rate, 0.3 ml/min. 1, Lysozyme; 2, peroxidase; 3, cytochrome *c*; 4, myoglobin; 5, β-lactoglobulin; 6, ovotransferrin; 7, ovalbumin; 8, human serum albumin. [From W. Müller, *GIT Fachz. Lab.* **32,** 233 (1988), with permission.]

(Pharmacia-LKB, Uppsala, Sweden) and 5.2% PVP 30 (BASF, Ludwigshafen/Rhein, FRG) becomes monophasic when 50 m*M* $NaClO_4$, 50 m*M* KBr, or 100 m*M* NaCl is incorporated. This means that larger potential differences between the phases, with a negatively charged upper phase, cannot be obtained. Salts known to give rise to positively charged PEG phases (50 m*M* $Na_2SO_4$, 33 m*M* lithium citrate) do not affect the PVP–Dx system, however. Of special interest is the fact that incorporation of 50 m*M* sodium octanoate leaves the system biphasic, whereas it becomes monophasic on centrifugation above 1000 *g*. The two phases are restored on standing at 1 *g* for about 10 min.[4] A separation of standard proteins using a PVP–Dx system is shown in Fig. 4.

The use of poly(vinyl alcohol) (PVA, Fluka, Buchs, Switzerland)–Dx systems does not present any clear advantage over the commonly employed PEG–Dx system when used to separate smaller proteins. If

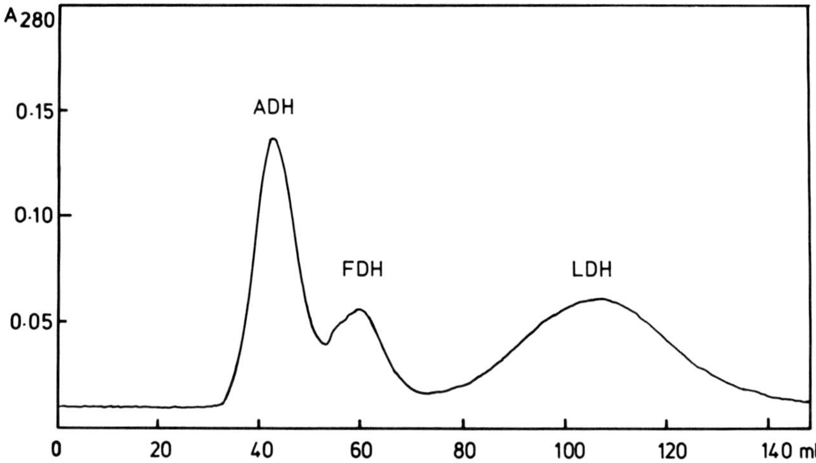


FIG. 3. Separation of three dehydrogenases on a LiParGel 750 column, 30 × 1.6 cm. The system contained 9.8% (w/w) dextran T40 and 5.4% (w/w) PEG 6000. Gradient elution was accomplished by running from 15 to 5 m*M* phosphate for 100 ml and followed by 5 m*M* phosphate (pH 8.4). Flow rate, 1 ml/min. ADH, Alcohol dehydrogenase (4 mg); FDH, formate dehydrogenase (8 mg); LDH, lactate dehydrogenase (2 mg). [From W. Müller, *GIT Fachz. Lab.* **32,** 233 (1988), with permission.]

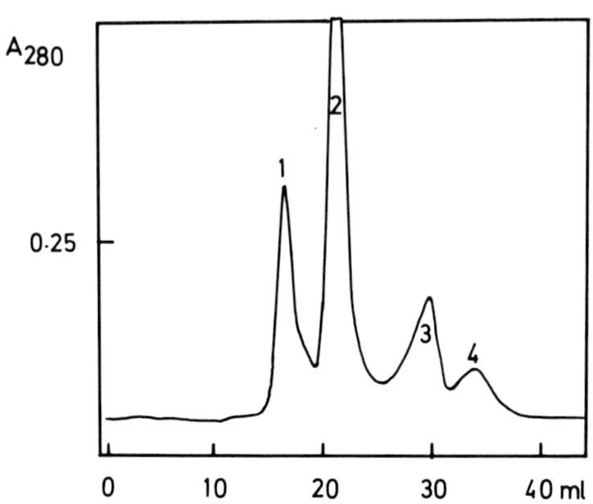


FIG. 4. Separation of four standard proteins on a PAA-coated silica column (1000 Å/10 $\mu$m, 30 × 1 cm). The system contained 10.0% (w/w) dextran T40 and 10.0% (w/w) PVP 30, 10 m*M* Tris-HCl, pH 8.0, 80 m*M* KCl, and 1 m*M* $NaN_3$. Flow rate, 0.5 ml/min; temperature, 23°. 1, Peroxidase (1.5 mg); 2, myoglobin (1.3 mg); 3, $\beta$-lactoglobulin (1 mg); 4, ovalbumin (1 mg). (From F. Engelhard, Diplomarbeit, Fachhochschule Darmstadt, Germany, 1987.)

combined with Dx T500 (molecular weight 500,000, Pharmacia-LKB) the high phase viscosity presents unnecessary technical problems, including back-pressures up to 100 bar at a flow rate of 0.25 ml/min through a 300 × 10 mm column packed with a 10 $\mu$m support. Replacing Dx T500 by Dx T40 (molecular weight 40,000) reduces the back-pressure by 33 to 50%. The selectivity of these systems is somewhat less than that of PEG–Dx systems of similar tie-line length, but the fact that immunoglobulins behave differently in these systems justifies the study of PVA systems in more detail.

The wide application PEG–salt systems have found in biotechnology due to low price and acceptable selectivity has, of course, suggested the use of these systems in LLPC. The results of a thorough study[5] revealed that separations are possible in such systems using Superose 12 (Pharmacia-LKB) or porous, regenerated cellulose as supports. LiParGel or PAA-coated silica–diols can also be used, but for none of these supports is there an acceptable correlation between the measured $K$ values and the elution volumes observed.[5]

### *Fractionation of Native Protein Mixtures*

The LLPC technique has been used widely for fractionating protein mixtures from tissues, sera, fermentation broths, etc. Figure 5 shows the elution profiles obtained for sera from five different mammals using a PEG–Dx system. Aside from the fact that they differ appreciably from one another, the immunoglobulins do not form visible bands but rather a broad background underlying all bands with the exception of the most prominent one appearing at the end of each profile, which contains the albumins. This elution position is typical for acid proteins like albumins (p*I* 4.6) in a PEG–Dx system containing KBr as the dominant electrolyte.

The fractionation of immunoglobulins has been possible in the PEG–Dx systems despite their low solubilities in the presence of PEG. Working at pH 9.0 in a system containing 0.1 *M* glycine yielded fractionation patterns for human IgG, IgA, and IgM as depicted in Fig. 6.[6] LLPC has gained special recognition in serum fractionation by providing the first pure corticosteroid-binding globulins and sex hormone-binding globulins.[7,8]

[5] S. Boese, Diplomarbeit, Fachhochschule Darmstadt, Germany, 1991.

[6] U.-B. Hansson, K. Andersson, Y. Liu, and P.-Å. Albertsson, *Anal. Biochem.* **183,** 305 (1989).

[7] A. Heubner, O. Belovsky, W. Müller, H.-J. Grill, B. Manz, M. Juchem, and K. Pollow, *J. Chromatogr.* **397,** 419 (1987).

[8] W. Müller, *Eur. J. Biochem.* **155,** 213 (1986).

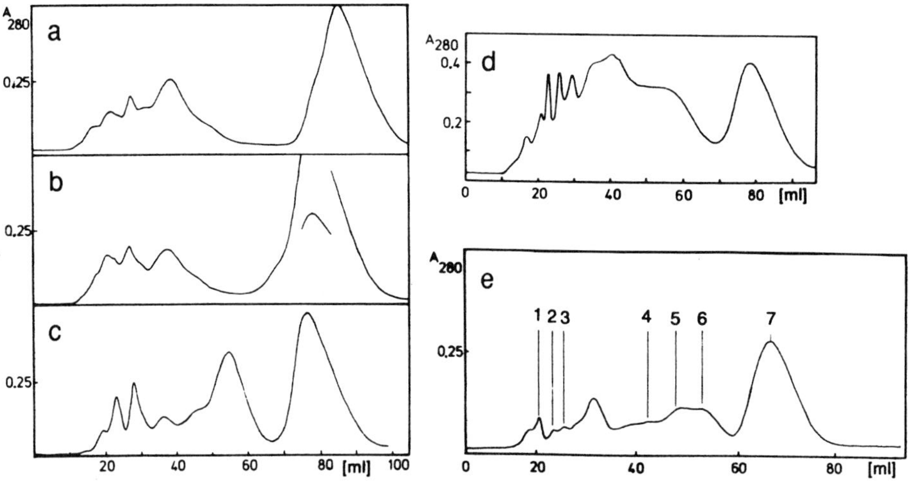


FIG. 5. Fractionation of five different sera by LLPC [(a) goat, (b) sheep, (c) dog, (d) horse, (e) human] on LiParGel 750 (30 × 1 cm). The system contained 9.8% (w/w) dextran T40, 5.4% (w/w) PEG 6000, 75 m*M* KBr, 10 m*M* sodium phosphate, pH 7.2, and 1 m*M* $NaN_3$. Flow rate, 0.3 ml/min; temperature, 23°. The numbers in (e) indicate the elution positions of 1, $\alpha_1$-acid glycoprotein; 2, sex hormone-binding globulin; 3, corticosteroid binding globulin; 4, $\alpha_1$-antitrypsin; 5, ceruloplasmin; 6, transferrin; 7, albumin. [From W. Müller, *GIT Fachz. Lab.* **32,** 233 (1988), with permission.]

An interesting application of LLPC in biotechnology is the isolation of mutarotase from a bacterial extract. The basic enzyme may be obtained in a highly enriched state by using a PEG–Dx system containing 50 m*M* lithium sulfate as the main electrolyte. Under these conditions nearly all other proteins present in the crude extract elute ahead of the mutarotase.

Using a PVP–Dx system for serum fractionation, the immunoglobulins are eluted close to the front.[4] When PVA is used instead of PEG or PVP as the phase-forming polymer, the elution pattern changes dramatically inasmuch as the albumin is eluted ahead of the immunoglobulins.[9]

## Fractionation of Nucleic Acids

### *Introduction*

In contrast to proteins, incompatibilities between nucleic acids and PEG or Dx have not been reported. Consequently, there was no need to test other two-phase systems for separating these biopolymers. Actually,

[9] M. Ruppel, Diplomarbeit, Fachhochschule Darmstadt, Germany, 1989.

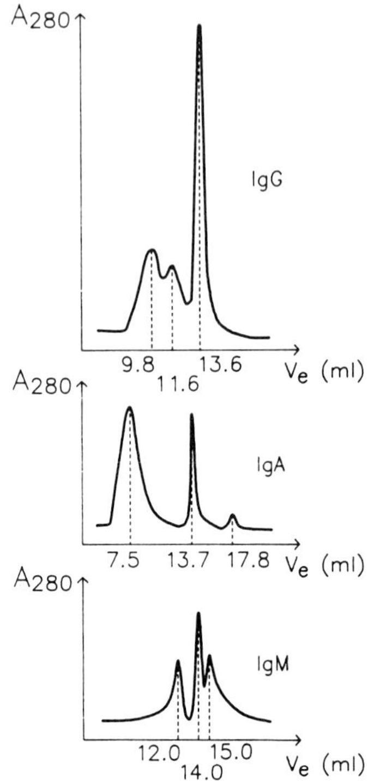


FIG. 6. Fractionation of human immunoglobulins on LiParGel 650 (35 × 1 cm) in a 6.2% Dx T500–4.4% PEG 8000 system at pH 9.0 in the presence of 75 m*M* Tris and 0.1 *M* glycine. Flow rate, 0.2 ml/min; temperature, 22°. IgG (1.5 mg), IgA (0.8 mg), and IgM (0.3 mg) were loaded in 1 ml of mobile phase. [From U.-B. Hansson, K. Andersson, Y. Liu, and P.-Å. Albertsson, *Anal. Biochem.* **183,** 305 (1989), with permission.]

substitution of another polymer for Dx might be of interest because the low diffusion rate of nucleic acids is decreased further by the high viscosity of the Dx-rich stationary phase. Whether the starch derivatives proposed as replacements for Dx improve nucleic acid fractionation in LLPC has not yet been examined.

All nucleic acid fractionations could be achieved using a single two-phase system composed of PEG 8000 (Union Carbide, New York, NY) and Dx T500. Use of a system composed of smaller phase-forming polymers, such as Dx T40 or PEG 6000 (Merck), always diminished the resolution. This can be explained by the fact that the molecular weights of the phase-forming polymers influence the partitioning of nucleic acids when they exceed a critical molecular weight;[10] another explanation given below seems more plausible.

The $K$ values of nucleic acids follow the semiempirical law of Brønstedt[11] which states that the $K$ values of "isochemical" substances (i.e., those having the same surface chemistry, as in a homologous series) decrease exponentially with increasing surface area. Because the surface area of nucleic acids is directly proportional to molecular weight, an isocratic elution of a set of typical DNA fragments (as obtained by isolation of DNA from tissues or from restriction digests) or of natural RNAs is barely possible. For this reason nucleic acids are always chromatographed using elution gradients within which their $K$ values are continuously decreased, thus promoting the elution of larger fragments. The gradients applied differ from the common gradients used in chromatography inasmuch as the eluting component or parameter in the system, for example, an organic solvent or the ionic strength, is kept constant. The effect of the gradient is achieved by a continuous exchange of one kind of salt for another. In general, potassium or sodium chlorides or bromides are replaced by lithium sulfate or citrate (taking into account ion valence), which are well known to render the mobile, PEG-rich phase positive with respect to the Dx-rich stationary phase. The resulting continuous change of the phase potential difference counteracts the size dependence of the $K$ value and thus promotes a size-dependent elution. Nucleic acids may be separated in principle according to the following parameters: size, secondary structure and morphology, base composition, and sequence. Separations based on these criteria can be carried out by LLPC.

### *Size Fractionation of DNA and RNA*

The first successful size fractionations with DNA fragments were carried out with LLPC columns prepared from cellulose (CC31, Whatman, Midstone, UK) coated with a stationary phase by swelling the dry cellulose in the mobile phase at 80°.[12] The cellulose binds water, which changes the composition of the system and causes the formation of a stationary phase in the form of fine droplets immediately "binding" to the cellulose. The very thin film of the stationary phase formed offers the advantage of minimizing all diffusion problems and thus increases the performance of

[10] P.-Å. Albertsson, A. Cajarville, D. E. Brooks, and F. Tjerneld, *Biochim. Biophys. Acta* **926,** 87 (1987).

[11] J. N. Brønstedt, *Z. Phys. Chem.* **157a,** 257 (1931).

[12] W. Müller, H.-J. Schütz, C. Guerrier, P. E. Cole, and R. Potts, *Nucleic Acids Res.* **7,** 2483 (1979).

the column at the expense, however, of capacity. DNA fragments with a molecular weight over $10^6$ tend to aggregate in the film of stationary phase owing to their high local concentration and are therefore eluted in rather broad peaks because the disaggregation becomes rate limiting for the mass transfer. In a subsequent study[13] it could be shown that careful coating in general provides the best balance in alleviating diffusion and aggregation problems.

The first fractionation of a restriction digest according to fragment size succeeded on a cellulose support.[13] The resolution obtained was high, even by today's standards, but the time required for a run to resolve fragments between 10 and 4000 base pairs could easily be several days. Changing the support to LiParGel or PAA-coated silica–diols improved the resolution and capacity, but not the speed of separation. Figure 7 shows the separation of restriction fragments generated from plasmid DNA.[14]

### *Separations According to Secondary Structure and Morphology*

Separations of DNA according to secondary structure are not easy to quantify because of the poorly defined state of denatured DNA. Heat denaturation studies performed by Albertsson a long time ago[13] have shown that single-stranded DNA has an appreciably higher $K$ than the double-stranded form and that it is possible to separate the two species simply by partitioning. The isolation of ribosomal genes from sea urchins was achieved by making use of the difference in their $K$ values.[15] The same partitioning behavior was found for single-stranded DNA from phage MS13. Fragments produced from this species could easily be separated according to size by LLPC (W. Müller and D. W. Stafford, unpublished results, 1984).

The separation of RNAs is possible under quite similar conditions. The isolation of a viroid RNA is shown in Fig. 8. Total RNA from yeast has also been fractionated.[16] The separation of covalently closed circular RNA from the nicked species of a pUC19 plasmid preparation has also been achieved by LLPC. Besides the two main forms the preparation contains several morphologically undefined species that are also separated.[14]

[13] P.-Å. Albertsson, *Biochim. Biophys. Acta* **103,** 1 (1965).

[14] W. Müller, "Liquid–Liquid Partition Chromatography of Biopolymers." GIT-Verlag GmbH, Darmstadt, Germany, 1988.

[15] J. B. Patterson and D. W. Stafford, *Biochemistry* **9,** 1278 (1970).

[16] W. Müller, *GIT Fachz. Lab.* **32,** 223 (1988).

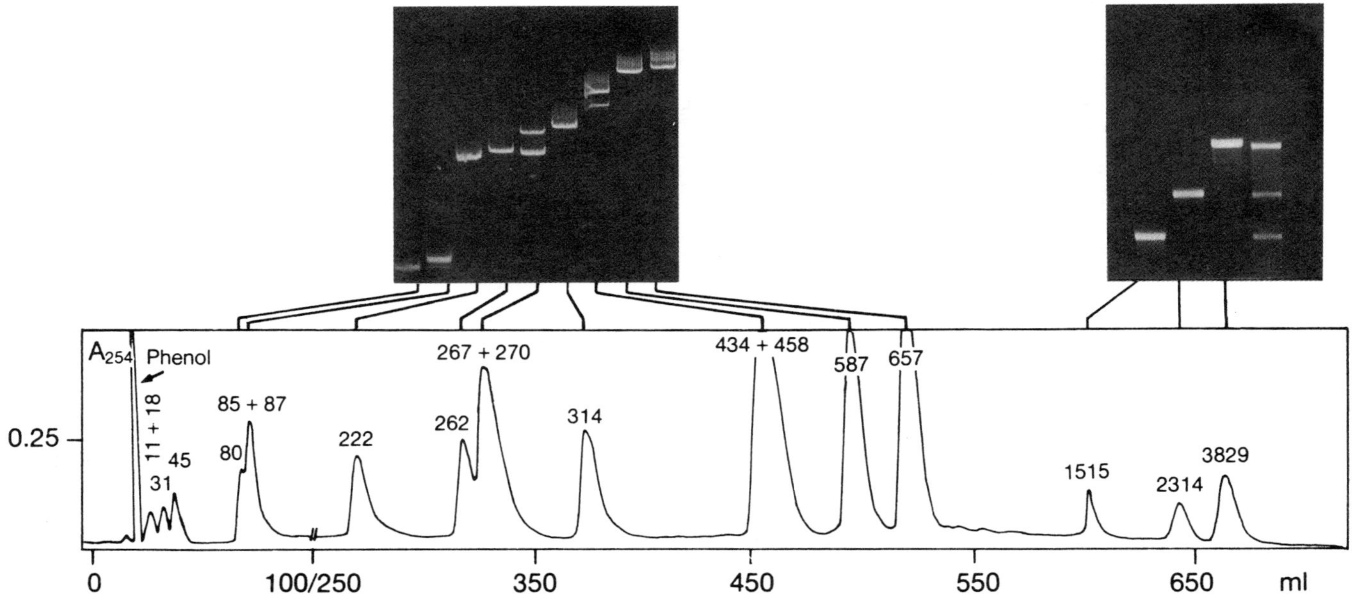


FIG. 7. Chromatographic fractionation of DNA restriction fragments obtained from *Bam*HI-linearized pDS1 plasmid (3829 bp) by digestion with *Hae*III and *Xba*I restriction endonucleases. The column was PAA-coated 10 $\mu$m silica with 4000 Å pore size (17 × 1.5 cm). The system contained 6.2% (w/w) dextran T500, 4.4% (w/w) PEG 8000. Temperature, 37°; flow rate, 0.33 ml/min. Gradient elution was accomplished using mixtures of 0.2 *M* sodium acetate (A) and 0.2 *M* lithium acetate (B) in the mobile phase, with the following gradient forms: 0–80 ml, 0% B; 80–160 ml, 0–20% B; 160–600 ml, 20–85% B; 600–800 ml, 85–100% B. The sample contained 2 mg of *Hae*III digest, 0.12 mg of *Xba*I digest, and 0.1 mg of linearized plasmid in 2 ml of mobile phase A. The numbers above and within the peaks indicate the fragment sizes in base pairs. The two gel photographs show the fragments present in the peaks as indicated: left, 5% polyacrylamide; right, 1% agarose (ethidium bromide stained). (From W. Müller, "Liquid–Liquid Partition Chromatography of Biopolymers." GIT-Verlag GmbH, Darmstadt, Germany, 1988, with permission.)

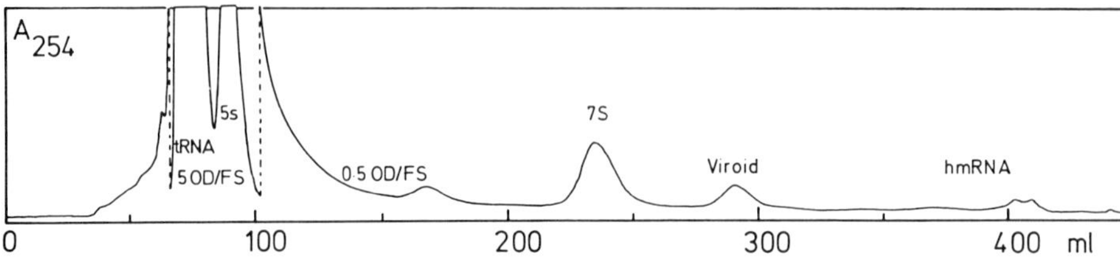


FIG. 8. Fractionation of soluble RNA from viroid-infected tomato plants on LiParGel 650 (43 × 1.5 cm) using the system 6.2% (w/w) dextran T500 and 4.4% (w/w) PEG 8000 by gradient elution. Component A was 0.2 *M* sodium acetate plus 10 m*M* cacodylate, pH 6.0; component B was 67 m*M* lithium citrate buffered as in A, pH 6.0. The gradient volume was 500 ml, linear. Flow rate, 0.25 ml/min; temperature, 37°. The sample was 42.5 mg of RNA applied in 4 ml mobile phase. (From W. Müller, "Liquid–Liquid Partition Chromatography of Biopolymers." GIT-Verlag GmbH, Darmstadt, Germany, 1988, with permission.)

TABLE II
STRUCTURE AND PHYSICAL PROPERTIES OF DNA-BINDING MACROLIGANDS[a]

| Macroligand | Chemical structure | Color | Specific for | Affinity *K* (liter/mol) *I* = 0.1, 37° | Molar extinction coefficient (nm) |
|---|---|---|---|---|---|
| **I** | 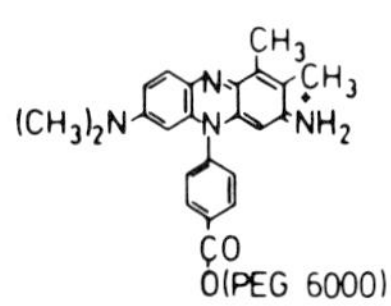 | Red | G:C / G:C and G:C / C:G | $7.4 \times 10^3$ | 44,700 (555 |
| **II** | 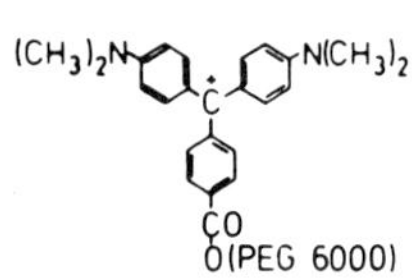 | Green | A:T / T:A | $6.2 \times 10^3$ | 89,000 (637 |
| **III** | 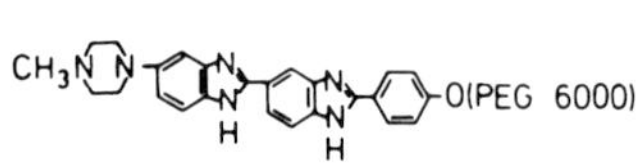 | Yellow | A:T / A:T / A:T | $2.5 \times 10^5$ | 38,000 (338 |

[a] From W. Müller *in* "Partitioning in Aqueous Two-Phase Systems" (H. Walter, D. E. Brooks, and D. Fisher, eds.), p. 235. Academic Press, Orlando, Florida, 1985, with permission.

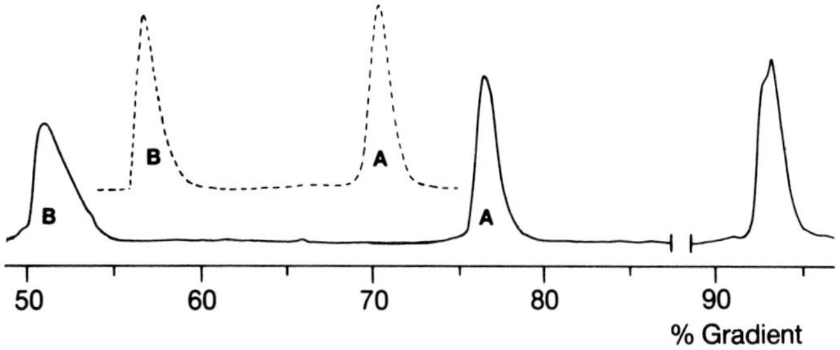


FIG. 9. Separation of two DNA restriction fragments of 2069 and 2293 bp differing by 7.7% in (A + T) content by application of base- and sequence-specific macroligands (Table II) on a cellulose column (Whatman CC31, 35 × 1.5 cm). The system contained 6.2% (w/w) dextran T500 and 4.4% (w/w) PEG 8000 and either 200 m*M* sodium acetate (phase A) or 200 m*M* lithium acetate (phase B). A linear gradient (630 ml) was applied with a flow rate of 0.25 ml/min, at a temperature of 37°. Approximately 1.5 mg of sample was loaded in each run. In the absence of macroligands both fragments were eluted in a peak at 93% of the gradient. In the presence of 13 $\mu M$ macroligand **III** (solid line) two peaks were eluted at 77% (peak A) and 52% (peak B) of the gradient containing the two fragments, 42.2% and 49.9% (A + T), respectively. The dashed line documents the same separation in the presence of 34 $\mu M$ macroligand **II**. [From W. Müller, *in* "Partitioning in Aqueous Two-Phase Systems" (H. Walter, D. E. Brooks, and D. Fisher, eds.), p. 227. Academic Press, Orlando, Florida, 1985, with permission.]

### *Separations According to Base Composition and Sequence*

The separation of nucleic acids according to base composition has been studied with DNA restriction fragments and calf thymus DNA known to contain several types of GC-rich satellite DNA sequences.[14,17,18] The fractionation represents one of the rare affinity chromatography processes not based on the denaturation–renaturation techniques normally used with DNA fragments. It uses highly selective base pair-specific synthetic dyes[19,20] that have been found to one terminal position of PEG. These PEG conjugates, called macroligands, retain their base pair specificity, on the one hand, and dramatically reduce the *K* value of the complexed DNA molecule, on the other. The latter effect is due to the PEG chain attached to the DNA molecule and can be used efficiently to shift the elution positions of complexed DNA molecules. Because the shift depends

[17] W. Müller, *in* "Affinity Chromatography and Related Techniques" (T. C. J. Gribnau, J. Visser, and R. J. F. Nivard, eds.), Analytical Chemistry Symposia Ser., Vol. 9, p. 437. Elsevier, Amsterdam, 1982.

[18] W. Müller, *in* "Partitioning in Aqueous Two-Phase Systems" (H. Walter, D. E. Brooks, and D. Fisher, eds.), p. 227. Academic Press, Orlando, Florida, 1985.

[19] W. Müller and D. M. Crothers, *Eur. J. Biochem.* **54,** 267 (1975).

[20] W. Müller, H. Bünemann, and N. Dattagupta, *Eur. J. Biochem.* **54,** 279 (1975).

on the number of PEG molecules fixed, a separation reflecting base composition results, which exceeds by far that obtainable by use of any other technique. The chemical structures and pertinent properties of the macroligands are given in Table II.

Figure 9 shows the separation of two DNA fragments both about 2000 bp in size, differing in A + T content by 7.7%. The elution profiles were obtained with macroligands specific for three and two adjacent AT pairs (yellow and green macroligand). The higher selectivity of the former, requiring three adjacent AT pairs, fortunately combined with a one order of magnitude higher affinity, provides the more efficient separation.

The separation of even larger DNA fragments obtained from λ DNA by digestion with *Eco*RI restriction enzyme with all three macroligands has been reported.[14] The fractionation of calf thymus DNA with respect to satellite DNAs has been very effective. Again all three macroligands (Table II) have been used.[14,17]

### *Conclusions*

When nucleic acids have to be fractionated in larger quantities, the technique described offers a reasonable size selectivity and unusual potential for preparative DNA separations according to base composition and topology. For larger ribonucleic acids like ribosomal RNAs the LLPC technique is unique because it provides the possibility of running gradients with increasing eluting strengths in aqueous media at constant ionic strength (i.e., by salt-exchange gradients). Because larger ribonucleic acids and single-stranded DNAs become sparingly soluble at salt concentrations above 0.5 *M*, a desorption from ion exchangers is, in general, impossible. Thus, for preparative nucleic acid separations LLPC may be the method of choice despite its more demanding technical requirements.

# [19] Purification of Nucleic Acid–Protein Complexes

*By* GILLIAN E. FRANCIS and DEREK FISHER

## Introduction

The analysis of DNA–protein interactions is central to the study of many processes including the regulation of gene expression, DNA replication, and chromatin organization. Although numerous techniques exist for the study of DNA–protein interactions *in vitro,* there are few that allow the examination of events taking place within cells.

The DNA topoisomerases are ubiquitous, important enzymes which regulate DNA topology and modulate the torsional stress of DNA (DNA supercoiling). Topoisomerase II achieves this by making a transient double-stranded break in DNA (attaching itself covalently to DNA in the process) and then changes the linking number in a maneuver whereby an adjacent part of the DNA strand passes through the enzyme-linked break between the dimeric subunits. Different strands can also be passed to catenate/decatenate circular DNA, and similar reactions modulate the topology of chromatin loop domains in eukaryotic cells, regulating supercoiling and disentangling DNA. Type I topoisomerases change the torsional stress in DNA by making a single-strand break (again with covalent attachment of the enzyme), but they cannot alone catenate DNA.

Almost all cellular functions involving DNA require the topoisomerases: DNA replication necessitates disentangling of daughter chromatids,[1] and transcription propagates waves of positive and negative supercoiling ahead of and behind RNA polymerase, respectively.[2] Torsional stress is also known to modulate the binding of transcriptional regulators,[3] and this may be the basis of the requirement for DNA topoisomerase II in some differentiation events[4]; it might act via the influence of DNA supercoiling on binding of transcriptional modulators and hence on transcription rates of key genes. Because changes in DNA supercoiling can only be influenced within individual, relatively short, chromatin loop domains (which are bounded by anchorage points), such "regulatory" topoisomerase binding sites potentially provide a means of identifying the genes being regulated.

Few methods allow purification of topoisomerase I or II while attached to DNA at their sites of action. Among these is the sodium dodecyl sulfate (SDS)/KCl precipitation procedure of Trask *et al.*[5] (where SDS traps the enzyme when it is covalently attached to DNA and the KCl precipitates the proteins and attached DNA). This is a relatively low-yield method requiring 10–20 molecules of enzyme to be attached to DNA fragments for efficient precipitation. We applied this method to differentiating hemopoietic cells in which topoisomerase II had been shown to mediate a supercoiling change that is necessary for differentiation.[4] We failed to recover more topoisomerase-linked DNA from differentiating than from

[1] W. G. Nelson, L. F. Liu, and D. S. Coffey, *Nature* (*London*) **322,** 187 (1986).

[2] L. F. Liu and J. C. Wang, *Proc. Natl. Acad. Sci. U.S.A.* **84,** 7024 (1987).

[3] B. Sanzey, *J. Bacteriol.* **138,** 40 (1979).

[4] G. E. Francis, J. J. Berney, P. S. North, Z. Khan, E. L. Wilson, P. Jacobs, and M. Ali, *Leukemia* **1,** 653 (1987).

[5] D. K. Trask, J. A. DiDonato, and M. T. Muller, *EMBO J.* **3,** 671 (1984).

control undifferentiated cells.[6] The method does, however, appear to be useful as a preliminary step to enrich protein-bound DNA with respect to free DNA, and we use it as such below.

The affinity partitioning procedure for complexes formed by DNA and topoisomerase II described here has several advantages over other available methods. It is less expensive than immunoprecipitation and requires no *a priori* assumptions about the isoform of the enzyme to be recovered. It should be noted that the same procedure is applicable to the recovery of topoisomerase I complexed to DNA and to proteins that normally bind DNA noncovalently (via the expedient of adding a DNA–protein cross-linking step).

## Principle of Method

When applied to topoisomerases I and II, the method exploits the fact that the enzyme links itself covalently to DNA, where it binds and can thus be the target for an affinity ligand. The availability of a mild method for coupling poly(ethylene glycol) (PEG) to proteins while *in situ* on DNA, provided us with a useful affinity ligand and, in conjunction with a PEG–phosphate phase system (in which DNA has affinity for the phosphate-rich phase while PEG-linked protein has affinity for the PEG-rich phase), gave a means for separating DNA–protein complexes from bulk DNA. The method is essentially an adaptation of conventional affinity phase partitioning. However, it contrasts to the use of, for example, PEG-bound antibodies as affinity ligands in phase systems containing PEG in which PEG-bound antibody provides affinity both for the PEG-rich phase via the PEG and for the target via the antibody.[7] Here, the ligand is PEG and is merely the source of differential affinity between phases and not of the specificity of the recovery (which is due to the DNA-binding protein itself which binds specific DNA sites before the ligand is attached).

## Experimental Procedures

### *Preparation of Cells for Recovery of Induced or Constitutive Covalent DNA–Protein Complexes*

DNA topoisomerases bind covalently to DNA as they act, but the complex is transient. If, at the moment of covalent attachment, the cell

[6] R. J. Anderson, C. Delgado, D. Fisher, J. M. Cunningham, and G. E. Francis, *Anal. Biochem.* **193,** 101 (1991).

[7] L. J. Karr, D. L. Donnelly, A. Kozlowski, and J. M. Harris, this volume [35].

is disrupted with SDS, it denatures the protein, the ligation step is prevented, and the complex is cleaved between the dimeric subunits. Thus "cleavable complexes" are trapped (normally the enzyme would religate DNA and dissociate). During this procedure the cell contents are disrupted, and proteolytic degradation of the attached enzyme is likely and must be prevented by the addition of protease inhibitors. The addition of some materials (e.g., EDTA) immediately prior to SDS should be avoided since they cause rapid ligation and dissociation of the enzyme.[5] There are numerous agents that trap the cleavable complexes (by formation of tertiary drug–DNA–enzyme complexes and inhibition of the DNA religation and enzyme dissociation steps) and can thus improve yield. These have to be used with caution because they alter the site specificity of cleavage and should be avoided altogether where physiological site specificity needs to be maintained.

As with all physical means of examining the genome, the recovered DNA is highly heterogeneous. With attempts to recover topoisomerase II differentiation-associated binding sites there will be sequences attached to topoisomerases from replication forks[8] and hence randomly sited in the genome, sequences from regions associated with the topoisomerase of the nuclear scaffold in undifferentiated cells, as well as the sequences of interest. A strategy is therefore needed to recognize, and preferably isolate, the sequences of interest. The differential hybridization procedure (outlined below) allows one to establish that DNA sequences of interest are being enriched by comparing hybridization signals from pairs of PEG and phosphate phases from treated and untreated cells (in our case cells induced to differentiate and appropriate controls). Caution is required in the preparation of such controls in that stressing cells can cause them to differentiate and may thus affect this differential. Sham-treated cell controls, handled exactly as test counterparts, are thus essential, and a third untreated control is desirable. Growth rate is also important because of marked fluctuations in the isoforms of topoisomerase II when log-phase growth slows at confluence.

A typical cell recovery scheme (see Fig. 1) is as follows: HL60 cells in log-phase growth are harvested by centrifugation at 400 *g* for 6 min and resuspended in serum-free culture medium. For the preparation of whole cell lysates (i.e., when using the new procedure without a prior SDS/KCl enrichment step), cells are resuspended at $5 \times 10^6$ cells/ml and lysed in the presence of protease inhibitors by the addition of aqueous SDS to a final concentration of 1%, with 1% (v/v) Triton X-100, 15 m*M*

[8] R. B. Alexander, W. G. Nelson, and D. S. Coffey, *Cancer Res.* **47,** 2403 (1987).

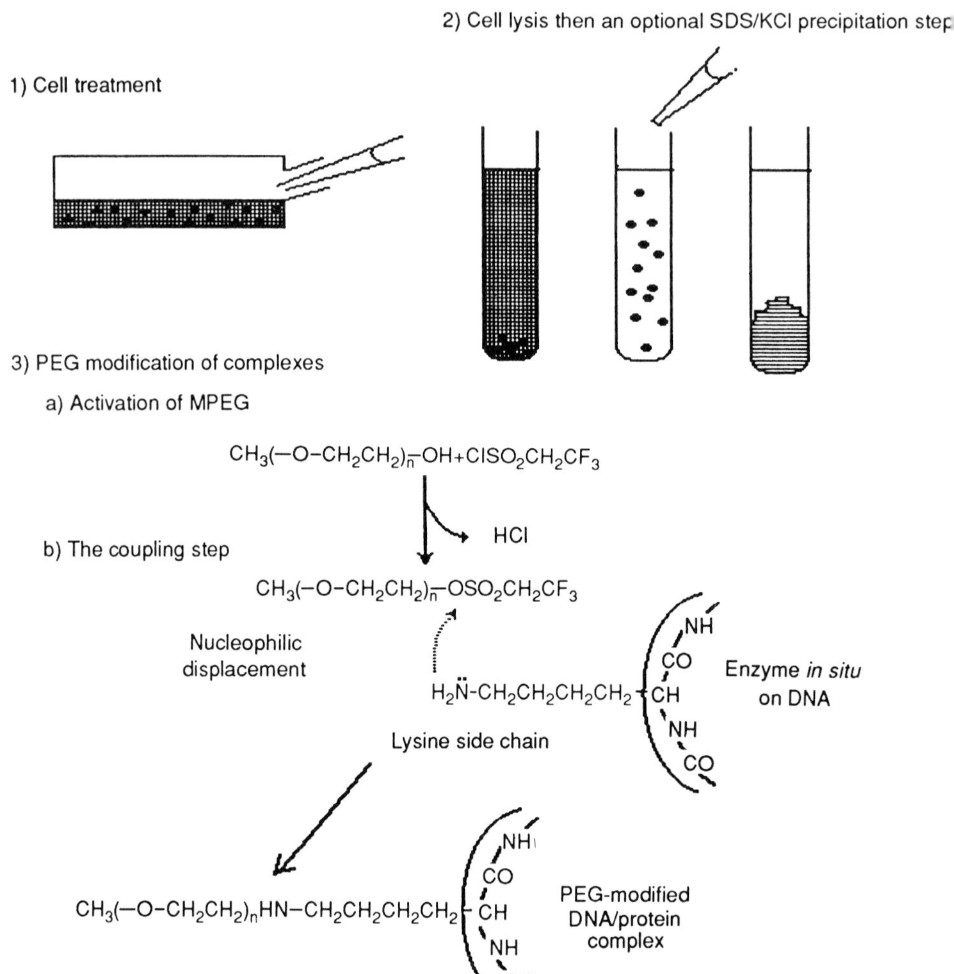


FIG. 1. Overview of the procedure. (1) Cells are first treated to initiate binding of the protein of interest (in our case a differentiation-associated binding of DNA topoisomerase II to DNA). (2) Cells are lysed with an SDS-containing mixture to denature the protein while conserving its covalent attachment to DNA (the cell lysate can be stored at this point). An optional SDS/KCl precipitation step for additional enrichment can be used at this stage. (3) The lysate or SDS/KCl-precipitated DNA–protein complex preparation is exposed to activated PEG, which couples PEG to all proteins including the enzyme. (4) The PEG-modified material is then loaded into the phase system, which is mixed and allowed to separate. Additional extractions with fresh phosphate phases may be used to improve separation, and then (5) DNA is recovered from the phases by proteinase K digestion followed by conventional extraction and desalting methods. Modifications of the basic procedure (not shown) include the addition of a cross-linking step to examine noncovalently bound proteins or the exposure of cells to drugs that modulate cleavable complex formation.

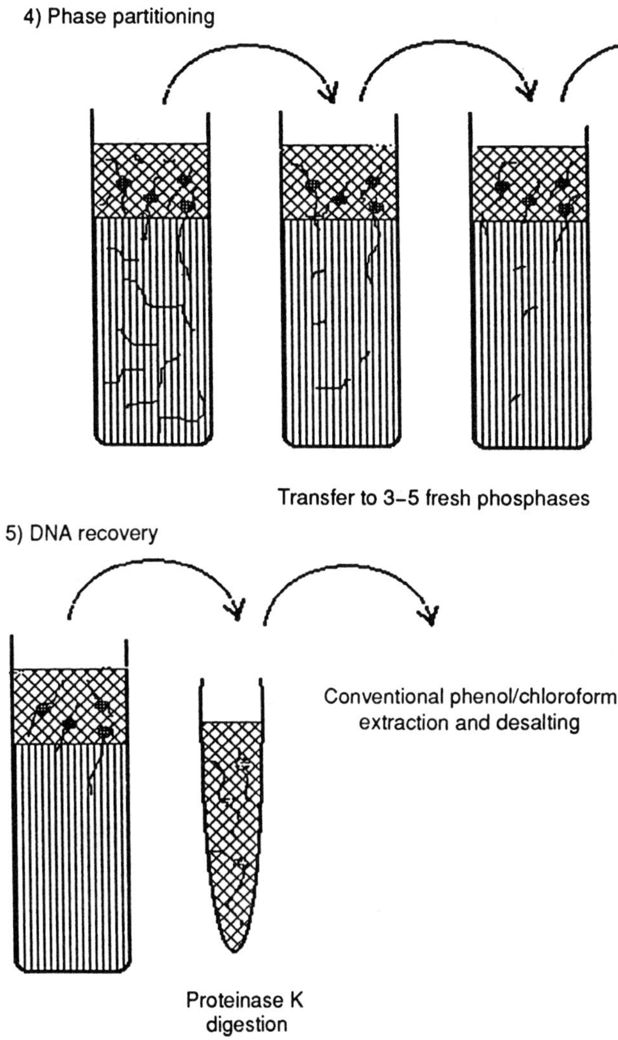


FIG. 1. (*Continued*)

EDTA, and 1 m*M* phenylmethylsulfonyl fluoride (PMSF). The cell lysate is vortexed, aliquoted, and stored at −70°.

Cell treatments known to induce topoisomerase-mediated DNA breakage/reunion reactions are applied using independent information to establish optimum timing (e.g., nucleoid sedimentation or alkaline elution experiments). In the case of hemopoietic cell differentiation induction, cells are pretreated for 70 min with either $10^{-6}$ *M* all-*trans*-retinoic acid or 2 ×

$10^{-8}$ $M$ phorbol 12-myristate 13-acetate (PMA). This timing is selected because protein-associated DNA breaks induced by retinoic acid during differentiation are present after this time interval.[4] Controls are sham-treated with the relevant diluent.

To establish that the sequences of interest are due to attachment to a particular protein some further controls are required. For topoisomerase II, to induce increased recovery of complexes (and compare that with the recovery of the DNA sequences of specific interest), HL60 cells are exposed to VP16-213 (etoposide, Sigma, St. Louis, MO), $10^{-9}$ to $10^{-5}$ $M$ for 15 min, prior to lysis (camptothecin, also from Sigma, can be substituted when topoisomerase I is studied). To reduce topoisomerase II complex formation, HL60 cells are preincubated with novobiocin (Sigma) or merbarone (N.C.I.) for 60 min at 37° under standard culture conditions. The dose required has to be established for each cell line since topoisomerase inhibitors can induce apoptosis. Should this occur, recovery of DNA will be influenced. Agarose gel electrophoresis of DNA samples should therefore be used to monitor the presence of apoptosis-induced nucleosomal "laddering."[9]

### *Preparation of DNA–Protein Complexes for Poly(ethylene Glycol) Modification*

In the phase system the fate of an individual complex depends on three forces (i.e., the relative affinities of the DNA, protein, and PEG for the two phases). The length of the DNA fragments should be short enough to allow a single attached protein, when PEG-modified, to carry it to the PEG-rich phase. DNA length has to be established for each preparation empirically since the $n$ in $PEG_n$–protein–DNA is not known. If recovery of very short strands of DNA at the break site will suffice, they can be obtained with a combination of sonication and DNase digestion (where the attached enzyme protects a short region of DNA). The DNase step can also follow PEG modification of the complexes, and, in this case, the DNA attached is somewhat longer.[6] The system described here can, using either approach, recover DNA to which a single molecule of topoisomerase II is attached.

Sonication and DNase treatment of DNA–protein complexes are performed as follows: aliquots (500 $\mu$l) of SDS/KCl DNA–protein precipitates or whole cell extracts are sonicated on ice, taking care to avoid frothing, for 5–60 sec at 12 $\mu$m peak to peak, 20 KHz, 150 W, in an ultrasonic disruptor (M.S.E. Instruments, Fisons Scientific Equipment, Loughborough, U.K.). For the preparation of short fragment DNA–protein com-

[9] S. J. Martin, S. V. Lennon, A. M. Bonham, and T. G. Cotter, *J. Immunol.* **145,** 1859 (1990).

plexes from which probes for hybridization are prepared, protein-associated DNA purified by the method of Trask *et al.*,[5] equivalent to 5 $\mu$g DNA, is exposed to 2.5 $\mu$g bovine pancreatic DNase I for a period of 5 min in a 250-$\mu$l reaction mixture [50 m*M* Tris-HCl, pH 8.0, 10% (v/v) glycerol, 0.1% (w/v) bovine serum albumin (BSA), 25 m*M* $MgCl_2$, 1 m*M* $CaCl_2$, and 20 m*M* KCl]. DNase digestion should be monitored in a representative sample, to determine that the incubation time and amount of DNase result in the appearance of a band of low molecular weight "protected" DNA (see Ref. 6).

*Attachment of Poly(ethylene Glycol) to Enzymes in Situ on DNA*

Coupling of PEG to proteins can be achieved by appropriate derivatization of the hydroxyl groups of PEG with a reagent that can be substituted by nucleophilic groups in the protein (the ε-amino groups of lysine side chains and terminal $NH_2$), and either few or many sites can be modified.[7,10] A PEG coupling method suitably mild for delicate proteins like cytokines and based on the procedure of Nilsson and Mosbach[11] for coupling enzymes and affinity ligands to solid phases bearing hydroxyl groups, using tresyl chloride (2,2,2-trifluoroethanesulfonyl chloride)-activated monomethoxy-PEG, was used. By activating monomethoxy-PEG (M-PEG) at its single free derivatizable hydroxyl group, the resultant tresylated M-PEG (TM-PEG) gives a reactive species capable of linking to only one protein molecule, thus eliminating the possibility of cross-linking.

The coupling of M-PEG to antibodies, albumin, and granulocyte–macrophage colony-stimulating factor (GM-CSF)[12–14] can occur under mild conditions (pH 7.5 phosphate buffer, room temperature). The coupling step is rapid and simple, leaves the protein–DNA complex intact, and fails to bind PEG directly to the DNA. We have not applied alternative PEG coupling methods to the recovery of DNA–protein complexes, but there appears to be no reason, *a priori,* why they should not be substituted (and this simplifies the method since several activated PEG derivatives are now available from Sigma Chemical Co., Poole, Dorset, UK). The important points to establish if substituting

[10] J. M. Harris, *Rev. Macromol. Chem. Phys.* **C25,** 325 (1985).

[11] K. Nilsson and K. Mosbach, *Biochem. Biophys. Res. Commun.* **102,** 449 (1981).

[12] C. Delgado, G. E. Francis, and D. Fisher, *in* "Separations Using Aqueous Phase Systems: Applications in Cell Biology and Biotechnology" (D. Fisher and I. A. Sutherland, eds.), p. 211. Plenum, New York and London, 1989.

[13] C. Delgado, J. N. Patel, G. E. Francis, and D. Fisher, *Biotechnol. Appl. Biochem.* **12,** 119 (1990).

[14] F. Malik, C. Delgado, C. Knusli, A. E. Irvine, D. Fisher, and G. E. Francis, *Exp. Hematol.* (*N.Y.*) **20,** 1028 (1992).

alternatives are that neither the activated PEG nor the coproducts formed in the coupling reaction damage the DNA–protein complex. Procedures should also be evaluated to establish that they do not pose problems in subsequent steps (such as phase partitioning, DNA recovery, DNA cloning).

Activation of M-PEG with tresyl chloride is performed as described elsewhere[13] using dried monomethoxy-PEG (Union Carbide Europe, Versoix, Switzerland) ($M_r$ 5000, 18 g, 3.5 mmol) dissolved in dry dichloromethane (45 ml) to which 1.125 ml (14 mmol) of pyridine and 1 ml (9 mmol) of ice-cold tresyl chloride are added dropwise. The reaction is allowed to continue at room temperature with constant stirring for 1.5 hr before the dichloromethane is removed by evaporation under reduced pressure, leaving the TM-PEG. Repeated precipitation from methanol–HCl at $-20°$ is used to remove pyridine,[13] and the TM-PEG is stored desiccated at 4°.

PEG modification of DNA–protein complexes is achieved by adding TM-PEG, 400 mg/ml in 50 m*M* sodium phosphate, 0.125 *M* NaCl buffer, pH 7.5, to the cell lysate (containing unpurified SDS-entrapped complexes) or SDS/KCl-precipitated DNA–protein complexes (i.e., complexes depleted of bulk DNA) at a volume ratio of 1 : 1 on a rotating mixer for 2 hr at room temperature. Because we do not know the total number of lysine residues on topoisomerase II and the other proteins present, TM-PEG is added in excess (i.e., in sufficient quantity to modify all proteins present). Experiments with albumin indicate maximum partitioning to the PEG phase with a TM-PEG : lysine molar ratio of 8 : 1, and significantly increased partitioning is achieved even at much lower values. The ratios of DNA–protein complexes to TM-PEG used therefore represent a gross excess of TM-PEG (0.8 mg of TM-PEG per microgram of protein), but this may have additional advantageous consequences, because of the effects of PEG concentration on DNA conformation.[15,16] The transition to the "psi" form will facilitate dissociation of noncovalently bound proteins, although many, but not all, will have been dissociated by prior exposure to 1% SDS in the cell lysate preparation or SDS/KCl procedure.[5] Although the presence of $\psi$ DNA has not been demonstrated in our procedure, the PEG-phase yield of DNA obtained experimentally[6] is consistent with the bulk of the histones and other noncovalently bound chromatin proteins having dissociated.

TM-PEG-treated material may contain unreacted TM-PEG capable of coupling to enzymes used in subsequent processing of the DNA. Lysine

[15] T. Maniatis, J. H. Venable, Jr., and L. S. Lerman, *J. Mol. Biol.* **84,** 37 (1974).

[16] C. F. Jordan, L. S. Lerman, and J. H. Venable, *Nature* (*London*) *New Biol.* **236,** 67 (1972).

or albumin can be used to prevent undesired reaction of free TM-PEG after the coupling step.[12–14] In practice, proteinase K is robust and, provided overnight incubations at 37° are used, is little affected by the addition of TM-PEG-treated material. When DNase is to be used after TM-PEG exposure, 1 *M* free base lysine dissolved in coupling buffer (1 volume) is added to the reaction mixture (6 volumes), and the latter is incubated at room temperature for an additional 1 hr.

*Phase Partitioning*

The phase system is constructed using 10% (w/w) PEG 6000 (BDH, Poole, UK), 14% phosphate (ratio of 0.124 mol $KH_2PO_4$ and 0.176 mol $K_2HPO_4 \cdot 3H_2O$, and 76% distilled deionized sterile water. The system is mixed and allowed to settle at 25°, the two phases are then separated, and the phosphate-rich lower phase is filtered through a 0.22-$\mu$m filter and the PEG-rich upper phase through a 0.45-$\mu$m filter. Both are aliquoted and stored at −20°.

The volume of the phases should be varied appropriate to favor extraction, generally using PEG : phosphate phase volume ratios of less than 1 : 1 to maximize the extraction of free DNA. This ratio needs to be reversed when the intention is to produce DNA effectively depleted of DNA–protein complexes. Although the phase system gives good extraction of high molecular weight DNA into the phosphate phase, samples containing small DNA fragments will show substantial contamination of the PEG phase by DNA.[6] This problem is solved by repeatedly extracting the PEG phase with fresh phosphate phases and monitoring that all the remaining PEG-phase DNA is protein associated.[6] The latter can be readily established by comparing an aliquot of a sample which was digested by proteinase K with one which was untreated. The former sample should not give any recovery of DNA from the PEG phase. The number of extractions required depends on the size distribution of DNA and thus has to be established empirically.[6]

Phase systems composed of various volume ratios of PEG : phosphate phases (for the reasons outlined above) are set up with no more than 15% the total volume comprising the TM-PEG-treated DNA–protein extract. For experiments in which small DNA fragments are fractionated, the ratio of PEG : phosphate phases is decreased to 250 $\mu$l : 750 $\mu$l, and multiple rounds of phosphate extraction are used. The phases are mixed by briefly vortexing and allowed to settle at 25° for 10 min. For multiple phosphate extractions the PEG phase is then transferred to a fresh phosphate phase and the procedure repeated (the number of extractions depends on the DNA size distribution). Typically 4–5 rounds of phosphate extraction clear the PEG phase of all but protein-associated DNA. Similarly, where DNA depleted of topoisomerase II attachment sites is required, phosphate phases are repeatedly extracted with fresh PEG phases.

*Recovery of DNA from Phases and Quantitation*

Many commonly used methods for estimating the amount of recovered DNA are not applicable while DNA is in the phases.[17] Chemiluminescence proved problematic for the use of [$^3$H]DNA and scintillation counting. The influence of PEG on fluorescence prevented the use of Hoechst 33258 to estimate DNA fluorimetrically. Therefore DNA is recovered prior to estimation by use of overnight proteinase K digestion followed by phenol/chloroform extraction. This procedure effectively removes PEG but not phosphate, and the latter is removed using Sephadex G-50 columns (110 $\mu$l of phase extract per 1.2-ml column). DNA is recovered from the eluate by precipitation in 2 volumes of absolute ethanol at $-70°$ for 1 hr followed by centrifugation at 11,000 *g* for 10 min at 4°. This DNA is either estimated by agarose gel electrophoresis and scanning densitometry (to examine the size distribution of the loaded and recovered DNA), or, for rapid quantitation, samples are spotted onto an agarose slab, photographed on a transilluminator, and compared to a set of DNA standards. An alternative rapid method [an adaptation of a filter binding method proposed by Pettijohn[17] substituting DE-81 positively charged paper for Whatman (Maidstone, UK) glass filters] is applicable where only quantitation of recovered DNA is required.[18]

*Determining Enrichment of Specific Sequences in Poly(ethylene Glycol) Phase by Differential Hybridization*

The principle used to determine whether any specific DNA sequences are being carried to the PEG phase via their attached protein(s), or whether the proteins are randomly attached throughout the genome, is illustrated in Fig. 2. Probes containing the putative specific sequences are prepared from differentiation-induced cells (or cells exposed to any other treatment under investigation) using the two-step procedure with 60 sec of sonication and DNase digestion to shorten attached DNA between the SDS/KCl and phase partitioning steps. Maximum shortening is important to reduce contamination by adjacent DNA (which may contain repetitive sequences) because, being heterogeneous, the probes will be prone to a low signal-to-noise ratio.

[17] D. E. Pettijohn, *Eur. J. Biochem.* **3,** 25 (1967).
[18] R. Anderson, Ph.D. Thesis, Council for National Academic Awards, London, UK.

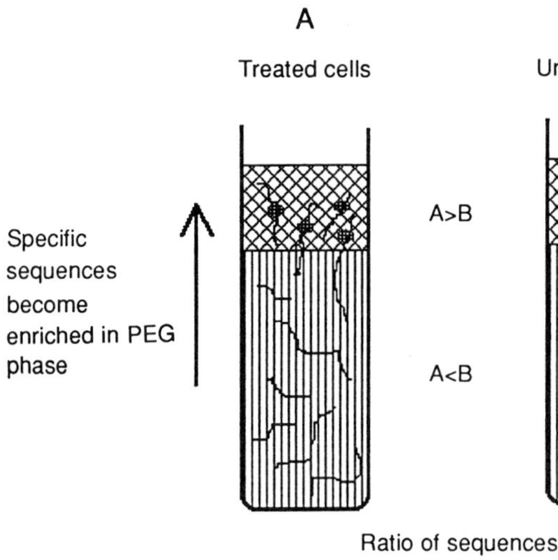


FIG. 2. Principle of the differential hybridization approach. Four samples of filter DNA are prepared (using 20 sec sonication, to give a high proportion of fragments of sizes between 0.5 and 1.0 kb). A pair of phase separations are performed for treated and untreated cells. Both PEG and phosphate phases are subjected to multiple extractions (typically 4–5 rounds) to improve any contrast in recovered DNA species between the phases. The DNA samples are recovered as described in the text and loaded onto nylon or nitrocellulose filters for hybridization. The probe DNA is either a known gene probe or very short DNA from the PEG-rich phase of treated cells. For probe preparation, the DNA of complexes is deliberately shortened prior to TM-PEG treatment and partitioning by using first 60 sec of sonication and then exposure to DNase (so that only DNA protected by the bound protein itself remains). This approach is adopted to reduce contamination of the probe with repetitive DNA. The probe is then labeled and hybridized to the filters and hybridization signals quantitated by scanning densitometry.

To examine differential hybridization with the above indicated probes, four target DNA samples are prepared per experiment: repeatedly extracted PEG phase recovered DNA from differentiating (or treated) and undifferentiated (or untreated) cells and their complementary phosphate phases (also repeatedly extracted to reduce cross-contamination). These filter samples are sonicated for 20 sec after the SDS/KCl precipitation step to produce longer fragments (this allows regions adjacent to the break sites to be examined with known gene probes and does not restrict samples to the immediate vicinity of the binding site). The filters, either nitrocellulose or nylon GeneScreen Plus (NEN Research Products, Boston, MA), are prepared by the dot-blot method of Kafatos *et al.*[19]

[19] F. C. Kafatos, C. W. Jones, and A. Efstratiadis, *Nucleic Acids Res.* **7,** 1541 (1979).

Four serial dilutions (usually 50, 25, 12.5, and 6.25 ng DNA) are loaded onto the nylon (or nitrocellulose) with a dot-blot device and allowed to stand at room temperature for at least 30 min prior to aspiration by vacuum. Filters are dried between Whatman 3MM paper and baked for 2 hr at 80° under vacuum (nitrocellulose filters). The filters are then prehybridized, for 3.5 hr with agitation, using 6× SSC (1× SSC is 0.15 *M* NaCl, 0.15 *M* sodium citrate, pH 7.0), 0.5% (w/v) SDS, 5× Denhardt's solution [1× Denhardt's solution is 0.1 g Ficoll, 0.1 g poly(vinylpyrrolidone), and 0.1 g BSA in 500 ml], and 100 $\mu$g/ml of highly sonicated salmon sperm DNA, allowing 1 ml/cm$^2$ of filter. This and all subsequent hybridization/washing steps are carried out at 68° with agitation. Hybridization fluid (6× SSC, 0.5% SDS, 5× Denhardt's solution, 100 $\mu$g/ml denatured salmon sperm DNA, and 0.1 *M* EDTA) is added, allowing 0.5 ml/cm$^2$ filter, with a probe labeled with [$\alpha^{32}$P] dCTP by the random primer method.[20] Hybridization is for 15–20 hr. Filters are washed twice for 45 min and twice for 30 min, in 1 ml/cm$^2$ 1× SSC, 1× Denhardt's solution, and 0.1% SDS (prewarmed to 68°). Finally, filters are rinsed twice for 20 min, in 1 ml/cm$^2$ of prewarmed 0.1× SSC, then blotted with Whatman 3MM paper and exposed to Fuji RX 100 X-ray film with two intensifying screens at −70°. The hybridization signal is estimated with a scanning densitometer using appropriate controls to ensure that exposures are in the linear range. It is important to perform replicate analyses on samples from different experiments since loading errors and differences in the average length of DNA will have a profound effect on signal strength. For most purposes analysis of more than 5 samples would be prudent.

## Optional Procedures and Adaptations of Basic Method

### *Use of Preliminary Sodium Dodecyl Sulfate/KCl Precipitation Step: Trask Procedure*

For many applications the phase partitioning method alone will not produce sufficient enrichment of the target DNA, and the DNA–protein complexes should be subjected to a preliminary purification step to deplete protein-free DNA. About 1 × 10$^8$ cells in log phase of growth are used, and the method is performed essentially as originally described[5] except that calf thymus DNA is omitted from the SDS buffer and vigorous vortexing is not used (so that long DNA fragments are not sheared). The yields of the Trask, new, and combined procedures are reported in Ref. 6.

[20] J. Sambrook, E. F. Fritsch, and T. Maniatis, "Molecular Cloning: A Laboratory Manual." Cold Spring Harbor Laboratory, Cold Spring Harbor, New York, 1989.

*Adapting Method as Topoisomerase Inhibitor Assay*

Many cytotoxic drugs act via inhibiting topoisomerase II in a manner that either traps it as a DNA–protein complex or prevents such complexes from forming. The method has proved to be a much more sensitive means of detecting the action of these chromotherapeutic agents than either conventional assays based on the detection of drug-induced complexes (alkaline elution, filter binding) or assays of topoisomerase enzyme activity, since this method detects etoposide at levels as low as $10^{-9}$ *M*, whereas the other methods require 1–2 orders of magnitude greater drug exposure. To use the method for the assay of topoisomerase inhibitors, DNA of standardized length, appropriately labeled for easy quantitation of recovered DNA, should be used. In practice even sonication, which produces a broad distribution of DNA fragment lengths, allows recovered DNA to serve as an index of DNA–protein complex formation and hence, of drug action,[6] but we would advise use of a standard substrate such as a plasmid.

*Studying Noncovalently Bound Proteins*

The authors have not fully explored the possibilities of studying noncovalently bound proteins, but we have demonstrated its feasibility. Our earlier studies on differentiation of hemopoietic cells using alkaline elution[4] had suggested that retinoic acid treatment, in addition to producing a small number of topoisomerase II-mediated DNA strand breaks, might also be producing an increased association of DNA with protein. We have used the present method in conjunction with UV cross-linking to examine the kinetics of this effect. A convenient means of cell exposure is to use a source emitting at 254 nm with an intensity of 5 W/cm$^2$ and to expose 1-ml aliquots of cell suspensions in 35-mm petri dishes (giving a maximum depth of 1 mm). Using this approach the temperature does not increase by more than 1° with up to 5 min of exposure. Two types of controls are necessary with this approach. First a protein-free control is needed: an identical sample treated with proteinase K (final concentration 25 $\mu$g/ml) at 56° for 1.5 hr before TM-PEG treatment. This control indicates the efficiency of the partitioning step for protein-associated DNA and should yield no PEG-phase DNA if sufficient rounds of phosphate extraction have been used. The second control (and in practice this is most important) is a sham-treatment control for each time point. It was evident that handling, per se introduces changes in DNA–protein associations in HL60 cells, and results of treatments, therefore, have to be expressed with respect to these controls.

Using the above UV exposure regimen a significant increment in PEG-phase recovery is obtained after 1 min of exposure with an approximately

3-fold increase in recovery after 5 min of exposure.[18] Using a constant UV exposure of 5 min we were able to detect an increment in PEG-phase recovery starting at 30 min after retinoic acid exposure, peaking at 70 min (about 3-fold control). This indicates that in addition to the changes in DNA topoisomerase II other, noncovalently bound proteins also change in their association with DNA after retinoic acid induction. They become so tightly associated that UV irradiation can cross-link them to DNA. In this preparation the change in PEG-phase recovery without UV cross-linking is insignificant, indicating that these proteins, unlike topoisomerases, are not covalently bound.

### *In Vitro Generation of DNA–Protein Complexes*

There are so many existing methods with which to examine DNA–protein complexes formed *in vitro* that it seems unlikely that the new method will have a major role. We have used labeled plasmid and examined the DNA protected by topoisomerase and its attached PEG on complexes formed *in vitro*,[6] but this is of little general interest and will not therefore be reviewed here. This method is very easy to scale up, however, and it should therefore be considered in applications where this is of importance. It should also be recognized that phase systems are a powerful yet simple tool for the analysis of equilibrium binding.[21] Phase systems in which high molecular weight DNA remains exclusively in one phase provide an alternative to physically restraining the target molecule with a dialysis membrane. They, in conjunction with multiple extraction procedures (e.g., countercurrent distribution), are also sensitive means for evaluating weak interactions.

## Acknowledgments

This work was supported by the Wellcome Trust. We acknowledge the excellent technical support of graduate student Robert Anderson in the development of the method.

[21] H. Walter, D. E. Brooks, and D. Fisher (eds.), "Partitioning in Aqueous Two-Phase Systems: Theory, Methods, Uses, and Applications to Biotechnology." Academic Press, Orlando, Florida, 1985.

# Section III

## Studies on Protein Interactions, Molecular Structure, Charge, Hydrophobicity, and Conformational Changes

# [20] Cross-Partitioning: Determination of Isoelectric Point by Partitioning

*By* HARRY WALTER and DANIEL FORCINITI

## Introduction

### *Usefulness*

Cross-partitioning is a simple method by which the isoelectric point of a protein, close to or identical with that obtained by electrophoretic means, can be determined.[1] The method has the additional advantage that, if a good biological or other assay is available for the specific protein of interest and the protein does not interact with contaminating materials in the solution, its isoelectric point can be established without prior protein purification.

The cross-partitioning method has also been used to determine an isoelectric point of membranes,[2] organelles,[3,4] and even cells.[5–8] In these cases cross-partitioning may, but need not, yield a value similar to the isoelectric point obtained by electrophoresis since the charge of particulates reflected by partitioning and that measured by electrophoresis can differ.[9]

### *Basic Phenomena*

Even though dextran (Dx) and poly(ethylene glycol) (PEG) are uncharged polymers, ions have unequal relative affinities for the top and bottom phases of aqueous two-phase systems made with these polymers.[10] A Donnan potential results between the phases, the relative magnitude

[1] P.-Å. Albertsson, S. Sasakawa, and H. Walter, *Nature (London)* **228,** 1329 (1970).

[2] H.-E. Åkerlund, B. Andersson, A. Persson, and P.-Å. Albertsson, *Biochim. Biophys. Acta* **552,** 238 (1979).

[3] I. Ericson, *Biochim. Biophys. Acta* **356,** 100 (1974).

[4] H. Westrin, V. P. Shanbhag, and P.-Å. Albertsson, *Biochim. Biophys. Acta* **732,** 83 (1983).

[5] H. Walter, *in* "Cell Separation: Methods and Selected Applications" (T. G. Pretlow and T. P. Pretlow, eds.), Vol. 1, p. 261. Academic Press, New York, 1982.

[6] H. Miörner, P.-Å. Albertsson, and G. Kronvall, *Infect. Immun.* **36,** 227 (1982).

[7] G. B. Ström, G. K. Blomquist, and B. Söderström, *J. Appl. Bacteriol.* **66,** 461 (1989).

[8] G. K. Blomquist and G. B. Ström, this volume [32].

[9] H. Walter, R. Tung, L. J. Jackson, and G. V. F. Seaman, *Biochem. Biophys. Res. Commun.* **48,** 565 (1972).

[10] G. Johansson, *Biochim. Biophys. Acta* **221,** 387 (1970).

and charge of which depend on the salt (or salts) chosen.[11] The partition coefficient, $K$, of the proteins (and of biomaterials in general) depends, among other things, on the salt chosen and on the charge of the protein. Thus, a negatively charged protein (i.e., one at pH values above its isoelectric point) will have higher $K$ values in $Na_2SO_4$ than in NaCl, whereas the reverse holds for a positively charged material (e.g., a protein at pH values below its isoelectric point).[1] By partitioning a protein at different pH values in two sets of phase systems, each containing one of two different neutral salts, two curves are obtained when the $K$ values with each salt are plotted against the respective pH. The point at which these curves cross is called the cross-point and is close to the isoelectric point of the protein.[1,12–14]

## Methods

### *Preparation of Proteins for Partitioning*

Proteins, whether purchased or prepared in the laboratory, should be dialyzed against distilled water before use.

### *Preparation of Phase Systems and Protein Partitioning*

Preparation of phase systems and their analysis is detailed by Brooks and Norris-Jones in this volume.[15] In short, the following stock solutions are prepared: 20% (w/w) Dx T500 (Pharmacia-LKB, Piscataway, NJ), 40% (w/w) PEG 8000 (Carbowax 8000, Union Carbide, New York, NY), a series of 40 m*M* buffers (glycine or sodium phosphate) spanning the pH range from 3.5 to 11.5, and alkali (i.e., lithium, sodium, potassium) chlorides and sulfates at four times the concentration needed in the final phase system (as indicated in the legends to the various figures).[14]

To simplify preparation of a series of phase systems having the same polymer concentrations but differing in other components, an aqueous mixture containing 14% (w/w) Dx T500 and 8.8% (w/w) PEG 8000 is first made ready. The following solutions are then weighed into 12 × 75 mm centrifuge tubes: 2 g of the well-shaken polymer mixture, 1 g of the sodium (or potassium or lithium) chloride stock solution or the analogous alkali sulfate stock solution containing between 0.5 and 1% of a given protein,

[11] R. Reitherman, S. D. Flanagan, and S. H. Barondes, *Biochim. Biophys. Acta* **297,** 193 (1973).

[12] H. Walter and S. Sasakawa, *Biochemistry* **10,** 108 (1971).

[13] S. Sasakawa and H. Walter, *Biochemistry* **11,** 2760 (1972).

[14] H. Walter, S. Sasakawa, and P.-Å. Albertsson, *Biochemistry* **11,** 3880 (1972).

[15] D. E. Brooks and R. Norris-Jones, this volume [2].

and 1 g of 40 m$M$ buffer. The components are well mixed. The final phase system composition is 7% (w/w) Dx, 4.4% (w/w) PEG, 0.1 $M$ alkali chloride *or* 50 m$M$ alkali sulfate, and 10 m$M$ glycine or sodium phosphate buffer. (The buffer concentration must be kept low so that the buffer serves only to maintain pH but its components are "drowned out" by the presence of the other salts and thus do not per se affect the $K$ value of the protein.) The phase systems are centrifuged at 1200 $g$ for 10 min to speed phase settling. The entire procedure is carried out at room temperature.

### *Determination of Protein Partition Coefficients*

To determine protein $K$ values, 0.5 ml of top phase and 0.5 ml of bottom phase are carefully pipetted from each partition tube and diluted to 2.5 ml with water. The absorbance is measured at 280 nm (nonhemoproteins) or 540 nm (hemoproteins) against a top or bottom phase blank similarly diluted. The pH is measured directly on the undiluted phases.

### *Cross-Partition Curves Obtained with Proteins*

A few representative cross-partition curves of proteins are shown in Fig. 1, and the general similarity between cross-points of a large number of different proteins[1] and their isoelectric points is shown in Table I. The cross-partition points for ovalbumin obtained by use of sodium, potassium, or lithium chlorides and sulfates have almost the same coordinates of $K$ value and pH (Fig. 2).[14] Thus the measured cross-point of a protein is virtually unaffected by the salts and salt combinations used. The slight difference in pH values observed with different salts has its counterpart in the electrophoretic determination of isoelectric points, which can also be slightly affected by the salt medium used.

## Discussion

### *Fundamentals*

The determination of isoelectric points by cross-partitioning is based on the insensitivity of the partition coefficient, $K$, of proteins at their isoelectric point to the type and concentration of salts. Away from the isoelectric pH, in contrast, the partition coefficient depends on these parameters. The observed $K$ can be considered the result of two contributions,[16] one is independent of protein charge, $K_0$, namely, the

[16] G. Johansson, *in* "Partitioning in Aqueous Two-Phase Systems: Theory, Methods, Uses, and Applications to Biotechnology" (H. Walter, D. E. Brooks, and D. Fisher, eds.), p. 161. Academic Press, Orlando, Florida, 1985.

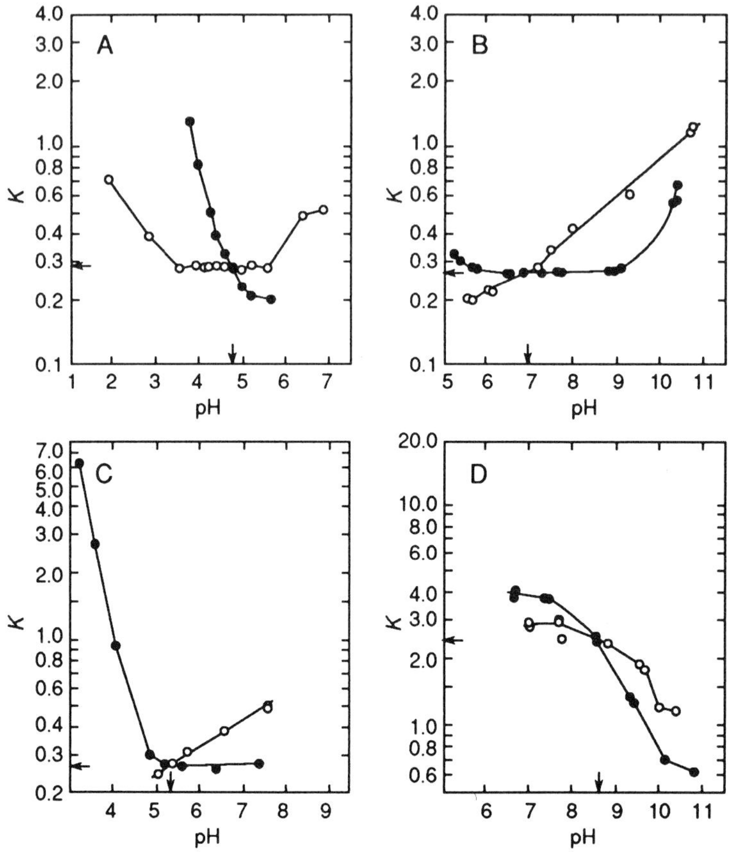


FIG. 1. Partition coefficient of proteins in two-polymer aqueous phase systems as a function of pH and two different salt compositions: (A) serum albumin, (B) hemoglobin, (C) transferrin, (D) papain. The phase system was composed of 7% (w/w) Dx, 4.4% (w/w) PEG, 10 m*M* buffer, and either 0.1 *M* NaCl (●) or 50 m*M* $Na_2SO_4$ (○). Partition coefficients (*K*) are defined as the protein absorbance at 280 nm (or 540 nm for hemoproteins) in the top phase divided by the protein absorbance in the bottom phase. [Data from Albertsson, Sasakawa, and Walter and reprinted by permission from NATURE, Vol. 228, pp. 1329–1330. Copyright © 1970 Macmillan Magazines Ltd.]

$K$ at the cross-point, and the other is protein charge-dependent, $K_z$, that is,

$$K = K_0 K_z \tag{1}$$

The $K_0$ value is a characteristic for each protein. The $K_0$ value is an indicator for the relative solvation of a protein in the two phases, that is, for the interactions between the protein and the phase-forming polymers, salts, and water. We expect $K_0$ to be a function of polymer molecular weight, type, and concentration, protein amino acid composition, protein concentration, and temperature.

The $K_0$ is generally less than 1 and decreases with increasing protein molecular weight.[13,17] Deviations from this rule may be explained by considering the association or aggregation of proteins in the phase systems. For example, in the partitioning of several nonhemoproteins, Sasakawa and Walter[13] found that $K_0$ of insulin and papain were respectively smaller and larger than expected. They suggested that insulin may be a dimer in the phase system and papain may aggregate at its isoelectric point. The same argument was used by Forciniti *et al.*[17] to explain the anomalous behavior of lysozyme in Dx–PEG systems. Hemoproteins do not follow the above rule, and their partitioning behavior is discussed below.

Forciniti *et al.*[17] partitioned five proteins at their isoelectric points in 64 different Dx–PEG phase systems to investigate the effect of polymer molecular weight and concentration on $K_0$. The $K_0$ of each protein was found to be affected differently by changes in the polymer molecular weight or concentration. Polymer concentration affects the $K_0$ value of large proteins (albumin, transferrin, and catalase) to a greater extent than that of smaller ones (lysozyme and chymotrypsinogen) (see Ref. 17 for details).

There is a small dependence of $K_0$ on the type of salt used. Light scattering experiments show that interaction coefficients between polymers and proteins depend on the type of salt present.[18] Salts may also interact differently with different proteins. The observed increase in $K_0$ with increasing amounts of charged PEG[19] supports the idea of specific interactions between ions and proteins. No general rule can, however, be given because of the uncertainty of the experimental data.

The pH at the cross-point usually agrees with the isoelectric point of the proteins (Table I). However, some discrepancies have been observed.[13] For example, trypsin gives a pH at the cross-point that differs from the electrophoretically determined isoelectric point by 1.4 pH units. Possibly, conformational changes of the protein in the phase systems or specific interactions between the polymers and protein cause the disagreement. Working at more than one polymer concentration or polymer molecular weight may help detect such dissimilarities.

[17] D. Forciniti, C. K. Hall, and M.-R. Kula, *Biotechnol. Bioeng.* **38,** 986 (1991).

[18] R. S. King, H. W. Blanch, and J. M. Prausnitz, *AIChE J.* **34,** 1585 (1988).

[19] G. Johansson, A. Hartman, and P.-Å. Albertsson, *Eur. J. Biochem.* **33,** 379 (1973).

TABLE I
CROSS-PARTITION POINTS AND ISOELECTRIC POINTS OF VARIOUS PROTEINS[a]

| Protein | pH units: Cross-point | pH units: Isoelectric point | Refs. |
|---|---|---|---|
| Albumin | | | |
| Beef serum | 4.7 | 4.71 | *b* |
| Human serum | 4.75 | 4.71 | *b* |
| Hen egg white (ovalbumin) | 4.75 | 4.6 | *c* |
| α-Amylase | | | |
| Bacterial | 4.7 | — | — |
| *Aspergillus oryzae* | — | 4.2 | *d* |
| α-Chymotrypsin | 9.0 | 8.1 | *b* |
| | | 8.6 | *e* |
| Cytochrome *c* | | | |
| Horse | 9.85 | 9.8 | *f* |
| | | 10.1 | *g* |
| | | 10.65 | *h* |
| Sheep | 9.85 | — | — |
| Hemoglobin (oxy) | | | |
| Human | | | |
| A | 7.0 | 6.87 | *i* |
| | | 7.07 | *f* |
| F | 7.1 | 7.0 | *j* |
| C | 7.5 | 7.3 | *k* |
| S | 7.2 | 7.09 | *i* |
| Horse | 6.95 | 6.92 | *l* |
| | | 6.8 | |
| Pig | 6.75 | — | — |
| Dog | 6.75 | — | — |
| Lamb | 6.9 | — | — |
| Rabbit | | | |
| Oxy | 6.95 | — | — |
| Met | 7.1 | — | — |
| Insulin (beef) | 5.4 | 5.35 | *f* |
| Lysozyme | | | |
| Hen | 11.0 | 10.5–11.0 | *m* |
| Turkey | 10.8 | — | — |
| Myoglobin | | | |
| Beef | 7.2 | — | — |
| Horse | 7.2 | 6.99 | *f* |
| Papain | 8.6 | 8.75 | *n* |
| Ribonuclease A | 7.2–9.7 | 7.8 | *o* |
| | | 8.2–9.6 | *p* |
| Transferrin (human) | 5.15 | 5.2 | *q* |
| Trypsin | 9.4 | 10.8 | *r* |

[a] Reprinted by permission from *Nature*, Vol. 228, pp. 1329–1330. 

It is widely believed that the partitioning of proteins away from the isoelectric point largely depends on the electrostatic potential difference, $\Delta\psi$, between the two phases which develops as a result of the partitioning of salts. This electrostatic potential difference can be measured following the protocol developed by Brooks et al.[20] or can be estimated by measuring the partition coefficient of a charged particle according to Eq. (2)[21]:

$$\ln K = \ln K_0 + \frac{Z\,\Delta\psi\,F}{RT} \tag{2}$$

where $Z$ is the net charge, $R$ is the gas constant, $T$ is the absolute temperature, and $F$ is Faraday's constant. A plot of ln $K$ versus the charge should be a straight line whose slope is proportional to $\Delta\psi$. However, a change in the pH may also affect $\Delta\psi$, the composition of the phases, the interaction between the ions of the salt and the protein, and the interaction between the polymers and the protein. Therefore, deviations from linear behavior are common. Still, Eq. (2) permits us to rationalize the curves ln $K$–pH as representations of the change in $K$ due to changes in the charge of the

[20] D. E. Brooks, K. A. Sharp, S. Bamberger, C. H. Tamblyn, G. V. F. Seaman, and H. Walter, *J. Colloid Interface Sci.* **102,** 1 (1984).

[21] G. Johansson, *Mol. Cell. Biochem.* **4,** 169 (1974).

---

[b] R. A. Alberty, *in* "The Proteins" (H. Neurath and K. Bailey, eds.), Vol. 1A, p. 541. Academic Press, New York, 1967.

[c] L. G. Longsworth, *Ann. N.Y. Acad. Sci.* **41,** 275 (1941).

[d] E. H. Fisher and E. A. Stein, *in* "The Enzymes" (P. D. Boyer, H. Lardy, and K. Myrbäck, eds.), Vol. 4, p. 313. Academic Press, New York, 1960.

[e] E. A. Anderson and R. A. Alberty, *J. Phys. Colloid Chem.* **52,** 1345 (1948).

[f] E. G. Young, *in* "Comprehensive Biochemistry" (S. Florkin and E. H. Stotz, eds.), Vol. 7, p. 22. Elsevier, Amsterdam, 1963.

[g] A. White, P. Handler, and E. L. Smith, "The Principles of Biochemistry," p. 131. McGraw-Hill, New York, 1964.

[h] H. Theorell, *J. Am. Chem. Soc.* **63,** 1820 (1941).

[i] L. Pauling, H. Itano, S. J. Singer, and I. C. Wells, *Science* **110,** 543 (1949).

[j] H. H. Zinsser, *Arch. Biochem. Biophys.* **38,** 195 (1952).

[k] E. C. Thomas, A. G. Motulsky, and D. H. Walters, *Am. J. Med.* **18,** 832 (1955).

[l] W. Grassmann and J. Trupke, *in* "Physiologische Chemie" (B. Flaschenträger and E. Lehnartz, eds.), Vol. 1, p. 623. Springer-Verlag, Berlin, 1951.

[m] G. Alderton, W. H. Ward, and H. L. Frevold, *J. Biol. Chem.* **157,** 43 (1945).

[n] E. L. Smith and J. R. Kimmel, *in* "The Enzymes" (P. D. Boyer, H. Lardy, and K. Myrbäck, eds.), Vol. 4, p. 137. Academic Press, New York, 1960.

[o] A. Rothen, *J. Gen. Phys.* **24,** 203 (1940).

[p] L. B. Barnett and H. B. Bull, *Arch. Biochem. Biophys.* **89,** 167 (1960).

[q] R. C. Roberts, D. G. Makey, and U. S. Seal, *J. Biol. Chem.* **241,** 4907 (1966).

[r] M. Bier and F. F. Nord, *Arch. Biochem. Biophys.* **33,** 320 (1950).

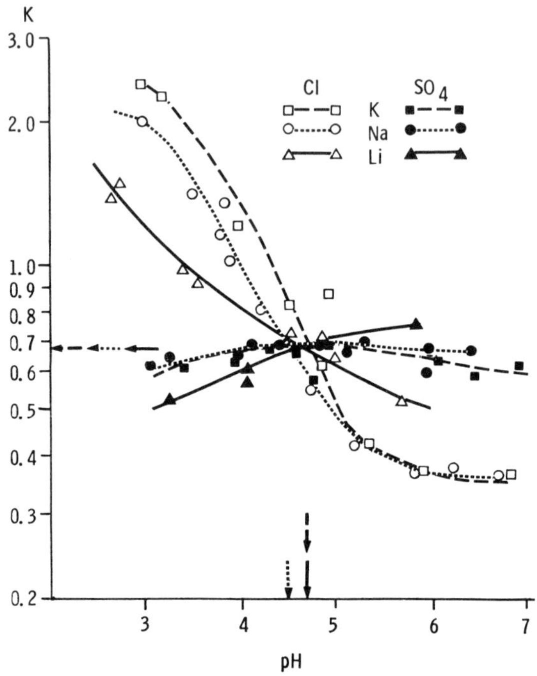


FIG. 2. Cross-partition curves of ovalbumin. Partition coefficients ($K$) in phase systems containing 7% (w/w) Dx, 4.4% (w/w) PEG, 0.1 $M$ alkali chloride or 50 m$M$ alkali sulfate, and 10 m$M$ buffer are plotted as a function of pH. Open symbols represent alkali chlorides, whereas closed symbols represent alkali sulfates. The arrows show the $K$ and pH of ovalbumin at the cross-point in each pair of corresponding alkali chloride–alkali sulfate phase systems. (Data from Walter, Sasakawa, and Albertsson and reprinted with permission from *Biochemistry,* Vol. 11, pp. 3880–3883. Copyright 1972 by the American Chemical Society.)

protein. Large discrepancies between measured and calculated electrostatic potential differences can be observed. For example, although Johansson[21] calculated that a Dx–PEG system containing KCl has a considerable potential and that the potential of a sytem containing $K_2SO_4$ is negligible, Brooks *et al.*[20] found, by direct measurements, the opposite. Nonetheless, it appears likely that specific interactions between salts and proteins in addition to the potential difference created by the salt are responsible for the effect of different salts on protein partitioning.

## Guidelines

The sensitivity of cross-partitioning depends on the angle at which the two lines intersect (Figs. 1 and 2). If the lines are perpendicular the sensitivity is at a maximum, whereas parallel lines yield no cross-point. The slope of the lines depends on the electrostatic potential difference created by the salt, the change in net charge of the protein with pH, the specific interactions between ions and proteins, and the salt-induced changes in interactions between polymer and protein. The sensitivity can be manipulated by altering the molecular weight of the polymers, the temperature, the concentration of polymers, and the type of salt and by adding ionic polymers.

Few reports on the effect of polymer molecular weight or concentration in partition experiments have been published.[17,22–24] An extensive data bank of $K$ values has been obtained at several pH values,[17,22] including that at the isoelectric point. Lysozyme, chymotrypsinogen, albumin, catalase, and human transferrin were examined in a number of Dx–PEG systems. It was found that the $K$ of proteins was more sensitive to changes in pH in two-phase systems containing low polymer molecular weights or concentrations. Thus, high polymer molecular weights should be avoided in cross-partition experiments.[22,23]

Protein partitioning is only weakly affected by changes in the temperature.[25] The $K$ increases slightly with increasing temperature in Dx–PEG systems between 4° and 40°. Because of this relative insensitivity to temperature, higher temperatures may be used to minimize protein association. Manipulation of temperature is especially attractive when determining the isoelectric point of proteins in protein mixtures.

The salts most commonly used in cross-partitioning experiments are $Na_2SO_4$ and NaCl. However, $Li_2SO_4$ and KCl have also been employed. The pH at the cross-point and the $K_0$ value are only marginally dependent on the combination of salts used (Fig. 2).[14] The concentration of salt generally has little or no influence on the cross-point pH obtained. For example, no influence of salt concentration on the cross-partitioning of trypsin, ovalbumin, horse heart cytochrome *c*, and hemoglobin has been found.[14] This independence of cross-partitioning on the type and concentration of salt makes cross-partitioning a viable option for determining the isoelectric points of proteins that are stable only at high salt concentrations. In contrast, the type and concentration of salts have a strong influence on the shape of the ln $K$–pH curves. Figure 2 illustrates the effect that different salts have on the cross-partition curves obtained with oval-

[22] D. Forciniti, C. K. Hall, and M.-R. Kula, *Chem. Eng. Sci.* **47,** 165 (1992).

[23] S. Sasakawa and H. Walter, *Biochemistry* **13,** 29 (1974).

[24] P.-Å. Albertsson, A. Cajarville, D. E. Brooks, and F. Tjerneld, *Biochim. Biophys. Acta* **926,** 87 (1987).

[25] D. Forciniti, C. K. Hall, and M.-R. Kula, *Bioseparation* **2,** 115 (1991).

bumin. The combination of KCl with either $Na_2SO_4$ or $K_2SO_4$ generates the most sensitive cross-partition plot.

PEG can be derivatized by replacing the hydroxyl groups with bromine and subsequent replacement of bromine with a charged group such as the trimethylamino group. Charged PEG has a profound effect on the partitioning of charged particles.[19] Charged PEG can partially replace salts to increase the sensitivity of cross-partitioning determinations. Blomquist[26] has used trimethylamino-PEG and PEG-sulfonate to determine the isoelectric points of the isoenzymes of enolase. The addition of salts is necessary to prevent protein association. The concentration of salt must be high enough to prevent protein association but low enough so as not to interfere appreciably in the charged polymer–protein interactions (i.e., around 5 m*M*).[19] An increase in temperature may also serve to alleviate protein associations at low ionic strengths.

## Examples and Illustrations

### *Cross-Partitioning of Ribonuclease*

When bovine pancreatic ribonuclease A is cross-partitioned in Dx–PEG systems,[13,14] a pH region of equal $K$ values rather than a cross-point is found. In addition the $K_0$ value is dependent on salt concentration. These results suggest that salt concentration-dependent conformational changes in the ribonuclease molecule affect the surface charge and/or hydrophobic–hydrophilic surface properties of the molecule.[14] This conclusion is in harmony with the wide range of isoelectric points that have been reported for ribonuclease.[1]

### *Cross-Partitioning of Hemoproteins*

Hemoproteins from cytochrome *c* (molecular weight 12,000) to catalase (molecular weight 240,000) give $K_0$ values in Dx–PEG systems that do not show the clear dependence on protein molecular weight found with nonhemoproteins.[13] Even though the molecular weights of human hemoglobin variants (A, F, S, C) and hemoglobins from different species are essentially the same, the $K_0$ values of hemoglobins A and F as well as the $K_0$ values of hemoglobins from different mammalian species show measurable differences.[12,13] Although the four human hemoglobin variants differ in charge, adult hemoglobins A, S, and C have the same $K_0$, whereas fetal hemoglobin (F) has a lower $K_0$. The partitioning behavior of hemo-

[26] G. Blomquist, *Biochim. Biophys. Acta* **420,** 81 (1976).

proteins most likely reflects subtle differences in the respective surface compositions.

*Back-Partitioning*

Another use of cross-partitioning is back-partitioning[12] (analogous to back-titration). Back-partitioning is carried out by mixing, in different ratios, the protein in the phase of the highest pH reached in a cross-partition experiment with protein-free phase of lower pH. Thus, the cross-partitioning curves are reconstructed from the highest to the lowest pH. Back-partitioning may be used to detect irreversible, pH-effectuated conformational changes. Walter and Sasakawa found that there are no irreversible changes in human hemoglobins F and A at alkaline pH values.[12] Back-partitioning has detected irreversible conformational changes in lysozyme at alkaline pH values.

*Cross-Partitioning of Amino Acids and Peptides*

Amino acids and peptides have been investigated in cross-partition experiments.[23] There appears to be a critical size below which forces other than electrostatic are responsible for partitioning behavior, since amino acids and small peptides do not yield a cross-point. Carnosine (molecular weight 226) is the smallest peptide studied that has a cross-point. For nonhemoproteins, increasing protein molecular weight is accompanied by a decreased $K_0$. This rule does not seem to apply to peptides, however, since oxytocin (molecular weight 1000) has a larger $K_0$ value than carnosine.

## Conclusions

Cross-partitioning is a simple and gentle method for the determination of the isoelectric point of a protein that, furthermore, requires no special equipment. It is particularly useful for the determination of isoelectric points of proteins that are stable only at high salt concentration because high salt concentrations obviate electrophoresis. Cross-partitioning also permits the determination of the isoelectric points of a protein in a mixture provided that a specific assay for the protein of interest exists and that components in the mixture do not interact. The analysis of the $K_0$ value may provide an indication of protein–protein interactions such as association, aggregation, or dissociation as well as protein conformational changes.

## Acknowledgment

Work in Harry Walter's laboratory was supported by the Medical Research Service of the U.S. Department of Veterans Affairs.

# [21] Charge Determination by Partitioning

*By* GÖTE JOHANSSON

## Introduction

The partitioning of a charged molecule or particle between the phases of a two-polymer biphasic system depends on the salt included in the system.[1] Partitioning such charged species in the presence of different salts permits the charge of small and large ions, including proteins, to be estimated.

## Theory

The relation between the partition coefficient, $K$, of a molecule with a net charge $Z$ and the specific salt used is given by Eq. (1):

$$\log K = \log K_0 + \gamma Z \tag{1}$$

where the constant $\gamma$ depends on the salt included in the system (at defined polymer composition and concentration and temperature). The $K_0$ value is essentially independent of salt.[2] If the sample ion has partition coefficients $K_1$ and $K_2$ with salt 1 and salt 2, respectively, the difference $\Delta \log K = \log K_1 - \log K_2$ will [from Eq. (1)] be proportional to the net charge, $Z$, of the ion. This dependence can be expressed as follows:

$$\Delta \log K = (\gamma_1 - \gamma_2)Z = \Delta\gamma Z \tag{2}$$

The $\Delta\gamma$ value can be obtained by partitioning an ion the charge of which is known (in solution). This permits the indirect determination of the net charge of the substance without knowing its molecular weight or nature.[3,4]

## Charge Determination of Low Molecular Weight Ions

The ions to be analyzed must be present at low concentrations compared with the concentration of the salt used to determine the partitioning. Usually 1–5 m$M$ of the ion to be studied is used when the concentration

[1] P.-Å. Albertsson, "Partition of Cell Particles and Macromolecules," 3rd Ed. Wiley, New York, 1986.

[2] H. Walter, S. Sasakawa, and P.-Å. Albertsson, *Biochemistry* **11,** 3880 (1972).

[3] G. Johansson, *J. Chromatogr.* **322,** 425 (1985).

[4] C. L. DeLigny and W. J. Gelsema, *Sep. Sci. Technol.* **17,** 375 (1982).

of the steering electrolyte is greater than 200 mmol/kg (1–1 electrolyte, e.g., NaCl) or over 100 mmol/kg (1–2 electrolyte, e.g., $Na_2SO_4$).

### *Experimental*

*Systems*. Two-phase systems containing 8% (w/w) dextran (Dx) T500 (Pharmacia LKB, Uppsala, Sweden) and 8% (w/w) poly(ethylene glycol) (PEG) 3400 (BP Chemicals, Hythe, UK) are prepared by mixing 8 g of 40% (w/w) PEG 3400 and 16 g of 20% (w/w) (determined by optical rotation) Dx T500 with 10 ml of either 0.5 *M* $Na_2SO_4$ or 1 *M* NaCl or 1 *M* $NaClO_4$. Water is added to give a total weight of 36 g, which leaves 4 g (i.e., 10% of the final weight of the system) for the sample. The system is stirred to permit the withdrawal of representative aliquots. Eight to nine 3.6-g systems are weighed into 10-ml centrifuge tubes as well as 0.4 g of sample ion (final concentration not higher than 5 m*M*). At least two different salts are used so that $\Delta \log K$ can be determined. The systems are mixed, equilibrated at 25° in a thermostatted water bath, and then mixed again. The phases are permitted to settle in the bath for 15–30 min, or they may be centrifuged for a few minutes. Aliquots of known volumes are withdrawn from each (i.e., top and bottom) phase[5] and analyzed for concentration of the partitioned ion of interest.

*Methods of Analysis*. Most analytical methods can be used to determine concentrations of the ion in the two phases. However, the influence on the analysis by polymers and salts in the system must be considered. Methods that have been used for small ions are photometric, acidometric, and argentometric titrations, absorbance measurements, and flame photometry.[3,4,6] Countercurrent distribution can be used to estimate the $K$ values[7] from the position of the concentration peak of the partitioned substance. The $K$ value can be determined accurately as long as the analytical method is adequate to indicate the maximum concentration.

*Results*. The determination of the net charge of a number of ions is shown in Table I. The $\Delta\gamma$ values are taken from the $\Delta \log K$ values of $H^+$ and $OH^-$. The high values for dyes indicate that they are present as dimers in solution.[3]

### *Dependence of Ion Charge on pH*

*Experimental*. To check the dependence of charge on pH, 4 g of sample is added to 36 g of initial system, and, with continuous stirring, the pH

[5] G. Johansson, this volume [3].

[6] W. J. Gelsema and C. L. DeLigny, *J. Chromatogr.* **498,** 325 (1990).

[7] H.-E. Åkerlund and P.-Å. Albertsson, this volume [8].

TABLE I
PARTITIONING OF IONS AND NEUTRAL MOLECULES[a]

| Partitioned species | $\Delta \log K = \log K_1 - \log K_2$ A | B | Calculated $Z$ A | B |
|---|---|---|---|---|
| $H^+$ | 0.05 | 0.10 | +1.0 | +1.0 |
| $OH^-$ | −0.05 | −0.10 | −1.0 | −1.0 |
| $Cl^-$ | n.d. | −0.10 | — | −1.0 |
| $Br^-$ | n.d. | −0.13 | — | −1.3 |
| $I^-$ | n.d. | −0.13 | — | −1.3 |
| $SCN^-$ | n.d. | −0.17 | — | −1.7 |
| $Cu^{2+}$ | 0.13 | 0.28 | +2.6 | +2.8 |
| $Fe^{3+}$ | 0.09 | 0.19 | +1.8 | +1.9 |
| $Sn^2+$ | 0.06 | 0.19 | +1.2 | +1.9 |
| $Fe(CN)_6{}^{3-}$ | −0.13 | −0.28 | −2.6 | −2.8 |
| $Ag^+$ | n.d. | 0.22 | — | +2.2 |
| $Ag(NH_3)_2{}^+$ | n.d. | 0.14 | — | +1.4 |
| Phthalic acid | n.d. | 0.00 | — | 0 |
| Hydrogen phthalate | n.d. | −0.08 | — | −0.8 |
| Phthalate | n.d. | −0.14 | — | −1.4 |
| Indigo disulfonate | −0.25 | −0.40 | −5.0 | −4.0 |
| Methyl red ion (−1) | n.d. | −0.17 | — | −1.7 |
| Methyl red (uncharged) | n.d. | −0.02 | — | −0.2 |

[a] Adapted from G. Johansson, *J. Chromatogr.* **322,** 425 (1985), with permission.
[b] For A, 1 is NaCl, 2 is $Na_2SO_4$; for B, 1 is $NaClO_4$, 2 is $Na_2SO_4$. n.d., Not determined.

is adjusted stepwise by addition of very small volumes (because of the low buffer capacity of the system) of 0.1–0.5 *M* HCl or NaOH. The pH is measured with a glass electrode, and samples of the phase system are withdrawn after each adjustment. The sampled two-phase systems are equilibrated in a 25° water bath, mixed, and the phases permitted to settle. The ion concentration of top and bottom phases is analyzed as above.

*Results.* The effect of pH changes on the partitioning of two charged molecules is shown in Fig. 1. The pyridinium ion (Fig. 1A) is deprotonized when the pH is increased and changes into electrically neutral pyridine. The $\Delta \log K$ at low pH is twice the expected value for one charge unit and may indicate that the pyridinium ion is present as a dimer. The chromate ion, $CrO_4{}^{2-}$, dimerizes to dichromate, $Cr_2O_7{}^{2-}$, at low pH, and this change is reflected in the $K$ of chromate with both NaCl and $Na_2SO_4$ (Fig. 1B). The $\Delta \log K$ value, 0.075–0.115, is, however, relatively constant, corresponding to $Z$ of 1.75–2.3. The transition between different forms of ionic species can, in both cases, be followed by spectral changes. Both

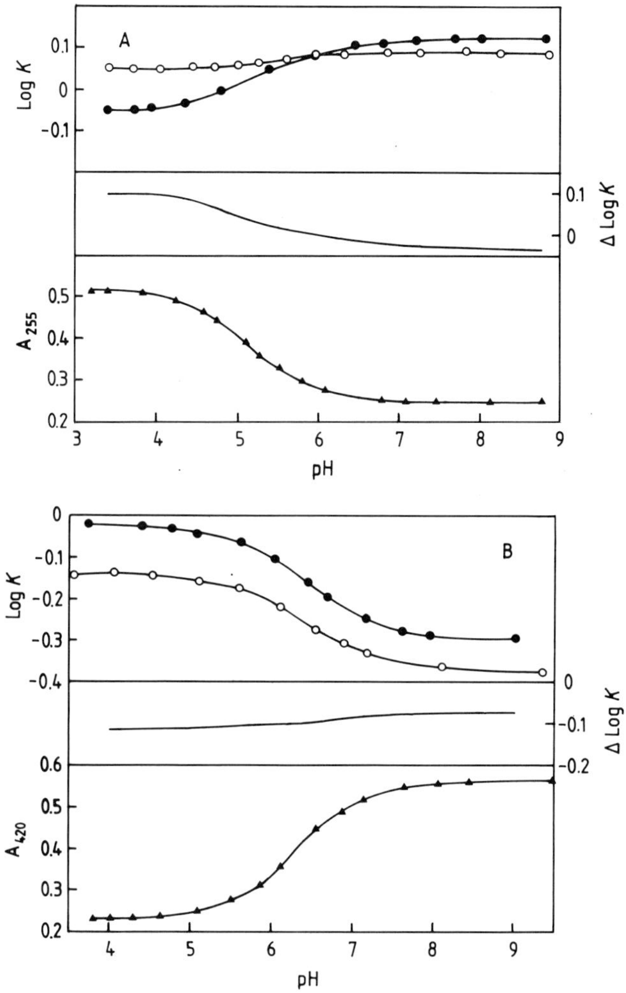


FIG. 1. Variation of log $K$ (●, ○) and Δ log $K$ (——) of (A) pyridine and (B) chromate with pH in systems containing $Na_2SO_4$ (●) or NaCl (○) and comparison with the spectral shifts (▲) with pH for (A) pyridine (0.1 m$M$) at 255 nm and (B) chromate (0.5 m$M$) at 420 nm and 25°. The system composition was 8% (w/w) dextran T500, 8% (w/w) PEG 3400, 1 mmol/kg pyridine or 5 mmol/kg potassium dichromate, and 125 mmol/kg $Na_2SO_4$ or 250 mmol/kg NaCl. Temperature, 25°.

the nonzero Δ log $K$ value for uncharged pyridine (Fig. 1A) and the variation in Δ log $K$ for chromate with pH (Fig. 1B) may be due to minor differences in $K_0$ as a function of the salt used, showing the limitations of Eq. (2).

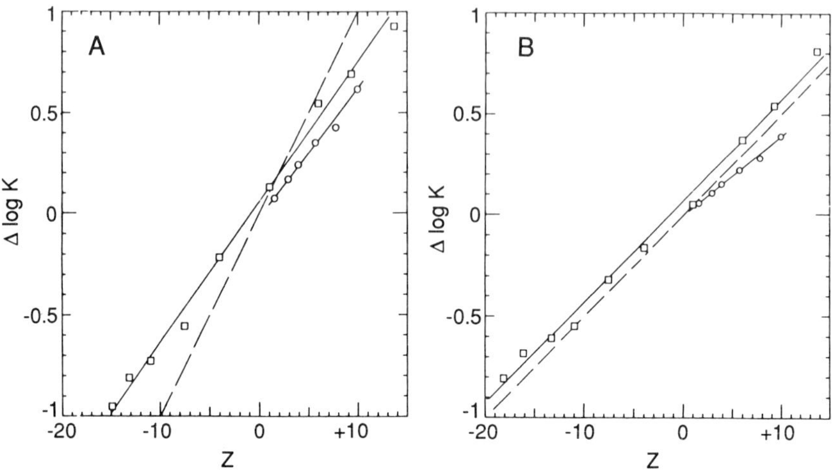


FIG. 2. Plot of Δ log *K* versus net charge, *Z* (determined from the titration curves of the proteins), for the proteins ribonuclease (○) and ovalbumin (□), for the salt pairs (A) $NaClO_4$–$Na_2SO_4$, and (B) NaCl–$Na_2SO_4$. The dashed line shows expected log *K* values calculated from *Z* and the $\gamma$ values obtained via partitioning of $H^+$. Phase systems were as in Fig. 1 but contained 5 g protein/kg. [From G. Johansson, *J. Chromatogr.* **322,** 425 (1985), with permission.]

## Charge Determination of Proteins

Because the charge on a protein changes with pH the latter must be adjusted (as described above). Proteins are partitioned in a manner similar to that described for small ions. The protein concentration can be up to 5 g per kilogram. Care should be taken when adding $H_2SO_4$ or NaOH to avoid protein denaturation. Stirring should be gentle so that the mixture does not foam. The concentration of protein, if pure, is measured at 280 nm or by staining with Coomassie Brilliant Blue according to Bradford.[8] In the case of enzymes the activity can be measured.

The Δ log *K* at different pH values establishes the relation between protein net charge, *Z*, and pH.[3] The *Z* value obtained by partitioning can be compared with the stoichiometric *Z* obtained via titration and the isoelectric point of the protein (Fig. 2). The Δ log *K* versus *Z* plot (from the titration curve), using two salts, gives a more linear dependence than when log *K* (with one salt) is plotted versus the same *Z* values. This is probably due to the observed small variations of $K_0$ with pH for proteins.[9] Proteins may interact especially, for example, with chaotropic ions such

[8] M. M. Bradford, *Anal. Biochem.* **72,** 248 (1976).

[9] G. Johansson, *Mol. Cell. Biochem.* **4,** 169 (1974).

TABLE II
$\Delta\gamma$ VALUES FOR VARIOUS SALTS IN RELATION TO SYSTEM CONTAINING SODIUM SULFATE[a]

| Salt | $\Delta\gamma$ |
|---|---|
| Tetrabutylammonium phosphate, pH 6.8 | −0.085 |
| Lithium phosphate buffer, pH 6.8 | −0.049 |
| Ammonium phosphate buffer, pH 6.8 | −0.039 |
| Sodium phosphate buffer, pH 6.8 | −0.031 |
| Potassium phosphate buffer, pH 6.8 | −0.026 |
| Tetrabutylammonium bromide | −0.025 |
| $Li_2SO_4$ | −0.015 |
| $(NH_4)_2SO_4$ | −0.010 |
| $K_2SO_4$ | 0.001 |
| LiCl | 0.024 |
| Tris-HCl | 0.026 |
| Potassium acetate | 0.041 |
| NaCl | 0.050 |
| KCl | 0.052 |
| NaBr | 0.062 |
| NaI | 0.074 |
| NaSCN | 0.080 |
| Potassium trichloroacetate | 0.088 |
| $NaClO_4$ | 0.10 |

[a] Systems contained 8% (w/w) dextran T500 and 8% (w/w) PEG 3400 at 25°. Values were determined as $\Delta\gamma = \gamma_{salt} - \gamma_{Na_2SO_4}$.

as perchlorate. This may explain the fact that the net charge, measured by $\Delta \log K$, using $NaClO_4$ as one of the salts, is lower than predicted from the titration curve.

## Optimizing Partitioning Conditions

Higher (or lower) $\Delta\gamma$ values can be obtained by using combinations of salts other than the ones mentioned above. As a guide $\Delta\gamma$ values for a number of salts in combination with $Na_2SO_4$ are presented in Table II. The $\Delta\gamma$ for any pair of salts is obtained as the difference between the two given, $Na_2SO_4$-related, $\Delta\gamma$ values. To obtain higher $\Delta\gamma$ values for a given pair of salts the concentrations of polymers are increased and/or polymers with higher molecular weights are substituted. In both cases the system is removed further from the critical point, resulting in more extreme $\gamma$ values. The $\Delta\gamma$ values are, as are the $\gamma$ values for the two salts, roughly proportional to the length of the tie line of the system.[10] This offers an easy way to estimate the increase in $\Delta \log K$ that results from raising polymer concentrations of a system.

To achieve greater $\Delta\gamma$ values charged PEG derivatives[11] can be used instead of the salts. The partitioning in a pair of systems one of which contains negatively charged PEG-sulfonate (Aqueous Affinity, Arlöv, Sweden) and the other positively charged trimethylamino-PEG (Aqueous Affinity) gives, under optimal conditions, $\Delta\gamma$ values of the order of 0.5. In this case low concentrations of proteins (<1 g/kg) and buffer (2–5 mmol/kg) must be used.[12]

## Net Charge of Biological Particles

How well the $Z$ determination works for biological particles (e.g., cells or cell organelles) is not established, but studies of the partition with two salts may give a relative net charge for different particles which are compared.[13,14] One problem is that particles partition in addition to the two bulk phases also to the interface, which complicates theoretical treatments. In addition, biological particles do not have a defined size and surface charge density, and a population of biologically "identical" particles may be inhomogeneous with regard to charge.

## Methodological Complications

Anions of the salts used to obtain the $\Delta\gamma$ value may, in some cases, interact (e.g., form complexes) with the ion to be analyzed. If so, no conclusion as to its charge can be drawn. Higher polymer concentrations may enhance ion pair formation as well as dimerization or formation of complex protein aggregates.

## Acknowledgments

I thank Monica Joelsson for excellent technical assistance. This work has been supported by grants from the Swedish Natural Science Research Council.

[10] P.-Å. Albertsson and F. Tjerneld, this volume [1].

[11] G. Johansson, this volume [5].

[12] G. Blomquist, *Biochim. Biophys. Acta* **420,** 81 (1976).

[13] H.-E. Åkerlund, B . Andersson, A. Persson, and P.-Å. Albertsson, *Biochim. Biophys. Acta* **522,** 238 (1979).

[14] H. Westrin, V. Shanbhag, and P.-Å. Albertsson, *Biochim. Biophys. Acta* **732,** 83 (1983).

# [22] Protein–Protein and Protein–Ligand Interactions

*By* SUSANNE LUNDBERG and LARS BACKMAN

Here we illustrate the use and applicability of aqueous two-phase partitioning to the study of protein–protein interactions by characterizing the association between brain spectrin and calmodulin. We also describe the use of the partitioning technique for analyzing interactions between small ligands and proteins, as exemplified by the binding of calcium ions to brain spectrin.

## Protein–Protein Interactions

Protein–protein interactions are involved in most biological processes. Complex cellular structures such as multienzyme complexes and cytoskeletons depend on high-affinity protein interactions for proper biological function. Reversible associations, though usually much weaker, are also involved in many regulatory mechanisms. The great importance of such protein interactions has led to the development of several tcchniques for defining the association in terms of binding and stoichiometry as well as dependence on different modulators. Mass transfer techniques, like sedimentation[1] and gel filtration,[2] have been widely used for quantitative studies, whereas solid-phase-based methods[3] and chemical cross-linking[4] have proved to be useful for identification of interacting proteins.

### *Protein–Protein Interactions Studied by Phase Partitioning*

A consequence of complex formation is that the surface properties (i.e., net charge, hydrophobicity, and solubility) of the complex most probably differ in some respect from those of the free proteins. It is therefore possible, at least in theory, to follow complex formation by any technique that separates molecules according to charge or hydrophobicity.

[1] J. R. Cann, "Interacting Molecules: The Theory and Practice of Their Electrophoresis, Ultracentrifugation, and Chromatography." Academic Press, New York, 1970.

[2] G. K. Ackers, *in* "The Proteins" (H. Neurath and R. L. Hill, eds.), 3rd Ed., p. 1. Academic Press, New York, 1975.

[3] M. L. Billingsley, K. R. Pennypacker, C. G. Hoover, D. J. Brigati, and R. L. Kincaid, *Proc. Natl. Acad. Sci. U.S.A.* **82,** 7585 (1985).

[4] K. Yamamoto, *Biochemistry* **28,** 5573 (1989).

One such technique is aqueous two-phase partitioning. Because the partitioning behavior is determined by the surface properties of the partitioned material, an association between two proteins should perturb the partition coefficient ($K$) of at least one, if not both, of the interacting proteins.[5] Several studies have shown that partitioning in aqueous two-phase systems is useful for detection of interacting proteins[6,7] as well as for quantification of interacting systems[8,9] (for a review, see Ref. 5).

### *Spectrin–Calmodulin Interactions*

The significance of protein interactions in regulation is probably best exemplified by calmodulin. On binding of calcium and concomitant activation, calmodulin binds to many different proteins and thereby regulates many target processes.[10] Many calmodulin-binding proteins have been identified; the myosin light chain kinase of smooth muscle[11] and erythrocyte $Ca^{2+}$,$Mg^{2+}$-ATPase[12] are two examples of several target proteins that have been characterized in detail. The cellular cytoskeleton also is an important target for calmodulin, and it has been suggested that calmodulin may regulate the cytoskeleton by binding to certain cytoskeletal proteins in a calcium-dependent manner.[13] Spectrin, a major component of actin-based cytoskeletons present in many different cell types, binds calmodulin in a cell-specific manner.[14] Erythrocyte spectrin has a relatively low affinity for calmodulin,[15] whereas brain spectrin appears to have a much higher affinity[14] and is believed to bind calmodulin *in vivo*.[13] However, there are other proteins involved in the cytoskeleton such as adducin that also have affinity for calmodulin.[16] It is therefore necessary to determine the affinities of these calmodulin-binding proteins before it is possible to identify pri-

[5] L. Backman, *in* "Partitioning in Aqueous Two-Phase Systems: Theory, Methods, Uses, and Applications in Biotechnology" (H. Walter, D. E. Brooks, and D. Fisher, eds.), p. 267. Academic Press, Orlando, Florida, 1985.

[6] J. Y. L. Chiang, *Arch. Biochem. Biophys.* **211,** 662 (1981).

[7] L. Backman and G. Johansson, *FEBS Lett.* **65,** 39 (1976).

[8] J. S. Patton, P.-Å. Albertsson, C. Erlanson, and B. Borgström, *J. Biol. Chem.* **253,** 4195 (1978).

[9] C. R. Middaugh and E. Q. Lawson, *Anal. Biochem.* **105,** 364 (1980).

[10] C. B. Klee, *in* "Molecular Aspects of Cellular Regulation" (P. Cohen and C. B. Klee, eds.), Vol. 5, p. 35. Elsevier, Amsterdam, 1988.

[11] I. C. Bagchi, Q. H. Huang, and A. R. Means, *J. Biol. Chem.* **267,** 3024 (1992).

[12] J. Krebs, T. Vorherr, P. James, E. Carafoli, T. A. Craig, and D. M. Watterson, *Adv. Exp. Med. Biol.* **269,** 163 (1990).

[13] A. S. Harris and J. S. Morrow, *Proc. Natl. Acad. Sci. U.S.A.* **87,** 3009 (1990).

[14] J. R. Glenney, Jr., P. Glenney, and K. Weber, *Eur. J. Biochem.* **144,** 529 (1982).

[15] Å. Berglund, L. Backman, and V. P. Shanbhag, *FEBS Lett.* **172,** 109 (1984).

[16] H. W. Kaiser, E. O'Keefe, and V. Bennett, *J. Cell Biol.* **109,** 557 (1989).

mary targets and to understand how calmodulin may exert its regulatory role on the cytoskeleton.

## Experimental Procedures for Studying Protein–Protein Interactions

### *Solutions*

20% (w/w) Dextran T-500: The dextran (Dx; Pharmacia LKB Biotechnology AB, Uppsala, Sweden) stock solution is prepared as described elsewhere.[17] The solution is stored in the cold.

50% (w/w) Ficoll 400: 100 g Ficoll (Pharmacia LKB) is dissolved in about 100 g water as described for Dx.[17] The optical rotation is measured, and the concentration of Ficoll in the stock is calculated using a specific optical rotation of +54.9°. The stock is diluted to 50.0% (w/w) and stored in the cold.

40% (w/w) Poly(ethylene glycol) 8000: 40 g poly(ethylene glycol) (PEG; Union Carbide, New York, NY) 8000 is dissolved in water to give 100 g solution.

200 m*M* HEPES, pH 7.0

2 *M* KCl

100 m*M* $MgCl_2$

100 m*M* $CaCl_2$

Test proteins: Brain spectrin is purified from calf brain using the procedure of Bennett *et al.*[18] Isolated calmodulin[19] is isotopically labeled by reductive [$^{14}$C]-methylation.[20]

Radiocalcium solutions: [$^{45}$Ca]-containing solutions having different concentrations are prepared by diluting a calcium chloride stock solution and adding a suitable amount of $^{45}CaCl_2$ (New England Nuclear, Stockholm, Sweden).

### *Partitioning Strategy*

In the phase system the *K* of calmodulin is defined by

$$K_{\text{Cam}}^{\text{free}} = \frac{[\text{Cam}]^{\text{upper}}}{[\text{Cam}]^{\text{lower}}} \tag{1}$$

where $[\text{Cam}]^{\text{upper}}$ and $[\text{Cam}]^{\text{lower}}$ denote the concentrations of calmodulin in the upper and lower phases, respectively. When a protein which binds

[17] D. E. Brooks and R. Norris-Jones, this volume [2].

[18] V. Bennett, A. J. Baines, and J. Davis, this series, Vol. 134, p. 55.

[19] R. Gopalakrishna and W. B. Anderson, *Biochem. Biophys. Res. Commun.* **104,** 830 (1982).

[20] D. Dottavio-Martin and J. M. Ravel, *Anal. Biochem.* **87,** 562 (1978).

calmodulin is present a certain fraction of calmodulin will be in the bound state, and the partitioning of calmodulin is therefore better described by an apparent $K$,[5] defined as

$$K_{\mathrm{Cam}}^{\mathrm{app}} = \frac{[\mathrm{Cam}]_{\mathrm{total}}^{\mathrm{upper}}}{[\mathrm{Cam}]_{\mathrm{total}}^{\mathrm{lower}}} = \frac{[\mathrm{Cam}]_{\mathrm{free}}^{\mathrm{upper}} + [\mathrm{Cam}]_{\mathrm{bound}}^{\mathrm{upper}}}{[\mathrm{Cam}]_{\mathrm{free}}^{\mathrm{lower}} + [\mathrm{Cam}]_{\mathrm{bound}}^{\mathrm{lower}}} \tag{2}$$

From Eq. (2) it follows that the partitioning of calmodulin will depend on the presence of a protein which binds calmodulin. Thus, when the concentration of such a protein is increased (at constant concentration of calmodulin) an increasing fraction of calmodulin will be bound, and the apparent $K$ will approach that of bound calmodulin. Similarly, if the concentration of the protein which binds calmodulin is kept constant, the partitioning will approach that of free calmodulin when the concentration of calmodulin is increased.

The partitioning behavior or isotherm is dependent not only on the partitioning properties of free and bound calmodulin but also on the number of binding sites and on the affinity of each site on the protein which binds calmodulin. It is possible, therefore, to extract binding characteristics (i.e., dissociation constant and stoichiometry), from the partitioning data. This usually requires fitting models describing different binding equilibria to the experimental data. An estimate of the binding may, however, be obtained directly from the partitioning isotherm since the transition in apparent partitioning corresponds approximately to the dissociation constant. If the binding occurs in only one of the phases (i.e., when one of the proteins partitions to only one phase), the data can also be analyzed according to Scatchard.[21]

To determine binding by partitioning, it is a prerequisite that the partitioning of free and bound forms of at least one of the proteins differs. It is also of advantage to have as large a difference as possible in $K$ between the two proteins, as this enhances sensitivity and facilitates data analysis. To obtain such a phase system we apply experimental design[22] (i.e., reduced factorial design). The design used includes the concentrations of Dx (5–7.5%), Ficoll (12–15%), NaCl (10–140 m$M$), and KCl (10–140 m$M$) within the defined limits. Using this approach we found an appropriate phase system after testing only about 20 different phase systems.

*Phase System*

To make up the stock phase system, a 125-ml separatory funnel is placed on a balance, and the following are weighed into it:

[21] G. Scatchard, *Ann. N.Y. Acad. Sci.* **51,** 660 (1949).

[22] G. E. P. Box, W. G. Hunter, and J. S. Hunter, "Statistics for Experimenters." Wiley, New York, 1978.

| | |
|---|---|
| 30.0 g | 50% Ficoll |
| 25.0 g | 20% Dx |
| 0.5 g | 40% PEG |
| 5.0 ml | 200 m*M* HEPES, pH 7.0 |
| 5.0 ml | 2 *M* KCl |
| 1.0 ml | 100 m*M* $MgCl_2$ |
| 1.0 ml | 100 m*M* $CaCl_2$ |

Water is added to give a final weight of 80.0 g. Note that the polymer solutions are weighed, whereas the buffer and salt solutions are measured by volume. This gives a phase system with the following composition: 18.75% Ficoll, 6.25% Dx, 0.25% PEG, 12.5 mmol/kg HEPES, pH 7.0, 125.0 mmol/kg KCl, 1.25 mmol/kg $MgCl_2$, and 1.25 mmol/kg $CaCl_2$.

The stock phase system is thoroughly mixed and left undisturbed overnight at 20° to let the phases settle. Next morning, or when a distinct interface develops, upper and lower phases are collected separately. First, the lower phase is slowly collected into a flask. When the interface approaches the stopcock, the flow is stopped to let the interface reform. The stopcock is then reopened, and the last drops of the lower phase are collected carefully, avoiding collection of the interface. The interface is removed by letting a small volume (1–2 ml) of upper phase flow slowly past the stopcock. After that the upper phase is poured (through the top of the funnel) directly into a flask. Finally, 0.4 ml of each phase is dispensed into polystyrene tubes. A large number of phase systems can easily be prepared by using a repetitive pipette (such as the Repetman from Gilson Villiers le Bel, France). The phase systems can be stored in the cold.

### *Partitioning Experiment*

For the binding studies, 0.2 ml of the spectrin and/or calmodulin (usually 0.15 ml of spectrin and 0.05 ml of [$^{14}$C]calmodulin) is added to the phase system, giving a 1-ml system. The phase systems are carefully mixed and incubated for 2 hr at 20° with end-over-end rotation. After incubation the phases are separated by centrifugation (2250 *g* for 15 min at 20°). During centrifugation it is very important to control the temperature (it should not increase more than a few degrees or the partitioning may change significantly).

To determine the partitioning of the protein, an aliquot of 0.1 ml from each phase is withdrawn and the protein concentration is measured, either by measuring the absorbance at 280 nm or by the Bradford method[23] (using appropriate blanks). The partitioning of [$^{14}$C]calmodulin is determined by

[23] M. M. Bradford, *Anal. Biochem.* **72,** 248 (1976).

scintillation counting of 0.1 ml upper and lower phases, respectively, in 1.8 ml water and 2 ml Lumagel scintillation fluid (Lumac, Landgraaf, The Netherlands).

*Analysis of Partitioning Data*

In the system used, spectrin partitions almost exclusively to the upper phase with a $K_{\text{spectrin}}$ value higher than 10, irrespective of the concentration. However, when concentrations of spectrin higher than about 0.4 $\mu M$ are used, spectrin aggregates and a precipitate could be observed in the phase system.

The $K_{\text{Cam}}^{\text{free}}$ determined for free calmodulin is 1.89 ± 0.07 and, like that of spectrin, is constant over the concentration range studied. Thus the equilibria occurring within and between the phases can be described as shown in Fig. 1.

The partitioning of calmodulin is dependent on the presence of spectrin, as Fig. 2 shows. When the concentration of spectrin is varied from about 1 n$M$ to 0.2 $\mu M$ the apparent $K$ of calmodulin increases from 1.9 to 2.7. The observed partitioning behavior of calmodulin must be caused by complex formation with spectrin. In addition, the partitioning isotherm implies that the binding strength lies somewhere in the low micromolar range.

To characterize the binding, models are fitted to the partitioning isotherm. A best fit for a single binding site is obtained with a dissociation constant ($K_d$) of 0.32 $\mu M$. When the model is extended to include a second independent calmodulin site on spectrin, regression analysis gives the same $K_d$ for the first site but indicates an exceedingly weak binding to the second site. Because inclusion of a second binding site does not improve the fit, it is apparent that the data are best described by the first model.

The binding strength determined compares favorably with earlier estimates of calmodulin binding to brain spectrin. Binding of calmodulin to brain spectrin is much stronger than that to erythrocyte spectrin,[14] and the affinity of erythrocyte spectrin for calmodulin, as determined by phase

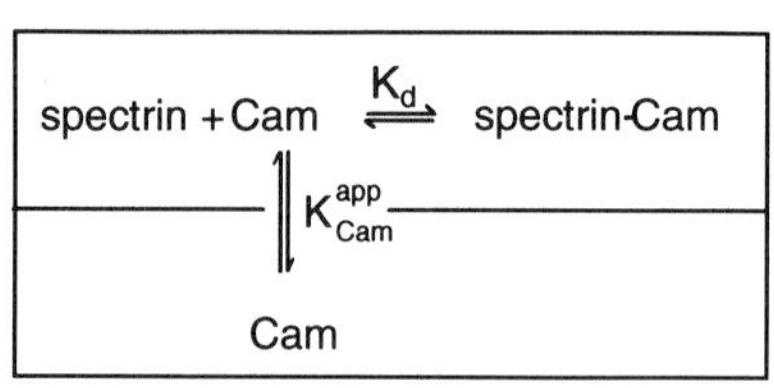


FIG. 1. Partitioning and dissociation equilibria of the interaction between calmodulin and spectrin.

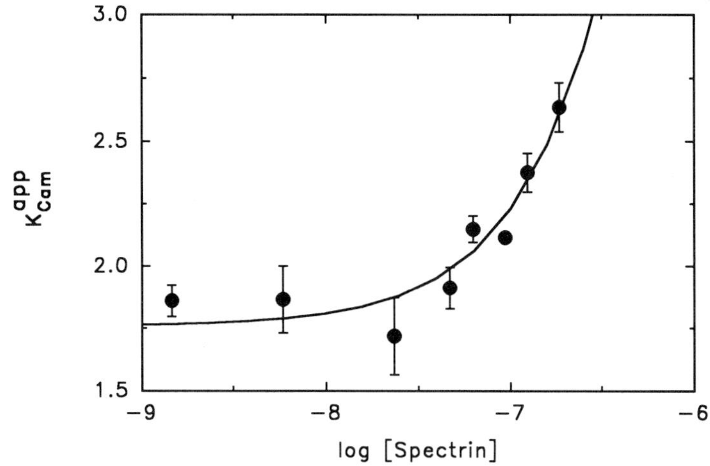


FIG. 2. Calmodulin binding to brain spectrin. The apparent partition coefficient of calmodulin (24.4 n$M$) was determined in the presence of brain spectrin at the indicted concentrations (●). The partitioning data were fitted to different models describing the binding. The solid line represents the best fit obtained with a model for one site; in the curve-fitting procedure the values obtained for the partition coefficients of spectrin ($K_{spectrin}$ = 10) and for free calmodulin ($K^{free}_{Cam}$ = 1.87) were used. It was assumed that the partitioning of all bound forms of calmodulin ($K^{bound}_{Cam}$) also was 10. The volumes of the upper and lower phases were 0.44 and 0.56 ml, respectively. A best fit was obtained with a dissociation constant of 0.32 $\mu M$.

partitioning,[15] is around 10 $\mu M$. Therefore, the value obtained for the dissociation constant appears reasonable.

Analysis of the primary structure of brain spectrin indicates that there are two plausible calmodulin-binding sequences in the molecule: one site close to the C terminus and the other in the middle of the protein.[11] It is possible that the calmodulin site in the middle of brain spectrin, which is not present in the erythrocyte form, constitutes the major (and perhaps only) site in brain spectrin. The large difference in affinity for calmodulin observed between these homologous proteins would thus be explained.

## Protein–Ligand Interactions

Calcium ions often act as an intracellular messenger, conveying an external signal to internal targets.[24,25] Inside the cell, transmission of the calcium signal may be by direct binding to and regulation of the target

[24] R. S. Adelstein, C. B. Klee, and M. Rodbells (eds.), *Adv. Second Messenger Phosphoprotein Res.* **21,** 1 (1988).

[25] H. Rasmussen, *Sci. Am.* **261,** 44 (1989).

protein. The signal may also be transmitted indirectly by binding to and concomitant activation of a regulatory protein which, in turn, regulates the target process.

To identify and characterize calcium-binding proteins there are several methods available. For identification and quick screening of a large number of samples a membrane overlay assay[26] is most suitable. This method is based on immobilization of protein on a nitrocellulose or nylon membrane, followed by incubation in a medium containing radioactive calcium. Calcium-binding proteins can then be identified by autoradiography. For more quantitative studies it is necessary to use techniques that are able not only to separate but also to quantify protein-bound calcium as well as free calcium. Of such methods, flow[27,28] and equilibrium[29] dialyses have been used for characterization of calcium-binding (as well as binding of other ligands). Spectrometric methods have also been used.[30-33] Gel filtration according to Hummel–Dryer[28,34] and phase partitioning[5] are well suited for the study of binding of ligands to proteins. In addition, methods based on sedimentation as well as filtration have been specially developed for particular proteins.[35]

## Experimental Procedures for Studying Protein–Ligand Interactions

### *Solutions*

All polymer, buffer, and salt solutions are prepared as described above.

### *Partitioning Strategy*

To simplify the analysis of the partitioning data[5] we would prefer a phase system in which spectrin is effectively excluded from one of the phases. In such a phase system, the binding equilibria will only occur in

[26] K. Maruyama, T. Mikawa, and S. Ebashi, *J. Biochem. (Tokyo)* **95,** 511 (1984).
[27] S. P. Colowick and F. C. Womak, *J. Biol. Chem.* **244,** 774 (1969).
[28] J. A. Cox, M. Milos, and J. P. McManus, *J. Biol. Chem.* **265,** 6633 (1990).
[29] T. Reinard and H.-J. Jacobsen, *Anal. Biochem.* **176,** 157 (1989).
[30] W. B. Gratzer and G. H. Beaven, *Anal. Biochem.* **81,** 118 (1977).
[31] S. Linse, P. Brodin, T. Drakenberg, E. Thulin, P. Sellers, K. Elmdén, T. Grundström, and S. Forsen, *Biochemistry* **26,** 6723 (1987).
[32] L. Serrano, J. C. Diez, F. Wandosell, J. Avila, and M. Little, *Biochem. Int.* **19,** 235 (1989).
[33] H. Hong, S. C. El-Saleh, and P. Johnson, *Int. J. Biochem.* **22,** 399 (1990).
[34] J. P. Hummel and W. J. Dryer, *Biochim. Biophys. Acta* **63,** 530 (1962).
[35] K. Suzuki, Y. Minami, Y. Emori, S. Imajoh, and H. Kawasaki, *Adv. Exp. Med. Biol.* **255,** 173 (1990).

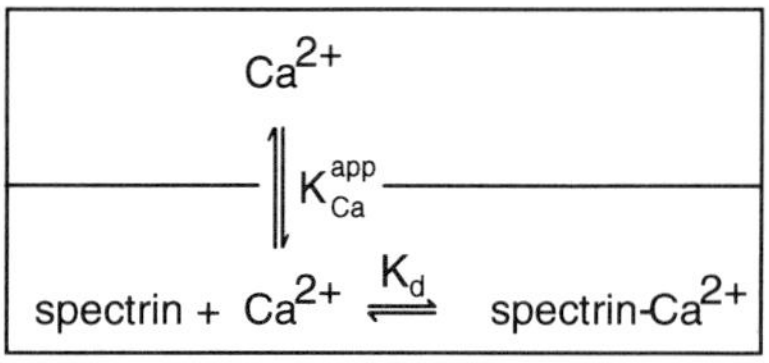


FIG. 3. Partitioning and dissociation equilibria of calcium and spectrin.

one of the phases, whereas the other phase will contain only free calcium, as depicted in Fig. 3.

To mimic intracellular conditions the calcium binding should be measured at a certain buffer and salt composition. Therefore, the partitioning of the protein can be adjusted only by varying the concentrations of the polymers. Because low molecular weight substances partition almost equally between the two phases,[36] independent of the phase composition, the desired phase system is obtained by simply adjusting the concentration of the polymers until the desired partitioning of the protein results.

The $K$ of calcium in the phase system (in the absence of protein) is defined by

$$K_{\mathrm{Ca}}^{\mathrm{free}} = \frac{[\mathrm{Ca}^{2+}]^{\mathrm{upper}}}{[\mathrm{Ca}^{2+}]^{\mathrm{lower}}} \tag{3}$$

where $[\mathrm{Ca}^{2+}]^{\mathrm{upper}}$ and $[\mathrm{Ca}^{2+}]^{\mathrm{lower}}$ denote the concentrations in the upper and lower phases, respectively. In the presence of a protein which binds calcium, a certain fraction of calcium will be bound. Therefore the partitioning of calcium is more appropriately described in terms of an apparent $K$:

$$K_{\mathrm{Ca}}^{\mathrm{app}} = \frac{[\mathrm{Ca}^{2+}]_{\mathrm{total}}^{\mathrm{upper}}}{[\mathrm{Ca}^{2+}]_{\mathrm{total}}^{\mathrm{lower}}} = \frac{[\mathrm{Ca}^{2+}]_{\mathrm{free}}^{\mathrm{upper}} + [\mathrm{Ca}^{2+}]_{\mathrm{bound}}^{\mathrm{upper}}}{[\mathrm{Ca}^{2+}]_{\mathrm{free}}^{\mathrm{lower}} + [\mathrm{Ca}^{2+}]_{\mathrm{bound}}^{\mathrm{lower}}} \tag{4}$$

In the absence of any binding protein, the apparent $K$ of calcium is equal to that of free calcium. As the concentration of the protein which binds calcium is increased (at constant concentration of calcium), an increasing fraction of calcium will be bound, and the apparent $K$ will approach that of bound calcium.

It is also possible to keep the concentration of the protein constant and vary that of calcium. In this case, when calcium is in excess, nearly

[36] G. Johansson, *in* "Partitioning in Aqueous Two-Phase Systems: Theory, Methods, Uses, and Applications in Biotechnology" (H. Walter, D. E. Brooks, and D. Fisher, eds.), p. 161. Academic Press, Orlando, Florida, 1985.

all ligand will be free and only a small fraction bound, and the apparent partitioning of calcium should be very similar to that of free calcium. When the concentration of calcium is decreased, the fraction of bound calcium increases owing to the law of mass action, and consequently the apparent partitioning of calcium is affected.

The shape of the partitioning isotherm of calcium depends not only on the partitioning properties of free and bound calcium, but also on the binding strength and the stoichiometry as well as the type of binding (i.e., cooperative or noncooperative). Therefore, the partitioning isotherm contains all pertinent data and allows the characterization of the binding equilibrium. Such an analysis requires fitting of different models describing binding equilibria to experimental data. It is, however, possible to obtain some information directly from the partitioning isotherm; for instance, the concentration at which the transition in apparent partitioning occurs corresponds approximately to the dissociation constant.

*Phase System*

To make up the stock phase system, a 125-ml separatory funnel is placed on a balance and filled with the following:

| | |
|---|---|
| 26.0 g | 50% Ficoll |
| 30.0 g | 20% Dx |
| 0.5 g | 40% PEG |
| 5.0 ml | 200 m*M* HEPES, pH 7.0 |
| 5.0 ml | 2 *M* KCl |
| 1.0 ml | 100 m*M* $MgCl_2$ |

Water is added to give a final weight of 80.0 g. Note that the polymer solutions are weighed whereas the buffer and salt solutions are measured by volume. This gives a phase system with the following composition: 16.25% Ficoll, 7.50% Dx, 0.25% PEG, 12.5 mmol/kg HEPES, pH 7.0, 125.0 mmol/kg KCl, and 1.25 mmol/kg $MgCl_2$.

The stock phase system is thoroughly mixed and left undisturbed overnight at 20° to let the phases settle. Next morning, or when a distinct interface develops, the upper and lower phases are collected separately, and 0.4 ml of each phase is dispensed into polystyrene tubes. A large number of phase systems can easily be prepared by using a repetitive pipette. The phase systems can be stored in the cold.

*Partitioning Experiment*

For the binding studies, 0.2 ml of the spectrin and/or radioactive calcium solution (usually 0.15 ml of the protein and 0.05 ml of the $^{45}Ca^{2+}$

solution) is added to the phase system, giving a 1-ml system having the following composition: 13% Ficoll, 6% Dx, 0.2% PEG, 10 m*M* HEPES, pH 7.0, 100 m*M* KCl, and 1 m*M* $MgCl_2$.

The phase systems are carefully mixed and incubated for 2 hr at 20° with end-over-end rotation. After the incubation, the phases are separated by centrifugation (2250 *g* for 15 min at 20°). During centrifugation it is very important to control the temperature (it should not increase by more than a few degrees, otherwise the partitioning may change significantly).

To determine the partitioning of the protein, aliquots of 0.1 ml from each phase are withdrawn and the protein concentration is measured either by the absorbance at 280 nm or by the Bradford method[23] (using appropriate blanks). The partitioning of $^{45}Ca^{2+}$ is determined by scintillation counting of 0.1 ml of upper and of lower phase each in 1.8 ml water and 2 ml Lumagel scintillation fluid (Lumac).

### *Partitioning Data*

To establish the partitioning behavior of free protein and free calcium, spectrin as well as $^{45}Ca^{2+}$ are partitioned separately, as described above. Irrespective of the concentration of added protein, most of the spectrin (>95%) is always recovered in the lower phase. This corresponds to a $K_{spectrin}$ of less than 0.05. Owing to the large difference in concentration between the two phases it is difficult to determine the value of $K_{spectrin}$ exactly. The partitioning of free calcium is also constant over the concentration range studied. In the phase system used the $K$ of free calcium ($K_{Ca}^{free}$) is 1.04 ± 0.01.

When calcium is partitioned together with spectrin, its partitioning is greatly affected. On the other hand, calcium does not affect the partitioning of spectrin. At constant concentrations of calcium, the apparent partition of $^{45}Ca^{2+}$ decreased when the concentration of spectrin is increased, indicating an increase in the fraction of calcium in the phase containing spectrin. When the calcium concentration is varied and that of spectrin kept constant, the apparent partitioning of calcium also follows the predicted behavior for binding equilibrium, as Fig. 4 shows. Because the transition in the partitioning isotherm occurs around 30 $\mu M$, this indicates that spectrin binds calcium with a binding strength somewhere around 30 $\mu M$.

### *Analysis of Partitioning Data*

To characterize the binding in more detail the fit of partitioning data to several models describing the partitioning behavior of different interacting systems is tested. The partitioning of spectrin is independent of calcium,

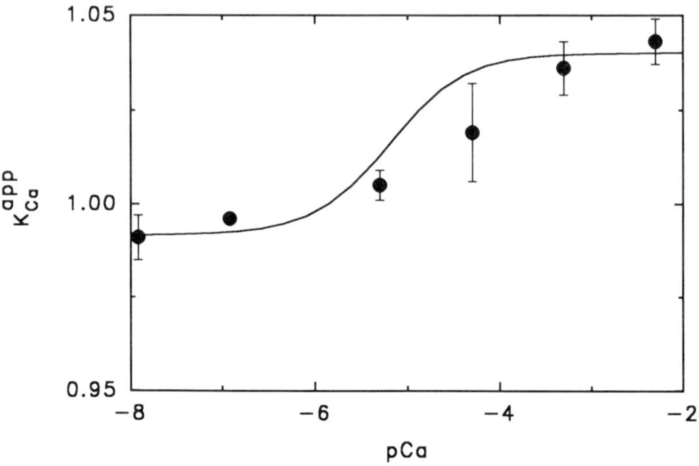


FIG. 4. Calcium binding to brain spectrin. The apparent partition coefficient of calcium was determined in the presence of 127 n*M* spectrin at the indicated total concentrations of calcium (●). The partitioning data were fitted to different models describing the binding. The solid line represents the best fit obtained with a model for two independent sites; in the curve-fitting procedure the values obtained for the partition coefficients of spectrin ($K_{spectrin}$ = 0.05) and of free calcium ($K_{Ca}^{free}$ = 1.04) were used. It was assumed that the partitioning of all bound forms of calcium ($K_{Ca}^{bound}$) also was 0.05. The volumes of the upper and lower phases were 0.42 and 0.58 ml, respectively. A best fit was obtained with a dissociation constant of 11 $\mu M$ for both sites.

so it is assumed that protein-bound calcium should partition similarly to free spectrin. With these partition coefficients (i.e., $K_{spectrin} = 0.05$, $K_{Ca}^{bound} = 0.05$, and $K_{Ca}^{free} = 1.04$), a best fit for a single binding site is obtained with a dissociation constant ($K_d$) of 6 $\mu M$. A better fit (i.e., a smaller least sum of squares) is obtained when two independent sites are allowed. In this case a $K_d$ of 11 $\mu M$ is determined for both sites. Increasing the number of binding sites improves the fit further; a model for four binding sites fits the data slightly better than the two-sites model and gives a $K_d$ of 23 $\mu M$ for all sites. Although the fit improves with increasing numbers of sites, it is not possible, from the partitioning data alone, to exclude any of the tested models. However, other experiments using synthetic peptides in conjunction with a membrane overlay assay have indicated that there are at least two binding sites for calcium in brain spectrin.[37] Therefore it seems reasonable to suggest that there are at least two sites for calcium on spectrin and that the binding strength is around 10 $\mu M$ or weaker.

[37] S. Lundberg, V.-P. Lehto, and L. Backman, *Biochemistry* **31,** 5665 (1992).

We have also analyzed the data according to Scatchard,[21] as has been done in several partitioning studies of the binding of small molecules to macromolecules (see Ref. 5). Because binding occurs only in the lower phase, the concentration of free and bound calcium in that phase could be calculated from the measured total concentration (i.e., free plus bound) in each phase, using the determined $K$ of free calcium. Although the curvilinear plot does not permit determination of binding strength and stoichiometry (too few data points), it is evident that there is more than one site for calcium on spectrin.

## Discussion

From this and many other studies in which phase partitioning in aqueous two-phase systems have been used (see Ref. 5), it is clear that the technique is useful for studying protein–protein interactions as well as protein–ligand interactions. The technique offers several advantages when compared to more traditional or classic methods, such as sedimentation velocity, gel filtration, or equilibrium and flow dialysis. Phase partitioning is a true equilibrium method in contrast to mass transfer methods such as gel filtration and flow dialysis. Equilibrium between the phases is attained in seconds compared to hours in equilibrium dialysis. A large number of samples can be handled simultaneously, and it is thus possible to collect much data rapidly. The technique does not require any sophisticated apparatus, and only ordinary labware is needed.

A major disadvantage of the technique is the fairly large amount of protein needed. It might be possible to use less protein by reducing the volume of the phase system. Alternatively, it is possible to recover the material after measurement, as the partitioning itself is nondestructive.

Depending on the purpose of the experiment, the analysis of partitioning data can be another obstacle. To prove that a binding occurs, it is sufficient to show that the partitioning of one of the components depends on the concentration of the other. The partitioning should be done in a phase system with excess of a neutral salt which, in the present case, is KCl. This is to avoid any significant influence on the interfacial potential by the partitioned components that could complicate the analysis.

The quantitative analysis is more complex because it requires fitting of defined binding models to the experimental data. That cannot be done without a dedicated computer program, and at present no such program is commercially available. The analysis can be simplified if a one-sided partitioning can be attained, because then the dissociation constant and the stoichiometry can be obtained from a Scatchard plot. It should be realized that data needed for the Scatchard plot are prone to errors because the calculation of free and bound component concentrations involves division. A small error in measurement of total concentration may lead to much larger errors in the calculated free and bound concentrations and in the precision of the determined binding characteristics.

## Acknowledgments

Research by the authors was supported by grants from the Swedish Natural Science Research Council (K-KU 1517-303) and Magn. Bergvalls Stiftelse. We thank Eva-Maj Hägglöv for excellent technical assistance.

# [23] Estimation of Surface Hydrophobicity of Proteins by Partitioning

*By* VITHALDAS P. SHANBHAG

## Introduction

The biological functions of many proteins including enzymes depend on the ability of a protein to interact with apolar groups, namely, their surface hydrophobicity or "effective hydrophobicity."[1] Hydrophobicity is also involved in stabilizing multimeric proteins or protein complexes. The surface hydrophobicity of proteins is often regulated through changes in conformation induced by binding of ions or other effectors. The determination of hydrophobicity is of importance for understanding the mechanism of interaction of proteins with small molecules or other macromolecules.

A method to determine surface hydrophobicity of proteins is to study the binding of hydrophobic molecules to them. This chapter describes the use of affinity partitioning in dextran (Dx)–poly(ethylene glycol) (PEG) aqueous two-phase systems containing nonpolar groups coupled to PEG[2–5] (cf. ch. 5 and ch. 26 this volume). Comparison of affiity partitioning of a protein in the presence of the nonpolar group with that in the presence of the same group carrying a charged (e.g., $COO^-$ or $NH_3^+$) or polar group

[1] A. Kato and S. Nakai, *Biochim. Biophys. Acta* **624,** 13 (1980).
[2] V. P. Shanbhag and C.-G. Axelsson, *Eur. J. Biochem.* **60,** 17 (1975).
[3] C.-G. Axelsson and V. P. Shanbhag, *Eur. J. Biochem.* **71,** 419 (1976).
[4] C.-G. Axelsson, *Biochim. Biophys. Acta* **533,** 34 (1978).
[5] G. Johansson and V. P. Shanbhag, *J. Chromatogr.* **284,** 63 (1984).

(e.g., OH) or unsaturated moieties permits further characterization of the binding site.[6]

## Relation between Hydrophobicity and Partition Coefficient, $K$

The theoretical basis of hydrophobic affinity partitioning has been described by Shanbhag and Axelsson.[2] In short, one measures the partition coefficient of a given protein in the absence ($K$) and presence ($K'$) of the hydrophobic ligand. The two types of $K$ are given by[2,7]

$$\log K = \log K_0 + \gamma Z \quad (1)$$

$$\log K' = \log K_0' + \gamma' Z \quad (2)$$

where $K_0$ and $K_0'$ are the charge-independent values of the partition coefficients, in the absence and presence of the hydrophobic ligand, respectively; $\gamma$ and $\gamma'$ are the corresponding values for the interfacial potential in the system; and $Z$ is the net charge of the protein molecule. At constant pH $Z$ is assumed to be constant, that is, there is no (or negligible) change in the charge of the protein caused by binding of the ligand. Further, if $\gamma$ is not affected by the presence of PEG-coupled ligand (i.e., $\gamma = \gamma'$), then the change in log $K$ to log $K'$ is given by the change in log $K_0$ to log $K_0'$. The latter change, $\Delta \log K_0$, reflects, under otherwise constant conditions of polymer concentrations, pH, and ionic strength, the affinity of the partitioned protein for the hydrophobic ligand included in the system. It is important to stress (see also section on choice and preparation of phase systems) that the concentration of the salt included should be on a molar basis at least 50 times that of the protein and of the charged ligand included in the system.[8] If the above conditions are fulfilled, then $\Delta \log K$ given by

$$\Delta \log K = \log K' - \log K \quad (3)$$

will be a measure of the affinity of the protein for the hydrophobic ligand and therefore the surface hydrophobicity of the protein in question. A plot of $\Delta \log K$ (or log $K$) versus the concentration of the PEG-coupled ligand, L–PEG, usually expressed as a fraction of total PEG in the system, will resemble a binding curve (Fig. 1).

Even though the partitioning curves (see Fig. 1) resemble a binding curve, it has not been possible to analyze the hydrophobic affinity parti-

[6] V. P. Shanbhag and G. Johansson, *Eur. J. Biochem.* **93,** 363 (1979).

[7] G. Johansson, *in* "Partitioning in Aqueous Two-Phase Systems" (H. Walter, D. E. Brooks, and D. Fisher, eds.), p. 175. Academic Press, London and Orlando, Florida, 1985.

[8] G. Johansson, *Acta Chem. Scand. Ser. B* **B28,** 873 (1974).

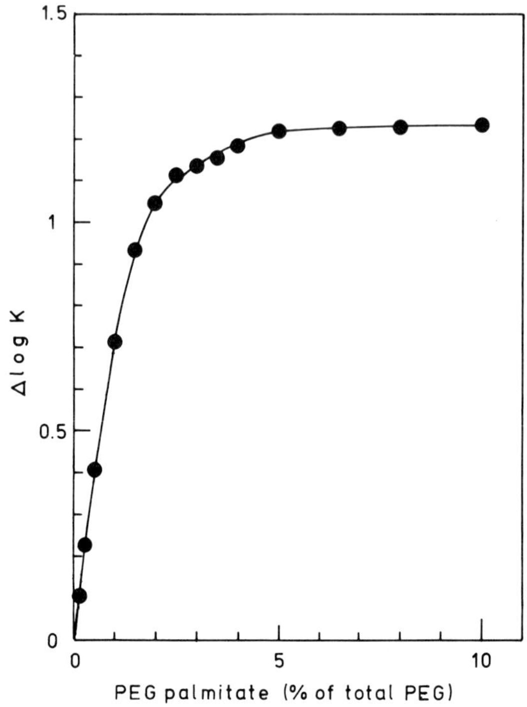


FIG. 1. Increase in Δ log $K$ of human serum albumin as a function of the concentration of PEG palmitate in the system. The system contained 4% (w/w) dextran T500, 4% (w/w) PEG 8000 (including PEG palmitate), 100 m$M$ KCl, 10 m$M$ potassium phosphate buffer, pH 7.1, and 2 g/liter albumin. Temperature, 23°. [Adapted by permission from G. Johansson and V. P. Shanbhag, *J. Chromatogr.* **284,** 63 (1984).]

tioning data in terms of affinity constants, $K_a$, or number of binding sites.[5,8] Maximum binding observed with human serum albumin has been two molecules of ligand bound per protein molecule.[5] Why partitioning does not provide the complete binding curve is discussed in Ref. 5. However, Δ log $K$ determined at the same molar ratio of protein to ligand still offers a relative measure for the affinity of a given protein for different hydrophobic ligands (or of the affinity of different proteins for the same ligand, e.g., the PEG-coupled palmitoyl group). This conclusion is borne out by the fact that Δ log $K$ obtained as described above is comparable with hydrophobicity determined by the use of other methods.[1,9]

[9] E. Keshavarz and S. Nakai, *Biochim. Biophys. Acta* **567,** 269 (1979).

## Information from Hydrophobic Affinity Partitioning

Table I shows that the affinity partitioning method is suitable for (a) obtaining a relative measure of the surface hydrophobicity of proteins; (b) studying the nature of hydrophobic sites; and (c) detecting and measuring

TABLE I
HYDROPHOBIC AFFINITY PARTITIONING

| Hydrophobic ligand | Proteins | Measured effect | Ref. |
|---|---|---|---|
| Saturated ($C_2$–$C_{18}$) and unsaturated (18 : 1) acyl groups | Human serum albumin, $\beta$-lactoglobulin, CO–hemoglobin, myoglobin, cytochrome *c* | Effect of chain length of ligand and concentration of L–PEG on $\Delta \log K$ | *a* |
| Saturated ($C_2$–$C_{18}$) and unsaturated (18 : 1, 18 : 2, 18 : 3) acyl group and benzoate | Serum albumins, lysozyme, ovalbumin, ribonuclease | Effect of chain length and type of ligand on $\log K$ | *b* |
| Saturated ($C_2$–$C_{18}$) and unsaturated (18 : 1, 18 : 2, 18 : 3) acyl group | Total histone, individual histones | Effect of chain length and type and concentration of salt on $\Delta \log K$ | *c* |
| Palmitoyl ($C_{16}$) | $\beta$-Lactoglobulin | Effect of pH on $\Delta \log K$ | *d* |
| Palmitoyl ($C_{16}$) | Human serum albumin, $\alpha$-lactalbumin | Variation of $\Delta \log K$ with polymer concentration | *e* |
| Saturated ($C_6$–$C_{18}$) acyl group and aromatic groups | Calmodulin | Effect of [$Ca^{2+}$] and chain length on $\Delta \log K$ | *f* |
| Saturated ($C_6$–$C_{18}$) acyl group | $\alpha$-Lactalbumin | Effect of [$Ca^{2+}$] and pH on $\Delta \log K$ | *g* |
| Saturated ($C_8$–$C_{18}$) acyl group | $\alpha_2$-macroglobulin | Dependence of $\Delta \log K$ on conformation | *h* |
| Saturated acyl groups ($C_6$–$C_{18}$) without and with terminal $COO^-$ group (fatty acid group) | Human serum albumin | Effect of terminal $COO^-$ group on $\Delta \log K$ | *i* |

[a] V. P. Shanbhag and C.-G. Axelsson, *Eur. J. Biochem.* **60,** 17 (1975).
[b] G. Johansson, *Biochim. Biophys. Acta* **451,** 517 (1976).
[c] C.-G. Axelsson and V. P. Shanbhag, *Eur. J. Biochem.* **71,** 419 (1976).
[d] C.-G. Axelsson, *Biochim. Biophys. Acta* **533,** 34 (1978).
[e] G. Johansson and V. P. Shanbhag, *J. Chromatogr.* **284,** 63 (1984).
[f] Å. Berglund, V. P. Shanbhag, and L. Backman, *Biochem.* (*Life Sci. Adv.*) **7,** 309 (1988).
[g] V. P. Shanbhag, G. Johansson, and A. Ortin, *Biochem. Int.* **24,** 439 (1991).
[h] G. Birkenmeier, L. Carlsson-Bostedt, V. Shanbhag, T. Kriegel, G. Kopperschläger, L. Söttrup-Jensen, and T. Stigbrand, *Eur. J. Biochem.* **183,** 239 (1989).
[i] V. P. Shanbhag and G. Johansson, *Eur. J. Biochem.* **93,** 363 (1979).

changes in surface hydrophobicity accompanying changes in protein conformation or state of aggregation. The following conclusions emanate from Table I: (1) Partitioning of some but not all proteins is affected by the presence of the PEG-coupled hydrophobic group ligand[2,10] (see also Fig. 2). (2) The value of $\Delta$ log $K$ increases with the amount of L–PEG (Fig. 1).[2,3,11,12] (3) At constant L–PEG concentration $\Delta$ log $K$ for a given protein increases with the chain length of the PEG-coupled acyl group, with maximum affinity exhibited for the palmitoyl ($C_{16}$) group. The increase differs for different proteins. All proteins studied have affinity only for acyl groups longer than eight carbon atoms[2,3,10,11] (see Fig. 2). (4) Introduction of a charged group at the free end of the acyl chain might increase (human serum albumin, HSA, for $COO^-$)[6] or decrease (in all other cases studied)[11] the affinity of the protein for the hydrophobic ligand. Similar effects have been observed when unsaturated moieties are introduced in acyl chains coupled to PEG[2,3,10] (see Fig. 2). (5) Changes in $\Delta$ log $K$ reflect changes in surface hydrophobicity arising from changes in conformation or state of aggregation.[3,4,11,13]

## Poly(ethylene Glycol)-Coupled Hydrophobic Ligands

Hydrophobic ligands have been linked to PEG via ester,[9,10,14] amide, or ether[15] bonds. The degree of substitution, $DS$, which denotes the fraction of the PEG hydroxyl groups reacted with the ligand, is usually expressed as a percentage, where $DS$ = 100% corresponds to fully substituted PEG. Because PEG contains two hydroxyl groups per molecule a $DS$ = 100% corresponds to 2 mol coupled ligand per mole PEG. Thus

$$DS = \frac{(\text{wt ligand})}{2(\text{wt L–PEG} - \text{wt ligand})} \times \frac{\text{mol wt PEG}}{\text{mol wt ligand}} \times 100 \tag{4}$$

$DS$ should be known to calculate the concentration of L–PEG and should therefore be measured and given. The $DS$ of PEG derivatives used in the work cited in Table I has ranged from 35 to 50%. Sometimes the $DS$ exceeds 50%, which indicates that a significant fraction of the PEG

[10] G. Johansson, *Biochim. Biophys. Acta* **451,** 517 (1976).

[11] Å. Berglund, V. P. Shanbhag, and L. Backman, *Biochem. (Life Sci. Adv.)* **7,** 309 (1988).

[12] V. P. Shanbhag, G. Johansson, and A. Ortin, *Biochem. Int.* **24,** 439 (1991).

[13] P. E. H. Jensen, E.-M. Hägglöf, L. F. Arbelaez, T. Stigbrand, and V. P. Shonbhag, *Biochim. Biophys. Acta.* **1164,** 152 (1993).

[14] V. P. Shanbhag and G. Johansson, *Biochem. Biophys. Res. Commun.* **61,** 1141 (1974).

[15] J. M. Harris, *J. Macromol. Sci. Res. Polym. Chem. Phys.* **C25,** 325 (1985).

molecules carry a ligand at each end. PEG derivatives with *DS* greater than 50% should be avoided, especially when studying proteins with a strong affinity for the ligand in question, so as to avoid the risk of a single PEG molecule binding to two sites on the same protein molecule or cross-linking two protein molecules.

### *Synthesis*

Esters of PEG with fatty acids or aromatic acids can be synthesized according to Shanbhag and Johansson[14] and Johansson.[10] PEG is reacted with the relevant acid chloride in dry toluene using triethylamine as catalyst. The ratio of the reactants should be about 1.2 mol acid chloride per mole PEG in order to obtain a high *DS* but one which does not exceed 50%. The PEG derivatives are purified by recrystallization from absolute ethanol, and traces of ethanol are removed by rotary evaporation under vacuum. The *DS* is determined by saponification with alkali followed by titration of the excess alkali with HCl.

The PEG-coupled hydrophobic ligands may also carry charged groups.[9] For a description of general methods for the synthesis PEG-coupled ligands, including hydrophobic ones, see Harris.[15]

## Two-Phase Systems

The phase systems used to evaluate protein hydrophobicity so far have all contained PEG 8000 (Union Carbide, New York, NY) and Dx T70 or Dx T500 (Pharmacia, Uppsala, Sweden). The choice of polymer concentrations is dictated by the ionic conditions selected, the pH, and the requirement for systems with or without an interfacial potential. Because the sensitivity of the method is directly proportional to the difference in log *K* in a system with and without the PEG-coupled ligands, it is best to work in a system in which the protein studied partitions to the lower, Dx-rich phase in the absence of ligand. A proper choice of system will be guided by reference to the chapter in this volume relating to the influence of salts, pH, and polymer concentration on the partitioning of proteins.[16] If the aim is to work in systems with charge-independent protein partitioning, then phase selection can be done according to Table II. The systems which give charge-independent protein partitioning are preferred when the change in log *K* (i.e., $\Delta$ log *K*) is measured over a wide range of pH values.[2,7]

[16] G. Johansson, this volume [3].

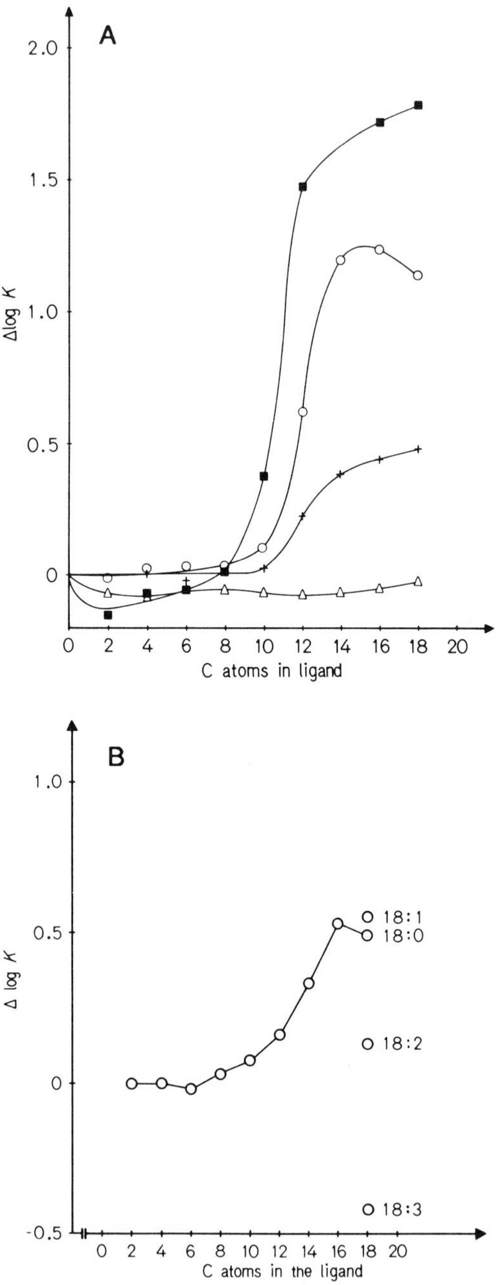
A
2.0
1.5
1.0
0.5
0
Δlog K
0 2 4 6 8 10 12 14 16 18 20
C atoms in ligand
B
1.0
0.5
0
-0.5
Δ log K
18:1
18:0
18:2
18:3
0 2 4 6 8 10 12 14 16 18 20
C atoms in the ligand

TABLE II
SYSTEMS FOR CHARGE-INDEPENDENT PROTEIN PARTITIONING

| | Dextran | | PEG | | Salt | |
|---|---|---|---|---|---|---|
| Temperature (°C) | Type | Concentration (%, w/w) | Type | Concentration (%, w/w) | Type | Concentration (m$M$) |
| 22 | T70 | 8 | 8000 | 8 | $K_2SO_4$ | 100 |
| 4 | T500 | 4 | 8000 | 4 | Sodium acetate | 100 |

The size of the system used depends on the sensitivity of the protein concentration measurement. Experimental methods using only 150–200 $\mu$l of phase system have been developed.[17] The phase systems can be made by mixing stock solutions of Dx, PEG, L–PEG, salt, buffer, and protein so that the concentrations of all components including total PEG (derivatized and underivatized PEG) is the same in all systems. The concentration of derivatized PEG is successively increased with a corresponding decrease in the concentration of the underivatized PEG. The concentration of the L–PEG in the system is usually expressed as a percentage of the total PEG and is given by

$$[\text{L–PEG}] = \frac{\text{weight of L–PEG}}{\text{total weight of PEG}} \times DS \tag{5}$$

Table III gives a protocol for making a 2-g phase system containing 8% Dx T70, 8% PEG 8000, 100 m$M$ $K_2SO_4$, 10 m$M$ sodium phosphate, pH , and 0.2 mg protein with different concentrations of L–PEG in the system.

[17] C. W. Gray and M. J. Chamberlin, *Anal. Biochem.* **41,** 83 (1971).

FIG. 2. Effect of chain length of the PEG-bound fatty acyl group on the partitioning of proteins expressed as $\Delta \log K$. (A) Partitioning of human serum albumin (■), $\beta$-lactoglobulin (O), carboxyhemoglobin (+), and cytochrome $c$ (Δ) at 23°. The system contained 7% (w/w) dextran T500, 4% (w/w) PEG 8000, 100 m$M$ $K_2SO_4$, and 2 m$M$ potassium phosphate buffer, pH 7.1. In systems with PEG-bound ligand, L–PEG constituted 10% of the total PEG. [Adapted by permission from V. P. Shanbhag and C.-G. Axelsson, *Eur. J. Biochem.* **60,** 17 (1975).] (B) Partitioning of total histone (2 g histone/liter system) in systems with and without 20% of the total PEG being replaced by PEG-bound acyl group. The system contained 6% (w/w) dextran T70, 6% (w/w) PEG 8000, 100 m$M$ KF, and 2 m$M$ potassium phosphate buffer, pH 7.1. For ligands containing 18 carbon atoms the number of double bonds is indicated. [Adapted by permission from C.-G. Axelsson and V. P. Shanbhag, *Eur. J. Biochem.* **71,** 419 (1976).]

TABLE III
PREPARATION OF TWO-PHASE SYSTEMS

| Concentration of L–PEG (%) | Dextran T70 (20%) | PEG 8000 (40%) | L–PEG (*DS* = 40%) (20%) | $K_2SO_4$ (500 m*M*) | Sodium phosphate, pH 7 (100 m*M*) | Protein[a] (2 mg/ml) |
|---|---|---|---|---|---|---|
| 0 | 0.8 | 0.400 | — | 0.4 | 0.2 | 0.2 |
| 2 | 0.8 | 0.384 | 0.016 | 0.4 | 0.2 | 0.2 |
| 5 | 0.8 | 0.368 | 0.032 | 0.4 | 0.2 | 0.2 |

[a] The protein here is dissolved in water. If the protein solution contains salt and buffer, correction has to be made for their contribution to the concentration of salt and buffer in the system, and the system is made up to the required weight (2 g above) by addition of water. For the blank the protein solution is replaced by an identical amount of the solvent used in making the protein solution.

For an unknown protein the concentration of L–PEG is varied in large intervals up to the maximum concentration possible, which is the *DS* of the L–PEG used. When the affinity of the ligand is strong, that is, when the $K'$ value reaches a plateau with only a small fraction of PEG replaced by L–PEG, it is advisable to change L–PEG in small steps in order to obtain an accurate extraction curve.[3,6,11,13]

To obtain a series of extraction curves (e.g., of different conformational states of the same protein), it is advisable to make two phase mixtures, one without and one with L–PEG. These are mixed in definite proportions and then the calculated amount of protein solution added to obtain systems with different concentrations of L–PEG. In the example given in Table IV, a 9-g phase mixture, PM I, without L–PEG is prepared by mixing all

TABLE IV
PREPARATION OF PHASE SYSTEMS FROM PHASE MIXTURES

| Concentration of L–PEG (%) | Weight (g) | |
|---|---|---|
| | PM I | PM II |
| 0 | 0.9 | — |
| 1 | 0.8 | 0.1 |
| 2 | 0.7 | 0.2 |
| 4 | 0.5 | 0.4 |
| 6 | 0.3 | 0.6 |
| 8 | 0.1 | 0.8 |

components except the protein solution in required proportions. Another phase mixture, PM II, is prepared likewise but with 9% of the PEG being replaced by L–PEG. The PM I and PM II preparations are mixed in the proportions given in Table IV to obtain 0.9 g of phase mixtures with the indicated concentration of L–PEG. Addition of 0.1 g protein solution yields systems (1 g) with different concentrations of L–PEG. The individual mixtures PM I and PM II should be thoroughly mixed while taking aliquots for making the phase systems.

## Determination of Partition Coefficients

The systems are mixed by about 40 inversions and equilibrated at the chosen temperature. They are mixed again as before and the phases separated by centrifugation at 2000 *g* for 5 min at the same temperature. Aliquots are withdrawn from each phase and diluted with water or buffer if necessary. Identically treated aliquots from the corresponding phase from a system not containing the protein are used as blanks. Absorbance at 280 nm, an enzyme-linked immunosorbent assay (ELISA), or the radioactivity of $^{3}H$-, $^{14}C$-, or $^{125}I$-labeled proteins can be used to determine the concentration of proteins in the two phases. It is necessary to dilute the phase aliquots at least 5 times for measurement of protein concentration via absorbance or ELISA. The scintillation cocktail used for measurement of radioactivity should be such that sensitivity is not affected by Dx, PEG, and L–PEG. The phases should be diluted if necessary to increase the sensitivity of measurement of radioactivity. The measurement of radioactivity of $^{125}I$-labeled proteins is not affected by the presence of Dx, PEG, and L–PEG. Even though measurement of radioactivity is a very sensitive method for determining protein concentration, it is advisable to also determine $K$ using absorbance or ELISA so as to confirm that the affinity for L–PEG is not significantly affected by isotope labeling. Determination of protein concentation in the presence of PEG-coupled hydrophobic ligands using the method of Bradford[18] or BAC (*N,N*-bisacrylylcystamine) reagent (Bio-Rad, Richmond, CA) is difficult because very high blanks are obtained.

## Concluding Remarks

The most popular method used to detect and characterize hydrophobic sites on native proteins is the binding of fluorescent probes.[19] All such

[18] M. M. Bradford, *Anal. Biochem.* **72,** 248 (1976).

[19] L. A. Sklar, B. S. Hudson, and R. D. Simoni, *Biochemistry* **16,** 5100 (1977).

probes used for studies in aqueous media have, of necessity, carried one or more polar or charged groups. Such groups could affect binding, and therefore the affinity might not reflect the true hydrophobicity (as has been pointed out for 1-anilinonaphthalene 8-sulfonate).[20] In contrast, affinity partitioning in systems with and without PEG-bound hydrocarbon chains yields a measure of the hydrophobicity of the surface of a protein and permits comparison of its different conformational states. Comparison of results obtained with ligands of different chain lengths or degrees of branching and those carrying polar or charged groups at specific points may shed additional light on the size and structure of a hydrophobic site.

[20] G. R. Penzer, *Eur. J. Biochem.* **25,** 218 (1972).

# [24] Detection of Conformational Changes in Proteins by Probing with Poly(ethylene Glycol)-Bound Ligands

*By* GERD BIRKENMEIER and GERHARD KOPPERSCHLÄGER

## Introduction

The biological function of many proteins is regulated by changes in the molecular conformation or by alterations of structural properties. Such surface alterations can be monitored by partitioning in aqueous two-phase systems. This method has proved to be sensitive in the detection of protein association–dissociation phenomena,[1] protein–protein interactions,[2] as well as of pH- or ligand-induced structural changes in proteins.[3,4]

In general, a conformation transition gives rise to an altered partitioning of a protein between the two phases if the surface area involved is large enough. When the structural change is restricted only to a small surface area exposed to the solvent, no changes in the partition coefficient, $K$, are to be expected unless such subtle changes are recognized by a specific ligand. We have used ligands bound to poly(ethylene glycol) (PEG) to detect and magnify, by affinity partitioning, conformational changes in human $\alpha_2$-macroglobulin and yeast phosphofructokinase. Use of such af-

[1] C. R. Middaugh and E. Q. Lawson, *Anal. Biochem.* **105,** 364 (1980).

[2] L. Backmann, *Eur. J. Biochem.* **120,** 257 (1981).

[3] K. Ichikawa, K. Hashizume, T. Miyamoto, Y. Nishii, K. Yamauchi, H. Ohtsuka, and T. Yamada, *J. Endocrinol.* **119,** 431 (1988).

[4] J. C. Hansen and J. Gorski, *J. Biol. Chem.* **261,** 13990 (1986).

finity ligands can yield considerable information on protein surface properties and structural dynamics.

## Materials and Methods

PEG 6000 and 5,5′-dithiobis(2-nitrobenzoic acid) (DTNB) are purchased from Serva (Heidelberg, Germany). Dextran (Dx) T70 is from Pharmacia (Uppsala, Sweden). Reactive dyes are obtained from I.C.I. Organic Divisions (Blackley, UK). The fatty acid ester of PEG is synthesized as described by Johansson.[5] The degree of substitution for palmitate is 41%, where 100% corresponds to substitution of 2 mol OH groups per mole PEG. The dye–PEG derivatives are prepared according to Johansson.[6]

### *Preparation of Two-Phase Systems for Partitioning of $\alpha_2$-Macroglobulin*

Two-phase systems of 3 g are prepared from stock solutions of PEG 6000 (40%, w/w) and Dx T70 (20%, w/w).[7] The polymer solutions are mixed with buffer, water, and sample to get the final concentrations indicated in Figs. 1–6. Hydrophobic affinity partitioning is carried out by replacing parts of the PEG by palmitate-PEG. The systems are equilibrated and mixed at 22°. Phase separation is speeded by brief centrifugation at 5000 *g* for 20 sec.

### *Preparation of Methylamine- and $\alpha$-Chymotrypsin-Treated $\alpha_2$-Macroglobulin*

$\alpha_2$-Macroglobulin from fresh blood is purified according to van Leuven *et al.*[8] Methylamine-treated $\alpha_2$-macroglobulin is prepared by incubation of native $\alpha_2$-macroglobulin (60 $\mu M$) with 0.2 $M$ methylamine (Aldrich, Steinheim, Germany) for 2 hr at room temperature and dialysis against 0.1 $M$ sodium phosphate buffer, pH 8.0. Complexation of $\alpha_2$-macroglobulin with $\alpha$-chymotrypsin (Sigma, St. Louis, MO) is performed by incubation of 60 $\mu M$ $\alpha_2$-macroglobulin with increasing concentrations of active $\alpha$-chymotrypsin (up to 4-fold molar excess) in 0.1 $M$ sodium phosphate buffer, pH 8.0, for 2 min. The reaction is stopped by the addition of 2 m$M$ phenylmethylsulfonyl fluoride (PMSF) (Serva) dissolved in dioxane.

[5] G. Johansson, *Biochim. Biophys. Acta* **451,** 517 (1976).
[6] G. Johansson, this series, Vol. 104, p. 356.
[7] D. E. Brooks and R. Norris-Jones, this volume [2].
[8] F. van Leuven, J.-J. Cassiman, and H. van den Berghe, *J. Biol. Chem.* **256,** 9016 (1981).

For preparation of $^{125}$I-labeled $\alpha$-chymotrypsin–$\alpha_2$-macroglobulin complexes, the incubation medium ($\alpha_2$-macroglobulin plus 4-fold molar excess of $\alpha$-chymotrypsin) is spiked with $^{125}$I-labeled $\alpha$-chymotrypsin. Complexed and free proteinase are separated by gel filtration on BioGel A-0.5 (Bio-Rad, Richmond, CA) in phosphate buffer (0.1 *M*, pH 8.0). The proteinase–inhibitor complex is eluted as the first peak. $\alpha$-Chymotrypsin is iodinated according to Hunter and Greenwood[9] with 0.5 mCi of $^{125}$I (Amersham Buchler, Braunschweig, Germany) using the Iodo-Beads technique (Pierce, Rockford, IL).

### *Determination of Partition Coefficient of $\alpha_2$-Macroglobulin*

The concentration of $\alpha_2$-macroglobulin in the top and the bottom phase is determined immunologically by rocket electroimmunoassay essentially as described in Ref. 10.

### *Preparation of Two-Phase Systems for Partitioning Phosphofructokinase*

Phosphofructokinse is prepared as described elsewhere.[11] Before use the enzyme is exhaustively dialyzed against the respective buffer as given in Figs. 5 and 6, followed by incubation with fructose 6-phosphate for 1 hr. Two-phase systems of 4 g are prepared from stock solutions of PEG 6000, Dx T70, sodium phosphate buffer, and different additives by weight as indicated in Figs. 5 and 6. The amount of dye–PEG given in percent refers to the total PEG present in the system which contained 5% (w/w) PEG, 7% (w/w) Dx, 50 m*M* sodium phosphate buffer, pH 7.0, and 5 m*M* 2-mercaptoethanol. The system is brought to 0°, and, after addition of 5 to 10 $\mu$l of the enzyme solution which correspond to approximately 1 unit, the tubes are gently shaken for about 15 sec and kept at 0° for 30 min in most cases. Then the systems are mixed once again, and phase separation is speeded by centrifugation (1500 *g*, 1 min at 4°).

For studying the time dependence of the affinity partitioning of phosphofructokinase, the systems are mixed after addition of the enzyme only for 10 sec followed by a short centrifugation at 0°. Exactly after 1 min, samples from the top and bottom phases are withdrawn for activity assays. At time intervals of 1 min the two-phase systems are mixed again, separated, and analyzed as described above.

[9] N. M. Hunter and F. C. Greenwood, *Nature* (*London*) **194,** 495 (1962).

[10] G. Birkenmeier, this volume [14].

[11] E. Hofmann and G. Kopperschläger, this series, Vol. 90, p. 49.

## Results and Discussion

### $\alpha_2$-*Macroglobulin*

*Conformational Changes.* Human $\alpha_2$-macroglobulin is an inhibitor of proteinases from all subclasses.[12] The unique steric mode by which $\alpha_2$-macroglobulin binds to proteinases is due to a sequence of several molecular events. It starts with the limited proteolysis of the "bait" region, which is followed by a cleavage of an internal $\beta$-cysteinyl-$\gamma$-glutamyl thiolester.[13–15] Activation of the internal thiolester leads to the appearance of nascent sulfhydryl groups. This triggers the labilization of the native structure of $\alpha_2$-macroglobulin. The resulting conformational change causes the proteinase to be trapped and leads to the exposure of previously hidden recognition sites for the receptor on several cell types.[16] An analogous, if not identical, conformational change as seen with proteinases is brought about by thiolester cleavage alone using a potent nucleophile like methylamine.[12]

*Partitioning.* The conformational transition can be detected by partitioning in aqueous two-phase systems using palmitate-PEG as a surface probe. $\alpha_2$-Macroglobulin was partitioned in a system in which increasing quantities of PEG were replaced by palmitate-PEG. The two-phase system was composed of 5% PEG 6000, 7.5% Dx T70, 10 m*M* sodium phosphate buffer, pH 7.0, and 0.1 *M* $Na_2SO_4$. The resulting change in the partitioning of $\alpha_2$-macroglobulin expressed as a $\Delta \log K$ value was taken as a measure of surface hydrophobicity.[17] Figure 1 indicates that the effect of palmitate-PEG on partitioning of native $\alpha_2$-macroglobulin was very small. The $\Delta \log K$ value obtained at saturation was only 0.2 units. This indicates that native $\alpha_2$-macroglobulin displays a low tendency for hydrophobic interaction. To verify that our sample was in the native state, we measured the free sulfhydryl groups in the protein by titration with DTNB. Figure 2 reveals that 0.5 of 4 mol sulfhydryl groups were already exposed in the native protein. This partial transition, which occurred during the isolation process, is most probably responsible for the small effect that palmitate-PEG has on native $\alpha_2$-macroglobulin. It suggests that native $\alpha_2$-macroglob-

[12] A. J. Barrett, this series, Vol. 80, p. 737.

[13] A. J. Barrett and P. N. Starkey, *Biochem. J.* **133,** 709 (1973).

[14] P. E. H. Jensen and L. Sottrup-Jensen, *J. Biol. Chem.* **261,** 15863 (1986).

[15] P. Gettings, A. H. Beth, and L. W. Cunningham, *Biochemistry* **27,** 2905 (1988).

[16] S. V. Pizzo and S. L. Gonias, *in* "The Receptors" (P. M. Conn, ed.), p. 177. Academic Press, New York, 1984.

[17] V. P. Shanbhag and C.-G. Axelsson, *Eur. J. Biol.* **93,** 363 (1975).

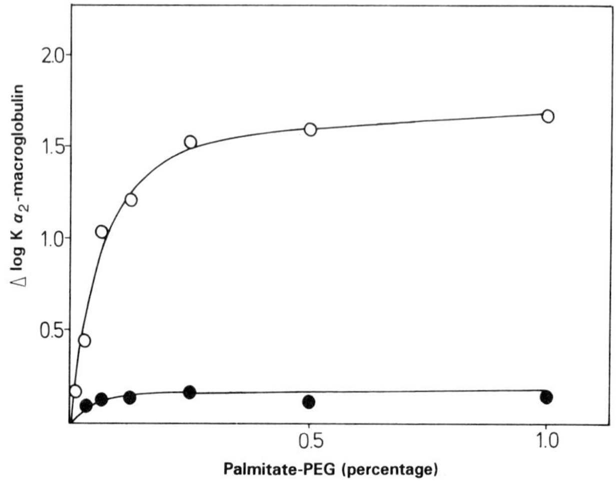


FIG. 1. Effect of palmitate-PEG on partitioning of native and methylamine-treated $\alpha_2$-macroglobulin. The two-phase systems (3 g) were composed of 5% PEG 6000, 7.5% dextran T70, 10 m$M$ sodium phosphate buffer, pH 7.0, 0.1 $M$ $Na_2SO_4$, protein, and increasing quantities of palmitate-PEG. The concentration of the PEG ester is expressed as a percentage of the total PEG in the system. (●) Native $\alpha_2$-macroglobulin (0.16 $\mu M$); (○) methylamine-treated $\alpha_2$-macroglobulin (0.16 $\mu M$).

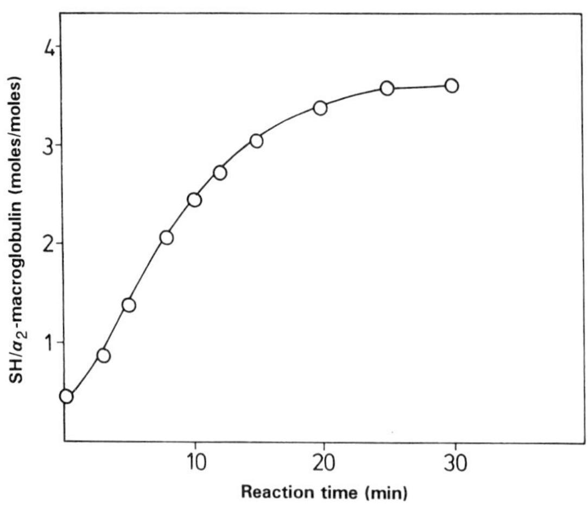


FIG. 2. Titration of thiolester cleavage in $\alpha_2$-macroglobulin during reaction with methylamine. Approximately 2 $\mu M$ native $\alpha_2$-macroglobulin was incubated with 0.24 m$M$ DTNB in 0.1 $M$ sodium phosphate buffer, pH 8.0, at 22°. After the initial absorbance at 412 nm was read, methylamine (0.2 $M$) was added and the absorbance change recorded with time against a blank containing the same solution without protein.

ulin with intact thiolesters displays negligible affinity for the hydrophobic ligand.

The low hydrophobicity of native $\alpha_2$-macroglobulin reflected by this study stands, to some extent, in contradiction to higher $\Delta \log K$ values reported.[18] The reason for this is most probably the direct iodination of $\alpha_2$-macroglobulin, which may partially alter the native protein.

Treatment of $\alpha_2$-macroglobulin with methylamine is known to cause cleavage of the thiolesters without "bait" site cleavage, and it results in increased electrophoretic mobility[19] and changes in the TNS [6-(*p*-toluidino)-2-naphthalenesulfonic acid] fluorescence,[20] antigenicity,[21] and several other physical properties.[22] This major conformational change is also characterized by exposure of hydrophobic sites on the protein surface (Fig. 1). Already at low palmitate-PEG concentration the methylamine-treated $\alpha_2$-macroglobulin was effectively transferred to the top phase. Thus, thiolester cleavage causes the transition of a hydrophilic protein into a largely hydrophobic form.

*Time-Dependent Events.* It is well known that the extent of conformational change is triggered by the rate of thiolester cleavage. The question was posed as to whether the cleavage and the exposure of hydrophobic sites are sequential events. We therefore compared the rate of sulfhydryl release with the partitioning behavior of $\alpha_2$-macroglobulin during treatment of native protein with methylamine. Figure 2 shows that 3.7 mol of SH groups per mole protein were released within 30 min. The reaction rate constant calculated from the curve by a first-order plot was $1.4 \times 10^{-3}$ $\text{sec}^{-1}$, which is in accord with data from the literature.[23]

The time dependence of the change in hydrophobicity of $\alpha_2$-macroglobulin during reaction with the nucleophile is depicted in Fig. 3. Native $\alpha_2$-macroglobulin was incubated with methylamine, and samples were withdrawn at certain intervals and immediately partitioned in two-phase systems in the absence and the presence of palmitate-PEG. As seen from the increase in the log $K$ value, $\alpha_2$-macroglobulin becomes increasingly hydrophobic with time of incubation. This change in hydrophobicity accompanying the change in conformation is also reflected in a partitioning

[18] G. Birkenmeier, L. Carlsson-Bostedt, V. P. Shanbhag, T. Kriegel, G. Kopperschläger, L. Sottrup-Jensen, and T. Stigbrand, *Eur. J. Biochem.* **183,** 239 (1989).

[19] F. van Leuven, P. Marynen, J.-J. Cassiman, and H. van den Berghe, *Biochem. J.* **203,** 405 (1982).

[20] D. K. Strickland and P. Bhattacharya, *Biochemistry* **23,** 3115 (1984).

[21] F. van Leuven, P. Marynen, J.-J. Cassiman, and H. van den Berghe, *J. Biol. Chem.* **261,** 6933 (1986).

[22] L. Sottrup-Jensen, *J. Biol. Chem.* **264,** 11539 (1989).

[23] L. J. Larsson and I. Björk, *Biochemistry* **23,** 2802 (1984).

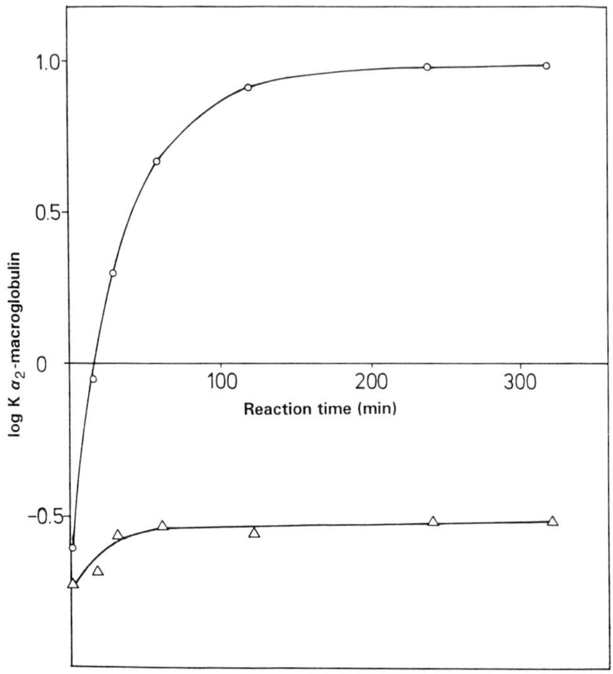


FIG. 3. Time course of transition of $\alpha_2$-macroglobulin by methylamine using palmitate-PEG as a surface probe. Two-phase systems were composed of 5% PEG 6000, 7.5% dextran T70, 10 m*M* sodium phosphate buffer, pH 7.0, and 0.1 *M* $Na_2SO_4$. Approximately 60 $\mu M$ $\alpha_2$-macroglobulin was incubated with 0.2 *M* methylamine in 0.1 *M* sodium phosphate buffer, pH 8.0, at 22°. At certain time intervals 25-$\mu$l aliquots were removed and added to the two-phase systems. After thorough mixing and brief centrifugation, 0.5 ml of each phase was removed and analyzed for protein. Partitioning of $\alpha_2$-macroglobulin in the absence (△) and in the presence (○) of 0.5% palmitate-PEG is shown.

change in the absence of palmitate-PEG. This was to be expected since hydrophobic proteins are known to have higher affinity to PEG than to Dx.[24] However, large magnification of this effect could be achieved by use of palmitate-PEG as a hydrophobic probe. In contrast to data in Fig. 2, the change in the hydrophobicity occurred on a longer time scale as compared to sulfhydryl group release. This is consistent with the significantly lower reaction rate constant of $0.17 \times 10^{-3}$ $sec^{-1}$ calculated from the partitioning curve and indicates that exposure of hydrophobic sites results from a conformational change initiated by thiolester cleavage.

[24] P.-Å. Albertsson, "Partition of Cell Particles and Macromolecules," 3rd ed. Wiley (Interscience), New York, 1986.

*Saturation with α-Chymotrypsin.* The reaction of $\alpha_2$-macroglobulin with proteinases ends in its conformational change, which is similar to that induced by methylamine as indicated by different physical and chemical methods. The inhibitor can accommodate and inhibit two molecules of small proteinases like trypsin or α-chymotrypsin.[22] Accordingly, we could observe maximum hydrophobicity of $\alpha_2$-macroglobulin under conditions in which about 2 molecules of α-chymotrypsin reacted with the inhibitor (Fig. 4). Similar to the effect of methylamine, complexation with proteinases also causes changes in the $K$ of $\alpha_2$-macroglobulin in systems without palmitate-PEG. The differences in the $K$ of free α-chymotrypsin and of α-chymotrypsin in the complex indicate that the proteinase itself has no affinity for the hydrophobic ligand and just follows the partitioning of $\alpha_2$-macroglobulin in two-phase systems with or without palmitate-PEG.

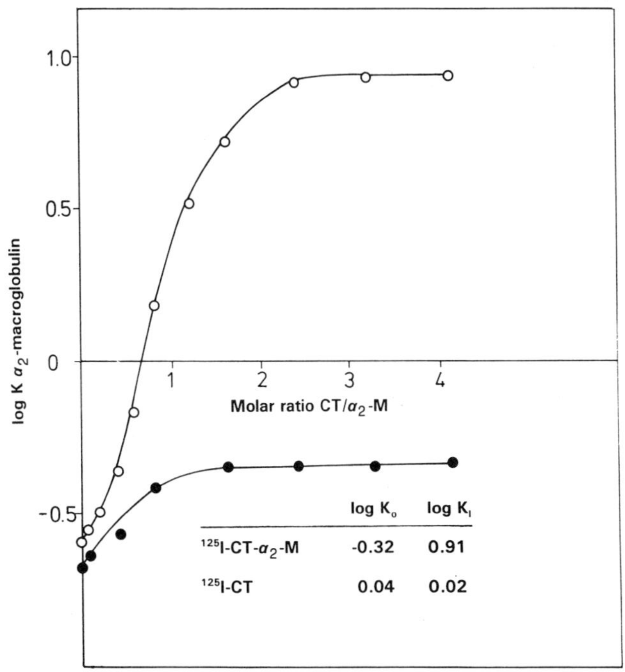


FIG. 4. Change in the partitioning of $\alpha_2$-macroglobulin in response to complexation with α-chymotrypsin. Two-phase systems used were as in Fig. 3. $\alpha_2$-Macroglobulin was incubated with increasing amounts of active chymotrypsin (CT) (see Materials and Methods). Aliquots (25 μl) were removed and added to two-phase systems containing no (●) or 0.5% (○) palmitate-PEG. After mixing and phase separation, the partitioning of $\alpha_2$-macroglobulin was determined. The partition coefficients of free $^{125}$I-labeled α-chymotrypsin and of $^{125}$I-labeled α-chymotrypsin–$\alpha_2$-macroglobulin complex in the absence (log $K_0$) and presence (log $K_1$) of the palmitate ligand are shown in the inset.

Thus, α-chymotrypsin does not contribute to the partitioning of the complex,[25] which supports the assumption that the proteinase is completely buried in the two cavities of the inhibitor.[13]

*Conclusions*. The results demonstrate that the differences in hydrophobicity parallel the different receptor-binding properties of native and transformed $\alpha_2$-macroglobulin.[16] Because the two forms of $\alpha_2$-macroglobulin are known to bind different proteins including cytokines,[26] we suggest that the observed "hydrophobic switch" which is triggered by a conformational change might well be involved in distinct biological functions of $\alpha_2$-macroglobulin.

*Phosphofructokinase*

*Affinity Partitioning with Dye-Poly(ethylene Glycol)*. Phosphofructokinase from yeast and other sources interacts with Cibacron Blue F3G-A and related dyes, as revealed by kinetic and chromatographic techniques.[27] Affinity partitioning in aqueous two-phase systems led to a 1000-fold increase of the $K$ value of the enzyme from baker's yeast when a dye–ligand with strong liganding properties was added to the system at saturating concentration.[28] This enormous response, caused by the dye–PEG complex formation, seemed to be a good prerequisite for detecting even small alterations of the enzyme–dye interaction caused, for example, by distinct changes in the dye-binding domain of the enzyme.

Phosphofructokinase from yeast and other sources is known to have two classes of binding sites for ATP: substrate and regulatory. The latter causes enzyme inhibition at higher concentrations of ATP.

*Allosteric Properties and Conformational Changes*. The structural flexibility of yeast phosphofructokinase was found by kinetic, physical, and chemical studies.[29–31] Owing to the allosteric properties of the enzyme, fructose 6-phosphate is capable of stabilizing a conformation in which

[25] P.-Å. Albertsson, *Methods Biochem. Anal.* **29,** 1 (1983).

[26] C. T. Chu and S. V. Pizzo, *Immunol. Today* **12,** 249 (1991).

[27] G. Kopperschläger, H.-J. Böhme, and E. Hofmann, *in* "Advances in Biochemical Engineering" (A. Fiechter, ed.), p. 101. Springer-Verlag, Berlin, Heidelberg, and New York, 1982.

[28] G. Kopperschläger, this volume [11].

[29] R. Reuter, K. Eschrich, W. Schellenberger, and E. Hofmann, *Acta Biol. Med. Germ.* **38,** 1068 (1979).

[30] M. Laurent, M. N. Tijane, C. Roucous, F. J. Seydoux, and A. Tardieu, *J. Biol. Chem.* **259,** 3124 (1984).

[31] T. Kriegel, W. Schellenberger, G. Kopperschläger, and E. Hofmann, *Biomed. Biochim. Acta* **50,** 1159 (1991).

the inhibition of ATP is diminished. If a certain dye mimics the natural nucleotide ATP by binding to the catalytic as well as the inhibitory sites of the enzyme, any alteration in the ATP-binding sites induced by fructose 6-phosphate should also be reflected in changes in the dye–protein interaction and, hence, in the partitioning behavior of the enzyme.

*Effect of Fructose 6-Phosphate*. In Fig. 5 the affinity partitioning of phosphofructokinase, with different dye–PEG derivatives and increasing amounts of fructose 6-phosphate, is depicted. Independent of the kind of dye used, a drastic change in the $K$ value of the enzyme was found in the range of 20–40 $\mu M$ of the substrate. The characteristics of the curves are similar to titration curves; the concentration range of the inflection points of the curves are of the same order of magnitude as the dissociation

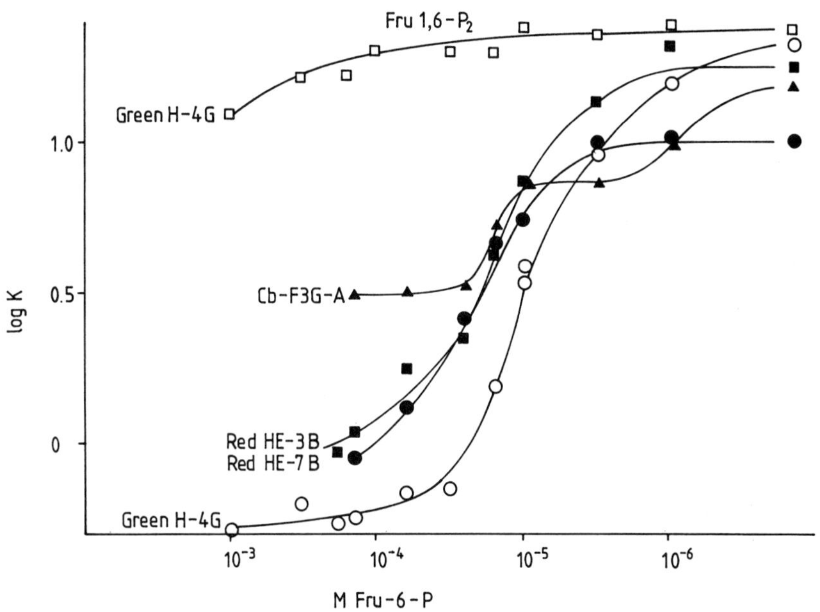


FIG. 5. Affinity partitioning of yeast phosphofructokinase as dependent on fructose 6-phosphate (Fru-6-P) and fructose 1,6-biphosphate (Fru 1,6-$P_2$). The systems contained 5% PEG 6000 (in which 1/2000 of the amount was replaced by the indicated dye–PEG), 7% dextran T70, and 50 m$M$ sodium phosphate buffer, pH 7.0. The dyes are Cibacron Blue F3G-A, Procion Green H-4G, Procion Red HE-3B, and Procion Red HE-7B. The dialyzed enzyme was incubated for at least 1 hr with the respective concentrations of the sugar phosphates before 1 unit of the enzyme was added to the system. The mixtures were allowed to stand for 30 min with occasional agitation before the partitioning of the enzyme was determined. [From G. Kopperschläger, J. Kirchberger, and T. Kriegel, *Makromol. Chem. Macromol. Symp.* **17,** 373 (1988), with permission.]

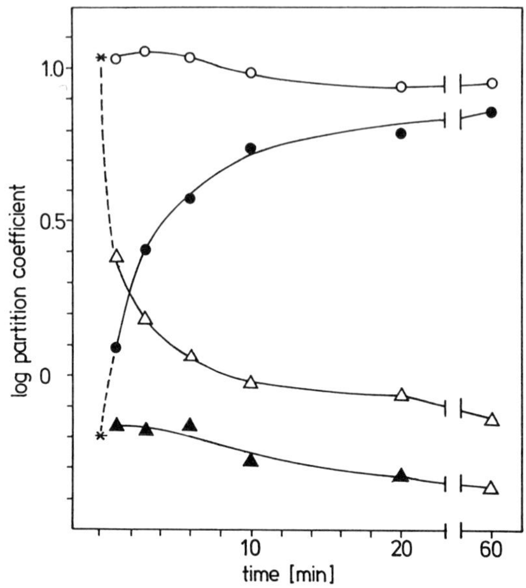


FIG. 6. Affinity partitioning of yeast phosphofructokinase as dependent on time in the presence and absence of fructose 6-phosphate. The basic systems contained 5% PEG 6000, 7% dextran T70, and 50 m*M* sodium phosphate buffer, pH 7.0. Approximately 0.05% of the total PEG was replaced by Procion Red HE-7B–PEG. The dialyzed enzyme was incubated with (●, ▲) or without (○, △) 1 m*M* fructose 6-phosphate. The systems contain no (●, ○) or 1 m*M* (▲, △) fructose 6-phosphate. On addition of 5–10 μl enzyme solution (~1 unit) to a 4-g system and mixing for 10 sec, phase separation was speeded by centrifugation. After exactly 1 min samples from the top and bottom phases were withdrawn. At intervals, as indicated, the systems were mixed again, separated, and analyzed. Asterisks denote extrapolated values at time = 0. [From G. Kopperschläger and G. Birkenmeier, *Bioseparation* **1,** 235 (1990), with permission.]

constant of the enzyme–fructose 6-phosphate complex.[32] Fructose 1,6-bisphosphate, on the other hand, is unable to change the *K* of the enzyme markedly over the same concentration range. There are two possible interpretations: either the binding of fructose 6-phosphate causes a steric hindrance for complexing enzyme and dye, or the substrate lowers the affinity of the dye to its binding site(s) as suggested from the kinetic analysis for the regulatory binding site of ATP.[29]

*Time-Dependent Events.* To obtain further insight on the effect of fructose 6-phosphate, the partitioning of phosphofructokinase was followed as a function of time in systems with and without the substrate. In

[32] K. Nissler, R. Kessler, W. Schellenberger, and E. Hofmann, *Biochem. Biophys. Res. Commun.* **79,** 973 (1973).

addition, in one series of experiments the enzyme was preincubated with 1 m$M$ fructose 6-phosphate. If 5–10 $\mu$l of this enzyme sample were put into a 4-g system free of fructose 6-phosphate, the original substrate concentration was diluted 400- to 800-fold. Thus, the enzyme goes from a high substrate concentration to a low substrate concentration on addition to the system. The experimental conditions are reversed when the enzyme, which is not preincubated with substrate, is added to a system containing fructose 6-phosphate.

The results of four experiments with different variables are shown in Fig. 6 and can be summarized as follows: (1) The $K$ of the enzyme preincubated with fructose 6-phosphate increases in a time-dependent manner in a two-phase system free of fructose 6-phosphate, whereas there is no change in the $K$ value in a system containing substrate. (2) The $K$ of phosphofructokinase not preincubated with fructose 6-phosphate decreases with time in a two-phase system containing fructose 6-phosphate, whereas it is unchanged in a system free of substrate. The rate constant, $k$, of the decrease of the log $K$ is in the range of 1/minute ($1/k = 0.21$ min). Because the rate constant for the formation of the enzyme–substrate complex is assumed to be in the range of 1/second, it is likely that the time-dependent change of the partitioning of phosphofructokinase is the result of a slow transition of the enzyme to a more ATP-sensitive conformational state and vice versa.

This interpretation was confirmed by cross-linking of the enzyme in the presence of fructose 6-phosphate to yield a substrate-imprinted enzyme in which the kinetic properties have been changed.[31] The reversible conversion of an ATP-sensitive form of phosphofructokinase to a less sensitive species, as earlier reported,[33] has also been observed by following the $K$ of phosphofructokinase in a two-phase system containing Procion Red HE-7B–PEG. This indicates that changes in the inhibitory binding sites for ATP are reflected in partitioning.[34]

*Conclusion.* Affinity partitioning of phosphofructokinase using biomimetic dyes as ligands was found to be a sensitive approach for recognition of small changes in binding properties of ligands that may be caused by local conformational changes of enzyme structure.

[33] E.-G. Afting and D. Ruppert, *Arch. Biochem. Biophys.* **156,** 720 (1973).

[34] G. Kopperschläger, J. Kirchberger, and T. Kriegel, *Makromol. Chem. Makromol. Symp.* **17,** 373 (1988).

# [25] Analysis of Structural Changes in Steroid Receptor Proteins by Partitioning

*By* JEFFREY C. HANSEN

## Introduction

The cellular action of steroid hormones is mediated by proteins known as steroid hormone receptors. Steroid hormone receptors, together with the receptor proteins for vitamin $D_3$, retinoic acid, and thyroid hormones, are members of a large structurally related superfamily of the gene regulatory proteins.[1] Steroid receptors function mainly as ligand-dependent transcription factors; binding of ligands to the unoccupied state converts the receptor to a form which interacts with the chromatin milieu of specifically targeted genes, thereby altering the levels of transcription.[1,2] The molecular mechanisms by which hormone binding to the receptors is coupled to signal transduction are the subject of extremely active investigation. In this chapter, I describe the applications of aqueous two-phase partitioning to the study of the structural dynamics of steroid hormone receptors. For a comprehensive review of recent progress in the field, see Ref. 1.

### *Factors Influencing Steroid Receptor Partitioning*

The usefulness of aqueous two-phase partitioning as a probe of steroid receptor dynamics is directly related to the thermodynamic principles that determine how a molecule distributes between the two aqueous phases. The treatment described below is simplistic, yet it accurately describes the way that estrogen receptors[3] and many other proteins[4,5] partition in aqueous two-phase systems. For more detailed discussions of and reference to partitioning theory, see Refs. 6 and 7.

The link between partitioning theory and empirical data is the partition

[1] M. A. Carson-Jurica, W. T. Schrader, and B. W. O'Malley, *Endocr. Rev.* **11,** 201 (1990).
[2] K. R. Yamamoto, *Annu. Rev. Genet.* **19,** 209 (1985).
[3] J. C. Hansen and J. Gorski, *Biochemistry* **24,** 6078 (1985).
[4] P.-Å. Albertsson, "Partitioning of Cells and Macromolecules," 3rd Ed. Wiley(Interscience), New York, 1986.
[5] G. Johansson, *in* "Partitioning in Aqueous Two-Phase Systems: Theory, Methods, Uses, and Applications to Biotechnology" (H. Walter, D. E. Brooks, and D. Fisher, eds.), p. 161. Academic Press, Orlando, Florida, 1985.
[6] N. L. Abbott, D. Blankschtein, and T. A. Hatton, *Bioseparation* **1,** 191 (1990).
[7] H. Walter, G. Johansson, and D. E. Brooks, *Anal. Biochem.* **197,** 1 (1991).

coefficient ($K_{obs}$), which is defined as the concentration of solute in the upper and lower phases, respectively, at partitioning equilibrium [Eq. (1)], and which is the experimentally determined parameter

$$K_{obs} = C_{top}/C_{bottom} \tag{1}$$

For macromolecules such as steroid receptors, there are both conformational ($K_0$) and electrostatic ($K_{elec}$) contributions to the $K_{obs}$ term, such that

$$\ln K_{obs} = \ln K_0 + \ln K_{elec} \tag{2}$$

where

$$\ln K_0 = \ln K_{hydrophobic/hydrophilic} + \ln K_{excluded\ volume} + \ln K_{other} \tag{3}$$

The $K_0$ term is dictated by receptor–solvent interactions, where the solvent in this treatment is the composite of the water, buffer, salts, and polymers in each phase. The contributions to the $K_0$ term are complex and not completely understood [hence the $K_{other}$ term in Eq. (3)] but can be generally traced to at least two features of the protein: surface hydrophobic/hydrophilic content[8] and excluded volume effects.[9] The $K_{elec}$ term arises because steroid receptors are charged and the phase systems selected contain an electrical potential difference across the phase boundary (see below). Importantly, as will be shown, it is possible to dissect $K_{obs}$ to determine whether $K_0$ or $K_{elec}$ is responsible for observed differences in partitioning behavior.

*Aqueous Two-Phase Partitioning of Steroid Receptors*

The application of aqueous two-phase partitioning to the study of steroid receptors was pioneered by Peter Andreasen. In a series of elegant papers, Andreasen used the partitioning technique to study the mechanism of salt-dependent "activation" of glucocorticoid receptors[10–16] and to char-

[8] B. Y. Zaslavsky, L. M. Miheeva, N. M. Mestechkina, and S. V. Rogozhin, *J. Chromatogr.* **253,** 139 (1982).

[9] S. B. Zimmerman and S. O. Trach, *Biopolymers* **30,** 703 (1990).

[10] P. A. Andreasen, *Biochim. Biophys. Acta* **428,** 792 (1976).

[11] P. A. Andreasen, *Biochim. Biophys. Acta* **540,** 484 (1978).

[12] P. A. Andreasen and W. I. P. Mainwaring, *Biochim. Biophys. Acta* **631,** 334 (1980).

[13] P. A. Andreasen, *Biochim. Biophys. Acta* **676,** 205 (1981).

[14] P. A. Andreasen, *Mol. Cell. Endocrinol.* **28,** 563 (1982).

[15] P. A. Andreasen, *Mol. Cell. Endocrinol.* **30,** 229 (1983).

[16] P. A. Andreasen and K. Junker, *in* "Molecular Mechanisms of Steroid Action" (V. K. Moudgil, ed.), p. 199. de Gruyter, Berlin, 1985.

acterize the properties of mutant S49.1 glucocorticoid receptors.[17] Details of this work have been reviewed.[18,19] Systematic partitioning studies of the effects of agonistic and antagonistic ligands and temperature on the structure of the estrogen receptor (ER) were performed by Hansen and Gorski.[3,20,21] The partitioning approach recently has been extended to the carboxyl-terminal ER steroid-binding domain by Fritsch *et al.*[22] The partitioning technique also has been applied to the progesterone[23] and glucocorticoid receptors[24] by Moudgil *et al.* and to the thyroid receptor[25] and the erbA receptor protein[26] by Ichikawa.

The results presented below are derived from studies that have been performed with the estrogen receptor (ER). Although this chapter is dedicated solely to steroid receptors, the general strategy outlined below should be equally applicable to many other dynamic proteins.

## Methods

Figure 1 shows the strategy used to study the unoccupied and liganded forms of the ER by partitioning. The sections below describe each of these steps in more detail. Supplementary information can be found in Refs. 3 and 22.

### *Receptor Terminology*

The steroid receptor literature is peppered with the terms "activation" and "transformation."[27] These terms generally describe the changes in hormone–receptor complexes that occur on exposure to high temperature

[17] P. A. Andreasen and U. Gehring, *Eur. J. Biochem.* **120,** 443 (1981).

[18] P. A. Andreasen, *in* "Partitioning in Aqueous Two-Phase Systems: Theory, Methods, Uses, and Applications to Biotechnology" (H. Walter, D. E. Brooks, and D. Fisher, eds.), p. 315. Academic Press, Orlando, Florida, 1985.

[19] P. A. Andreasen, *in* "Biochemical Actions of Hormones" (G. Litwack, ed.), p. 181. Academic Press, San Diego, 1987.

[20] J. C. Hansen and J. Gorski, *J. Biol. Chem.* **261,** 13990 (1986).

[21] J. C. Hansen and J. Gorski, *Biochemistry* **28,** 623 (1989).

[22] M. Fritsch, C. M. Leary, J. D. Furlow, H. Ahrens, T. J. Schuh, G. C. Mueller, and J. Gorski, *Biochemistry* **31,** 5303 (1992).

[23] V. K. Moudgil and C. Hurd, *Biochemistry* **26,** 4994 (1987).

[24] V. K. Moudgil, L. Vandenheede, C. Hurd, N. Eliezer, and G. Lombardo, *J. Biol. Chem.* **262,** 5180 (1987).

[25] K. Ichikawa, K. Hashizume, T. Miyamoto, Y. Nishii, K. Yamauchi, H. Ohtsuka, and T. Yamada, *J. Endocrinol.* **119,** 431 (1988).

[26] K. Ichikawa, K. Hashizume, S. Furuta, T. Osami, T. Miyamoto, K. Yamauchi, T. Takeda, and T. Yamada, *Mol. Cell. Endocrinol.* **70,** 175 (1990).

[27] W. W. Grody, W. T. Schrader, and B. W. O'Malley, *Endocr. Rev.* **3,** 141 (1982).

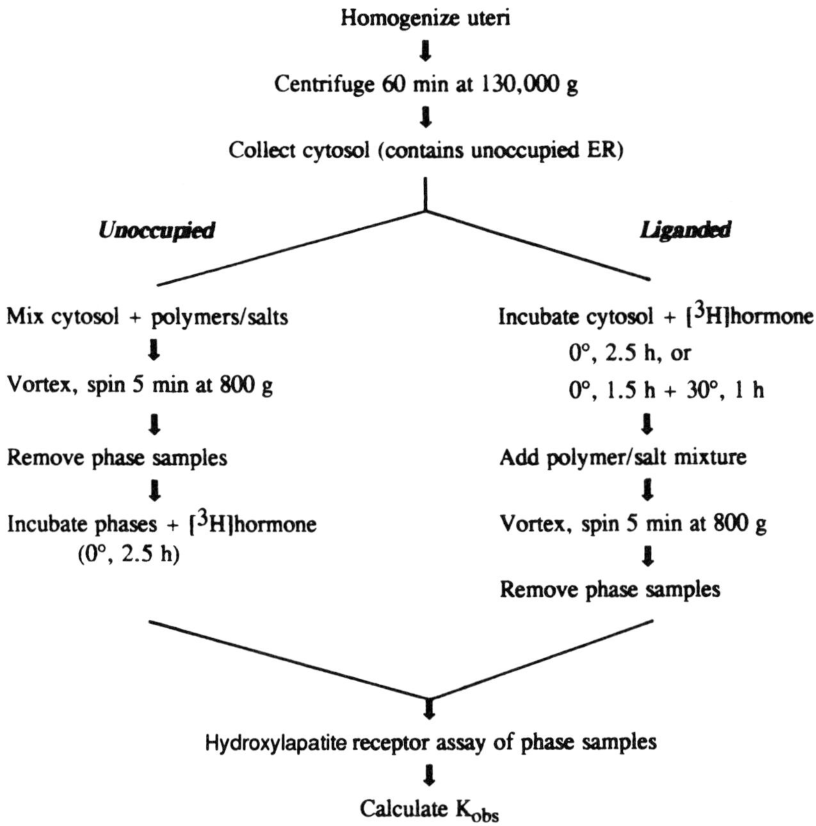


FIG. 1. Flow chart for aqueous two-phase partitioning of unoccupied and liganded estrogen receptors.

or high salt concentrations. Keeping with this convention, I refer to the estrogen receptor–ligand complex at 0° as the "nontransformed" ER. The ER–ligand complex at 30° is called the "transformed" ER. The temperature-dependent changes in ER structure are called "transformation."

*Preparation of Different Estrogen Receptor Forms*

Cytosol containing 4–6 n*M* unoccupied ER is prepared by homogenization of minced immature rat uteri in 10 m*M* Tris-HCl, 1.5 m*M* $Na_2EDTA$, 0.5 m*M* dithiothreitol (DTT) buffer (TED buffer), pH 8.0, followed by centrifugation of the homogenates at 130,000 *g* for 60 min at 4°.[3] The nontransformed ER is obtained by incubating cytosol with 5–10 n*M*

[$^3$H]estradiol ($E_2$) (New England Nuclear, Boston, MA) for 2.5–3 hr at 0°. The transformed ER is obtained by performing the last 60 min of the incubation at 30°.

*Phase Partitioning*

Dextran (Dx), $M_r$ approximately 500,000 (Sigma, St. Louis, MO), and poly(ethylene glycol) (PEG), $M_r$ 8000 (Sigma), stock solutions containing 25–35% (w/w) polymer are prepared in TED buffer, pH 8.0. In some experiments the Tris is replaced with an equivalent amount of HEPES or potassium phosphate. Polymer concentrations are determined by weighing a sample on a glass fiber filter before and after incubation for 24 hr at 37°.

Generally, phase partitioning is performed by combining an aliquot of a concentrated polymer–salt mixture with an aliquot of cytosol, followed by rapid vortexing. The concentrated polymer mixture is composed of 9% (w/w) each of Dx and PEG and various types and concentrations of salts, and it is mixed by rapid stirring prior to combination with cytosol. If desired (e.g., for the cross-partitioning experiments), the pH of the mixture is progressively lowered from 8.0 to approximately 4.5 by stepwise addition of concentrated HCl. Aliquots (0.9 ml) of the concentrated polymer mixture are combined with 0.6 ml uterine cytosol containing either unoccupied, nontransformed, or transformed ER. The final phase mixture thus contains 5.4% (w/w) of each polymer, with different salts and salt concentrations, and in some cases a different pH. The final phase mixture is incubated at 0° for several minutes, then vortexed at maximum setting for 15 sec (which is sufficient to reach partitioning equilibrium[3]). The equilibrated phase systems are then centrifuged for 5 min at 800 *g* to accelerate separation of the two-phase systems. All steps in the partition protocol are performed at 0°.

This protocol produces two-phase systems in which the upper phase is 90.4 ± 0.4% water (S.D., $n = 95$), 63% of the total volume, and PEG rich, whereas the bottom phase is 80.5 ± 0.5% water (S.D., $n = 95$), 37% of the total volume, and Dx rich. These characteristics of the phase systems are independent of the buffer, salt type, and pH used in the present experiments. It should be noted that an extremely high degree of phase system reproducibility (as judged by phase water content and volume) is achieved by handling all polymer stocks and phase system solutions with positive displacement pipettes (SMI, Emeryville, CA). The partitioning pH is determined by measuring the pH of the phase systems with a Radiometer (Copenhagen, Denmark) pH meter. The PEG palmitate used in the hydrophobic partitioning studies is obtained from Aqueous Affinity AB (Arlöv, Sweden).

*Determination of Estrogen Receptor $K_{obs}$*

To calculate the partition coefficient ($K_{obs}$) it is necessary to determine the concentration of ER present in each phase after partitioning. This is accomplished by first removing 0.15–0.20 ml of each phase using a positive displacement pipette, followed by a hydroxylapatite (HAP) receptor binding assay. The HAP assay involves binding of $[^3H]E_2$–receptor complexes to the hydroxylapatite matrix, followed by removal of the free $[^3H]E_2$ with several buffer washings.[3] If the unoccupied ER has been partitioned, phase samples are incubated with 5–10 n*M* $[^3H]E_2$ with or without 1–2 $\mu M$ unlabeled estradiol for 2.5–3 hr prior to the HAP assay. Each receptor $K_{obs}$ is derived from replicate partitionings of both total and nonspecific $[^3H]E_2$ binding. The counts per minute (cpm) $[^3H]E_2$ in each phase sample determined by the HAP assay are converted to disintegrations per minute (dpm) per milliliter of receptor-bound $[^3H]E_2$ in each phase. The $K_{obs}$ is calculated as the ratio of mean ER concentrations in the upper and lower phases. The percent standard error of each individual $K_{obs}$ is calculated as described.[3]

## Results and Discussion

*Analysis of Estrogen Receptor Forms in Phase Systems Containing Different Salts*

All studies of steroid receptor partitioning have begun with a determination of the $K_{obs}$ of different receptor forms as a function of the salt type in the phase system. Recall from Eq. (2) that the $K_{obs}$ is influenced by both $K_0$ and $K_{elec}$. The $K_{elec}$ term is equal to

$$\ln K_{elec} = Z\, \Delta\psi\, (F/RT) \tag{4}$$

where $Z$ is the molecular net charge, $\Delta\psi$ is the interfacial potential difference across the phase boundary, $T$ is the absolute temperature, $F$ is the Faraday constant, and $R$ is the gas constant. In aqueous two-phase systems, the magnitude of $\Delta\psi$ depends on the types of salts used.[28,29] Thus, if the molecule is charged, Eq. (4) predicts that the value of $K_{obs}$ should be a sensitive function of the salt type present in the phase system. In the studies described here, the ER $K_{obs}$ initially was determined in phase systems containing different salts (in each case at pH ~8). The values of

[28] G. Johansson, *Mol. Cell. Biochem.* **4,** 169 (1974).

[29] D. E. Brooks, K. A. Sharp, S. Bamberger, C. H. Tamblyn, G. V. F. Seaman, and H. Walter, *J. Colloid Interface Sci.* **102,** 1 (1984).

the ER $K_{obs}$ were found to vary between 0.01 (in NaCl) and 2.0 (in $Li_2SO_4$).[3] The relative effects of anions and cations on the ER $K_{obs}$ were the same as found previously by Albertsson for negatively charged proteins,[4] indicating that the effects of salts on ER partitioning were mediated through $K_{elec}$. The values of ER $K_{obs}$, although strongly dependent on salt type, were independent of ionic strengths up to at least 0.6 $M$.[3]

The $K_{obs}$ versus salt studies described above also indicated that both ligand binding and heating produce a change in the ER $K_{obs}$.[3] Exposure of the nontransformed ER to 30° for 60 min (i.e., ER transformation) lowered the $K_{obs}$ in all combinations of salts tested. This result was not surprising in light of Andreasen's previous partitioning studies of glucocorticoid receptor transformation.[10–19] A more unique observation was that the $K_{obs}$ for the unoccupied ER was different than that for the nontransformed ER under all partitioning conditions,[3] indicating that ligand binding at 0° causes a structural change on the ER surface.

Recently, the partitioning properties of the unoccupied and liganded forms of the progesterone receptor,[23] glucocorticoid receptor,[24] thyroid receptor,[25] and c-erbA protein[26] have been compared. In each case, binding of hormone to the unoccupied receptor induced a significant change in receptor $K_{obs}$. Thus, a hormone-induced structural change appears to be a general feature of all steroid receptors.

### *Characterization of $K_0$ and $K_{elec}$ Terms by Cross-Partitioning*

To determine whether the difference in $K_{obs}$ of the unoccupied and nontransformed ER seen in different salts at pH 8 was due to a difference in $K_0$ and/or $K_{elec}$, the "cross-partition" analysis was performed. In this approach,[30,31] one determines the $K_{obs}$ versus pH profiles in the presence of different salts. These profiles span 3–4 pH units and should intersect (the "cross-point") at the pH at which the net charge on the molecule is zero. The pH at the point of intersection is close to the isoelectric point of the molecule as determined by electrophoresis, and the $K_{obs}$ at the point of intersection is the $K_0$.

The cross-partition analysis of the unoccupied and nontransformed ER is shown in Fig. 2. Receptors were partitioned in three different salts (0.1 $M$ NaCl, 0.3 $M$ LiCl, and 0.1 $M$ $Li_2SO_4$) over the pH range of 4.5–7.2. Lowering the pH reduces the large salt-dependent differences in ER $K_{obs}$ present at pH 7.2, until the pH profiles intersect at discrete points on both axes. The values at the points of intersection indicate a cross-point pH of approximately 5.5 for both forms of the protein. However, the $K_0$ of

[30] P.-Å. Albertsson, S. Sasakawa, and H. Walter, *Nature (London)* **228,** 1329 (1970).
[31] H. Walter and D. Forciniti, this volume [20].

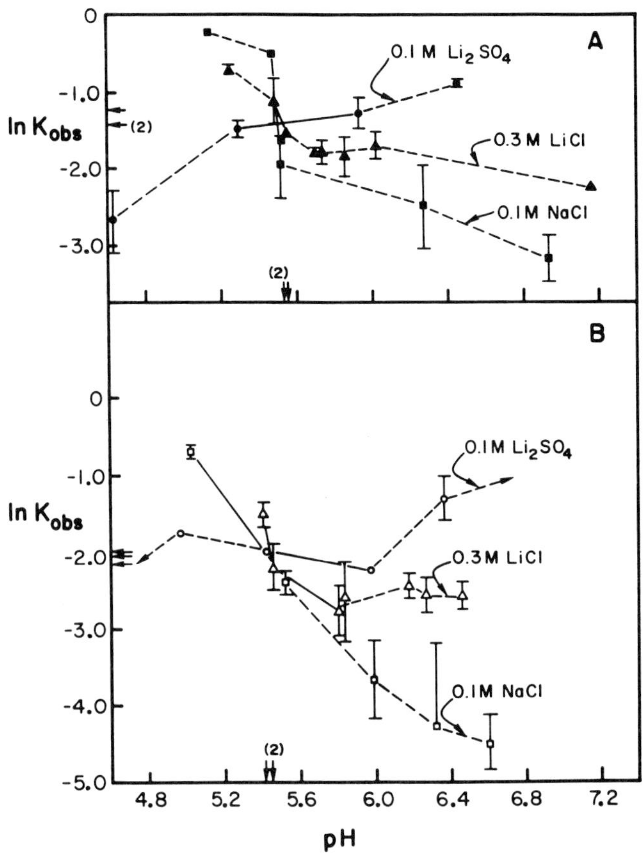


FIG. 2. Cross-partition analysis of unoccupied and nontransformed rat uterine estrogen receptor. Tubes containing either unoccupied (A) or nontransformed (B) ER, plus 5.4% of both dextran ($M_r$ 510,000) and poly(ethylene glycol) and either 0.1 $M$ NaCl, 0.3 $M$ LiCl, or 0.1 $M$ $Li_2SO_4$ were partitioned at the indicated pH and otherwise treated as described in the text. Manipulation and determination of partitioning pH are described in the text. All systems were buffered with 10 m$M$ Tris-HCl, 1.5 m$M$ EDTA, and 0.5 m$M$ DTT (TED). Each point represents the mean (±S.D.) of one $K_{obs}$ value derived from duplicate partitionings. Arrows indicate the points of intersection of each salt pair on each axis. The $K_0$ of the unoccupied ER under these conditions is 0.25 ± 0.03, and that of the nontransformed ER is 0.12 ± 0.02. The cross-point pH of the unoccupied ER is 5.55 ± 0.15, and that of the nontransformed ER is 5.45 ± 0.15. [Adapted with permission from J. C. Hansen and J. Gorski, *Biochemistry* **24,** 6078 (1985). Copyright 1985 American Chemical Society.]

the unoccupied receptor is 0.25, whereas the $K_0$ of the nontransformed ER is 0.12. These results establish that the hormone-dependent change in the ER $K_{obs}$ at pH 8 results from a change in the $K_0$ of the protein. Aggregates of steroid receptors (e.g., with heat-shock proteins) are disrupted by high salt concentrations and dilution,[1] yet the ER $K_{obs}$ values are independent of these parameters.[3] It therefore is most likely that the difference in ER $K_0$ reflects a conformational change rather than a change in aggregation. This ER conformational change has recently been localized to the C-terminal hormone-binding domain of the receptor.[22]

### *Hydrophobic Partitioning*

One possible explanation for the difference in unoccupied and nontransformed ER $K_0$ values is that the conformational change induced by ligand binding may reduce the surface hydrophobicity. This notion was tested using the method[32–34] of hydrophobic partitioning. In this approach one compares the $K_{obs}$ obtained in the absence and presence of PEG palmitate. The $K_{obs}$ of the unoccupied ER was found to increase significantly in the PEG palmitate-substituted phase systems, whereas the $K_{obs}$ of the nontransformed ER actually decreased in the presence of PEG palmitate.[20] Thus, at least part of the hormone-induced decrease in the $K_0$ term appears to be due to a loss of hydrophobic content on the ER surface after hormone binding. The hormone-induced decrease in ER hydrophobicity observed by hydrophobic partitioning has been localized to the hormone-binding domain of the receptor.[22]

### *Kinetic Analyses Using Partitioning*

As described above, ER transformation lowers the ER $K_{obs}$ beyond that caused by ligand binding. This temperature-dependent change in ER partitioning behavior was studied with kinetics,[20,21] as the $K_{obs}$ values of the transformed ER were too low in all salts to permit accurate cross-partition analysis. Andreasen was the first to show that the kinetics of steroid receptor transformation can be measured with aqueous two-phase partitioning.[14] Figure 3 shows a typical time course of the change in ER $K_{obs}$ after exposure to 30°. The intermediate $K_{obs}$ values obtained at exposures of less than 30 min (Fig. 3) are directly proportional to the fraction of receptors that have undergone thermal transformation and therefore

[32] V. P. Shanbhag and G. Johansson, *Biochem. Biophys. Res. Commun.* **61,** 1141 (1974).

[33] V. P. Shanbhag and C. G. Axelsson, *Eur. J. Biochem.* **60,** 17 (1975).

[34] V. P. Shanbhag, this volume [23].

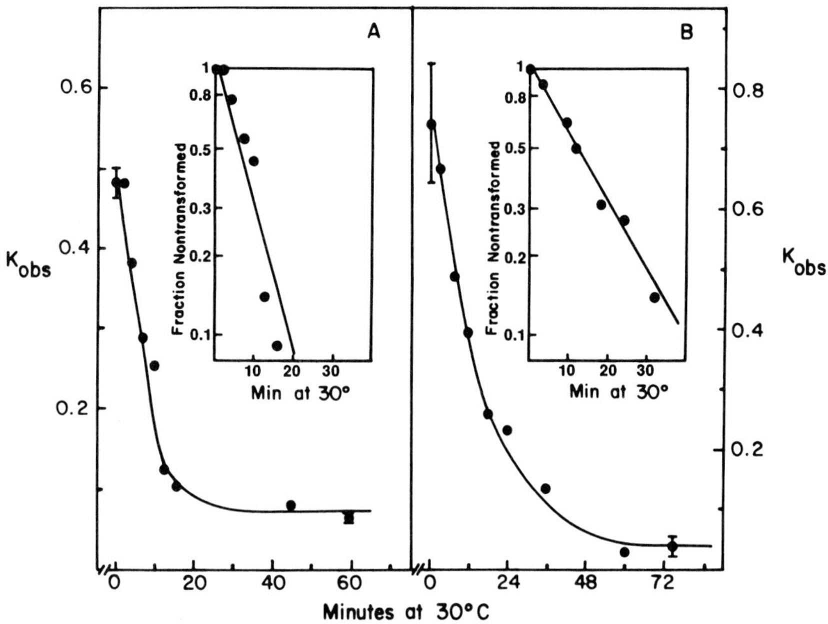


FIG. 3. Kinetics of transformation of estrogen–receptor and antiestrogen–receptor complexes at 30°. Nontransformed estrogen receptors bound to either 17β-estradiol (A) or 4-hydroxytamoxifen (B) were prepared by incubating cytosol with 5 n*M* [$^3$H]estradiol or 5 n*M* $^3$H-labeled 4-hydroxytamoxifen for 1.5–2.5 hr at 0°. Heating at 30° was for 0–75 min, as shown. Reactions were stopped by transfer to 0°, followed immediately by phase partitioning. Partitionings were performed in 5.4% dextran, 5.4% poly(ethylene glycol), 0.1 *M* $Li_2SO_4$ systems buffered with TED, pH 8.1. The $K_{obs}$ values of the nontransformed and transformed estradiol–receptor complexes were 0.48 ± 0.02 and 0.07 ± 0.01, respectively. The $K_{obs}$ values of the nontransformed and transformed 4-hydroxytamoxifen–receptor complexes were 0.74 ± 0.10 and 0.04 ± 0.01, respectively. The correlation coefficients of the linear regression of the semilogarithmic plots (insets) were −0.96 for A and −0.99 for B. From the slopes of the linear regressions, the observed rate constants were 0.068 min$^{-1}$ for receptors bound to 17β-estradiol and 0.026 min$^{-1}$ for receptors bound to 4-hydroxytamoxifen. [Adapted from J. C. Hansen and J. Gorski, *J. Biol. Chem.* **261,** 13990 (1986), with permission.]

can be used to derive kinetic information. A detailed kinetic analysis indicted that the temperature-dependent change in ER $K_{obs}$ values results from an irreversible, hormone-independent conformational change having an activation energy of 25 kcal/mol.[21] It was also observed that this conformational change occurs more slowly when receptors are bound to the antagonist 4-hydroxytamoxifen (Fig. 3 and Ref. 20).

## Conclusions

This chapter describes an integrated approach for studying the structural changes of macromolecules by partitioning in aqueous two-phase polymer systems. The approach works best in a comparative context (i.e., when different structural states of the same molecule are partitioned under identical conditions). This allows one to assign differences in $K_{obs}$ to either the $K_0$ and/or $K_{elec}$ terms, and it avoids the difficulties involved in interpreting their absolute values. The recommended sequence of partitioning experiments is as follows. First, determine the $K_{obs}$ as a function of salt type. This will indicate whether the anion and cation effects follow previously determined relationships.[4] If so, it is likely that the salt effects on $K_{obs}$ are mediated through the $K_{elec}$ term. At this point, partition different forms (e.g., conformers, aggregates) of the molecule in several different salts to determine the conditions under which different forms of the molecule have different $K_{obs}$ values. The next step is to determine the pH dependence of the $K_{obs}$ values in different salts (cross-partitioning). If the points of intersection of the pH profiles are independent of salt type, cross-partitioning will indicate both the $K_0$ and the pH at which the net molecular charge is zero. This will indicate whether any differences in $K_{obs}$ of the different molecular forms are due to differences in $K_0$ or $K_{elec}$. To further explore the molecular basis for differences in $K_{obs}$, perform hydrophobic and other types of affinity partitioning. Finally, if possible, determine the kinetic mechanism of the structural transition by measuring the change in partition coefficients with time.

How useful has the partitioning technique been for studying steroid receptors? The partitioning studies remain the most rigorous physicochemical analyses of the effects of ligand binding on steroid receptor structure, largely because of the unprecedented ability of the technique to characterize the unoccupied receptor state. Andreasen's conclusion over a decade ago that the mutant S49.1 nt$^i$ glucocorticoid receptor was essentially identical to the chymotrypsinized wild-type receptor[17] is another impressive illustration of the power of the partitioning technique. The studies of Fritsch *et al.*[22] further demonstrate the great potential of the partitioning technique when used in combination with contemporary molecular genetic tools. Although it should be cautioned that not every macromolecule will be as "well-behaved" as steroid receptors, aqueous two-phase partitioning clearly in many cases will be able to yield a large amount of unique and interesting information. As such, it is a recommended technique for anyone interested in macromolecular structure–function relationships.

# [26] Use of Partitioning in Predicting Mild Hydrophobic Interaction Chromatography Behavior

*By* PATRICK HUBERT

## Introduction

For the purification of biochemicals that display strong hydrophobic interactions the use of ligands having intermediate hydrophobic character should provide sufficient binding strength without causing the irreversible adsorption, denaturation during elution owing to harsh conditions, etc., sometimes associated with the ligands (phenyl groups or alkyl chains) generally employed in traditional hydrophobic chromatography. Such moderate hydrophobic interactions are involved in partitioning between aqueous two-phase systems. In this technique the system is composed of two mutually incompatible polymers or a polymer and a salt; for example, poly(ethylene glycol) (PEG)–dextran (Dx), poly(vinyl alcohol) (PVA)–Dx, and PEG–$MgSO_4$ are among the most widely used biphasic systems. It seems logical that the principles of this batch partitioning technique could be applied to a chromatographic procedure if one of the phase-forming polymers were immobilized on a chromatographic support.

Such immobilization of a polymer can be effected by its mere adsorption on the chromatographic matrix. This approach, explored as early as 1963 by Morris,[1] was subsequently exploited by Müller[2,3] when suitable supports, based on polymer materials or silica, and which exhibited sufficient selectivity and affinity for one of the two aqueous phase polymers, became available. This new type of chromatography, named liquid–liquid partition chromatography (LLPC) by analogy to its parent batch procedure, is the subject of a chapter elsewhere in this volume.[4] An alternative approach is to bind covalently to the chromatographic support one of the phase-forming polymers used in the batch partitioning procedure. Thus far polyether- and PVA-bound stationary phases are the only polymers that have been used in this manner.[5–10]

[1] C. J. O. R. Morris, *Protides Biol. Fluids* **10,** 325 (1963).
[2] W. Müller, *Eur. J. Biochem.* **155,** 213 (1986).
[3] W. Müller, *Bioseparation* **1,** 265 (1990).
[4] W. Müller, this volume [9].
[5] S. H. Chang, K. M. Gooding, and F. E. Regnier, *J. Chromatogr.* **120,** 321 (1976).
[6] T. G. I. Ling and B. Mattiasson, *J. Chromatogr.* **254,** 83 (1983).
[7] N. T. Miller and B. L. Karger, *J. Chromatogr.* **326,** 45 (1985).
[8] J. P. Chang, Z. El Rassi, and Cs. Horvath, *J. Chromatogr.* **319,** 396 (1985).

In contrast with other groups who focused on standard protein mixtures or simple extracts for physicochemical or thermodynamic determinations, we tested mild hydrophobic interaction chromatography in actual fractionations of complex mixtures. Amino derivatives of various polyethers of increasing hydrophobicity were coupled to Sepharose 6B. The resulting mild hydrophobic stationary phases were tested in the purification of enzymes acting on hydrophobic substrates (lipids or steroids).[11–16] This approach is illustrated by the extraction from *Pseudomonas testosteroni* of three closely related enzymes [3-oxosteroid $\Delta^5 \rightarrow \Delta^4$-isomerase (isomerase), $3\alpha$-hydroxysteroid dehydrogenase ($\alpha$-HSD), and $3\beta,17\beta$-hydroxysteroid dehydrogenase ($\beta$-HSD)] with the additional challenge to obtain the purification by stepwise elution of the same crude extract.

Excellent purification ratios as well as satisfactory recoveries of enzyme activities can be obtained, provided that the chromatographic conditions are optimized. In fact, not only is the nature of the stationary phase important but, since hydrophobic interactions are highly dependent on ionic strength, the strategy of mild hydrophobic interaction chromatography implies a judicious selection of the mobile phase as well during both the adsorption and desorption steps. For instance, high ionic strength (yet low enough to avoid retention of slightly hydrophobic contaminants) is required for the adsorption step, whereas low ionic strength (yet high enough to minimize release of strongly adsorbed contaminants) is necessary during desorption.

In this respect, preliminary studies of the crude enzyme extract by partitioning in aqueous two-phase systems prepared with the same chemicals as those employed in the chromatographic experiments clearly affords valuable information. Results from such studies can be used to predict the chromatographic behavior of the enzymes in question and to help to select conditions that are likely to afford optimal chromatographic fractionation.

## Synthesis of Stationary Phases

Because of their solubility in water the polymers involved in the batch partitioning technique are generally regarded as being strictly hydrophilic.

---

[9] Y. Shibusawa, U. Matsumoto, and M. Takatori, *J. Chromatogr.* **398,** 153 (1987).
[10] T. Suzuki, N. Muroi, and T. Tomono, *J. Biomater. Sci. Polym. Ed.* **1,** 3 (1989).
[11] R. Mathis, P. Hubert, and E. Dellacherie, *J. Chromatogr.* **319,** 43 (1985).
[12] R. Mathis, P. Hubert, and E. Dellacherie, *J. Chromatogr.* **347,** 291 (1985).
[13] R. Mathis, A. Mourey, and P. Hubert, *Appl. Environ. Microbiol.* **54,** 1307 (1988).
[14] R. Mathis, P. Hubert, and E. Dellacherie, *J. Chromatogr.* **474,** 396 (1989).
[15] R. Mathis, P. Hubert, and E. Dellacherie, *J. Chromatogr.* **538,** 456 (1991).
[16] P. Hubert, R. Mathis, and E. Dellacherie, *J. Chromatogr.* **539,** 297 (1991).

However, if one considers thermodynamic parameters such as surface tension, this hydrophilicity concept appears somewhat more relative. For instance, the surface tension of PEG molecules ($\gamma_{sv} \approx 42$–$46$ mJ/m$^2$)[17] is midway between those of water ($\gamma_{sv} \approx 70$ mJ/m$^2$) and low-energy materials such as $C_6$–$C_{16}$ alkanes ($\gamma_{sv} \approx 18$–$28$ mJ/m$^2$).[18] From this point of view, PEG polymers as well as the other water-soluble polymers most often used in partitioning [PVA, poly(vinylpyrrolidone) (PVP), or Dx], which all fall within a narrow range on the hydrophobicity scale, may be considered as hydrophobic. At the same time, when regarded from the other extremity of the hydrophobicity scale ($C_6$–$C_{16}$ alkanes and beyond), their designation as "mild hydrophobic ligands," first suggested by Ling and Mattiasson,[6] also appears to be fully justified.

The stationary phases that we used have been prepared by covalent immobilization on Sepharose 6B of various polyethers: PEG HO—$(CH_2CH_2$—$O)_n$—H; polypropylene glycol (PPG) HO—$[CH_2CH(CH_3)$—$O]_n$—H; polytetramethylene glycol (PTMG) HO—$(CH_2CH_2CH_2CH_2$—$O)_n$—H; as well as various ethylene oxide–propylene oxide (PEO/PO) copolymers, commercially available as Pluronics (Serva, Heidelberg, Germany) or Jeffamines (Fluka, Mulhouse, France). The hydrophobic character of these polymers increases with increasing number of carbon atoms in the repeating unit: PEG < PEO/PO < PPG < PTMG.

Covalent immobilization of polymers requires preliminary activation of the matrix. Strongly interacting spacer arms should be avoided since these would mask the mild hydrophobic interactions otherwise involved. This problem was, in effect, experienced when the activation reaction was carried out with certain commonly used bisepoxides such as 1,4-butanediol diglycidyl ether or ethylene glycol diglycidyl ether. In fact, these activation groups remained on the gel in large amounts after the coupling step and, owing to the presence of $(CH_2)_n$ units in the structure ($n = 2$ or 4), acted as hydrophobic ligands, to afford strongly adsorbing stationary phases, even in the absence of any immobilized polyether.[11]

Unambiguous immobilization of polyethers on chromatographic supports, by means of spacer arms not apt to interfere in the hydrophobic interaction, can be obtained by using carbonyldiimidazole (CDI) or *p*-nitrophenyl chloroformate as activation reagents. When using these, however, the polyethers should first be transformed into their amino derivatives so as to obtain, after coupling, a urethane group, which is chemically more resistant to hydrolysis.[12]

[17] G. Körösi and E. Kovats, *J. Chem. Eng. Data* **26,** 323 (1981).

[18] N. Jarvis, R. B. Fox, and W. A. Zesnian, *in* "Contact Angle, Wettability and Adhesion" (Advances in Chemistry Series), p. 323. American Chemical Society, Washington, D.C., 1979.

### Batch Partitioning

Mild hydrophobic chromatography experiments are carried out with either Sepharose–PEG or Sepharose–Pluronic as the stationary phase and different concentrations of potassium phosphate in the mobile phase. To mimic these conditions, batch partitioning studies are conducted with different two-phase systems composed of a potassium phosphate-rich bottom phase and a top phase rich in a mixture of PEG and Pluronic.

The partition coefficient, $K$, of a particular enzyme, which is the ratio of its concentration in the upper phase to that in the lower one, is taken as an indication of the interaction of the enzyme with the polyethers. When chromatography on Sepharose–polyether is carried out with potassium phosphate as eluent, enzymes exhibiting batch $K$ values above 1 should be easily retained in or retarded by the stationary phases. In contrast, species that favor the phosphate-rich bottom phase ($K < 1$) should display only little, if any, retardation.

Batch partitioning of the crude extract from *Pseudomonas testosteroni* is studied in a number of biphasic systems in which both the hydrophobic character of the upper phase (presence of different amounts of Pluronic) and the ionic strength of the lower phase differ. The biphasic systems are prepared as follows. Desired amounts of PEG 6000 (Fluka), Pluronic L 64, and potassium phosphate [potassium phosphate relates to a mixture of $K_2HPO_4$ and $KH_2PO_4$ in the ratio 306.9/168.6 (w/w)] are dissolved in 3 ml water in hemolysis tubes. After addition of *P. testosteroni* crude extract (0.25 ml),[11] the mixtures are adjusted to 5 g with water.

The final concentration of PEG plus Pluronic is 10% (w/w) in all biphasic systems. Only the PEG : Pluronic ratio (10 : 0, 9.5 : 0.5, 9 : 1, and 8 : 2) is varied. The final concentration of potassium phosphate is 10, 15, or 20% (w/w). After 30 inversions, the mixtures are allowed to stand at room temperature for 1 hr. Aliquots of both phases are collected by pipette and assayed for the various enzyme activities according to described procedures.[19,20]

### Partition Coefficients versus Chromatographic Behavior

Figure 1 shows variation of the $K$ values of the three indicated enzymes versus the content of Pluronic in the biphasic system, at three different potassium phosphate concentrations. All the $K$ values increase with increasing amounts of Pluronic in the upper phase (i.e., as the hydrophobic character of this layer becomes stronger). In addition, at any given percent-

[19] P. Talalay and F. S. Wang, *Biochim. Biophys. Acta* **18,** 300 (1955).

[20] P. I. Marcus and P. Talalay, *J. Biol. Chem.* **218,** 661 (1956).

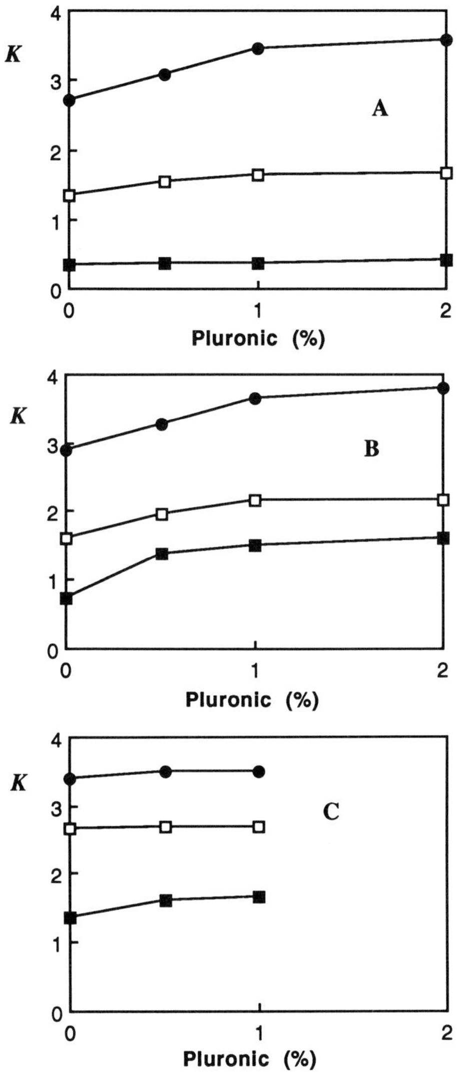


FIG. 1. Partition coefficients ($K$) of (●) 3-oxosteroid $\Delta^5 \rightarrow \Delta^4$-isomerase (isomerase), (□) $3\beta,17\beta$-hydroxysteroid dehydrogenase ($\beta$-HSD), and (■) $3\alpha$-hydroxysteroid dehydrogenase ($\alpha$-HSD) versus percentage of Pluronic in the biphasic system at (A) 10, (B) 15, and (C) 20% (w/w) potassium phosphate concentrations. [From R. Mathis, P. Hubert, and E. Dellacherie, *J. Chromatogr.* **538,** 456 (1991), with permission.]

age of Pluronic, all enzymes partition increasingly toward the upper phase with increasing concentration of potassium phosphate. These facts support the hypothesis that the interaction between enzymes and polyethers is of hydrophobic nature. The data in Fig. 1 correlate with the chromatographic elution profiles shown in Figs. 2 and 3.

The most immediate information given by Fig. 1 is that, irrespective of both potassium phosphate concentration and Pluronic content, all biphasic systems give the following: $K_{\text{isomerase}} > K_{\beta\text{-HSD}} \gg K_{\alpha\text{-HSD}}$. In other words, the strength of the interaction between the polyether(s) and the respective enzymes follows the order isomerase > $\beta$-HSD $\gg$ $\alpha$-HSD. The chromatography of the three enzymes on Sepharose–PEG or Sepharose–Pluronic with potassium phosphate as the eluent should thus result in an elution order of $\alpha$-HSD, $\beta$-HSD, then isomerase, whichever stationary phase is used and whatever the initial ionic strength of eluent. This is in agreement with the data shown in Figs. 2 and 3.

The second observation relates to $\alpha$-HSD. This enzyme, in contrast to $\beta$-HSD, and also isomerase, interacts weakly with the polyethers ($K \ll 1$ at 10% potassium phosphate). However, the fact that its $K$ can

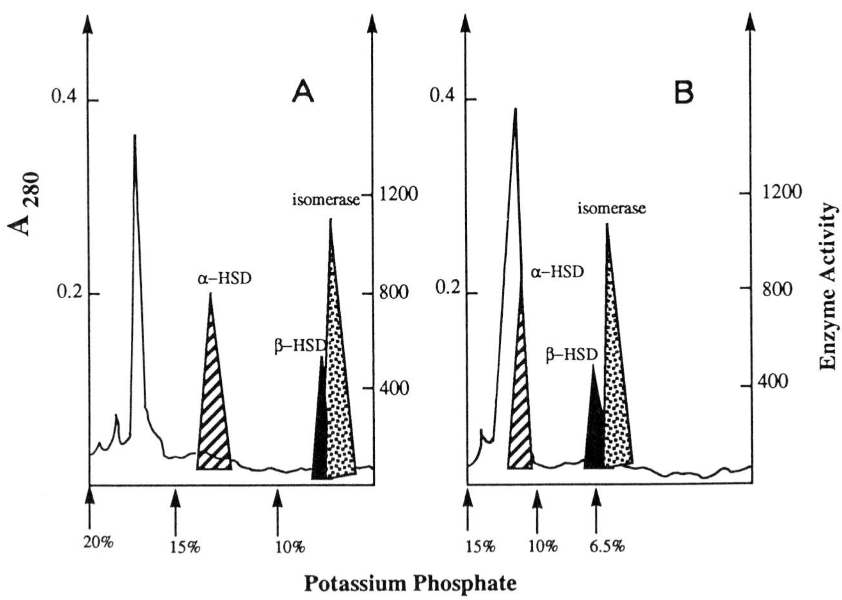


FIG. 2. Sepharose–PEG stationary phase. Elution profile of *Pseudomonas testosteroni* crude extract, starting with (A) 20% or (B) 15% potassium phosphate in the mobile phase. [From R. Mathis, P. Hubert, and E. Dellacherie, *J. Chromatogr.* **474,** 396 (1989), with permission.]

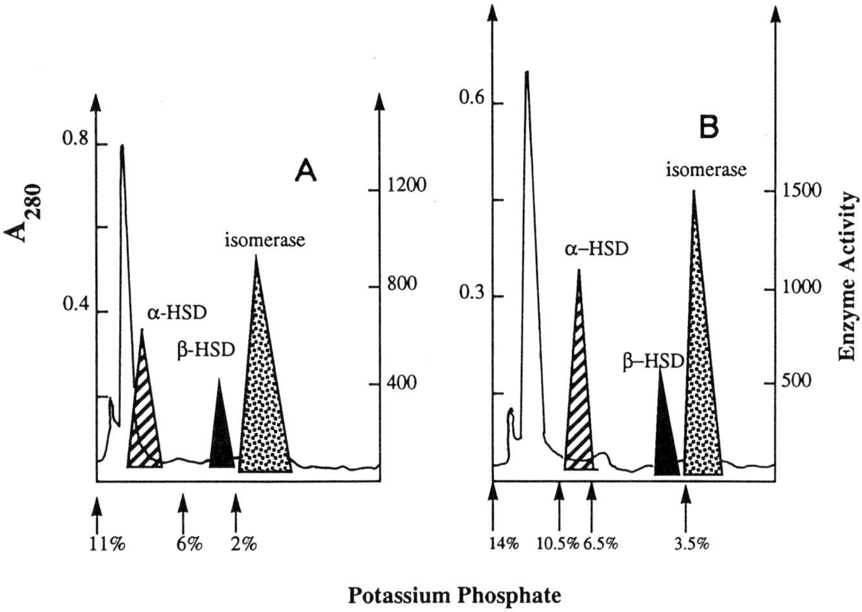


FIG. 3. Sepharose–Pluronic stationary phase. Elution profile of *Pseudomonas testosteroni* crude extract, starting with (A) 11% or (B) 14% potassium phosphate in the mobile phase. [From R. Mathis, P. Hubert, and E. Dellacherie, *J. Chromatogr.* **474,** 396 (1989), with permission.]

be shifted above 1, by use of either a higher ionic strength or an increase in the hydrophobicity of the top phase, or both (Fig. 4), suggests that chromatographic conditions permitting its extraction can be found.

On Sepharose–PEG, retention of $\alpha$-HSD requires a 20% potassium phosphate mobile phase (compare Fig. 2A,B). Thus, in batch partitioning systems containing only PEG (Pluronic content = 0), $K_{\alpha\text{-HSD}}$ is less than 1 for potassium phosphate concentrations of 10 and 15%; $K_{\alpha\text{-HSD}}$ exceeds 1 for a potassium phosphate concentration of 20% (Fig. 4). The slightly stronger hydrophobic character of Sepharose–Pluronic causes retention of this enzyme at only 14% potassium phosphate concentration in the mobile phase (compare Fig. 3A,B), which is in agreement with the partitioning experiment [e.g., with 0.5% Pluronic, $K_{\alpha\text{-HSD}} < 1$ at 10% potassium phosphate and $K_{\alpha\text{-HSD}} > 1$ at 15% potassium phosphate (Fig. 4)].

Finally, $\beta$-HSD and isomerase have $K$ values greater than 1 in all cases described in Fig. 1. As a result both should be retarded on Sepharose–PEG or Sepharose–Pluronic, provided that the mobile phase phosphate concentration is at least 10%. Furthermore, as $\Delta K$ ($K_{\text{isomerase}} - K_{\beta\text{-HSD}}$) increases

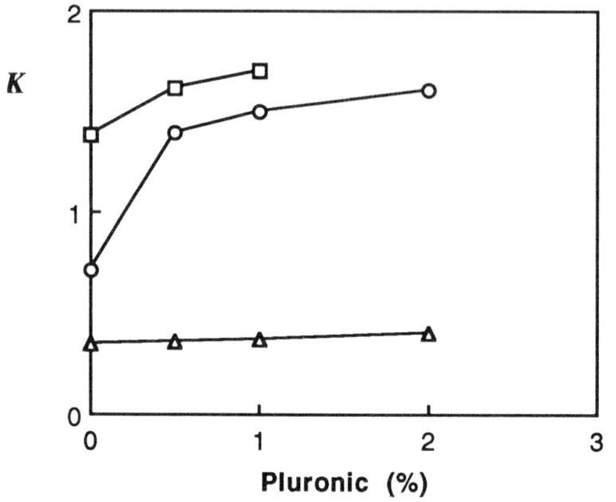


FIG. 4. Partition coefficients ($K$) of $\alpha$-HSD versus percentage of Pluronic in the biphasic system, at (△) 10, (○) 15, and (□) 20% (w/w) potassium phosphate concentrations. [From R. Mathis, P. Hubert, and E. Dellacherie, *J. Chromatogr.* **538,** 456 (1991), with permission.]

with higher Pluronic content, a better separation of these enzymes is predicted for chromatography on Sepharose–Pluronic than on Sepharose–PEG. The chromatographic profiles obtained with these two stationary phases correspond with this expectation (compare Figs. 2 and 3).

Figure 3B illustrates the strategy developed from the information derived from batch partitioning. The chromatography of the crude extract was carried out, starting with a potassium phosphate concentration (14%) sufficient for total retention of the three enzymes and on Sepharose–Pluronic, rather than on Sepharose–PEG, for a better discrimination between isomerase and $\beta$-HSD. Once the nonretained contaminants were removed, the potassium phosphate concentration was decreased to 10.5%, resulting, as expected, in the elution of $\alpha$-HSD. A further decrease in the potassium phosphate concentration to 6.5% caused the elution of $\beta$-HSD and finally of isomerase at 3.5% potassium phosphate. Under these conditions, good recoveries of the enzyme activities are obtained ($\alpha$-HSD, 60%; $\beta$-HSD, 40%; isomerase, 65%) together with the almost quantitative removal of contaminants from each fraction (residual contaminants expressed as a percentage of the total proteins in the starting crude extract: $\alpha$-HSD, 1.5%; $\beta$-HSD, 2%; isomerase, not detectable).

## Conclusions

The results reported illustrate how preliminary simple studies using batch partitioning yield data which correlate with mild hydrophobic interaction chromatography experiments and which can readily be exploited to predict conditions for an optimal fractionation. Such a batch partitioning approach is synonymous with time saving, particularly in situations in which special problems (e.g., complexity of starting extract, similarity of the species to be purified, and necessity to perform stepwise elution) are encountered.

# Section IV

## Partitioning and Affinity Partitioning of Particulates

A. Cells
Separation and Fractionation of Cell Populations and Obtaining Information on Their Surface Properties
*Articles 27 through 32*
Affinity Partitioning
*Articles 33 through 37*

B. Organelles: Separation and Subfractionation
*Articles 38 through 41*

C. Membranes
Separation and Subfractionation
*Articles 42 through 46*
Affinity Partitioning
*Articles 47 and 48*
Membrane Domain Analysis
*Articles 49 and 50*

# [27] Analytical Applications of Partitioning: Detection of Differences or Changes in Surface Properties of Mammalian Cell Populations

*By* HARRY WALTER

## Introduction

Partitioning in dextran (Dx)–poly(ethylene glycol) (PEG) aqueous two-phase systems can be used to separate distinct cell populations based on differences in their surface properties.[1] Its most powerful application, however, and one in which unique information is often obtained, lies in the ability to fractionate cells into subpopulations and to trace cell surface alterations that occur as a function of normal and abnormal *in vivo* processes (e.g., differentiation, maturation, aging) or *in vitro* treatments (e.g., cell storage, incubation with enzyme, aldehyde fixation).[2,3] Here I illustrate, by use of selected examples, the analytical applications of partitioning to the detection of subtle differences between closely related cell populations and/or of cell membrane surface changes that accompany biochemical processes. For reviews of this area, see Refs. 2 and 3.

## General Approach to Analytical Detection of Cell Surface Differences

The procedures used in the selection of Dx–PEG phase systems having different physical properties (charge-sensitive, non-charge-sensitive, affinity) and for single-tube partitioning or countercurrent distribution (CCD) of cells have been outlined in the chapter by Walter and Larsson.[1] When the partitions, $P$ values (defined as the quantity of cells in the top phase as a percentage of total cells added[1]), of two cell populations differ significantly the populations can be separated in a single extraction step (i.e., by partitioning in a single tube and physically separating cells in the top phase from those at the interface plus bottom phase). Smaller differences in $P$ values between two (or among more) cell populations may still permit

[1] H. Walter and C. Larsson, this volume [4].

[2] H. Walter and D. Fisher, *in* "Partitioning in Aqueous Two-Phase Systems: Theory, Methods, Uses, and Applications to Biotechnology" (H. Walter, D. E. Brooks, and D. Fisher, eds.), p. 377. Academic Press, Orlando, Florida, 1985.

[3] H. Walter, *in* "Partitioning in Aqueous Two-Phase Systems: Theory, Methods, Uses, and Applications to Biotechnology" (H. Walter, D. E. Brooks, and D. Fisher, eds.), p. 327. Academic Press, Orlando, Florida, 1985.

cell resolution but only if a multiextraction procedure (e.g., CCD[4]) is employed. A successful preparatory separation results in distinct peaks on CCD, each representing one of the cell populations. Also feasible is the detection, for analytical purposes, of differences in surface properties between closely related cell populations by use of a combination of isotopic labeling and CCD techniques[3,5] as described below.

Fractionation of a single cell population by CCD, in contrast, often (though not always) results in a single, even symmetrical, distribution curve where cells with selected distinct properties are found under different parts of the curve.[2,3] Ways of determining whether cells under a CCD curve have been fractionated into subpopulations are given in Ref. 1.

Under some circumstances surface differences between cell subpopulations that are not detectable by partitioning can be rendered detectable if a differential reactivity of a subpopulation with certain reagents exists. When such a reaction also results in a change in the $P$ value of that cell subpopulation, a difference can be detected and, if large enough, a separation effected [see example of erythrocytes from patients with paroxysmal nocturnal hemoglobinuria (PNH)[6] below].

The tracing of surface alterations as a function of different *in vivo* or *in vitro* processes involves a combination of cell sampling at selected times during the studied process, CCD, and cell subpopulation identification (by isotopic labeling, microscopic recognition, marker enzyme analysis, or other means) through the CCD curve.

### *Method for Detection of Surface Differences between Closely Related Cell Populations*

To determine whether closely related cell populations (e.g., cultured cell sublines, red blood cells from different individuals) differ in surface properties, a simple method based on a combination of isotopic labeling and CCD can be used.[5,7] Aliquots of two cell populations (A and B) to be compared are labeled with an isotope that does not affect the surface properties of the cells, that is, the $P$ value is not affected. [$^{51}$Cr]Chromate (ICN, Irvine, CA) has been found suitable for this purpose. Cells are incubated for 30 min to 1 hr with [$^{51}$Cr]chromate (e.g., 10 $\mu$Ci of $^{51}$Cr per milliliter of anticoagulated blood or 150 $\mu$Ci of $^{51}$Cr per $5 \times 10^6$ leukocytes are suitable quantities of isotope). The labeled cells are washed several

[4] H.-E. Åkerlund and P.-Å. Albertsson, this volume [8].

[5] H. Walter and E. J. Krob, *Biochem. Biophys. Res. Commun.* **120,** 250 (1984).

[6] M. K. Pangburn and H. Walter, *Biochim. Biophys. Acta* **902,** 278 (1987).

[7] H. Walter, E. J. Krob, F. A. Al-Romaihi, D. Johnson, and C. B. Lozzio, *Cell Biophys.* **13,** 173 (1988).

times with saline or culture medium to remove excess isotope; the unlabeled cells are washed similarly. To compare the surface properties of A and B one part of $^{51}$Cr-labeled cells are pipetted into a centrifuge tube containing 4 or 5 ml of phosphate-buffered saline. Four parts of the unlabeled cells to which the labeled cells are to be compared are pipetted into the same tube. Cells in the tubes are gently mixed, centrifuged, and used to make a load mix for CCD. The following four mixtures are examined in each experiment: $^{51}$Cr-labeled cells A plus unlabeled cells B (A + B); $^{51}$Cr-labeled cells B plus unlabeled cells A (B + A); $^{51}$Cr-labeled cells A plus unlabeled cells A (A + A); and $^{51}$Cr-labeled cells B plus unlabeled cells B (B + B). Mixing labeled and unlabeled cells of the same population is an essential control to indicate that the $^{51}$Cr-labeling procedure does not affect the surface properties of the labeled cells as reflected by partitioning (Fig. 1).

A CCD experiment is carried out separately but simultaneously on the four mixtures (in a charge-sensitive, non-charge-sensitive, or affinity system as desired). The total cell distribution is determined[1] and gives, primarily, the distribution of unlabeled cells in each case since these cells are present in great excess. The distribution of labeled cells is obtained as counts/minute (cpm). Deviation of the relative specific activity (see Fig. 1) from 1.00 (the relative specific activity of the original, unfractionated cell mixture) through the distribution curve reflects the degree of displacement and of differences between any two such cell populations examined.

## Tracing Surface Alterations as Function of Normal or Abnormal *in Vivo* Processes or *in Vitro* Treatments: Specific Examples

### *Surface Changes in Cell Differentiation, Maturation, and Aging*

*Preparation, Fractionation, and Analysis of Rat Intestinal Epithelial Cells.*

AIM. To trace possible surface alterations associated with cell differentiation, maturation, and aging of rat intestinal epithelial cells.[8]

APPROACH. Rats are injected intraperitoneally with [$^{14}$C]methylthymidine and sacrificed at different times (4–48 hr) thereafter. The intestinal epithelial cells are prepared by the method of Weiser.[9] The cells are

[8] H. Walter and E. J. Krob, *Exp. Cell Res.* **91,** 6 (1975).
[9] M. M. Weiser, *J. Biol. Chem.* **248,** 2536, 2542 (1973).

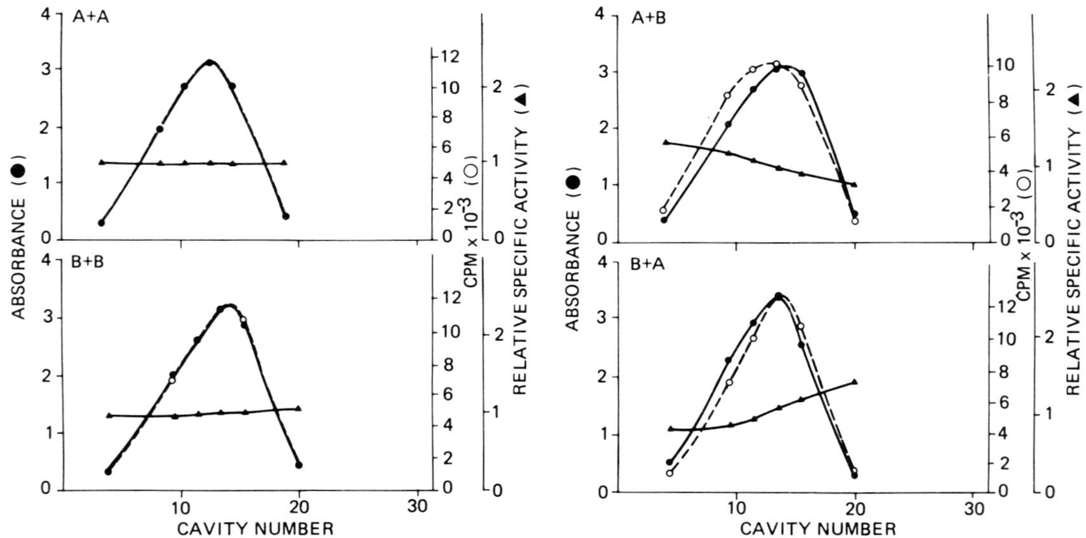


FIG. 1. Comparison of surface properties of red blood cell populations obtained from two arbitrarily selected, presumably hematologically normal, individuals. Blood was drawn from two individuals (A and B). Aliquots of red cells from each were labeled with [$^{51}$Cr]chromate. Labeled cells were washed and mixed with an excess of unlabeled red cells from the other individual (A + B, B + A). As controls, aliquots of labeled cells were also mixed with an excess of unlabeled cells from the same person (A + A, B + B). The four mixtures were subjected to countercurrent distribution in a phase system composed of 5% (w/w) dextran T500 (Pharmacia LKB, Piscataway, NJ), 4% (w/w) PEG 8000 (Union Carbide, Long Beach, CA), and 0.11 *M* sodium phosphate, pH 6.8, which is charge sensitive. Thirty transfers were completed at 4° with a settling time of 6 min and a shaking time of 22 sec. ●, Total cell distribution in terms of hemoglobin absorbance at 540 nm; ○, distribution of labeled cells (in terms of counts/minute); ▲, relative specific activities (with 1.0 being the specific activity of the original, unfractionated cell mixture in each case). [From H. Walter and E. J. Krob, *Biochem. Biophys. Res. Commun.* **120,** 250 (1984).]

subjected to CCD in a charge-sensitive Dx–PEG phase system (for composition and conditions of CCD, see Fig. 2). Typically 1.0 ml of packed cells, previously washed in top phase, is suspended in 8.8 ml of top phase. Cavities 0–7 receive 0.5 ml of bottom phase while cavities 8–119 receive 0.6 ml of bottom phase. Cavities 0–7 receive 0.9 ml of the indicated intestinal cell suspension in top phase, and cavities 8–119 receive 0.8 ml top phase. Because the bottom phase capacity of the CCD unit is 0.7 ml, loading in the manner indicated results in a 0.1 ml stationary top phase.[2,4] Cells are collected after CCD directly into plastic centrifuge tubes, and 0.5 ml of isotonic saline is added to each tube. Tubes are then pooled by groups of 3 or 5 along the extraction train, and the cells are centrifuged at 200 *g* for 10 min. The supernatant solution is discarded and the cells are suspended in a known volume of isotonic saline. The cell distribution is obtained by measuring the absorbance of the cell suspensions at 500 nm. Radioactivity (in counts/minute) is determined using a liquid scintillation counter. The relative specific activity of each fraction is also calculated, with 1.00 being defined as the specific activity of the original, unfractionated cell population.

RESULTS AND DISCUSSION. Rat intestinal epithelial cells, prepared by the method of Weiser, are obtained, sequentially, as cell fractions from villus tips to crypt base. Cells from these fractions were subjected to CCD in a charge-sensitive Dx–PEG phase system. CCD (data not shown) on the first fractions obtained by Weiser's method gave a larger peak to the left and a smaller peak to the right, indicating heterogeneity of the upper villus cells; the last fractions gave a peak only to the right. When all fractions were pooled before CCD two well-separated peaks were obtained, with the right peak sometimes showing additional heterogeneities (Figs. 2 and 3).

Some rats were injected with [$^{14}$C]methylthymidine, which is incorporated into the differentiating crypt cells and which, with time, is associated with maturing and aged villus cells to which the crypt cells give rise. Injected rats were sacrificed at different times after injection. Intestinal epithelial cells were prepared, subjected to CCD and the fractionation patterns of labeled cells analyzed (Fig. 2). Results indicate that the, presumably charge-associated, partitioning of crypt base cells increases during differentiation and that of the villus cells diminishes during maturation and aging. The partitioning of the upper (oldest) villus cells is the lowest of any in the intestinal cell population (see Fig. 3). The phenomenon of changing membrane surface properties as a function of cell differentiation, maturation, and aging appears to be general, having been found and traced by partitioning in a number of different cell populations.[2,3]

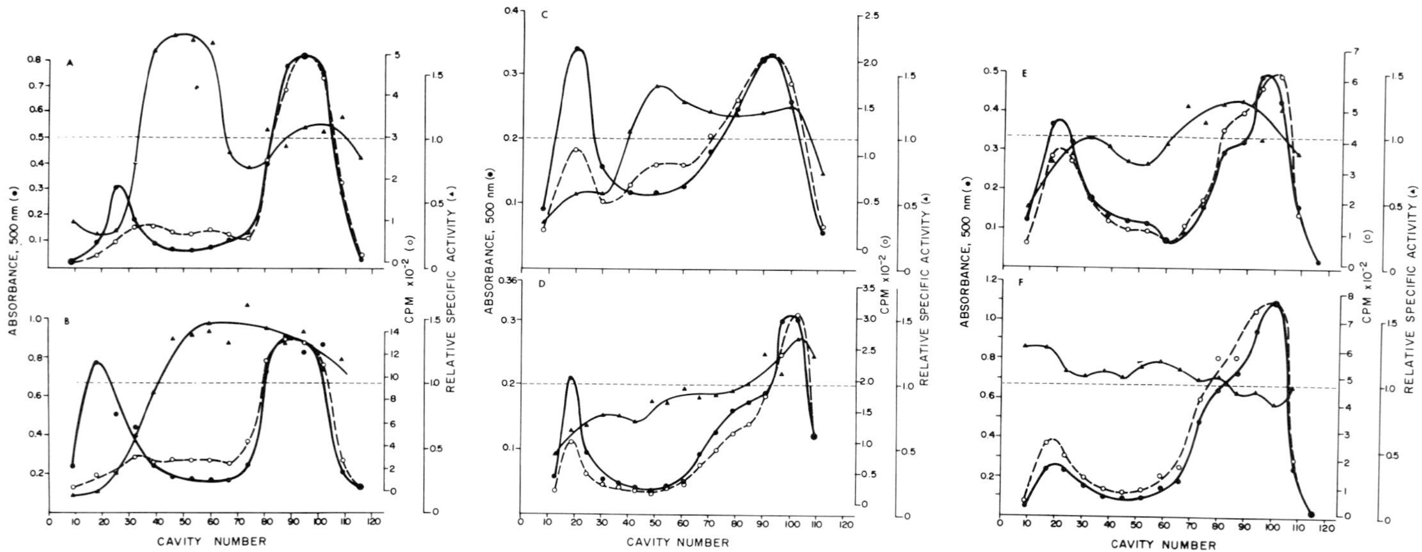


FIG. 2. Rats were injected with [$^{14}$C]methylthymidine and sacrificed at different times thereafter. Intestinal epithelial cells were prepared and subjected to countercurrent distribution in a charge-sensitive system composed of 5% dextran T500, 3.8% PEG 8000, 100 mOs*M* sodium phosphate buffer, pH 7.4, and 170 mOs*M* sucrose. One hundred twenty transfers were carried out at 4–5° using a settling time of 7 min and a shaking time of 25 sec. Cell distribution is given in terms of absorbance at 500 nm (●) and radioisotope distribution in counts/minute (○). A relative specific activity is also shown (▲) in which 1.00 is always the specific activity of the whole, unfractionated cell population. (A) Four hours after injection of isotope; (B) 11 hr; (C) 24 hr; (D) 23 hr; (E) 30 hr; (F) 48 hr. [From H. Walter and E. J. Krob, *Exp. Cell Res.* **91,** 6 (1975), with permission.]

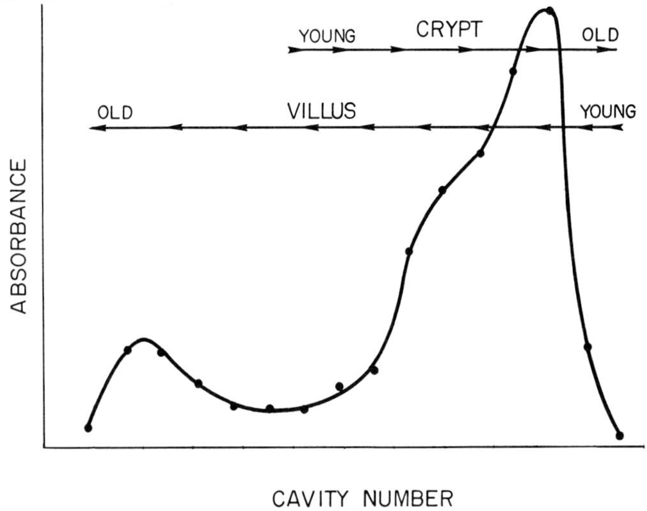


FIG. 3. Schematic presentation (summarizing data in Fig. 2) of the usual countercurrent distribution pattern obtained with pooled intestinal epithelial cells obtained by Weiser's method. The distribution is given in terms of cell absorbance (at 500 nm). Lines indicate increase in crypt and decrease in villus cell partition ($P$ values) as a function of cell differentiation, maturation, and aging. Such changes in partitioning behavior are a sensitive reflection of surface alterations accompanying these *in vivo* processes. [From H. Walter and E. J. Krob, *Exp. Cell Res.* **91,** 6 (1975), with permission.]

## *Surface Differences among Cell Subpopulations*

*Preparation, Fractionation, and Analysis of Human Peripheral Blood Monocytes.*

AIM. To test for correlations between surface properties and selected biological functions (i.e., phagocytosis) of human peripheral blood monocytes.[10]

APPROACH. All procedures are carried out under sterile conditions. Human peripheral blood mononuclear cells are obtained from 120 ml defibrinated blood, diluted 1:2 with isotonic saline, and layered on Hypaque-M (Winthrop Labs., New York, NY)–Ficoll 400 (Pharmacia LKB, Piscataway, NJ) using standard procedures.[11] The cells are collected from the Hypaque cushion and washed with Hanks' balanced salt solution

[10] H. Walter, L. L. Graham, E. J. Krob, and M. Hill, *Biochim. Biophys. Acta* **602,** 309 (1980).

[11] H. Walter, T. J. Webber, J. P. Michalski, C. C. McCombs, B. J. Moncla, E. J. Krob, and L. L. Graham, *J. Immunol.* **123,** 1687 (1979).

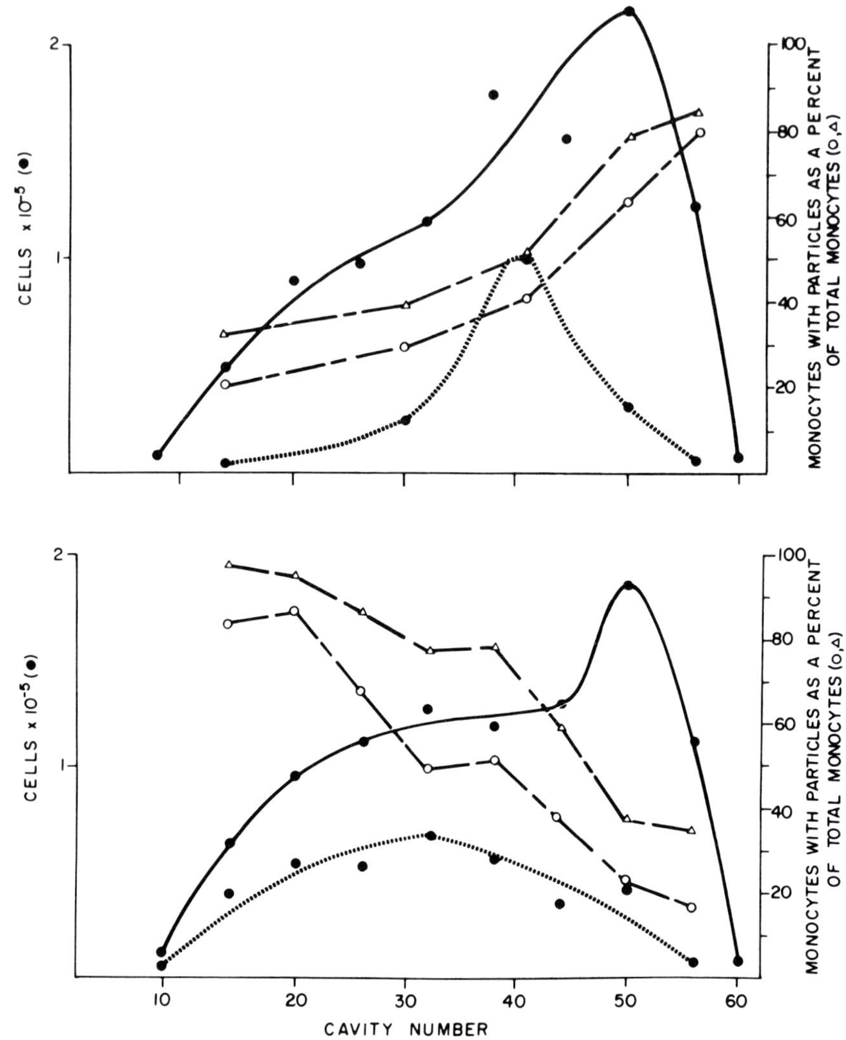


FIG. 4. (Top) Countercurrent distribution patterns of aliquots of a monocyte-enriched cell population, obtained by velocity sedimentation at unit gravity, in a charge-sensitive phase system containing 5% (w/w) dextran T500, 4% (w/w) PEG 8000, 240 mOs*M* sodium phosphate buffer, pH 7.4, 30 mOs*M* NaCl, and 5% (w/w) fetal bovine serum (heat inactivated). Sixty transfers were completed at 4–5° using a settling time of 6 min and a shaking time of 25 sec. After countercurrent distribution cells from cavities along the extraction train were permitted to phagocytose latex particles. Slides of cells were prepared, stained for myeloperoxidase activity, and counted. ●, Distribution of entire cell population (solid line) or monocytes (dotted line); △, monocytes with any number of phagocytosed particles as a percentage of total monocytes; ○, monocytes with two or more phagocytosed particles

(HBSS, GIBCO, Grand Island, NY) containing 2% (w/w) fetal bovine serum (FBS), heat-inactivated. An aliquot is electronically counted on an Electrozone Celloscope (Particle Data, Elmhurst, IL). A volume (of HBSS plus 2% heat-inactivated FBS) containing 1.2–1.5 × $10^8$ cells is then layered on a Ficoll 70 (Pharmacia LKB) gradient in a Bont velocity–sedimentation apparatus (Netherlands Cancer Institute, Amsterdam), at 4–5°.[11] Cell separation at unit gravity is primarily a function of cell size and is carried out as previously described in detail.[11] Fractions, 16 in all, are collected from the velocity sedimentation apparatus and slides prepared from aliquots of each. Cells are stained for myeloperoxidase activity with 3-amino-9-ethylcarbazole (Sigma Chemical Co., St. Louis, MO), a marker for monocytes, and counted microscopically.

Upper gradient fractions contain virtually monocyte-free lymphocytes, whereas lower ones contain increasing quantities of monocytes (up to 80%). Monocyte-enriched fractions from the velocity–sedimentation apparatus are pooled so as to have not less than 40% and often as much as 70% monocytes. Monocytes, washed 2 or 3 times with HBSS plus 2% heat-inactivated FBS, are suspended in Eagle's modified medium (GIBCO) containing 20% heat-inactivated FBS at a final concentration of $10^6$ cells/ml. A latex particle suspension (Dow Chemical Co., Midland, MI; 1.1 μm, 10% solids), diluted 1 : 10 with the same medium as used to suspend monocytes, is added to the monocytes (5 μl/ml cell suspension). The cells are incubated at 37° for 35 min with agitation. After incubation, cells are washed again with HBSS plus 2% heat-inactivated FBS solution.

Between 1.5 and 2.5 × $10^7$ monocytes which have phagocytized latex particles (see above) or which have not phagocytized latex particles (but are treated identically except that the incubation is without latex particles) are washed with top phase (of the system indicated in Fig. 4), and each preparation is finally suspended in 4 ml of top phase ("load mix"). Note that the phase system also contains FBS, which is beneficial for maintaining cell viability, and has been filtered through a 0.45-μm filter unit (Nalgene Labware Div., Rochester, NY). Cavities 0–3 and 60–63 of the 120-cavity CCD apparatus[4] each receive 0.5 ml of bottom phase and 0.9 ml of one of the load mixes, whereas cavities 4–59 and 64–119 each receive 0.6 ml of bottom phase and 0.8 ml of top phase. Both cell populations

as a percentage of total monocytes. (Bottom) CCD of aliquot of same monocyte-enriched cell population as in top and run at the same time. Prior to countercurrent distribution, monocytes were permitted to phagocytose latex particles. [From H. Walter, L. L. Graham, E. J. Krob, and M. Hill, *Biochim. Biophys. Acta* **602,** 309 (1980), with permission.]

could thus be subjected to CCD simultaneously and without overlap. Other conditions for CCD are given in Fig. 4.

Aliquots of the cell suspensions from different cavities along the extraction train are washed with HBSS plus 2% heat-inactivated FBS and electronically counted. They are then suspended in Eagle's medium plus 20% heat-inactivated FBS ($10^6$ cells/ml). Cells from different parts of the extraction train which have not phagocytized latex particles are now permitted to phagocytize the particles while cells which belong to the population that phagocytized latex particles prior to CCD are again incubated but without latex particles. Slides are prepared from all fractions, and cells are stained for myeloperoxidase. Cells are counted on each slide for relative percentages of lymphocytes and monocytes, monocytes containing latex particles as a percentage of total monocytes, and monocytes containing two or more latex particles as a percentage of total monocytes (Fig. 4).

Results and Discussion. Monocyte viability (trypan blue exclusion test) after CCD was of the order of 90%. Cells from different parts of the extraction train were examined for their ability to phagocytize latex particles. The ratio of monocytes which phagocytized particles to monocytes which did not phagocytize particles increased appreciably with increasing *P* values, presumably reflecting their higher charge-associated surface properties (Fig. 4, top). In contrast, when monocytes were permitted to phagocytize latex particles prior to CCD, an increase in *P* value was associated with an appreciable decrease in the above-indicated ratio (Fig. 4, bottom). Thus a correlation is evident between the surface properties of monocytes and their ability to phagocytize latex particles. Furthermore, phagocytosis of particles appears to alter the surface properties of monocytes which have ingested particles.[10]

*Preparation, Fractionation, and Analysis of Human T Lymphocytes.*

Aim. To examine the relative surface properties, as reflected by partitioning, of T lymphocytes having different functions and bearing different markers.[12]

Approach. All procedures are carried out under sterile conditions. Human peripheral blood mononuclear cells are obtained and counted as described in the previous section. The phase system used (for composition, see caption to Fig. 5) also contains FBS. The cells, of the order of $1.5 \times 10^8$, are suspended in 4 ml top phase, and a CCD experiment is set up in a manner analogous to that described in the previous section except that only one sample is loaded in cavities 0–3. Tubes, along the extraction train, are then pooled in groups of 5, and aliquots of the cell suspensions are counted electronically to obtain the cell distribution curve. Cells are washed 3 times with HBSS plus 2% FBS (heat inactivated) solution.

[12] J. P. Michalski, B. Bozelka, and H. Walter, *Clin. Exp. Immunol.* **67**, 565 (1987).

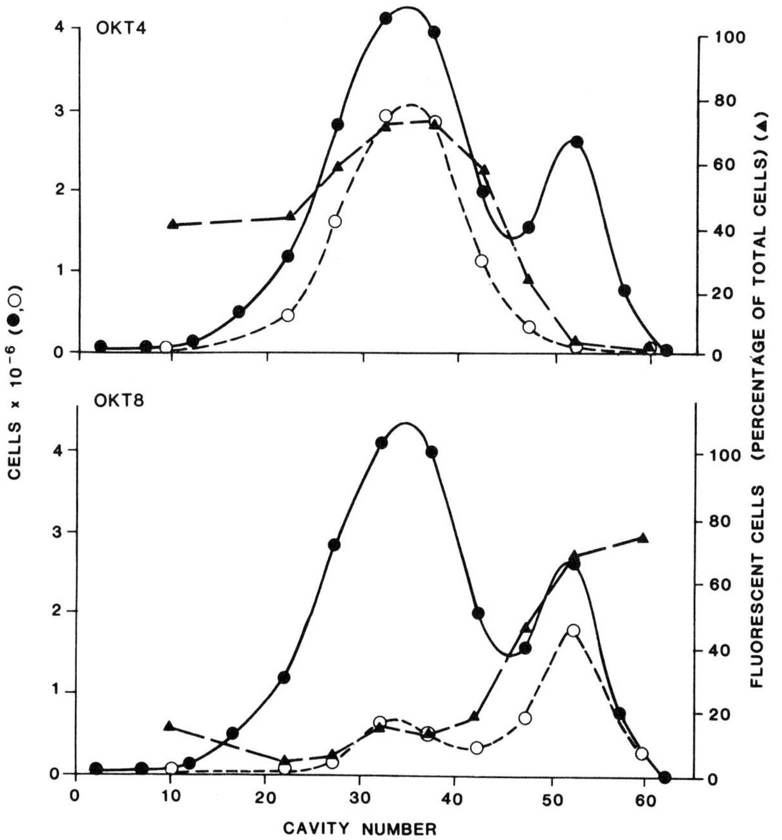


FIG. 5. Countercurrent distribution pattern of human peripheral blood mononuclear cells in a charge-sensitive aqueous two-phase system composed of 5% (w/w) dextran T500, 4% (w/w) PEG 8000, 200 mOs*M* sodium phosphate buffer, pH 7.4, 70 mOs*M* NaCl, and 5% (w/w) fetal bovine serum (heat-inactivated). Fifty-nine transfers were completed at 5° using a settling time of 6 min and a shaking time of 22 sec. Distribution of total mononuclear cells is given in cell counts (●). Cells were also analyzed, using fluorescent monoclonal antibodies, for the OKT4 (top) and OKT8 (bottom) markers. Percentages of these cells in different cavities along with extraction train are indicated (▲), as is the distribution, in terms of counts, of cells bearing the specific marker indicated (○). [From J. P. Michalski, B. Bozelka, and H. Walter, *Clin. Exp. Immunol.* **67,** 565 (1987), with permission.]

They are then analyzed using fluorescent monoclonal antibodies (Ortho Diagnostics, Raritan, NJ) to the OKT4 (T helper/inducer) or OKT8 (T suppressor/cytotoxic) markers. The fluorescent cells are counted on a fluorescence-activated cell sorter (FACS III, Becton Dickinson, Sunnyvale, CA). In each analysis 25,000 cells are assessed by means of the 488 nm laser line with the gain settings of scatter (4 × 5) and fluorescence

(2 × 1) at 500 mV. The percentage of positive cells is calculated after correction for background staining, with daily results being confirmed by fluorescence microscopy. Absolute numbers of lymphocyte subpopulations are calculated (and plotted, Fig. 5) by multiplying the percentage of that subpopulation (from FACS analysis) by the total cell count.

RESULTS AND DISCUSSION. Human peripheral blood mononuclear cells were subjected to CCD and were thereby subfractionated, presumably on the basis of their charge-associated surface properties (Fig. 5). Selected markers of cells along the extraction train were analyzed by means of monoclonal fluorescent antibodies. The helper/inducer T-cell subset (OKT4) was found to give a single distribution curve with a low partition (*P* value), that is, was to the left (Fig. 5, top). The suppressor/cytotoxic (OKT8) subset was found to have a bimodal distribution (Fig. 5, bottom) and, hence, to be heterogeneous. The OKT8 cells with the higher *P* value constitute part of a minor subpopulation of cells enriched with natural killer ($Leu7^+$) cells (data not shown). For additional separations and analyses (including double-labeling studies) of human peripheral blood lymphocyte subpopulations, see Refs. 11 and 12.

### *Fractionation of Tissue Cultured Cells*

*Fractionation and Analysis of Murine Lymphosarcoma Cell Line.*

AIM. To determine whether the known surface alterations associated with enhanced malignant characteristics of RAW117 sublines permit fractionation of the cells according to their metastatic potential.

APPROACH. The lymphosarcoma parental line RAW117-P is induced *in vitro* from spleen cultures of BALB/c mice infected with Abelson leukemia virus as previously described.[13] The *in vivo* selection for a highly malignant liver colonizing variant subline, RAW117-H10, has also been described.[13] All procedures are carried out under sterile conditions. Cells are grown in suspension using Dulbecco's modified Eagle's medium (DMEM, GIBCO) supplemened with 15% FBS without antibiotics. RAW117-P cells are sampled at different times during continuous culture (from 1 to 7 weeks) at a density of 2–3 × $10^6$ cells/ml. They are harvested by centrifugation, washed once, and resuspended in 4 ml of top phase ("load mix"; for composition, see Fig. 6). Cavities 0–3 of the 120-cavity CCD unit used[4] receive 0.5 ml of bottom phase and 0.9 ml of the load mix. All other cavities receive 0.6 ml bottom phase and 0.8 ml top phase. The CCD experiment is then run at 4–5° (for details of run, see Fig. 6). Cells are collected and kept at 4°. Cold isotonic saline is added to each tube and

[13] K. M. Miner, H. Walter, and G. L. Nicolson, *Biochemistry* **20,** 6244 (1981).

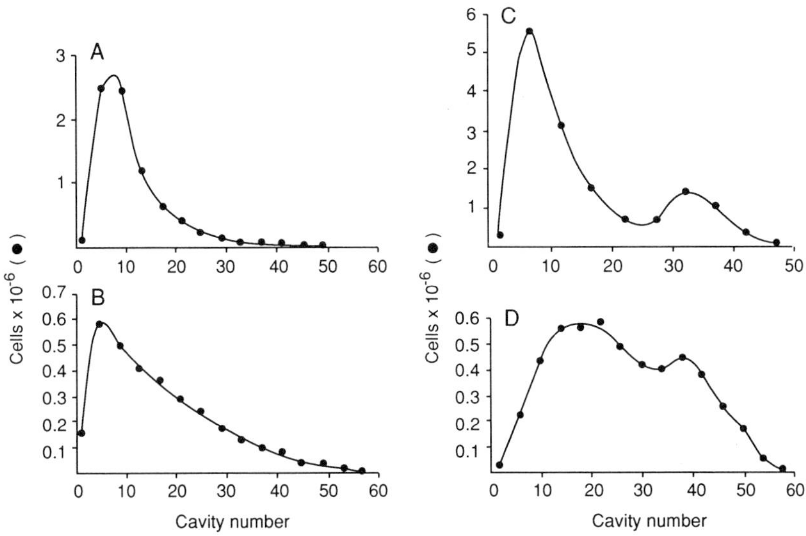


FIG. 6. Countercurrent distribution patterns obtained when murine lymphosarcoma cells (RAW 117-P) were sampled at different times during continuous culture. (A) Cells grown from a frozen sample were subjected to countercurrent distribution within 1 week; (B) cells as described in (A) sampled after an additional 1 week; (C) cells as in (A) sampled 3 weeks later; (D) cells as described in (A) sampled 6 weeks later. A charge-sensitive phase system was used containing 5% (w/w) dextran T500, 4% (w/w) PEG 8000, 160 mOs*M* sodium phosphate buffer, pH 7.4, 120 mOs*M* NaCl, and 5% (w/w) fetal bovine serum. Fifty-nine transfers were carried out (except in C, where it was 50 transfers) at 4–5° with a settling time of 6 min and a shaking time of 25 sec. [From K. M. Miner, H. Walter, and G. L. Nicolson, *Biochemistry* **20,** 6244 (1981). Copyright 1981 American Chemical Society.]

adjacent tubes are pooled (usually by groups of 4). The cells are washed once and finally suspended in DMEM plus 10% FBS and 12 m*M* HEPES, pH 7.4. Cell distribution curves are obtained by electronic cell counting of aliquots of the cell suspensions. Cell viability ranges from 60 to 96% (trypan blue exclusion test).

In the experiment depicted in Fig. 6D the cells are assayed *in vivo* for metastatic potential. The cells are incubated in DMEM plus 10% FBS and gentamicin sulfate (50 $\mu$g/ml) for 12 hr at 37°. Because the pooled CCD fractions at the ends of the distribution contain fewer cells compared to the other pooled fractions, they require a 36-hr incubation to obtain enough cells for injection into BALB/c mice. Cells are assayed for organ colonization after intravenous injection of 5000 viable cells into at least 10 mice per group. Twenty-three days after injection, organs are removed from all live animals and the number of tumor colonies determined visually (Table I).

TABLE I

TUMORS FORMED AFTER INTRAVENOUS INJECTION OF RAW117 LYMPHOSARCOMA CELLS WITH OR WITHOUT FRACTIONATION BY COUNTERCURRENT DISTRIBUTION[a]

| Cells | Number of liver tumor colonies in each animal | Median | Range | Tumor colonies at other sites |
|---|---|---|---|---|
| RAW117-P | 0, 0, 0, 0, 0, 0, 0, 50, >200[b] | 0 | 0, >200 | Lymph nodes (1)[c] |
| RAW117-P LM[d] | 0, 0, 0, 0, 0, 10, 12, >200,[b] >200,[b] >200[b] | 0 | 0, >200 | Lymph nodes (2), kidney (2) |
| RAW117-P CCD 3[d] | 0, 0, 0, 0, 0, 0, 0, 0, >200,[b] >200[b] | 0 | 0, >200 | Lymph nodes (1), spleen (1) |
| RAW117-P CCD 9 | 0, 0, 0, 0, 0, 0, 51, >200,[b] >200,[b] >200[b] | 0 | 0, >200 | Ovary (1), kidney (1) |
| RAW117-P CCD 11 | 0, >200,[b] >200,[b] >200,[b] >200,[b] >200,[b] >200,[b] >200,[b] >200[b] | >200 | 0, >200 | Lymph nodes (3), spleen (3) |
| RAW117-P CCD 12 | 34, >200,[b] >200,[b] >200,[b] >200,[b] >200,[b] >200,[b] >200,[b] >200,[b] >200[b] | >200 | 34, >200 | Spleen (4), kidney (4) |
| RAW117-P CCD 13–15 | 0, 0, 0, 0, 0, >200,[b] >200,[b] >200,[b] >200,[b] >200[b] | 0 | 0, >200 | Spleen (2), ovary (2) |
| RAW117-H10 | >200,[b] >200,[b] >200,[b] >200,[b] >200,[b] >200,[b] >200,[b] >200,[b] >200,[b] >200[b] | >200 | >200 | Lymph nodes (4) |

[a] A total of 500 viable RAW117 cells were injected intravenously in 0.2 ml of DMEM. All live animals were sacrificed at 23 days. From K. M. Miner, H. Walter, and G. L. Nicolson, *Biochemistry* **20,** 6244 (1981). Copyright 1981 American Chemical Society.

[b] Animal died before assay date. Postmortem examination indicated over 200 liver tumors.

[c] The number of animals with metastases to site are given in parentheses.

[d] LM, Load mix; CCD 3 = CCD fraction 3. Cells with a low fraction number are those having a lower partition.

RESULTS AND DISCUSSION. A low metastatic murine lymphosarcoma parental cell line (RAW117-P) and a high metastatic subline (RAW117-H10), selected for liver colonization, were grown in suspension culture and subjected to CCD in a charge-sensitive phase system. Both cell lines gave broad distribution curves (not shown), with the mean $P$ value of the metastatic line being higher than that of the parental line.[13] RAW117-P was kept in culture for almost two months, during which time the CCD curves shifted to higher mean $P$ values and revealed increasing cell population heterogeneity (Fig. 6). Cells from different cavities along the extraction train (as in Fig. 6D) were placed in tissue culture, allowed to double overnight, and assayed by *in vivo* injection in BALB/c mice. After 23 days, cells which had had higher $P$ values (corresponding more closely to the $P$ values associated with the malignant cell subline RAW117-H10) formed significantly more liver tumor colonies than did cells with lower $P$ values (Table I). This indicates that malignant cell variants exist in the parental (RAW117-P) cell population and that these can be separated by CCD from the majority of cells having low metastatic potential.

*Cell Cycle Analysis of Countercurrent Distribution-Fractionated Cultured Cell Population.*

AIM. To examine surface properties of tissue cultured cells associated with different parts of the cell cycle.

APPROACH. A K-562 subline (from Ronald Herberman, Pittsburgh Cancer Inst.) is grown in suspension culture under standard conditions[14] and harvested in the log phase of growth. The suspension is centrifuged at 500 $g$ for 15 min at room temperature, the supernatant solution is discarded, and the cell pellet is washed once with 10 ml RPMI-1640 (GIBCO). The cells are suspended in 2–5 ml of the same medium, and an aliquot is counted electronically and microscopically on a hemocytometer to determine cell viability (trypan blue exclusion). The cells ($\sim 3 \times 10^7$) are centrifuged again, and the supernatant solution is discarded; the cells are suspended in 4 ml top phase (for phase composition, see Fig. 7) and subjected to CCD.

All cavities receive 0.5 ml bottom phase. Cavities 0–2 receive 0.7 ml of the indicated load mix, whereas cavities 3–119 receive 0.7 ml top phase (see Fig. 7 for other details of the CCD run). After CCD 0.7 ml saline is added to each tube and the contents gently mixed. Adjacent tubes are pooled in groups of 4 or 5. The cells are centrifuged, washed, and finally suspended in 0.6 ml RPMI-1640. Aliquots of the suspensions (as well as an aliquot of the original, unfractionated cells) are electronically counted and cell viability determined. Remaining aliquots are additionally washed, and an adequate number of cells (not fewer than $1.5 \times 10^6$ with at least 80% of cells viable) are processed for analysis by flow cytometry using

[14] H. Walter, F. A. Al-Romaihi, E. J. Krob, and G. V. F. Seaman, *Cell Biophys.* **10,** 217 (1987).

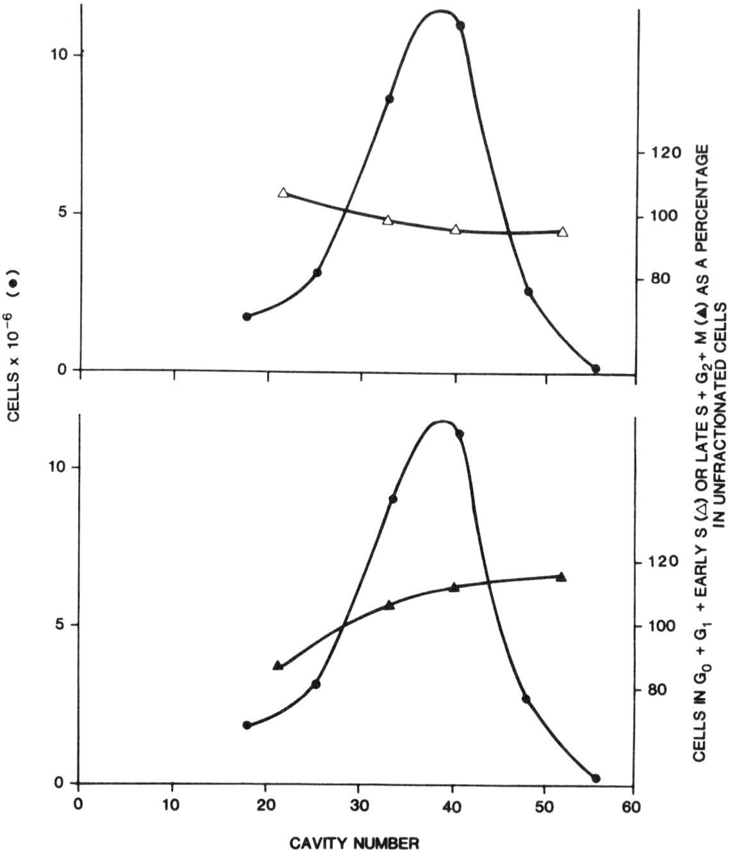


FIG. 7. Countercurrent distribution pattern of K-562 (RH) cells, harvested in log phase of growth, in a charge-sensitive phase system composed of 5% (w/w) dextran T500, 4% (w/w) PEG 8000, 220 mOs*M* sodium phosphate buffer, pH 7.4, 46 mOs*M* NaCl, and 5% (w/w) fetal bovine serum (heat-inactivated). Fifty-nine transfers were completed at 4–5° using a 6-min settling time and a 22-sec shaking time. Cell distribution is given in terms of cell counts (●). Cell cycle analysis by flow cytometry (FACS) was carried out on cells in selected fractions along the extraction train. The top graph shows the cells in $G_0$ plus $G_1$ plus early S (△) through the distribution, whereas the bottom part indicates the cells in late S plus $G_2$ plus M (▲), all expressed as a percentage in the unfractionated cell population. [From H. Walter, F. A. Al-Romaihi, E. J. Krob, and G. V. F. Seaman, *Cell Biophys.* **10,** 217 (1987), with permission.]

propidium iodide.[14] A Becton Dickinson FACS IV flow cytometer with excitation from the 488 nm line of an argon laser and emission filters from Ditric Optics of 590D(2) and 620LP are used for the analysis. The mathematical program used to calculate DNA distributions has been described.[15] The CCD curve obtained is shown in Fig. 7 as are the results from the DNA distribution analysis.

RESULTS AND DISCUSSION. K-562 cells (a culture of human leukemia stem cells originally derived from a patient with chronic myelogenous leukemia in blast crisis) were harvested in the log phase of growth and fractionated by CCD in a charge-sensitive Dx–PEG phase system. Aliquots of unfractionated cells and of cells along the extraction train were subjected to cell-cycle analysis by flow cytometry. DNA distributions were calculated. The unfractionated cells had 41% in $G_0$ plus $G_1$, 57% in S, and 13% in $G_2$ plus M. The distribution of these cells is shown in Fig. 7, as is the quantity of cells in $G_0$ plus $G_1$ plus early S (Fig. 7, top) or late S plus $G_2$ plus M (Fig. 7, bottom) each as a percentage in the original, unfractionated cells. There is a tendency for the cells in $G_0$ plus $G_1$ plus early S to decrease and those in late S plus $G_2$ plus M to increase through the distribution curve with increasing cell $P$ value (i.e., from left to right through the curve in Fig. 7). Thus, part of the heterogeneity under the distribution curve of K-562 cells[14] is related to differences in surface properties of cells in different phases of the cell cycle.

*Detection and Study of in Vitro Modification of Cell Membranes*

*Neuraminidase Treatment of Rat Red Blood Cell Population.*

AIM. To test the relative effects of an enzyme (neuraminidase) on two subpopulations (reticulocytes, mature erythrocytes) present in the rat red blood cell population.[16]

APPROACH. Rats are injected intravenously with [$^{59}$Fe]ferrous citrate (Mallinckrodt, St. Louis, MO) and bled 16.5 hr later by heart puncture using acid–citrate–dextrose as anticoagulant. Red cells are then washed 3 times with isotonic cacodylate buffer, pH 6.4. Two aliquots of 0.5 ml of washed, packed red cells are suspended in 3.5 ml of cacodylate buffer. One aliquot is incubated without any other additions, and the other receives 200 $\mu$l (0.2 IU) of neuraminidase (*Vibrio cholerae,* Calbiochem, La Jolla, CA). Both samples are incubated with gentle shaking at 37° for 90 min. After incubation the samples are again washed with cacodylate buffer 3 times, and 0.3 ml of each of the washed, packed cells is suspended, in the cold (4°), in 3.3 ml of top phase ("load mix"; see phase composition in Fig. 8). Cavities 0–2 and 60–62 each receive 0.5 ml of bottom phase and 0.7 ml of one of the load mixes. All other cavities receive 0.5 ml bottom phase and 0.7 ml top phase (other details in Fig. 8). After CCD, the cell distributions are obtained by washing the cells, lysing them, and

[15] P. N. Dean and J. H. Jett, *J. Cell Biol.* **60,** 523 (1974).

[16] H. Walter, E. J. Krob, and A. Pedram, *Biochim. Biophys. Acta* **731,** 487 (1983).

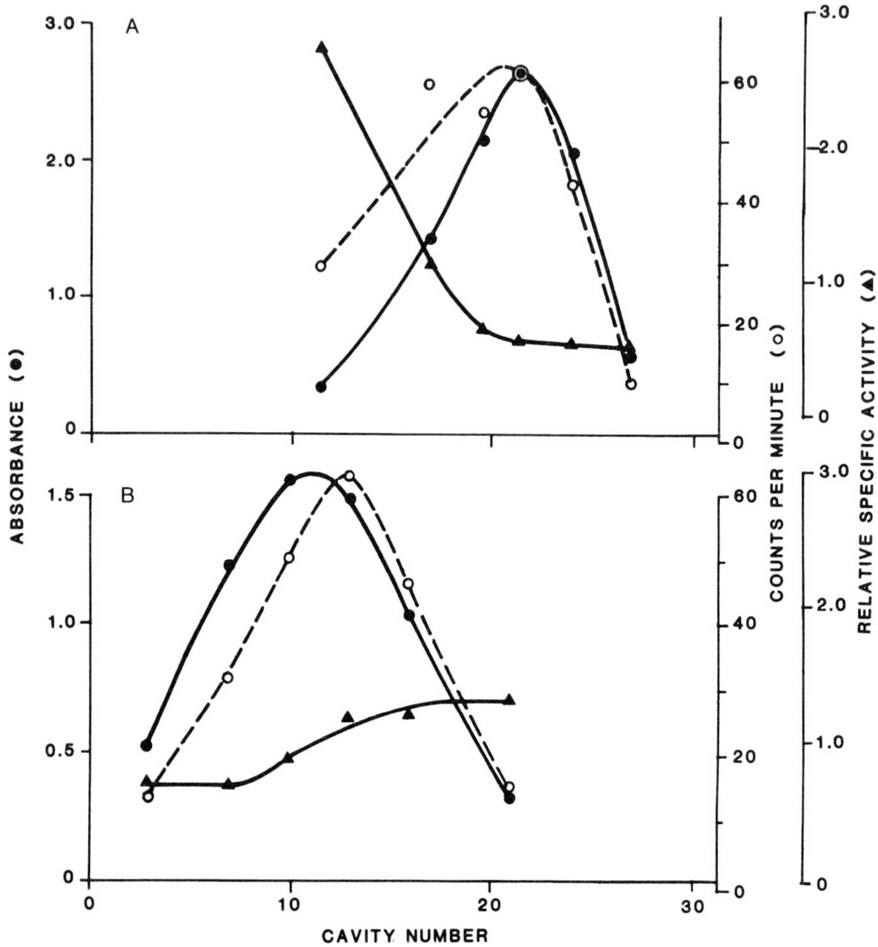


FIG. 8. Countercurrent distribution patterns of red blood cells from a rat injected with [$^{59}$Fe]ferrous citrate and bled 16.5 hr later, yielding a cell population in which the reticulocytes are labeled. Red cells were subjected to countercurrent distribution in a charge-sensitive phase system containing 5% (w/w) dextran T500, 4% (w/w) PEG 8000, 90 m*M* sodium phosphate buffer, pH 6.8, and 30 m*M* NaCl. Thirty transfers were completed at 4–5° using a settling time of 6 min and a shaking time of 22 sec. Whole red cell distribution is given in terms of hemoglobin absorbance at 540 nm (●), distribution of the labeled reticulocytes is given in terms of counts/minute (○), and a relative specific activity is also shown (▲). (A) Original, incubated red blood cell population; (B) cell population as in (A) but incubated with neuraminidase. [From H. Walter, E. J. Krob, and A. Pedram, *Biochim. Biophys. Acta* **731,** 487 (1983), with permission.]

measuring the hemoglobin absorbance at 540 nm.[1] The distribution of reticulocytes (the cell population labeled) is obtained by plotting the counts/minute in the different cavities. A relative specific activity is also calculated which relates the specific activities (counts/minute/hemoglobin absorbance) obtained in each sample to that of the original, unfractionated cell population (Fig. 8).

RESULTS AND DISCUSSION. When rats are injected with $^{59}$Fe and bled 16.5 hr later, a red cell population is obtained in which the reticulocytes are labeled.[3] These have a lower *P* value in charge-sensitive Dx–PEG phase systems than do rat mature erythrocytes. Sialic acid is a major charge-bearing group on mammalian red blood cells and, in the case of the rat erythrocytes, 70% of the sialic acid can be removed by treatment of the cells with neuraminidase. By subjecting a labeled rat red cell population, as described above, to CCD, the relative effect of this enzyme treatment on rat reticulocytes and mature erythrocytes can be studied. Although neuraminidase treatment does not eliminate the surface differences between reticulocytes and erythrocytes, it reduces the *P* value of mature erythrocytes to a greater extent than that of reticulocytes (Fig. 8). A differential effect of the enzyme on the two cell populations is thereby indicated.

## Making Use of Differences in Membrane Surface Reactivities of Different Cell Subpopulations to Alter P Values of Subpopulations: Specific Examples

### *Testing for Cell Surface Changes in Selected Disease State*

*Preparation and Separation of Red Blood Cell Populations from Patients Having Paroxysmal Nocturnal Hemoglobinuria.*

AIM. To separate two cell subpopulations ("normal" and abnormal erythrocytes in PNH) by a selective surface modification of one of the subpopulations (the abnormal cells), resulting in an altered *P* value.[6]

APPROACH. A large number of molecules of the complement protein C3b is attached to red blood cells lacking the regulatory activity of decay accelerating factor (DAF), namely, abnormal erythrocytes present in patients with PNH. In short, washed erythrocytes ($\sim 5 \times 10^9$) from a PNH patient are resuspended in 150 $\mu$l of a solution containing 1.8 mg C3 (prepared as described in Ref. 6). Trypsin (5 $\mu$l of 1 mg trypsin/ml) is added to activate the C3, producing metastable C3b, a small proportion of which attaches randomly to cells. After 6 min of incubation at 22° the cells are washed five times in 5 m*M* veronal buffer, pH 7.4, with 150 m*M* NaCl and 0.1% gelatin. This procedure deposits between 100 and 500 C3b

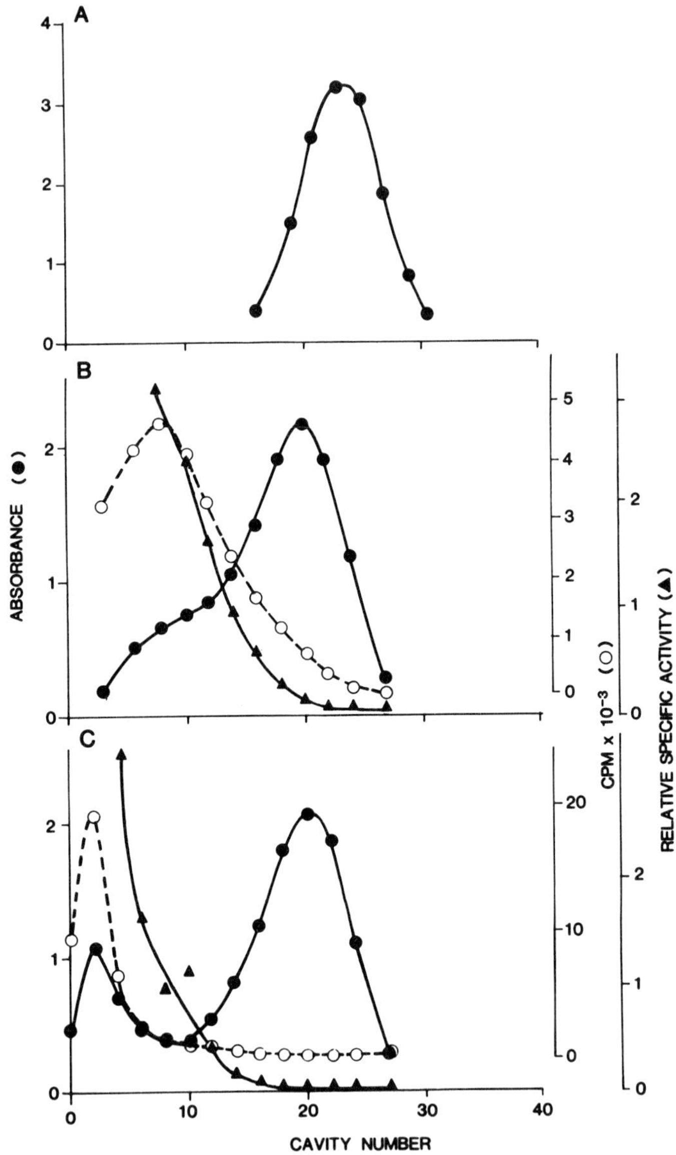


FIG. 9. Countercurrent distribution patterns of (A) erythrocytes from a patient (DM) having paroxysmal nocturnal hemoglobinuria (PNH); (B) cells as in (A) except that 24,000 $^{125}$I-labeled C3b molecules had been specifically deposited on the abnormal red cells (as referenced in the text); and (C) experiment as in (B) except that 43,000 $^{125}$I-labeled C3b molecules had been deposited on the abnormal cells. A charge-sensitive phase system containing 5% (w/w) dextran T500, 4% (w/w) PEG 8000, and 0.11 $M$ sodium phosphate

molecules/cell. A mixture (50 $\mu$l) containing 18 $\mu$g Factor B, 0.5 $\mu$g Factor D (both prepared as in Ref. 6), and 25 n*M* $NiCl_2$ is added to the cell pellet, mixed, and allowed to incubate 3 min at 22°. This permits formation of cell-bound C3 convertase. Enzyme formation is much more efficient on the abnormal cells. Formation is stopped by addition of 20 $\mu$l of 0.2 *M* EDTA, pH 7.4, and the C3 convertase is allowed to decay for 4 min at 22°, during which time most of the enzymes formed on normal cells are inactivated by DAF, but PNH erythrocytes, which lack DAF, retain more than 80% of the surface-bound enzyme. The surviving enzymes are stabilized by addition of an excess of nephritic factor (100 $\mu$g in 5 $\mu$l, prepared as in Ref. 6). Addition of 0.3 to 2.3 mg C3 containing 5 $\mu$Ci $^{125}$I-labeled C3 (labeled as in Ref. 6) results in deposition of large numbers of C3 molecules onto cells bearing active enzyme. After 2 hr of incubation at 37° the cells are washed four times in veronal buffer, pH 7.4, with 150 m*M* NaCl, and 0.1% gelatin. The number of C3b molecules bound per cell is determined by using Factor H (prepared as in Ref. 6) as previously described.[6]

Erythrocytes from a PNH patient on which no or 24,000 or 43,000 C3b molecules/cell have been deposited are subjected to CCD (Fig. 9). Three load mixes, each containing 0.2 ml of washed, packed cells suspended in 2.2 ml of top phase (see caption to Fig. 9 for composition), are prepared. These samples are run separately but simultaneously on a 120-cavity CCD unit.[4] All cavities receive 0.5 ml of bottom phase. Cavities 0–2, 40–42, and 80–82 receive 0.7 ml of one of the load mixes, while cavities 3–39, 43–79, and 83–119 each receive 0.7 ml of top phase (additional details in Fig. 9). Cells are analyzed, after CCD, for distribution (in terms of hemoglobin absorbance) and radioactivity (in terms of counts/min).

RESULTS AND DISCUSSION. The deposition of complement proteins on the surface of a cell reduces the *P* value of the cell in charge-sensitive Dx–PEG phase systems.[3] This phenomenon was used to separate, by CCD, subpopulations of erythrocytes from patients with PNH. Purified components of the complement system were used to deposit specifically $^{125}$I-labeled C3b molecules on the abnormal erythrocytes, which lack the complement regulatory protein DAF (decay accelerating factor). Around 24,000 $^{125}$I-labeled C3b molecules/cell resulted in resolution of C3b-bearing and C3b-nonbearing red cells (Fig. 9).

---

buffer, pH 6.8, was used. Forty transfers were completed at 4–5° using a 6-min settling time and a 22-sec shaking time. ●, Total cell distribution (in terms of hemoglobin absorbance at 540 nm); ○, distribution of labeled cells (in counts/minute); ▲, relative specific activity. [From M. K. Pangburn and H. Walter, *Biochim. Biophys. Acta* **902,** 278 (1987), with permission.]

## Conclusion

In this chapter I have illustrated, with specific examples, how partitioning methodology can be used to detect subtle surface differences between closely related cell populations; to trace surface changes accompanying tissue cell differentiation, maturation, and aging; to establish correlations between a cell biological function and its surface properties (as well as alteration of the surface properties of a cell subsequent to exercising its biological function); to fractionate cultured cells into subpopulations with different biological properties; and to study the effect of *in vitro* reactions on the surface properties of cells (or even the relative effect of the reactions on cell subpopulations). Finally, I have shown how, with certain modifications directed specifically to the surface of a cell subpopulation, its separation becomes feasible. Many of these analyses are not feasible by use of any other method. Additional applications have been presented elsewhere.[2,3,17,18]

## Acknowledgment

This work was supported by the Medical Research Service of the U.S. Department of Veterans Affairs.

[17] H. Walter and G. Johansson, *Anal. Biochem.* **155,** 215 (1986).
[18] H. Walter, G. Johansson, and D. E. Brooks, *Anal. Biochem.* **197,** 1 (1991).

# [28] Cell–Cell Affinity

*By* HARRY WALTER

## Introduction

### *Usefulness*

This chapter describes a highly sensitive method for the detection of affinity between cell populations. Such cell–cell interactions may fulfill a requirement in the initiation of a biological function in a cell population or may be a specific, albeit fortuitous, occurrence.

### *Rationale*

In a given phase system, components of a mixture partition independently of one another in accord with their respective characteristic partition coefficients, $K$[1], or, in the case of particulates, $P$ values,[2] except when an interaction between constituents takes place. The $K$ (or $P$) value of a biomaterial measured in the absence and presence of a second material thus indicates, by the constancy or shift in its $K$ (or $P$) value, whether an interaction has taken place. The method, which is rapid and relatively simple, has been widely applied primarily in the study of protein interactions and binding. It has led, among other things, to the demonstration of weak enzyme–enzyme complexes[3,4] and has been the subject of a number of reviews.[5-7]

### *Model System*

In this chapter, I illustrate by use of a model system, consisting of human peripheral blood mononuclear cells (PBMNC) and sheep erythrocytes (SRBC, which form spontaneous "E rosettes" with human T lymphocytes), how cell–cell affinity can be probed by countercurrent distribution (CCD) in aqueous phases.[8]

## Methods

### *General*

The primary requirement for testing the effect of one cell population on the apparent partition ratio, $G$,[9] of another cell population is that the two cell populations initially have adequately different $P$ values. Because these depend on the relative surface properties of the cells examined, no general rule can be given as to how to select a useful phase system. If an

[1] The partition coefficient $K$, used to describe the partitioning behavior of soluble materials, is defined as the concentration of a material in the top phase divided by its concentration in the bottom phase.

[2] The $P$ value, used to describe the partitioning behavior of particulates (e.g., cells), is defined as the quantity of the particulate in a bulk phase (usually the top phase) as a percentage of the total particulate added.

[3] L. Backman and G. Johansson, *FEBS Lett.* **65,** 39 (1976).

[4] L.-O. Persson and G. Johansson, *Arch. Biochem. Biophys.* **276,** 227 (1990).

[5] P.-Å. Albertsson, *Methods Biochem. Anal.* **29,** 1 (1983).

[6] L. Backman, *in* "Partitioning in Aqueous Two-Phase Systems: Theory, Methods, Uses, and Applications to Biotechnology" (H. Walter, D. E. Brooks, and D. Fisher, eds.), p. 267. Academic Press, Orlando, Florida, 1985.

[7] S. Lundberg and L. Backman, this volume [22].

[8] H. Walter, E. J. Krob, and B. J. Moncla, *Exp. Cell Res.* **115,** 379 (1978).

[9] The apparent partition ratio, $G$, is obtained from the location of the peak, $r_{max}$, in a CCD curve using the empirical equation $G = r_{max}/(n - r_{max})$, where $n$ is the number of transfers.

initially chosen phase system[10,11] turns out to be inappropriate, one can alter phase composition parameters as outlined by Walter and Larsson[10] and examine whether such manipulation results in a suitable system by affecting the relative *P* values of the two populations in a differential manner.

### *Specific Procedures Used with Model System*

Blood is collected from presumably hematologically normal individuals or from patients having chronic lymphocytic leukemia (CLL), using acid–citrate–dextrose (ACD) as anticoagulant. PBMNC are prepared by the centrifugation method using Hypaque (Winthrop Labs., New York, NY)–Ficoll (Pharmacia-LKB, Piscataway, NJ) according to Böyum.[12] Five milliliters of fresh, washed SRBC are suspended in three times their volume of cacodylate buffer (ionic strength of 0.15), pH 6.4, and incubated with 50 U of neuraminidase (*Vibrio cholerae,* Calbiochem, San Diego, CA) for 1.5 hr at 37°. The cells are then washed several times with aqueous isotonic salt solution. Rabbit RBC are washed three times prior to use in experiments.

An electronic cell counter (Particle Data, Elmhurst, IL) operating on the Coulter principle is used. With the settings employed both red and white blood cells could be counted. When only PBMNC are to be counted in a mixture of these and RBC, a red cell lysing agent (American Hospital Supply, McGaw Park, IL) is employed. Thus total cell counts and PBMNC counts could be obtained and RBC counts in a mixture of these and PBMNC by difference.

The phase system used is prepared as described elsewhere in this volume[10] and contains 5% (w/w) dextran T500 (Pharmacia-LKB), 4% (w/w) poly(ethylene glycol) 8000 (Union Carbide, New York, NY), 240 mOs*M* sodium phosphate buffer, pH 7.4, 30 mOs*M* NaCl, and 5% (w/w) fetal bovine serum (FBS). The FBS employed in the phase system is previously absorbed with SRBC (or rabbit RBC depending on the experiment in which it is to be used).[8] Its presence is considered to be beneficial to the PBMNC.

The CCD experiment is carried out on a thin-layer unit having 120 concentric cavities with a lower phase capacity of 0.7 ml as described elsewhere in this volume.[10,13] In each experiment two distributions are carried out with PBMNC, one in the presence and the other in the absence

[10] H. Walter and C. Larsson, this volume [4].

[11] D. E. Brooks and R. Norris-Jones, this volume [2].

[12] A. Böyum, *Scand. J. Clin. Lab. Invest.* **21,** (Suppl. 97), 77 (1968).

[13] H.-E. Åkerlund and P.-Å. Albertsson, this volume [8].

of erythrocytes. This can be done separately but simultaneously and without overlap as follows[8,10]: Cavities 0 and 60 each receive 0.5 ml of bottom phase and 0.9 ml of load mix (i.e., PBMNC suspended in top phase); cavities 61–63 each receive 0.5 ml of bottom phase and 0.9 ml of top phase containing neuraminidase-treated SRBC or rabbit RBC. The quantity of cells used differ in the experiments so as to yield the ratios of red cells to PBMNC indicated in Figs. 1 and 2. Cavities 1–59 and 64–119 each receive 0.6 ml bottom phase and 0.8 ml top phase. Plates are loaded in this manner to give a stationary interface during the extraction procedure.[10] Forty or sixty transfers are completed at 5° with a shaking time of 25 sec and a settling time of 7 min.

Cells are collected after CCD as described elsewhere in this volume.[10] Seven-tenths isotonic aqueous salt solution is added to each tube to convert the two-phase system to a single homogeneous suspending medium. Tubes are pooled in groups of 6, the cells centrifuged, the supernatant solution discarded, and the cells suspended in a known volume of isotonic medium. Aliquots of the cell suspensions are electronically counted and the quantity of PBMNC and red cells determined as described earlier. Figures 1 and 2 show the CCD curves for the PBMNC alone and in the presence of SRBC (Fig. 1) or rabbit RBC (Fig. 2).[8]

## Discussion

When human PBMNC are subjected to CCD in a charge-sensitive[10] phase system, B lymphocytes have the lowest, T lymphocytes an intermediate, and null lymphocytes the highest *G* value,[14] whereas monocytes, which themselves are also fractionated,[15] are spread under the entire curve. T cells constitute the largest lymphocyte subpopulation and are found predominantly under and slightly to the left of the main (i.e., right-hand) peak.[14] To obtain adequately different *G* values for PBMNC and SRBC, SRBC were treated with neuraminidase (a treatment which actually enhances E-rosette formation[16]). SRBC so treated have a lower surface charge (and lower partition ratio in charge-sensitive phases) than the untreated red cells.[16] Their *G* value is also lower than the mean *G* value of PBMNC.[8]

Figure 1 shows the distribution patterns of human PBMNC (solid lines) in the absence (top half, Fig. 1A–D) and presence (bottom half) of SRBC

[14] H. Walter, T. J. Webber, J. P. Michalski, C. C. McCombs, B. J. Moncla, E. J. Krob, and L. L. Graham, *J. Immunol.* **123,** 1687 (1979).

[15] H. Walter, L. L. Graham, E. J. Krob, and M. Hill, *Biochim. Biophys. Acta* **602,** 309 (1980).

[16] H. Walter and E. J. Krob, *Cell Biophys.* **1,** 149 (1979).

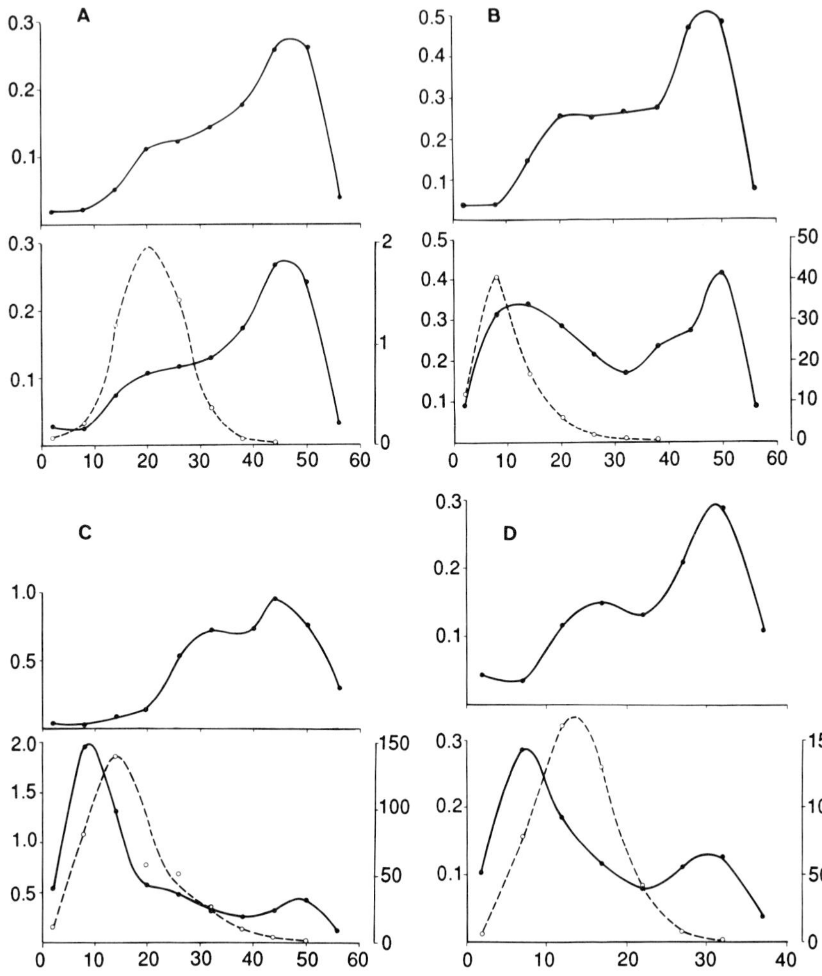


FIG. 1. *Abscissa:* Cavity number; *ordinate:* (*left*) mononuclear cells × $10^{-6}$ (●), (*right*) red blood cells × $10^{-6}$ (○). Countercurrent distribution (CCD) patterns of human normal peripheral blood mononuclear cells in either the absence (*top*) or presence (*bottom*) of different quantities of sheep red blood cells (SRBC). Mononuclear cells were obtained by use of Hypaque–Ficoll and subjected to CCD in a phase system composed of 5% (w/w) Dx T500, 4% (w/w) poly(ethylene glycol) 8000, 240 mOs*M* sodium phosphate buffer, pH 7.4, 30 mOsM NaCl, and 5% (w/w) fetal bovine serum. Either 40 or 60 transfers (as indicated) were completed. Red and white cell counts were obtained with an electronic cell counter. Ratio of SRBC to mononuclear cells: (A) 4.4, (B) 32, (C) 61, (D) 407. [From H. Walter, E. J. Krob, and B. J. Moncla, *Exp. Cell Res.* **115,** 379 (1978).]

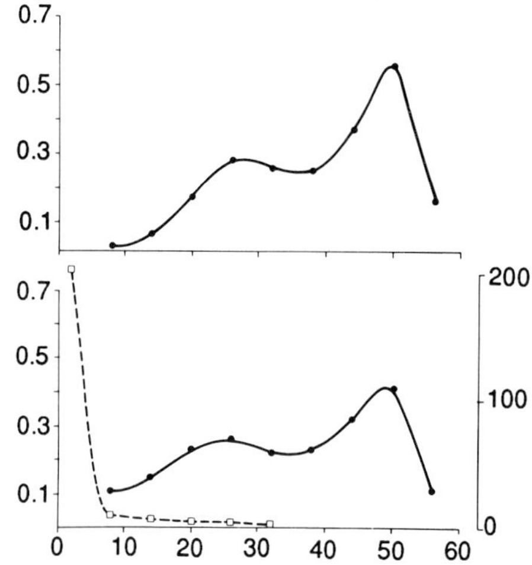


FIG. 2. *Abscissa:* Cavity number; *ordinate:* (*left*) mononuclear cells × $10^{-6}$ (●), (*right*) red blood cells × $10^{-6}$ (□). CCD pattern of human normal peripheral blood mononuclear cells in either the absence (*top*) or presence (*bottom*) of rabbit red blood cells. Ratio of rabbit RBC to mononuclear cells: 97. [From H. Walter, E. J. Krob, and B. J. Moncla, *Exp. Cell Res.* **115,** 379 (1978).]

(dashed lines). The ratio of SRBC to PBMNC was increased from 4.4 (A) to 32 (B), 61 (C), and, finally, 407 (D). With a low ratio of SRBC to PBMNC (Fig. 1A) no effect on the distribution of the latter is in evidence. With an increase in the ratio (Fig. 1B–D) there is a systematic shift of increasing quantities of T lymphocytes to lower $G$ values (i.e., to the left, where the bulk of SRBC is distributed). Lymphocytes which shift with smaller quantities of SRBC (Fig. 1B) are those with greatest binding to the SRBC, suggesting that the method yields a subfractionation of T lymphocytes based on their relative affinity for SRBC. The PBMNC peak in evidence at the far righthand end after the bulk of T lymphocytes have shifted to lower $G$ values (Fig. 1D) represents the null cells.[14,17] The reason that the SRBC population itself retains an essentially symmetrical distribution and appears not to be shifted is that the red cells are present in great excess.

In Fig. 2 an experiment was run using rabbit instead of SRBC (at a ratio of 97 : 1). Note that there is a small (but reproducible[8]) shift of PBMNC which occurs in the presence of rabbit RBC. A very small percentage of PBMNC has been reported to rosette with rabbit RBC, and the data thus appear to reflect the sensitivity of the method.

[17] J. P. Michalski, S. Zanki, J. L. Anderson, and H. Walter, *J. Clin. Lab. Immunol.* **21,** 43 (1986).

Control experiments with a human peripheral blood lymphocyte population from a patient with chronic lymphocytic leukemia (CLL, virtually all B lymphocytes) revealed that no shift in the lymphocyte population on CCD takes place in the presence of SRBC.[8]

Although the experiments described here were carried out in the cold, some cell–cell affinity studies may require higher temperatures. Preincubation of the cell populations may, in some cases, be of benefit.

In conclusion, because CCD is a stepwise procedure in which cell populations are, in essence, brought into contact and shaken apart repeatedly,[13] the detection of and information on cell–cell affinity is enhanced. The method should be especially useful in those cases in which interactions are weak, transitory, or reversible in nature.

## Acknowledgment

Work in the author's laboratory was supported by the Medical Research Service of the U.S. Department of Veterans Affairs.

# [29] Testing for Charge and Hydrophobicity Correlates in Cell–Cell Adhesion

*By* KARL-ERIC I. MAGNUSSON

## Introduction

Cell–cell adhesion is a multifactorial event governed by specific protein–protein and protein–carbohydrate interactions,[1] as well as by nonspecific physicochemical forces that depend on surface charge and hydrophobic interactions.[2–5] Specific and nonspecific factors determine the cell–cell contact at a molecular level, that is, at close proximities. Nonspecific physicochemical characteristics can also act at longer distances, and, in this case, theories developed to describe the stability of colloidal particles help to explain, as one example, aggregation phenomena of particulates

[1] H. Lis and N. Sharon, *in* "The Lectins: Properties, Functions, and Applications in Biology and Medicine" (I. E. Liener, N. Sharon, and I. J. Goldstein, eds.), p. 265. Academic Press, Orlando, Florida, 1986.

[2] P. Bongrand, C. Capo, and R. Depieds, *Prog. Surf. Sci.* **12,** 217 (1982).

[3] K.-E. Magnusson, *Scand. J. Infect. Dis. Suppl.* **33,** 32 (1982).

[4] K.-E. Magnusson, *Biochem. Soc. Trans.* **17,** 453 (1989).

[5] C. J. Van Oss, C. F. Gillman, and A. W. Neumann, "Phagocytic Engulfment and Cell Adhesiveness as Cellular Surface Phenomena." Dekker, New York, 1975.

including bacteria and mammalian cells.[6–8] Cell–cell contact is a dynamic process crucial also for the establishment of complex organisms and maintenance of their integrity. The following examples serve to elucidate some of the complex interactions involved in these processes.

During differentiation, embryogenesis, and development of specialized tissues and whole organisms, gradients of information exist that are expressed on the cell surface and on extracellular matrices, and which govern individual cells until they find their final compartment(s). Some cells, like polymorphonuclear leukocytes (PMNL), are instrumental in the defense against intruding bacteria.[9,10] They can fulfill this task because they maintain very primitive functions such as found in free-living amebas. Like the original predators they ingest almost any particulate matter such as microorganisms which may have invaded the host.[10] They also sense soluble and surface-bound chemoattractants, the concentration gradients of which help them find the direction to their prey and the sites of invasion and inflammation.[11,12]

Before reaching the site of infection, activated PMNL adhere strongly to endothelial cells of vessel walls, marginate, and move along the chemoattractant gradients. At the site, adhesion is so strong that motility stops, and phagocytosis may follow. During the crusade from the blood to the inflammatory site, cell–cell adhesion and detachment and finally adhesion and ingestion are governed by an ensemble of specific interactions, being dependent on cell–adhesion molecules and ligand–receptor-mediated intracellular signals ($Ca^{2+}$, pH, phospholipid turnover) as well as nonspecific interactions.[10,13] The effects that charge and hydrophobicity exert, both at the cellular and molecular levels and especially in the contact between bacteria and animal cells, are well documented. Both kinds of attractive forces act at long and at short distances.[2,5–8]

Cells that have a net negative charge should, according to colloidal theories, repel one another, but repulsion forces are easily negated by,

[6] B. A. Pethica, *in* "Microbial Adhesion to Surfaces" (R. C. W. Berkeley, J. M. Lynch, J. Melling, P. R. Rutter, and B. Vincent, eds.), p. 19. Society for Chemistry and Industry, Ellis Harwood, Chichester, 1980.

[7] W. C. Duncan-Hewitt, *in* "Microbial Cell Surface Hydrophobicity" (R. J. Doyle and M. Rosenberg, eds.), p. 39. American Society for Microbiology, Washington, D.C., 1990.

[8] C. Tanford, "The Hydrophobic Effect: Formation of Micelles and Biological Membranes," 2nd Ed. Wiley, New York, 1980.

[9] S. Mudd, M. McCutcheon, and B. Lucke, *Physiol. Rev.* **14,** 210 (1934).

[10] L. Öhman, G. Maluszynska, K.-E. Magnusson, and O. Stendahl, *in* "Progress in Drug Research" (E. Jucker, ed.), p. 131. Birkhauser Verlag, Basel, 1988.

[11] C. Dahlgren, H. Elwing, and K.-E. Magnusson, *Colloids Surf.* **17,** 295 (1986).

[12] J. M. Lackie, "Cell Movements and Cell Behaviour." Allen & Unwin, London, 1986.

[13] M. Patarroyo, *Clin. Immunol. Immunopathol.* **60,** 333 (1991).

for example, $Ca^{2+}$ bridging. Different cell adhesion molecules also promote cell–cell contact in a $Ca^{2+}$-dependent manner.[13] For instance, in enterocyte–enterocyte associations all molecules binding adjacent cells require $Ca^{2+}$.[11,12] Moreover, membrane proteins are usually mobile in the plane of the membrane,[14] which may change the repulsion–attraction force as distinct from stable colloidal particles like latex beads. Lateral segregation of charged proteins on either of two interacting cells may cause positive–negative charge coupling and may also permit long- and short-distance hydrophobic interactions to occur. A mosaic of negative and positive domains on cell membranes, on the other hand, might offset the overall net negative charge and the putative repulsion of two or more particles.

## Aqueous Phase Partitioning for Probing Cell Surface Charge and Hydrophobicity

### *Background*

For individual proteins, peptide (amino acid)-associated plots of hydrophobicity–hydrophilicity and of positive–negative charge can be used to describe the topography of hydrophobic and charge properties. However, because neither the total, integrated membrane distribution of lipids (glycolipids, neutral, positively, and negatively charge lipids) and proteins (glycoproteins, negatively and positively charged) nor their relative distributions are known, this approach cannot be used for whole-cell surface analysis. Aqueous two-phase partitioning has proved to be a powerful tool in probing the surfaces of prokaryotic and eukaryotic cells.

### *Aqueous Two-Phase Systems*

The system used contains 5% (w/w) poly(ethylene glycol) (PEG) 4000 or 6000 (Carbowax, Union Carbide, New York, NY; molecular weight 4000 or 6000–8000) and 5% (w/w) dextran (Dx) T500 (Pharmacia Biotechnology, Uppsala, Sweden; molecular weight around 470,000), which gives a PEG-rich top and a Dx-rich bottom phase. Some ions have different affinities for the two phases, and, in those cases, an electrostatic potential difference is created between the two phases.[15,16] With a 10 m*M* sodium phosphate buffer (pH 7), the top phase is positive relative to the bottom phase. The electrostatic potential can be virtually eliminated by choosing a salt having ions that have equal affinities for the two phases (e.g., NaCl,

[14] K. Jacobson, A. Ishihara, and R. Inman, *Annu. Rev. Physiol.* **49,** 163 (1987).

[15] D. E. Brooks and R. Norris-Jones, this volume [2].

[16] H. Walter and C. Larsson, this volume [4].

50–100 m*M*). In this way a non-charge-sensitive system is created, where partitioning depends more on van der Waals interactions (e.g., surface hydrophobicity).[6–8]

The introduction of hydrophobic PEG–ligands, such as PEG palmitate (P-PEG) or other fatty acid esters of PEG,[17,18] has greatly facilitated the interpretation of partitioning results in the study of hydrophobic surface properties. Moreover, positively charged trimethylamino-PEG (TMA-PEG) and negatively charged PEG sulfonate (S-PEG)[19,20] have proved useful in assessing surface charge of particles by partitioning.

### *Single-Step Partitioning Using Hydrophobic or Charged Ligand–Poly(ethylene Glycol)*

The following procedure is used to assess the surface hydrophobicity and charge of cell surfaces. A system composed of PEG 6000 (4.4%) and Dx T500 (6.2%) in a 30 m*M* trishydroxyaminomethane (Tris) buffer is prepared from sterile stock solutions. It is allowed to settle in a separatory funnel overnight at 4°. The bottom and top phases are separated and stored at 4° until used.

A single-tube system is prepared by mixing 2 ml bottom phase, 2 ml top phase, 0.2 ml hydrophobic (or charged) ligand–PEG (in distilled water), and 0.2 ml cell suspension in a graduated tube. The concentration of bacteria can be varied considerably, that is, between $10^6$ and $10^8$/ml, without affecting their partitioning, and the amount is determined primarily by the sensitivity of the detection system. The system is mixed by inverting the tubes 10–20 times, then allowed to separate for 30–60 min at 4°. After phase settling the volumes of the phases are read and recorded, and 0.5- to 1-ml samples are withdrawn from the top and bottom phases. After vortex-mixing an aliquot is withdrawn from the total system, which also contains the interface. The relative volumes of the components in the tube are chosen from a practical point of view. For instance, the addition of 0.2 ml of 2% (w/w) P-PEG (stock concentration) yields an approximate relative concentration of 2% P-PEG ($100 \times 0.2 \times 2/(4 \times 4.4 + 0.2 \times 2) \approx 2.2$) compared to the total amount of PEG. The effects of other hydrophobic or charged PEG–ligands on cell partitioning can be probed in the same manner.

[17] G. Johansson, *Biochim. Biophys. Acta* **451,** 517 (1976).

[18] G. Johansson, this volume [5].

[19] P.-Å. Albertsson, "Partition of Cell Particles and Macromolecules," 3rd Ed. Wiley(Interscience), New York, 1986.

[20] G. Johansson, *Biochim. Biophys. Acta* **222,** 381 (1970).

### Measurements of Distribution of Microorganisms and Animal Cells

The distribution of particles in the phase system can be measured in a number of ways: (1) by particle counting under the microscope or with an electronic particle counter; (2) by turbidity measurements, using a standard spectrophotometer (550–650 nm) or a Klett colorimeter; (3) by using radiolabeling techniques [when possible, the cells should be labeled metabolically by adding, for instance, $^{3}H$- or $^{14}C$-labeled amino acids or nucleotides (adenine, thymidine) to the culture medium; *in vitro* labeling with $^{125}I$ or $^{51}Cr$ has also been used, but one must ascertain that surface properties are not affected by the label]; and (4) cell plating, where the number of viable bacteria can readily be determined by the standard agar plate method and biochemical marker analysis (enzyme activity, ATP, DNA, etc.).

### *Quantification of Effect of Hydrophobic or Charged Ligands on Cell Partitioning*

When the distribution of particles between the bottom phase, the interface, and the top phase is known, a partition ratio $P_t/P_b$ can be calculated. $P_t$ and $P_b$ are the percentages of cells partitioning into the top or bottom phase, respectively. However, because significant amounts of material may adsorb at the interface, changes in partitioning in single-step analysis after incorporation of a PEG–hydrophobic (or charged) ligand cannot simply be interpreted in terms of alterations of $P_t/P_b$. We therefore have introduced another empirical measure, namely, the change of partition $(CP)$[21,22]:

$$CP = P_{\text{t,after}} - P_{\text{t,before}} + P_{\text{b,before}} - P_{\text{b,after}} = (P_{\text{t,after}} - P_{\text{b,after}}) - (P_{\text{t,before}} - P_{\text{b,before}})$$

Thus, $CP$ takes into account the transfer of material from the interface to the top phase and from the bottom phase to the interface. For example, if the particles were 100% in the bottom phase before addition of the PEG–ligand ($P_{\text{b,before}} = 100$) and were 100% transferred to the top phase by the PEG–ligand ($P_{\text{t,after}} = 100$), then $CP = 100 - 0 + 100 - 0 = 200$.

This measure has been used to assess a relative hydrophobicity of bacteria, as reflected by their interaction with P-PEG.[4,21–24] In each parti-

[21] E. Kihlström and K.-E. Magnusson, *Cell Biophys.* **2,** 177 (1980).

[22] K.-E. Magnusson and O. Stendahl, *in* "Partitioning in Aqueous Two-Phase Systems" (H. Walter, D. E. Brooks, and D. Fisher, eds.), p. 415. Academic Press, Orlando, Florida, 1985.

[23] C. Svanborg-Eden, L.-M. Bjursten, R. Hull, S. Hull, K.-E. Magnusson, Z. Moldovano, and H. Leffler, *Infect. Immun.* **44,** 672 (1984).

[24] C. Svanborg Edén, L. Hagberg, R. Hull, S. Hull, K.-E. Magnusson, and L. Öhman, *Infect. Immun.* **55,** 1224 (1987).

tioning situation and for calibration purposes, *CP* can be compared to other measures of cell surface hydrophobicity (or of charge), such as contact angles and electrophoretic mobilities.[5,11,25]

## *Other Studies on Surface Hydrophobicity or on Charge of Bacteria*

Stendahl *et al.* studied lipopolysaccharide (LPS) mutants of *Salmonella typhimurium* by a combination of aqueous biphasic partitioning and assessment of phagocytosis by neutrophils.[26] The partitioning was determined primarily by the LPS structure, as evidenced by the distribution between the top and bottom phase of the UDPgalactose-4-epimerase-less mutant LT2-M1. The bacteria consisted of two subpopulations when examined on countercurrent distribution[16] as a function of the gradual elongation of the lipopolysaccharide.

With the thin-layer countercurrent distribution apparatus described by Åkerlund and Albertsson,[27] the partitioning behavior of *S. typhimurium* bacteria was examined in a phase system composed of PEG 6000 (4.4%), Dx T500 (6.2%), and 30 m*M* Tris buffer (pH 7.3). The rough (R) bacteria and LPS-defective mutants of *S. typhimurium*[22,26] accumulated in the Dx-rich bottom phase, whereas the smooth (S) parent strain with a long complete LPS was recovered from the PEG-rich top phase. Mutant strains with a shorter polysaccharide (PS) showed a graduated partitioning depending on the size of the PS. The Tris buffer creates a potential difference between the Dx and PEG phases, top phase negative, and the partitioning between the phases was reversed when 0.01 *M* sodium phosphate buffer (pH 7)[15,16] was substituted for the Tris buffer.[15,16] The use of positively and negatively charged PEG[18] (trimethylamino-PEG and PEG sulfonate) and of hydrophobic ligand–PEG[17,18] indicate that the R-type LPS mutants were negatively charged and relatively more hydrophobic than the smooth (S) parent strains.[21,22]

Studies of gram-negative bacterial surface charge and hydrophobicity have been extended to partitioning of different strains of *Escherichia coli,*[28–34] *Neisseria gonorrhoeae,*[35,36] *Klebsiella,*[37] *Shigella,*[38] and *Yer-*

[25] G. F. V. Seaman and G. Uhlenbruck, *Arch. Biochem. Biophys.* **100,** 493 (1963).

[26] O. Stendahl, K.-E. Magnusson, C. Tagesson, R. Cunningham, and L. Edebo, *Infect. Immun.* **7,** 573 (1973).

[27] H.-E. Åkerlund and P.-Å. Albertsson, this volume [8].

[28] K.-E. Magnusson, O. Stendahl, I. Stjernström, and L. Edebo, *Acta Pathol. Microbiol. Scand. Sect. B* **86,** 113 (1978).

[29] T. Grundström, S. Normark, and K.-E. Magnusson, *J. Bacteriol.* **144,** 884 (1980).

[30] K.-E. Magnusson, J. Davies, T. Grundström, E. Kihlström, and S. Normark, *Scand. J. Infect. Dis. Suppl.* **24,** 135 (1980).

[31] J. Huixiu, K.-E. Magnusson, O. Stendahl, and L. Edebo, *J. Gen. Microbiol.* **129,** 3075 (1983).

*sinia*.[39] Effects of lipopolysaccharide, capsule, and fimbriation on the overall surface properties have been evaluated. The partitioning data have also been compared to cell affinity for positively charged (DEAE-Sepharose) and hydrophobic (octyl- or phenyl-Sepharose) materials.[21] The results obtained also correlate with contact angle measurements.[40] In all instances increased surface hydrophobicity of the bacteria correlated positively with increased interaction with phagocytic[10] and epithelial cells.[21,41] By varying the length of the hydrophobic ligand coupled to PEG, the depth of the hydrophobic moiety from the outermost surface of *S. typhimurium* and *S. minnesota* bacteria could be estimated.[21,42]

### *Effects of Immunoglobulins and Other Serum Proteins on Surface Properties and Cell–Cell Adhesion*

Coating *E. coli* and *S. typhimurium* bacteria with the immunoglobulins IgG and/or secretory IgA (SIgA) had opposite effects on the surface characteristics of the bacteria.[4,28,43–49] Thus, IgG and SIgA rendered the

---

[32] L. Öhman, K.-E. Magnusson, and O. Stendahl, *Acta Pathol. Microbiol. Immun. Sect. B* **93,** 133 (1985).

[33] L. Öhman, K.-E. Magnusson, and O. Stendahl, *Acta Pathol. Microbiol. Immunol. Scand. Sect. B,* **93,** 125 (1985).

[34] G. Maluszynska, O. Stendahl, and K.-E. Magnusson, *Acta Pathol. Microbiol. Immunol. Scand. Sect. B* **93,** 139 (1985).

[35] K.-E. Magnusson, E. Kihlström, L. Norlander, A. Nordqvist, J. Davies, and S. Normark, *Infect. Immun.* **26,** 402 (1979).

[36] K.-E. Magnusson, E. Kihlström, L. Norlander, A. Nordqvist, J. Davies, and S. Normark, *Infect. Immun.* **26,** 397 (1979).

[37] P. M. Allen, D. Fischer, J. R. Saunders, and C. A. Hart, *J. Med. Microbiol.* **24,** 363 (1987).

[38] L. Edebo, K.-E. Magnusson, and O. Stendahl, *Acta Pathol. Microbiol. Scand. Sect. B* **91,** 101 (1983).

[39] E. Kihlström and K.-E. Magnusson, *Acta Pathol. Microbiol. Scand. Sect. B* **91,** 113 (1983).

[40] R. K. Cunningham, T. O. Söderström, C. F. Gillman, and C. J. Van Oss, *Immunol. Commun.* **4,** 429 (1975).

[41] M. Rosenberg and R. J. Doyle, *in* "Microbial Cell Surface Hydrophobicity" (R. J. Doyle and M. Rosenberg, eds.), p. 1. American Society for Microbiology, Washington, D.C., 1990.

[42] K.-E. Magnusson and G. Johansson, *FEMS Microbiol. Lett.* **2,** 225 (1977).

[43] L. Edebo, E. Kihlström, K.-E. Magnusson, and O. Stendahl, *in* "Cell Adhesion and Motility" (A. S. G. Curtis and J. D. Pitts, eds.), 3rd Symposium of the British Society for Cell Biology, p. 65. Cambridge Univ. Press, Cambridge, 1980.

[44] O. Stendahl, C. Tagesson, K.-E. Magnusson, and L. Edebo, *Immunology* **32,** 11 (1977).

[45] C. Tagesson, K.-E. Magnusson, and O. Stendahl, *J. Immunol.* **119,** 609 (1977).

[46] I. Stjernström, K.-E. Magnusson, O. Stendahl, and C. Tagesson, *Infect. Immun.* **18,** 261 (1977).

[47] K.-E. Magnusson, O. Stendahl, I. Stjernström, and L. Edebo, *Immunology* **36,** 439 (1979).

bacteria more hydrophobic and more hydrophilic, respectively. The IgG effect was furthermore potentiated by serum complement activation (C3b).[4,44] These effects were also accompanied by increased and decreased affinities for PMNL.[48] Thus, IgG-specific, Fc receptor-mediated interactions are accompanied by increased hydrophobicity, promoting bacteria–cell and bacteria–liposome[45,50] interactions, possibly via the hydrophobic effect.[3,5,8] Thus, overall hydrophobic properties favor cell–cell contact, whereas hydrophilic characteristics counteract cell–cell adhesion.[4,51,52]

## Conclusions

The positive role of surface hydrophobicity in bacteria–animal cell interactions has been confirmed in a number of independent studies.[41,53–56] However, no single method can characterize cell surfaces to the point where information can predict cell–cell interactions. Although there are correlations, few direct comparisons have been made between aqueous two-phase partitioning and results obtained by use of other methods. Differences in the interaction between different hydrophobic probes and the same cell surface is a serious shortcoming of all methods used to assess surface hydrophobicity, namely, aqueous biphasic partition,[21,57]

[48] L. Edebo, M. Edebo, E. Kihlström, K.-E. Magnusson, L. Öhman, T. Skogh, and O. Stendahl, *in* "Antibodies: Protective, Destructive and Regulatory Role," Proceedings of the Ninth International Convocation on Immunology, Buffalo, New York, 1984, p. 82. Karger, Basel, 1985.

[49] K.-E. Magnusson and I. Stjernström, *Immunology* **45,** 239 (1982).

[50] C. Tagesson, O. Stendahl, and K.-E. Magnusson, *Stud. Biophys.* **64,** 151 (1977).

[51] M. C. M. van Loodrecht, J. Lyklema, W. Norde, G. Schraa, and A. J. B. Zehnder, *Appl. Environ. Microbiol.* **53,** 1893 (1987).

[52] R. Steadman, N. Topley, J. M. Knowlden, R. K. Mackenzie, and J. D. Williams, *Biochim. Biophys. Acta* **1013,** 21 (1989).

[53] R. V. Lachica, *in* "Microbial Cell Surface Hydrophobicity" (R. J. Doyle and M. Rosenberg, eds.), p. 297. American Society for Microbiology, Washington, D.C., 1990.

[54] H. J. Busscher, J. Sjollema, and H. C. van der Mei, *in* "Microbial Cell Surface Hydrophobicity" (R. J. Doyle and M. Rosenberg, eds.), p. 335. American Society for Microbiology, Washington, D.C., 1990.

[55] H. S. Courtney, D. L. Hasty, and I. Ofek, *in* "Microbial Cell Surface Hydrophobicity" (R. J. Doyle and M. Rosenberg, eds.), p. 361, American Society for Microbiology, Washington, D.C., 1990.

[56] T. Wadström, *in* "Microbial Cell Surface Hydrophobicity" (R. J. Doyle and M. Rosenberg, eds.), p. 315. American Society for Microbiology, Washington, D.C., 1990.

[57] D. Gerson and J. Akit, *Biochim. Biophys. Acta* **602,** 281 (1980).

hydrophobic interaction chromatography (HIC),[46] salt aggregation test (SAT),[58] hydrocarbon affinity,[59] and fatty acid binding.[60-62]

With regard to surface charge, there appears to be no direct correlation to adhesion, as indicated by the following examples, When the surface of bacteria is negatively charged and hydrophilic, as for capsulated *E. coli,* the interaction with animal cells is counteracted.[23,24] However, when the negative charge is presented on a hydrophobic background, as for deep-rough mutants of *S. typhimurium* and *S. minnesota,* the bacteria adhere avidly to different mammalian cells.[4,10,22] Therefore, surface hydrophobicity of microorganisms, as reflected by partitioning in aqueous two-phase systems, is a very useful correlate for cell–cell adhesion.

[58] M. A. Lindahl, A. Faris, T. Wadström, and S. Hjertén, *Biochim. Biophys. Acta* **677,** 471 (1981).
[59] M. Rosenberg, D. Gutnick, and E. Rosenberg, *FEMS Microbiol. Lett.* **9,** 29 (1980).
[60] M. Hermansson, S. Kjelleberg, and B. Norkrans, *FEMS Microbiol. Lett.* **6,** 129 (1979).
[61] S. Kjelleberg, C. Lagercrantz, and T. Larsson, *FEMS Microbiol. Lett.* **7,** 41 (1980).
[62] T. Malmqvist, *Acta Pathol. Microbiol. Scand. Sect. B* **91,** 69 (1983).

# [30] Surface Changes during Growth and Development of Slime Molds

*By* PAUL T. SHARPE

## Introduction

The best characterized and most studied slime molds are the cellular slime mold *Dictyostelium discoideum* (*Dd*) and the acellular slime mold *Physarum polycephalum* (*Pp*). Both organisms exist as haploid, free-living soil amebas, feeding on bacteria and reproducing by binary fission. When their food supply is exhausted, a survival response is initiated in which members of each species undergo uniquely distinct sequences of developmental changes that result in the formation of resistant spores. Whereas *Dd* has been extensively studied for many years by developmental biologists, *Pp* is more of a geneticist's organism, largely because of the multiallelic nature of its sexual cycle.

The free-living single-cell nature of *Dd* and *Pp* growth means that they are ideal cell populations for study by aqueous two-phase partitioning. Surface changes during the growth of both slime molds, and also during development of *Dd*, have been extensively studied by this method, the original impetus for the work being different in each case. With *Dd*, synchronously developing cells (amebas) have been subjected to thin-layer countercurrent distribution (TLCCD) in order to identify the earliest stages of differentiation into the precursors of the two basic cell types that form during development, prespore and prestalk cells. With *Pp*, single-tube aqueous two-phase partitioning studies have concentrated on growth, in particular addressing the question of whether the multiallelic nature of mating type loci *mat-A* and *mat-B* are reflected in different cell surface properties.

## Methods

### *Dictyostelium discoideum* Partitioning

The composition of aqueous two-phase systems to partition *Dd* amebas at various developmental stages can be selected experimentally in the manner outlined in Ref. 1. Initial studies showed that charge-sensitive phase systems are not suitable for *Dd* cells, probably because the plasma membrane contains little sialic acid.[2] Non-charge-sensitive phase systems were found to be suitable, and most studies have used a 60-chamber Bioshef TLCCD apparatus (Department of Biochemistry, University of Sheffield, UK). *Dd* (Ax-2 strain) amebas are grown in axenic culture (HL5 medium plus 86 mmol glucose) at 22° and harvested in the mid-exponential phase of growth (cell density between 1 and $5 \times 10^6$ cells/ml). Amebas are washed once in cold distilled water, once in the upper phase of the selected system, and are resuspended in 1 ml of upper phase at a density of $5 \times 10^7$/ml. They are then loaded onto the TLCCD rotors immediately. The bottom rotor chambers are filled to 85% of their capacity with lower phase, approximately 0.77 ml, and the same volume of upper phase is added to 59 chambers. Amebas suspended in upper phase (~0.77 ml) are added to the "load" chamber, which contains no upper phase. A total of 59 transfers are carried out with a settling time of 10 min and a shaking time of 30 sec. Phases are "broken" by the injection of 3 ml water into each rotor chamber, and the rotors are unloaded by emptying the contents of each chamber into a 4-ml tube. The number of cells in each fraction is determined either by counting aliquots on a Coulter counter or by lysis in Empigen detergent and measuring total cell protein by absorbance at 280 nm.

Phase systems (200 or 500 g) are prepared essentially as outlined by Brooks and Norris-Jones.[3] Dextran (Dx) T500 (Pharmacia LKB, Uppsala,

[1] H. Walter and C. Larsson, this volume [4].

[2] P. T. Sharpe, T. E. Treffry and D. J. Watts, *J. Embryol. Exp. Morphol.* **67,** 181 (1982).

[3] D. E. Brooks and R. Norris-Jones, this volume [2].

Sweden) and poly(ethylene glycol) (PEG) 4000 (BDH, London) are used. Stock salt solutions of 1 *M* sodium chloride, 0.2 *M* sodium sulfate, and 0.2 *M* potassium phosphate buffer, pH 7.8, are stored at room temperature. The salt composition of all phase systems used is 0.5 mol/kg NaCl (10 ml of 1 *M* stock to 200 g total phase), 10 mmol/kg $Na_2SO_4$ (10 ml of 0.2 *M* stock to 200 g), and 1 mmol/kg $KH_2PO_4$–$K_2HPO_4$ buffer, pH 7.8 (1 ml of 0.2 *M* stock to 200 g). The phase systems are equilibrated at 4° overnight, then vigorously shaken and transferred to 500-ml separatory funnels. Phase separation is allowed to take place at 4° until both upper and lower phases are transparent, at which time the phases are separated and stored in the cold.

The Dx–PEG concentrations of the phases are varied until the peak of the distribution curve is located in the center. For amebas harvested at $5 \times 10^6$ cells/ml, a Dx–PEG concentration of 5.5% (w/w) Dx–5.5% (w/w) PEG gives this desired result with the particular batches of Dx and PEG used. Different lots of Dx and PEG can result in slight shifts in the position of the peak.

*Growth.* From the earliest experiments it was evident that axenically growing amebas produce broad single-peak distribution curves. The position of the peak was found to depend on the phase of growth.[4] Thus, amebas harvested in early exponential growth ($10^6$/ml) gave a peak in fraction 20, amebas in late exponential growth ($5 \times 10^6$/ml) in fraction 28, amebas in early stationary phase ($10^7$/ml) fraction 32, and late stationary phase amebas ($2 \times 10^7$/ml) in fraction 36 (Fig. 1). Similar changes in partitioning are observed for amebas grown nonaxenically with bacteria as a food source. Because partitioning is carried out in a non-charge-sensitive phase system, the results may reflect a progressive increase in cell surface hydrophobicity (preference for the PEG-rich upper phase) as cells proceed through their growth curve.

The growth-associated surface changes may be due to a loss of carbohydrate residues from cell surface molecules. Such surface changes have also been shown to have possible developmental significance. When amebas in exponential phase and amebas in stationary phase are mixed and allowed to develop, they sort out during aggregation; moreover, spores are formed predominantly from exponential phase amebas, whereas stalk cells form predominantly from stationary phase amebas.[4] Thus, cell surface changes detected by TLCCD can be correlated with developmental potential. Furthermore, when amebas, harvested in exponential growth, are collected from the leading edge of the TLCCD (low fraction numbers) and are mixed with amebas obtained from the trailing edge (high fraction

[4] P. T. Sharpe and D. J. Watts, *J. Cell Sci.* **75,** 339 (1985).

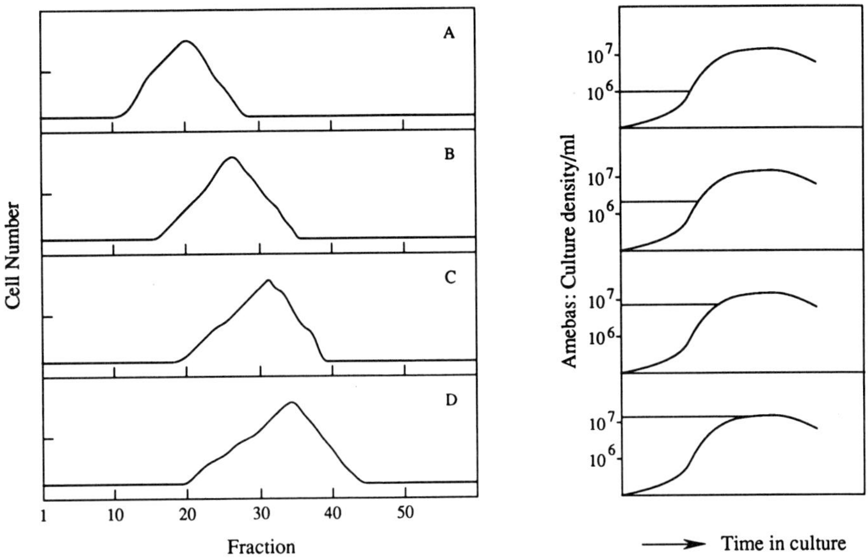


FIG. 1. Thin-layer countercurrent distribution patterns of *D. discoideum* axenically grown amebas obtained where indicated on the accompanying growth curves. (A) Cell density $10^6$/ml, (B) $5 \times 10^6$/ml, (C) $9 \times 10^6$/ml, (D) $2 \times 10^7$/ml. [From P. T. Sharpe and D. J. Watts, *J. Cell Sci.* **75,** 339 (1985).]

numbers) of the same distribution, then allowed to develop, spores again form predominantly from amebas with the less hydrophobic surface (i.e., those from the leading edge).[4]

To study possible correlates of surface heterogeneity in the exponentially growing amebas, the cell cycle distribution was analyzed in different fractions of the TLCCD. Amebas in early exponential phase were subjected to TLCCD and cells collected from different parts of the distribution (Fig. 2). These were analyzed for phase of cell cycle by measurement of DNA content per cell. Growing cultures were also pulsed with [$^3$H]thymidine (1 mCi) for 30 min and the incorporation measured in the different fractions recovered from the TLCCD apparatus. The results suggest that amebas in the leading edge (least hydrophobic and greater spore forming potential) are predominantly in $G_1$ phase. Amebas in the center fractions are in $G_2$ phase, while amebas in the trailing edge (most hydrophobic, least spore forming potential) are predominantly in S phase.[5,6] Thus, partitioning of *Dd* amebas indicates that not only do surface molecules change

[5] P. T. Sharpe and D. J. Watts, *FEBS Lett.* **168,** 89 (1983).

[6] P. T. Sharpe, G. M. Knight, and D. J. Watts, *Biochem. J.* **217,** 839 (1987).

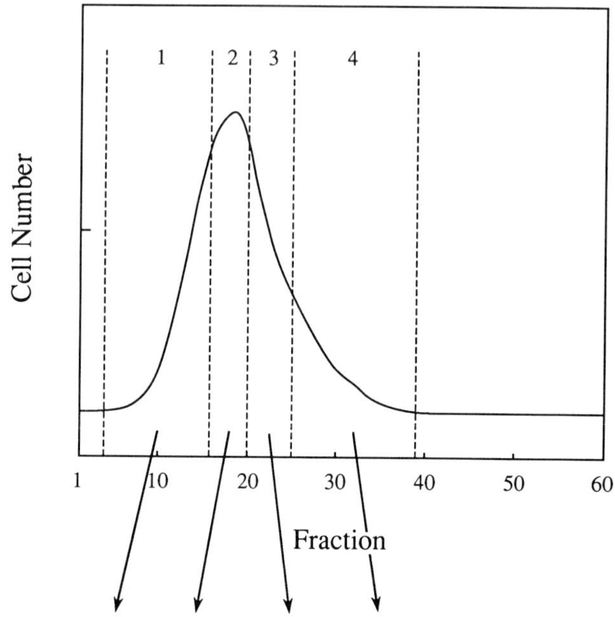


| | | | | |
|---|---|---|---|---|
| DNA pg/ameba | 0.28 | 0.38 | 0.36 | 0.28 |
| [$^3$H]Thymidine incorporation cpm/$10^7$ amebas | 2040 | 1140 | 1050 | 750 |
| Cell cycle phase | $G_1$ | $G_2$ | | S |

FIG. 2. Cell cycle analysis of cells under different parts of the thin-layer countercurrent distribution curve with *D. discoideum* amebas harvested in the exponential phase of growth. [From P. T. Sharpe and D. J. Watts, *FEBS Lett.* **168,** 89 (1983).]

during growth of amebae, but the changes correlate with subsequent developmental decisions. Analogous partitioning changes occur during the normal cell cycle, reflecting surface changes, and again these appear to influence developmental fate of the amebas.

*Development.* The same partitioning conditions as described above for growth are used on amebas harvested at different states of synchronous development.[2] Ax2 strain *Dd* amebas are grown axenically, harvested in the exponential phase of growth, and washed in cold distilled water. Synchronous development is initiated by spreading 0.5 ml of cell suspension in water at a concentration of $1 \times 10^8$ cells/ml onto 4.7-cm-diameter Millipore (Bedford, MA) filters (average pore size 0.8 $\mu$m). Each filter is placed on a prefilter saturated with 1.8 ml of pad diluted fluid containing 1.5 g KCl, 0.5 g $MgCl_2$, and 0.5 g streptomycin sulfate in 1 liter of 50 m$M$ $NaH_2PO_4$–$K_2HPO_4$ buffer, pH 6.5. The filters and prefilters sit inside 50-mm-diameter petri dishes kept humid in the dark at 22°. Under these conditions development is synchronous, with the tipped aggregate stage reached at 8 to 10 hr, the slug stage at 15 hr, and culmination at 24 hr. Fruiting bodies reach full maturity after approximately 30 hr. Six Millipore filters provide ample cells for each experiment. Routinely, cells are harvested from five filters at the required time and prepared for TLCCD. The remaining filter is usually left for 24 hr as a check that development has proceeded normally. Cells at the required time of development are harvested from the filters by washing in cold distilled water, followed by vigorous vortex-mixing to create a single cell suspension. Cells are washed in upper phase and resuspended at a density of $7 \times 10^7$ cells/ml for loading onto the TLCCD rotors.

Cells, between 0 and 11 hr of development, can be partitioned in this manner. At stages after 11 hr, however, it is difficult to disaggregate cells easily and produce single cell suspensions. The resulting cell clumps can make interpretation of the TLCCD difficult. TLCCD of cells, between 0 and 11 hr of development, in a 5.5% Dx–5.5% PEG non-charge-sensitive phase system is shown in Fig. 3. The results reveal a progressive increase in cell partition ratio (shift of distribution peak to higher fraction numbers) between 0 and 8 hr of development. At 8 hr the distribution curve broadens considerably, indicating an increase in cell surface heterogeneity. Even more striking is the distribution curve obtained at 9 hr of development, when two clearly discernible peaks are visible. The number of cells in the first small peak (peak I) apparently increases between 9 and 11 hr with respect to the large peak (peak II). Analysis of the nature of the cells in these populations showed that cells in peak I are predominantly prespore and cells in peak II are predominantly prestalk. Genetic markers of prestalk cell differentiation have been identified and indicate that the distribution is not as simple as the original analysis suggested, since the cell population comprises more than one type of stalk cell.[7] The location of these different cell types in the TLCCD is being investigated.

Evidence that the differentially acquired cell surface molecules detectable between 8 and 10 hr of development are glycoproteins has come from inhibition experiments using tunicamycin.[8] Addition of tumicamycin at 15

[7] M. J. Gaskell, K. A. Jermyn, D. J. Watts, T. E. Treffry, and J. G. Williams, *Differentiation (Berlin)*, **51,** 171 (1992).

[8] P. T. Sharpe and D. J. Watts, *Biosci. Rep.* **4,** 589 (1984).

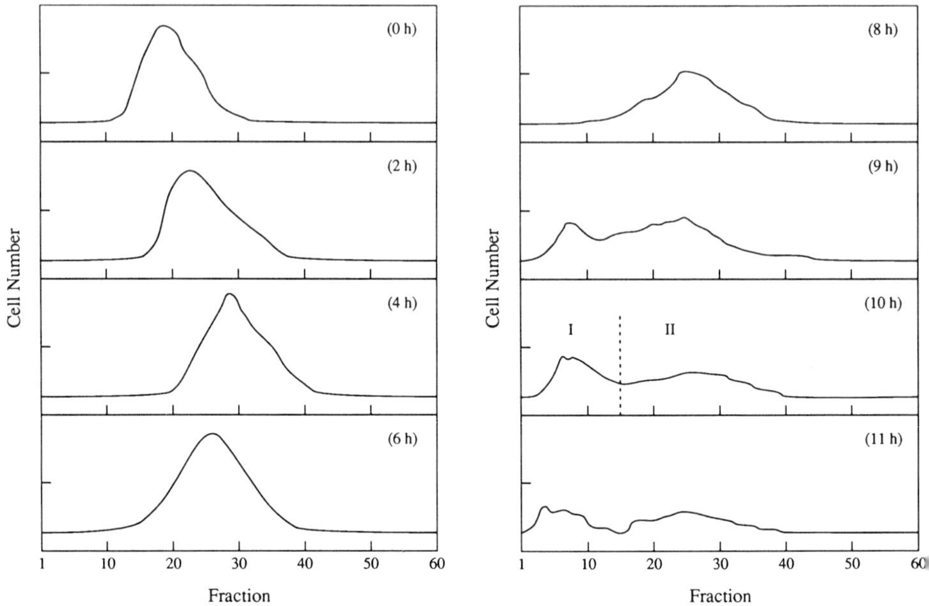


FIG. 3. Thin-layer countercurrent distribution patterns of *D. discoideum* amebas during development. [From P. T. Sharpe, T. E. Treffry, and D. J. Watts, *J. Embryol. Exp. Morphol.* **67,** 181 (1982).]

$\mu M$/ml to developing *Dd* cells fails to inhibit any of the surface changes between 0 and 6 hr detected by TLCCD, but it prevents the broadening of the peak at 8 hr and results in the absence of formation of peak I between 9 and 10 hr (Fig. 4).

*Cell Recovery and Viability.* Recovery of *Dd* cells after TLCCD was greater than 80%. Viability was estimated by dilution plating of TLCCD recovered cells with *A. aerogenes* (0.2 ml) as a food source on SM nutrient agar plates and followed by incubation at 22°. The number of colonies, as a percentage of cells plated, provides a measure of viability. Typical viabilities ranged from 90 to 95%.

### *Physarum polycephalum Partitioning*

Development of *Pp* involves the formation of large syncytial diploid plasmodia by fusion of haploid amebas. Fusion is controlled by multiallelic mating-type loci. Two of these loci, *mat-A* and *mat-B*, must be different for plasmodia to form. Differences at the *mat-B* locus are required for fusion, whereas differences at the *mat-A* locus facilitate the amebal–plas-

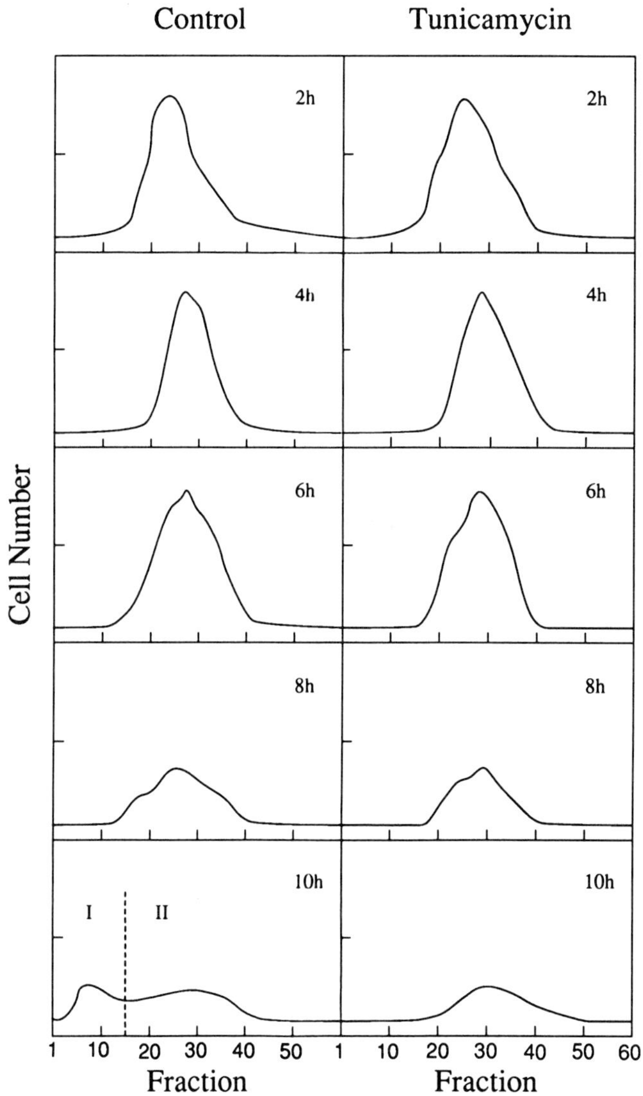


FIG. 4. Effect of tunicamycin on the thin-layer countercurrent distribution of developing *D. discoideum* amebas. [From P. T. Sharpe and D. J. Watts, *Biosci. Rep.* **4,** 589 (1984).]

modian transition. The molecular nature of these loci is not known, but their multiallelic nature is intriguing. One possibility is that different mating type loci alleles involve different cell surface molecules and provide the basis for compatibility.

Aqueous two-phase partitioning provides a simple, rapid method for selecting surface differences between amebas with different mating-type alleles.[9,10] Amebas are grown axenically, harvested at different densities, and washed in water, and $2.0 \times 10^7$ amebas are resuspended in 1 ml water. Then 1 ml of amebas is added to a 9-ml (9-g) phase system in a 10-ml disposable test tube. Non-charge-sensitive and charge-sensitive phase systems can be used for *Pp* partitioning, but non-charge-sensitive phases have been used primarily. PEG 8000 and Dx T500 concentrations are manipulated to produce a desired partition ratio for amebas in exponential growth ($5 \times 10^6$ cells/ml) as described above. The salt concentrations used are 0.5 mol/kg NaCl and 0.01 mol/kg potassium phosphate, pH 7.0. Fifty-gram phase systems are made in 45-g aliquots by direct weighing as described for *Dd* (5 g being left for addition of ameba suspension). Phases and amebas are mixed by inversion 30 times, and phases are left to settle upright at room temperature for 30 min. One milliliter is withdrawn from the center of the settled upper phase and the number of amebas counted. Partition is expressed as the percentage of added amebas recovered in the upper phase. In common with *Dd*, *Pp* amebas show progressive surface changes with cell density during growth. However, unlike *Dd*, *Pp* amebal partitioning decreases with growth such that stationary phase amebas have the lowest partition ratio (Fig. 5).

When comparing partitioning of amebas with different mating-type alleles, it is important to harvest the amebas at equivalent growth densities. Strains MA185 (*matA3 matB3*) and RSD4 (*matA1 matB1*) display markedly different relative partitioning in non-charge-sensitive phase systems, with MA185 showing higher partition ratios at nearly all equivalent stages of growth (Fig. 5). Investigation of surface differences reflected by partitioning implicates a low molecular weight soluble factor (SMF, surface-modifying factor) secreted into the axenic growth medium; incubation of RSD4 amebas in MA185-conditioned medium produces an increase in partitioning (Table I). No decrease of MA185 partitioning with increasing time of growth in RSD4-conditioned medium is observed. These experiments illustrate the potential of partitioning as a method for comparing surface changes on cells with genetic differences and also for tracing changes in response to soluble factors.

Aqueous two-phase partitioning can provide valuable information on the role of cell surface changes in slime mold growth and development. The beauty of the technique is that it does not require that any specific information on the nature of the cell surface changes be known. Indeed,

[9] P. T. Sharpe and E. M. Goodman, *Eur. J. Cell Biol.* **40,** 248 (1986).

[10] P. T. Sharpe and E. M. Goodman, *J. Gen. Microbiol.* **132,** 3491 (1986).

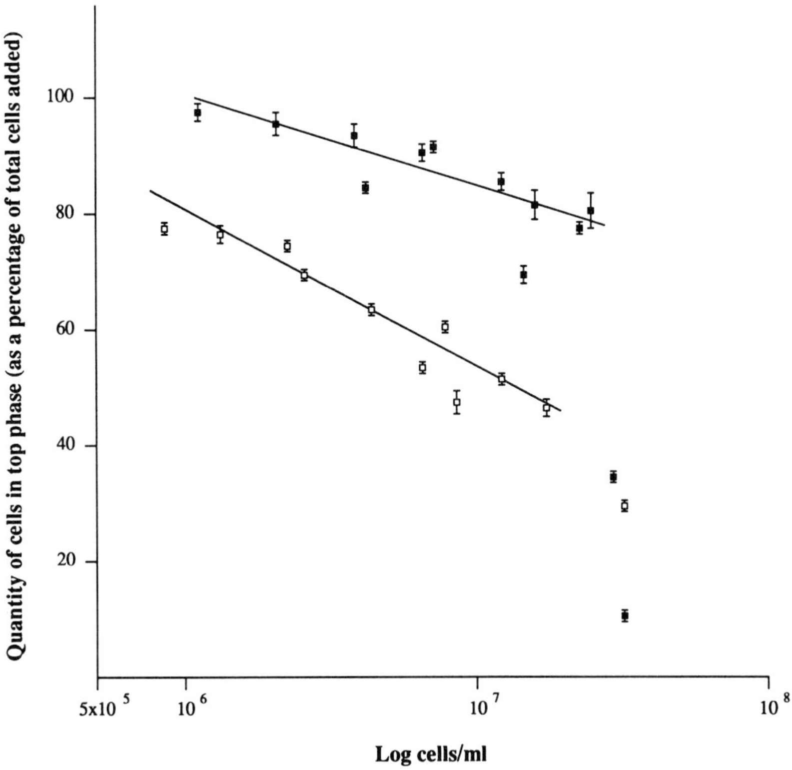


FIG. 5. Changes in the partition ratio of *P. polycephalum* amebas during growth (culture density). ■ MA185; □ RSD4. [From P. T. Sharpe and E. M. Goodman, *Eur. J. Cell Biol.* **40,** 248 (1986).]

TABLE I

EFFECT OF INCUBATION FOR DIFFERENT TIMES IN CONDITIONED MEDIUM ON PARTITIONING OF COMPATIBLE MATING TYPES OF *Physarum polycephalum* AMEBAS[a]

| Amebas | Partition fraction[b] | | |
|---|---|---|---|
| | 1 hr | 2 hr | 4 hr |
| RSD4 (+ MA185 medium) | 20 ± 2.3 | 36 ± 1.7 | 40 ± 2.1 |
| MA185 (+ RSD4 medium) | 36 ± 1.9 | 39 ± 3.1 | 38 ± 3.8 |

[a] From P. T. Sharpe and E. M. Goodman, *J. Gen. Microbiol.* **132,** 3491 (1986).

[b] Quantity of amebas in top phase (±SEM) as a percentage of total cells added.

despite the fact that none of the molecules involved in any of the surface changes detected during growth and development of slime molds have been identified, the partitioning studies have allowed us to learn a great deal about the surface changes that accompany these processes.


### Acknowledgments

I acknowledge the contributions of Donald Watts, Tim Treffy, and Gene Goodman in the work described, and Donald Watts for also critically reading the manuscript.


# [31] Separation of Pollen from *Nicotiana tabacum*

*By* BOEL MÅRTENSSON and SUSANNE WIDELL

## Introduction

Haploid plants have long been important in plant breeding because homozygotic diploid plants can be produced from them. One way to produce haploids is to influence the male gamete to develop into a plant. In principle, this can be done in two ways: (1) by cultivating isolated pollen/microspores (below we use the term pollen to include both microspores, i.e., pollen that has not undergone the first mitotic division, and true pollen) and (2) by cultivating anthers. With both methods the yield is quite low and, in order to increase yield, efforts have been made to determine from which pollen types haploid plants originate.

Microscopic analyses have revealed that the pollen population within an anther can be divided into different classes. The classification is based on the developmental stages of the same type of pollen and the fact that different types of pollen exist, since pollen from some plants are dimorphic (e.g., *Nicotiana tabacum*).[1] The predominant pollen is mainly gametophytic, that is, can fertilize stigmas, whereas their sporophytic (nongametophytic) capacity is restricted to growth of the pollen tube. Another, and much smaller, fraction of the pollen has increased sporophytic and decreased gametophytic capacities. Under favorable conditions the sporophytic pollen develops into embryos, and the term embryogenic pollen or P-grain is therefore often used for such pollen.[2] The sporophytic capacity develops during early flower induction and depends on the growth conditions of the mother plant. Short days, low temperature, and nitrogen

[1] M. Horner and H. E. Street, *Ann. Bot.* **42,** 763 (1978).
[2] E. Heberle-Bors, *Planta* **156,** 396 (1982).

starvation during flower induction are all factors that increase the amount of embryogenic pollen. Subsequent treatments, for example, cold storage of buds and preculture of anthers in liquid medium, do not influence the quantity but increase the viability of the embryogenic pollen,[3] and thereby the yield of embryos.

It is advantageous to separate embryogenic pollen from normal gametophytic pollen because the latter usually inhibits early embryo development.[3] One way of doing this is by Percoll gradient centrifugation,[4] but in 1987 phase partitioning was introduced as an alternative method for separating pollen of different types. A fraction enriched in embryogenic pollen showed a different partitioning behavior from that of gametophytic pollen.[5] Advantages of this method are both biological (the spindle formation is disturbed by strong gravitational forces[6]) and methodological (only tubes are needed for the separation but no ultracentrifuge).

As mentioned above the pollen population in a bud is not synchronized, that is, the individual grains are at different developmental stages. During this developmental process changes also occur in the pollen cell wall, which explains why pollen can be separated by phase partitioning. Therefore, a third advantage with phase separation is that it can be used not only as a means of purifying pollen that has the potential to form embryos, but also in studies on the developmental processes in sporophytic as well as gametophytic pollen.

## Procedure

### *Chemicals*

Dextran (Dx) T500 is from Pharmacia LKB (Uppsala, Sweden), and poly(ethylene) glycol (PEG) 3350 is from Union Carbide (New York, NY).

### *Plant Material*

Plants of *Nicotiana tabacum* L. var. Wisconsin are grown in the greenhouse (20°, 16 hr fluorescent white light, 75 W/m$^2$, Philips HPLHP, Eindhoven, The Netherlands). Alternatively, they are first grown under short-day conditions (15°, 8 hr light, 32 W/m$^2$, OSRAM Powerstar HQI-E 400W/D, München, Germany) until flower induction is complete and then moved to the greenhouse. Short-day conditions during flower induction has been shown to increase the yield of sporophytic pollen.[3]

[3] E. Heberle-Bors, *Theor. Appl. Genet.* **71,** 361 (1985).

[4] E. Heberle-Bors and J. Reinert, *Naturwissenschaften* **67,** 311 (1980).

[5] B. K. Mårtensson and S. Widell, *Plant Cell Tissue Organ Cult.* **8,** 27 (1987).

[6] B. S. Sangwan-Norreel, *J. Exp. Bot.* **28,** 843 (1977).

TABLE I
COMPOSITION OF NITSCH H MEDIUM USED FOR POLLEN CULTURE[a]

| Macronutrients | m$M$ | Micronutrients | $\mu M$ | Other | $\mu M$ |
|---|---|---|---|---|---|
| $KNO_3$ | 9.4 | $MnSO_4$ | 150 | *myo*-Inositol | 550 |
| $NH_4NO_3$ | 9.0 | $H_3BO_3$ | 160 | Nicotinic acid | 40 |
| $MgSO_4 \cdot 7H_2O$ | 0.75 | $FeSO_4 \cdot 7H_2O$ | 100 | Glycine | 27 |
| $CaCl_2$ | 1.5 | $Na_2EDTA \cdot 2H_2O$ | 100 | Thiamin hydrochloride | 1.5 |
| $KH_2PO_4$ | 0.5 | $ZnSO_4 \cdot 7H_2O$ | 35 | Pyridoxine hydrochloride | 2 |
| | | $NaMoO_4 \cdot 2H_2O$ | 1 | Biotin | 0.2 |
| | | $CuSO_4 \cdot 5H_2O$ | 0.1 | Folic acid | 1 |
| Sucrose 58 m$M$ | | | | | |

[a] Recalculated from J. P. Nitsch, *Phytomorphology* **19,** 389 (1969).

### *Bud Harvest*

Flower buds in which the sepals and petals are of about equal length are gently isolated and incubated in darkness at 5° to 10° for 0 to 10 days. Anthers are isolated from the buds and either used directly for the preparation of pollen suspensions or precultured before this preparation in a liquid medium (macronutrients and Fe–Na–EDTA from Nitsch's H medium,[7] see Table I) for 7 days at 25° in darkness.[8]

### *Preparation of Pollen Suspensions*

Dissected anthers from 30–40 buds are gently crushed in 5 m$M$ potassium phosphate, pH 7.8, and the pollen suspension thus obtained is filtered through two nylon nets (mesh 80 and 45 $\mu$m, respectively). The pollen is then collected by centrifugation (350 g, 7 min), washed twice, and suspended in 5 m$M$ potassium phosphate, pH 7.8. If the quantity of material is inadequate for partitioning, the pollen suspension can be frozen in liquid nitrogen and suspensions from several preparations pooled prior to phase partitioning.

The filtration is time consuming but can be speeded up by shaking the suspension during filtration through a nylon net attached to a wide tube. A simple filtration device is made by mounting the net onto a 100-ml plastic bottle (bottomless), where the net is held in place by the bottle lid which has a hole cut in the center.

[7] J. P. Nitsch, *Phytomorphology* **19,** 389 (1969).
[8] N. Sunderland and M. Roberts, *Ann. Bot.* **43,** 405 (1979).

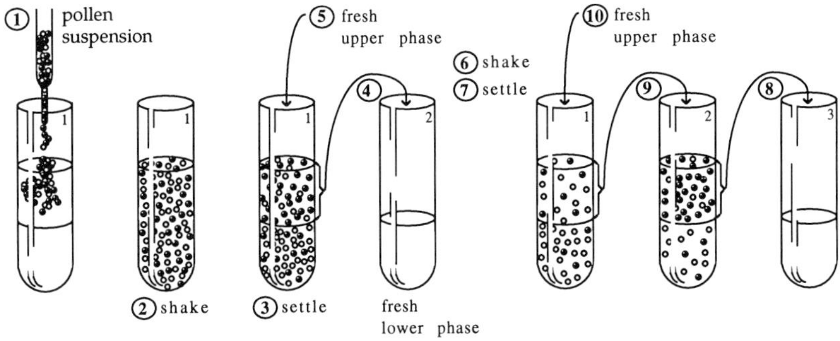


FIG. 1. Procedure for fractionation of pollen using countercurrent distribution. [From B. K. Mårtensson and S. Widell, *Plant Cell Tissue Organ Cult.* **8**, 27 (1987).]

### *Countercurrent Distribution*

As indicated, the suspension of pollen to be separated contains gametophytic pollen in different developmental stages as well as sporophytic pollen (the pollen species are identified by light microscopy, using acetocarmine staining; see below). Even in routine preparations, countercurrent distribution (CCD) is preferred[9] to a batch extraction procedure because shifts in partition ratios occur owing to the history of the pollen. The sequence in which different pollen types partition, however, is usually the same. Irrespective of whether anthers used for isolation of pollen have been precultured, the separation is carried out at room temperature as follows.

Three grams (3.0 g) of pollen suspension in 5 m*M* potassium phosphate, pH 7.8, is transferred to 21.0 g of a two-phase mixture to become the 24.0-g phase system used in the separation (see Fig. 1 and Table II). The contents of the tubes are thoroughly mixed and the phases permitted to settle (about 10–30 min) at unit gravity. The upper phase is transferred to tube 2 containing 12 ml fresh lower phase obtained from a bulk phase system (see Table II), and 12 ml fresh upper phase is added to the lower phase plus interface in the first tube. The tubes are shaken and permitted to settle; the phases are transferred as before until a total of 8 tubes is obtained (Fig. 1). The contents of the tubes are diluted three times with 5 m*M* potassium phosphate, pH 7.8, and the pollen is pelleted at 350 *g* for around 10 min. Repeated washes of the pollen are sometimes necessary.

[9] H.-E. Åkerlund and P.-Å. Albertsson, this volume [8].

TABLE II
COMPOSITIONS OF PHASE MIXTURE[a] AND BULK PHASE SYSTEM[b] FOR SEPARATION OF POLLEN INTO DIFFERENT CLASSES[c]

| Step | Phase mixture | Phase system |
|---|---|---|
| A. System useful for pollen from cold-treated buds [5.9/5.9% (w/w) of dextran and of PEG][d] | | |
| 20% (w/w) Dextran T-500 | 7.08 g | 70.8 g |
| 40% (w/w) PEG 3350 | 3.54 g | 35.4 g |
| 0.2 *M* Potassium phosphate, pH 7.8 | 5.25 g | 60.0 g |
| Sucrose (solid) | 2.05 g | 20.5 g |
| Water | 3.08 g | 53.3 g |
| | 21.0 g | 240.0 g |
| Pollen suspension (in 5 m*M* potassium phosphate) | 3.00 g | 0.0 g |
| | 24.0 g | |
| B. System useful for pollen from precultured anthers of cold-treated buds [6.1/6.1% (w/w) of dextran and of PEG][e] | | |
| 20% (w/w) Dextran T-500 | 7.32 g | 73.2 g |
| 40% (w/w) PEG 3350 | 3.66 g | 36.6 g |
| 0.2 *M* potassium phosphate, pH 7.8 | 5.25 g | 60.0 g |
| Sucrose (solid) | 2.05 g | 20.5 g |
| Water | 2.72 g | 49.7 g |
| | 21.0 g | 240.0 g |
| Pollen suspension (in 5 m*M* potassium phosphate) | 3.00 g | 0.0 g |
| | 24.0 g | |

[a] Becomes the phase system used on addition of the pollen suspension.
[b] From which upper and lower phases used in the subsequent transfers are obtained.
[c] The final composition of the phase systems are as follows: (A) 5.9% (w/w) dextran T500, 5.9% (w/w) PEG 3350, 0.25 *M* sucrose, and 5 m*M* potassium phosphate, pH 7.8; (B) 6.1% (w/w) dextran T500, 6.1% (w/w) PEG 3350, 0.25 *M* sucrose, and 5 m*M* potassium phosphate, pH 7.8.
[d] From B. K. Mårtensson and S. Widell, *Plant Cell Tissue Organ Cult.* **35,** 141 (1993).
[e] From B. K. Mårtensson and S. Widell, *Plant Cell Tissue Organ Cult.* **8,** 27 (1987).

## *Finding Optimal Phase Composition*

The optimal concentrations[10] of Dx and PEG vary, depending on how the anthers and buds are pretreated. In principle, a longer pretreatment of buds results in pollen with a higher partition ratio, provided the treatment is not too long. As shown in Fig. 2, cold pretreatments of buds for up to 8 days gave increasingly higher partition ratios for the isolated pollen, whereas 10 days of pretreatment resulted in a phase partitioning behavior similar to

[10] H. Walter and C. Larsson, this volume [4].

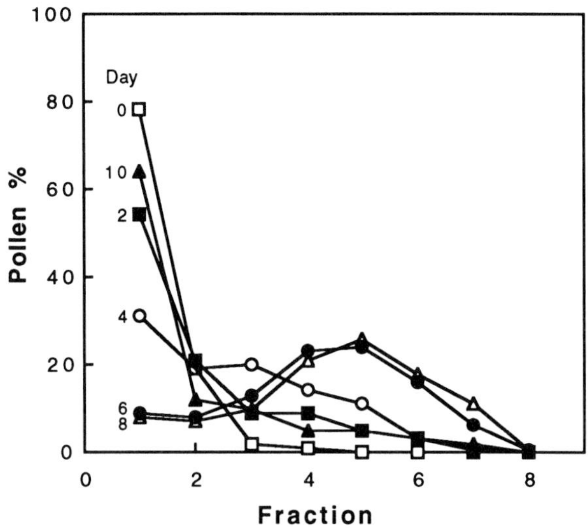


FIG. 2. Countercurrent distribution of pollen suspensions isolated from buds that have been subjected to cold for different lengths of time (0–10 days). The phase system used was composed of 5.9% (w/w) dextran T500, 5.9% (w/w) PEG 3350, 0.25 *M* sucrose, and 5 m*M* potassium phosphate, pH 7.8, and the upper phase was transferred 7 times until 8 fractions were obtained. Note that the material partitions higher (i.e., is further to the right) as the buds are cold-treated for longer times, in this case up to 8 days, but that pollen from buds kept too long in the cold (10 days) show low partition ratios again. [Redrawn from B. K. Mårtensson and S. Widell, *Plant Cell Tissue Organ Cult.* **35,** 141 (1993).]

pollen isolated from fresh buds. The effect of length of cold treatment on pollen partitioning has to be tested because sometimes the "switch point" appears earlier. Usually embryogenic pollen have a higher partition ratio than the bulk, but this also needs to be established before a suitable phase system can be chosen. In Table II a phase system with final concentrations of 5.9% (w/w) each of Dx and PEG is described. This system is useful for the separation of pollen isolated from cold-treated buds (0–10 days in darkness, see above), without preculture of the anthers in liquid medium prior to pollen isolation.

The isolated pollen population from precultured anthers (from cold-treated buds as above) has a different composition, and higher polymer concentrations are needed to get the best separation (Table II). With such pollen suspensions 6.1% (w/w) each of Dx and PEG should be used. In Fig. 3 a separation of this kind of pollen suspension is depicted as well as the number of embryos formed in each fraction after 8 weeks of culture (see below for procedure for culture of phase-separated material).

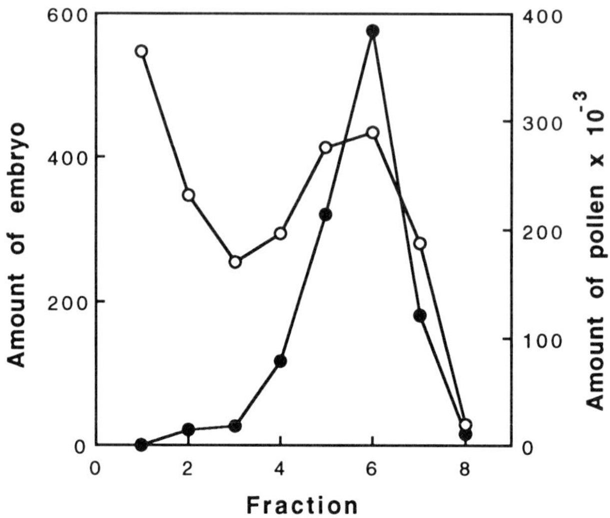


FIG. 3. Countercurrent distribution of pollen suspensions isolated from precultured anthers taken from buds that have been subjected to cold for 5 days. The phase system used was composed of 6.1% (w/w) dextran T500, 6.1% (w/w) PEG 3350, 0.25 *M* sucrose, and 5 m*M* potassium phosphate, and the upper phase was transferred 7 times until 8 fractions were obtained. ○, Total pollen; ●, total embryos developed after 8 weeks of culture. Note that the embryo-forming pollen are enriched in fractions with an intermediate to high partition ratio. [From B. K. Mårtensson and S. Widell, *Plant Cell Tissue Organ Cult.* **8,** 27 (1987).]

All manipulations are at room temperature (20°) since the pollen to be separated are intact entities. The phase systems are also equilibrated at this temperature.

*Properties of Fractions*

As shown in Fig. 3 pollen isolated from precultured anthers of cold-treated buds, and capable of forming embryos, have a higher partition ratio than does the bulk. This difference is not due to the fact that the anthers are precultured, since even without preculture the partition ratio is higher for the embryogenic pollen (Fig. 4). These embryogenic pollen are also characterized by higher specific peroxidase activity compared to nonembryogenic ones (approximate 5 times higher). A higher specific enzyme activity, although not so large, can also be seen with nonseparated pollen suspension.[11] Certain isozymes of peroxidase, especially those with

[11] B. K. Mårtensson and S. Widell, unpublished (1992).

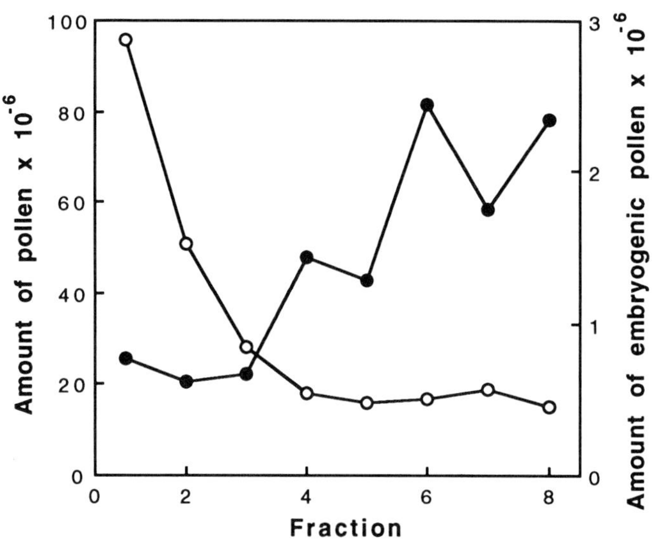


FIG. 4. Distribution of pollen in fractions obtained by countercurrent distribution of pollen suspensions isolated from cold-treated buds (7 days) without preculture of the anthers. The phase system used was composed of 5.9% (w/w) dextran T 500, 5.9% (w/w) PEG 3350, 0.25 *M* sucrose, and 5 m*M* potassium phosphate, pH 7.8, and the upper phase was transferred 7 times until 8 fractions were obtained. ○, Total pollen; ●, embryogenic pollen. [Redrawn from B. K. Mårtensson and S. Widell, *Plant Cell Tissue Organ Cult.* **35,** 141 (1993).]

more alkaline isoelectric points, are associated with embryogenesis.[11] This has also been found with embryogenic carrot suspensions and could therefore be a general phenomenon.[12] Therefore, in the search for pretreatments of buds and anthers that could result in better embryo yields, data on peroxidase activities might be useful (see below for assay of peroxidase activity).

The cold-treated and precultured pollen suspension (as in Fig. 3) was also analyzed with respect to pollen type. Different pollen classes can be distinguished as indicated in Fig. 5.[5] The embryos formed on culture originate from pollen with an intermediate to high partition ratio, and the distribution pattern correlates best with small, densely stained pollen, containing two or more nuclei (IV in Fig. 5). Mature pollen are also enriched in this fraction but are nonembryogenic.

[12] M. Joersbo, J. M. Andersen, F. T. Okkels, and R. Rajagopal, *Physiol. Plant.* **76,** 10 (1989).

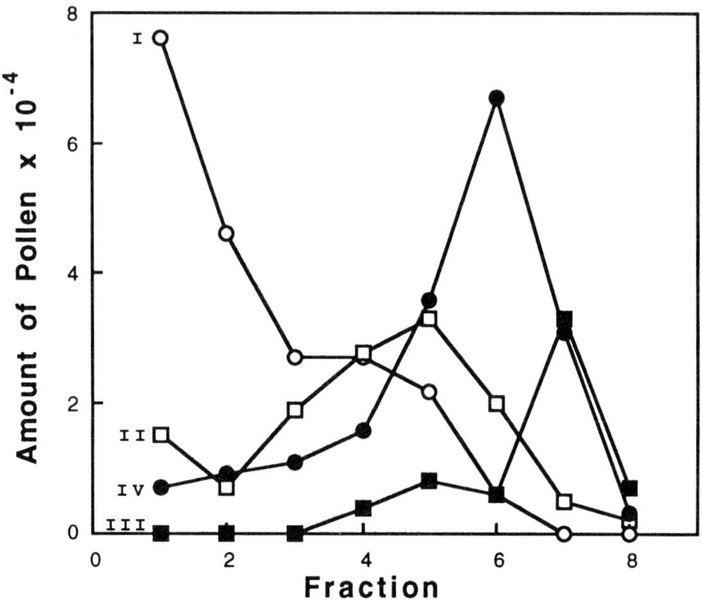


FIG. 5. Different distribution of some of the pollen classes within a pollen suspension isolated from precultured anthers from cold-treated buds (5 days). For clarity only 4 (out of 8) different pollen classes are shown. I, Small lightly stained pollen with large nucleus; II, pollen with a pollen tube; III, giant pollen with amyloplasts; IV, small, densely stained pollen, with two or more nuclei (judged to be embryogenic). The original pollen suspension was composed of 8% I pollen, 7% II pollen, 8% III pollen, 9% IV pollen, 21% dead pollen, and 47% other pollen (i.e., 16% were normal sized pollen with no clear nucleus, having a low partition ratio; 25% were normal-sized, had a lightly stained nucleus, and showed an intermediate partition ratio; and 6% were mature pollen with a thick wall). [From B. K. Mårtensson and S. Widell, *Plant Cell Tissue Organ Cult.* **8,** 27 (1987).]

### *Assays*

For microscopic observations, two drops of pollen suspension (density $10^4$ pollen/ml) from each fraction are stained with 4% (w/v) acetocarmine in 45% (v/v) acetic acid for 12 hr and mounted in glycerol.

Peroxidase activity is assayed on pollen extracts obtained by grinding 1 volume of pollen in a glass mortar together with washed Carborundum (400–500 grains/$mm^2$) in three times the volume of an extraction medium (0.1 *M* sodium phosphate, pH 6.0, 0.4 m*M* $Na_2SO_3$, 0.1 m*M* $CaCl_2$, 10.3 m*M* NaCl, and 0.11 *M* sorbitol) and 6% (w/v) polyvinylpolypyrrolidone (PVPP).[11] Carborundum and cell wall fragments are pelleted by 10 min of centrifugation at 3000 *g* and soluble protein is separated from membrane-bound protein by centrifugation at 80,000 *g* at 4° for 60 min. The supernatant solution used for peroxidase activity measurements is concentrated using Millipore (Bedford, MA) ultrafiltration or an Amicon (Danvers, MA) microconcentrator before enzyme assay. The activity is measured spectrophotometrically at 25° using $H_2O_2$ and 3,3′,5,5′-tetramethylbenzidine as substrate according to Bos *et al.*[13] All extraction steps are done at 5°.

Usually it is difficult to grind the pollen in the presence of PVPP. If phenolic compounds are not a problem, PVPP could probably be omitted from the extraction medium. Protein determination is measured spectrophotometrically according to Bearden.[14]

## Comparison of Phase Partitioning with Other Methods

As indicated, embryogenic pollen can also be isolated using Percoll gradient centrifugation.[4] We find the following advantages with phase partitioning: the avoidance of high-speed centrifugations, the easier handling of material, and more qualitative analysis that is obtained, which is sufficiently attractive to outweigh the larger embryo yield using Percoll gradient centrifugation (30 versus 15% of embryogenic pollen within a purified fraction).

## Practical Advice

Before beginning the isolation of embryogenic pollen by partitioning, determine how many days the buds should be cold-treated to yield the highest specific peroxidase activity in extracted pollen. Using this pretreatment, with or without preculture of the anthers, isolate pollen in 5 m*M* potassium phosphate, pH 7.8, and add such pollen suspensions to a series of phase mixtures to make 4.0-g phase systems, each composed of identical concentrations of Dx T500 and PEG 3350: 5.7% (w/w), 5.9% (w/w), 6.1% (w/w), or 6.3% (w/w) in 0.25 *M* sucrose and 5 m*M* potassium phosphate, pH 7.8. Mix the contents in the tubes, let the phases separate, and choose the phase system in which the material is fairly evenly distributed between upper and lower phases.

If the embryo-forming capacity is to be studied, preculture of anthers is advantageous in some systems (e.g., the one used here). The phase separations must be done under sterile conditions, using autoclaved phases (they are too viscous to be Millipore-filtered) and sterile tubes. Some water loss from the phases occurs, which must be compensated by addition

[13] E. S. Bos, A. A. van der Doelen, N. van Roy, and A. H. W. M. Schuurs, *J. Immunoassay* **2,** 187 (1981).

[14] J. C. Bearden, Jr., *Biochim. Biophys. Acta* **533,** 525 (1978).

of sterile water. After phase partitioning the fractions are washed twice with 5 m*M* sterile potassium phosphate, pH 7.8, followed by at least one wash in the medium chosen for cultivation (e.g., Nitsch H medium, appropriate for cultivation of tobacco pollen, Table I). The pollen density in each fraction is determined, and the fractions are diluted with culture medium to the same density (for tobacco about $10^4$ pollen/ml). The fractions are transferred to 5-cm petri dishes (3 ml per dish) and incubated for 3 weeks at 28° with 16 hr dim fluorescent light, 6 W/$m^2$ (Osram Powerstar HA1-E 400W/D, München, Germany). After another 5 weeks at 42 W/$m^2$, the number of embryos formed in the cultures is counted under the microscope.

# [32] Classification of *Penicillium* Fungi by Cross-Partition Analysis

*By* GÖRAN BLOMQUIST and GUNNAR B. STRÖM

## Introduction

*Penicillium* species are normally classified according to morphological criteria. This has caused confusion among mycologists since certain species, especially within the *P. viridicatum* group, have very similar microstructure characteristics. To solve these problems chemical analytical methods have been developed. Some examples are fatty acid analysis of fungal conidia,[1] pyrolysis–mass spectrometry,[2] mycotoxin production profile studies,[3] pyrolysis–gas chromatography,[4] and pyrolysis–gas chromatography combined with multivariate data analysis.[5] The disadvantage of many of these methods is that they are time consuming, and some demand highly sophisticated instrumentation. As an alternative approach, aqueous polymer two-phase systems have been used analytically for classification purposes.[6] The simplicity of this technique and the fact that the conidia retain their viability are clear advantages.

[1] G. Blomquist, B. Andersson, K. Anderson, and I. Brondz, *J. Microbiol. Methods.* in press (1992).

[2] A. C. M. Weijman, *in* "Analytical Pyrolysis" (C. E. R. Jones and C. A. Cramers, eds.), Proceedings International Symposium, 3rd Meeting, p. 225. Elsevier, Amsterdam, 1977.

[3] J. C. Frisvad, *Appl. Environ. Microbiol.* **41,** 569 (1981).

[4] T. D. Burns, R. J. Stretton, and S. D. A. K. Jayatilake, *J. Chromatogr.* **116,** 271 (1973).

[5] G. Blomquist, E. Johansson, B. Söderström, and S. Wold, *J. Anal. Appl. Pyrolysis* **1,** 53 (1979).

[6] G. Ström, G. Blomquist, and B. Söderström, *J. Appl. Bacteriol.* **66,** 461 (1989).

Aqueous two-phase systems have been used not only for analysis but also for separation in microbiological systems. It has been shown that, in a one-step procedure, conidia of some *Penicillium* species and sporangiospores of some phycomycetes can be separated.[7] With a multistep procedure (ten extractions in a countercurrent distribution) it was possible to separate mixtures of spores of different genera as well as within genera.[8] The partition behavior of conidia of different species within the *Penicillium* genera indicated that two-phase partitioning could be used to characterize *Penicillium* species. To increase the sensitivity in characterizing closely related fungi, cross-point determination[9] of *Penicillium* conidia harvested from colonies growing on different nutritional media was performed.[10] By incorporating other parameters such as the interactions between charged polymers in the systems and charged groups on the cell surfaces, hydrophobic–hydrophilic interactions, and the growth rate on different cultivation media, it was possible to classify closely related *Penicillium* differing primarily in mycotoxin production profiles.[6]

## Materials and Methods

### *Organisms*

The following strains of *Penicillium* are used: (1) *Penicillium brevicompactum* (CBS 629.66), (2) *P. verrucosum* var. *cyclopium* (CBS 483.75), (3) *P. hirsutum* (IV ATCC 32000), (4) *P. hirsutum* (IV IMI 285513), (5) *P. camembertii* II (MRA 49 b2), (6) *P. camembertii* III (PEA 50), (7) *P. crustosum* (NRRL1983), (8) *P. crustosum* (VIA 4), (9) *P. solitum* (M2/p19), (10) *P. solitum* (C127), (11) *P. solitum* (MAISY8), (12) *P. crustosum* (IIMI 285510), and (13) *P. hirsutum* (IV ATCC 32002). *Penicillium crustosum* and *P. camembertii* groups II and III were grouped by Dr. Jens Frisvad, Food Technology Laboratory, The Technical University of Denmark. CBS Strains are obtained from the Centraalbureau voor Schimmelcultures, Baarn, The Netherlands. All other species are obtained from Professor Bengt Söderström, Microbial Ecology, University of Lund, Sweden.

### *Cultivation Substrates*

The following nutritional media are used:

Corn meal agar (CMA; Difco Laboratories, Detroit, MI)

[7] G. Blomquist, G. Ström, and B. Söderström, *Appl. Environ. Microbiol.* **47,** 1316 (1984).

[8] G. Ström and G. Blomquist, *Appl. Environ. Microbiol.* **52,** 723 (1986).

[9] H. Walter and D. Forciniti, this volume [20].

[10] G. Ström, *J. Appl. Bacteriol.* **60,** 557 (1986).

TABLE I
DATA SET OF 42 VARIABLES OBTAINED FROM CROSS-PARTITION EXPERIMENTS AND GROWTH RATE MEASUREMENTS WITH *Penicillium brevi-compactum* USED IN SIMCA ANALYSIS

| Medium | Cross-point | $y'$ | Growth rate | $y_1$ | $y_2$ | $y_3$ | $y_4$ |
|---|---|---|---|---|---|---|---|
| CMA | 4.52 | 25 | 18 | 31 | 12 | 24 | 37 |
| MEA | 4.2 | 24 | 12 | 24 | 24 | 20 | 29 |
| MEA (1 : 10) | 4.73 | 38 | 16 | 66 | 31 | 36 | 41 |
| PDA | 4.3 | 30 | 13 | 42 | 29 | 24 | 33 |
| PDA (1 : 10) | 4.65 | 22 | 24 | 61 | 48 | 19 | 23 |
| SDA | 4.35 | 21 | 10 | 28 | 22 | 19 | 22 |

Malt extract agar (MEA), full strength, containing 2% malt extract (Oxoid L38) and 1.5% (w/v) purified agar (Oxoid L28, Oxoid Ltd., Basingstoke, UK)
MEA (1 : 10) containing 0.2% (w/v) malt extract and 1.5% (w/v) purified agar
Potato dextrose agar (PDA; full strength; Oxoid CM13)
PDA (1 : 10) containing 1.35% (w/v) purified agar and PDA diluted 1/10.
Saboraud dextrose agar (SDA; Oxoid CM 41)

### *Cultivation Conditions*

For the production of conidial suspensions for the two-phase partitioning experiments, the nutritional substrates are spread/inoculated with a suspension of conidia in a 0.01% (w/v) Tween 80 solution. The cultivation plates are incubated at room temperature (22°) for 10 days.

### *Growth Rate Determination*

For growth rate determinations the center of 9-cm cultivation plates are inoculated with a needle. After 10 days of cultivation the colony diameter on the six different culture media is determined (Table I).

### *Preparation of Conidial Suspensions*

To each cultivation plate 10 ml of an aqueous particle-free solution containing 0.01% Tween 80 is added. The plates are gently shaken for a couple of minutes, and the suspension is transferred to test tubes. The test tubes are placed into an ultrasonic bath for 3 min to disrupt aggregates. The conidia are sedimented by centrifugation at 3000 $g$ for 2 min. The supernatant solution is discarded, and the conidia are resuspended in 10 ml of 0.01% Tween 80 particle-free solution. This procedure is repeated.

The final pellet of conidia is suspended in 50 ml particle-free 0.01% Tween 80 aqueous solution. The concentration of conidia in the final suspension (sample suspension) is approximately $10^8$ conidia/ml.

*Preparation of Stock Solutions of Polymers, Buffers, and Salts*

An aqueous 22% (w/w) dextran (Dx) T500 (molecular weight 500,000; Pharmacia-LKB, Uppsala, Sweden) solution (500 ml) is prepared. The concentration is determined by polarimetry and adjusted to 20% (w/w).

An approximately 42% (w/w) Ficoll 400 (molecular weight 400,000; Pharmacia-LKB) solution (500 ml) is prepared. The concentration is also determined by polarimetry and adjusted to 40% (w/w).

Aqueous solutions of DEAE-Dx T500 (20%, w/w) and of sodium dextran sulfate (20%, w/w) are prepared (50 ml each). The substituted dextrans are obtained from Pharmacia-LKB.

A 40% (w/w) aqueous solution of poly(ethylene glycol) (PEG) (Carbowax 8000, molecular weight 8000; Union Carbide, New York, NY) is prepared (500 ml).

The stock solutions of PEG, Dx, and Ficoll are stored in a refrigerator to prevent microbial growth. To make weighing of the phase system components easier, stock solutions are stored in plastic drop bottles. Stock solutions of citrate–phosphate buffers of different pH are made in the following way: 161.5 g of 1 *M* disodium phosphate (A) is mixed with 425 g of 0.5 *M* citric acid (B) to give pH 3.5. To obtain pH 4.0, 200 g of A is mixed with 415 g of B and, finally, 233.5 g of A and 425 g of B gave pH 4.5. Stock solutions (100 ml) of 2 *M* sodium chloride and of 2 *M* lithium sulfate are prepared.

*Composition of Phase Systems*

The phase systems used for the partition experiments have the following compositions:

System I: From the stock solutions, 1.2 g Dx, 0.5 g Ficoll, 0.4 g PEG, 0.1 g DEAE-Dx, 0.1 ml of 2 *M* $Li_2SO_4$, 0.1 ml buffer solution, and 0.6 ml water are weighed into 10-ml test tubes. After adding 1 ml of sample suspension, the phase systems thus are composed of 6% (w/w) Dx, 5% (w/w) Ficoll, 4% (w/w) PEG, 0.5% (w/w) DEAE-Dx, 50 m*M* $Li_2SO_4$, and 7–10 m*M* sodium citrate–phosphate buffer (pH 3.5, 4.0, 4.5). The contents of each test tube are thoroughly mixed by inversion. The pH of each system is then measured directly: pH 3.8, 4.3, and 4.8, respectively.

System II: From the stock solutions 1.2 g Dx, 0.5 g Ficoll, 0.4 g PEG, 0.1 g sodium dextran sulfate, 0.1 ml of 2 *M* NaCl, 0.1 ml buffer solution, and 0.6 ml water are weighed into test tubes. After adding 1 ml of sample suspension, the phase systems thus are composed of 6% (w/w) Dx, 5% (w/w) Ficoll, 4% (w/w) PEG, 0.5% (w/w) sodium dextran sulfate, 50 m*M* NaCl, and 7–10 m*M* sodium citrate–phosphate buffer (pH 3.5, 4.0, 4.5). The contents of each test tube are thoroughly mixed by inversion. The pH of each system is measured and found to be pH 4.0, 4.5, and 5.0, respectively. The buffer concentrations given are based on phosphate.

### *Partition Experiments*

The conidia suspension from each of the six cultivation plates is added to the phase system series I and II. This means that for each fungus 6 phase systems are obtained for each cultivation medium. Thus a total of 36 phase systems resulted. For each fungal strain this procedure is repeated five times.

The contents of the test tubes are mixed by 30 inversions and the phases then allowed to settle at room temperature (22°) for 90 min. Then 1.0 ml of the upper phase of each tube is withdrawn into a separate tube and diluted with 2 ml of sterile, particle-free water containing 0.01% Tween

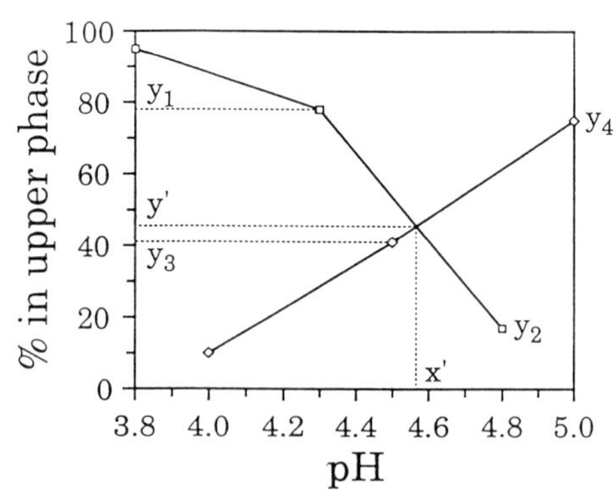


FIG. 1. Variables used in the SIMCA analysis obtained from cross-point determination. The variables are represented by the cross-point ($x'$), percentage of conidia in the upper phase at the cross-point ($y'$) and other measures as follows: from phase system I $y_1$ represents the percentage of conidia in the upper phase at pH 4.3 and $y_2$ the percentage of conidia in the upper phase at pH 4.8, and $y_3$ and $y_4$ represent the percentage in upper phase at pH 4.5 and 5.0, respectively, in phase system II.

80. The same amount of water is added to the remaining part of the system for phase dissolution (i.e., to obtain a single, homogeneous suspending medium). From a knowledge of the relative phase volumes, the partition ratio (i.e., the percentage of conidia in the upper phase relative to the interface plus lower phase) is determined. This is done by measuring the absorbance of the diluted upper phases and the diluted remaining parts of the systems separately at 560 nm.

*Evaluation of Cross-Partition Experiments*

The percentage of conidia in the upper phase was plotted in a diagram against the pH of the phase system (Fig. 1). As shown in Fig. 1 the cross-

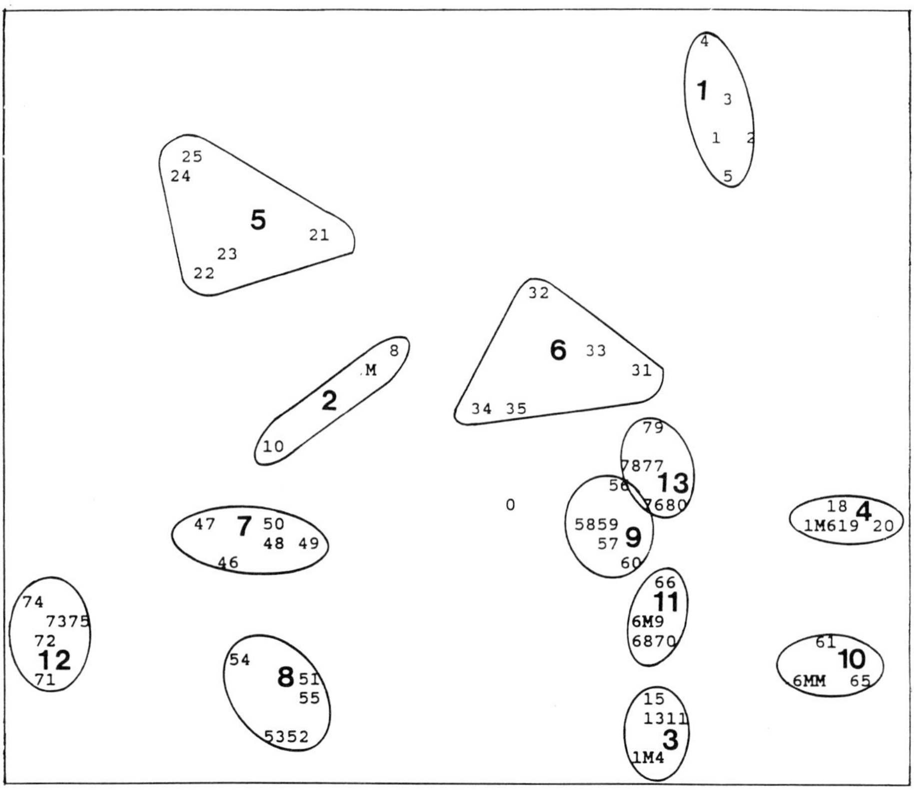


FIG. 2. The first and second components of a principal component analysis based on 42 variables for 13 strains of *Penicillium* species. Five replicate experiments were run with each strain. Bold numbers represent the different strains of *Penicillium* used. M means that two or more objects overlap.

partition of one strain cultivated on one medium resulted in six variables ($x'$, $y'$, $y_1$, $y_2$, $y_3$, $y_4$). Thus, for six cultivation media 36 variables relating to each fungal species were obtained.

### *Data Analysis*

For each of the 13 strains five replicate experiments (objects), each with 42 variables (six $x'$ values + six $y'$ values + six $y_1$ values + six $y_2$ values + six $y_3$ values + six $y_4$ values + six growth rate values), were used in the statistical treatment. The entire data set thus consisted of 65 objects and 42 variables. All data were processed with an IBM-PC-

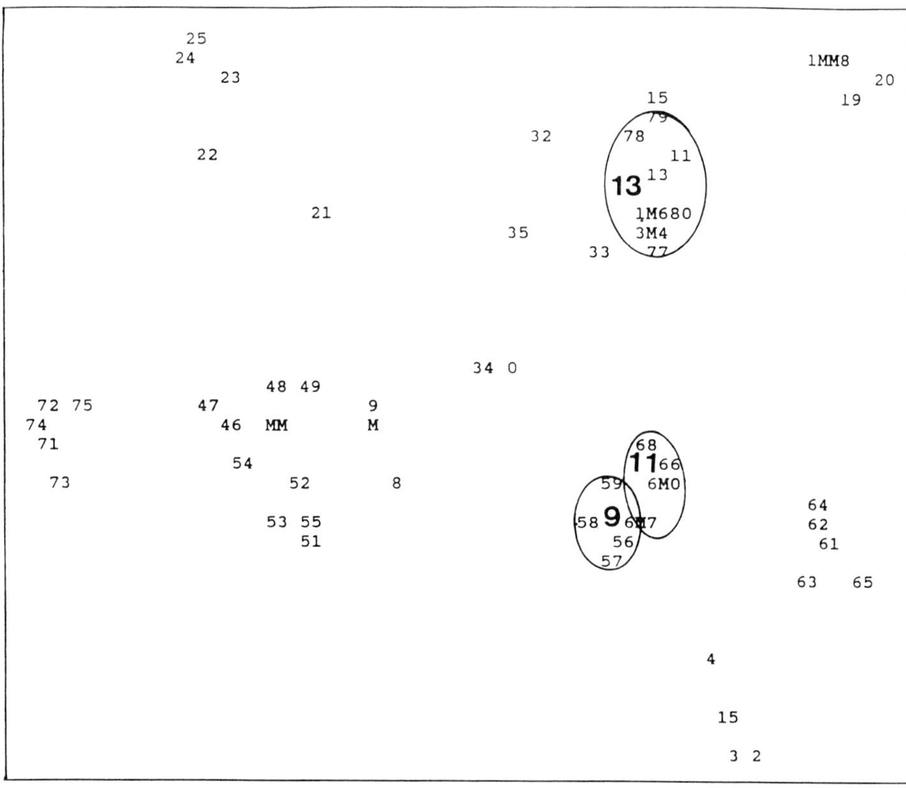


FIG. 3. The first and third components of a principal component analysis based on 42 variables for 13 strains of *Penicillium* species. Bold numbers represent the different strains of *Penicillium* species. M means that two or more objects overlap.

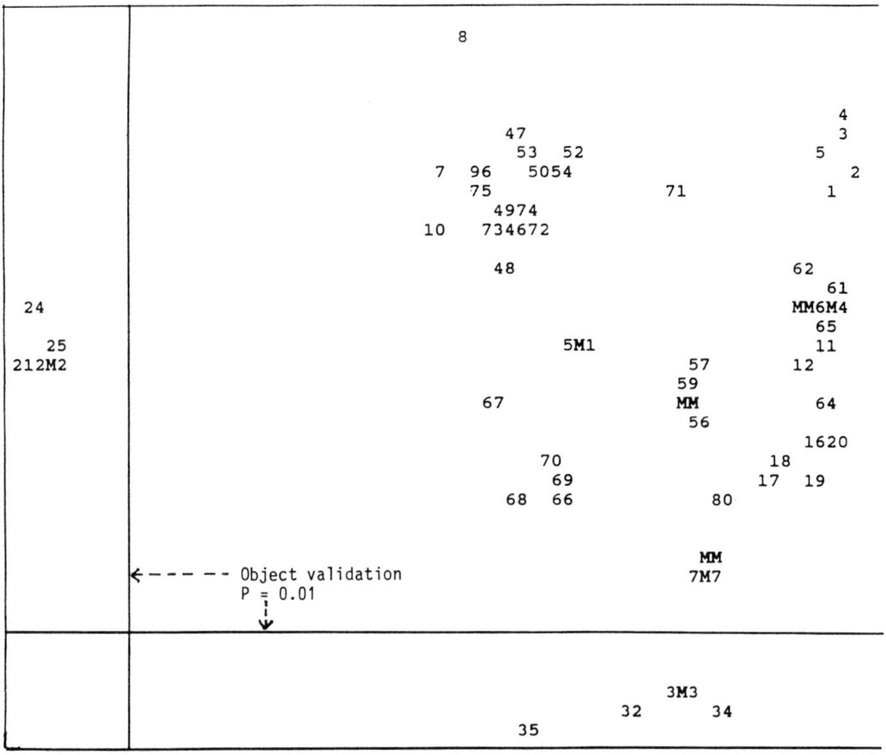


FIG. 4. Classification analysis of *P. camembertii* II (numbers 21–25) and *P. camembertii* III (numbers 31–35). The other numbers represent alien objects. M means that two or more objects overlap.

compatible computer with a classification program, SIMCA[11] (soft independent modeling of class analogies), obtained from Professor S. Wold (Chemometric Research Group, University of Umeå, Sweden). A typical data set for *Penicillium brevi-compactum* is shown in Table I.

In the SIMCA analysis the different variables are first normalized so that the sum over all 42 variables in each experiment is equal to 42 (mean peak value 1.0).[5] The objects are then subjected to a principal component analysis to study clustering tendency (Figs. 2 and 3). Figures 2 and 3 show that the different strains are separated from one another. Sometimes it appears that different strains overlap or are closely located in the principal

[11] S. Wold and M. Sjöström, *in* "Chemometric Theory and Application" (B. R. Kowalski, ed.), ACS Symp. Ser., No. 52. p. 243. American Chemical Society, Washington, D.C., 1977.

component plot. In such cases it is necessary to plot another dimension. For example, strains 9, 11, and 13 are clustered (Fig. 2), and another projection of the principal component analysis was carried out. Figure 3 shows that strain 13 is well separated from the strains 9 and 11. Strain 13 represents *P. hirsutum,* whereas strains 9 and 11 represent *P. solitum.*

To show a difference between two strains the fitness of different classes (strains) to one another was tested using the classification program in the SIMCA package. Figure 4 illustrates the classification of *P. camembertii* II and *P. camembertii* III, showing that the two strains of *Penicillium,* at a 99% confidence level, are classified as separate strains. The numbers outside the confidence interval represent other *Penicillium* strains, which are classified as alien objects.

## Discussion

### *Advantages and Disadvantages of Method*

A major advantage of the technique is that it gives highly reproducible results, even with conidia harvested on different occasions. For example, five replicate experiments with *P. crustosum,* grown on malt extract agar, gave a cross-point of 4.52 with a standard error of 0.04. A further advantage is that conidia used in the cross-partition experiments can be recultivated for other studies. Because of the slow growth of conidia, the experiments can be done at room temperature. The conidial suspensions can be stored overnight in a refrigerator without any growth occurring. The major drawback of the method is that nonsporulating fungi cannot be identified.

### *Possible Problems*

If nominally identical polymers supplied by other manufacturers are used problems may arise in phase formation owing to minor differences in polymer molecular weight. In such cases slightly higher polymer concentrations than those given above may be necessary for constituting phase systems. A good strategy under such circumstances is to make a phase diagram[12] for selection of the phase system. It is of course necessary to avoid contamination of culture plates with other fungi. The time for phase separation should not exceed 90 min, owing to the risk of settling of conidia to the interface, resulting in a misleading concentration of conidia in the upper phase.

[12] P.-Å. Albertsson and F. Tjerneld, this volume [1].

# [33] Charge-Directed Affinity Partitioning of Cells

*By* JESUS MENDIETA

## Introduction

An alternative approach to the usual affinity partitioning of cells in which polymer ligands are used is to modify surface properties (such as net charge) of the specific cells of interest and thereby the partition ratio of the cells. The use of two-phase systems highly sensitive to net charge permits the selective extraction of cells by means of coupling a ligand (specific for a surface receptor or property of the membrane) to a polyelectrolyte. When done at a low level of substitution, the affinity of the ligand to its specific receptor is preserved.

## Experimental Model

### *Ligand and Cell Population*

The ligand that has been used to study the feasibility of this approach is transferrin (purified and iron-saturated human transferrin, Sigma, St. Louis, MO), the plasma $\beta$-globulin which delivers iron to cells. Polylysine ($M_r$ 4000–15,000, Sigma) is coupled to transferrin in order to obtain a positively charged derivative which retains its affinity for the receptor. The reaction was carried out by means of *N*-succinimidyl-3-(2-pyridyldithio)propionate (SPDP, Pharmacia-LKB, Uppsala, Sweden), using the method of Carlsson *et al.*[1]

The uptake of iron by cells from transferrin is mediated by a specific surface receptor.[2] Transferrin receptors abound in the immature precursors of erythroid cells,[3,4] decrease during erythroid differentiation,[5] and are absent in erythrocytes. The transferrin receptor is thus a suitable marker for erythroid cell differentiation and can be used for their fractionation.

*Erythroid Cells.* The erythroid cell population is isolated from rat bone marrow cells on a self-generated Percoll (Pharmacia-LKB) gradient.[6] The

[1] J. Carlsson, H. Drevin, and R. Axén, *Biochem. J.* **173,** 723 (1978).

[2] I. S. Trowbridge and M. B. Omary, *Proc. Natl. Acad. Sci. U.S.A.* **78,** 3039 (1981).

[3] B. J. Iacopetta, E. H. Morgan, and G. C. T. Yeoh, *Biochim. Biophys. Acta* **887,** 204 (1982).

[4] M. T. Nunez, J. Glass, S. Fisher, L. M. Lavidor, E. M. Lenk, and S. H. Robinson, *Br. J. Haematol.* **36,** 519 (1977).

[5] J. Mendieta, A. Herraez, P. Sancho, and J. Luque, *Biosci. Rep.* **9,** 541 (1989).

[6] M. Pineda, M. Pinilla, and J. Luque, *Cell Biochem. Funct.* **2,** 254 (1984).

cells in the density range of 1.08–1.10 g/ml are collected and washed free of Percoll with isotonic Hanks' buffered salt solution (HBSS). The cell population contains more than 80% erythroid cells, mainly erythroblasts in different stages of differentiation, and reticulocytes.

### *Two-Phase System*

The two-phase system used contains 5% (w/w) dextran (Dx) T500 (Pharmacia-LKB), 4% (w/w) poly(ethylene glycol) (PEG) 8000 (Union Carbide, New York, NY), 90 m*M* sodium phosphate buffer, pH 7.4, and 30 m*M* NaCl. The presence of phosphate ions in the system gives rise to an interfacial potential difference between the phases (top phase positive) because of the affinity of this ion for the Dx-rich bottom phase.[7] In this so-called charge-sensitive system one of the determinant factors influencing the partitioning is the charge-associated properties of the biological material. The partitioning behavior of erythroid cells in this kind of system has been extensively studied.[8]

### *Transferrin–Polylysine Conjugation*

The coupling procedure[9] is outlined in Fig. 1. Between 50 and 60% of transferrin moleclues are activated by SPDP when mixed in a molecular ratio of 1 : 2. For activation of polylysine with SPDP, a molar ratio of 1 : 1 is used, which yields 80–90% substitution. To generate -SH groups the activated polylysine is treated with 0.1 *M* dithiothreitol (DTT). The excess of low molecular weight reagents is removed by gel filtration on Sephadex G-25 (Pharmacia-LKB).

Thiolated polylysine and 2-pyridyl disulfide propioimide–transferrin solutions are mixed (molar ratio 3 : 1), and the coupling reaction goes to completion in 1 hr. The release of pyridine-2-thione can be followed by the increase in absorbance at 343 nm. The reaction mixture is dialyzed overnight against 0.3 *M* glycine hydrochloride, pH 6.2, and purified on a DEAE-cellulose column (DE-52, Whatman, Maidstone, UK) in the same solvent. Unmodified transferrin is adsorbed while the transferrin–polylysine conjugate is eluted and collected. The conjugate is then dialyzed against HBSS.

[7] G. Johansson, *Biochim. Biophys. Acta* **221,** 387 (1970).

[8] H. Walter, *in* "Partitioning in Aqueous Two-Phase Systems: Theory, Methods, Uses, and Applications to Biotechnology" (H. Walter, D. E. Brooks, and D. Fisher, eds.), p. 328. Academic Press, Orlando, Florida, 1985.

[9] J. Mendieta and G. Johansson, *Anal. Biochem.* **200,** 280 (1992).

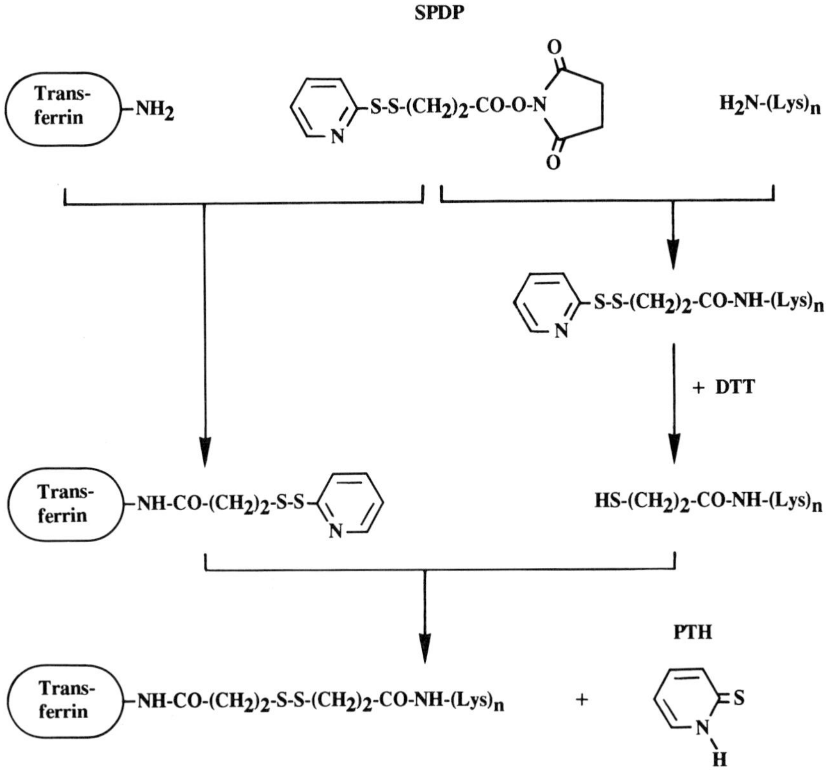


FIG. 1. Scheme of the coupling procedure. $(Lys)_n$, Polylysine; SPDP, *N*-succinimidyl-3-(2-pyridyldithio)propionate; DTT, dithiothreitol; PTH, pyridine-2-thione. [From J. Mendieta and G. Johansson, *Anal. Biochem.* **200,** 280 (1992), with permission.]

## Results

### *Partition Coefficient of Transferrin/Polylysine Conjugate*

Partitioning of soluble molecules (such as proteins) is expressed by a partition coefficient ($K$) defined as the ratio between the concentration in the upper and lower phases. The progress of the coupling reaction can be followed by the decrease of $K$ for transferrin in the charge-sensitive two-phase system described above. The decrease in log $K$ value with time indicates a change in charge of the transferrin owing to its reaction with the positively charged polylysine.

The $K$ for the transferrin–polylysine conjugate ($K = 0.24 \pm 0.06$) is 3.4 times lower than the value for native transferrin ($K = 0.82 \pm 0.04$). Because the bond between transferrin and polylysine is a disulfide linkage, the polyelectrolyte can be removed from the conjugate by reduction (e.g., 0.1 *M* DTT). After reduction, the $K$ for transferrin ($K = 0.91 \pm 0.06$) is close to the $K$ value for the unmodified protein.

### *Characteristics of the Conjugate*

Biochemical properties of the transferrin are preserved after reaction with polylysine. The iron-binding capacity of the transferrin–polylysine conjugate is unaltered by the conjugation procedure. The degree of saturation, measured by means of the absorbance ratio $A_{470}/A_{280}$, gives values between 0.040 and 0.045, corresponding to complete saturation, for both transferrin and the conjugate. No significant difference in affinity constants to the specific receptor is evident for transferrin–polylysine versus native transferrin.

### *Effect of Transferrin–Polylysine on Cell Partitioning*

The erythroid cell population from bone marrow ($6 \times 10^7$ cells/ml) is incubated with the transferrin–polylysine conjugate (500 $\mu$g/ml) in a final volume of 0.4 ml HBSS. After 90 min at 4° the cells are washed 3 times with HBSS.

The system containing 5% (w/w) Dx T500, 8% PEG 8000, 90 m*M* sodium phosphate buffer, pH 7.4, and 30 m*M* NaCl is prepared and equilibrated at 4°. A small aliquot of incubated cells is added to 0.9 ml of top phase and resuspended, then 0.7 ml of bottom phase is added. The system is carefully mixed by repeated inversion, and an aliquot (50 $\mu$l) is immediately removed. The rest of the system is allowed to settle for 5 min. An aliquot (50 $\mu$l) of the top phase is then removed, and the quantity of cells is determined in both the top phase and the total mixed system by measurement of the absorbance at 400 nm after a 4-fold dilution with HBSS.

Incubation of erythroid cells with the conjugate caused a 15-fold decrease in the partition ratio ($G$ = ratio of number of cells in the upper phase and in lower phase plus interface) owing to an increase in positive charge on the cell surface. Incubation with the split conjugate (obtained by reduction) does not restore the partitioning of cells to their original $G$ value (Table I), reflecting a superimposed nonspecific interaction between the positively charged polylysine and the negatively charged cell surface.[10] In addition, the partitioning of erythrocytes, which have lost the transferrin receptor during erythroid maturation,[5] is nonspecifically influenced by the

[10] Y. Marikovsky, D. Danon, and A. Katchalsky, *Biochim. Biophys. Acta* **124,** 154 (1966).

TABLE I
EFFECT OF SIALIC ACID AND TRANSFERRIN–POLYLYSINE CONJUGATE ON PARTITION RATIO, $G$,[a] OF ERYTHROCYTES AND ERYTHROBLASTS PARTITIONED IN CHARGE-SENSITIVE TWO-PHASE SYSTEM[b,c]

| | | $G$ | | |
|---|---|---|---|---|
| | | | Transferrin–polylysine incubation | |
| Cells | Sialic acid[d] | No incubation | Untreated | DTT-treated[e] |
| Erythroblasts | – | 2.7 ± 0.1 | 0.18 ± 0.05 | 0.9 ± 0.2 |
| | + | — | 0.50 ± 0.09 | 2.1 ± 0.3 |
| Erythrocytes | – | 4.0 ± 0.1 | 1.1 ± 0.1 | 2.4 ± 0.4 |
| | + | — | 3.9 ± 0.2 | 4.1 ± 0.3 |

[a] The partition ratio, $G$, is defined as the ratio between the number of cells in the top phase and the cells in bottom phase plus interface.
[b] System composition: 5% (w/w) dextran T500, 4% (w/w) PEG 8000, 30 m$M$ NaCl, and 90 m$M$ sodium phosphate buffer, pH 7.4. Temperature, 4°.
[c] From J. Mendieta and G. Johansson, *Anal. Biochem.* **200,** 280 (1992), with permission.
[d] Solution of sialic acid, adjusted to pH 7.4, was added to the incubation medium (final concentration 2 mg/ml).
[e] Cleavage of the conjugate by use of dithiothreitol (DTT).

incubation with transferrin–polylysine conjugate, as reflected by a change in $G$ (Table I).

Sialic acid residues on glycoproteins and glycolipids are the most prevalent negatively charged groups on the red cell surface.[11] It is possible to reduce nonspecific binding of conjugate to cells by including free sialic acid (sodium salt) in the incubation medium. When erythrocytes are incubated in the presence of 2 mg/ml sialic acid (Sigma), which represents a 50-fold molar excess compared with the number of lysine residues coupled to transferrin, the $G$ value is restored to its original value (Table I). However, after incubation of erythroblasts in the presence of sialic acid the $G$ value is as low as 0.5, in the presence of transferrin–polylysine, whereas it is 2.1 with the transferrin part of the DTT-cleaved conjugate.

After incubation of a heterogeneous mixture of cells at saturating concentrations of transferrin–polylysine conjugate, the cells change their log $G$ value in proportion to the number of transferrin receptors on the cell surface.[9] The use of multistep extraction procedures, such as countercurrent distribution, permits the fractionation of cells according to the number

[11] R. C. Hughes, "Membranes and Glycoproteins." Butterworth, London, 1975.

of receptors expressed and, thus, according to the stage of cell differentiation as reflected by specific enzymatic markers.[12]

## Discussion

Cell charge-directed affinity partitioning provides good specificity in cell extraction with the capacity to handle a large number of cells as shown by the fractionation of erythroid cells. By using ligand-bound, negatively charged polyelectrolytes (such as polyglutamate), selected cells can be made more negatively charged and can be extracted into the oppositely charged phase.

## Acknowledgments

This work has been supported by a fellowship from the foundation Wenner-Grenska Samfundet.

[12] J. Mendieta and G. Johansson, *Mol. Cell. Biochem.* **121,** 93 (1993).

# [34] Factors in the Affinity Extraction of Red Blood Cells Using Poly(ethylene Glycol)–Metal Chelate

*By* GERD BIRKENMEIER, HARRY WALTER, and KIM E. WIDEN

## Introduction

Partitioning of cells generally occurs between one bulk phase and the interface of dextran (Dx)–poly(ethylene glycol) (PEG) aqueous two-phase systems and is governed by the interaction of the physical properties of the phases and the physical and biochemical surface properties of the cells.[1,2] Thus, in charge-sensitive phase systems (i.e., those having an electrical potential difference between the phases) the surface charge of the cells is a determinant in their partitioning behavior. Non-charge-sensitive phase systems (i.e., systems which have essentially no potential difference

[1] H. Walter, *in* "Partitioning in Aqueous Two-Phase Systems: Theory, Methods, Uses, and Applications to Biotechnology" (H. Walter, D. E. Brooks, and D. Fisher, eds.), p. 327. Academic Press, Orlando, Florida, 1985.

[2] H. Walter and C. Larsson, this volume [4].

between the phases) recognize cell surface properties other than charge in the partitioning process.[1,2]

Affinity partitioning, utilizing the interaction between surface biomolecules and affinity ligands confined primarily to one of the phases by covalently linking them to one of the polymers, usually PEG, has also been exploited in cell separation technology, as evidenced by a number of chapters in this volume.[3–5] Thus, the partitioning of cells can be steered by incorporation of charged or hydrophobic ligands into the phase system,[6] while immunoaffinity partitioning can be used for the extraction of specific cell populations with polymer-linked antibodies.[7,8]

Another approach to increase the selectivity of separation is based on the interaction of macromolecules with chelated transition metal ions.[9] In this case accessible histidine residues on the surface of a protein are the predominant metal chelate binding sites.[9a] Experiments with metal chelate-derivatized PEG in Dx–PEG and PEG–salt phase systems have shown that extraction of proteins by use of these ligands is a histidine-mediated process.[9a,10]

The feasibility of extracting and separating cells by metal chelate [i.e., PEG iminodiacetate charged with Cu(II), PEG–IDA–Cu] affinity partitioning has been established.[11] Additional study on the influence of chelated metal ions on the partitioning of erythrocytes from different species in both charge-sensitive and non-charge-sensitive Dx–PEG phase systems has revealed aspects of the extractions that can be ascribed to "specific" and "nonspecific" parameters.[12] One of the uses of PEG–IDA–Cu (and, most likely, of other PEG–ligands) is that of extending the partitioning range of Dx–PEG phase systems to cells having such low partition ratios that the traditional manipulations of phase system composition[2] are inadequate to result in a usable partition.

[3] D. E. Brooks and S. J. Stocks, this volume [36].

[4] C. Delgado, G. E. Francis, and D. Fisher, this volume [37].

[5] J. Mendieta, this volume [33].

[6] G. Johansson, this volume [5].

[7] K. A. Sharp, M. Yalpani, S. J. Howard, and D. E. Brooks, *Anal. Biochem.* **154,** 110 (1986).

[8] L. J. Karr, D. L. Donnelly, A. Kozlowski, and J. M. Harris, this volume [35].

[9] J. Porath, J. Carlsson, I. Olsson, and G. Belfrage, *Nature (London)* **258,** 598 (1975).

[9a] B. H. Chung, D. Bailey, and F. H. Arnold, this volume [15].

[10] G. Birkenmeier, M. A. Vijayalakshmi, T. Stigbrand, and G. Kopperschläger, *J. Chromatogr.* **539,** 267 (1991).

[11] H. Goubran-Botros, G. Birkenmeier, A. Otto, G. Kopperschläger, and M. A. Vijayalakshmi, *Biochim. Biophys. Acta* **1074,** 69 (1991).

[12] H. Walter, K. E. Widen, and G. Birkenmeier, *J. Chromatogr.* **641,** 279 (1993).

## Methods

### *Materials*

Dextran T500 (Lot No. 0106905) is obtained from Pharmacia LKB (Piscataway, NJ). PEG 8000 is supplied by Union Carbide (Long Beach, CA). Monomethoxy-PEG 5000, imidazole and trypsin are purchased from Sigma Chemical Co. (St. Louis, MO). Neuraminidase (*Vibrio cholerae*) is from Calbiochem (San Diego, CA). All other reagents are of analytical reagent grade.

### *Preparation of Metal Chelate–Poly(ethylene Glycol)*

PEG iminodiacetate (PEG–IDA) is synthesized as previously described.[9a,10] Charging with Cu(II) is performed in 50 m*M* sodium acetate buffer, pH 4.0. The content of Cu(II) per mole of PEG–IDA is 0.64 to 0.83 mol.

### *Blood Specimen*

Blood specimens from beef, dog, horse, human, pig, rabbit, and sheep are collected in acid–citrate–dextrose (ACD, Squibb, New Brunswick, NJ) solution (10 ml of blood plus 3 ml of ACD).

### *Treatment of Human Erythrocytes with Neuraminidase or Trypsin*

One-half milliliter of washed and packed human erythrocytes (RBC, red blood cells) are suspended in 3.5 ml of phosphate-buffered saline (PBS), pH 7.0, and incubated with neuraminidase (200 $\mu$l, 1 IU/ml) or with trypsin (200 $\mu$l, 1 mg/ml) at 37° for 60 min. Aliquots of cells treated under the same conditions but without enzyme are taken as control. The cells are washed 3 times with PBS and used in the partitioning experiments.

### *Preparation of Aqueous Two-Phase Systems*

The different two-phase systems are made up by weighing required quantities from stock solutions of Dx T500 (20%, w/w), PEG 8000 (40%, w/w), PEG–IDA–Cu (5%, w/w), 0.44 *M* sodium phosphate buffer, pH 6.8, and 0.60 *M* NaCl. The composition of the phase system used is given in a shorthand as follows: The first number is the percentage Dx, followed by a colon and a number giving the percentage PEG, followed again by a colon and a number giving the percentage PEG–IDA–Cu. The salt composition is designated by number (#): #1 is 0.11 *M* sodium phosphate buffer, pH 6.8; #2 is 90 m*M* sodium phosphate buffer, pH 6.8, plus 30 m*M* NaCl; and #5 is 10 m*M* sodium phosphate buffer, pH 6.8, plus 0.15 *M* NaCl.

*Partitioning of Erythrocytes*

Phase systems selected for partitioning are separated by centrifugation, and the top and bottom phase volumes are adjusted to be equal. Aliquots of the mixed phases (3 ml) are pipetted into 12 × 75 mm tubes, and 50 μl of a suspension of erythrocytes corresponding to 6–17 × $10^7$ cells is added to the mixed phase system, which is mixed again. After permitting the phases to settle for 15 min in a vertical position, an 0.8-ml aliquot is withdrawn from the middle of the top phase. The quantity of cells in the top phase and in the sample is determined by measuring the hemoglobin absorbance at 540 nm after lysing the cells.[2] For experiments in which phase systems are close to the critical point (i.e., lower polymer concentrations), 10-g systems are used. Top and bottom phase volumes are adjusted to 5 ml, and a 30-μl aliquot of washed packed erythrocytes is added and the phases mixed. The phase systems, in capped tubes (125 × 16 mm), are allowed to settle in a horizontal position for 7 min. The tubes are then gently raised to vertical position (without agitating the contents), and the phases are permitted to settle for an additional 1 min. A 1-ml aliquot of top phase is then withdrawn for analysis.

The partition, *P*, value of cells is defined as the quantity of cells in the top phase, at time of sampling, as a percentage of the total cells added. Cell partitioning can also be expressed as a partition ratio, which equals the quantity of cells in the top phase divided by that at the interface plus bottom phase.

For the experiments shown in Figs. 1 and 2 the partitioning of cells in the presence of PEG–IDA–Cu was so adjusted that the *P* value in the presence of imidazole (1 m*M*) was quite small ($<10$). This was done to establish that the cell partitioning ratios obtained in phases containing PEG–IDA–Cu were predominantly due to the affinity ligand and not due to other physical properties of the phases. Imidazole is known to interfere with binding of copper chelate to cell surface accessible histidine residues, but it does not affect *P* values of cells in systems devoid of PEG–IDA–Cu.[11]

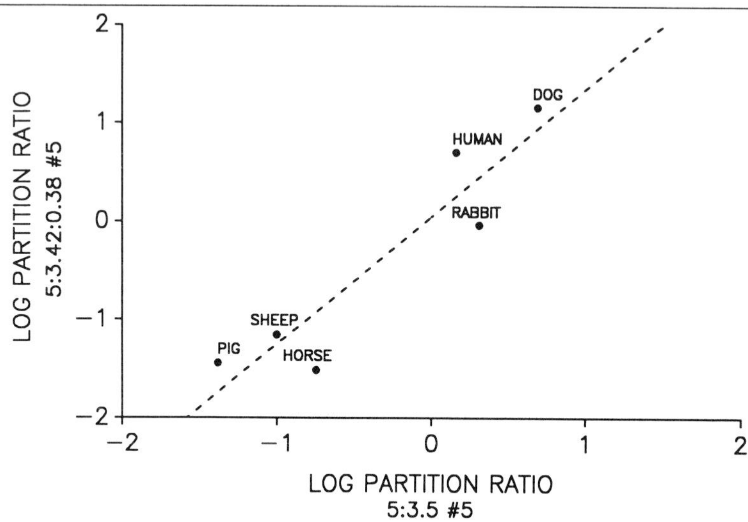


FIG. 1. Log partition ratios (i.e., the ratio of the number of cells in the top phase divided by the number of cells at the interface plus bottom phase) of erythrocytes from different species in a non-charge-sensitive phase system containing PEG–IDA–Cu (ordinate) plotted versus those obtained in a non-charge-sensitive system with lower polymer concentration and without chelate (abscissa). Note that because of the higher polymer concentration cells would not partition in the non-charge-sensitive phase system containing 5% dextran and 3.8% total PEG (ordinate) if it was without PEG–IDA–Cu but rather would remain at the interface. [From H. Walter, K. E. Widen, and G. Birkenmeier, *J. Chromatogr.* **641,** 279 (1993). Reprinted by permission of Elsevier.]

## Discussion

### *Metal Chelate Affinity Partitioning of Erythrocytes from Different Species in Charge- and Non-Charge-Sensitive Dextran–Poly(ethylene Glycol) Phase Systems*

Erythrocytes are known to partition between the top and the interface of systems composed of Dx T500–PEG 8000.[1,2] In non-charge-sensitive phase systems with adequately low polymer concentrations, a species-specific partitioning of erythrocytes has been found.[1,2] Erythrocytes with high *P* values (e.g., dog erythrocytes) have been shown to adhere more weakly to the interface than do erythrocytes with low *P* values (e.g., sheep cells).[13] Thus, to evaluate the effect of an affinity ligand (e.g., PEG–IDA–Cu) on the partitioning of cells (e.g., erythrocytes from different species) the phase system selected should be one in which virtually all cells are at the interface in the absence of the PEG–ligand. Even under

[13] H. Walter, F. D. Raymond, and D. Fisher, *J. Chromatogr.* **609,** 219 (1992).

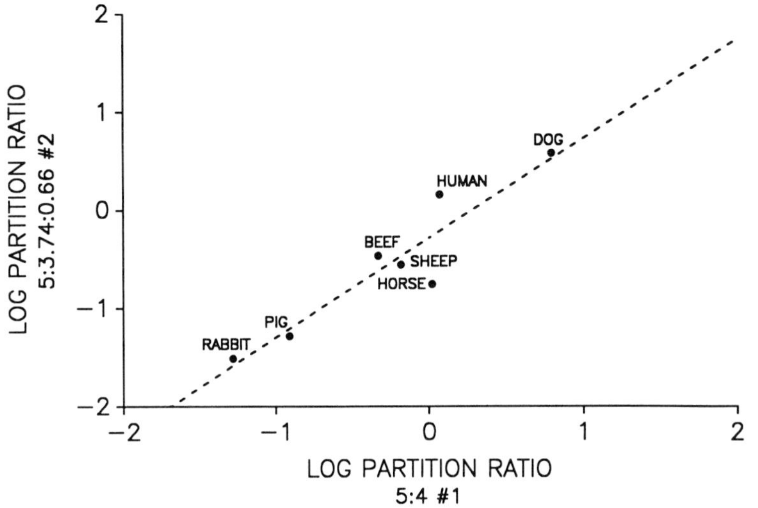


FIG. 2. Log partition ratios (i.e., the ratio of the number of cells in the top phase divided by the number of cells at the interface plus bottom phase) of erythrocytes from different species in a charge-sensitive phase system containing PEG–IDA–Cu (ordinate) plotted versus those obtained in a charge-sensitive system with lower polymer concentration and without chelate (abscissa). Note that because of the higher polymer concentration cells would not partition in the charge-sensitive phase system containing 5% dextran and 4.4% total PEG (ordinate) if it was without PEG–IDA–Cu but rather would remain at the interface. [From H. Walter, K. E. Widen, and G. Birkenmeier, *J. Chromatogr.* **641,** 279 (1993). Reprinted by permission of Elsevier.]

these conditions one would expect that the affinity ligand extracts cells of some species (e.g., those from dog) more easily into the top phase than cells from other species (e.g., sheep).

Figure 1 indicates the partition ratios of erythrocytes from different species in a non-charge-sensitive phase system in which 10% of the PEG has been substituted by PEG–IDA–Cu. Dx–PEG phase systems in which NaCl is the predominant salt are considered to have virtually no Donnan potential between the phases and are, thus, non-charge-sensitive for cells.[1,2] The composition of the phases chosen, 5 : 3.42 : 0.38 #5 (ordinate), contains polymer concentrations (i.e., 5 : 3.8) at which all species examined are almost completely adsorbed at the interface in the absence of PEG–IDA–Cu. Hence, the fact that erythrocytes partition in this system must be attributed to the PEG–IDA–Cu, which "pulls" cells out of the interface and into the top phase in a seemingly species-specific manner. The grouping of distribution ratios of erythrocytes from different species in the phase system resembles one obtained in phase systems at lower polymer concentration but devoid of the affinity ligand[14] (5 : 3.5 #5, abscissa in Fig. 1).

Partitioning experiments were also performed in charge-sensitive phase systems and yielded analogous results (Fig. 2). Charge-sensitive phases can be obtained by including sodium phosphate as the main salt in the phase system. A Donnan potential between the phases is created, with the top phase being positive.[1,2] Again, the phase system with the lowest polymer concentration at which the erythrocytes of different species are almost completely adsorbed at the interface was selected, namely, 5 : 4.4 #2, and 15% of the PEG was substituted by PEG–IDA–Cu (ordinate in Fig. 2). As Fig. 2 shows, a close correlation exists between the partition ratios of erythrocytes from different species in charge-sensitive phase systems at lower polymer concentrations without PEG–IDA–Cu (5 : 4 #1) and at higher polymer concentrations containing PEG–IDA–Cu (5 : 3.74 : 0.66 #2).[12]

It is known that in charge-sensitive phase systems at lower polymer concentrations the $P$ values of erythrocytes have some correlation with the relative electrophoretic mobilities of the cells.[1,2] A similar tendency exists in charge-sensitive phases at higher polymer concentrations in the presence of PEG–IDA–Cu.[12]

The relationship between the partition ratios of erythrocytes from different species in non-charge-sensitive and charge-sensitive phase systems with and without PEG–IDA–Cu indicates that a major effect of the PEG–IDA–Cu is the facilitation of partitioning of cells at polymer concentrations higher than those at which they can partition in the absence of chelate. Thus, the use of PEG–IDA–Cu may extend the partitioning range of Dx–PEG phase systems to cells having such low non-charge-associated *or* charge-related surface properties as to preclude their partitioning in Dx–PEG systems solely by reducing the polymer concentrations of the phase systems. Further examination of the partioning of the cells reveals that, although there are similarities in the grouping of the partition ratios in systems with and without PEG–IDA–Cu, the extraction sequence of cells is not identical (e.g., note the higher ratio of human compared to rabbit erythrocytes in non-charge-sensitive phases containing PEG–IDA–Cu,[12] Fig. 1).

Accessible histidine residues are assumed to be the predominant metal ion-combining sites in proteins.[9,9a] The findings that red blood cell partitioning in charge-sensitive or non-charge-sensitive phase systems, at higher polymer concentrations, gives essentially the same sequences in the presence of the affinity ligand as, at lower polymer concentrations,

[14] H. Walter, E. J. Krob, and D. E. Brooks, *Biochemistry* **15,** 2959 (1976).

in its absence, implies that the histidine per unit surface area of erythrocytes from different species is about the same or that the other cell surface properties outweigh the differences in metal chelate binding sites and effect the observed correlation.[12]

*Procedures for Selecting Appropriate Phase Systems and Probing of Cell Surfaces by Means of Metal Chelates*

To gauge the relative affinity of red blood cells for the PEG-bound ligand the appropriate selection of phase system is important. To this end one has to determine the lowest polymer concentration in a charge- or non-charge-sensitive phase system (as desired) at which virtually all of the cells are at the interface. Some polymer-bound ligands (e.g., PEG palmitate[1]) extract cells at very low polymer–ligand concentrations (e.g., 0.001%, w/w). Incorporation of a PEG–ligand at such low concentrations will not appreciably alter the phase composition of a system. Extraction of cells by use of less effective ligands (e.g., PEG–IDA–Cu) requires a considerably higher PEG–ligand concentration (3–15%, w/w). Ideally, in the latter case, the molecular weight of the polymer (i.e., PEG) constituting a bulk phase and that used in carrying the ligand is the same. If, however, the molecular weights differ, an appropriate "compensation" for this difference must be made. The molecular weights of the bulk PEG (PEG 8000) in the phases and that coupled to the ligand (the PEG in PEG–IDA–Cu has a molecular weight of 5000) in the examples presented here differ. Substitution of larger amounts of PEG–IDA–Cu (for the bulk PEG) in a phase system will cause cells to have higher $P$ values irrespective of any binding to the affinity ligand because, as a general rule, reducing the molecular weight of a polymer constituting one of the phases causes more of a material to partition into that phase.[15,16] To minimize this effect and to optimize that of the PEG–ligand, $P$ values of the cells were measured in phase systems containing PEG–IDA–Cu in the presence and absence of a competing ligand (i.e., imidazole).[12] The PEG concentration (PEG plus PEG–ligand) of the phase system selected should be high enough so that, tested in the presence of imidazole ("control"), cells partition more than 1% but less than 10% into the top phase. If these "base" values are higher, cells that partition into the top phase in the absence of the PEG–ligand may be part of the population that is extractable with the metal chelate but will not be measured as such. Hence, expressing

[15] D. E. Brooks, K. A. Sharp, and D. Fisher, *in* "Partitioning in Aqueous Two-Phase Systems: Theory, Methods, Uses, and Applications to Biotechnology" (H. Walter, D. E. Brooks, and D. Fisher, eds), p. 11. Academic Press, Orlando, Florida, 1985.

[16] D. E. Brooks and R. Norris-Jones, this volume [2].

a net $P$ value as the difference between a $P$ value obtained in the absence and a $P$ value obtained in the presence of PEG–IDA–Cu may be misleading.

The procedures just outlined were used in selecting phase systems for studies on the relative extractions of different cell populations by metal chelate (Figs. 1 and 2) and for examining the partitioning of normal and *in vitro* treated cells (Table I). Treatment of human erythrocytes with neuraminidase or trypsin reduces the $P$ value in charge-sensitive phases (e.g., 5 : 3.9 #1) and increases them in non-charge-sensitive phases (e.g., 5 : 3.4 #5).[1,2] In a non-charge-sensitive phase system having a higher polymer concentration and containing PEG–IDA–Cu (5 : 3.78 : 0.42 #5), selected so that in the presence of imidazole the $P$ values of cells are low (Table I), the $P$ value of neuraminidase-treated cells, but not of trypsin-treated cells, is also found to increase. On the other hand, in a charge-sensitive phase system (5 : 3.74 : 0.66 #2), the $P$ value of neuraminidase-treated cells is not only not reduced but is actually increased.

Perhaps removal of sialic acid permits better penetration of the PEG–IDA–Cu into ligand site-harboring regions of the cell surface. Reduction of the negative charge (sialic acid) may also cause local perturbations of the p$K$ values of histidine residues, leading to altered binding affinities. It is known that the binding of chelated transition metal ions to proteins is very much affected by the degree of protonation of accessible

TABLE I
PARTITION, $P$, OF HUMAN NORMAL ERYTHROCYTES AND ERYTHROCYTES TREATED WITH NEURAMINIDASE OR TRYPSIN[a]

| Phase system | Normal RBC | Neuraminidase-treated RBC | Trypsin-treated RBC |
|---|---|---|---|
| 5:3.9 #1 | 63 ± 3 (12) | 7 ± 3 (7) | 23 ± 1 (5) |
| 5:3.74:0.66 #2 | 61 ± 2 (12) | 80 ± 6 (8) | 20 ± 2 (5) |
| With imidazole | 9 ± 2 (7) | 10 ± 0 (3) | 10 ± 1 (5) |
| 5:3.4 #5 | 44 ± 3 (12) | 89 ± 3 (7) | 82 ± 2 (5) |
| 5:3.78:0.42 #5 | 21 ± 2 (5) | 81 ± 1 (3) | 13 ± 1 (3) |
| With imidazole | 3 ± 0 (5) | 8 ± 1 (3) | 7 ± 1 (3) |

[a] In charge-sensitive or non-charge-sensitive phase systems with and without PEG–IDA–Cu. $P$, the partition, is defined as the quantity of cells in the top phase as a percentage of total cells added. Values given are means ±S.D., with the number of experiments in parentheses. RBCs (30 $\mu$l packed cells) were partitioned in a 10-g phase system having equal volumes of top and bottom phase. Settling time was 7 min in the horizontal position plus 1 min in the vertical position. [From H. Walter, K. E. Widen, and G. Birkenmeier, *J. Chromatogr.* **641,** 279 (1993). Reprinted by permission of Elsevier.]

histidine residues.[9a,17] The proteolytic degradation of membrane proteins apparently removes histidine residues as reflected by the reduced *P* values.

## Conclusion

PEG–IDA–Cu binds to erythrocytes from different species and causes them to partition in a sequence reminiscent of that obtained in either charge-sensitive or non-charge-sensitive phase systems at lower polymer concentrations without the affinity ligand. This effect of PEG–IDA–Cu on cells may be useful in extending the partitioning range of Dx–PEG phase systems to cells having such low partitions in non-charge-sensitive or charge-sensitive phase systems that the traditional manipulations[2] of phase system composition are inadequate to result in a useful cell partition ratio or *P* value. PEG–IDA–Cu has also been found to be a sensitive probe for certain changes or differences in surface properties of erythrocytes (from different sources or subsequent to *in vitro* treatments) which are not observed in the absence of the ligand.

## Acknowledgments

This work was supported by the German Ministry of Science and Technology and by the Medical Research Service of the U.S. Department of Veterans Affairs.

[17] F. H. Arnold, *Bio/Technology* **9,** 151 (1991).

# [35] Use of Poly(ethylene Glycol)-Modified Antibody in Cell Extraction

*By* LAUREL J. KARR, DONNA L. DONNELLY, ANTONI KOZLOWSKI, and J. MILTON HARRIS

## Introduction

Immunoaffinity partitioning makes use of the specificity of antigen–antibody reactions to provide cell separations within the benign environment of aqueous polymer two-phase systems prepared from poly(ethylene glycol) (PEG) and dextran (Dx). This chapter deals with the methodology of utilizing PEG-derivatized antibodies or PEG-derivatized secondary li-

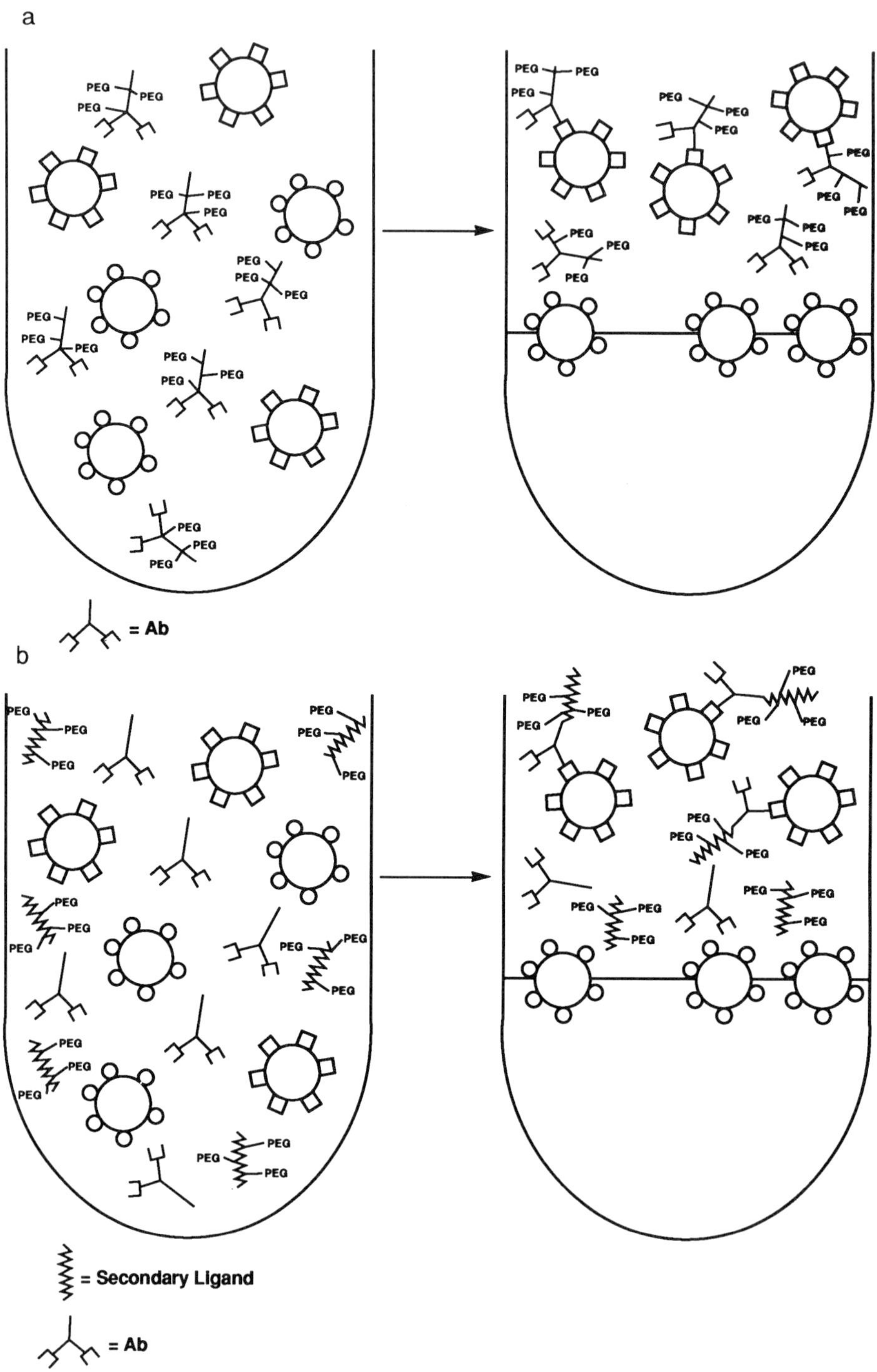
a
PEG
= Ab
b
PEG
= Secondary Ligand
= Ab

gands (such as protein A) to shift partitioning of target cells into the PEG-rich phase. Affinity partitioning of small quantities of cells is described elsewhere in this volume.[1] A related method that utilizes polyacrylamide-derivatized antibodies to direct cell partitioning into the Dx-rich phase is presented in another chapter.[2]

The methodology for immunoaffinity partitioning of cells is relatively new, and many of the techniques have been developed with model systems. Derivatives of antibodies (Ab) with PEG (to produce PEG–Ab) have been used to separate mixtures of red blood cells (RBCs) from different species.[3–5] Depending on the percentage of available lysine groups coupled to PEG and the molecular weight of PEG used, PEG–Ab can be made to partition to the PEG-rich phase while the ability of the antibody to agglutinate cells is reduced. Following incubation of PEG–Ab with RBC mixtures, partitioning of target cells into the PEG-rich phase is also increased (Fig. 1a), and two populations of cells can be separated, depending on their relative partition ratios, by a single partition step or by using a multiple extraction procedure such as countercurrent distribution (CCD).[6,7]

Methods that use PEG-derivatized ligands that bind to antibodies (i.e., "secondary" ligands) have also been described.[8–10] These ligands include

[1] C. Delgado, G. E. Francis, and D. Fisher, this volume [37].
[2] D. E. Brooks and S. J. Stocks, this volume [36].
[3] L. J. Karr, S. G. Shafer, J. M. Harris, J. M. Van Alstine, and R. S. Snyder, *J. Chromatogr.* **354,** 269 (1986).
[4] K. A. Sharp, M. Yalpani, S. J. Howard, and D. E. Brooks, *Anal. Biochem.* **154,** 110 (1986).
[5] C. Delgado, R. J. Anderson, G. E. Francis, and D. Fisher, *Anal. Biochem.* **192,** 322 (1991).
[6] H.-E. Åkerlund and P.-Å. Albertsson, this volume [8].
[7] H. Walter and C. Larsson, this volume [4].
[8] L. J. Karr, J. M. Van Alstine, R. S. Snyder, S. G. Shafer, and J. M. Harris, *J. Chromatogr.* **442,** 219 (1988).
[9] S. J. Stocks and D. E. Brooks, *Anal. Biochem.* **173,** 86 (1988).
[10] L. J. Karr, D. L. Donnelly, A. Kozlowski, and J. M. Harris, unpublished results.

FIG. 1. (a) Depiction of immunoaffinity partitioning using PEG-derivatized primary antibody (PEG–Ab). A mixture of cells is incubated with PEG–Ab, washed, and then introduced into a PEG–dextran two-phase system chosen so that most cells will partition between the lower, dextran-rich phase and the interface. Following mixing, the phase system is allowed to settle. The target cells partition into the PEG-rich upper phase while the rest of the cells remain at the interface. (b) Depiction of immunoaffinity partitioning utilizing a PEG-derivatized secondary ligand. Secondary ligands that have been used include PEG–protein A, PEG–protein G, PEG–avidin with biotinylated antibody, and PEG–anti-Fc. [From L. J. Karr, D. L. Donnelly, A. Kozlowski, and J. M. Harris, unpublished results.

staphylococcal protein A (SpA) and recombinant streptococcal protein G (SpG, produced in *Escherichia coli*) that react with the Fc portion of immunoglobulin G (IgG) antibody, avidin that reacts with biotinylated antibody, and derivatized antibody that reacts with the Fc portion of the primary antibody (anti-Fc). Incubation of PEG–secondary ligand, primary antibody, and cells leads to formation of compound ligands that react with surface antigens inherent to the target cells and in turn cause an increase in partitioning of these cells to the PEG-rich phase (Fig. 1b).

Derivatization of antibodies and secondary ligands with PEG and subsequent application to cell extraction by phase partitioning is described.

## Methods

### *Poly(ethylene Glycol) Activation*

Before PEG (Union Carbide, Danbury, CT) can be attached to a protein or other ligand it is necessary to prepare an active form of PEG, usually one that will couple to lysine groups. Several such active derivatives of PEG have been prepared in our laboratories,[11,12] and they are also available from a number of commercial sources including Sigma (St. Louis, MO), Polysciences (Philadelphia, PA), and Shearwater Polymers (Huntsville, AL). For cell separations we have used the cyanuric chloride derivative of PEG and PEG tresylate. Similar results are obtained with both derivatives in terms of ease of protein modification and efficiency of subsequent cell separations, although it has been shown with PEG–enzymes that the cyanuric chloride derivative produces modified protein with reduced activity.[13] In the case of immunoaffinity partitioning, some loss of antibody activity is desirable to reduce cell aggregation. To obtain reproducible protein modifications it is best to store activated PEG derivatives frozen under nitrogen in aliquots appropriate for one-time use.

*Cyanuric Chloride Method.* The following is a typical preparation of the cyanuric chloride derivative.[14] A 5-ml aliquot of 2 *M* butyllithium in hexane is withdrawn under nitrogen from a bottle sealed with a septum and added to 100 ml of benzene (stored over sodium) containing approximately

[11] J. M. Harris, *J. Macromol. Sci. Rev. Macromol. Chem. Phys.* **C25,** 325 (1985).

[12] J. M. Harris, J. M. Dust, R. A. McGill, P. A. Harris, M. J. Edgell, M. R. Sedaghat-Herati, L. J. Karr, and D. L. Donnelly, *in* "Water-Soluble Polymers" (S. W. Shalaby, C. L. McCormick, and G. B. Butler, eds.), ACS Symp. Ser. 467, Chapt. 27. Washington, D.C., 1991.

[13] K. Yoshinaga and J. M. Harris, *J. Bioact. Compat. Polym.* **4,** 17 (1989).

[14] S. G. Shafer and J. M. Harris, *J. Polym. Sci. Polym. Chem. Ed.* **24,** 375 (1986).

1 mg of triphenylmethane as indicator. The resulting colorless solution is titrated with 10 m$M$ monomethoxy-PEG (M-PEG) 1900 in benzene to reach the red end point, indicating a slight excess of M-PEG (~20 g of M-PEG added). This solution is added dropwise, under nitrogen, to 9.25 g cyanuric chloride (50 mmol, recrystallized 3 times from toluene, stored at −50°) in 100 ml benzene. The solution is stirred for 1 hr. Product is collected by addition to 300 ml dry hexane, and it is reprecipitated twice by solution in toluene and addition to two volumes hexane. The yield is approximately 50%, with 100% end-group conversion. Product is stored under nitrogen at −50° as some decomposition is noted for storage at room temperature.

The product can be analyzed by elemental analysis for nitrogen or by chloride titration, as follows: The dried product is carefully weighed and dissolved in an aqueous buffer of pH 10 (10.7 ml of 0.1 $M$ sodium hydroxide plus 50 ml of 50 m$M$ sodium bicarbonate diluted to 100 ml). After heating to reflux for 1 hr, the solution is acidified by adding 2 ml of concentrated nitric acid. The chloride solution is then titrated, with stirring, with 10 m$M$ silver nitrate using a silver billet electrode.[15]

*Tresylate Method.* The following is a modification of the original procedure of Nilsson and Mosbach.[13,16,17] In a typical preparation, 10 g of PEG monomethyl ether (or M-PEG) is dissolved in 50 ml toluene, and water is removed by azeotropic distillation of toluene on a Dean–Stark apparatus (~5 ml of toluene removed). Dichloromethane [25 ml, dried with 3 Å molecular sieves (Sigma, St. Louis, MO) activated at 330° for 12 hr] and 1.2 ml of triethylamine (distilled over $CaH_2$) are added, and the flask is embedded in an ice bath. A nitrogen atmosphere is maintained. Tresyl chloride (0.600 ml) is added during 5 min, and the reaction is stirred overnight while permitting the ice to melt. Most of the dichloromethane is distilled off under reduced pressure (using an aspirator for 30 min, temperature less than 30°). The residue is filtered to remove amine salt, and the filtrate is added to 400 ml of anhydrous ether in an ice bath. The precipitated product is filtered and dried under vacuum. The yield is 9.8 g. This product is stored under nitrogen at −50°.

The tresylate can be analyzed by elemental analysis for sulfur or by hydrolysis and titration of the resulting acid. Similarly, $^1H$ nuclear magnetic resonance can be used to compare the areas of the following peaks (in $d_6$-dimethyl sulfoxide): tresylate methylene at 4.94 ppm, ether methyl

[15] *Standard Methods,* 13th Ed., p. 376 (1971).

[16] K. Nilsson and K. Mosbach, this series, Vol. 104, p. 56.

[17] J. M. Dust, Z.-H. Fang, and J. M. Harris, *Macromolecules* **23,** 3742 (1990).

at 3.26 ppm, ethylene oxide backbone at 3.51 ppm, and unreacted M-PEG hydroxyl at 3.57 ppm.[17] We typically find greater than 90% end-group conversion.

*Protein Derivatization*

Polyclonal antibodies have been used for separating cells in model systems (human and sheep RBCs). For isolation of cell populations with small surface differences, it will be necessary to use preparations of monoclonal antibodies that consist of identical copies of only one antibody molecule. In polyclonal antibody preparations properties such as antigen specificity, binding affinity, and resistance to denaturation are average values for many slightly different antibody molecules. For monoclonal antibody preparations, unlike polyclonal antibody preparations, any eccentricities inherent in the molecule are greatly magnified, so that each preparation has its own unique characteristics. Although this makes experiments using a monoclonal antibody preparation highly reproducible, it can cause frustration when trying to devise techniques that will work for all monoclonal antibody preparations.

Purification of cells and proteins by immunoaffinity partitioning using monoclonal antibodies can be enhanced by having an arsenal of reagents from which to choose. To this end we have used various molecular weights of PEG and branched as well as linear PEG types.

*General Method for Preparing Poly(ethylene Glycol)-Protein Derivatives.* For consistent results we have found it necessary to either filter the antibody preparation or to centrifuge it at high speed to remove aggregates prior to modification. In a typical modification we use 12 mg of antibody (or protein A, protein G, avidin, etc.) dissolved in 2.0 ml of 0.10 *M* $Na_2B_4O_7$, pH 9.2. The appropriate molar ratio of activated PEG (stored frozen under nitrogen) is dissolved in 1.0 ml of the above borate buffer and added dropwise to the stirred protein at 4°. Gentle stirring is continued for 1 hr at 4°. Unreacted PEG is removed by diafiltration using an Amicon YM or PM filter (Amicon, Beverly, MA) of the appropriate pore size, and PEG–protein is concentrated to 7–8 ml. These preparations are stable for at least a month at 4°, but frozen storage in aliquots may be safer if application of the preparation is delayed.

*Comments on Method: Removal of Unreacted Polymer.* Although it is not critical to have all unreacted PEG removed prior to determining percentage lysine modification with the fluorescamine method, it is critical for the trinitrobenzenesulfonic acid (TNBS) method (see below). Because free PEG could interfere in the subsequent partitioning experiment, it is generally best to remove PEG before proceeding with the fluorescamine

TABLE I
PERMEABILITY OF AMICON ULTRAFILTRATION MEMBRANES TO $^{14}$C-LABELED MONOMETHOXYPOLY(ETHYLENE GLYCOL) 5000[a]

| Membrane | Volume buffer through membrane (ml) | $^{14}$C counts removed (%) |
|---|---|---|
| PM30 | 60 | 67.9 |
| | 170 | 91.7 |
| | 277 | 97.1 |
| | 327 | 99.2 |
| YM30 | 50 | 70.2 |
| | 155 | 85.9 |
| | 235 | 92.4 |
| | 320 | 98.7 |
| XM50 | 50 | 18.8 |
| | 165 | 55.4 |
| | 270 | 70.4 |
| | 360 | 85.1 |
| YM10 | 75 | 19.0 |
| | 175 | 34.0 |
| | 305 | 46.8 |
| | 355 | 50.6 |

[a] Contained in 10 ml and maintained at this volume throughout the procedure. From L. J. Karr, D. L. Donnelly, A. Kozlowski, and J. M. Harris, unpublished results.

analysis as well. We have found that the Amicon XM series of membranes retain an unacceptable amount of unreacted PEG. A comparison of the amounts of $^{14}$C-labeled PEG 5000 (Shearwater Polymers) retained by YM, PM, and XM membranes is given in Table I.

*Variation with Antibody Lots.* We have used several lots of some antibodies and only once have we found significant differences in derivatization results. For that particular lot it was necessary to use twice as much activated PEG to get the same percentage of lysines modified as with the other lots of the same antibody.

*Molar Ratio of Activated Polymer to Protein.* It may be necessary to try several PEG : lysine molar ratios to determine which is best for a particular system. For PEG of molecular weight 5000, it is generally observed that a 1 : 1 molar ratio is optimum to provide a percentage lysine modification that gives a large increase in cell partitioning and little cell agglutination. For larger PEGs less of the activated PEG is required. Table II shows some results from experiments in which the PEG–lysine ratio and percentage lysine modification have been optimized.

TABLE II
AMOUNT OF POLY(ETHYLENE GLYCOL) TRESYLATE USED TO MODIFY PROTEINS

| Protein | Lysines modified (%)[a] | PEG molecular weight | PEG : lysine molar ratio[b] | Partition (%) | | Number of lysines seen by fluorescence |
|---|---|---|---|---|---|---|
| | | | | Prior to modification | After modification | |
| IgG | 45–55 | 5000 | 1 : 1 | 48.2 | 84.7 | 39 of 90 |
| | | 14,000 | 1 : 4 | 48.2 | — | |
| | | 20,000 | 1 : 4 | 48.2 | 83.4 | |
| SpG | 70–75 | 5000 | 2 : 3 | 42.7 | 88.7 | 7 of 27 |
| | | 14,000 | 1 : 6 | 42.7 | — | |
| | | 20,000 | 1 : 6 | 42.7 | 91.2 | |
| SpA | 60–70 | 5000 | 1 : 1 | 33.6 | 72.5 | 16 of 49 |
| | | 14,000 | 1 : 4 | 33.6 | — | |
| | | 20,000 | 1 : 4 | 33.6 | 84.2 | |
| Avidin | 55–60 | 5000 | 2.5 : 1 | 42.7 | 82.6 | 31 of 36 |

[a] Percentage modification of the lysines available as detected by the fluorescamine method.
[b] Calculated from total lysines in the protein, not available lysines.

*Molecular Weight of Activated Polymer.* Using a series of single-tube partitioning experiments with PEG-modified antibodies, protein A, protein G, and avidin, we have found that it is best to use PEG with a molecular weight greater than 1900,[3] although differences between PEG 5000 (Union Carbide) and PEG 20,000 (Fluka, Ronkonkoma, NY) are minimal, indicating a plateau in effectiveness.[10] PEG 20,000 reagents can be used at somewhat lower concentrations with minimal agglutination. PEG 14,000 (Berol Nobel, Stenungsund, Sweden), which is branched and has four arms, causes extensive agglutination of cells and so was not extensively characterized.

### *Determination of Percentage of Lysines Modified with Poly(ethylene Glycol)*

The percentage of lysine groups to which PEG has been attached is frequently determined by using a colorimetric or fluorescence method for measuring primary amine concentration before and after the reaction. Two methods that have been used extensively are the fluorescamine assay of Brooks and co-workers[18] and the TNBS assay of Habeeb.[19] We prefer the fluorescamine method because less PEG–protein is required and because small amounts of free PEG do not interfere with the assay. Veronese

[18] S. J. Stocks, A. J. M. Jones, C. W. Ramey, and D. E. Brooks, *Anal. Biochem.* **154,** 232 (1986).
[19] A. F. S. A. Habeeb, *Anal. Biochem.* **14,** 328 (1966).

and co-workers have discussed the attributes of a variety of methods for determining percent lysine modification for proteins.[20] The fluorescamine method, as modified in our laboratory, is performed as follows:

1. The protein sample (either unmodified or PEG-modified) is diluted to make a solution (~2 ml) with a concentration of approximately 200 $\mu$g/ml. The concentration of the solution is determined by optical density measurement. Solution is then added in portions to a set of five test tubes: normally 0.5 ml in the first, 0.4 ml in the second, 0.3 ml in the third, 0.2 ml in the fourth, and 0.1 ml in the fifth. Each tube is diluted to 2.0 ml with 0.1 *M* sodium phosphate buffer, pH 8.0. A sixth tube containing only 2.0 ml buffer is included as a blank. These tubes are taken to step 2 below. Smaller amounts of protein can be utilized, but it is important that sufficient quantities be present in the initial solution to permit concentration measurement by optical density.

2. One milliliter of fluorescamine solution (0.3 mg/ml in acetone) is added to the first sample during vigorous vortexing. After 5 min the sample is assayed on a fluorescence spectrometer with excitation set at 390 nm and emission set at 475 nm. The slit width should be set as narrow as possible while still giving a usable reading. This procedure is followed for each of the samples and the blank.

3. Results are plotted as fluorescence units versus concentration, with the slope of the line being determined by linear regression. The experiment is duplicated and the slopes averaged. The percentage modification is determined by the following formula:

$$\%\ \text{Modification} = 1 - (\text{slope modified protein}/\text{slope unmodified protein}) \times 100$$

*Comments*. It is necessary to keep the fluorescamine solution in a dark container, and the cap must be replaced promptly after sample removal. For optimal results we make up no more than 15 ml of fluorescamine solution per vial. Fluorescamine is weighed out in each vial to be used, and acetone is added just prior to use. It is very important to vortex the dilution sample while adding fluorescamine solution.

The fluorescamine method is extremely useful for monitoring relative amounts of lysine modification and for determining reproducibility of preparations. However, it is important to note that the percentage lysine modification derived by this method is inherently inaccurate. As Veronese and co-workers have pointed out,[20] fluorescence yields from the same fluorophore in different environments vary because of local environment

[20] L. Sartore, P. Caliceti, O. Schiavon, C. Monfardini, and F. M. Veronese, *Appl. Biochem. Biotechnol.* **31,** 213 (1991).

effects. Fluorescamine molecules attached to a PEG–protein will give variable fluorescence yields because most lysines are environmentally different and because substitution of PEG changes local environments. Our view is that the local environment effects are small. After extensive use of the method for several years, we have found that fluorescamine results correlate well with observations on subsequent partitioning experiments.

Despite this inherent limitation of the fluorescamine method, it is important to emphasize that the method is highly reproducible and easy to apply. It thus provides an accurate measure of relative modification and is reproducible.

*Single-Tube Partitioning Methods*

Small-scale experiments with single test tubes are useful for determining optimum values for incubation concentrations, time, and temperature as well as for choice of phase systems and proportions of reagents for modifying immunoaffinity ligands. A significant advantage of partitioning in aqueous polymer two-phase systems relative to other separation methods is the ease of scale-up once experimental parameters have been determined. Single-tube experiments are performed as follows:

1. Between 1 and $2 \times 10^7$ total cells are washed twice in buffer, by centrifugation at low speed and removal of supernatant solution, and then once in upper phase. Following centrifugation and removal of the upper-phase wash, the pelleted cells are suspended with 0.2 ml of appropriately diluted PEG–antibody or PEG–ligand (SpA, SpG, avidin, etc.) plus antibody in a $12 \times 75$ mm polystyrene tube. PEG–ligand must be added to cells prior to antibody to avoid cell agglutination, and plastic tubes reduce cell adherence to tube walls. Generally, incubation is for 1 hr at room temperature if the cells under consideration are not labile.

2. Cells are washed once in fresh upper phase and, following removal of supernatant solution, are suspended in 1.0 ml upper phase and quantified by impedance counting. The purpose of quantifying cells at this point is to determine the extent of cell agglutination (as compared to controls with no antibody or secondary ligand present) and to provide a total count of cells for each tube.

3. Then 1.0 ml of lower phase is added to each tube. The suspension is mixed by inversion 20 times and allowed to settle for 15 min. Mixing is repeated and the system is allowed to settle again, at which time a portion is removed and quantified. Partition of cells is expressed either as a percentage of total cells in a bulk (usually upper) phase or as a partition ratio, which is the ratio of cells in the top phase to the remainder

of cells. Generally a plot of log partition ratio versus incubation concentration of PEG–Ab yields a straight line, thus emphasizing the exponential relationship for which partitioning is noted. The increase in partitioning produced by the immunoaffinity ligand is determined by comparison to control, single-tube partitioning in which no PEG–antibody or PEG–ligand is used.

*Comments.* In general, the best results (i.e., greatest increase in target cell partitioning into the PEG-rich phase with little cell agglutination) have been achieved using ratios of SpA or SpG to antibody of 30–50 : 1. PEG–avidin seems to do best at a 1 : 1 ratio with biotinylated antibody. We have used a succinimide-activated biotin [sulfosuccinimidyl 6-(biotinamido)hexanoate, Pierce, Rockford, IL) and have achieved biotinylation of antibodies using the manufacturer's recommended protocol for attaching two biotin molecules per molecule of IgG.

Depending on the extent of purification and the number of cells required, single-tube partitioning using larger volumes may be sufficient. More complete separations of model cell systems can also be obtained by use of CCD.[6] Figure 2 shows CCD separations of mixtures of sheep and human RBCs using PEG–SpA and antibody directed to either cell type (using the phase system described below).

### *Choice of Phase Systems for Immunoaffinity Partitioning*

We have examined the effects of various phase system parameters on immunoaffinity partitioning (Table III). In brief, we observed that the most dramatic shifts in cell partitioning occur in systems having a low interfacial tension and a low electrostatic potential difference between the phases. The phase system generally used in our laboratories for immunoaffinity partitioning consists of 4.6% (w/w) Dx T500 (Pharmacia LKB, Piscataway, NJ), 3.9% (w/w) PEG 8000 (Union Carbide), 150 m*M* NaCl, 7.3 m*M* $Na_2HPO_4$, and 2.3 m*M* $NaH_2PO_4$, pH 7.2 (buffer V in Table III).

Although we have not observed differences in partitioning for different lots of Dx T500, differences obtained with different lots of PEG 8000 are considerable. Because PEG 8000 is inexpensive and because it is difficult to obtain the same lot number in a subsequent order, we generally order several lot numbers in 10-pound quantities. Each lot is tested for immunoaffinity partitioning of human RBCs with PEG-derivatized antibody directed against human RBCs using the phase system described above. We look for low partitioning into the upper phase in single-tube control experiments without PEG–antibody (usually on the order of 3–5%) and a large increase in partitioning to the upper phase following incubation of the cells with PEG–antibody. Normally two out of three lots ordered are

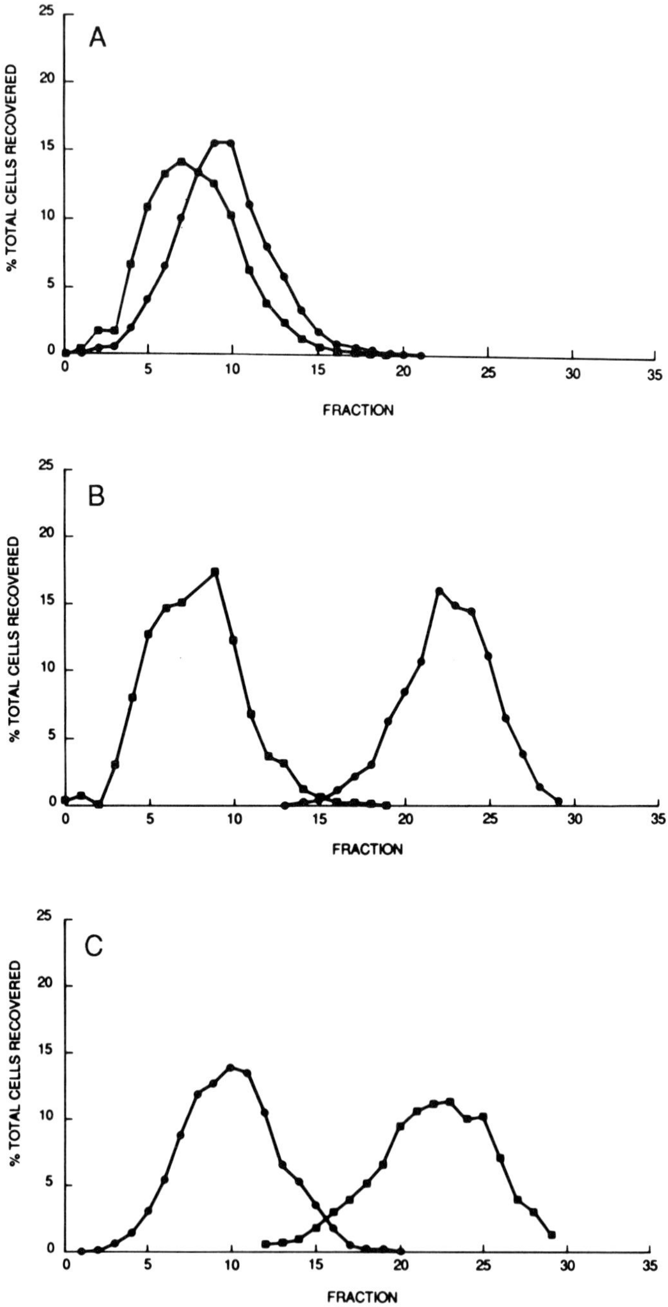
A
B
C
% TOTAL CELLS RECOVERED
FRACTION
0
5
10
15
20
25
30
35

TABLE III
EFFECT OF PHASE SYSTEM COMPOSITION ON ABILITY OF POLY(ETHYLENE GLYCOL)–ANTIBODY TO INDUCE ALTERATION OF ERYTHROCYTE PARTITIONING[a]

| Two-phase system[b] | Interfacial tension ($\mu$N/m) | Electrostatic bulk phase potential (mV) | Cell location (phase) | Partition (%) | |
|---|---|---|---|---|---|
| | | | | No PEG–Ab | PEG–Ab |
| (5, 4) I | 11.0 | 2.2 | Upper | 14 ± 1 | 58 ± 4 |
| (5, 4) II | 8 | 1.0 | Upper | 6 ± 1 | 64 ± 1 |
| (5, 3.5) V | 0.7 | 0.2 | Upper | 2 ± 1 | 77 ± 1 |
| (4.6, 3.9) V | 3 | 0.2 | Upper | 2 ± 1 | 69 ± 3 |
| (5, 4) V | 4.9 | 0.3 | Upper | 0 ± 1 | 30 ± 1 |
| (7, 4) V | 17.2 | 0.2 | Upper | 0 ± 1 | 4 ± 2 |
| (7, 5) V | 5.0 | 0.2 | Upper | 0 ± 1 | 1 ± 1 |
| | | | Interface | 45 ± 3 | 97 ± 3 |
| | | | Lower | 55 ± 3 | 3 ± 1 |

[a] From L. J. Karr, S. G. Shafer, J. M. Harris, J. M. Van Alstine, and R. S. Snyder, *J. Chromatogr.* **354,** 269 (1986).

[b] All systems were composed of PEG 8000 and dextran T500 except (7, 5) V which was prepared with dextran T40. Buffers (all pH 7.2) contain the following amounts of NaCl, $Na_2HPO_4$, and $NaH_2PO_4$: I = 0, 109, and 35 m*M*; II = 50, 73, and 23 m*M*; and V = 150, 7.3, and 2.3 m*M*.

good, so a quantity of usable PEG sufficient to last several years is obtained. This material is stored in sealed plastic bags at 4°.

## Conclusions

The immunoaffinity techniques described provide effective, reproducible extraction of large numbers of apparently undamaged cells. The PEG derivatives for antibody and ligand modification are commercially available so that only minor chemical manipulation is required. PEG–SpA and PEG–SpG use considerably less antibody than direct derivatization, and

FIG. 2. (A) Countercurrent distribution pattern of a mixture of $3 \times 10^7$ untreated human (●) and sheep (■) RBCs (~1 : 1). Thirty transfers were completed in a phase system containing 4.6% dextran, 3.9% PEG, and buffer V (Table III) with a 12-sec shaking and a 3-min settling time at room temperature. (B) Identical to (A) except that, prior to CCD, the RBC mixture was incubated with 0.8 mg/ml PEG–protein A and 0.12 mg/ml unmodified antibody directed against human RBCs. (C) Identical to (A) except that, prior to CCD, the RBC mixture was incubated with 0.8 mg/ml PEG–protein A and 0.08 mg/ml unmodified antibody directed against sheep RBCs. [From L. J. Karr, J. M. Van Alstine, R. S. Snyder, S. G. Shafer, and J. M. Harris, *J. Chromatogr.* **442,** 219 (1988).]

this approach eliminates the need for a different antibody preparation for each cell type. This approach may also be useful for those monoclonal antibodies that lose an unacceptable level of antigen-binding capability through direct PEG modification. Biotin–avidin binding is stronger than antibody–antigen binding or binding of SpG or SpA to Fc fragments, and this approach may be of use with cells having few binding sites.

# [36] Use of Polyacrylamide-Derivatized Antibody in Dextran–Poly(ethylene Glycol) Systems

*By* DONALD E. BROOKS and S. JILL STOCKS

## Introduction

As discussed in the chapter by Karr *et al.*,[1] antibodies (Abs) provide a powerful and versatile tool for the isolation of cells that are specified by the presence of a particular surface antigen (Ag). In immunoaffinity partitioning the principle is to modify the Ab in such a way as to cause it to partition strongly in favor of one phase while leaving it reactive with its target antigen. Appropriately modified Abs have been produced by a variety of means including derivatization with poly(ethylene glycol) (PEG),[2–4] trypan blue,[5] and biotin,[5] all of which cause them to partition to varying degrees into the top, PEG-rich phase of a dextran (Dx)–PEG two-phase system. Primary Abs,[2,3] secondary Abs,[4,5] and PEG-derivatized protein A[6] have been used as immunoaffinity ligands for the separation of erythrocytes from different species.

Although the work referred to above has amply demonstrated the principle of immunoaffinity partitioning, the actual separations that have been carried out have not been particularly challenging, largely because they have dealt with cell surface antigens present in abundance on the

[1] L. J. Karr, D. L. Donnelly, A. Kozlowski, and J. M. Harris, this volume [35].

[2] K. A. Sharp, M. Yalpani, S. J. Howard, and D. E. Brooks, *Anal. Biochem.* **154,** 110 (1986).

[3] L. J. Karr, J. M. Harris, S. G. Shafer, J. M. Van Alstine, and R. S. Snyder, *J. Chromatogr.* **354,** 269 (1986).

[4] S. J. Stocks and D. E. Brooks, *Anal. Biochem.* **173,** 86 (1988).

[5] S. J. Stocks and D. E. Brooks, *in* "Advances in Separations Using Aqueous Phase Systems in Biology and Biotechnology" (D. Fisher and I. A. Sutherland, eds.), p. 183. Plenum, London, 1989.

[6] L. J. Karr, J. M. Harris, S. G. Shafer, J. M. Van Alstine, and R. S. Snyder, *in* "Advances in Separations Using Aqueous Phase Systems in Biology and Biotechnology" (D. Fisher and I. A. Sutherland, eds.), p. 193. Plenum, London, 1989.

cell type of interest and absent on all other cells present. To solve more stringent separation problems, in which the antigenic differences among cells in the parent mixture are based on antigens present at low density or present on all cells but at different densities, the procedure needs to be optimized to effect the largest partitioning change per bound Ab.

Derivatization to produce partitioning into the PEG-rich phase can be expected to make the derivatized Ab behave like PEG and approach similar partitioning characteristics in Dx–PEG systems. The reason for choosing PEG is largely historical, however, and there are advantages to developing affinity reagents which cause partitioning into the Dx-rich phase. For instance, if the concentration ratio between the phases for PEG is compared to that for Dx, as determined from the phase diagram,[7] it is seen that Dx partitions from 2 to 10 times more strongly in favor of the bottom phase than PEG does into the top phase. Hence, basing an affinity ligand on Dx affinity or PEG incompatibility would seem to hold promise.

## Polyacrylamide as Dextran Affinity Ligand

It is known from work by Müller[8] on bead derivatization that grafting hydroxylated gel beads with polyacrylamide (PAAm) produces a solid phase which can immobilize the Dx-rich phase of a Dx–PEG system. This allows partition chromatography to be performed on macromolecules[8]; cell separations have been performed on such columns as well.[9,10] The reason for the immobilization is not the affinity of PAAm for Dx but its strong incompatibility with PEG.[11] We therefore have developed a method for derivatizing protein with PAAm which has been shown to be extremely effective, when applied to Abs, in producing immunoaffinity separations on the basis of antigen density differences.[12]

## Polyacrylamide Derivatization of Proteins

Grafting and polymerization of acrylamide is carried out via a free radical mechanism. Some ceric salts, for instance, nitrate and sulfate, form effective redox systems in the presence of organic reducing agents

[7] P.-Å. Albertsson and F. Tjerneld, this volume [1].

[8] W. Müller, *Eur. J. Biochem.* **155,** 213 (1986).

[9] D. R. Skuse and D. E. Brooks, *J. Chromatogr.* **432,** 127 (1988).

[10] D. R. Skuse, W. Müller, and D. E. Brooks, *Anal. Biochem.* **174,** 628 (1988).

[11] W. Müller, this volume [9].

[12] S. J. Stocks, Ph.D. thesis, Department of Chemistry, University of British Columbia, Vancouver, Canada, 1989.

such as alcohols and aldehydes.[13] This produces cerous ions and free radicals capable of initiating vinyl polymerization. If the reducing agent is a macromolecule and the reduction is carried out in the presence of a vinyl monomer, the free radical created on the macromolecule initiates polymerization to give a polymer graft. The reactions result in predominantly grafted polymers since the free radicals are formed preferentially on the backbone. In the absence of a reducing agent, ceric salts will initiate acrylamide polymerization at a very slow rate, producing free, water-soluble PAAm.

Optimization of the reaction conditions for grafting PAAm onto protein was done using bovine serum albumin (BSA) and nonspecific bovine immunoglobulin G (IgG). One problem with this reaction is that almost all ceric salts are insoluble in water; hence, buffering has not been possible since all normal buffers precipitate the initiator. All reactions are carried out in water, therefore, the only buffering being supplied by the protein itself.

Protein solution in water (25 ml, ~2 mg/ml) containing acrylamide (Bio-Rad, Richmond, CA; 0.02–0.2 mg/ml) is placed in a three-necked reaction vessel. Neck 1 contains a nitrogen bubbler inlet, neck 2 a reflux condenser topped with a nitrogen outlet via a paraffin bubbler neck, and neck 3 a stoppered separating funnel with a pressure equalizing arm containing degassed $(NH_4)_2[Ce(NO_3)_6]$ (BDH, Darmstadt, Germany; 54 mg/ml) in 1.0 *M* $HNO_3$. The solution is degassed by bubbling $N_2$ through for 3 hr under reflux at room temperature. The reaction is initiated by adding 0.1 ml of the Ce(IV) solution and stirred minimally, and the reaction is continued for 3 hr without stirring and with constant $N_2$ bubbling. The reaction is terminated by admitting air to the reaction vessel. Some estimate of the extent of reaction is provided by the increase in solution viscosity as the grafting and solution reactions proceed. The derivatives are concentrated and free acrylamide removed by ultrafiltration through an Amicon (Danvers, MA) XM100A filter.

## Partitioning Behavior of Polyacrylamide-Derivatized Proteins

Polyacrylamide was grafted onto BSA, bovine IgG, and rabbit anti-human red blood cell (RBC) IgG by the above procedure. The partition coefficients for these derivatives are given in Table I. As expected, in all cases PAAm derivatization strongly increased protein partitioning into the bottom, Dx-rich phase. The derivatized anti-human RBC antibodies were also tested for specific binding capacity by hemagglutination assays on human and rabbit RBC. They showed no decrease in agglutinating

[13] G. Mino and S. Kaizerman, *J. Polym. Sci.* **31,** 242 (1958).

TABLE I
PARTITIONING OF POLYACRYLAMIDE-DERIVATIZED PROTEINS

| Protein | Hydroxyl : acrylamide reaction ratio[a] (mol : mol) | Viscosity (mPa sec) | | $K$[b] | |
|---|---|---|---|---|---|
| | | Unreacted | Reacted | Native protein | PAAm–protein |
| BSA[c] | 1 : 32 | 0.94 | 1.34 | 2.1 | 0.38 |
| B IgG[d] | 1 : 5 | 0.98 | 2.31 | 1.5 | 0.16 |
| B IgG | 1 : 3 | 0.92 | 1.23 | 1.5 | 0.25 |
| B IgG | 1 : 1.6 | 0.91 | 0.99 | 1.5 | 0.28 |
| R α-h RBC[e] | 1 : 100 | 0.91 | 1.36 | 1.1 | 0.18 |
| R α-h RBC | 1 : 25 | 0.90 | 1.04 | 1.1 | 0.21 |
| R α-h RBC | 1 : 20 | 0.90 | 0.99 | 1.1 | 0.19 |

[a] Based on serine plus threonine content only.
[b] $K$ is the partition coefficient (upper/lower) of the indicated protein in 5% (w/w) dextran T500 (Pharmacia, Uppsala, Sweden), 4% (w/w) PEG 8000 (Union Carbide, New York, NY), 110 m$M$ sodium phosphate buffer, pH 7.5.
[c] Bovine serum albumin.
[d] Bovine IgG with no specificity for red blood cells.
[e] Rabbit anti-human RBC IgG.

ability due to PAAm derivatization, nor was nonspecific agglutination of control rabbit RBC observed (Table II). The most highly modified antibody (1 : 100) showed a slight increase in agglutinating ability, probably owing to the presence of free PAAm molecules. High molecular weight PAAm is known to agglutinate cells nonspecifically, and neutral polymers such as these are able to potentiate antibody-induced agglutination.[14]

These results indicate that because the antibodies retain their capacity to cross-link cells following PAAm derivatization, it might be expected that agglutination would lead to problems during partitioning. The relatively high polymer concentrations in phase systems have a disaggregating effect on cells, however,[15] and few aggregation problems were encountered when using the PAAm–anti-RBC antibodies.

## Immunoaffinity Partitioning of Cells Using Polyacrylamide–Antibodies

The ability of PAAm-derivatized Ab to increase immunospecifically cell partitioning into the Dx-rich lower phase is illustrated with data on

[14] D. E. Brooks, *in* "Human Blood Groups," 5th International Convocation on Immunology, Buffalo, New York, 1976, p. 27. Karger, Basel, 1977.

[15] D. E. Brooks, K. A. Sharp, and D. Fisher, *in* "Partitioning in Aqueous Two-Phase Systems: Theory, Methods, Uses, and Applications to Biotechnology" (H. Walter, D. E. Brooks, and D. Fisher, eds.), p. 11. Academic Press, New York, 1985.

TABLE II
HEMAGGLUTINATION ACTIVITY OF POLYACRYLAMIDE-DERIVATIZED ANTIBODIES

| Antibody | Hydroxyl : acrylamide[a] reaction ratio (mol : mol) | Cell | Minimum agglutinating concentration ($\mu$g/ml) |
|---|---|---|---|
| R $\alpha$-h RBC[b] | C[c] | Human RBC | 2.0 |
| PAAm–R $\alpha$-h RBC[d] | 1 : 25 | Human RBC | 2.0 |
| PAAm–R $\alpha$-h RBC | 1 : 100 | Human RBC | 0.4 |
| R $\alpha$-h RBC | C | Rabbit RBC | None[e] |
| PAAm–R $\alpha$-h RBC | 1 : 25 | Rabbit RBC | None |
| PAAm–R $\alpha$-h RBC | 1 : 100 | Rabbit RBC | None |

[a] Estimated from the numbers of serine and threonine residues.
[b] Rabbit anti-human RBC IgG.
[c] Control: underivatized antibody.
[d] Polyacrylamide-derivatized rabbit anti-human RBC IgG.
[e] No agglutination at the highest concentration of antibody (0.5 mg/ml).

human and rabbit RBC and PAAm–anti-human RBC Ab in Table III. In both phase systems used, only the immunologically specific cells partition into the lower phase, and only when the Ab is derivatized. The effect was also observed at PAAm–Ab concentrations lower than those shown in Table III, 80 $\mu$g/ml of PAAm–anti-human RBC being sufficient to move about 90% of the human cells into the bottom phase (data not shown).

TABLE III
IMMUNOAFFINITY CELL PARTITIONING WITH POLYACRYLAMIDE-DERIVATIZED ANTIBODIES

| Ligand | Cell type | Lower phase partition %[a] | |
|---|---|---|---|
| | | System 1[b] | System 2[c] |
| None | Human RBC | 12 | 33 |
| PAAm–R $\alpha$-h RBC[d] | Human RBC | 93 | 99 |
| None | Rabbit RBC | 16 | 22 |
| PAAm–R $\alpha$-h RBC[d] | Rabbit RBC | 19 | 39 |

[a] % Lower phase partition = [(no. cells in lower phase) × 100] ÷ (no. in upper phase + no. at interface).
[b] System I: 5% (w/w) dextran T500, 3.4% (w/w) PEG 8000, 110 m*M* sodium phosphate buffer, pH 7.5.
[c] System II: 5% (w/w) dextran T500, 3.4% (w/w) PEG 8000, 130 m*M* sodium chloride, 10 m*M* sodium phosphate buffer, pH 7.5.
[d] Polyacrylamide-derivatized rabbit anti-RBC (0.2 mg/ml).

## Conclusions

Polyacrylamide derivatization can be very effective in directing protein partitioning into the Dx-rich phase. The disadvantage of this approach is the lack of added buffer and the harsh conditions under which the derivatization chemistry takes place. It is possible that not all proteins will withstand the acidity required for the grafting reaction and will lose activity. Successful derivatization of both polyclonal and monoclonal Abs is feasible, however, using the technique described.[12]

Collectively, the results in Table III represent the most efficient immunoaffinity separation of RBC reported to date. Clearly, under appropriate conditions, PAAm derivatization provides very effective immunoaffinity ligands. These reagents carry the additional interesting feature that they react predominantly with the hydroxylated amino acids serine and threonine and with the carbohydrate on glycoproteins, thus offering an alternative to the usual amine-based coupling chemistry common in modification reactions. They also open the possibility of performing two immunospecific separations simultaneously, by incorporating in a phase system two Abs with specificities for two different cell types, one derivatized with PAAm and the other modified to partition into the PEG-rich phase. It is likely that conditions could be found under which the two antigenically defined subpopulations could be made to separate into the two bulk phases with the remaining cells left in the interface, an interesting prospect.

# [37] Bioextraction of Low Abundance Cells by Affinity Partitioning

*By* Cristina Delgado, Gillian E. Francis, and Derek Fisher

## Introduction

Progress in methods for coupling polymers to proteins has provided the opportunity for developing affinity partitioning techniques for the separation of cells based on protein–ligand/cell surface receptor interactions. The mildness of aqueous biphasic systems, together with the specificity of affinity methods, has generated a powerful method for the isolation of cells in biological mixtures. The partitioning of a selected cell type is directed toward the top, poly(ethylene glycol) (PEG)-rich phase by coating the cell with a specific antibody–ligand that has been covalently linked to PEG.[1-5] Ideally, the contributions of other surface properties of the cell (e.g., charge, hydrophobicity), which can also cause cells to partition into the top phase, are minimized in order for the affinity ligand to have a maximal effect on top phase partitioning. For PEG–ligands this can be achieved by selecting non-charge-sensitive phase systems (i.e., those with a low content of phosphate and a high content of sodium chloride) in which partitioning of the bulk cells is easily directed toward the interface by manipulation of the polymer concentrations.

To select the experimental conditions leading to a low partition of the contaminating cells and a high partition of the PEG–ligand coated target cells, the following parameters have to be controlled: (1) the degree of PEG modification of the antibody/ligand, (2) the concentration of the PEG-modified antibody/ligand, (3) the concentration of the cells, and (4) the composition of the biphasic system. The degree of PEG modification of the antibody–ligand influences the partition coefficient of the PEG conjugate in the biphasic system and also the affinity of the PEG conjugate for the target cell: the former increases with the degree of PEG modification whereas the latter might decrease. The concentration of the PEG–antibody/ligand and the concentration of cells influence the number of PEG–ligand molecules attached per cell and therefore the degree of coverage of the cell surface with PEG. As the latter increases, the partition coefficient of the PEG-coated cells also increases. The composition of the biphasic system influences the partitioning of the cells in the absence and presence of PEG–ligand and also the partitioning of the PEG–antibody/ligand. Increasing the concentration of the polymers directs the partitioning of the cells toward the interface (owing to the increased interfacial tension) and increases the partition coefficient of the PEG–ligand. However, the PEG–ligand is less efficient in pulling the cells up from the interface and into the top phase as the concentration of polymers increases.[1]

The following constitute steps used for cell affinity partitioning: (1) production of the PEG-modified antibody/ligand, (2) incubation of the PEG–antibody/ligand with the cell mixture, (3) positive selection of the target cells into the PEG-rich phase by partitioning in an appropriate biphasic system, and (4) recovery and concentration of the cells.

[1] L. J. Karr, S. G. Shafer, J. M. Harris, J. M. van Alstine, and R. S. Snyder, *J. Chromatogr.* **354,** 269 (1986).

[2] L. J. Karr, J. M. van Alstine, R. S. Snyder, S. G. Shafer, and J. M. Harris, *J. Chromatogr.* **442,** 219 (1988).

[3] K. A. Sharp, M. Yalpani, S. J. Howard, and D. E. Brooks, *Anal. Biochem.* **154,** 110 (1986).

[4] S. J. Stocks and D. E. Brooks, *Anal. Biochem.* **173,** 86 (1988).

[5] C. Delgado, R. J. Anderson, G. E. Francis, and D. Fisher, *Anal. Biochem.* **192,** 322 (1991).

## Methods

### *Chemicals*

- Poly(ethylene glycol) (PEG) (usually $M_r$ 6000, BDH, Poole, UK)
- Monomethoxy-PEG (M-PEG) ($M_r$ 5000, Union Carbide Europe, Versoix, Switzerland)
- Dextran (Dx) T-500 (Pharmacia, Uppsala, Sweden)
- Activated M-PEG (Shearwater, Huntsville, AL)
- Ligand [examples given are immunoglobulin G (IgG) and transferrin]

### *Stock Solutions*[6]

- PEG, 40% (w/w)
- Dx, 20% (w/w)
- Sodium phosphate buffer, 0.44 *M* pH 6.8
- Sodium chloride, 0.60 *M*
- Phosphate-buffered saline (PBS, stock solution from GIBCO, Paisley, UK)

### *Modification of Proteins with Monomethoxypoly(ethylene Glycol)*

Covalent attachment of PEG to proteins is achieved by nucleophilic attack of the ε-amino groups of lysine residues and N-terminal amino group at a suitable leaving group that has been introduced at the hydroxyl terminus of the polymer (activated polymer). Generally, it is preferred to activate monomethoxy-PEG to restrict the reactive groups to one per molecule, thereby avoiding the potential formation of cross-linked PEG-modified products. The reaction between the activated PEG and the protein can take place in any buffer free of strong nucleophiles (e.g., phosphate, HEPES, carbonate). The extent of the reaction depends on the pH (reaction is favored by alkaline conditions) and the concentration of both the protein and the activated PEG. In general, a molar excess of the activated PEG is required. Examples given are with treslyated M-PEG (TM-PEG).[7] Other activated PEG derivatives can be used similarly to produce PEG conjugates, but the reaction time, coupling buffer, and molar excess of the polymer might need optimization.

All coupling reactions are performed in PBS (50 m*M* sodium phosphate buffer, pH 7.5, containing 125 m*M* sodium chloride) for 2 hr at room temperature, with gentle mixing on a rotary mixer. As a precaution the activated PEG should be made up in PBS just before use to minimize any

[6] D. E. Brooks and R. Norris-Jones, this volume [2].

[7] C. Delgado, J. N. Patel, G. E. Francis, and D. Fisher, *Biotechnol. Appl. Biochem.* **12,** 119 (1990).

possible hydrolysis. Some typical reactions are summarized in Table I. It is important that with dilute protein solutions the molar excess of activated PEG required be substantially increased compared to conditions employed with more concentrated protein solutions.

*Characterization of Poly(ethylene Glycol)–Ligand–Antibody Conjugate*

The partition coefficient of the PEG–ligand is measured in an appropriate biphasic system (1 g) prepared from the stock solutions of PEG, Dx, sodium phosphate, and sodium chloride, incorporating 0.1 g of the reaction mixture containing PEG–ligand and adjusting the final weight to 1 g with distilled water. The excess activated PEG in the reaction mixture contributes to the total weight of PEG in the biphasic system and has to be taken into account when preparing the phase system. This is particularly important when the partition coefficients of PEG–ligands prepared at increasing molar ratios of activated PEG to lysine residues are to be compared. The phase system components are equilibrated to 25° and weighed out, and the phase system is then well mixed and allowed to stand at 25° in a water bath until complete separation of the phases is obtained (usually 20 min for the biphasic systems disclosed in Table I). Alternatively, phase settling is speeded by centrifugation at 25° using a bench-top centrifuge. Samples from top and bottom phases are analyzed for protein content [Coomassie blue assay is suitable, (Pierce, Rockford, U.S.A.)]. Gilson pipettes with positive displacement or Hamilton syringes are recommended to sample the phases (the Dx bottom phase is often viscous, and sampling with ordinary Gilson pipettes is not reliable).

If $^{125}$I-labeled ligands are used, the experimental approach can be modified since the $^{125}$I-labeled ligands can be detected even after addition of only 1 to 5 $\mu$l of the reaction mixture to the biphasic system. The biphasic system is prepared at 25° in large quantities in advance and the top and bottom phases stored separately at 4° until required. The biphasic system is reconstituted by mixing 1 ml of bottom phase and 1 ml of top phase after the phases have been equilibrated to 25°. Then 1 to 5 $\mu$l of the reaction mixture is added and the biphasic system mixed and allowed to settle at 25° in a water bath as before. After separation of the phases, the protein contents in the top and bottom phases are analyzed with a gamma counter.

Further characterization of the PEG–ligand/antibody involves determination of the number of PEG chains attached per molecule of protein. This is usually done with a fluorometric assay in which fluorescamine reacts with the unmodified $\varepsilon$-amino groups to form fluorophores, the result-

TABLE I
PREPARATION OF POLY(ETHYLENE GLYCOL)–LIGANDS FOR AFFINITY PARTITIONING

| | Protein | | TM-PEG stock | | BSA for quenching | | Partitioning of conjugates | |
|---|---|---|---|---|---|---|---|---|
| Ligand | Concentration (mg/ml) | Volume (ml) | Concentration (mg/ml) | Volume (ml) | Concentration (mg/ml) | Volume (ml) | Biphasic system[a] | Partition coefficient |
| IgG[b] | 6.5 | 0.5 | 172 | 0.5 | 84 | 1 | 4.60% | 5.9[c] |
| | | | | | | | 4.75% | 7.9[c] |
| IgG[d] | 0.2 | 0.5 | 21 | 0.5 | — | — | 4.75% | 2.1 |
| IgG[d] | 0.2 | 0.5 | 45 | 0.5 | — | — | 4.75% | 3.3 |
| IgG[d] | 0.2 | 0.5 | 90 | 0.5 | — | — | 4.75% | 4.9 |
| Tf[e] | 3.0 | 0.5 | 216 | 0.5 | 90 | 1 | 4.75% | 6.1[f] |

[a] The percentages indicate the concentrations for both PEG and dextran. In all the systems the concentration of salts was 10 m*M* sodium phosphate, pH 6.8, and 150 m*M* sodium chloride.

[b] Anti-human red blood cell IgG (α-HRBC).

[c] C. Delgado, R. J. Anderson, G. E. Francis, and D. Fisher, *Anal. Biochem.* **192,** 322 (1991).

[d] Anticarcinoembryonic antigen.

[e] Transferrin.

[f] C. Delgado, P. Sancho, J. Mendieta, and J. Luque, *J. Chromatogr.* **594,** 97 (1992).

ing fluorescence being proportional to the primary amine concentration.[8] The affinity of the PEG–ligand toward the target can be measured using standard binding assays.[9]

### *Preparation of Poly(ethylene Glycol)–Ligand/Antibody Conjugate for Cell Affinity Partitioning*

To use the PEG–ligand/antibody conjugate in cell affinity partitioning, the excess of activated PEG present in the preparation must either be removed or its activity quenched (by addition of a nucleophile). Removal of the activated PEG can be accomplished by gel filtration in a fast protein liquid chromatography (FPLC) system using, for instance, a Superose 12 HR 10/30 column (Pharmacia LKB). Selection of the column depends on the protein in question. However, the activated PEG above certain concentrations interferes with the resolution of gel permeation chromatography.[10]

"Quenching" the excess activated PEG is an attractive alternative when labile ligands are being attached to PEG or if dilution of the sample for FPLC leads to protein concentrations below the limit of detection. Bovine serum albumin (BSA) has been used as a quenching agent. BSA is added to the PEG–ligand/antibody preparation to provide a molar excess of 4 lysine residues per molecule of activated PEG, and the reaction mixture is incubated for an additional 2 hr at room temperature in a rotary mixer (the concentration of TM-PEG in the reaction mixture also imposes a practical limit for the use of this strategy). The preparation is then diluted with buffer to the required conjugate concentration. This procedure is especially attractive when addition of the BSA to the reaction mixture produces a solution of physiological osmolarity which can then be used directly for incubation with the cell mixture. Table I summarizes quenching of TM-PEG with BSA.

### *Application of Poly(ethylene Glycol)–Antibody for Affinity Extraction of Cells Present at Low Abundance*

The following illustrates the extraction of red blood cells present at 1% abundance in a mixture with HL-60 cells using as the affinity ligand PEG modified anti-human red blood cell membrane IgG ($\alpha$-HRBC) (Table I). A non-charge-sensitive phase system containing 4.75% PEG 6000/

[8] S. J. Stocks, A. J. Jones, C. W. Ramey, and D. E. Brooks, *Anal. Biochem.* **154,** 232 (1986).

[9] C. Delgado, P. Sancho, J. Mendieta, and J. Luque, *J. Chromatogr.* **594,** 97 (1992).

[10] F. Malik, C. Delgado, C. Knusli, A. E. Irvine, D. Fisher, and G. E. Francis, *Exp. Hematol. (N.Y.)* **20,** 1028 (1992).

TABLE II
EXTRACTION OF ERYTHROCYTES[a]

| Partitioning step | First: Total cells | First: Cells in top phase | Second: Total cells | Second: Cells in top phase | Third: Total cells | Third: Cells in top phase |
|---|---|---|---|---|---|---|
| 1 | 9900 HL-60 | 297 | 864 | 26 | 75 | 2.3 |
| | 100 RBC | 50 | 88 | 44 | 77 | 38.3 |
| 2 | 9603 HL-60 | 288 | 838 | 25 | 73.2 | 2.2 |
| | 50 RBC | 25 | 44 | 22 | 38.3 | 19.1 |
| 3 | 9315 HL-60 | 279 | 813 | 24 | 71 | 2.1 |
| | 25 RBC | 12.5 | 22 | 11 | 19.1 | 9.5 |

Adapted from C. Delgado, R. J. Anderson, G. E. Francis, and D. Fisher, *Anal. Biochem.* **192,** 322 (1991).

[a] Extraction of erythrocytes (RBC) at 1% abundance from a mixture with HL-60 cells using M-PEG–α-HRBC as the immunoaffinity ligand in a combination of bulk extractions. The partitioning of the RBC and the HL-60 cells into the top phase were 50 and 3%, respectively. A total of 10,000 cells were subjected to a first round of three partitions of the bottom phase + interface with fresh top phases. The combined top phases were then subjected to the second round of extractions. The combined top phases obtained in the second round were subjected to the third round of extractions.

4.75% Dx T500 is selected because the partitioning of the contaminating cells, HL-60 cells, is about 3% and the partitioning of the target cells, after interaction with the PEG–antibody, is greater than 50%.[6] The biphasic system is prepared in advance, usually at 250 g total, from the stock solutions of PEG, Dx, sodium phosphate buffer, sodium chloride, and distilled water.

For the extraction, $2.75 \times 10^7$ cells are resuspended in 1 ml of PBS containing $5.5 \times 10^{-9}$ mol of M-PEG–α-HRBC conjugate (this is prepared by adding 500 μl of the M-PEG–α-HRBC quenched with BSA to 500 μl of PBS) and incubated for 30 min at 37°. Cells are collected by centrifugation (400 *g*, 10 min, 4°), resuspended in 5 ml of top phase, and mixed in a Universal tube with 5 ml of bottom phase. After 60 gentle inversions the biphasic system is allowed to phase separate for 20 min in a water bath at 25°. The top phase (containing ~50% of the total red blood cells and 3% of the total HL-60 cells) is removed and retained. The bottom phase is subjected to two additional partitioning steps with fresh top phase to recover more target cells (in each partition step 50% of the remaining target cells are recovered in the top phase together with 3% of contaminant cells) (Table II). Cells in the combined top phases are collected by centrifugation. Two more rounds of bulk extractions (each comprising three partitioning steps) using, as a loading mixture, the cells contained in the combined top phases from the previous extraction, yields about 67% of the target cells with a purity of 91%. This is summarized in Table II.

## Discussion

Affinity partitioning methods can be applied for the positive selection of cells present in a biological mixture at 1% abundance. Having designed strategies of immunoaffinity partitioning for the bioextraction of cells present at low abundance, it remains to be established whether the method will work with cells with low expression of the target antigen or receptor. The majority of the examples reported to date use as the target a surface antigen of relatively high expression (in our case 500,000 copies per cell). A recent report has indicated 80,000 as the minimum number of receptors that need to interact with the PEG–ligand to provide a sufficient PEG coat with which to increase the partitioning of the cells.[9] Because for many biological applications the target antigen has lower expression than this, new affinity ligands capable of loading more PEG to the cell surface need to be developed. Covalent attachment of polylysine to the candidate protein will increase markedly the number of positions that can be PEG modified (for methods to link polylysine to proteins, see Mendieta[11]), and thereby the load of PEG to the cell surface per molecule will increase. This might reduce the number of antigens/receptors required to allow efficient extractions. Such a strategy will also extend the range of possible affinity ligands to small proteins and peptides (e.g., cytokines and growth factors), which have relatively few (lysine) sites for PEG derivatization.

## Acknowledgments

This project was supported by the British Council and the Spanish Ministry of Education (C.D.) and the Wellcome Trust (G.E.F. and D.F.).

[11] J. Mendieta, this volume [33].

# [38] Preparation of Synaptosomes and Mitochondria from Mammalian Brain

*By* M. J. López-Pérez

## Introduction

Synaptosomes are "pinched-off" nerve terminals containing synaptic mitochondria, cytoplasm, and vesicles which cosediment with free mitochondria when brain homonogenates are subfractionated by conventional differential centrifugation.[1] Hence, the preparation of brain synaptosomes and free mitochondria, metabolically active and free of contaminants, has been a recognized problem.[2,3] Countercurrent distribution of synaptosomes in aqueous two-phase systems has been used for analytical purposes.[4,5] However, this procedure requires special equipment often not available, and the separation process is long and not always suitable for preparative purposes. Two-phase systems containing 6.4% (w/w) dextran (Dx) T500 and 6.4% (w/w) poly(ethylene glycol) (PEG) 4000 were found suitable for achieving the separation of brain synaptosomes and mitochondria by simple batch extraction.[6] In this system mitochondria show a high affinity for the lower, Dx-rich phase, whereas synaptosomes strongly partition to the upper, PEG-rich phase when (approximately) 0.32 *M* sorbitol, 0.1 m*M* potassium EDTA, and 5 m*M* potassium phosphate, pH 7.4–7.8, are included in the system.

## Phase System Components

Dextran T500 is obtained from Pharmacia-LKB (Uppsala, Sweden). A 20% (w/w) solution of deionized Dx T500 in doubly distilled water is prepared as described by Brooks in this volume.[7] PEG 4000 (Av. Mol.

[1] E. G. Gray and V. P. Whittaker, *J. Anat.* **96,** 79 (1962).

[2] V. P. Whittaker, *in* "Handbook of Neurochemistry" (A. Lajtha, ed.), Vol. 7, p. 1. Plenum, New York, 1984.

[3] J. B. Clark and W. J. Nicklas, *J. Biol. Chem.* **245,** 4724 (1970).

[4] G. Johansson and B. Olde, *J. Biochem. Biophys. Methods* **17,** 285 (1988).

[5] B. Olde and G. Johansson, *Mol. Cell. Biochem.* **87,** 153 (1989).

[6] M. J. López-Pérez, G. Paris, and C. Larsson, *Biochim. Biophys. Acta* **635,** 359 (1981).

[7] D. E. Brooks and R. Norris-Jones, this volume [2].

Wt. 3350) is purchased from Sigma Chemical Co. (St. Louis, MO), and a 40% (w/w) solution is also prepared in doubly distilled water.

## Method I: Preparation of Rat Brain Synaptosomes

### *Solutions*

The isolation medium (medium A) contains 0.32 *M* sucrose, 1 m*M* potassium EDTA, and 10 m*M* Tris-HCl, pH 7.4. The resuspension medium (medium B) contains 0.32 *M* sorbitol, 0.1 m*M* potassium EDTA, and 5 m*M* potassium phosphate, pH 7.4. The solutions needed for preparation of aqueous two-phase systems are as follows:

20% (w/w) Dx T500
40% (w/w) PEG 4000
1 *M* Sorbitol
0.2 *M* Potassium phosphate, pH 7.4
10 m*M* Potassium EDTA

### *Procedure*

The following procedure is carried out at 4°.

*Preparation of Phase Systems.* Two different phase systems are prepared: one (24 g) used to obtain a very enriched synaptosomal upper phase (purification phase) and another (30 g) devoted to the preparation of a fresh lower phase (washing phase). This fresh lower phase will be used for repartitioning the synaptosome-enriched upper phase (Fig. 1). Both phases are prepared by precise weighing of their respective components as described by Brooks and Norris-Jones.[7]

Purification phase: Weigh the following into a transparent plastic 50-ml centrifuge tube:
7.68 g of 20% (w/w) Dx T500
3.84 g of 40% (w/w) PEG 4000
6.72 g of 1 *M* sorbitol
0.6 g of 0.2 *M* potassium phosphate, pH 7.4
0.3 g of 10 m*M* potassium EDTA
Water to a weight of 21 g

The remaining 3 g to give the final 24-g phase system will be added later as a crude synaptosomal preparation (P3 fraction) (see below). Mix well by 20 inversions and store at 4°.

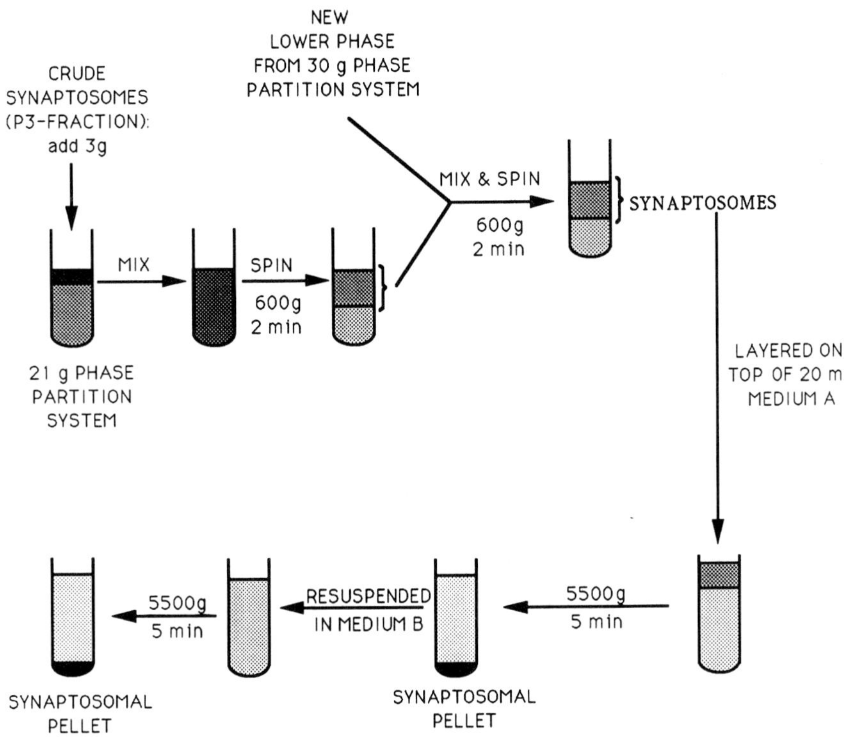


FIG. 1. Preparation of rat brain synaptosomes. [From J. A. Enríquez, J. Sànchez-Prieto, M. T. Muiño-Blanco, J. Hernández-Yago, and M. J. López-Pérez, *J. Neurochem.* **55,** 1841 (1990).]

Washing phase: Weigh the following into a flask:

- 9.6 g of 20% (w/w) Dx T500
- 4.8 g of 40% (w/w) PEG 4000
- 9.6 g of 1 *M* sorbitol
- 0.75 g of 0.2 *M* potassium phosphate, pH 7.4
- 0.3 g of 10 m*M* potassium EDTA
- Water to a final weight of 30 g

After temperature equilibration at 4°, mix well by 20 inversions, place into a separatory funnel, and allow to settle overnight. Collect the lower phase and store at 4°. Alternatively, the phase can be weighed into a transparent plastic centrifuge tube and, after temperature equilibration at 4°, mixed well by 20 inversions. The mixture is centrifuged at 600 *g* for 2 min, the upper phase and interface discarded, and the lower phase stored at 4°.

*Preparation of Brain Homogenate.* In each experiment four forebrains from male Wistar rats (150–200 g) are finely cut with a pair of scissors. The chopped material is washed twice with cold isolation medium to remove as much blood and debris as possible. The minced tissue obtained is homogenized in 30 ml of isolation medium in a Dounce-type homogenizer with a glass pestle (total clearance 0.1 mm) with 12 up-and-down strokes. The homogenate is then diluted to 60 ml with cold isolation medium.

*Preparation of Crude Synaptosomal Fraction.* The diluted homogenate (in two tubes) is centrifuged at 1300 *g* for 3 min. The supernatant solutions obtained are further centrifuged at 17,000 *g* for 10 min, producing a crude synaptosomal pellet (P2 fraction). This pellet is carefully resuspended in 30 ml of medium B, then centrifuged at 12,000 *g* for 10 min. The pellet obtained ("crude synaptosomal fraction," P3 fraction) is resuspended in 3 ml of medium B.

*Purification of Synaptosomes by Phase Partitioning.* Three grams of the suspended P3 fraction is added to the 21 g of purification phase. The system is mixed by 20 inversions of the tube and centrifuged at 600 *g* for 2 min. This centrifugation shortens the time of phase settling. The upper phase is carefully removed with a Pasteur pipette, and 9 ml of fresh lower phase obtained from the washing phase is added. The new phase system is again mixed by 20 inversions and centrifuged at 600 *g* for 2 min. The upper phase thus obtained is carefully removed and layered on top of a cushion of 20 ml of medium A (the upper phase is only slightly less dense than medium A, and layering must be done with care). The tube is centrifuged at 5500 *g* for 5 min, and the purified synaptosomes are pelleted.[8–11]

## Properties of Brain Synaptosomes Prepared by Phase Partitioning

Brain synaptosomal preparations contain about 30–40% of contaminating material, mostly free mitochondria, myelin, and membrane vesicles or "ghosts" of unidentified origin.[2,9,12] The best assessment of synaptosomal enrichment of preparations is the morphometric estimation of particles

[8] V. Corbatón, P. Fernández-Silva, M. J. López-Pérez, and J. Montoya, *Neurochem. Res.* **15,** 711 (1990).

[9] J. A. Enríquez, J. Sánchez-Príeto, M. T. Muiño Blanco, J. Hernández-Yago, and M. J. López-Pérez, *J. Neurochem.* **55,** 1841 (1990).

[10] J. Sánchez-Prieto and M. J. López-Pérez, *Biochim. Biophys. Acta* **778,** 81 (1984).

[11] M. P. Arregui, A. Pérez-Martos, and M. J. López-Pérez, *Biosci. Rep.* **11,** 181 (1991).

[12] A. Nagy and A. V. Delgado-Escueta, *J. Neurochem.* **43,** 1119 (1984).

TABLE I
MORPHOLOGICAL ASSESSMENT OF SYNAPTOSOMAL PREPARATIONS

| Preparation | Synaptosomes (%) | Mitochondria (%) | Others (%) | Ref. |
|---|---|---|---|---|
| Percoll/sucrose gradient | 70.2 | 3.3 | 26.4 | *a* |
| Discontinuous sucrose gradient | 48 | 4 | 48 | *b* |
| Discontinuous Ficoll gradient | 60–70 | 10 | 20 | *c* |
| Phase partitioning | 62.9 | 1.8 | 35.3 | *d* |

[a] A. Nagy and A. V. Delgado-Escueta, *J. Neurochem.* **43,** 1119 (1984).
[b] P. R. Dodd, J. A. Hardy, A. E. Oakley, J. A. Edwardson, E. K. Perry, and J. P. Delannoy, *Brain Res.* **226,** 107 (1981).
[c] C. W. Cotman and D. A. Matthews, *Biochim. Biophys. Acta* **249,** 380 (1971).
[d] J. A. Enríquez, J. Sánchez-Prieto, M. T. Muiño-Blanco, J. Hernández-Yago, and M.J. López Pérez, *J. Neurochem.* **55,** 1841 (1990).

observed in electron micrographs of purified fractions.[9,12] Table I shows the morphometric assessment of brain synaptosomal preparations obtained by some of the most commonly used methods. "Other" contaminants refer to particles of unidentified origin and also includes minor quantities of myelin present in the preparations. It must be emphasized that the criteria used for this assessment are variable. Therefore, the comparative aspects of these figures should be only considered on a relative basis. It can be concluded that synaptosomal enrichment of preparations obtained by phase partitioning is similar to that of fractions purified by the most efficient density gradients (Ficoll or Percoll). Phase partitioning has proved to be a very efficient method for separation of free mitochondria and synaptosomes as shown by the low value of mitochondrial contamination of the brain synaptosomes (Table I).

Brain synaptosomes prepared by phase partitioning show good metabolic performance as assayed by respiratory activities and ATP levels in the absence and presence of depolarizing agents,[9] and they release the neurotransmitter glutamate in a $Ca^{2+}$-dependent manner.[9]

The main advantage of purifying synaptosomes by phase partitioning is the short duration of the isolation process (<1 hr). The most laborious part of the method is the weighing of phase system components. However, phase systems can be prepared in advance and stored overnight at 4°. An ultracentrifuge is not required, yet the method can be applied for large-scale preparations of synaptosomes. This last advantage can be of use in the purification of proteins or nucleic acids from nerve-ending particles.[8]

## Method II: Preparation of Rat Brain Mitochondria

### *Solutions*

The isolation medium (BSA–medium A) contains 0.32 *M* sucrose, 1 m*M* potassium EDTA, 0.1% fatty acid-free bovine serum albumin (BSA), and 10 m*M* Tris-HCl, pH 7.4. The resuspension medium (BSA–medium B) contains 0.32 *M* sorbitol, 0.1 m*M* potassium EDTA, 0.1% fatty acid-free BSA, and 5 m*M* potassium phosphate, pH 7.8. The 6% Ficoll medium is a solution of 6% Ficoll 400 in BSA–medium A. The solutions needed for preparation of aqueous two-phase systems are as follows:

20% (w/w) Dx T500
40% (w/w) PEG 4000
1 *M* Sorbitol
0.2 *M* Potassium phosphate, pH 7.8
10 m*M* Potassium EDTA
1% Fatty acid-free BSA

### *Procedure*

The following procedure is carried out at 4°.

*Preparation of Phase Systems.* As in the case of the preparation of synaptosomes, two different systems are used: a purification and a washing phase (Fig. 2). These are prepared as follows.

Purification phase: Weigh the following into a transparent 50-ml centrifuge tube:

2.56 g of 20% (w/w) Dx T500
1.28 g of 40% (w/w) PEG 4000
2.24 g of 1 *M* sorbitol
0.07 g of 10 m*M* potassium EDTA
0.7 g of 1% BSA
0.2 g of 0.2 *M* potassium phosphate, pH 7.8
Water to a weight of 7.0 g

Mix well by 20 inversions and store at 4°.

Washing phase: Weigh the following into a flask:

9.6 g of 20% (w/w) Dx T500
4.8 g of 40% (w/w) PEG 4000
9.6 g of 1 *M* sorbitol
0.75 g of 0.2 *M* potassium phosphate, pH 7.8
0.3 g of 10 m*M* potassium EDTA

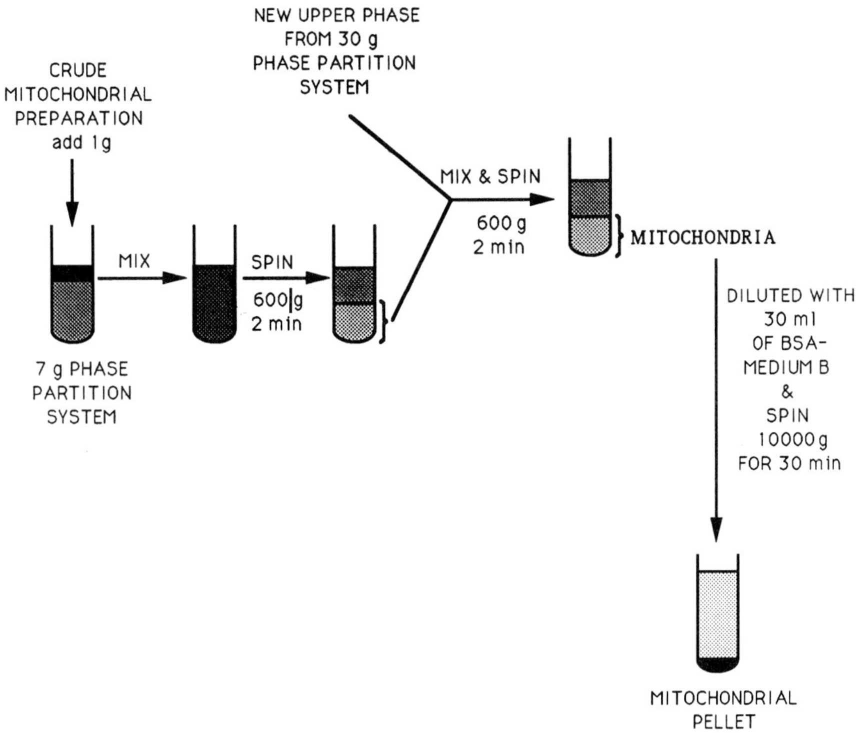


FIG. 2. Preparation of brain mitochondria by phase partitioning. [From M. J. López-Pérez, G. París, and C. Larsson, *Biochim. Biophys. Acta* **635,** 359 (1981).]

3.0 g of 1% BSA
Water to a final weight of 30 g

The washing phase is as described for the isolation of synaptosomes, the upper phase being collected and stored at 4°.

*Preparation of Brain Homogenate.* Twelve forebrains from rats are cut and washed with BSA–medium A as described for the preparation of synaptosomes. The minced tissue thus obtained is homogenized with 10 volumes of BSA–medium A in a Dounce-type homogenizer fitted with a Teflon pestle at 400 rpm with five up-and-down strokes.

*Preparation of Crude Mitochondrial Fraction.* The homogenate is centrifuged at 1100 *g* for 10 min. The supernatant solution obtained is further centrifuged at 19,000 *g* for 20 min. The pellet is resuspended in 5 ml of BSA–medium A, and the resulting suspension is layered on top of a cushion of 6% Ficoll and centrifuged at 19,000 *g* for 30 min. The superna-

TABLE II
ENZYME SPECIFIC ACTIVITIES OF MITOCHONDRIAL PREPARATIONS

| Preparation | Activity (μmol/min/mg protein) | | | Ref. |
|---|---|---|---|---|
| | Fumarase | Lactate dehydrogenase[a] | Acetylcholinesterase | |
| Discontinuous Ficoll gradient[b] | 0.364 | 0.105 | 0.012 | *c* |
| Clark and Nicklas method | 0.372 | 0.079 | 0.025 | *c* |
| Phase partitioning | 0.340 | 0.014 | 0.010 | *d* |

[a] Lactate dehydrogenase activity was determined in all cases in the presence of Triton.
[b] Discontinous Ficoll gradient data refer the values obtained from the free mitochondria population
[c] J. C. K. Lai and J. B. Clark, this series, Vol. 55, p. 51 Vol. LV.
[d] M. J. López-Pérez, G. París, and C. Larsson, *Biochim. Biophys. Acta* **635,** 359 (1981).

tant solution is discarded. The sediment consists of an upper fluffy layer and a yellowish-brown pellet. Loosen the fluffy layer, by gently shaking the tube, and discard it. The pellet is resuspended in 1 ml of BSA–medium B ("crude mitochondrial fraction").

*Purification of Brain Mitochondria by Phase Partitioning.* One gram of the crude mitochondrial fraction is added to the 7 g purification phase, mixed by 20 inversions, and centrifuged at 600 *g* for 2 min in a swinging-bucket centrifuge. The upper phase is carefully removed with a Pasteur pipette. To the remaining lower phase and interface add 3 ml of washing upper phase. After centrifugation at 600 *g* for 2 min, the upper phase is removed and the lower phase (containing the interface) is diluted with 30

TABLE III
RESPIRATORY ACTIVITIES AND SWELLING RATES OF BRAIN MITOCHONDRIA PURIFIED BY PHASE PARTITIONING[a]

| Purification method | KCl (m*M*) | Activity with 2.5 m*M* malate + 5 m*M* glutamate | | | Activity with 10 m*M* succinate | |
|---|---|---|---|---|---|---|
| | | State 3 | RCR | Swelling rate | State 3 | RCR |
| K$^+$ containing | 5 | 85 | 7.0 | 2.1 | 103 | 5.1 |
| | 100 | 87 | 3.7 | 5.6 | 95 | 2.2 |
| Na$^+$ containing | 5 | 77 | 4.8 | 1.7 | 114 | 3.9 |
| | 100 | 106 | 3.5 | 4.0 | 143 | 2.6 |

[a] State 3 respiratory activities are expressed as nanoatoms of $O_2$/min/mg protein. RCR, Respiratory control ratio. Swelling rates are expressd as the percent change in $A_{600\ nm/min}$. From J. Sanchez-Prieto and M. J. López-Pérez, *Rev. Esp. Fisiol.* **42,** 427 (1986).

ml of BSA–medium B and centrifuged at 19,000 $g$ for 30 min. The final pellet is resuspended in 0.5–1 ml of BSA–medium B.

## Properties of Brain Mitochondria Prepared by Phase Partitioning

Table II shows some enzyme specific activities of different brain mitochondrial preparations. The presence in these preparations of contaminating synaptosomes and synaptic membranes was assessed by the estimation of lactate dehydrogenase and acetylcholinesterase activities, respectively. The specific activities of both enzymes were very low in brain mitochondria isolated by phase partitioning, reflecting the high purity of the preparations.

Highly purified brain mitochondria prepared by phase partitioning have good respiratory control rates (RCR) and state 3 respiratory and swelling rates when the organelles are purified using media containing BSA and low temperature (4°) is maintained throughout the isolation procedure (Table III). The absence of BSA during phase partitioning of the organelles or carrying out some isolation steps at room temperature may cause uncoupling of mitochondria.

A particular feature of brain mitochondria consists in their stimulation of maximal respiratory activities (state 3 rates) by exogenous $K^+$.[3] Table III shows that increasing exogenous $K^+$ concentrations from 5 to 100 m$M$ does not cause enhancement of the state 3 rate of brain mitochondria purified by the above described method when $K^+$-containing media were used. If this $K^+$-stimulated maximal oxidation rate is to be studied, mitochondria must be isolated using $K^+$-free media (Table III). In this case, the same isolation method should be employed but using sodium EDTA or sodium phosphate (at the same concentrations).

The method can be also applied to the large-scale preparation of mitochondria for use in the purification of specific macromolecules.[3] Intrasynaptosomal (synaptic) mitochondria show the same affinity for the lower phase as do free mitochondria. Therefore, hypoosmotically released mitochondria from synaptosomes have also been separated from the synaptic membranes by phase partitioning in a two-phase system of the same composition as described.[3]

# [39] Isolation of all Major Organelles and Membranous Cell Components from a Single Homogenate of Green Leaves

*By* D. JAMES MORRÉ and BERTIL ANDERSSON

## Introduction

The isolation and purification of intracellular membranes and organelles are a prerequisite for identification of components, for enzyme analyses of various subfractions, and for obtaining information about organizational aspects of cellular functions. Moreover, analyses of subfractionated components permit studies on metabolic cooperation among different organelles and on the mechanisms that underlie intracellular trafficking of proteins, nucleic acids, and lipids.

Cellular subfractionation of plant tissues, in particular of green leaves, has been quite difficult to effect when compared to etiolated plant parts or animal tissues. A large portion of the intracellular membranes of green leaves are chloroplast thylakoid membranes, and the isolation of "non-green" fractions of plasma membranes, mitochondria, and membranes of the Golgi apparatus or the endoplasmic reticulum is not trivial. For plants in general the existence of a cell wall requires a procedure to break the cell in a manner that does not lead to functional impairment or to uncontrolled destruction of the intracellular membrane components. The large central vacuole occupies a significant portion of the cellular volume of mature plant cells and contains inhibitory substances such as tannins.

Today there are established procedures, some based on aqueous polymer two-phase partitioning, for the purification of most organelles or biomembranes from leaf tissues.[1] However, for many molecular studies of metabolic cooperation among organelles or on intracellular trafficking, more than one type of subcellular fraction and, ultimately, the isolation of all intracellular membranes from the same homogenate may be required.

Here we summarize a method to obtain all major membrane and organelle fractions from a single homogenate of spinach leaves.[2] The procedure is based on a combination of centrifugation steps and aqueous phase partitioning. It thus involves a "two-dimensional" resolution of the complex mixture of the leaf cell membranes, namely, size and density differences in one dimension and surface differences in the other. The procedure

[1] This series, Vols. 118 and 148.

[2] D. J. Morré, C. Penel, D. M. Morré, A. S. Sandelius, P. Moreau, and B. Andersson, *Protoplasma* **160**, 49 (1990).

given below is for spinach leaves but has been successfully applied, and without modifications, to green leaves of pea and soybean.

## Methods

### *Preparation and Crude Fractionation of Leaf Cell Homogenate*

Spinach plants are grown for 4 weeks in a chamber with 8 hr of light daily at an average temperature of 20°. Leaves (50 g) are harvested and cut coarsely with a pair of scissors and finely chopped into a small volume of medium with razor blades. Homogenization is by use of a mortar and pestle at 4°, using 1 to 2 ml medium to 1 g fresh weight of leaf material. The homogenization medium, modified from Gibeaut and Carpita,[3] is 50 m*M* HEPES (pH 7.4), 10 m*M* KCl, 1 m*M* EDTA, 10 m*M* ascorbate, 0.1% bovine serum albumin (BSA), 20 m*M* dithiothreitol, and 400 m*M* sucrose. The homogenate is squeezed through a single layer of Miracloth to remove unbroken cells, cell walls, and debris. The homogenate is transferred to 35- or 50-ml tubes and centrifuged at 1000 *g* for 10 min using a swing-out rotor. The resulting pellet, the supernatant solution, as well as the loose layer covering the pellet are collected and used to isolate various intracellular membrane and organelle fractions (see below and Fig. 1).

For the isolation of all intracellular membranes except nuclei and intact chloroplasts, the 1000 *g*–10 min supernatant solution is used as a starting material. In an initial fractionation step this supernatant solution is layered onto a discontinuous sucrose gradient consisting of 4 ml of 37% (w/v) sucrose and 8 ml of 21.5% (w/v) sucrose in 30-ml tubes. After centrifugation at 65,000 *g* for 30 min the membranes at the sample–21.5% sucrose interface and at the 21.5–37% sucrose interface are collected and used to isolate endoplasmic reticulum (ER) and Golgi apparatus, respectively. The pellet is collected and used to isolate mitochondria, plasma membranes (PM), and tonoplasts.

### *Nuclei*

The 1000 *g*–10 min pellet is resuspended in the homogenization medium after removal of the overlaying loose layer containing primarily chloroplast material. The suspension is layered on top of a 2.0 *M* sucrose cushion prepared in 37.5 m*M* Tris–maleate (pH 6.5), 5 m*M* $MgCl_2$, and 1% dextran (Dx) T500 (Pharmacia, Uppsala, Sweden) and centrifuged at 50,000 *g* for 30 min at 0 to 4°. The pellet contains nuclei.

[3] D. M. Gibeaut and N. C. Carpita, *Protoplasma* **156,** 82 (1990).

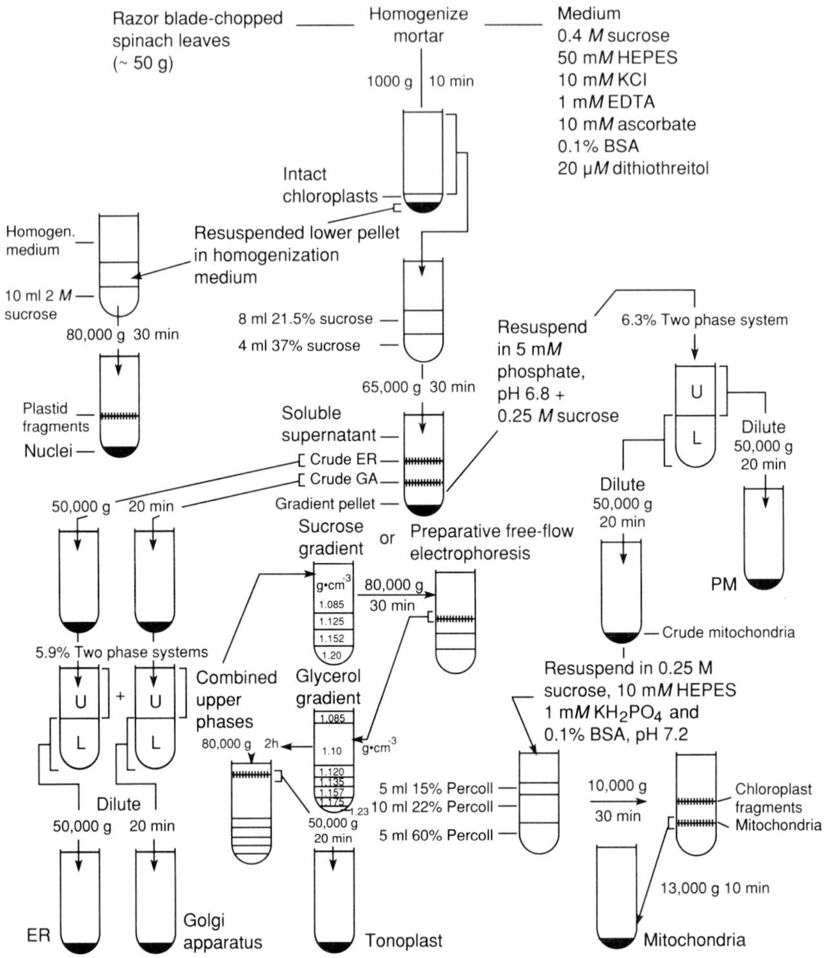


FIG. 1. Single homogenate fractionation procedure for plants yielding purified fractions of endoplasmic reticulum, Golgi apparatus, nuclei, plasma membranes, tonoplast, mitochondria, and chloroplasts. [From D. J. Morré, C. Penel, D. M. Morré, A. S. Sandelius, P. Moreau, and B. Andersson, *Protoplasma* **160,** 49 (1990).]

## *Intact Chloroplasts*

The loose upper layer overlaying the 1000 *g*–10 min pellet may be used directly as the source for intact chloroplasts. Contaminating thylakoids and other organelle membranes can be removed by phase partitioning.[4,5]

[4] C. Larsson and P.-Å. Albertsson, *Biochim. Biophys. Acta* **354,** 412 (1974).

[5] C. Larsson, this volume [40].

*Mitochondria*

The material which passes through the 37% sucrose layer of the initial gradient centrifugation of the 1000 *g*–10 min supernatant solution and pelleted (Fig. 1) is resuspended in 2 ml of 5 m*M* sodium phosphate (pH 6.8), 5 m*M* KCl, and 0.25 *M* sucrose. This suspension is added to a polymer mixture to give a 16 g phase system composed of 6.3% (w/w) Dx and 6.3% (w/w) poly(ethylene glycol) (PEG) 3350 (Union Carbide, New York, NY), 5 m*M* potassium phosphate (pH 6.8), and 0.25 *M* sucrose. After reaching ice bath temperature the tubes are mixed by 40 inversions in the cold. Phase settling is facilitated by a low-speed centrifugation at 1000 *g* for 5 min in the cold. The upper phase (plus interface) and lower phase are collected separately and diluted with homogenization medium. The lower phase, which contains mitochondria contaminated with thylakoid fragments, is centrifuged at 50,000 *g* for 20 min to yield a pellet of crude mitochondria. After resuspension in 1–2 ml of 10 m*M* HEPES (pH 7.2), 1 m*M* $KH_2PO_4$, 0.1% BSA, and 0.25 *M* sucrose using a Teflon–glass homogenizer, the mitochondria are purified by Percoll gradient centrifugation.[6]

The suspension is layered on top of a discontinuous gradient composed of 5 ml of 15% (w/v) Percoll, 10 ml of 22% (w/v) Percoll, and 5 ml of 60% (w/v) Percoll contained in a 30- or 35-ml tube. The gradient solutions contain 10 m*M* HEPES (pH 7.2), 1 m*M* $KH_2PO_4$, 0.1% BSA, and 0.25 *M* sucrose. The Percoll gradient is centrifuged at 10,000 *g* for 30 min at 0 to 4°. The mitochondria are recovered from the overlaying 22% (w/v) Percoll layer (Fig. 1). This fraction is diluted with 10 volumes of homogenization medium and collected by centrifugation at 13,000 *g* for 10 min. The supernatant solution is removed carefully, and the pellet, including overlying loose material, is resuspended and diluted to 35 ml with homogenization buffer and repelleted at 13,000 *g* for 10 min at 0 to 4°.

*Plasma Membranes*

Among cell membranes the plasma membrane from both animal and plant tissues has the greatest affinity for the upper phase in aqueous polymer two-phase systems used for membrane purification.[7–9] Indeed, plasma membranes from the spinach homogenate are obtained from the upper phase of the two-phase system used to isolate the mitochondria (Fig. 1). The plasma membranes are collected by centrifugation of 50,000 *g*

[6] G. E. Edwards and P. Gardeström, this series, Vol. 148, p. 421.

[7] C. Larsson, S. Widell, and P. Kjellbom, this series, Vol. 148, p. 558.

[8] A. S. Sandelius and D. J. Morré, *in* "Plasma Membrane Isolation" (C. Larsson and I. M. Møller, eds.), p. 44. Springer-Verlag, Berlin, 1990.

[9] C. Larsson, M. Sommarin, and S. Widell, this volume [44].

for 20 min after dilution of the upper phase in a solution containing 5 m*M* potassium phosphate (pH 6.8) and 0.25 *M* sucrose.

The aqueous polymer two-phase partitioning technique has been extremely useful for the isolation of plasma membranes from a variety of biological material[8] and has become the standard procedure for preparation of plant plasma membranes.[7-9] The phase partitioning technique has been applied to cyanobacterial cells.[10] In contrast to plasma membranes from eukaryotic cells, the corresponding membranes from the prokaryotic cyanobacteria partition to the lower phase.

## *Endoplasmic Reticulum and Golgi Apparatus*

Crude fractions of endoplasmic reticulum and Golgi apparatus are obtained after discontinuous sucrose gradient centrifugation of the 1000 *g*–10 min supernatant solution (Fig. 1). The endoplasmic reticulum is collected from the homogenate–21.5% sucrose interface, whereas the Golgi membranes are obtained at the 21.5–37% sucrose interface. After separation the two membrane fractions are resuspended in the clear yellow supernatant solution from the top of the gradient and pelleted by centrifugation at 50,000 *g* for 20 min. The two crude pellets are each resuspended in 0.5 ml of 5 m*M* potassium phosphate (pH 6.8) and 0.25 *M* sucrose and further purified by phase partitioning. Each of the two suspensions (0.5 ml) are added to polymer mixtures to yield two-polymer phase systems with a final composition of 5.9% (w/w) Dx T500, 5.9% (w/w) PEG 3350, 5 m*M* potassium phosphate (pH 6.8), and 0.25 m*M* sucrose (Fig. 1). After reaching ice bath temperature, the two phase systems are mixed by 40 inversions in the cold. Phase settling is facilitated by centrifugation at 1000 *g* for 5 min. The upper phases (plus the interfaces) and lower phases are collected separately. The endoplasmic reticulum and the Golgi membranes are recovered in the respective lower phases while contaminating chloroplast membranes and tonoplasts partition to the upper phases (Fig. 1). The highly purified membranes of the endoplasmic reticulum and the Golgi apparatus are collected by centrifugation at 50,000 *g* for 20 min after dilution of the two lower phases with homogenization buffer devoid of sucrose.

The preparation of pure endoplasmic reticulum and Golgi membranes together with the other separation steps described above results in a nearly complete subfractionation of the leaf cell homogenate. The procedure affords the possibility for a direct and specific isolation of these two organelle membranes.

[10] B. Norling, V. Mirzakhanian, F. Nilsson, D. J. Morré, and B. Andersson, *Arch. Biochem. Biophys.* in press (1994).

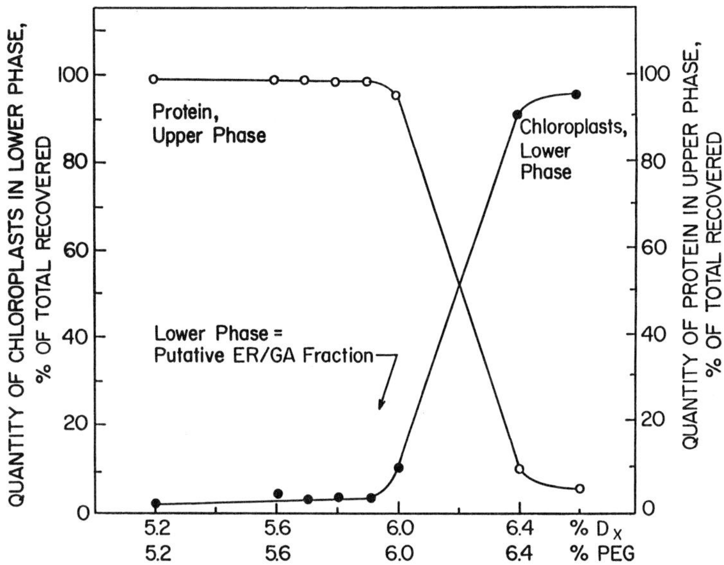


FIG. 2. Comparison of phase systems consisting of different proportions of dextran T500 and PEG 3350 for retention of chloroplasts in the upper phase and the choice of a phase system containing 5.9% of each polymer to purify endoplasmic reticulum and Golgi apparatus in the lower phase. The protein content (○) is given for the upper phase as a percentage of total recovered. That amount subtracted from 100 gives the percentage in the lower phase. The quantity of chloroplasts determined from total chlorophyll (●) is given for the lower phase as the percentage of total recovered. That amount subtracted from 100 gives the percentage in the upper phase. Phase composition in percent by weight of dextran and PEG is given on the abscissa. At 5.9% of each polymer and below, chloroplasts are largely excluded from the lower phase. The purified endoplasmic reticulum (ER) or Golgi apparatus (GA) fractions are prepared using a polymer concentration of 5.9% (arrow). As the polymer concentrations approached 6.6%, the bulk of the chloroplasts entered the lower phase together with most of the protein. When the chloroplasts and chloroplast fragments were shifted to the lower phase over the range of polymer concentrations of 6.2 to 6.6%, principally plasma membrane vesicles remained in the upper phase (D. J. Morré and B. Andersson, unpublished).

To obtain membranes free of chloroplast contamination it is necessary to optimize the phase composition by an examination of partitioning behavior in phase systems of increasing polymer concentrations (Fig. 2). At the chosen polymer concentration of 5.9% (w/w) (for both Dx and PEG) the chloroplast membranes partition into the upper phase while the membranes of the endoplasmic reticulum and the Golgi apparatus partition into the lower phase. At higher polymer concentrations the risk of chloroplast contamination increases dramatically. At lower polymer concentrations, on the other hand, there is a risk of reducing the yield of the desired membranes owing to their increased partitioning into the upper phase. Such a partitioning experiment should be made with each new batch of polymer to determine the optimal polymer concentrations for obtaining pure endoplasmic reticulum and Golgi membranes since polymer batches may differ with respect to average molecular weight.

### *Tonoplasts*

The two upper phases from the partitioning step used for isolation of the endoplasmic reticulum and the Golgi apparatus are combined and used for isolation of tonoplast vesicles (vacuole membranes). The tonoplasts are purified either by preparative free-flow electrophoresis[11] or by a glycerol gradient procedure[12] (Fig. 1).

In the latter approach, the combined upper phases are layered onto a discontinuous sucrose gradient with density steps as indicated in Fig. 1. After centrifugation at 80,000 *g* for 30 min, the membrane material collected at the 1.085–1.125 g/cm$^3$ interface is reloaded onto a glycerol gradient. The second gradient is centrifuged at the same speed for 2 hr. The tonoplast membranes are obtained at the 1.085–1.10 g/cm$^3$ interface and finally collected by dilution in homogenization buffer and centrifugation at 50,000 *g* for 20 min.

## Final Comments

Except for nuclei, the preparations are highly enriched. Plasma membranes, tonoplasts, mitochondria, and chloroplasts were found to be more than 90% pure as judged by morphometry and were virtually free of thylakoid fragments. The endoplasmic reticulum fraction consisted of 80–85% endoplasmic reticulum (plus minor amounts of Golgi apparatus), 10–15% thylakoids plus tonoplasts, and 5 to 10% plasma membranes as indicated by staining with phosphotungstic acid at low pH.[13] The Golgi apparatus fraction consisted of 50 to 65% stacks of cisternae or flattened cisternae of Golgi apparatus origin. Contaminants included plasma membrane vesicles (15–20%), thylakoids (10–15%), and unidentified vesicles (10–15%). Fraction identify and composition as judged by electron microscope morphometry were confirmed by analyses of marker enzymes.[2,14]

[11] G. Auderset, A. S. Sandelius, C. Penel, A. Brightman, H. Greppin, and D. J. Morré, *Physiol. Plant.* **68,** 1 (1986).

[12] G. F. E. Scherer and G. Fischer, *Protoplasma* **129,** 109 (1985).

[13] J. C. Roland, C. A. Lembi, and D. J. Morré, *Stain Technol.* **47,** 195 (1972).

[14] D. J. Morré and T. J. Buckhout, *in* "Plant Organelles" (E. Reid, ed.), p. 117. Ellis Horwood, Chichester, 1979.

## Acknowledgments

This work was supported by the Bank of Sweden Tercentenary Research Foundation. D. J. Morré was the 1989–1990 Tage Erlander professor of the Swedish Natural Science Research Council.

# [40] Isolation of Highly Purified Intact Chloroplasts and of Multiorganelle Complexes Containing Chloroplasts

*By* CHRISTER LARSSON

## Introduction

For studies on major chloroplast reactions, such as those of the Calvin cycle, intact chloroplasts prepared solely by differential centrifugation are preferred because of the rapidity and simplicity of that purification process. However, for studies on compartmentation of metabolism between the chloroplast and other cellular compartments more highly purified chloroplasts are needed, since contamination (by cytoplasm, mitochondria, peroxisomes, etc.) would otherwise obscure the results. Such highly purified chloroplasts can be prepared by phase partitioning.[1] As a by-product a fraction containing "multiorganelle complexes" is obtained. These complexes consist of one or two chloroplasts surrounded by cytoplasm, mitochondria, and peroxisomes all enclosed by some plasma membrane. Multiorganelle complexes are formed during homogenization of leaf tissue, and they are one of the contaminants removed during the purification process. The multiorganelle complexes can, however, be used to advantage as a reference point representing "highly contaminated chloroplasts" with a metabolic capacity close to that of the intact cell.[1]

## Procedure

A phase system (of, e.g., 50 g), to provide the fresh upper and lower phases indicated below, is prepared in a separatory funnel by weighing and pipetting from stock solutions of polymers [20% (w/w) dextran (Dx) T500 (Pharmacia LKB, Uppsala, Sweden), 40% (w/w) poly(ethylene glycol) (PEG) 3350 (Union Carbide, New York, NY)] and salts.[2,3] After

[1] C. Larsson and P.-Å. Albertsson, *Biochim. Biophys. Acta* **357,** 412 (1974).

[2] D. E. Brooks and R. Norris-Jones, this volume [2].

[3] H. Walter and C. Larsson, this volume [4].

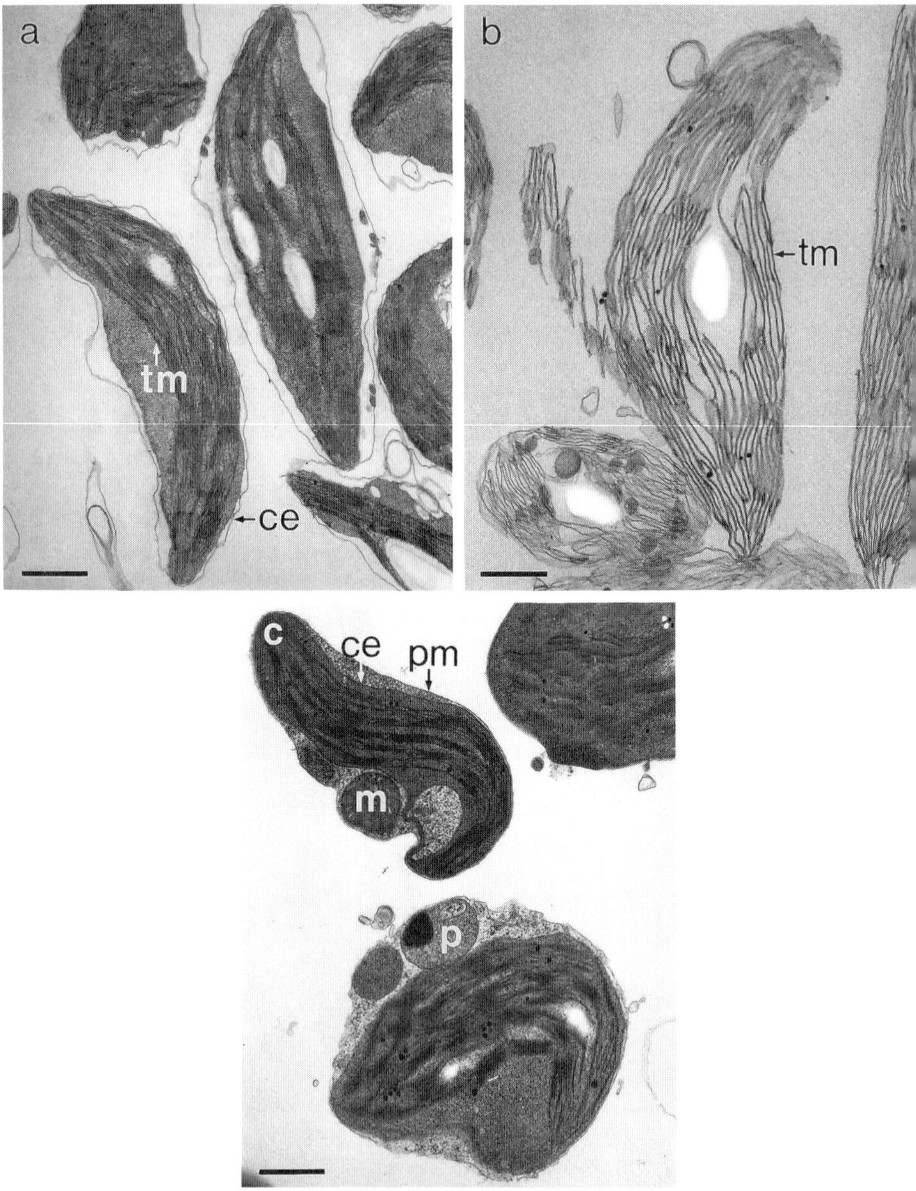


FIG. 1. Electron micrographs of (a) intact chloroplasts, (b) broken chloroplasts (thylakoids), and (c) multiorganelle complexes, all obtained by phase partitioning. Note that the multiorganelle complexes are surrounded by only a thin layer of cytoplasm and are, therefore, similar to free intact chloroplasts (a) in size and density. Observe also that the different

equilibration at 4°, the phase system is well shaken and left standing for phase settling to occur. Upper and lower phases are collected and stored separately in the cold or are frozen.

### *Intact Chloroplasts*

A well-drained chloroplast pellet, obtained from 50 g of spinach leaves by differential centrifugation,[4] is suspended in 1 ml of 0.3 *M* sorbitol, 5 m*M* potassium phosphate, pH 7.8. The chloroplast suspension (1.0 ml) is added to a phase mixture (7.0 g) to yield a phase system (8.0 g) composed of 5.8% (w/w) Dx T500, 5.8% (w/w) PEG 3350, 0.3 *M* sorbitol, and 5 m*M* potassium phosphate, pH 7.8, at a temperature of 4°. The phase system is mixed by 10–20 inversions of the tube and is either permitted to stand at unit gravity for about 10 min or subjected to low-speed centrifugation (swing-out buckets) to decrease the time required for phase settling. If centrifugation is used care must be taken that material partitioning into the upper phase not be sedimented onto the interface. The upper phase, containing mainly thylakoids (Fig. 1b) but also multiorganelle complexes and other contaminants, is removed (Pasteur pipette) and the lower phase plus interface is reextracted twice with fresh upper phase. The final lower phase plus interface is diluted 10-fold with a suitable medium, and the intact chloroplasts (Fig. 1a) are sedimented by centrifugation (500 *g* for 5 min, preferably using swing-out buckets) and resuspended.

### *Multiorganelle Complexes*

A batch preparation of multiorganelle complexes (Fig. 1c) is obtained in a similar way. In contrast to that of multiorganelle complexes, the partitioning of thylakoids is very sensitive to changes in the interfacial potential. Thus, if 5 m*M* KCl is included in the phase system (by pipetting from 2 *M* KCl) the chloroplast thylakoids will collect at the

[4] R. G. Jensen and J. A. Bassham, *Proc. Natl. Acad. Sci. U.S.A.* **56,** 1095 (1966).

particulates in a, b, and c expose different membrane surfaces to the phase system, which provides the basis for their separation by phase partitioning. ce, Chloroplast envelope, outer membrane; tm, thylakoid membrane; pm, plasma membrane; m, mitochondrion; p, peroxisome. Bars, 1 $\mu$m. [From P.-Å. Albertsson, B. Andersson, C. Larsson, and H.-E. Åkerlund, *Methods Biochem. Anal.* **28,** 115 (1982). Copyright © 1982 John Wiley & Sons, Inc. Reprinted by permission of John Wiley & Sons, Inc.]

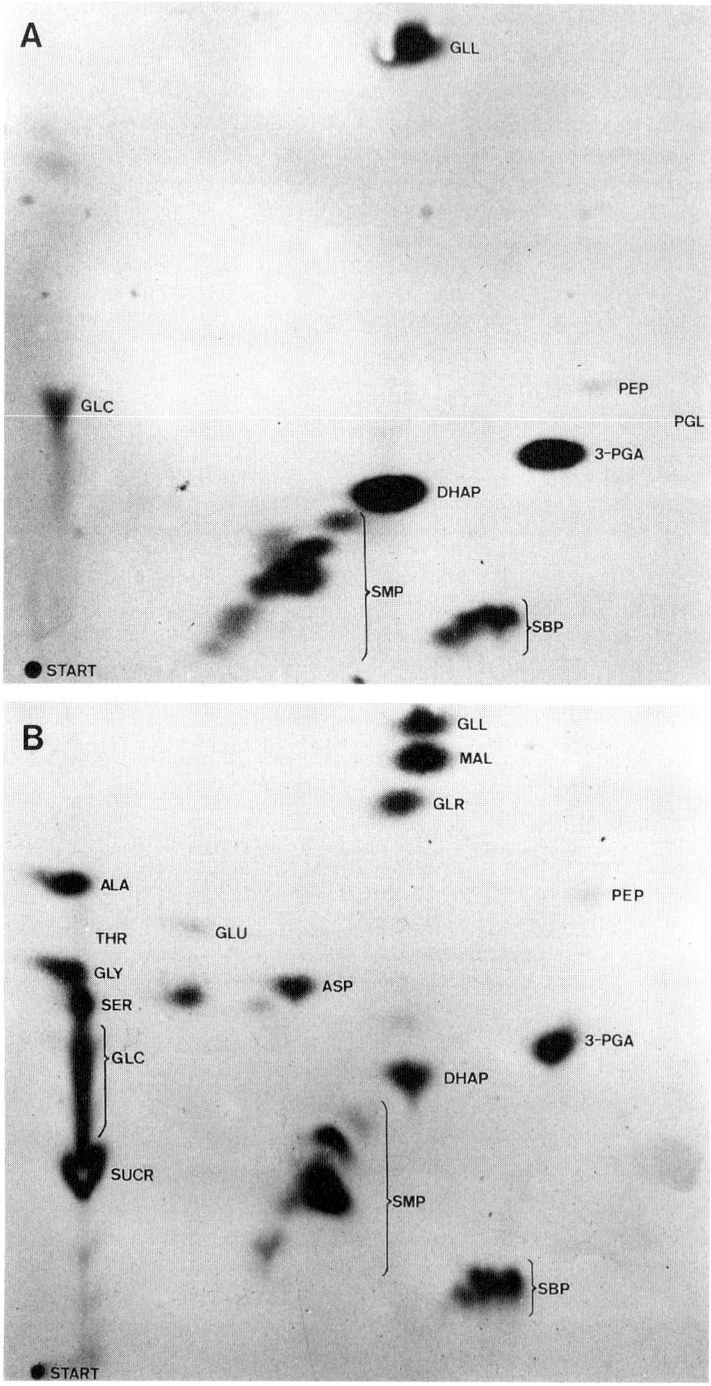
A
GLL
GLC
PEP
PGL
3-PGA
DHAP
SMP
SBP
START
B
GLL
MAL
GLR
ALA
PEP
THR
GLU
GLY
ASP
SER
3-PGA
GLC
DHAP
SUCR
SMP
SBP
START

interface together with the intact chloroplasts, while the multiorganelle complexes will remain in the upper phase (compare Refs. 3 and 5). The first upper phase from the isolation of intact chloroplasts (above) can be used as a source for multiorganelle complexes. However, because the amounts of multiorganelle complexes in chloroplast preparations are low (0.5–5% of total chlorophyll) it is usually necessary to scale up the process. The best yields are obtained from cotyledons.[1,6] A chloroplast preparation from 200–300 g of leaves is partitioned in a phase system of 24–36 g using a separatory funnel. After phase settling for 15–20 min, the lower phase plus interface is discarded and the upper phase is reextracted twice with fresh lower phase (also containing 5 m*M* KCl). The final upper phase is diluted at least once with a suitable medium, and the multiorganelle complexes are collected by centrifugation at 500 *g* for 5 min.

## Properties of Fractions

Intact chloroplasts purified by phase partitioning incorporate $^{14}CO_2$ almost exclusively into glycolate, starch, and intermediates of the Calvin cycle and starch synthesis (Fig. 2A). By contrast, multiorganelle complexes give a much broader spectrum of $^{14}C$-labeled products including sucrose, malate, glycerate, aspartate, alanine, glycine, serine, and other amino acids (Fig. 2B).

Intact chloroplasts can be separated from broken chloroplasts (thylakoids, Fig. 1b) by differential[4] or by Percoll gradient centrifugation.[7] However, intact chloroplasts and multiorganelle complexes, which are readily separated by phase partitioning because of differences in their surface properties (Fig. 1), are separated poorly by methods which depend on

[5] C. Larsson, M. Sommarin, and S. Widell, this volume [44].
[6] C. Larsson, C. Collin, and P.-Å. Albertsson, *Biochim. Biophys. Acta* **245,** 425 (1971).
[7] K. W. Joy and W. R. Mills, this series, Vol. 148, p. 179.

FIG. 2. Radioautographs of $^{14}C$-labeled products after 6 min of photosynthetic $^{14}CO_2$ fixation by (A) intact chloroplasts and (B) multiorganelle complexes, purified from spinach leaves by phase partitioning. Products were separated by thin-layer electrophoresis (horizontal) and chromatography (vertical) [P. Schürmann, *J. Chromatogr.* **39,** 507 (1969)]. ALA, Alanine; ASP, aspartate; DHAP, dihydroxyacetone phosphate; GLC, glucose; GLL, glycolate; GLR, glycerate; GLU, glutamate GLY, glycine; MAL, malate; PEP, phosphoenolpyruvate; 3-PGA, 3-phosphoglycerate; PGL, phosphoglycolate; SBP, sugar bisphosphates; SER, serine; SMP, sugar monophosphates; SUCR, sucrose; THR, threonine. [Modified from C. Larsson and P.-Å. Albertsson, *Biochim. Biophys. Acta* **357,** 412 (1974)].

size and density. Because multiorganelle complexes have a metabolic capacity close to that of the intact cell (Fig. 2B), they may constitute a serious problem in studies on chloroplast metabolism. This is well illustrated by Bickel *et al.*[8] in a study on the biosynthesis of prenylquinones. After incubation of a standard chloroplast preparation with [$^{14}$C]tyrosine they found about 3% of total label in plastoquinone and $\alpha$-tocopherol. If instead chloroplasts purified by phase partitioning were used, only 0.01% of the $^{14}$C was in prenylquinones. Bickel et al.[8] were also able to show (by reconstitution experiments) that peroxisomes were necessary for the conversion of tyrosine to homogentisate, a precursor for the prenylquinones. Thus, the high activity of prenylquinone synthesis observed with the unpurified chloroplasts was due to contamination by peroxisomes. It should be noted that the $^{14}CO_2$ fixation rate of the chloroplasts is not changed by the purification process. Chloroplasts purified by phase partitioning are therefore well suited for studies on compartmentation of metabolic pathways.

## Acknowledgements

Work in the author's laboratory was supported by the Swedish Natural Science Research Council.

[8] H. Bickel, B. Buchholz, and G. Schultz, *in* "Advances in the Biochemistry and Physiology of Plant Lipids" (L. Å. Appelqvist and C. Liljenberg, eds.), p. 369. Elsevier, Amsterdam, 1979.

# [41] Purification and Characterization of Plant Mitochondria and Submitochondrial Particles

*By* PER GARDESTRÖM, PATRICE X. PETIT, and IAN M. MØLLER

## Introduction

To study plant respiration it is important to isolate functionally intact mitochondria from different tissues. Functionally intact plant mitochondria can be obtained by differential centrifugation, yielding a crude fraction.[1] For many types of studies it is also necessary to have pure mitochondria. Different methods have been used to remove contaminating material,

[1] R. Douce, J. Bourguignon, R. Brouquisse, and M. Neuburger, this series, Vol. 148, p. 403.

including density gradient centrifugation in different media[2] and phase partitioning.[3] In the study of the transverse and lateral distribution of components in the inner mitochondrial membrane it is useful to have access to submitochondrial particles (SMP) of well-defined sidedness or polarity. Most biochemistry textbooks give the impression that inside-out (IO) SMP are the only vesicles formed when mitochondria are disrupted, but that is far from being the case. Depending on the disruption method and the composition of the disruption medium, predominantly IO-SMP, predominantly right-side-out (RO) SMP, or a mixture of these can be produced.[4-9] Because phase partitioning separates membrane particles based on surface properties, the method is very useful to separate and analyze RO- and IO-SMP.

Here we describe how phase partitioning can (1) purify mitochondria from green leaves, (2) separate IO-SMP and RO-SMP from a population of mixed sidedness, and (3) determine the isopartition point of intact mitochondria and submitochondrial particles.

## Materials and Methods

### *Reagents*

Poly(ethylene glycol) 3350 (PEG 3350) is obtained from Union Carbide (New York, NY) and dextran (Dx) T500 from Pharmacia LKB (Uppsala, Sweden). The substituted PEG types used are obtained from Aqueous Affinity (Arlöv, Sweden), namely, trimethylaminopoly(ethylene glycol) (TMA-PEG), poly(ethylene glycol) sulfonate (S-PEG), and hexamethyldiaminopoly(ethylene glycol) (HMDA-PEG).

### *Purification of Spinach Leaf Mitochondria*

The mitochondrial fraction obtained by differential centrifugation of a spinach leaf extract is heavily contaminated with thylakoid fragments.[1]

[2] R. Douce, "Mitochondria in Higher Plants: Structure, Function, and Biogenesis." Academic Press, Orlando, Florida, 1985.

[3] P. Gardeström, I. Ericson, and C. Larsson, *Plant Sci. Lett.* **13,** 231 (1978).

[4] I. M. Møller, A. Bergman, P. Gardeström, I. Ericson, and J. M. Palmer, *FEBS Lett.* **126,** 13 (1981).

[5] C. J. Kay, I. Ericson, P. Gardeström, J. M. Palmer, and I. M. Møller, *FEBS Lett.* **193,** 169 (1985).

[6] I. M. Møller, A. C. Lidén, I. Ericson, and P. Gardeström, this series, Vol. 148, p. 442.

[7] P. X. Petit, K. A. Edman, P. Gardeström, and I. Ericson, *Biochim. Biophys. Acta* **890,** 377 (1987).

[8] P. X. Petit, P. Gardeström, A. G. Rasmusson, and I. M. Møller, *Plant Sci.* **78,** 177 (1991).

[9] A. G. Rasmusson and I. M. Møller, *Physiol. Plant.* **83,** 357 (1991).

TABLE I
COMPOSITION OF PHASE SYSTEM USED FOR BATCH PURIFICATION OF LEAF MITOCHONDRIA[a]

| Amount (g) | Stock solution | Final concentration |
|---|---|---|
| 31.0 | 20% (w/w) Dextran T500 | 6.2% (w/w) |
| 15.5 | 40% (w/w) PEG 3350 | 6.2% (w/w) |
| 25.0 | 1.2 mol/kg Sucrose | 0.3 mol/kg |
| 5.0 | 0.1 mol/kg Potassium phosphate, pH 7.8 | 5 mmol/kg |
| 2.0 | 0.1 mol/kg KCl | 2 mmol/kg |
| 21.5 | Water | |
| 100.0 | | |

[a] The phase system (100 g) is prepared in a separatory funnel, cooled to 4°, and then carefully mixed. After settling for several hours (overnight is convenient), the phases are separated and the interface discarded. The crude mitochondrial pellets obtained after differential centrifugation of homogenate (from about 120 g spinach leaves[3]) is directly resuspended in 10 ml top phase, and 8 ml bottom phase is added. After mixing by inversions of the tube, the phases are separated by a low-speed centrifugation in a swing-out rotor (600 *g* for 3–4 min). About 80% of the thylakoid fragments are found in the top phase and about 80% of the mitochondria in the bottom phase/interface. The green top phase is removed and the bottom phase plus interface is washed twice with fresh top phase. Finally the bottom phase plus interface is diluted about 10 times with 0.3 *M* sucrose, and the mitochondria are collected by centrifugation at approximately 10,000 *g* for 10 min. All of the steps are made at 4°. From P. Gardeström and I. Ericson, this series, Vol. 148, p. 434.

Phase partitioning in a Dx–PEG phase system can be used to remove most of the contaminating thylakoid membranes.[3]

Spinach leaf mitochondria partition more to the bottom phase/interface than thylakoid fragments when the polymer concentration is increased as well as when the ratio of $Cl^-/HPO_4^{2-}$ is increased.[10] The phase system shown in Table I is used for batch purification of spinach leaf mitochondria. The enrichment of mitochondria relative to thylakoid fragments is 10–20 times in the phase partitioning step.[10] The mitochondria are functionally intact as shown by good respiratory control and ADP/O rates with different substrates.[3]

For leaf mitochondria the first chlorophyll-free preparation was obtained by combining differential centrifugation, phase partitioning, and density gradient centrifugation using Percoll.[11] Percoll gradient centrifugation in combination with differential centrifugation has been developed as the standard method for the purification of plant mitochondria.[2,12] This

[10] P. Gardeström and I. Ericson, this series, Vol. 148, p. 434.

[11] A. Bergman, P. Gardeström, and I. Ericson, *Plant Physiol.* **66,** 442 (1980).

[12] M. Neuburger, *Encycl. Plant Physiol. New Ser.* **18,** 7 (1985).

approach requires that contaminating membranes differ in density from mitochondria. This often appears to be so, but in some cases the method is likely to fail. Phase partitioning in these latter cases is a good alternative since membrane separation reflects differences in surface properties.

### *Separation of Right-Side-Out and Inside-Out Submitochondrial Particles*

Potato tuber mitochondria can be purified using either a step[7] or a continuous[12] Percoll gradient. Crude and purified Jerusalem artichoke mitochondria are obtained as described by Møller *et al.*[6] and Lidén and Møller.[13]

The mitochondria (20–30 mg protein/ml) suspended in 2–4 ml low-salt medium (0.3 *M* sucrose, 5 m*M* MOPS, pH 7.2) are sonicated on ice 4–6 times for 5 sec with 1 min between bursts to permit cooling. We have used either a Branson model B-30 sonifier or an MSE sonicator, but any standard sonicator can be used. Longer sonication gives better disruption and better yield of SMP but carries the risk of inactivating vital enzymes such as cytochrome-*c* oxidase. After sonication the sample is diluted to 24 ml with low-salt medium, and intact mitochondria and larger fragments are pelleted at 48,000 $g_{max}$ for 16 min at 4°. The resulting supernatant solution is centrifuged at 105,000 $g_{max}$ for 65 min to pellet the SMP, which are resuspended in 0.3 *M* sucrose to give a final volume of 1.1–1.3 ml. When the mitochondria are disrupted by sonication the 105,000 *g* supernatant solution hardly contains any cytochrome-*c* oxidase. This indicates that very few small SMP are produced and is in contrast to what is obtained after French press treatment.[7,8]

The SMP produced by sonication in a low-salt medium have a cytochrome-*c* oxidase latency of 60–80% depending on the species and the preparation.[5,7] Accepting the latency as a polarity assay (for discussion, see Ref. 6), the preparation contains 60–80% IO-SMP and 40–20% RO-SMP, which can subsequently be separated by phase partitioning.

To find the optimal phase system different polymer concentrations should be tested (Fig. 1). The phase systems contain Dx T500/PEG 3350 at the concentrations indicated plus 300 mmol/kg sucrose and 10 mmol/kg potassium phosphate, pH 7.0. As the polymer concentrations are increased the amount of SMP in the top phase decreases and that in the interface and bottom phase increases (Fig. 1A). At the same time the polarity changes such that at 6.6% polymers the bottom phase contains mostly IO-SMP and the top phase mostly RO-SMP (Fig. 1B). A slightly better

[13] A. C. Lidén and I. M. Møller, *Physiol. Plant.* **72,** 265 (1988).

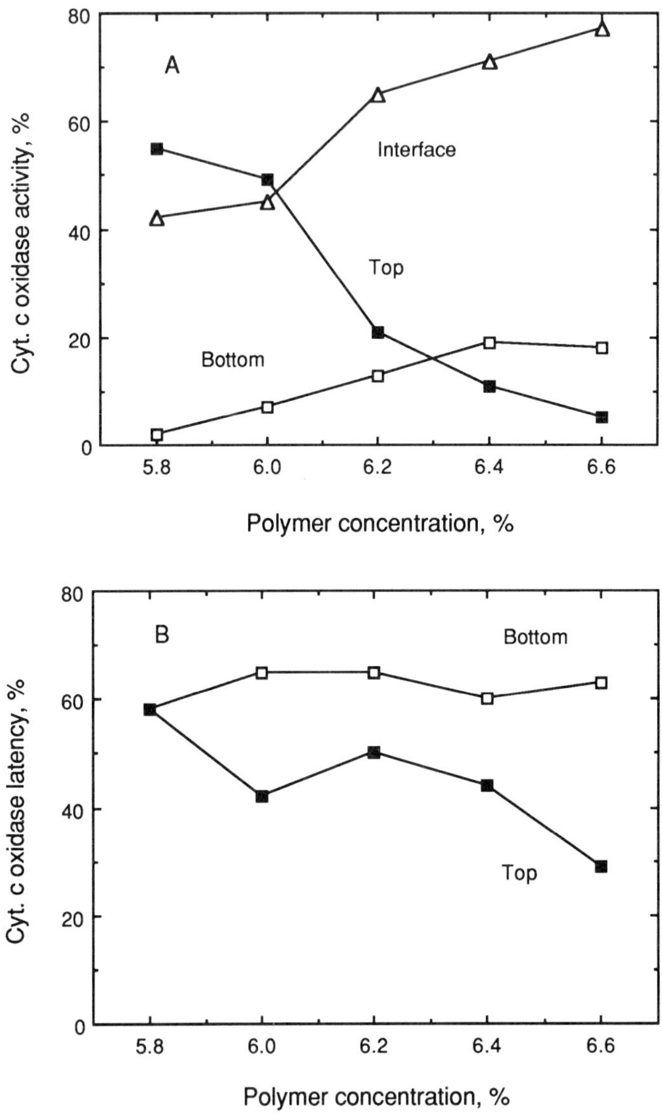


FIG. 1. Optimization of the aqueous polymer two-phase system for discriminating between IO- and RO-SMP from Jerusalem artichoke tuber mitochondria. In (A) the activity at the interface was calculated by subtracting those in the top and bottom phases from 100%. [Modified from C. J. Kay, I. Ericson, P. Gardeström, J. M. Palmer, and I. M. Møller, *FEBS Lett.* **193,** 169 (1985).]

TABLE II
SEPARATION OF RIGHT-SIDE-OUT AND INSIDE-OUT SUBMITOCHONDRIAL PARTICLES FROM POPULATION WITH MIXED SIDEDNESS FROM CRUDE JERUSALEM ARTICHOKE MITOCHONDRIA[a]

| Fraction | Latency of cytochrome-*c* oxidase (% IO-SMP) | Enzyme activity ($\mu$mol/mg protein/min) | | |
|---|---|---|---|---|
| | | Cytochrome-*c* oxidase | Succinate dehydrogenase | Latent malate dehydrogenase |
| Mitochondria | — | 2.0 ± 0.1 | 0.07 ± 0.02 | 16.2 ± 2.6 |
| SMP | 55 ± 3 | 2.3 ± 0.7 | 0.14 ± 0.03 | 4.5 ± 0.9 |
| Bottom phase | 74 ± 1 | 3.3 ± 0.3 | 0.19 ± 0.04 | 3.7 ± 0.6 |
| Top phase | 24 ± 7 | 2.2 ± 0.7 | 0.12 ± 0.04 | 9.7 ± 3.8 |

[a] A phase system containing 6.6% (w/w) Dx T500, 5.94% (w/w) PEG 3350, 0.66% (w/w) HMDA-PEG 6000, 300 mmol/kg sucrose, and 10 mmol/kg potassium phosphate, pH 8.0, was used to separate SMP, produced by sonication in a low-salt medium, in a one-step procedure. The data are means ± S.D.) of three independent preparations. Modified from C. J. Kay, I. Ericson, P. Gardeström, J. M. Palmer, and I. M. Møller, *FEBS Lett.* **193,** 169 (1985).

separation can be obtained at pH 8.0 and when 10% of the PEG is substituted with positively charged HMDA-PEG.[5] However, it is not necessary to use substituted PEG. A lower polymer concentration and some KCl (1–2 m*M*) give similar good results with potato tuber SMP.[14]

This purification of RO-SMP from mitochondria sonicated in low-salt medium has the advantage of avoiding French press treatment, which can damage the mitochondrial inner membrane. The RO-SMP produced by French press treatment are not able to generate a membrane potential.[8]

It is possible to separate, in a single extraction step, two populations of SMP with appreciably different sidedness (Table II). When SMP from Jerusalem artichoke mitochondria are used, the IO-SMP have a significantly higher specific activity of the membrane-bound enzymes cytochrome-*c* oxidase and succinate dehydrogenase, whereas the RO-SMP contain more latent malate dehydrogenase (Table II). This is consistent with a model in which IO-SMP are formed by neighboring cristae membranes resealing around the intermembrane space and RO-SMP are formed when inner membrane fragments reseal around part of the matrix.[5,6]

In general, RO-SMP always have a higher affinity for the upper phase than IO-SMP. This has also been observed for SMP isolated from mitochondria of potato tubers,[14] *Arum* spadices,[4] and beef heart.[15]

[14] I. M. Møller (1992), unpublished.

[15] I. M. Møller, E. Glaser, and B. Andersson, (1987), unpublished.

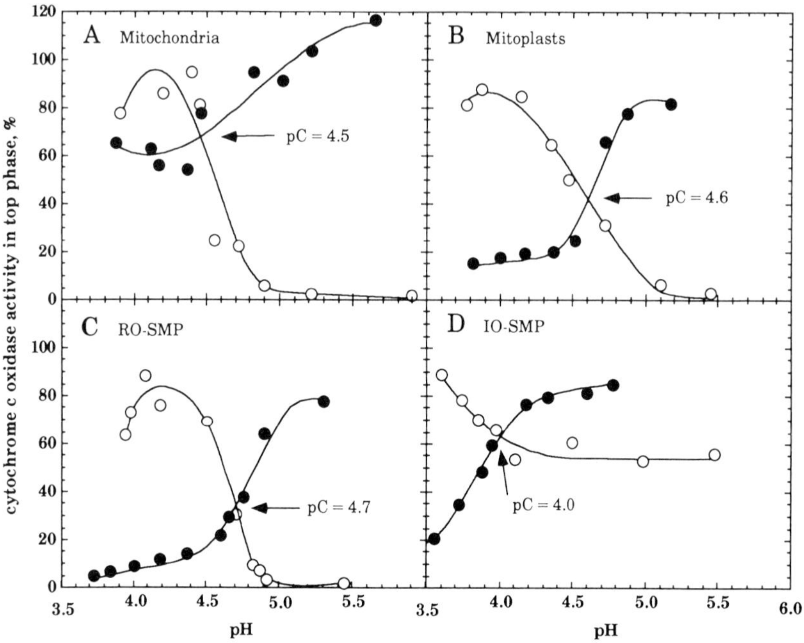


FIG. 2. Cross-partitioning of intact mitochondria (A), mitoplasts (B), RO-SMP (C), and IO-SMP (D) from potato tuber mitochondria. A final two-phase system weighed 2.0 g and contained 6.8% (w/w) Dx T500, 5.1% (w/w) PEG 3350, 1.7% (w/w) TMA-PEG 3350 or S-PEG 3350, 0.3 *M* sucrose, 0.7–2.5 m*M* citrate, water to a final weight of 1.9 g, and 100 $\mu$l of sample in 0.3 *M* sucrose and 50 m*M* potassium phosphate, pH 7.2 at 4°. Filled circles represent the series of phase systems with TMA-PEG (positively charged) and the open circles the series of phase systems with S-PEG (negatively charged). The cross-point values (pC) given are means from two or three independent determinations, and the deviation was not more than ±0.05 pH unit. [Modified from P. X. Petit, K. A. Edman, P. Gardeström, and I. Ericson, *Biochim. Biophys. Acta* **890,** 377 (1987).]

## *Characterization by Cross-Partition Analysis*

The cross-partition points[16] (which may be close to the isoelectric points) of Percoll-purified potato mitochondria, mitoplasts, RO-SMP, and IO-SMP are determined by cross-partitioning at 4°.[7] A relatively large electrostatic potential difference between the bulk phases is produced by including either a cationic or an anionic PEG derivative in the phase systems. In one series of phase systems at different pH values, TMA-PEG (cationic) is used, whereas in the other it is S-PEG (anionic) (see Fig. 2).

At low pH the activity of cytochrome-*c* oxidase decreases rapidly with

[16] H. Walter and D. Forciniti, this volume [20].

time. To correct for this a sample is withdrawn from the phase system immediately after mixing, and the phase separation is accelerated by centrifugation at 150 *g* for 5 min. A sample is withdrawn from the top phase. This sample and the sample from the mixed system are neutralized by mixing with an equal volume of 0.1 *M* sucrose, 50 m*M* potassium phosphate, pH 7.2, and the enzyme activity is measured. The remaining phase system is mixed with an equal volume of 0.3 m*M* sucrose and the pH determined at room temperature.

The pH at the cross-partition point, pC, is similar for intact potato tuber mitochondria and for mitoplasts and RO-SMP isolated therefrom (4.5–4.7). However, the amount of SMP partitioning into the top phase at this point differs markedly, being 70% for the mitochondria but only 30–40% for mitoplasts and RO-SMP (Fig. 2A–C). This indicates that although the outer surface of the outer mitochondrial membrane has about the same pC as the outer surface of the inner membrane, other surface properties differ. The pC of the IO-SMP, that is, of the inner surface of the inner mitochondrial membrane, is significantly lower than that of the outer surface (Fig. 2C, D).

The cross-partition point of intact spinach mitochondria,[10] intact chloroplasts, and the outer surface of the thylakoid membrane[17] all fall in the range of pH 4–5. The cross-partition or isoelectric point of the outer apoplastic surface of the plasma membrane appears to be below pH 4 when determined on plasma membrane vesicles by cross-partitioning[18] or by isoelectric focusing,[19] respectively. Thus, the membranes have a net negative charge at physiological pH. This negative charge gives rise to a negative surface potential that can affect membrane–membrane interactions, protein–membrane interactions, protein–protein interactions in the membrane, and the properties of membrane-bound enzymes.[20,21]

## Acknowledgments

We are grateful to Professor Christer Larsson for useful suggestions in the preparation of this chapter. Work done in the authors' laboratories has been supported by the Swedish Natural Science Research Council (to P.G. and I.M.M.) and the CNRS (to P.X.P.), and by a grant jointly from CNRS, France, and NFR, Sweden (P.X.P. and P.G.).

[17] A. Westrin, V. P. Shanbhag, and P.-Å. Albertsson, *Biochim. Biophys. Acta* **732,** 83 (1983).
[18] L. E. Körner, P. Kjellbom, C. Larsson, and I. M. Møller, *Plant Physiol.* **79,** 72 (1985).
[19] L. R. Griffing and R. S. Quatrano, *Proc. Natl. Acad. Sci. U.S.A.* **81,** 4804 (1984).
[20] J. Barber, *Biochim. Biophys. Acta* **594,** 253 (1980).
[21] P. Douzou and P. Maurel, *Trends Biochem. Sci.* **2,** 14 (1977).

# [42] Isolation and Identification of Plasma Membrane Populations

*By* MICHAEL E. BRADLEY, ROSS W. LAMBERT, and AUSTIN K. MIRCHEFF

## Introduction

To isolate a plasma membrane sample from virtually any nucleated cell, one must address the dual problems of identifying the plasma membrane fragments in a cell lysate and of separating them from intact organelles and other membrane fragments. A topologically continuous plasma membrane may be organized into microdomains that give rise to multiple membrane populations with distinct physical properties.[1] Epithelial cells, whose plasma membranes are organized into topologically distinct domains, each of which may be further organized into microdomains, pose an even more complex problem. Plasma membrane constituents usually turn over and recycle, and in some cases the recycling traffic may be quite rapid. Consequently, constituents traditionally employed as intrinsic plasma membrane markers are expressed both in surface and in intracellular pools. In extreme cases the intracellular pools may be manyfold larger than the surface pools.

Our strategy for solving these problems[2] is based on the premise that one can infer the distributions of populations of membrane particles from the frequency distributions of membrane markers along continua related to physical separation procedures.[3] We combine isopycnic centrifugation and partitioning in dextran (Dx)–poly(ethylene glycol) (PEG) aqueous two-phase systems to generate a two-dimensional separation. Distinct membrane samples are delineated, and the samples most likely to have been derived from the plasma membranes are identified on the basis of available information. $Na^+,K^+$-ATPase is expressed in the plasma membranes of most vertebrate cells and in the basal–lateral membranes of most vertebrate epithelia. Alkaline phosphatase frequently has an oppositely polarized surface distribution. Galactosyltransferase is expressed in the trans compartments of the Golgi complex. NADPH–cytochrome-*c* reductase is expressed in the endoplasmic reticulum. Succinate dehydrogenase is expressed in the mitochondria. Thus, membrane populations highly

[1] J. P. Gierow, this volume [50].

[2] A. K. Mircheff, *Am. J. Physiol.* **244,** G347 (1983).

[3] C. DeDuve, *J. Theor. Biol.* **6,** 35 (1964).

enriched in $Na^+$,$K^+$-ATPase and alkaline phosphatase with respect to the other markers are good candidates for identification as plasma membranes. Such identifications should be regarded as working hypotheses, to be tested against additional criteria.

Detailed instructions for cell lysis, differential sedimentation, isopycnic centrifugation, and phase partitioning have been presented.[4,5] We focus here on interpretation of the resulting marker distribution data, illustrated with the subcellular fractionation analysis of rat exorbital lacrimal gland. References to applications of this approach to other epithelial tissues are also provided.

## Methods

### *Tissue Preparation*

All experiments are performed with exorbital lacrimal glands from male Sprague-Dawley rats. The glands are cut into fragments, which are equilibrated in a modified Krebs improved Ringer I bicarbonate buffer (KRB); descriptions of the *in vitro* incubation may be found elsewhere.[6,7] The protease inhibitors phenylmethylsulfonyl fluoride (PMSF, 0.2 m*M*) and aprotinin (9 $\mu$g/ml) are added to KRB and to the media used for cell lysis and differential and density gradient centrifugation. It is also feasible to begin the isolation procedures with freshly harvested tissue, cultured cells, or, in some cases, cultured cells that have been snap-frozen and stored at $-70°$.[8]

### *Subcellular Fractionation*

After the equilibration period the gland fragments are rinsed rapidly with ice-cold isolation buffer, which consists of 5% (w/v) sorbitol (analyzed reagent grade, J. T. Baker, Phillipsburg, NJ), 0.5 m*M* NaEDTA, and 5 m*M* histidine–imidazole buffer, pH 7.5.[4,5] They are resuspended in isolation buffer and homogenized with a Tissumiser (Tekmar, Cincinnati OH); the homogenization vessel is immersed in an ice–water bath.[4–7]

[4] A. K. Mircheff, this series, Vol. 172, p. 18.

[5] A. K. Mircheff and E. J. J. M. van Corven, this series, Vol. 192, p. 341.

[6] S. C. Yiu, R. W. Lambert, M. E. Bradley, C. E. Ingham, K. L. Hales, R. L. Wood, and A. K. Mircheff, *J. Membr. Biol.* **102,** 185 (1988).

[7] M. E. Bradley, C. L. Peters, R. W. Lambert, S. C. Yiu, and A. K. Mircheff, *Invest. Ophthalmol. Visual Sci.* **31,** 977 (1990).

[8] A. K. Mircheff, J. W. Bowen, S. C. Yiu, and A. A. McDonough, *Am. J. Physiol.* **262,** C470 (1992).

The homogenate is centrifuged at 2000 *g* for 10 min at 4°; the resulting pellet is resuspended, homogenized, and again centrifuged. The supernatant fractions pooled from four cycles of homogenization and centrifugation, designated $S_0$, are brought to a final sorbitol concentration of 55% and subjected to isopycnic centrifugation in a Beckman Z-60 zonal rotor (Fullerton, CA) loaded with a 2-stage (35–55% and 55–70%) hyperbolic sorbitol density gradient.[4,5] After the run at 235,000 *g* for 70 min at 4°, the rotor contents are collected in 24 equal-volume fractions, the fractions are diluted with isolation buffer, and membranes are harvested at 250,000 *g* for 75 min at 4°. The resulting series of pellets, designated $\Sigma P_i$, are resuspended in 2.4 ml isolation buffer, and 21.9 ml samples are snap-frozen and stored at −70° until used for phase partitioning. The remainder may be kept at 4° for a few days until the marker determinations are completed.

Membranes from certain pooled density gradient fractions are subjected to differential rate centrifugation through columns of 17.5% sorbitol.[4] The resulting mitochondrial pellet is resuspended in isolation buffer. Micrososomal membranes are harvested from the entire supernatant fraction by dilution and centrifugation at 250,000 *g* for 75 min at 4°, then resuspended in isolation buffer.

Phase partitioning is carried out in a thin-layer countercurrent distribution (CCD) apparatus.[9,10] The two-phase systems contain 5% (w/w) Dx T500 (Pharmacia LKB, Piscataway, NJ), 3.5% (w/w) PEG 8000 (Carbowax 8000, Union Carbide, New York, NY), 5% sorbitol, and 10 $\mu M$ NaEDTA. Systems with pH 7.3 contain 8.33 m$M$ imidazole; systems with pH 8.0 contain 16.7 m$M$ imidazole. The pH is adjusted with HCl. Membranes are harvested by centrifugation at 250,000 g for 75 min at 4°, then resuspended in upper phase to a final protein concentration of either 0.4 or 1.2 mg/ml. The circular CCD plates have 120 cavities and are divided into three equal sectors. Each cavity is loaded with 0.65 ml lower phase; cavities 0–3 of each sector are loaded with 0.65-ml aliquots of sample suspended in upper phase, and cavities 4–39 of each sector are loaded with 0.65-ml aliquots of upper phase without sample. After 36 transfers 1.0-ml aliquots of isolation buffer are added to each cavity, the contents are mixed, collected, pooled in groups of 4, and diluted a further 4-fold with isolation buffer. Membranes are harvested by centrifugation at 250,000 *g* for 75 min at 4°, then resuspended in 2.0-ml aliquots of isolation buffer.

[9] P.-Å. Albertsson, B. Andersson, C. Larsson, H.-E. Åkerlund, *Methods Biochem. Anal.* **28,** 115 (1982).

[10] H.-E. Åkerlund and P.-Å. Albertsson, this volume [8].

*Biochemical Marker Determinations*

References and notes to procedures for most marker determinations have been described previously.[4,5] Distributions of markers among the fractions generated by an individual separation procedure are initially expressed in terms of percentage of the activity recovered from that separation. The cumulated products for a sequence of separation steps then represent marker yields as percentages of total activities recovered. Cumulative enrichment factors are calculated as ratios of marker yields to protein yield.[11]

## Results

*Differential Sedimentation*

The low- and high-speed centrifugation steps separate the homogenate into three broad sedimentation windows, with marker contents as summarized in Table I. The low-speed pellet, $P_0$, contains connective tissue fragments, intact nuclei, intact secretory vesicles, and some mitochondria. The $\Sigma P_i$ group contains most of the microsomal membranes and mitochondria, whereas the high-speed supernatant fraction, $\Sigma S_i$, contains soluble constituents and microsomal populations which fail to sediment with high-speed centrifugation. Most of the protein in $\Sigma S_i$ is soluble, whereas most of the $Na^+,K^+$-ATPase is associated with the nonsedimenting microsomes.[6]

*Density Gradient Centrifugation*

Distributions of markers within $\Sigma P_i$ are summarized in Fig. 1. The distribution of galactosyltransferase overlaps the distributions of $Na^+,K^+$-ATPase and alkaline phosphatase within density window II, and its shape approximately parallels that of the $Na^+,K^+$-ATPase distribution within windows III, IV, and V. The $Na^+,K^+$-ATPase and alkaline phosphatase distributions diverge markedly within density window IV, in which $Na^+,K^+$,ATPase exhibits a major peak and alkaline phosphatase exhibits only a shoulder or minor peak. The density distributions of other plasma membrane constituents, including $Na^+/H^+$ antiporters,[12] $Cl^-/HCO_3^-$ antiporters,[13] and muscarinic cholinergic receptors,[7] more closely resemble

[11] A. K. Mircheff and C. C. Lu, *Am. J. Physiol.* **247,** G651 (1984).

[12] A. K. Mircheff, C. E. Ingham, R. W. Lambert, K. L. Hales, C. B. Hensley, and S. C. Yiu, *Invest. Ophthalmol. Visual Sci.* **28,** 1726, 1987.

[13] R. W. Lambert, M. E. Bradley, and A. K. Mircheff, *Am. J. Physiol.* **255,** G367 (1988).

TABLE I

MARKER CONTENTS OF FRACTIONS GENERATED BY DIFFERENTIAL SEDIMENTATION[a]

| Marker ($n$) | Total activity | $P_0$ | Percent recovered activity $\Sigma P_i$ | $\Sigma S_i$ | Recovery |
|---|---|---|---|---|---|
| $Na^+$,$K^+$-ATPase (11) | 12.3 ± 4.3 | 5.2 ± 3.5 | 71.1 ± 6.6 | 23.7 ± 8.2 | 97.3 ± 20.0 |
| Alkaline-phosphatase (11) | 89.2 ± 12.3 | 4.1 ± 1.4 | 51.9 ± 8.6 | 44.0 ± 8.5 | 87.8 ± 12.3 |
| Acid phosphatase (11) | 425.1 ± 24.8 | 2.9 ± 1.7 | 53.4 ± 3.5 | 43.7 ± 4.5 | 90.4 ± 15.8 |
| Succinate dehydrogenase (10) | 21.3 ± 6.2 | 9.9 ± 8.3 | 77.5 ± 16.8 | 10.1 ± 8.5 | 151.2 ± 15.9 |
| Galactosyltransferase (5) | 107.2 ± 67.9 | 3.3 ± 2.0 | 86.0 ± 6.4 | 10.7 ± 7.3 | 76.1 ± 19.3 |
| NADPH–cytochrome-*c* reductase (4) | 1.6 ± 0.4 | 1.4 ± 0.6 | 44.6 ± 8.6 | 54.0 ± 8.9 | 130.4 ± 22.6 |
| Protein (11) | 339.3 ± 48.8 | 2.0 ± 0.9 | 31.0 ± 6.9 | 67.0 ± 7.1 | 98.6 ± 11.7 |

[a] Values presented are percentages of total activities recovered from the initial homogenate. Also presented are percentages of the markers measured in the initial low-speed pellet, $S_0$, recovered in $\Sigma P_i$ plus $\Sigma S_i$. Units for galactosyltransferase are nmol galactose transferred/min. Units for NADPH–cytochrome-*c* reductase are $\Delta$ absorbance/min. Units for all other activities are $\mu$mol/hr. Total protein is expressed in mg. All values are means ± S.D.; $n$ is the number of separate preparations in which markers were determined.

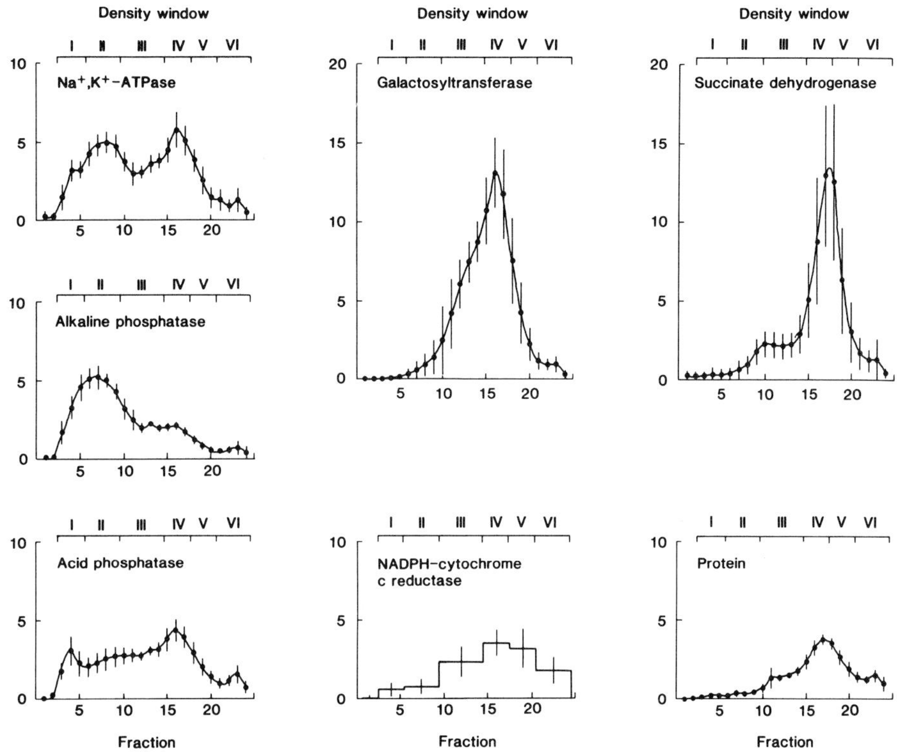


FIG. 1. Distributions of membrane-associated biochemical markers. Low-speed supernatant fractions were subjected to isopycnic centrifugation on sorbitol density gradients. Membranes were then harvested by high-speed centrifugation. After biochemical determinations, gradient fractions were pooled in density windows capturing salient features of the marker distribution patterns. Values given are mean percentages of total activities recovered (±S.D.). Numbers of separate preparations in which markers were determined are given in Table I. [Data presented are a compilation of new replicates and data previously published in A. K. Mircheff, M. E. Bradley, C. B. Hensley, E. J. J. M. van Corven, S. C. Yiu, and R. W. Lambert, *in* "Separations Using Aqueous Phase Systems" (D. Fisher and I. A. Sutherland, eds.), p. 67. Plenum, New York and London, 1989, included with permission.]

that of alkaline phosphatase. Thus, density window IV contains a pool of $Na^+,K^+$-ATPase in excess of other plasma membrane activities.

*Phase Partitioning*

To resolve all of the $Na^+,K^+$-ATPase- and alkaline phosphatase-containing membrane populations adequately it is necessary to analyze most of the density windows under two or three conditions: phase system pH

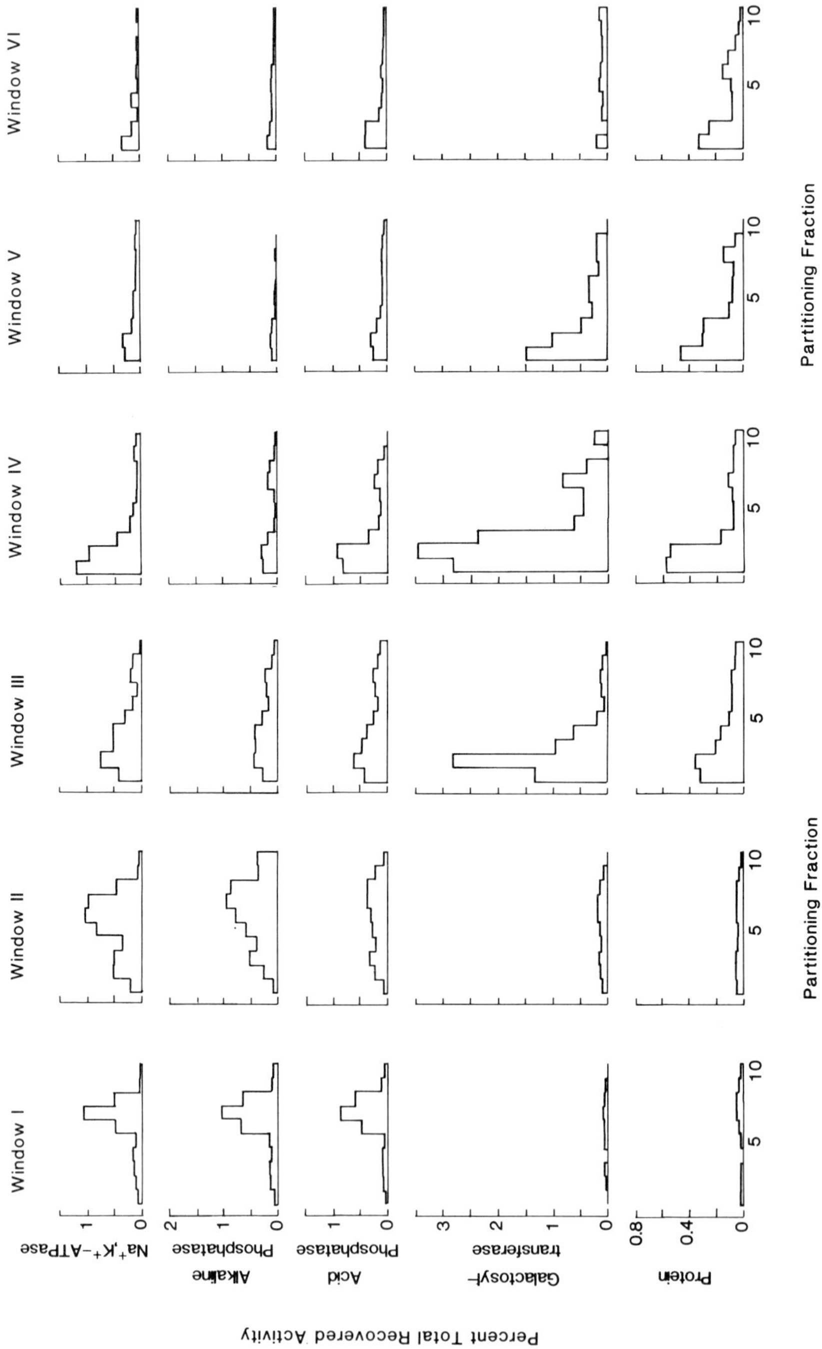
Window I
Window II
Window III
Window IV
Window V
Window VI
Na+,K+-ATPase
Alkaline Phosphatase
Acid Phosphatase
Galactosyl-transferase
Protein
Percent Total Recovered Activity
Partitioning Fraction
Partitioning Fraction

7.3 and initial sample protein concentration 0.4 mg/ml (Fig. 2); phase system pH 8.0 and initial sample protein concentration 0.4 mg/ml (Fig. 3); and phase system pH 8.0 and initial sample protein concentration 1.2 mg/ml (Figs. 4 and 5).

A minimum of 20 distinct microsomal membrane samples, which have been given arbitrary letter designations, can account for the marker distributions in Figs. 2–5. Their approximate positions on the three density gradient–phase partitioning planes are summarized in Fig. 6, and their marker cumulative enrichment factors are summarized in Table II. Adjacent phase partitioning fractions are considered to contain similar membrane samples when they exhibit similar cumulative factors for all of the markers. Nearby fractions are considered to contain different membrane samples when they exhibit distinct modes or consistently observed shoulders in the distribution of at least one marker, or when they contain significantly different cumulative enrichment factors for at least one marker. Sample j is broadly distributed with respect to density but is homogeneous in all other respects. For this reason it is considered to comprise a distinct population of membranes. In other cases, for example, f, k, and o, samples with distinct sets of marker cumulative enrichment factors appear to represent a continuous two-dimensional distribution, with the suggestion that they may all belong to a single membrane population in which marker cumulative factors vary with density.

Sample a accounts for the coinciding peaks of $Na^+$,$K^+$-ATPase, alkaline phosphatase, and acid phosphatase centered at fraction 7, window I (Fig. 2). Sample b accounts for the minor overlapping peaks of the same markers spanning partitioning fractions 1–5. Sample c accounts for the excesses of alkaline phosphatase and acid phosphatase activities over $Na^+$,$K^+$-ATPase activity in fractions 8–10, window II (Fig. 2). It can be distinguished from sample a on the basis of its cumulative enrichment factors for each of the markers in Table II. Sample d can be visualized from the peak of $Na^+$,$K^+$-ATPase activity centered at fraction 6, window

FIG. 2. Phase partitioning analyses of density windows I through VI from Fig. 1. (For density windows IV and V, only the microsomal fractions, obtained by differential rate centrifugation, were analyzed). Phase system pH was 7.3, and the initial sample protein concentration was 0.4 mg/ml upper phase. Values given are percentages of total recovered activity per partitioning fraction divided by the number of density gradient fractions per density window from a representative analysis; each density window was analyzed at least 3 times. [Data for density windows I and II have been published in A. K. Mircheff, M. E. Bradley, C. B. Hensley, E. J. J. M. van Corven, S. C. Yiu, and R. W. Lambert, *in* "Separations Using Aqueous Phase Systems" (D. Fisher and I. A. Sutherland, eds.), p. 67. Plenum, New York and London, 1989, and are included with permission.]

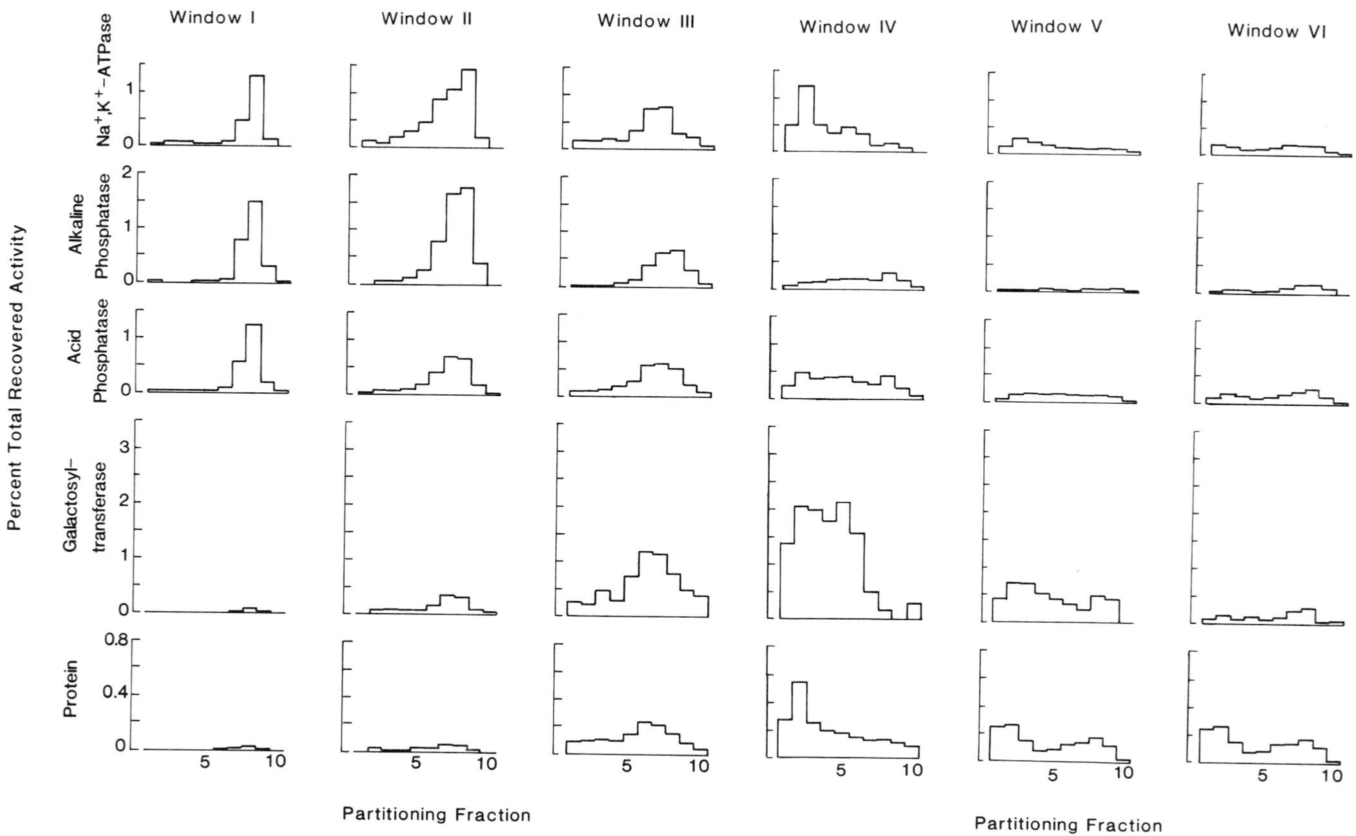


FIG. 3. Phase partitioning analyses of density windows I through VI. Phase system pH was 8.0, and the initial sample protein concentration was 0.4 mg/ml upper phase. Other details are as in Fig. 2. [Data for density windows I and II have been published in A. K. Mircheff, M. E. Bradley, C. B. Hensley, E. J. J. M. van Corven, S. C. Yiu, and R. W. Lambert, *in* "Separations Using Aqueous Phase Systems" (D. Fisher and I. A. Sutherland, eds.), p. 67, Plenum, New York and London, 1989, and are included with permission.]

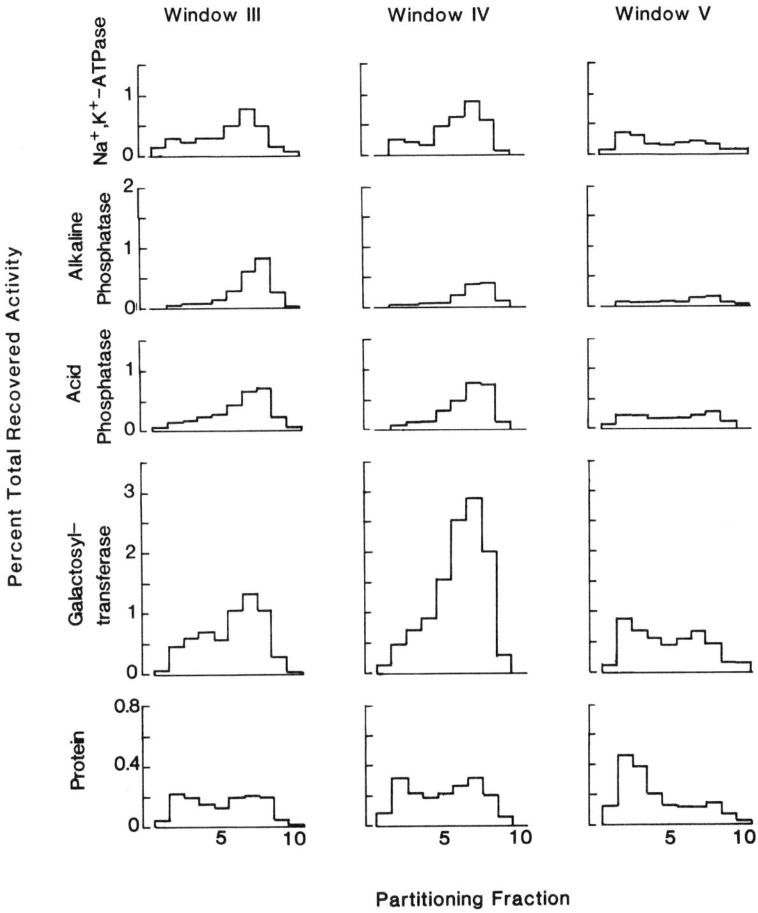


FIG. 4. Phase partitioning analyses of density windows III through V. Phase system pH was 7.3, and the initial sample protein concentration was 1.2 mg/ml upper phase. Other details are as in Fig. 2.

II (Fig. 2). Sample e accounts for the overlapping peaks of $Na^+,K^+$-ATPase, alkaline phosphatase, acid phosphatase, and galactosyltransferase spanning fractions 1–4 after the same analysis.

Sample j accounts for the peaks of protein and NADPH–cytochrome-*c* reductase spanning partitioning fractions 1–4, windows IV and V (Fig. 5). Sample f accounts for the overlapping peaks of $Na^+,K^+$-ATPase, alkaline phosphatase, acid phosphatase, and galactosyltransferase spanning fractions 7–10, window III (Fig. 2). Sample h accounts for the striking peak of galactosyltransferase and the smaller peaks of $Na^+,K^+$-ATPase,

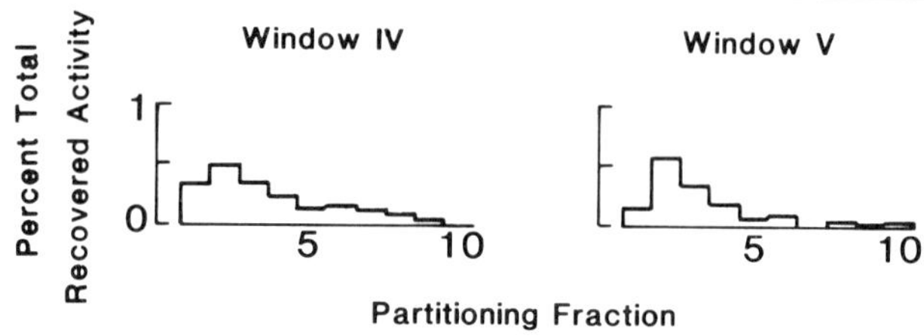


FIG. 5. Distributions of NADPH–cytochrome-*c* reductase activity after partitioning analyses under the same conditions as in Fig. 4.

alkaline phosphatase, and acid phosphatase centered at fraction 2, window III (Fig. 2). Sample g accounts for the mode of the alkaline phosphatase distribution and the shoulders of the $Na^+,K^+$-ATPase and acid phosphatase distributions at fraction 4, window III (Fig. 2). Increasing the pH of the phase system to 8.0 (Fig. 3) preferentially shifts samples g and h to the upper phase, displacing them to the right with respect to samples i and j. Increasing pH also partially separates sample j from sample i, which accounts for the peak of galactosyltransferase at partitioning fraction 4.

Sample k accounts for modes in the $Na^+,K^+$-ATPase, alkaline phosphatase, and acid phosphatase distributions in fractions 7 and 8, density window IV (Fig. 2). Samples l and m overlap one another, as well as samples i, j, and n, after analyses in the pH 7.3 phase system. Increasing the phase system pH to 8.0 causes rightward shifts in the distributions of samples l and m, window IV (Fig. 3). Although samples l and k overlap in the pH 8.0 phase system, they can be distinguished based on differences in their marker cumulative enrichment factors (Table II). Sample m accounts for the major peak of galactosyltransferase and minor peak of $Na^+,K^+$-ATPase at partitioning fraction 5, window IV (Fig. 3). Sample n accounts for the overlapping peaks of $Na^+,K^+$-ATPase and galactosyltransferase at partitioning fraction 2, window V (Fig. 3), where it overlaps samples i and j. Increasing the initial sample protein concentration from 0.4 to 1.2 mg/ml shifts the distributions of m and n further toward the right (Fig. 4) and separates n from i and j. Elements of sample i account for the shoulder of the galactosyltransferase distribution in fractions 3 and 4, windows III and IV (Fig. 4).

In addition to elements of samples j and n, density window V contains the minor samples o, p, and q. Density window VI contains samples r, s, and t.

## Discussion

Having delineated a series of membrane samples, one must now attempt to infer the subcellular origins of the samples containing the major

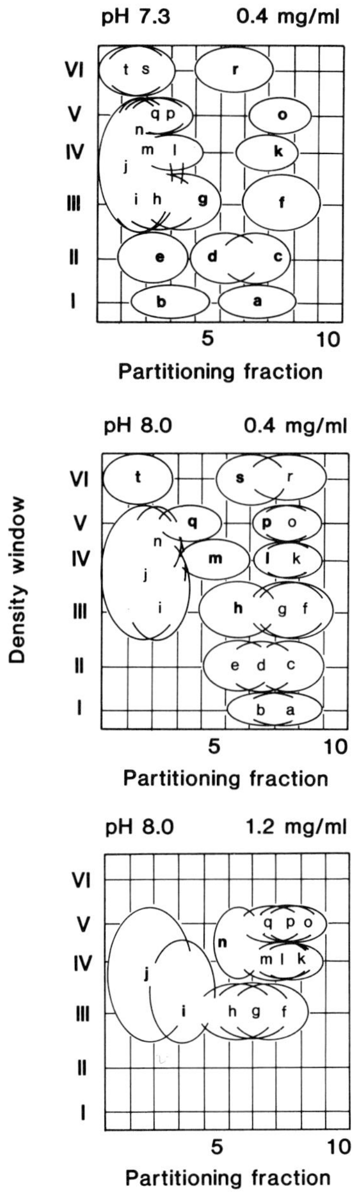


FIG. 6. Density–phase partitioning diagrams compiled from the analyses in Figs. 3–5 and indicating approximate positions of membrane populations that have been delineated from the biochemical marker distribution patterns.

TABLE II
MARKER CUMULATIVE ENRICHMENT FACTORS IN MEMBRANE SAMPLES DELINEATED BY SEQUENTIAL DENSITY–PHASE PARTITIONING ANALYSES[a]

| Membrane sample | $Na^+,K^+$-ATPase | Alkaline phosphatase | Acid phosphatase | Galactosyl-transferase |
|---|---|---|---|---|
| a | 23.3 ± 3.1 | 29.1 ± 4.7 | 21.9 ± 4.7 | 0.9 ± 0.0 |
| b | 16.9 ± 1.7 | 19.7 ± 2.4 | 12.3 ± 2.6 | 0.5 ± 0.0 |
| c | 11.4 ± 1.7 | 19.8 ± 3.5 | 8.6 ± 1.4 | 2.5 ± 0.4 |
| d | 23.9 ± 0.8 | 18.7 ± 1.5 | 8.0 ± 0.7 | 4.0 ± 1.0 |
| e | 12.2 ± 3.5 | 11.6 ± 1.6 | 7.6 ± 0.2 | 2.7 ± 0.1 |
| f | 3.8 ± 1.2 | 4.3 ± 1.2 | 4.3 ± 1.1 | 6.3 ± 3.5 |
| g | 4.4 ± 0.7 | 3.1 ± 0.5 | 3.3 ± 0.6 | 4.3 ± 0.9 |
| h | 3.8 ± 0.2 | 2.5 ± 0.7 | 3.0 ± 0.2 | 6.5 ± 0.6 |
| i | 2.0 ± 0.6 | 0.8 ± 0.4 | 1.8 ± 0.5 | 5.4 ± 1.5 |
| j | 0.7 ± 0.0 | 0.1 ± 0.0 | 0.4 ± 0.0 | 1.7 ± 0.2 |
| k | 1.8 ± 0.9 | 1.4 ± 0.0 | 2.5 ± 0.2 | 0.1 ± N.D. |
| l | 2.2 ± 1.2 | 2.5 ± 0.7 | 3.9 ± 1.0 | 1.3 ± N.D. |
| m | 2.2 ± 0.4 | 1.2 ± 0.7 | 2.7 ± 0.7 | 7.2 ± N.D. |
| n | 1.8 ± 0.5 | 0.5 ± 0.3 | 1.3 ± 0.3 | 4.3 ± 1.1 |
| o | 0.9 ± 0.6 | 0.5 ± 0.1 | 1.1 ± 0.2 | 2.7 ± N.D. |
| p | 0.6 ± 0.6 | 0.6 ± 0.0 | 1.2 ± 0.1 | 2.9 ± 2.9 |
| q | 0.9 ± 0.4 | 0.5 ± 0.1 | 1.1 ± 0.1 | 3.1 ± 2.1 |
| r | 0.6 ± 0.0 | 0.2 ± 0.0 | 0.5 ± 0.3 | 1.6 ± N.D. |
| s | 1.1 ± 0.1 | 0.8 ± 0.2 | 1.3 ± 0.2 | 1.6 ± N.D. |
| t | 0.6 ± 0.2 | 0.2 ± 0.1 | 0.7 ± 0.2 | 0.5 ± N.D. |

[a] Values given are mean cumulative enrichment factors ± S.D. Positions at which marker activity and protein data were sampled are indicated by boldface designation in Fig. 6. N.D., Not determined.

$Na^+,K^+$-ATPase and alkaline phosphatase activities. Because the analysis has been performed with lacrimal gland fragments, it is possible that different samples were derived from different cell types, or that a given sample consisted of a mixture of membranes derived from one or more different cell types. However, the acinar cells account for 80% of the mass of the lacrimal gland, and the remainder is apportioned to a number of minor cell types. Moreover, similar analyses have been performed with preparations of isolated lacrimal acini,[14] with results similar to those described here. Thus, we conclude most samples are dominated by membranes derived from acinar cells.

Sample a, which accounts for, roughly, 10% of the total $Na^+,K^+$-ATPase activity, can be identified as having been derived from the basal–lateral plasma membranes on the basis of its large enrichment fac-

[14] S. C. Yiu, R. L. Wood, and A. K. Mircheff, *Invest. Ophthalmol. Visual Sci.* **31,** 2437 (1990).

tors for $Na^+,K^+$-ATPase and alkaline phosphatase and the high ratios of these to the enrichment factor for galactosyltransferase. Because samples g, h, m, and n account for such large components of the total recovered galactosyltransferase activity, they can be attributed most plausibly to distinct domains of the Golgi complex. The observation that significant components of the $Na^+,K^+$-ATPase within density windows III–V distribute in parallel with components of the galactosyltransferase under three different conditions of phase partitioning analysis supports the conclusion that $Na^+,K^+$-ATPase is associated with these Golgi samples. Because sample j is the major locus of protein and NADPH–cytochrome-*c* reductase, it appears to represent the major population of endoplasmic reticulum membranes; its $Na^+,K^+$-ATPase activity may be due to newly synthesized pump units.[8]

Identification of the other samples containing major $Na^+,K^+$-ATPase activities, namely, c, d, and e, is somewhat problematic. Because it is the locus of the highest ratio of alkaline phosphatase to $Na^+,K^+$-ATPase cumulative enrichment, sample c is the most plausible candidate for identification as the major population of apical membrane vesicles. That corresponding samples from previous studies were major loci of $Na^+$-dependent amino acid transport activities characteristic of epithelial cell apical membranes[15] is consistent with this inference, as is the observation that $Na^+,K^+$-ATPase immunoreactivity is present in both the apical and the basal–lateral membranes of lacrimal acinar cells.[16]

The relatively large cumulative enrichment factors for $Na^+,K^+$-ATPase, alkaline phosphatase, and galactosyltransferase in samples d and e seem equally consistent with two alternative hypotheses: (1) that they originated in microdomains of the basal–lateral membranes which had relatively large contents of galactosyltransferase and (2) that they originated in intracellular compartments involved in segregating newly assembled and recycling basal–lateral membrane-expressed constituents from constituents retained intracellularly.

The hypothesis that samples d and e are derived from intracellular compartments was not obvious when the first subcellular fractionation analyses of lacrimal and salivary glands were performed, and fractions corresponding to sample d were originally identified as the major population of acinar cell basal–lateral membranes.[2,10,14] If we accept the hypothesis that sample a contains the acinar cell basal–lateral membrane population, whereas d and e contain membranes from intracellular compartments, we are left to conclude that only a modest fraction of the total $Na^+,K^+$-

[15] A. K. Mircheff, C. C. Lu, and C. N. Conteas, *Am. J. Physiol.* **245,** G661 (1983).
[16] R. L. Wood and A. K. Mircheff, *Invest. Ophthalmol. Visual Sci.* **27,** 1293 (1986).

TABLE III
TISSUES IN WHICH PLASMA MEMBRANE AND INTRACELLULAR MEMBRANE SAMPLES HAVE BEEN DELINEATED WITH SEQUENTIAL DENSITY–PHASE PARTITIONING ANALYSES

| Tissue | Species | Preparation | Refs. |
|---|---|---|---|
| Lacrimal gland | Rat | Freshly obtained fragments | *a, b* |
| | Rat | *In vitro* incubated fragments | *c, d* |
| | Rat | Isolated acini | *e* |
| | Rabbit | Isolated acini | *f* |
| Parotid gland | Rat | Freshly obtained fragments | *g* |
| Small intestine | Rat | Fresh mucosal scrappings | *h* |
| | Rat | Isolated enterocytes | *i* |
| Retinal pigment epithelium | Human | Cultured | *j* |
| Kidney | Rat | Freshly obtained fragments | *k* |
| | Rat | Isolated proximal tubules | *l* |
| | Rabbit | Freshly obtained fragments | *m* |
| | Rabbit | Partially purified brush borde | *n* |
| | Dog | MDCK cells | *o* |
| | Pig | LLC-$PK_1$ cells | *o* |

[a] A. K. Mircheff, C. C. Lu, and C. N. Conteas, *Am. J. Physiol.* **245,** G661 (1983).
[b] A. K. Mircheff and C. C. Lu, *Am. J. Physiol.* **247,** G651 (1984).
[c] S. C. Yiu, R. W. Lambert, M. E. Bradley, C. E. Ingham, K. L. Hales, R. L. Wood, and A. K. Mircheff, *J. Membr. Biol.* **102,** 185 (1988).
[d] M. E. Bradley, C. L. Peters, R. W. Lambert, S. C. Yiu, and A. K. Mircheff, *Invest. Ophthalmol. Visual Sci.* **31,** 977 (1990).
[e] S. C. Yiu, R. L. Wood, and A. K. Mircheff, *Invest. Ophthalmol. Visual Sci.* **31,** 2437 (1990).
[f] M. E. Bradley, R. W. Lambert, L. M. Lee, and A. K. Mircheff, *Invest. Ophthalmol. Visual Sci.* **33,** 2951 (1992).
[g] C. N. Conteas, A. A. McDonough, T. R. Kozlowski, C. B. Hensley, R. L. Wood, and A. K. Mircheff, *Am. J. Physiol.* **250,** C430 (1986).
[h] A. K. Mircheff, D. J. Ahnen, A. Islam, N. A. Santiago, and G. M. Gray, *J. Membr. Biol.* **83,** 95 (1985).
[i] E. J. J. M. Van Corven, C. H. van Os, and A. K. Mircheff, *Biochim. Biophys. Acta* **861,** 267 (1986).
[j] A. K. Mircheff, S. S. Miller, D. B. Farber, M. E. Bradley, W. T. O'Day, and D. Bok, *Invest. Ophthalmol. Visual Sci.* **31,** 863 (1990).
[k] C. B. Hensley, M. E. Bradley, and A. K. Mircheff, *Kidney Int.* **37,** 707 (1990).
[l] C. B. Hensley, M. E. Bradley, and A. K. Mircheff, *Am. J. Physiol.* **257,** C637 (1989).
[m] C. B. Hensley and A. K. Mircheff, *Kidney Int.* in press (1994).
[n] A. K. Mircheff, H. E. Ives, V. J. Yee, and D. J. Warnock, *Am. J. Physiol.* **246,** F853 (1984).
[o] A. K. Mircheff, J. W. Bowen, S. C. Yiu, and A. A. McDonough, *Am. J. Physiol.* **262,** C470 (1992).

ATPase activity of the lacrimal acinar cell is expressed in the basal–lateral plasma membrane. This hypothesis is supported by several different experimental observations. First, immunofluorescence analyses[15,17] indicate that lacrimal acinar cells contain very large intracellular pools of material reactive to antibodies against the $Na^+,K^+$-ATPase holoenzyme, $\beta_1$-subunit isoform and $\alpha_1$-subunit isoform. Second, when lacrimal gland fragments[6] and isolated lacrimal acini[18] are stimulated with a cholinergic agonist which accelerates $Na^+$ influx, $Na^+,K^+$-ATPase activity is removed from membrane samples spanning density windows IV and V and preferentially inserted into a membrane sample equilibrating in density window I. This result would suggest that a portion of the total $Na^+,K^+$-ATPase pump units of the acinar cell are expressed in an intracellular pool from which they can be rapidly mobilized and translocated to the basal–lateral membrane. Finally, preliminary experiments with the membrane surface labeling reagent sulfo-*N*-hydroxysuccinimidylbiotin[19] indicate that surface-expressed labeled groups rapidly equilibrate between a small surface-expressed pool and a 10- to 20-fold larger intracellular pool.

The existence of large intracellular pools of $Na^+,K^+$-ATPase and other plasma membrane constituents appears to be a common, but not a universal, phenomenon. As enumerated in Table III, the fractionation strategy described in this chapter has been applied to a variety of other epithelial cell types. In renal cortex, retinal pigment epithelium, and freshly harvested small intestinal mucosa, most (roughly, 75%) of the total $Na^+,K^+$-ATPase activities are associated with membrane samples that appear to consist largely of plasma membrane vesicle populations. Isolated enterocytes, in contrast, appear to contain a membrane sample which corresponds to sample d from exorbital lacrimal glands. This sample may include basal–lateral membrane constituents which are internalized as attachments between the enterocyte and extracellular matrix are disrupted.[20] The subcellular distribution of $Na^+,K^+$-ATPase in parotid gland acinar cells appears to be similar to that in lacrimal acinar cells. Madin–Darby canine kidney (MDCK) cells and porcine kidney proximal tubule (LLC-$PK_1$) cells both appear to have intracellular $Na^+,K^+$-ATPase pools which account for half their total $Na^+,K^+$-ATPase activities.

[17] K. K. Azuma, M. E. Bradley, R. L. Wood, A. A. McDonough, and A. K. Mircheff, *J. Cell Biol.* **111,** 190a (1990).

[18] S. C. Yiu, R. W. Lambert, P. J. Tortoriello, and A. K. Mircheff, *Invest. Ophthalmol. Visual Sci.* **32,** 2976 (1991).

[19] R. W. Lambert, R. L. Wood, and A. K. Mircheff, *Invest. Ophthalmol. Visual Sci.* **33s,** 1290 (1992).

[20] H. M. Amerongen, J. A. Mack, J. M. Wilson, and M. R. Neutra, *J. Cell Biol.* **109,** 2129 (1989).

## Acknowledgments

Work described in this chapter was supported by National Institutes of Health Grant RO1 EY 05801. M.E.B. and R.W.L. were recipients of Sigma Xi Grants-in-Aid of Research. The authors thank Drs. Kay K. Azuma, Charles B. Hensley, Robert W. Lambert, Carol A. Maves, Alicia A. McDonough, and Richard L. Wood for helpful comments during the course of this work.


# [43] Plasma and Internal Membranes from Cultured Mammalian Cells

*By* D. JAMES MORRÉ, TIMOTHY REUST, and DOROTHY M. MORRÉ

## Introduction

Decisive fractionation of tissue culture cells, especially the preparation of plasma membranes free of internal membranes and of internal membranes free of plasma membranes is often difficult by conventional gradient methods. A major contributory factor to this problem is the extensive cytoskeletal system that tends to result in poor dispersion of membranes. Here we describe a method for fractionation of mammalian tissue culture cells into nuclei, plasma membranes, and an internal membrane fraction each largely uncontaminated by the other.

## Methods

Attached cells are released by scraping and are collected by centrifugation for 6 min at 1000 *g* or 15 min at 120 *g*. Cells grown in suspension culture are collected by centrifugation.

Cell pellets are resuspended in 0.2 m*M* EDTA in 1 m*M* $NaHCO_3$ in an approximate ratio of 1 ml per $10^8$ cells. The resuspended cells are incubated on ice for 10 to 30 min to permit the cells to swell. Homogenization is with a Polytron homogenizer (Kinematica, Lucerne, Switzerland) for 30 to 40 sec at 10,500 rpm using a PT-PA 3012/23 probe (Kinematica), ST-10 probe (Kinematica), or equivalent. The cells are homogenized in 7- to 8-ml aliquots contained in a 10-ml glass mortar. To estimate breakage, the cells are monitored by light microscopy before and after homogenization. At least 90% cell breakage without breakage of nuclei is achieved routinely. If fewer than 90% of the cells are broken, the time of homogenization or the speed of the Polytron is increased. If all cells are broken as well as nuclei, the time of homogenization or the speed of the Polytron

TABLE I
PHASE SYSTEM USED IN ISOLATION OF PLASMA MEMBRANE FRACTION FROM CULTURED MAMMALIAN CELLS[a]

| Stock solution | Phase system: With sample | Phase system: Without sample |
|---|---|---|
| 20% (w/w) Dextran T500 | 5.28 g | 5.28 g |
| 40% (w/w) Poly(ethylene glycol) 3350 | 2.64 g | 2.64 g |
| 200 m*M* Potassium phosphate, pH 7.2 | 0.32 ml | 2.32 ml |
| Add distilled water to weight | 14.00 g | 16.00 g |
| Membranes, suspended in 0.2 *M* potassium phosphate, pH 7.2 | 2.00 g | — |

[a] The phase system without sample is used for reextraction of material (see text). The final composition of the system is 6.6% (w/w) dextran T500 (Pharmacia, Piscataway, NJ), 6.6% (w/w) poly(ethylene glycol) 3350 (Fisher Chemical Co., Chicago, IL), and 5.0 m*M* potassium phosphate, pH 7.2. From D. J. Morré and D. M. Morré, *BioTechniques* **7,** 946 (1989).

is decreased. The homogenate is centrifuged for 10 min at 175 *g* to remove unbroken cells and nuclei.

The supernatant solution from the first centrifugation is recentrifuged at $1.4 \times 10^6$ *g* min (e.g., 1 hr at 23,500 *g*) to prepare a microsome fraction enriched in plasma membranes. The supernatant solution is discarded, and the pellets are resuspended in 0.2 *M* potassium phosphate buffer in a ratio of approximately 1 ml per pellet from $5 \times 10^8$ cells. The resuspended membranes are then incorporated in the two-phase system described in Table I (see also Refs. 1 and 2).[3]

The contents of the two-phase system are mixed by 30–40 inversions of the tubes in the cold (4°). Care should be taken to maintain the temperature of the system during mixing. Partitioning is temperature dependent, so it is best to work consistently either from an ice bucket or in a cold room. It is essential that the contents of the tube do mix and that an unmixed cushion of dextran (Dx) does not remain along the wall of the tube. Phase separation occurs slowly at unit gravity, and the settling is usually accelerated by centrifugation (swinging bucket rotor) at 150–200 *g* for 5 (3–10) min in the cold. The upper phase, enriched in plasma membranes, is diluted 5-fold with 1 m*M* sodium bicarbonate and the membranes are collected by centrifugation (30,000 *g* for 30 min).

[1] C. Larsson, M. Sommarin, and S. Widell, this volume [44].
[2] P. Gierow, B. Jergil, and P.-Å. Albertsson, *Biochem. J.* **249,** 369 (1988).
[3] D. J. Morré and D. M. Morré, *BioTechniques* **7,** 946 (1989).

The purity of plasma membranes may be increased by a second partitioning step. Here, the upper phase enriched in plasma membranes from the first partitioning is reextracted with the fresh lower phase taken from a second tube, followed by 30–40 inversions and centrifugation to separate the phases. Simultaneously, a fresh upper phase may be used to reextract the original lower phase, followed by extraction of the lower phase that was used to extract the original upper phase. Finally, the two upper phases containing plasma membranes are combined and diluted with 1 m*M* bicarbonate and the membranes collected by centrifugation. The number of partitioning steps need not be limited to two. For cultured cells, however, increasing the number of partitions to more than two normally decreases yield without substantial improvement in purity.

The upper phase should be diluted 5-fold before collecting the plasma membrane by centrifugation (30,000 g for 30 min). To remove membranes from the Dx-containing lower phase, a 5-fold dilution is used followed by a wash in buffered isotonic sucrose to further reduce the polymer content.

## Comments

It is practical to prepare stock solutions of the phase system and store them frozen as aliquots. The stock solutions of Dx and poly(ethylene glycol) (PEG) are prepared on a weight basis, whereas sucrose, buffer, and salt stock solutions are prepared on a molar basis.[4]

The membranes in the lower phase obtained after two-phase partitioning steps provide plasma membrane-free membranes. After several washes to remove the Dx, the membranes may be further fractionated by preparative free-flow electrophoresis or sucrose gradient centrifugation to yield fractions enriched in mitochondria, endoplasmic reticulum, or Golgi apparatus.[5]

[4] D. E. Brooks and R. Norris-Jones, this volume [2].

[5] P. Navas, D. D. Nowack, and D. J. Morré, *Cancer Res.* **49**, 2146 (1989).

# [44] Isolation of Highly Purified Plant Plasma Membranes and Separation of Inside-Out and Right-Side-Out Vesicles

*By* CHRISTER LARSSON, MARIANNE SOMMARIN, and SUSANNE WIDELL

## Introduction

Partitioning in aqueous two-phase systems as a method for plant plasma membrane purification was introduced in 1981,[1] and it has since been successfully applied to a large number of species and tissues (for a compilation, see Ref. 2). The plasma membrane preparations obtained are of high purity (~95%) and consist mainly of sealed right-side-out (apoplastic side out) vesicles.[3,4] Some of these vesicles can be turned inside-out (cytoplasmic side out) by freezing and thawing, and sealed inside-out and right-side-out vesicles can subsequently be separated by repeating the phase partitioning step.[4] The preparations of inside-out and right-side-out vesicles (each ~80% pure) are ideal for studies on vectorial properties of the plasma membrane, such as transport, signal transduction mechanisms, and enzyme topology.

## Isolation of Plasma Membranes from a Microsomal Fraction

### *Preparation of the Microsomal Fraction*

Plant tissue (125 g of, e.g., oat roots or sugar beet leaves) is homogenized in 275 ml of 0.33 *M* sucrose, 50 m*M* 3-(*N*-morpholino)propanesulfonic acid (MOPS)–KOH, pH 7.5, 5 m*M* EDTA, 5 m*M* dithiothreitol (DTT), 5 m*M* ascorbate, 0.5 m*M* phenylmethylsulfonyl fluoride (PMSF), 0.2% (w/v) bovine serum albumin (BSA, protease free; Sigma, St. Louis, MO, A-3294), 0.2% casein (enzymatic hydrolyzate boiled for 10 min; Sigma, C-0626), and 0.6% insoluble poly(vinylpyrrolidone) (PVP), using a kitchen blender. The homogenate is filtered (240-$\mu$m nylon net), and plastids, mitochondria, etc., are pelleted by centrifugation at 10,000 *g* for 10 min. Microsomal membranes are pelleted from the supernatant

[1] S. Widell and C. Larsson, *Physiol. Plant.* **51,** 368 (1981).

[2] A. S. Sandelius and D. J. Morré, *in* "The Plant Plasma Membrane—Structure, Function and Molecular Biology" (C. Larsson and I. M. Møller, eds.), p. 44. Springer-Verlag, Berlin and Heidelberg, 1990.

[3] C. Larsson, P. Kjellbom, S. Widell, and T. Lundborg, *FEBS Lett.* **171,** 271 (1984).

[4] M. G. Palmgren, P. Askerlund, K. Fredrikson, S. Widell, M. Sommarin, and C. Larsson, *Plant Physiol.* **92,** 871 (1990).

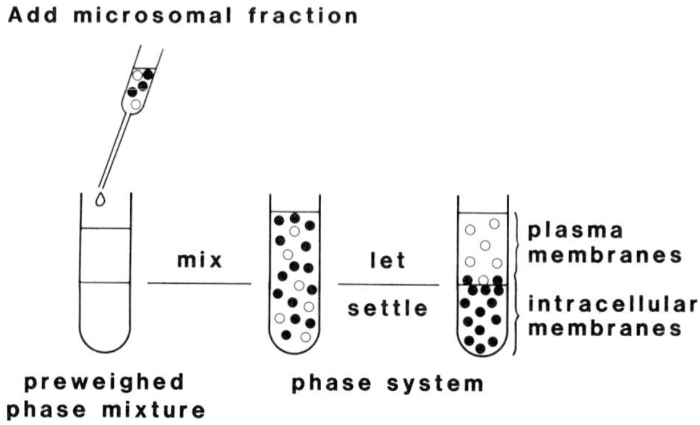


FIG. 1. Partitioning of membrane vesicles in a two-phase system. The microsomal fraction is added to a preweighed phase mixture to give a phase system of desired final composition (compare Table I). The phase system is mixed and left to settle (phase settling may be accelerated by centrifugation). Plasma membranes partition to the poly(ethylene glycol)-rich upper phase, whereas intracellular membranes partition to the interface and to the dextran-rich lower phase. [From C. Larsson, *in* "Modern Methods in Plant Analysis, New Series" (H. F. Linskens and J. F. Jackson, eds.), Vol. 1, p. 85. Springer-Verlag, Berlin and Heidelberg, 1985.]

suspension by centrifugation at 50,000 *g* for 30 min. The pellets ($\leq$100 mg of protein) are resuspended to approximately 10 ml in 0.33 *M* sucrose, 5 m*M* potassium phosphate, pH 7.8, 0–5 m*M* KCl, 1 m*M* DTT, and 0.1 m*M* EDTA and constitute the microsomal fraction.

### *Selection of Phase System*

When the procedure is optimized about 90% of the plasma membranes partition to the upper phase, whereas about 90% of the intracellular membranes partition to the interface plus lower phase (Fig. 1). Optimal conditions are found by first partitioning the microsomal fraction in a series of phase systems with increasing polymer concentrations.

Different phase mixtures (3.0 g) are prepared in identical tubes by weighing and pipetting from stock solutions of polymers and salts (for preparation of polymer stock solutions, see Ref. 5) such that addition of 1.0 ml of microsomal fraction to each tube gives a 4.0-g phase system with the desired final composition: 5.5–6.5% (w/w) each of dextran (Dx) T500 (Pharmacia, Uppsala, Sweden) and poly(ethylene glycol) (PEG) 3350 (Union Carbide, New York, NY) in 0.33 *M* sucrose, 5 m*M* potassium

[5] D. E. Brooks and R. Norris-Jones, this volume [2].

TABLE I
PHASE MIXTURES AND PHASE SYSTEMS USED IN PLASMA MEMBRANE PURIFICATION

| Component | Phase mixtures: 3.0 g | Phase mixtures: 27.00 g | Bulk phase system[a]: 300.0 g |
|---|---|---|---|
| 20% (w/w) Dextran T500[b] | 1.10–1.30 g | 11.16 | 93.0 g |
| 40% (w/w) Poly(ethylene glycol) 3350[b] | 0.55–0.65 g | 5.58 g | 46.5 g |
| Sucrose (solid) | 0.339 g | 3.05 g | 33.9 g |
| 0.2 *M* Potassium phosphate, pH 7.8 | 0.075 ml | 0.675 ml | 7.50 ml |
| 2 *M* KCl | 0–0.0075 ml | 0.068 ml | 0.45 ml |
| 0.5 *M* DTT in 50 m*M* EDTA[c] | 0.008 ml | 0.072 ml | $x$[d] |
| Add water to a final weight of | 3.00 g | 27.00 g | 300.0 g |
| Sample | 1.0 ml | 9.0 ml | — |
| Final weight of phase system | 4.0 g | 36.0 g | 300.0 g |

[a] Prepared 1 day before.
[b] Preparation of polymer stock solutions, see Ref. 5.
[c] DTT is added fresh from a concentrated stock solution in EDTA, pH ~ 7.
[d] DTT in EDTA is added to the "wash" phases during the phase separation to final concentrations of 1 m*M* DTT and 0.1 m*M* EDTA.

phosphate, pH 7.8, 0–5 m*M* KCl, 1 m*M* DTT, and 0.1 m*M* EDTA (Table I). Each tube is capped (Parafilm), and the contents are thoroughly mixed by around 20 inversions of the tube. Phase settling is achieved in 10–20 min at unit gravity but may be speeded by low speed centrifugation (1500 *g* for 3–5 min). Once a distinct interface has formed, about 90% of the upper phase is carefully removed with a Pasteur pipette (wide tip), without disturbing the interface, and added to a new tube. The upper phase and the rest of the phase system are diluted with a suitable medium to the same final volume (or, more conveniently, to the same final height in all tubes); samples are withdrawn and markers are assayed to determine the partitioning of different membrane vesicles. Alternatively, the membranes are pelleted and resuspended in a small volume before marker assays are performed. In the latter case, the lower phase must be diluted about 10-fold before pelleting owing to the high viscosity.

The partitioning of a particular marker is calculated as the percentage recovered in the upper phase relative to the total amount recovered in both phases. Based on the results shown in Fig. 2, a polymer concentration of 6.2–6.5% would be chosen since in that range most of the plasma membrane marker is in the upper phase and the markers for intracellular membranes are mainly in the lower phase plus interface.

Separation conditions are finally optimized by preparing a series of phase systems with increasing KCl concentrations (0–5 ml KCl) using the

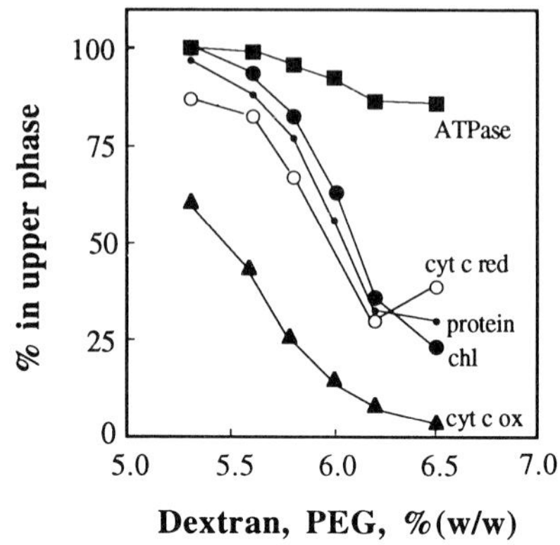


FIG. 2. Effects of increasing polymer concentrations on the partitioning of different membrane markers in a microsomal fraction from oat leaves. Markers used were cytochrome-*c* oxidase for the mitochondrial inner membrane, chlorophyll for the chloroplast thylakoid, antimycin A-resistant NADPH–cytochrome-*c* reductase for the endoplasmic reticulum (but see text), and total $K^+$,$Mg^{2+}$-ATPase for the plasma membrane. In addition to the polymers, the phase systems contained 0.33 *M* sucrose and 5 m*M* potassium phosphate, pH 7.8. Temperature, 4°. (From C. Larsson, S. Widell, and P. Kjellbom, this series, Vol. 148, p. 558.)

optimal polymer concentration from the first series. The effect of increasing $Cl^-$ concentration is similar and synergistic to that of increasing polymer concentration, that is, intracellular membranes are selectively drawn to the interface plus lower phase. Usually a KCl concentration of 2–5 m*M* is used in combination with 5 m*M* potassium phosphate, pH 7.8 (for a discussion of the effects of polymer concentrations and salts on partitioning, see Ref. 6). All procedures are run in the cold (4–5°, or on ice in a bucket), and once a temperature is chosen it should be maintained since the partitioning varies with temperature. The phase composition deemed most useful from the KCl series would be tested in the batch procedure outlined below.

## *Batch Procedure*

Plasma membranes are purified from a microsomal fraction (10,000–50,000 *g* pellet) of sugar beet leaves as follows. The microsomal fraction (9.0 ml, ~100 mg of protein) is added to a phase mixture (27.0 g) to give a phase system (36.0 g) of 6.5% (w/w) Dx T500, 6.5% (w/w) PEG 3350, 0.33 *M* sucrose, 5 m*M* potassium phosphate, pH 7.8, 5 m*M* KCl, 1 m*M* DTT, and 0.1 m*M* EDTA (4°) (Table I). We routinely start with

[6] H. Walter and C. Larsson, this volume [4].

0.5 to 1 kg of leaves and process four to eight 36-g systems in parallel using the three-step batch procedure outlined in Fig. 3. The final upper phases containing the plasma membranes are diluted at least 2-fold with a suitable medium, and the membranes are pelleted at 100,000 *g* for 1 hr. The yield is 15 to 18 mg of protein per 125 g of leaves, and the preparations are free of chlorophyll and are also otherwise of high purity as determined by standard marker assays (Table II). The plasma membranes (usually >90% right-side-out vesicles) are resuspended to 15 to 20 mg/ml protein in a suitable medium (without DTT in aliquots intended for oxidoreductase assays) and are used either fresh or stored in liquid $N_2$ until used.

The fresh upper and lower phases used in the batch procedure are obtained from a bulk phase system prepared in a separatory funnel (Table I). After reaching the desired temperature (4°), the phase system is well shaken and left standing for phase settling to occur. Upper and lower phases are collected separately and stored in the cold or frozen.

The yield and purity of the plasma membranes at each step of the procedure can be calculated, provided the partitioning of markers is known. Assuming that 90% of the plasma membranes and 10% of the intracellular membranes partition to the upper phase, and that 90% of the upper phase is transferred in each step, we calculate the data presented in Table III. There is only a slight decrease in the yield of plasma membranes at each step so long as the lower phases are reextracted with a second upper phase (giving $U_{1'}$, $U_{2'}$, and $U_{3'}$). In contrast, the degree of contamination (by intracellular membranes) decreases by one order of magnitude for each step. The calculated yield after three steps is, in each case, close to that obtained experimentally (compare Tables II and III): 83% of the plasma membrane marker vanadate-inhibited $K^+$, $Mg^{2+}$-ATPase was obtained in the plasma membrane fraction ($U_3 + U_{3'}$), which is identical to the calculated yield, and 0.4% of cytochrome-*c* oxidase and 0.0% of chlorophyll, markers for the main contaminants in leaf microsomal fractions (mitochondrial inner membrane and thylakoids, respectively), were obtained (compared to a calculated yield of 0.2%).

## Separation of Inside-Out and Right-Side-Out Vesicles

### *Formation of Inside-Out Vesicles from Right-Side-Out Vesicles*

The highly purified, right-side-out plasma membrane vesicles (obtained by the batch procedure) are frozen and thawed to produce a mixture of inside-out and right-side-out vesicles. Aliquots of 0.8 ml (15–20 mg protein) are repeatedly (2–4 times) frozen in liquid $N_2$ and thawed in water at 20°. The formation of sealed inside-out vesicles can be traced by monitoring

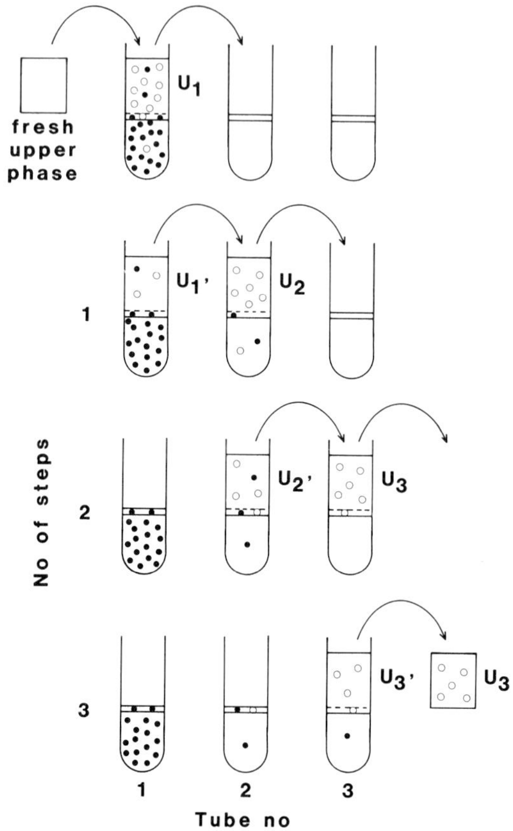


FIG. 3. Separation of plasma membranes (○) and intracellular membranes (●) by a batch procedure consisting of three steps (1, 2, 3). The microsomal fraction, suspended in the medium used in the phase system, is added to a preweighed phase mixture to give a phase system of desired final composition (Table I). The phase system is well mixed (by 20–30 inversions of the tube), and phase settling is speeded by centrifugation in a swinging-bucket rotor (5–10 min at 1500 *g*). Then, 90% of the upper phase ($U_1$) is removed (dashed line) and reextracted twice with fresh lower phase (tubes 2 and 3) to increase purity (giving $U_3$). To increase the yield of plasma membranes the lower phases may be reextracted, in sequence, with fresh upper phase (giving $U_{3'}$). The upper phases ($U_3$ and $U_{3'}$) are pooled and diluted, and the plasma membranes are collected by ultracentrifugation. In addition to fresh lower phase, tubes 2 and 3 contain a small volume of upper phase corresponding to the 10% left at each step. [From C. Larsson, *in* "Modern Methods in Plant Analysis, New Series" (H. F. Linskens and J. F. Jackson, eds.), Vol. 1, p. 85. Springer-Verlag, Berlin and Heidelberg, 1985.]

TABLE II
DISTRIBUTION OF PROTEIN, CHLOROPHYLL, AND TOTAL ACTIVITIES OF MARKER ENZYMES[a]

| Fraction | Protein (mg) | Chlorophyll (mg) | Activity (μmol/min) Glucan synthase | Vanadate-inhibited ATPase | Cytochrome-*c* oxidase |
|---|---|---|---|---|---|
| Plasma membranes[b] | 33<br>(17) | 0.0<br>(0.0) | 56<br>(230) | 15<br>(83) | 0.17<br>(0.4) |
| Intracellular membranes[c] | 130<br>(65) | 8.7<br>(77) | 0.9<br>(2) | 2.6<br>(14) | 34<br>(94) |
| Microsomal fraction | 200 | 11.3 | 24 | 18 | 36 |

[a] The plasma membrane, intracellular membrane, and microsomal fractions were obtained from 250 g of sugar beet leaves. Yield in percent is given in parentheses (microsomal fraction = 100%). Data from O. Gallet, R. Lemoine, C. Larsson, and S. Delrot, *Biochim. Biophys. Acta* **978,** 56 (1989).

[b] $U_3 + U_{3'}$ in Fig. 3.

[c] Lower phase plus interface of tube 1 after 3 steps (Fig. 3).

appropriate markers (see below), such as $H^+$ pumping and nonlatent ATPase activity (Fig. 4). The composition of the freeze–thaw medium is of importance: addition of 50 m*M* KCl to the basic medium (0.33 *M* sucrose, 5 m*M* potassium phosphate, pH 7.8, 1 m*M* DTT, and 0.1 m*M* EDTA) increases significantly the yield of inside-out vesicles. Usually about 40% of the vesicles are oriented inside-out after two to four freeze–thaw cycles.

TABLE III
CALCULATED YIELD OF PLASMA MEMBRANES AND INTRACELLULAR MEMBRANES IN BATCH PROCEDURE[a]

| Step | Phase | % Plasma membranes | (sum) | % Intracellular membranes | (sum) |
|---|---|---|---|---|---|
| 1 | $U_1$ | 81 | | 9 | |
| | $U_{1'}$ | 15 | (96) | 8 | (17) |
| 2 | $U_2$ | 66 | | 1 | |
| | $U_{2'}$ | 24 | (90) | 1 | (2) |
| 3 | $U_3$ | 53 | | 0.1 | |
| | $U_{3'}$ | 30 | (83) | 0.1 | (0.2) |

[a] The batch procedure is outlined in Fig. 3. It is assumed that 90% of the plasma membranes and 10% of the intracellular membranes partition to the upper phase at each step.

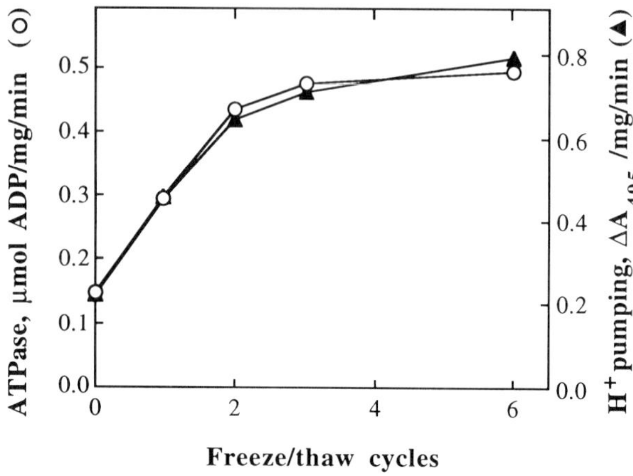


FIG. 4. Effect of the number of freeze–thaw cycles on the formation of inside-out plasma membrane vesicles (sugar beet leaf). $H^+$ pumping and nonlatent ATPase activity were used as markers for inside-out vesicles. [From M. G. Palmgren, P. Askerlund, K. Fredrikson, S. Widell, M. Sommarin, and C. Larsson, *Plant Physiol.* **92,** 871 (1990); reproduced with permission of the American Society of Plant Physiologists.]

*Separation of Vesicles of Opposite Sidedness*

The optimal composition of the phase system for separation of inside-out and right-side-out vesicles is determined by partitioning the freeze–thawed plasma membranes in a series of phase systems with increasing polymer concentrations (as above). From the experiment in Fig. 5, a concentration of 6.2% (w/w) of each polymer is chosen for the separation of inside-out and right-side-out plasma membrane vesicles.

The freeze–thawed plasma membranes are subfractionated using the countercurrent distribution procedure (CCD)[7] to produce one fraction enriched in inside-out vesicles, another fraction enriched in right-side-out vesicles, and two intermediate fractions (Fig. 6). The freeze–thawed plasma membranes (0.8 ml) are added to a phase mixture (7.2 g) to give a phase system (8.0 g) with a final composition of 6.2% (w/w) Dx T500, 6.2% (w/w) PEG 3350, 0.33 *M* sucrose, 5 m*M* potassium phosphate, pH 7.8, 5 m*M* KCl, 1 m*M* DTT, and 0.1 m*M* EDTA (4°). The phase system is shaken and centrifuged for about 5 min at 1500 *g* (swinging-bucket rotor) to speed phase separation. About 90% of the upper phase is removed without disturbing the interface and is added to a second tube containing fresh lower phase plus upper phase corresponding to the 10% upper phase not removed from tube 1. Fresh upper phase is added to tube 1, and

[7] H.-E. Åkerlund and P.-Å. Albertsson, this volume [8].

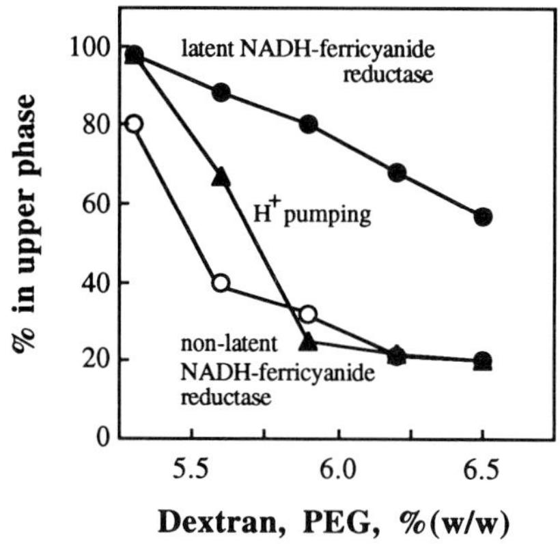


FIG. 5. Effect of polymer concentration on the partitioning of inside-out and right-side-out plasma membrane vesicles (sugar beet leaf). The phase system contained 0.33 *M* sucrose, 5 m*M* potassium phosphate, pH 7.8, 5 m*M* KCl, 0.1 m*M* EDTA, and equal concentrations of dextran T500 and poly(ethylene glycol) 3350 as indicated. Markers for inside-out vesicles were $H^+$ pumping and nonlatent NADH–ferricyanide reductase, and that for right-side-out vesicles was latent NADH–ferricyanide reductase. Assays were performed on aliquots withdrawn from the phases after separation and dilution. [From M. G. Palmgren, P. Askerlund, K. Fredrikson, S. Widell, M. Sommarin, and C. Larsson, *Plant Physiol.* **92,** 871 (1990); reproduced with permission of the American Society of Plant Physiologists.]

mixing and centrifugation are repeated. Then, 90% of the upper phase in tube 2 is transferred to a third tube containing fresh lower phase, the upper phase in tube 1 is transferred to tube 2, and fresh upper phase is added to tube 1. The procedure is repeated once more to produce four tubes containing complete phase systems and plasma membrane vesicles. The contents of each tube (fractions 1–4 in Fig. 6) are diluted about 10-fold with 0.33 *M* sucrose, 10 m*M* MOPS–KOH, pH 7.5, 0.1 m*M* EDTA, and 1 m*M* DTT, and the plasma membranes are pelleted at 100,000 *g* for 1 hr. The pellets are gently resuspended in the same medium (without DTT for oxidoreductase assays) and immediately used in the different assays. The procedure is scaled up either by processing two or more 8-g phase systems in parallel or by using larger phase systems.

Once the procedure is optimized one protein peak is found in fraction 1 and another in fraction 4, with minimum protein in fraction 2 (Fig. 7A). To determine the proportions of inside-out and right-side-out vesicles in fractions 1–4, the activity of markers for the cytoplasmic surface [$H^+$-ATPase, 1,3-$\beta$-glucan synthase, NADH-acceptor oxidoreductase (see below)] are assayed with and without detergent. Nonlatent activities (markers for inside-out vesicles) are enriched in fraction 1, whereas the activities in fractions 3 and 4 are highly latent, indicating a high proportion of right-

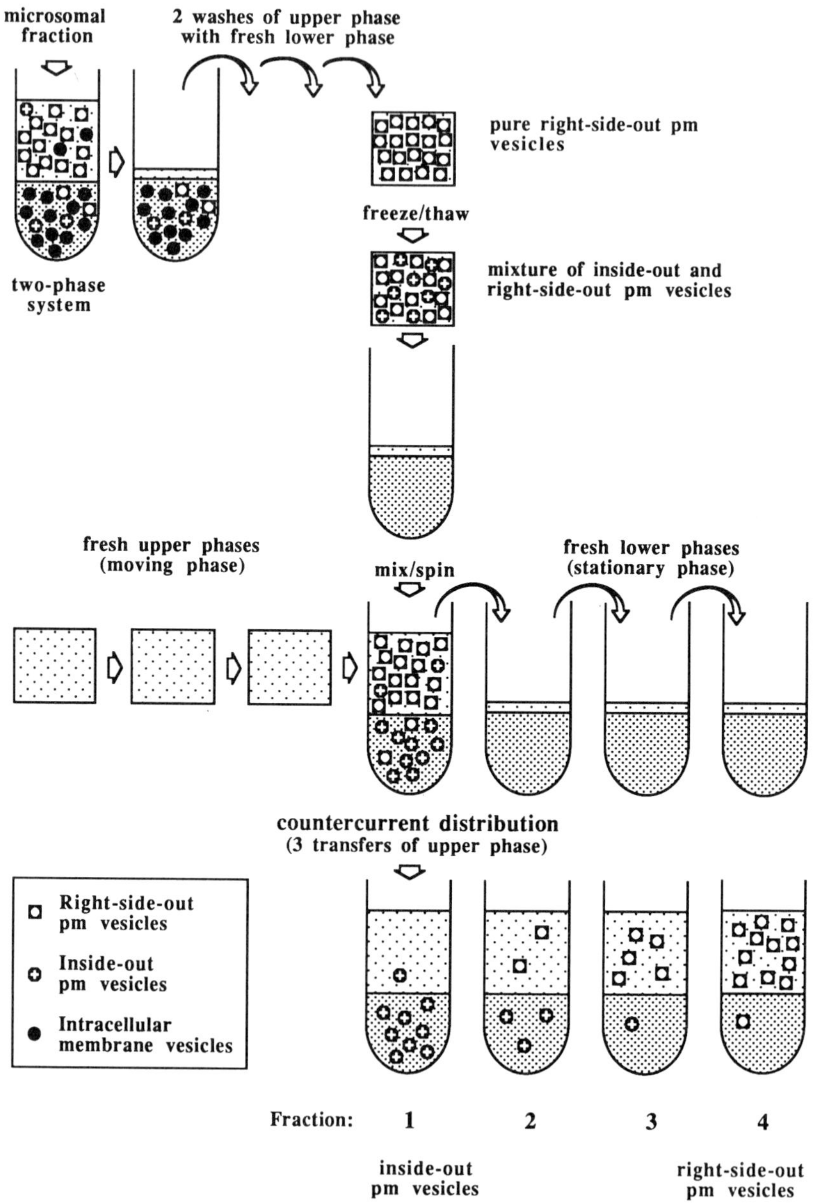


FIG. 6. Flow scheme for preparation of inside-out and right-side-out plasma membrane (pm) vesicles. Plasma membranes are purified from a microsomal fraction using a batch procedure. The highly purified, right-side-out plasma membrane vesicles are frozen and thawed to produce a mixture of right-side-out and inside-out vesicles. These are then separated by further phase partitioning steps using a countercurrent distribution procedure.

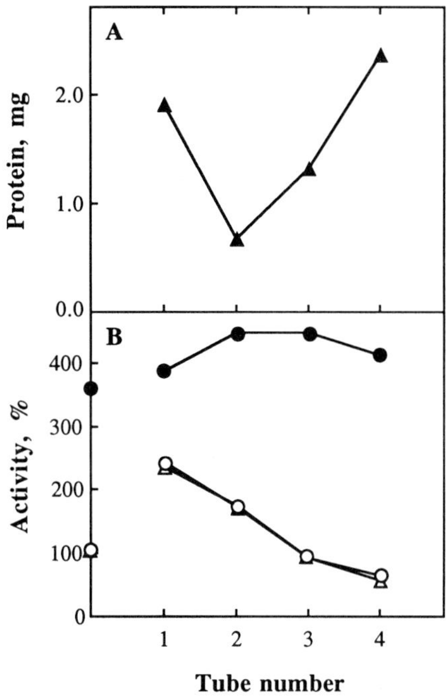


FIG. 7. (A) Distribution of protein among the four fractions obtained after countercurrent distribution of freeze–thawed plasma membranes (wheat root; see Fig. 6). (B) Relative specific activity of markers for the cytoplasmic surface of the plasma membrane in fractions 1–4. ATPase activity was assayed both in the absence (○) and presence (●) of 0.02% (w/v) Triton X-100. $H^+$ pumping (△) was assayed with the ΔpH probe acridine orange. Data on the *y* axis are for the unfractionated material. For marker enzymes, 100% activity was 0.10 μmol/mg protein/min for nonlatent ATPase, and 0.52 $\Delta A_{495}$/mg protein/min for $H^+$ pumping. [From M. Olbe and M. Sommarin, *Physiol. Plant.* **83,** 535 (1991).]

side-out vesicles. Intermediate values are found in fraction 2. $H^+$ pumping, a more definitive marker for sealed, inside-out vesicles, correlates with the nonlatent ATPase activity (Fig. 7B). According to the data for ATPase latency shown in Fig. 7B, fraction 1 would contain 62% inside-out vesicles and fraction 4 would contain 86% right-side-out vesicles. However, the detergent Triton X-100 used has a slightly stimulatory effect on the ATPase activity (besides permeabilizing the vesicles), which results in overestima-

---

Inside-out vesicles are enriched in fraction 1, and right-side-out vesicles are enriched in fraction 4 (see text for details). [From M. G. Palmgren, P. Askerlund, K. Fredrikson, S. Widell, M. Sommarin, and C. Larsson, *Plant Physiol.* **92,** 871 (1990); reproduced with permission of the American Society of Plant Physiologists.]

tion of latency and hence of the calculated percentage of right-side-out vesicles. Use of Brij 58 instead of Triton X-100 circumvents this problem. Another relative measure of the proportions of inside-out vesicles in fractions 1 and 4 is given by the ratio of $H^+$ pumping between the two fractions. The observed ratio of about 5 (Fig. 7B) fits well with the values of approximately 80% inside-out and 80% right-side-out vesicles obtained for fractions 1 and 4, respectively, when a larger number of markers was analyzed.[4]

In the CCD procedure, the inside-out and right-side-out vesicles are separated by repeating the phase partitioning step used to isolate the plasma membranes as right-side-out vesicles (Fig. 3). The same phase composition may often be used, but the phase composition should again be optimized (Fig. 5), because with some materials a much higher polymer and/or $Cl^-$ concentration is needed to obtain a good separation of inside-out and right-side-out vesicles.[8] As in the batch procedure, the right-side-out vesicles partition to the upper phase, whereas the inside-out plasma membrane vesicles behave like intracellular membranes and are therefore recovered in the lower phase plus interface (compare Figs. 3 and 6). The main difference between the two procedures is that in CCD the "wash" phases are saved to produce the inside-out fraction and the two intermediate fractions, whereas the corresponding fractions are simply discarded in the batch procedure. The CCD procedure is recommended for initial work since all material is saved and accounted for. This makes it simpler to optimize the separation.

## Markers

The plasma membrane preparations obtained by two-phase partitioning have a purity of greater than 90% as judged by biochemical and morphological markers. However, precise estimation of purity is at present not feasible because most membrane markers are not confined to a single membrane type and, even when they are, are not uniformly distributed in that membrane. In addition, there are no known markers for some membranes (e.g., the peroxisomal membrane). The most reliable estimations of fraction purity are currently those obtained by morphometric analysis of electron micrographs using silico- or phosphotungstic acid (STA or PTA) to stain the plasma membrane specifically.

Isolated plasma membranes typically appear as vesicles exposing either the apoplastic surface or the cytoplasmic surface to the bulk medium. The former may be termed right-side-out and the latter inside-out vesicles.

[8] M. Olbe and M. Sommarin, *Physiol. Plant.* **83,** 535 (1991).

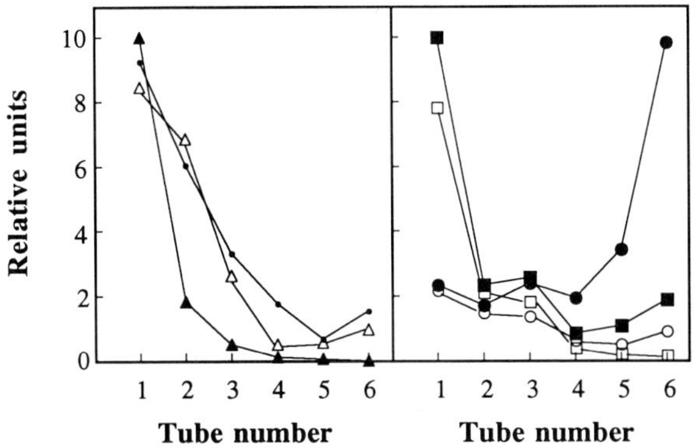


FIG. 8. Distribution of protein (●) and marker enzymes after countercurrent distribution of a microsomal fraction from wheat roots. The microsomal fraction was loaded in tube 1, and five transfers were made of the upper phase, leaving the interface plus lower phase as the stationary phase (compare Fig. 6), resulting in six tubes with complete phase systems. The countercurrent distribution was run by hand using a swinging-bucket rotor to speed phase separation. Markers: Cytochrome-*c* oxidase (▲), cytochrome-*c* reductase (△), vanadate-inhibited ATPase assayed in the presence (■) and absence (□) of Triton X-100, and 1,3-$\beta$-glucan synthase assayed in the presence (●) and absence (○) of digitonin. [From A. Bérczi, C. Larsson, S. Widell, and I. M. Møller, *Physiol. Plant.* **77,** 12 (1989).]

Because most marker assays involve water-soluble substrates that do not penetrate membranes, vesicle orientation is of major importance for the outcome of an assay (compare Figs. 7B and 8). Hence, the orientation of isolated vesicles can be established by assaying for enzyme latency.

Below, we outline the methodology to assay the most useful markers. We also include some procedures to determine plasma membrane vesicle orientation, i.e., the proportions of inside-out and right-side-out plasma membrane vesicles in a preparation.

### *Enzyme Latency*

The orientation of plasma membrane vesicles is usually determined by assaying a marker for the cytoplasmic surface (i.e., an enzyme with its active site located at the cytoplasmic surface, e.g., the $H^+$-ATPase) in the absence and presence of a suitable detergent. Nonlatent activity (activity observed in the absence of a detergent) indicates the presence of inside-out vesicles, whereas latent activity (the increase in activity on addition of detergent) indicates the presence of right-side-out vesicles. The percentage of inside-out vesicles is calculated as

$$\%\ \text{Inside-out vesicles} = \frac{\text{activity in the absence of detergent}}{\text{activity in the presence of detergent}} \times 100$$

and the percentage of right-side-out vesicles (percent latent activity) is calculated as:

$$\frac{\text{(activity in presence of detergent)} - \text{(activity in absence of detergent)}}{\text{activity in presence of detergent}} \times 100$$

The ideal detergent exposes all hidden active sites without otherwise inhibiting or stimulating the assayed enzyme activity. Few detergents meet these criteria completely, but the method is still useful provided that the above artifacts are accounted for. An obligatory test is to titrate the enzyme activity against increasing concentration of detergent in the assay medium. If an increase in activity is obtained which ends in a plateau above the critical micellar concentration (CMC), this is a good sign. The plateau value is taken as 100% activity, and in future determinations the assay is simply run with and without this detergent concentration, provided the same amount of protein and also otherwise identical conditions are used.

### *$H^+$-ATPase: Vanadate-Inhibited ATP Hydrolysis*

Vanadate-inhibited $H^+$-ATPase is the classic marker for the plant plasma membrane.[9] Because plant cells contain several different soluble and membrane-bound phosphohydrolases using MgATP as substrate, the plasma membrane $H^+$-ATPase can only be distinguished from other phosphohydrolases in crude fractions by detailed analysis. Thus, properties relatively specific to the plasma membrane-bound $H^+$-ATPase (EC 3.1.6.35) have been exploited to obtain a specific assay for the enzyme.[10]

The vanadate-inhibited ATPase activity is determined with 2–25 $\mu$g protein in 120 $\mu$l of 0.33 *M* sucrose, 50 m*M* 2-(*N*-morpholino)ethanesulfonic acid (MES)–KOH, pH 6.5, 3 m*M* ATP, 4 m*M* $MgSO_4$, 50 m*M* $KNO_3$, 1 m*M* sodium azide, 0.1 m*M* sodium molybdate, 0.1 m*M* EDTA, and 1 m*M* DTT. The assay is run in the presence and absence of 0.1 m*M* orthovanadate, freshly prepared and boiled in buffer prior to addition, and with and without detergent, namely, 0.015–0.05% (w/v) Triton X-100 or 0.05% Brij 58 (Note that lysophosphatidylcholine should not be used since it is a potent activator of the $H^+$-ATPase[11]; see Ref.

[9] T. K. Hodges, R. T. Leonard, C. E. Bracker, and T. W. Keenan, *Proc. Natl. Acad. Sci. U.S.A.* **69,** 3307 (1972).

[10] S. R. Gallagher and R. T. Leonard, *Plant Physiol.* **70,** 1335 (1982).

[11] M. G. Palmgren and M. Sommarin, *Plant Physiol.* **90,** 1009 (1989).

12 for a comparison of 42 different detergents.) The reaction is initiated by adding ATP and is run at 20–38° for 5–30 min. Blanks lacking $MgSO_4$ are run in parallel. The enzyme reaction is stopped, and released inorganic phosphate is determined by a method[13] where excess molybdate is complexed with arsenite to form a colorless product to prevent complexation with phosphate slowly liberated from the biological material and ATP. Indeed, it is not possible to measure ATPase activity with some materials (e.g., spruce) without using this modified procedure.

Following appropriate procedures, vanadate-inhibited ATPase is a useful marker with many materials, although there are exceptions (Fig. 8). With a few materials (e.g., oat root and leaf; Fig. 2) even total $K^+,Mg^{2+}$-ATPase activity is a useful marker.

### *$H^+$-ATPase: $H^+$ Pumping and ATP Hydrolysis*

At least theoretically, nonlatent activities of enzyme markers for the cytoplasmic surface may also arise from leaky right-side-out vesicles or membrane sheets. Thus a positive marker for sealed, inside-out vesicles is needed. ATP-dependent $H^+$-accumulation is such a marker.

Nonlatent ATPase activity and $H^+$ pumping can be measured simultaneously.[11,14] In this coupled assay the ATP hydrolyzed is regenerated, and phosphorylation of ADP is coupled to oxidation of NADH. Thus, formation of ADP is recorded as the absorbance decrease at 340 nm, and $H^+$ uptake into the vesicles is monitored as the absorbance decrease at 495 nm of the ΔpH probe acridine orange.

For the assay, 50–100 $\mu$g protein (in $\leq$80 $\mu$l) is added to 850 $\mu$l of medium (in a cuvette) to give the following final concentrations in the assay system (1 ml): 140 m*M* KCl, 10 m*M* MOPS–KOH, pH 7.0, 2 m*M* ATP, 1 m*M* EDTA, 1 m*M* DTT, 20 $\mu M$ acridine orange, and 1 mg/ml BSA (fatty acid free). Then 5 $\mu$l of valinomycin (0.1 mg/ml in ethanol) is added with vigorous mixing, followed by 42.5 $\mu$l substrate mixture (6 m*M* NADH, 24 m*M* phosphoenolpyruvate, and 10 m*M* MOPS–KOH, pH 7.0) and 8.5 $\mu$l enzyme mixture (3 mg/ml pyruvate kinase, 1.5 mg/ml lactate dehydrogenase, and 10 m*M* MOPS–KOH, pH 7.0; enzymes in glycerol rather than in ammonium sulfate should be used because $NH_4^+$ acts as ionophore). The reaction mixture is incubated at 20° for 5 min, then the baseline is recorded for 1 min; the reaction is started by addition of 15 $\mu$l of 0.27 *M* $MgCl_2$ (final concentration 4 m*M*). If a spectrophotometer which

[12] M. G. Palmgren, M. Sommarin, P. Ulvskov, and C. Larsson, *Biochim. Biophys. Acta* **1021,** 133 (1990).

[13] E. S. Baginski, P. P. Foa, and B. Zak, *Clin. Chim. Acta* **15,** 155 (1967).

[14] M. G. Palmgren, *Plant Physiol.* **94,** 882 (1990).

can shift automatically between two wavelengths is available, it is set to measure the absorbance at 495 and 340 nm in cycles of 10 sec. Alternatively, the initial rate of $H^+$ pumping is first recorded for 1 min at 495 nm, and ATPase activity then recorded at 340 nm.

This assay permits ATPase measurements even with fractions in media containing phosphate. It cannot be used in the presence of vanadate, however, and it is therefore not so useful with crude fractions. BSA binds free fatty acids, which can act as uncouplers in the reaction.[12] For measurements of $H^+$ pumping or ATPase activity alone, appropriate deletions in the assay medium are made and continuous recording used. To measure ATP latency, Brij 58 [0.05% (w/v), final concentration] is added during the course of a continuous recording.

### *1,3-β-Glucan Synthase*

The most reliable enzyme marker for the plant plasma membrane is 1,3-β-glucan synthase (EC 2.4.1.34). It catalyzes the incorporation of glucose units from UDPglucose (UDPG) into 1,3-β-polyglucan. The assay[4] is run in a final volume of 100 μl containing 0.1–10 μg protein, 50 m*M* *N*-(2-hydroxyethyl)piperazine-*N*′-2-ethanesulfonic acid (HEPES)–KOH, pH 7.25, 0.33 *M* sucrose, 2 m*M* spermine, 20 m*M* cellobiose, 0.2 m*M* $CaCl_2$, and 2 m*M* DTT, with or without 0.006–0.015% (w/v) digitonin to determine enzyme latency (use disposable plastic test tubes). The reaction is started by addition of UDP[$^3$H]glucose (20 GBq/mol) to a final concentration of 2 m*M*. The reaction is run for 15–30 min at 25°) and is terminated by immersing the test tubes in boiling water. The samples are transferred to paper filters (25-mm circles, Whatman, Clifton, NJ, 3MM), dried, and then washed 2 times for 1 hr each time in 0.35 *M* ammonium acetate, pH 3.6, and 30% (v/v) ethanol (60 filters per 0.5 liter). The filter background is determined by running blanks without protein. A radioactivity standard is prepared by adding 10 μl of [$^3$H]UDPG (half the amount used in the assay) to a wetted glass filter (Whatman GF/F). Paper filters cannot be used for the standard, since the low molecular weight UDPG will be absorbed into the paper filter and the radioactivity quenched. The filters are dried and the radioactivity measured by liquid scintillation counting.

### *Silicotungstic/Phosphotungstic Acid Staining*

STA or PTA at low pH can be used to stain specifically the plasma membrane in thin sections for electron microscopy.[15] Thus, the actual

[15] J. C. Roland, *in* "Electron Microscopy and Cytochemistry of Plant Cells" (J. L. Hall, ed.), p. 1. Elsevier, Amsterdam, 1978.

percentage of plasma membrane in a fraction can be determined by counting stained and unstained membranes. This is in contrast to enzyme markers, which only give relative values of enrichment. We use the following modified procedure. Thin sections (Epon) of glutardialdehyde- and $OsO_4$-fixed material are collected on nickel grids and destained by placing a droplet of 1% (w/v) periodate, pH 3.0, on the grid and leaving it there for 15 min at 20°. The grid is rinsed for 5 min with distilled water: first a droplet, then a stream. The grid is then incubated (droplet) with 1% (w/v) STA, pH 3.0, for 10 min at 37°, and rinsed with 1 m*M* HCl (droplet, then stream) for 10 min. The specificity of the stain should be checked on sections of whole tissue. If the staining is insufficient or if nonspecific staining is obtained, the time and/or temperature for the STA (PTA) step should be varied.

### *NADH–Acceptor Oxidoreductase*

A rapid and simple assay for vesicle orientation is to measure NADH–acceptor oxidoreductase in the absence and presence of detergent. NADH–acceptor oxidoreductase is assayed with either ferricyanide or cytochrome-*c* as acceptor.[16]

NADH–ferricyanide reductase activity is measured as $\Delta(A_{420} - A_{500})$ using a spectrophotometer operated in the dual-beam mode. The assay is run at 25° in 1 ml of 0.33 *M* sucrose, 25 m*M* HEPES–KOH, pH 7.3, 0.25 m*M* NADH, 0.2 m*M* $K_3[Fe(CN)_6]$, and 40 $\mu$g protein, with or without 0.025% (w/v) Triton X-100. The reaction is initiated by addition of NADH. Correction is made for nonenzymatic reduction of ferricyanide.

NADH–cytochrome-*c* reductase activity is measured similarly using 40 $\mu M$ cytochrome-*c* as acceptor instead of 0.2 m*M* $K_3[Fe(CN)_6]$, and with 0.4 $\mu M$ antimycin A and 1 m*M* KCN present in the assay medium. The activity is recorded as $\Delta(A_{550} - A_{600})$, and is determined with and without 0.015% (w/v) Triton X-100 (the reaction is strongly inhibited by higher concentrations of Triton X-100). The extinction coefficients are 1 and 19 $mM^{-1}$ $cm^{-1}$ for ferricyanide and cytochrome-*c*, respectively.

Cytochrome-*c* reductase is often used as a marker for the endoplasmic reticulum, but since it is also a plasma membrane activity[16] it cannot be used to determine contamination by endoplasmic reticulum in plasma membrane preparations. Assayed in the absence of detergent it is a useful marker for the endoplasmic reticulum in crude fractions, however, since

[16] P. Askerlund, C. Larsson, and S. Widell, *FEBS Lett.* **239,** 23 (1988).

the plasma membrane-bound activity is located at the cytoplasmic surface and therefore mainly latent in crude fractions.

## Practical Advice

To begin the study of phase partitioning, choose some green material, such as leaves of spinach, oat, maize, tobacco, or sugar beet. This will enable you to follow the partitioning of the intracellular membranes by eye, and it will give you the necessary experience to attack more difficult material.

If the separation of interest generates data on marker distribution that are contradictory or otherwise difficult to interpret, it is generally advantageous to use the CCD procedure (Fig. 6) instead of the batch procedure (Fig. 3) for the separation of the membranes in the microsomal fraction as well. In the CCD procedure no phases are discarded. Thus, a complete picture is obtained of the distribution of all membrane markers assayed.

This is illustrated in Fig. 8 for a wheat root microsomal fraction. Most of the protein is found in tubes 1–4, which means it partitions to the interface plus lower phase. This protein peak corresponds to intracellular membranes as confirmed by the distribution of cytochrome-*c* oxidase (mitochondrial inner membranes) and by the distribution of most of the cytochrome-*c* reductase (endoplasmic reticulum). The small protein peak in tube 6, corresponding to material partitioning in the upper phase, is due to plasma membranes, as confirmed by the distribution of 1,3-$\beta$-glucan synthase assayed in the presence of detergent. Note that the distribution of glucan synthase assayed without detergent present gives the erroneous impression that most of the plasma membranes are in tubes 1–3; however, this activity is probably due to inside-out plasma membrane vesicles and Golgi. Furthermore, vanadate-inhibited ATPase activity is not a useful marker for this material since most of the activity is recovered in tube 1 and is thus not located in the plasma membrane.[17] Note also that the small peak of cytochrome-*c* reductase in tube 6, arising from plasma membrane-bound activity,[16] shows that cytochrome-*c* reductase cannot be used to determine contamination by endoplasmic reticulum in plasma membrane preparations. The wheat root data in Fig. 8 emphasize the need for using more than one marker for the plasma membrane and the importance of performing assays in

[17] A. Bérczi, C. Larsson, S. Widell, and I. M. Møller, *Physiol. Plant.* **77,** 12 (1989).

the presence and absence of detergent. (For a critical discussion of markers, see Ref. 18).

Acknowledgments

Work in the authors' laboratories was supported by the Swedish Natural Science Research Council and the Swedish Council for Forestry and Agricultural Research.

[18] S. Widell and C. Larsson, *in* "The Plant Plasma Membrane—Structure, Function and Molecular Biology" (C. Larsson and I. M. Møller, eds.), p. 16. Springer-Verlag, Berlin and Heidelberg, 1990.

# [45] Fractionation of Thylakoid Membrane

*By* PER-ÅKE ALBERTSSON, EVA ANDREASSON, HREINN STEFÁNSSON, and LOUIE WOLLENBERGER

## Introduction

The structure of the photosynthetic membrane from plants, the thylakoid, is complex because it carries out diverse functions under varying conditions of light and temperature. The thylakoid membrane consists essentially of two compartments: the grana and the stroma lamellae interconnecting the grana. One can also distinguish between appressed and nonappressed (stroma exposed) membranes and between different membrane domains such as the partitions (appressed), the margins of both grana and stroma lamellae, the grana end membranes (stroma exposed), and the zone at the border between the grana and the stroma lamellae. The different domains are schematically illustrated in Fig. 1. To separate these different domains of the thylakoid membrane so as to understand its structure and function is of great importance. Here we describe how fragmentation combined with centrifugation and aqueous two-phase partitioning can be used to isolate different parts of the thylakoid membrane. The fragments are in membrane vesicular form (no detergents are used) and maintain photochemical activity.

## Fractionation Strategy

Leaves are homogenized in a sucrose medium, and chloroplasts, more or less intact, are prepared by differential centrifugation. Thylakoids are obtained after lysis of chloroplasts in a solution of 5 m$M$ $MgCl_2$. The

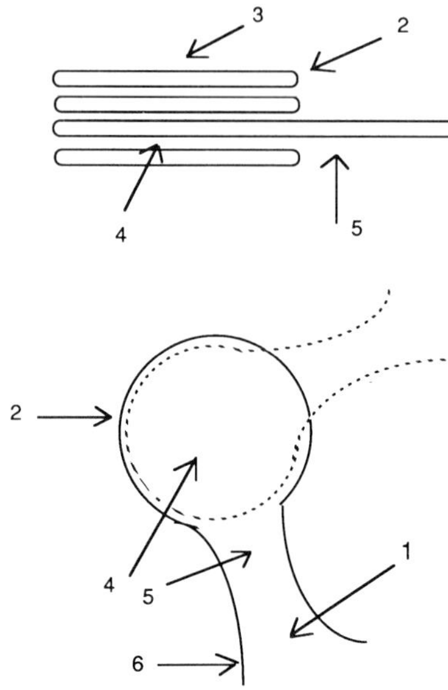


FIG. 1. Domains of the thylakoid membrane. (Upper) Cross section; (lower) top view. 1, Flat stroma lamellae; 2, margin of grana; 3, end membrane; 4, partition region; 5, zone between grana and stroma lamellae; 6, margin of stroma lamellae. [From P.-Å. Albertsson, E. Andreasson, and P. Svensson, *FEBS Lett.* **273,** 36 (1990).]

thylakoids are washed, suspended in a medium containing 1 m*M* $MgCl_2$, which preserves the stacking of the grana, and fragmented by sonication. For separation of grana and stroma lamellae vesicles countercurrent distribution (Fig. 2a) or batchwise extraction (Fig. 2b) may be chosen. Countercurrent distribution (CCD) yields two well-separated peaks: $\alpha$, consisting of grana vesicles, and $\beta$, consisting of stroma lamellae vesicles (Fig. 2a). Batchwise extraction (three steps) results in a bottom phase, B3, containing grana vesicles and a top phase, T3, containing stroma lamellae vesicles (Fig. 2b). Countercurrent distribution is the method of choice for analysis of the heterogeneity of a vesicle population and for the quantitative estimation of different vesicle subpopulations. The batch procedure is a preparative method and requires less time.

The grana vesicles or the stroma lamellae vesicles can be further fragmented by additional sonication to yield subgrana vesicles or substroma lamellae vesicles. Countercurrent distribution (analysis, Fig. 2a) or batch extraction (preparative, Fig. 2b) can again be applied.

The different procedures outlined in the scheme can be varied and combined at will. Batch procedures for isolating grana vesicles, stroma lamellae vesicles, margin vesicles, and photosystem II vesicles, the latter originating from the grana partitions, are given below. A few fractionations by use of countercurrent distributions are also shown. The experimental procedure for these is described in the original literature.[1,2]

## Selection of Phase System

Cell organelles and membrane vesicles partition between the bulk phases and the interface. For practical purposes the particles in the top phase are collected as one fraction while those in the bottom phase and at the interface are collected together as another (named bottom phase or B). The quantity of particles in each fraction is expressed as a percentage of total particles added to the phase system.

The distribution of cell organelles and membrane vesicles depends on several factors, such as type and molecular weight of polymer, polymer concentration, ionic composition, and temperature.[3] For the dextran–poly(ethylene glycol) (PEG) system the following hold. (1) An increase in molecular weight of PEG will decrease the percentage of material partitioning into the top phase. An increase in the molecular weight of dextran will decrease the percentage of material partitioning into the bottom phase. (2) An increase in polymer concentration decreases the amount of material in the top phase fraction and increases the amount in the bottom phase fraction owing to the increased adsorption of particles to the interface. (3) Different ions have different chemical affinity for the two phases, and, as a result, an electrical potential is created between the two phases when a salt is included in the phase system. Because cell particles carry a net negative charge their partitioning depends strongly on the ionic composition of the phase system. For example, for the dextran T500–PEG 3350 system, a sodium phosphate buffer at pH 7–8 will make the bottom phase negative relative to the top phase and tend to push the particles into the top phase. In contrast, sodium chloride will make the top phase negative relative to the bottom phase (as revealed by cross-partition[4,5]) and tend to push the particles into the bottom phase.[3–5] Generally, $Li^+$ is more effective

[1] E. Andreasson, P. Svensson, C. Weibull, and P.-Å. Albertsson, *Biochim. Biophys. Acta* **936,** 339 (1988).

[2] P. Svensson and P.-Å. Albertsson, *Photosynth. Res.* **20,** 249 (1989).

[3] P.-Å. Albertsson, "Partition of Cell Particles and Macromolecules." Wiley(Interscience), New York, 1986.

[4] H. Westrin, V. P. Shanbhag, and P.-Å. Albertsson, *Biochim. Biophys. Acta* **732,** 83 (1983).

[5] H. Walter and D. Forciniti, this volume [20].

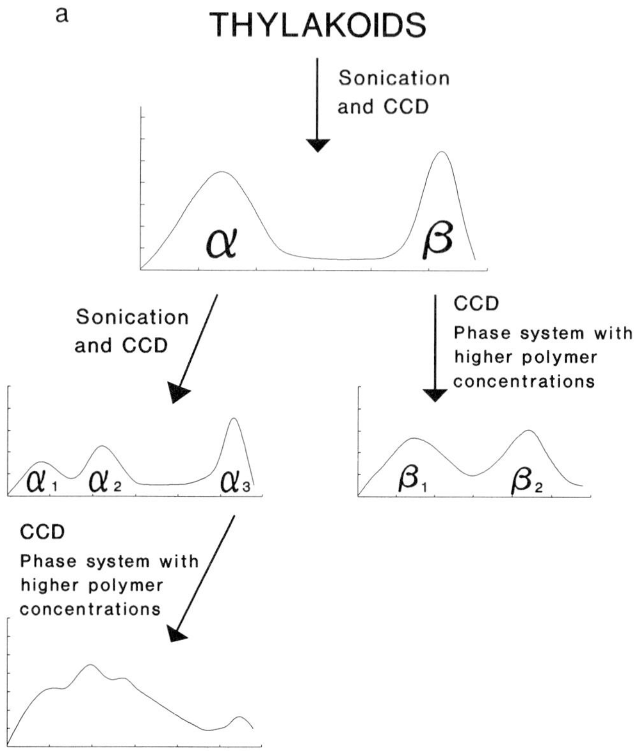


FIG. 2. (a) Scheme for fragmentation of thylakoids by sonication and separation of membrane vesicles by countercurrent distribution. (b) Scheme for fragmentation of thylakoids by sonication and separation of membrane vesicles by batch partitioning.

than $Na^+$ which is more effective than $K^+$ in pushing particles into the top phase. For negative ions, $HPO_4^{2-}$ $SO_4^{2-}$, and citrate are more effective than $Cl^-$ or $H_2PO_4^-$ in pushing particles into the top phase. The effect of the different salts is additive.[3] (4) An increase in temperature has an opposite effect to an increase in polymer concentration, that is, it will increase the percentage of material in the top phase.

For selection of a suitable system one can therefore vary the parameters mentioned above. The phase system composition described below was selected through systematic study of the indicated factors. Additional considerations for the preservation of biological activity dictated the temperature (4°), which must be strictly adhered to, and the concentration of ions and sucrose.

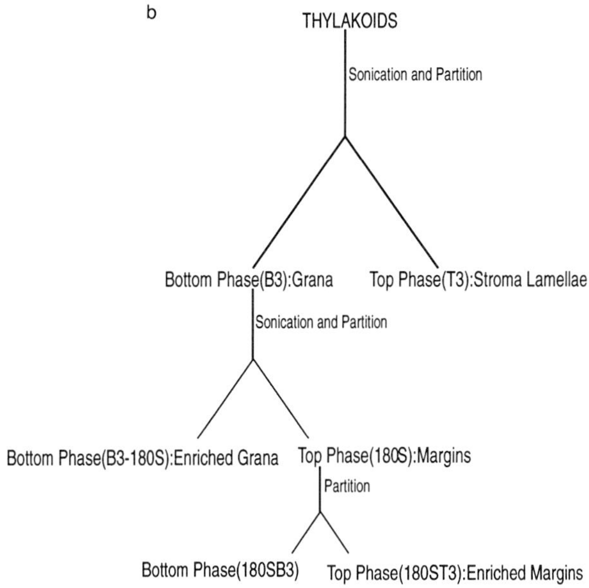


FIG. 2. (*continued*)

## Countercurrent Distribution of Sonicated Thylakoids

### *Reagents*

For reagents, see "Preparation of Different Vesicle Fractions from Thylakoid Membrane by Sonication and Batch Extraction."

### *Procedure*

*Chloroplast Isolation and Sonication.* See section on "Preparation of Different Vesicle Fractions from Thylakoid Membrane by Sonication and Batch Extraction."

*Countercurrent Distribution*

1. A countercurrent distribution apparatus in which the phase separation is speeded by centrifugation is used.[6] It contains 60 cavities, numbered 0–59.
2. Equal amounts, 0.90 ml, of pure top phase and bottom phase are added to each of the cavities 5–59.

[6] H.-E. Åkerlund and P.-Å. Albertsson, this volume [8].

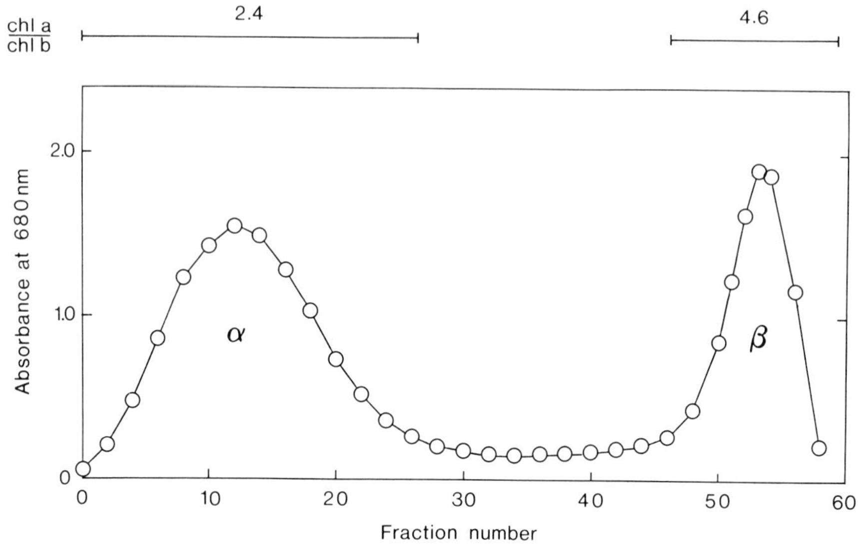


FIG. 3. Countercurrent distribution of a sonicate of stacked thylakoids from spinach. The phase system contained 5.7% (w/w) dextran T500, 5.7% (w/w) PEG 3350, 20 m*M* sucrose, 10 m*M* sodium phosphate buffer, pH 7.4, and 5 m*M* NaCl. The vesicles of the α peak originate in the grana and the vseicles in the β peak in the stroma lamellae. [From E. Andreasson, P. Svensson, C. Weibull, and P.-Å. Albertsson, *Biochim. Biophys. Acta* **936,** 339 (1988).]

3. Each of the cavities 0–4 is loaded with the sample system (1.80 ml).
4. Each cycle in the countercurrent distribution comprises 30 sec of mixing and 90 sec of centrifugation, including one transfer.
5. After 53 transfers the fractions are diluted with diluting medium in order to break the phase systems.
6. The diluted fractions are collected, and the relative chlorophyll content for of fraction is determined by measuring the absorbance at 680 nm.
7. Fractions 0–26 and 46–59 are pooled, diluted 1 : 1 with diluting medium, pelleted at 100,000 *g* for 90 min, and resuspended in the same medium.

Countercurrent distribution of sonicated, stacked thylakoids results in two well-separated peaks,[1] α and β (Fig. 3). The α vesicles, originating in the grana, are enriched in photosystem II (PSII; PSIIα) but also contain phytosystem I (PSI), which is in the form of PSIα. The α vesicles are inside-out; that is, they expose the lumen side to the medium. The β vesicles originate in the stroma lamellae, are enriched in PSI (PSIβ), depleted in PSII, which is in the form of PSIIβ, and are right-side-out.

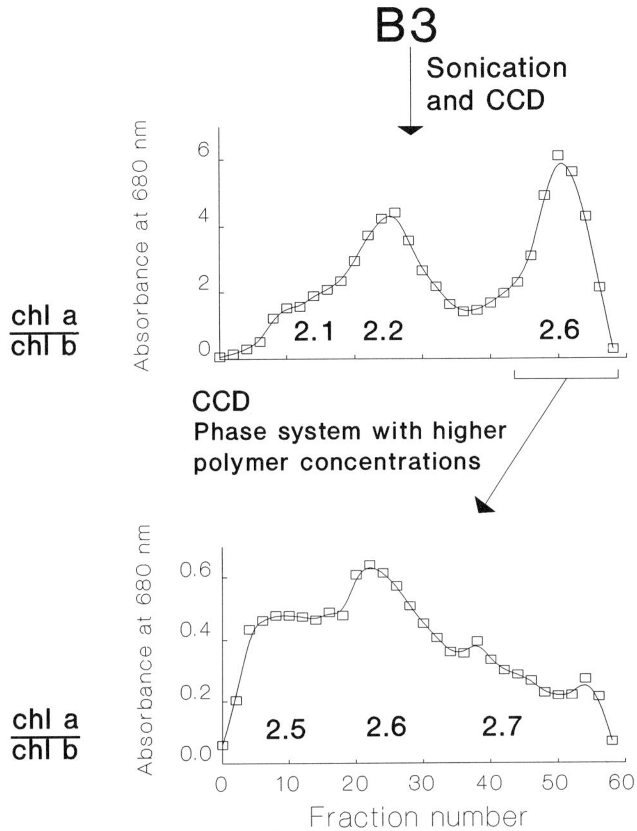


FIG. 4. Countercurrent distribution (CCD) of sonicated grana vesicles (B3 or $\alpha$ vesicles). The upper diagram shows separation into at least three different populations [phase system: 5.7% (w/w) dextran T500, 5.7% (w/w) PEG 3350]. The two populations to the left (Chl $a/b$ = 2.1 and 2.2) originate in the center of the grana partitions and are essentially composed of photosystem II and cytochrome $b/f$ complex. The peak to the right (Chl $a/b$ = 2.6) represents a population which originates from the periphery of the grana (margins) and is enriched in photosystem I. The margin vesicles can be further separated into different subpopulations by a second countercurrent distribution (lower diagram) with a phase system having higher polymer concentrations [6.0% (w/w) dextran T500, 6.0% (w/w) PEG 3350]. Salt and sucrose concentrations in both phase systems were 20 m*M* sucrose, 10 m*M* sodium phosphate buffer, pH 7.4, and 5 m*M* NaCl.

When the grana vesicles ($\alpha$ or B3) are further sonicated and subjected to countercurrent distribution, results depicted in Fig. 4 are obtained. Two or three separate peaks are apparent that represent vesicles of different origin and composition. The peak to the left has the lowest chlorophyll

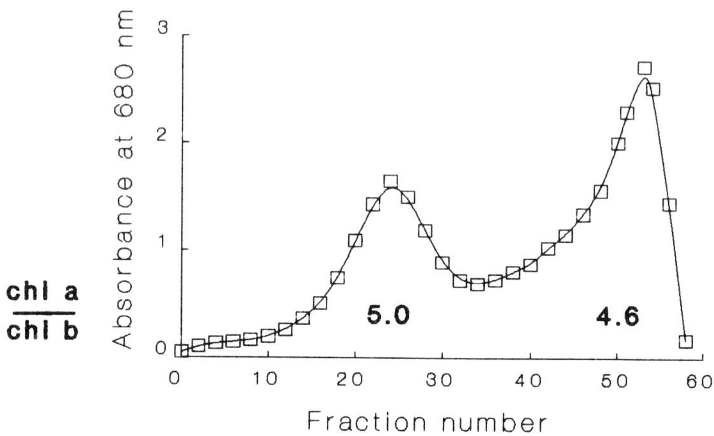


FIG. 5. Countercurrent distribution of stroma lamellae vesicles (T3 or $\beta$ vesicles). The phase system contained 6.1% (w/w) dextran T500, 6.1% (w/w) PEG 3350, 20 m*M* sucrose, 10 m*M* sodium phosphate buffer, pH 7.4, and 5 m*M* NaCl.

*a*/*b* ratio, the highest PSII activity, and the lowest PSI content. These vesicles originate in the core of the stacked grana. The peak to the right has a higher chlorophyll *a*/*b* ratio, relatively more PSI, and originates most probably from the grana margins. The middle peak has an intermediate composition. The grana margin vesicles can be further separated into subpopulations by a second countercurrent distribution using a phase system with higher polymer concentration (Fig. 4).

An experiment with stroma lamellae vesicles ($\beta$ or T3) is shown in Fig. 5. The countercurrent distribution diagram reveals considerable heterogeneity among the vesicles with respect to their partitioning (surface) properties, whereas the chlorophyll *a*/*b* ratio differs little among the fractions.

## Preparation of Different Vesicle Fractions from Thylakoid Membrane by Sonication and Batch Extraction

### *Grana and Stroma Lamellae Vesicles*

#### *Reagents*

Preparation medium: 50 m*M* sodium phosphate buffer (pH 7.4), 5 m*M* $MgCl_2$, 300 m*M* sucrose

5 m*M* $MgCl_2$

Washing medium I: 10 m$M$ tricine (pH 7.4), 5 m$M$ $MgCl_2$, 300 m$M$ sucrose

Washing medium II: 10 m$M$ sodium phosphate buffer, pH 7.4, 5 m$M$ NaCl, 1 m$M$ $MgCl_2$, 100 m$M$ sucrose

Polymer mixture: 3.32 g of 20% (w/w) dextran T500 (Pharmacia, Uppsala, Sweden), 1.66 g of 40% PEG 3350 (Union Carbide, New York, NY), 0.53 g of 0.2 $M$ sodium phosphate buffer, pH 7.4, 0.30 g of 0.1 $M$ NaCl, 1.07 g of 10 m$M$ $MgCl_2$, 1.33 g of 0.1 $M$ sucrose, and 2.45 g of distilled water

Pure top and bottom phases from the aqueous two-phase system containing 5.7% (w/w) dextran T500, 5.7% (w/w) PEG 3350, 10 m$M$ sodium phosphate buffer, pH 7.4, 5 m$M$ NaCl, and 20 m$M$ sucrose are prepared by weighing out 57 g of 20% (w/w) dextran T500, 28.5 g of 40% (w/w) PEG 3350, 10 g of 0.2 $M$ sodium phosphate buffer, pH 7.4, 10 g 0.1 $M$ NaCl, 4 g of 1 $M$ sucrose, and 90.5 g of distilled water. The components are thoroughly mixed at 4°, and the phase system so obtained is allowed to settle overnight at the same temperature. The top and bottom phases are subsequently separated.

Diluting medium: 10 m$M$ sodium phosphate buffer, pH 7.4, 5 m$M$ NaCl, 100 m$M$ sucrose

*Chloroplast Isolation*

1. Spinach leaves (30 g) are homogenized for 5 sec at maximum speed in a blender containing 100 ml of preparation medium.
2. The slurry is filtered through four layers of nylon mesh (25 $\mu$m) and centrifuged for 2 min at 1000 $g$.
3. The pellets are resuspended in preparation medium and centrifuged at 2000 $g$ for 5 min.
4. The chloroplasts are osmotically broken in 5 m$M$ $MgCl_2$ followed by centrifugation at 2000 $g$ for 5 min.
5. The thylakoids are washed twice in washing medium I and centrifuged as before.
6. The membranes are washed once in washing medium II, centrifuged as before, and resuspended in the same medium to give a chlorophyll concentration of about 4 mg/ml.

*Sonication*

1. An aliquot of 1 ml of the thylakoid suspension is added to the polymer mixture (see reagents list above) so as to give a phase system containing 5.7% (w/w) dextran T500, 5.7% (w/w) PEG 3350, 10 m$M$ so-

dium phosphate buffer, pH 7.4, 3 m$M$ NaCl, 1 m$M$ $MgCl_2$, and 20 m$M$ sucrose.

2. About 45 min after suspension in washing medium II the phase system containing thylakoids is sonicated in a Vibra-cell ultrasonic processor, Model VC500 (Sonics and Materials Inc., Danbury, CT) equipped with a 1/2-inch horn (intensity output setting 7, 20% duty pulses) for 6 pulses of 30 sec each with 60-sec resting intervals with continuous cooling.

3. To the sonicate is added 1.43 g of pure bottom phase to obtain equal volumes of the phases in the so-called sample system.

### *Batch Extraction*

1. Pure bottom and top phase (5 ml of each phase) are added to the sample system.

2. The phase system is mixed thoroughly and separated at 4° using low-speed centrifugation (1000 $g$ for 3 min).

3. The top phase (T1) and the bottom phase (B1) are collected and reeextracted twice with 10 ml of pure bottom and pure top phase, respectively, to obtain the T3 ($\beta$ vesicles) and B3 fractions ($\alpha$ vesicles), respectively.[1]

4. The vesicles are removed from the polymers by 3-fold dilution with the diluting medium and pelleting at 100,000 $g$ for 90 min, and they are resuspended in the same medium.

Data relating to the T3 and B3 vesicles are given in Tables I, II, and III.

## *Photosystem II Membrane Vesicles*

Grana vesicles are prepared according to the procedure detailed above. Further sonication of the grana vesicles, while still present in the phase

TABLE I
CHLOROPHYLL YIELD, CHLOROPHYLL *a*/*b* RATIO, PROTEIN, P700, AND CYTOCHROME *f* CONTENT OF THYLAKOID MEMBRANE FRACTIONS[a]

| Fraction | Chlorophyll % | Chlorophyll *a*/*b* | Protein/chlorophyll (g/g) | P700/chlorophyll (mmol/mol) | Cytochrome *f*/chlorophyll (mmol/mol) |
|---|---|---|---|---|---|
| Thylakoids | 100 | 3.1 | 2.9 | 2.6 | 1.3 |
| B3 ($\alpha$ vesicles) | 37 | 2.3 | 2.7 | 1.5 | 1.5 |
| T3 ($\beta$ vesicles) | 21 | 5.0 | 3.4 | 4.7 | 1.2 |

[a] From E. Andreasson, P. Svensson, C. Weibull, and P.-Å. Albertsson, *Biochim. Biophys. Acta* **936,** 339 (1988).

TABLE II
PROPERTIES OF DIFFERENT FRACTIONS OF THYLAKOID MEMBRANE[a]

| Fraction | Chlorophyll *a*/*b* (mol/mol) | P700/ chlorophyll (mmol/mol) | PSII activity | Cytochrome *f*/chlorophyll (mmol/mol) | Sidedness | Form of PSII | Form of PSI |
|---|---|---|---|---|---|---|---|
| Thylakoids | 2.8–3.1 | 2.6 | 120 | 1.3 | — | $\alpha + \beta$ | $\alpha + \beta$ |
| Stroma lamellae (T3) vesicles | 4.5–5.0 | 4.7 | 80 | 1.2 | Right-side-out | $\beta$ | $\beta$ |
| Grana vesicles (B3) | 2.2–2.4 | 1.5 | 257 | 1.5 | Inside-out | $\alpha$ | $\alpha$ |
| PSII vesicles (BS) | 1.8–2.0 | 0.5–0.9 | 300 | 1.7 | Inside-out | $\alpha$ | — |
| Margin vesicles | 3.0–3.3 | 2.5 | — | <1.2 | Right-side-out | $\beta$ | $\alpha$ |

[a] From E. Andreasson, P. Svensson, C. Weibull, and P.-Å. Albertsson, *Biochim. Biophys. Acta* **936,** 339 (1988); P. Svensson and P.-Å. Albertsson, *Photosynth. Res.* **20,** 249 (1989); and P. Svensson, S.-G. Yu, U. K. Larsson, E. Andreasson, and P.-Å. Albertsson, *in* "Current Research in Photosynthesis" (M. Baltscheffsky, ed.), Vol. 2, p. 839. Kluwer Academic Publishers, Dordrecht, The Netherlands, 1990. PSII activity is expressed as oxygen evolution ($\mu$mol $O_2$/mg chl · h) with phenyl-p-benzoquinone as electron acceptor. Chl = chlorophyll (a + b).

TABLE III
PIGMENT COMPOSITION OF DIFFERENT FRACTIONS OF SPINACH THYLAKOID MEMBRANE[a]

| Fraction | Percentage of total pigment on molar basis | | | | | |
|---|---|---|---|---|---|---|
| | Neoxanthin | Violaxanthin | Lutein | β-Carotene | Chl *a* | Chl *b* |
| Thylakoids | 2.5 | 3.7 | 8.1 | 7.1 | 57.5 | 21.2 |
| Stroma lamellae (T3) vesicles | 1.4 | 3.9 | 6.6 | 9.6 | 63.0 | 15.6 |
| Grana vesicles (B3) | 3.4 | 3.9 | 9.3 | 5.2 | 55.3 | 22.9 |
| PSII vesicles (BS) | 3.6 | 3.9 | 9.5 | 5.5 | 52.8 | 24.8 |
| Margin vesicles | 2.6 | 4.2 | 7.7 | 7.0 | 57.6 | 20.9 |

[a] From R. K. Juhler, E. Andreasson, S.-G. Yu, and P.-Å. Albertsson, *Photosynth. Res.* **35,** 171 (1993).

system, detaches the margins from the membranes rich in photosystem II, which remains in the bottom phase, while the margin vesicles partition to the top phase.[2] See above for reagents.

*Procedure*

1. To 15 g of the bottom phase (B3), obtained after the batch procedure for making grana vesicles, is added 15 g of fresh top phase.
2. The mixture is sonicated 20 times for 30 sec each with resting intervals of 60 sec while continuously cooling (0–4°).
3. After mixing the two phases at 4°, the phases are separated by centrifugation at 1000 *g* for 3 min.
4. The top phase is discarded and the bottom phase is washed once with new top phase.
5. The vesicles in the bottom phase are collected, after 3-fold dilution, by centrifugation at 100,000 *g* for 90 min. (Alternatively, they can be collected at 40,000 *g* for 10 min, if MgCl is added to a concentration of 10 m*M*).

Some properties of the photosystem II membranes are indicated in Table II.

*Margins*

Grana vesicles (B3, α vesicles) are first prepared according to the procedure described above. These are sonicated while still present in the two-phase system. Margin vesicles are then obtained by further batch extraction. See above for reagents.

*Procedure*

1. To the bottom phase (B3) obtained after the batch procedure for making grana vesicles (~12 ml), 10 ml of fresh top phase is added, and the mixture is sonicated 6 times for 30 sec each with resting intervals of 60 sec under continuous cooling.

2. Ten milliliters of fresh top phase is added, and, after shaking the two phases (at 4°) they are separated by centrifugation at 1000 *g* for 3 min.

3. The resulting top phase fraction, named 180 s, is enriched in margins. Contaminating inside-out vesicles are removed by means of three additional extraction steps with pure bottom phase to obtain a margin fraction named 180s T3.

4. The vesicles are collected by centrifugation at 100,000 *g* for 90 min from the top phase which had been diluted 3-fold with diluting medium. Vesicles are then resuspended in the same medium.

Some of the properties of the margin fraction are described in Table II.

## Conclusions

The procedures described allow the isolation of membranes vesicles originating in different regions of the thylakoid membrane. Some of their properties are listed in Tables I–III. From these several conclusions can be reached. (1) The stroma lamellae vesicles are right-side-out and are enriched in photosystem I, chlorophyll *a*, and $\beta$-carotene. Their photosystems, I and II, are both of the $\beta$ type, that is, they have an antenna size which is smaller than the respective $\alpha$ forms which reside in the grana. In addition, the stroma lamellae vesicles are enriched in the ATP synthase complex.[1] (2) The grana vesicles are inside-out and are enriched in photosystem II, chlorophyll *b*, and lutein. Their photosystems are both of the $\alpha$ type. (3) The photosystem II vesicles (BS), which originate in the central core of the grana, are highly enriched in photosystem II$\alpha$ and depleted in photosystem I. They have the lowest chlorophyll *a*/*b* ratio, and they have the highest content of lutein. This is due to the high content of the light-harvesting chlorophyll *a*/*b* protein complex associated with photosystem II. They also have the highest concentration of cytochrome *f* (about 2 per 1000 chlorophylls). Because the antenna size of PSII$\alpha$ is about 250 chlorophylls (about 4 reaction centers per 1000 chlorophylls), one can conclude that these vesicles mainly consist of photosystem II and the cytochrome *b*/*f* complex. (4) The margin vesicles are right-side-out and contain photosystem I of the $\alpha$ type and photosystem II of the $\beta$ type. Their pigment composition is between those of the stroma lamellae and photosystem II vesicles. The margin vesicles also differ from the other types of vesicles in that they have the lowest cytochrome *f* content (Table II) and are enriched in the 67-kDa membrane-bound protein kinase of the thylakoid.[7] The margin vesicles are also, like the stroma lamellae vesicles, enriched in the ATP synthase complex. From the data in Table III one can also conclude that the molar ratio between total chlorophyll ($a + b$) and total carotenoid is the same (about 4) for all fractions.

The different vesicle preparations are very useful for studies on the localization of different components in the thylakoid and of the function of the different thylakoid domains. With these preparations one can also examine dynamic aspects of the thylakoid membrane, for example, if and how different components move laterally in the membrane as a result of environmental factors.

[7] S.-G. Yu, H. Stefánsson, and P.-Å. Albertsson, Photosynthesis Congress, Nagoya, Japan 1992.

# [46] Isolation of Smooth Endoplasmic Reticulum and Tonoplast from Etiolated Mung Bean Hypocotyls

*By* SHIZUO YOSHIDA

## Introduction

The endoplasmic reticulum (ER) is the most abundant membrane in eukaryotic cells. The ER plays an important role in the biogenesis of several types of single-membrane organelles such as protein bodies, vacuoles, Golgi bodies, and plasma membranes.[1] The ER is also actively involved in the glycosylation of proteins, in the synthesis of membrane-associated proteins, lipids, storage, or secretory proteins,[1] and in the regulation of cellular calcium concentration.[2] To clarify the role of the ER in these related biochemical processes, it is important to isolate the ER in high purity. Hitherto, the ER-enriched fraction has been isolated from plant tissues using a discontinuous or linear sucrose gradient system.[3] However, it is sometimes difficult to avoid ER contamination with cellular membranes of lighter density, especially tonoplasts and plastid envelopes.

[1] M. J. Chrispeels, *in* "The Biochemistry of Plants" (N. E. Tolbert, ed.), Vol. 1, p. 389. Academic Press, New York, 1980.
[2] T. J. Buckhout, *Plant Physiol.* **76,** 926 (1984).
[3] J. M. Lord, *in* "Isolation of Membranes and Organelles from Plant Cells" (J. L. Halland and A. L. Moore, eds.), p. 119. Academic Press, New York, 1983.

Etiolated mung bean hypocotyls are a suitable plant material for biochemical studies of the ER, because of their great abundance.[4] This chapter describes the method we have used to isolate smooth ER from etiolated mung bean seedlings. Incidentally, tonoplast vesicles can be simultaneously isolated from the same tissue.

## Plant Materials

The seeds of mung bean (*Vigna radiata* L. Wilczek) are soaked and germinated at 26° in darkness on a stainless steel mesh plate installed at one-third the depth of a plastic box (20 × 30 × 16 cm). $CaSO_4$ solution (1 m*M*) is added up to the level of the seeds on the stainless steel mesh plate (about 3 liters). After 3.5 days of germination, hypocotyls (~6 cm in length) are excised and used for the preparation of the microsomal fraction.

## Stock Solutions

Homogenizing buffer solution: 250 m*M* sorbitol, 50 m*M* 3-(*N*-morpholino)propanesulfonic acid (MOPS)–KOH, pH 7.6, 5 m*M* EDTA, 5 m*M* EGTA, 0.5% (w/v) defatted bovine serum albumin (BSA), 1 m*M* phenylmethylsulfonyl fluoride (PMSF), 10 μg/ml tertiary-butylated hydroxytoluene (BHT), and 2.5 m*M* dithiothreitol (DTT)

Sorbitol suspending solution: 250 m*M* sorbitol, 5 m*M* MOPS–KOH, pH 7.3, 1 m*M* EDTA, 0.1 m*M* PMSF, 10 μg/ml BHT, 10 m*M* KCl, and 2 m*M* DTT

Ficoll solution (5.5%, w/w): 5.5 g of Ficoll (Pharmacia LKB, Piscataway, NJ) is dissolved in 94.5 ml of 15% (w/w) sucrose solution made up in the same buffer system as used for the sorbitol suspending solution

Two-polymer phase mixture (50 g system): 2.8 g each of poly(ethylene glycol) (PEG) (Sigma, St. Louis, MO, MW 3350) and dextran (Dx) T500 (Pharmacia LKB) and 28.4 mg of NaCl are dissolved in 33.3 ml of 10 m*M* potassium phosphate buffer, pH 7.8, containing 250 m*M* sucrose; when 10 ml of membrane suspension is added into this polymer mixture, the final concentrations of polymers and NaCl are 5.6% (w/w) and 10 m*M*, respectively

[4] S. Yoshida and T. Kawata, *Plant Cell Physiol.* **29,** 1391 (1988).

## Methods

### *Enzyme Assays*

ATPase is assayed in a reaction mixture containing 3 m$M$ ATP, 3 m$M$ $MgSO_4$, 30 m$M$ Tris/2-($N$-morpholino)ethanesulfonic acid (MES), pH 7.0, 50 m$M$ KCl, 1 m$M$ sodium molybdate, 5 m$M$ sodium azide, plus 0.03% Triton X-100. Potassium nitrate- and vanadate-sensitive ATPase activities are assayed in the presence of 100 m$M$ $KNO_3$ and 100 $\mu M$ sodium vanadate, respectively. Pyrophosphatase (PPase) is assayed in a reaction mixture that consists of 1 m$M$ sodium pyrophosphate, 1 m$M$ $MgSO_4$, 30 m$M$ Tris/MES, pH 7.2, 50 m$M$ KCl, 1 m$M$ sodium molybdate, and 0.03% Triton X-100. ATPase and PPase assays are performed at 30°, and liberated $P_i$ is measured by the modified method of Hodges and Leonard.[5] UDPase activity is assayed according to Nagahashi and Kane.[6] Antimycin A-insensitive NADH–cytochrome-$c$ reductase activity is assayed as reported elsewhere.[5] Protein is analyzed by the method of Bradford.[7]

### *Preparation of Crude Microsomal Fraction*

Excised hypocotyl tissues (100 g fresh weight) are chilled at 5° and chopped into ice-cold homogenizing buffer solution, then homogenized with Polytron PT 30 (Kinematica GmbH, Lucerne, Switzerland) at medium speed (speed dial at 5) for 30 sec. After homogenization, the slurry is passed through 4 layers of gauze and centrifuged at 10,000 $g$ for 20 min. The supernatant solution is then centrifuged at 150,000 $g$ for 20 min. The 10,000 and 150,000 $g$ pellets are suspended in 17 ml of the sorbitol suspending solution and kept on ice.

### *Separation of Smooth Endoplasmic Reticulum and Tonoplast on Ficoll Gradient*

Eight milliliters of Ficoll solution (5.5%, w/w) is put into an ultracentrifuge tube (25 ml), and the microsomal suspension is then overlayered to the top of the tube. The sample is centrifuged at 200,000 $g$ for 2.5 hr in a fixed-angle rotor (Hitachi RP 50-2). At the beginning and end of centrifugation, acceleration and deceleration rates should be low so as not to disturb the interface between membrane sample and Ficoll solution. After centrifugation the fraction at this interface, which is enriched in smooth ER and tonoplasts, is collected using a Pasteur pipette and diluted

[5] T. K. Hodges and R. T. Leonard, this series, Vol. 32, p. 392.

[6] J. Nagahashi and A. P. Kane, *Protoplasma* **112,** 167 (1982).

[7] M. M. Bradford, *Anal. Biochem.* **72,** 248 (1976).

TABLE I
SEPARATION OF ENDOPLASMIC RETICULUM AND TONOPLAST-ENRICHED FRACTIONS FROM MICROSOMAL MEMBRANES AFTER CENTRIFUGATION ON FICOLL GRADIENT[a]

| Fraction | Protein (mg) | PPase ($\mu$mol $P_i$/hr) | ATPase ($\mu$mol $P_i$/hr) | | UDPase ($\mu$mol $P_i$/hr) | NADH–cytochrome-*c* reductase ($\mu$mol/10 min) |
|---|---|---|---|---|---|---|
| | | | $\Delta KNO_3$ | $\Delta VO_4$ | | |
| Interface | 22.1 | 2517.0 | 252.7 | 76.3 | 499.2 | 78.4 |
| | (50.5) | (89.1) | (72.7) | (20.5) | (21.4) | (80.7) |
| Pellet | 21.7 | 309.8 | 95.0 | 296.2 | 1,831.6 | 18.7 |
| | (49.5) | (10.9) | (27.3) | (79.5) | (78.6) | (19.3) |

[a] $\Delta KNO_3$ and $\Delta VO_4$ indicate $KNO_3$- and vanadate-sensitive ATPase activities, respectively. Membrane materials retained in the Ficoll layer were combined into the pellet. Values in parentheses show the distribution as a percentage of activities between the two fractions. From S. Yoshida and T. Kawata, *Plant Cell Physiol.* **29,** 1391 (1988), with permission.

with 2 volumes of sorbitol suspending solution, then pelleted by centrifugation at 150,000 *g* for 30 min. The membrane pellet is suspended in 10 ml of 250 m*M* sucrose solution made up in 10 m*M* potassium phosphate buffer, pH 7.8, which is used in the two-polymer aqueous phase system (see below).

Table I shows that after the Ficoll gradient centrifugation about 90% of total pyrophosphatase activity (tonoplast marker) and 80% of total antimycin A-insensitive NADH–cytochrome-*c* reductase activity (ER marker) are recovered at the interface. Most of the carotenoids (marker for plastid envelopes), UDPase activity (Golgi marker), and vanadate-sensitive ATPase (plasma membrane marker) are sedimented through the Ficoll layer.

### *Separation of Endoplasmic Reticulum and Tonoplast Vesicles by Two-Polymer Phase Partitioning*

The suspension of the interfacial fraction (10 ml) is put into the two-polymer phase solution (which has been cooled to 0°), mixed by several inversions, then centrifuged at 4500 *g* for 5 min at 0°. The upper and lower phases are separated and mixed, respectively, with fresh lower and upper phases (which have been prepared using sucrose–phosphate buffer alone without membrane sample) and subjected to a second partitioning step. The upper and lower phases after the second partitioning are separated, and each is diluted with 4 volumes of the sorbitol suspending solution and centrifuged at 150,000 *g* for 30 min. The membrane pellets are resuspended in sorbitol suspending solution. As shown in Table II, the upper and lower phases are highly enriched in tonoplasts and ER, respectively. More than 90% of total pyrophosphatase activity is found to partition into the upper phase, whereas more than 95% of total antimycin A-insensitive NADH–cytochrome-*c* reductase activity partitions into the lower phase. It is noteworthy that the ER-enriched fraction (lower phase) contains an ATPase that is sensitive to Triton X-100 (0.03%) and vanadate (100 $\mu M$), but not to $NO_3^-$. This ATPase seems to be unique to the ER because other ATPase associated with tonoplasts, plasma membranes, and Golgi membranes are generally stimulated by the addition of the detergent to the reaction mixture.

Table III shows the specific activities of marker enzymes in the different fractions obtained during the separatory procedure. The specific activities of pyrophosphatase and $NO_3^-$-sensitive ATPase in the lower phase are fairly low, whereas NADH–cytochrome-*c* reductase is about 25 times higher in the lower phase than in the upper phase fraction. The specific activity of UDPase in the ER-enriched fraction is about one-third of that in the Ficoll–pellet fraction, indicating that the former fraction is still contaminated with a small amount of Golgi membranes. One can remove this contamination by use of a linear (15–50%, w/w) or a discontinuous (26/32%, w/w) sucrose density gradient. The tonoplast-enriched fraction (upper phase) can, if necessary, be further purified by centrifugation over a cushion of 32% (w/w) sucrose. After centrifugation at 170,000 g for 90 min in a fixed-angle rotor (RP 50-2, Hitachi), most of the contaminated plasma membranes are sedimented.

For the separation of a large amount of tonoplast vesicles with a high purity from mung bean hypocotyls and, perhaps, other materials with soft texture, we are currently using a floating centrifugation method.[8] The microsomal fraction prepared in homogenizing buffer solution which contained 250 m*M* sorbitol as an osmoticum is suspended in 300 m*M* sucrose buffer solution. The membrane–sucrose suspension is placed in centrifuge tubes and overlayered with sorbitol buffer solution to the top of the tubes and centrifuged at 123,000 *g* for 30 min in a fixed-angle rotor (Hitachi RP 50-2). After centrifugation, tonoplast vesicles are found banded at the interface of the sorbitol–sucrose solutions.

The protocol for the isolation of smooth ER is principally based on the fact that tonoplast vesicles from mung bean hypocotyls specifically partition to the upper phase.[4] This property of tonoplast vesicles is, however, unique to mung bean and some other kinds of leguminous plants (e.g., pea). With other plant materials such as orchard grass and winter

[8] M. Maeshima and S. Yoshida, *J. Biol. Chem.* **264,** 20068 (1989).

TABLE II

SEPARATION OF ENDOPLASMIC RETICULUM FROM FICOLL INTERFACE FRACTION BY TWO-POLYMER PHASE PARTITIONING[a]

| Phase | Protein (mg) | NADH–cytochrome-*c* reductase ($\mu$mol/10 min) | PPase ($\mu$mol $P_i$/hr) | UDPase ($\mu$mol $P_i$/hr) | ATPase ($\mu$mol $P_i$/hr) | | | | |
|---|---|---|---|---|---|---|---|---|---|
| | | | | | −TX | +TX | +TX/−TX | $\Delta KNO_3$ | $\Delta VO_4$ |
| Upper | 7.15 | 2.0 | 730.8 | 39.5 | 75.3 | 164.0 | 2.18 | 95.7 | 34.8 |
| | (57.6) | (4.7) | (93.5) | (20.6) | (47.9) | (84.0) | | (86.3) | (79.8) |
| Lower | 5.26 | 40.0 | 50.4 | 152.2 | 81.8 | 31.2 | 0.38 | 15.2 | 8.8 |
| | (42.4) | (95.3) | (6.5) | (79.4) | (52.1) | (16.0) | | (13.7) | (20.2) |

[a] Ficoll interface fraction was resuspended in 0.25 *M* sucrose, 10 m*M* potassium phosphate buffer (pH 7.8) that contained 2 m*M* DTT and subjected to phase partitioning. Values in parentheses show the distribution as a percentage in the upper and lower phases. The ATPase activity was analyzed in the presence (+TX) and in the absence (−TX) of 0.03% Triton X-100. The $KNO_3$-sensitive ATPase activities were measured in the presence of 0.03% Triton X-100. From S. Yoshida and T. Kawata, *Plant Cell Physiol.* **29,** (1988), with permission.

TABLE III
DISTRIBUTION OF SPECIFIC ACTIVITIES OF MARKER ENZYMES AMONG VARIOUS FRACTIONS[a]

| Fraction | NADH–cytochrome-*c* reductase ($\mu$mol/mg/hr) | UDPase ($\mu$mol $P_i$/mg/hr) | PPase ($\mu$mol $P_i$/mg/hr) | ATPase ($\mu$mol $P_i$/mg/hr) | | | | | |
|---|---|---|---|---|---|---|---|---|---|
| | | | | | | | | $\Delta VO_4$ | |
| | | | | −TX | +TX | ΔTX | $\Delta KNO_3$(+TX) | (−TX) | (+TX) |
| Ficoll interface | 3.5 | 22.5 | 113.7 | 14.7 | 19.3 | — | 11.4 | 9.4 | 3.4 |
| Ficoll pellet | 0.9 | 84.4 | 14.3 | 12.8 | 20.8 | — | 4.4 | 8.9 | 13.6 |
| Upper phase | 0.3 | 5.5 | 102.2 | 10.5 | 22.9 | — | 13.4 | 4.6 | 4.8 |
| Lower phase | 7.6 | 28.9 | 9.6 | 15.6 | 5.9 | 9.7 | 2.9 | 12.9 | 1.7 |

[a] From S. Yoshida and T. Kawata, *Plant Cell Physiol.* **29,** 1391 (1988), with permission.

rye, tonoplast vesicles, as well as ER, preferentially partition into the lower phase as the NaCl concentration is increased up to 30 m*M* in the phase system.[9]

## Concluding Remarks

The combination of Ficoll gradient centrifugation and two-polymer aqueous phase partitioning is useful for the isolation of smooth ER and tonoplast from etiolated mung bean hypocotyls. In particular, this method is effective for the isolation of these membranes from mung bean seedlings at early germination stages, stages at which the recovery of tonoplast vesicles is relatively poor if a sorbitol–sucrose density gradient or a floating centrifugation method is used,[8] because of the small size of vacuoles in the cells. Several types of ATPases have been unequivocally located in various cellular membranes of plant cells.[10] Vanadate-sensitive ATPase is not restricted to the plasma membrane but is also distributed in other membranes such as plastid envelopes.[11] It has been suggested that the vanadate-sensitive ATPase is located on the ER and may participate in the primary calcium pump. In this context the finding of a specific association of ATPase with the mung bean ER, sensitive both to neutral detergent and vanadate, may throw a light on the molecular properties as well as the physiological functions of the enzyme for the regulation of cellular calcium levels.

[9] M. Uemura and S. Yoshida, *Plant Physiol.* **73,** 586 (1983).
[10] T. Kawata and S. Yoshida, *Plant Cell Physiol.* **29,** 1399 (1988).
[11] W. Wu and G. A. Berkowitz, *Plant Physiol.* **99,** 553 (1992).

# [47] Rat Liver Plasma Membranes

*By* ANDERS PERSSON and BENGT JERGIL

## Introduction

An important factor when isolating biological membranes is to find separation protocols that are sufficiently selective for the membrane of interest. Selectivity is particularly important when less abundant membranes are to be isolated in order to minimize contamination by other membranes and to avoid undue losses during preparation. The use of biospecific affinity ligands coupled to one of the phase polymers in an aqueous two-phase partitioning system offers selectivity combined with gentle separation conditions for membranes.

Affinity partitioning has been used for isolating cholinergic receptors from the electroplax of *Torpedo californica*,[1,2] opiate receptors from calf brain cortex,[3] and plasma membranes from rat liver and hepatocytes.[4,5] The ligand used for affinity partitioning of plasma membranes was the lectin wheat germ agglutinin, which specifically binds *N*-acetylglucosamine and sialic acid residues. In this case the ligand was coupled to dextran (Dx) [rather than to poly(ethylene glycol) (PEG)] in a PEG–Dx polymer system. When attached to large phase polymers, ligands are expected to pull large components, such as membranes, more efficiently into the phase in which that polymer predominates. In addition, the bulk of membranes can be forced into the PEG-rich top phase by including borate buffer in the system. This facilitates the selective ligand-dependent extraction of membrane fractions into the Dx-rich bottom phase.

Two procedures are presented here: affinity partitioning of plasma membranes using wheat-germ agglutinin coupled to Dx as a ligand and the rapid purification of rat liver plasma membranes by combining traditional two-phase partitioning with the affinity partitioning technique.

## Experimental

### *Chemicals*

Stock solutions of 20% (w/w) Dx T500 (Pharmacia LKB Biotechnology, Uppsala, Sweden) and 40% (w/w) PEG 3350 (Carbowax 3350; Union Carbide, New York, NY) in water are prepared.[6] These can be stored at −20°. It is convenient to make the Dx stock solution from newly freeze-dried Dx. This dissolves readily, and the commonly used optical rotation measurement for determining the Dx concentration[7] can then be omitted. 2,2,2-Trifluoroethanesulfonyl chloride (tresyl chloride) is from Synthelec (Lund, Sweden), and the affinity ligand, wheat germ agglutinin, is from Boehringer (Mannheim, Germany).

[1] S. D. Flanagan, P. Taylor, and S. H. Barondes, *Nature (London)* **254,** 441 (1975).
[2] G. Johansson, R. Gysin, and S. D. Flanagan, *J. Biol. Chem.* **256,** 9126 (1981).
[3] B. Olde and G. Johansson, *Neuroscience (Oxford)* **15,** 1247 (1985).
[4] A. Persson, B. Johansson, H. Olsson, and B. Jergil, *Biochem. J.* **273,** 173 (1991).
[5] A. Persson and B. Jergil, *Anal. Biochem.* **204,** 131 (1992).
[6] D. E. Brooks and R. Norris-Jones, this volume [2].
[7] M. J. López-Pérez, G. Paris, and C. Larsson, *Biochim. Biophys. Acta* **635,** 359 (1981).

*Activation of Dextran*

Freeze-dried Dx T500 is activated with tresyl chloride essentially as previously described.[8] Five grams of Dx is dissolved in 25 ml dimethyl sulfoxide (DMSO), then 1 ml triethylamine is added slowly, followed by 5 ml dichloromethane. After chilling on ice, 0.35 g tresyl chloride is added dropwise while stirring vigorously. The mixture is stirred on ice for 30 min, for 60 min at 4°, and overnight at 20°. The reaction is terminated by precipitating the Dx with 50 ml of dichloromethane. After washing and thorough kneading with several portions of dichloromethane, the precipitate is dissolved in water, dialyzed against water overnight, and freeze-dried. The activated Dx is stable at −20° for at least 2 months. The recovery of Dx is about 95%.

When freeze-dried Dx is used for activation the time required for it to dissolve is reduced from several hours to a few minutes, and, also, the risk of hydrolyzing the tresyl chloride during activation is minimized. For this reason all organic solvents should be dried over a molecular sieve overnight before use.

*Coupling of Affinity Ligand to Tresyl Dextran*

Two grams of tresyl-Dx are dissolved in 10 ml coupling buffer (0.1 *M* $NaH_2PO_4/Na_2HPO_4$/0.5 *M* NaCl, pH 7.5) before addition of 10 mg of wheat-germ agglutinin dissolved in 1 ml of the same buffer. The coupling is allowed to proceed overnight at 4° with gentle agitation. The reaction is terminated by adding 10 ml of 0.1 *M* Tris-HCl, pH 7.5, which inactivates excess tresyl groups.

Uncoupled ligand and salts are removed by ultrafiltration of the Dx–ligand mixture in a Filtron Omegacell (150 ml size, cutoff 100K; Filtron Technology Corp., Clinton, MA). After freeze-drying the amount of ligand coupled is determined by protein analysis[9] using wheat germ agglutinin as standard. Routinely 4 mg are bound per gram freeze-dried product, and the recovery of Dx is more than 90%.

*Composition of Two-Phase Affinity System*

The two-phase polymer system is prepared by weighing the required amount of polymer stock solutions, Dx-bound ligand, and buffer into a suitable test tube. The final composition of the routinely used system is 6.0% (w/w) of each polymer, including a suitable amount of Dx-bound ligand, and 20 m*M* boric acid, adjusted to pH 7.8 with Tris base (see

[8] L.-O. Persson and B. Olde, *J. Chromatogr.* **457,** 183 (1988).

[9] M. M. Bradford, *Anal. Biochem.* **72,** 248 (1976).

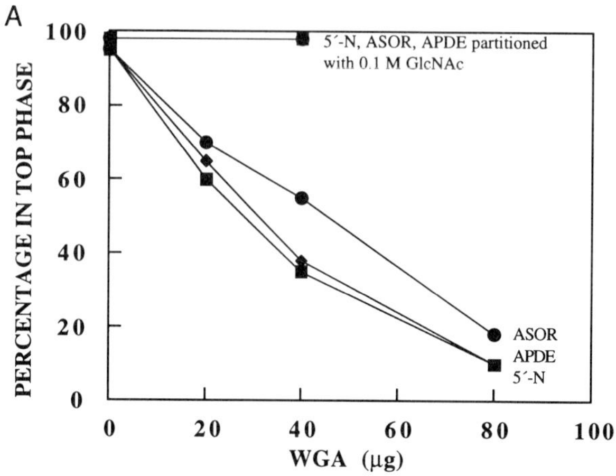


FIG. 1. Affinity partitioning of rat liver membranes. Membranes were partitioned in the presence of increasing amounts of Dx-bound wheat germ agglutinin. The systems used had a total weight of 2 g and contained 5.6% (w/w) dextran and 5.6% (w/w) PEG in 16.5 m$M$ $H_3BO_3$–Tris, pH 7.8. About 1 mg of membrane protein was added to each system. [From A. Persson, B. Johansson, H. Olsson, and B. Jergil, *Biochem. J.* **273,** 173 (1991).] (A) Partitioning of purified plasma membranes enriched in the apical domain [N. N. Aronson and O. Touster, Jr., this series, Vol. 31, p. 90]. The experiment was performed in the absence or presence of 0.1 *M N*-acetylglucosamine (GlcNAc). The plasma membrane markers measured after phase separation were the basolateral marker asialoorosomucoid binding [A. L. Hubbard, D. A. Wall, and A. Ma, *J. Cell Biol.* **96,** 217 (1983); P. Gierow, M. Sommarin, C. Larsson, and B. Jergil, *Biochem. J.* **235,** 685 (1986)] (ASOR, ●) and the apical markers alkaline phosphodiesterase [G. D. Smith and T. J. Peters, *Eur. J. Biochem.* **104,** 305 (1980)] (APDE, ◆) and 5′-nucleotidase [J. Avruch and D. F. H. Wallach, *Biochim. Biophys. Acta* **233,** 334 (1971)] (5′-N, ■). (B) Partitioning of a microsomal fraction [N. N. Aronson and O. Touster, Jr., this series, Vol. 31, p. 90]. The components measured after phase separation were 5′-nucleotidase (5′-N, ■), the endoplasmic reticulum marker arylesterase [E. H. Shephard and G. Hübscher, *Biochem. J.* **113,** 429 (1969)] (AE, ○), the Golgi marker galactosyltransferase [B. Fleischer and M. Smigel, *J. Biol. Chem.* **253,** 1632 (1978)] (Gal, □), and protein [M. M. Bradford, *Anal. Biochem.* **72,** 248 (1976)] (◇).

below). It should be noted that the polymer concentration is not so critical as in nonaffinity two-phase partitioning.[10]

To establish the amount of Dx-linked wheat germ agglutinin required to pull plasma membranes into the bottom phase, and also to check on the selectivity of the extraction, it is useful to partition the membrane fraction of interest with increasing amounts of Dx-linked wheat germ agglutinin and to determine appropriate marker components in the resulting top and bottom phases (Fig. 1).

[10] H. Walter and C. Larsson, this volume [4].

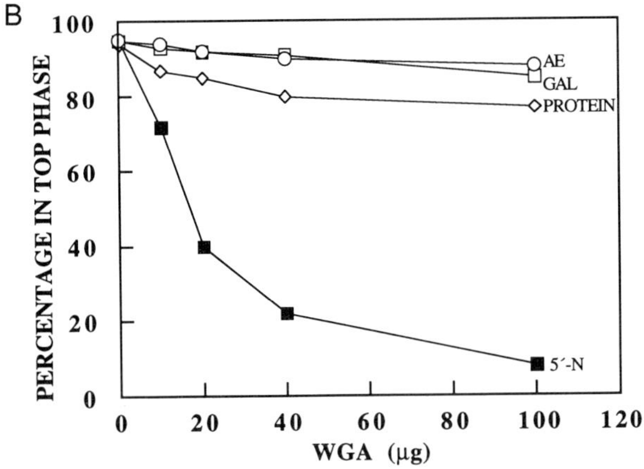


FIG. 1. (*continued*)

## *Affinity Partitioning*

Crude membrane fractions can be prepared by any common procedure. After pelleting, the membranes are resuspended in the homogenization medium or the same buffer as in the affinity system. The membrane fraction obtained is added to the affinity system. The system is inverted 20 times, vortexed, and inverted another 20 times to ensure thorough mixing. The phases are separated by gentle centrifugation (150 *g*) for 5 min, and the top phase is siphoned off. Depending on the crudeness of the starting material the purity of the plasma membranes may be increased by reextracting the bottom phase with an equal amount of fresh preequilibrated top phase. After the final phase separation the bottom phase is diluted 10-fold in a solution containing 0.25 *M* sucrose and 0.1 *M* *N*-acetylglucosamine in 5 m*M* Tris-HCl, pH 8.0. The membranes are pelleted by centrifugation for 60 min at 100,000 *g*, whereas the Dx-bound ligand, dissociated from the membranes by *N*-acetylglucosamine, remains in the supernatant solution. The membrane pellet is then resuspended in a suitable buffer. All phase separations should be done with solutions carefully maintained at 4°.

## *Rapid Purification of Rat Liver Plasma Membranes*

By combining homogenization in a conventional aqueous polymer two-phase system followed by phase separation with affinity partitioning, rat liver plasma membranes can be prepared rapidly to a high degree of purity and with a high yield.[5]

*Composition of the Phase Systems*

Homogenization system: The final mixture contains 5.7% (w/w) of each polymer, 15 m*M* Tris–$H_2SO_4$, pH 7.8, diluted from stock solutions, total weight 15 g, and is prepared by mixing 4.28 g of Dx T500 stock solution (20%), 2.14 g of PEG 3350 stock solution (40%), 1.12 ml of 0.20 *M* Tris–$H_2SO_4$, pH 7.8, 5.46 ml water, and 2.00 g squashed liver

Affinity system: The final mixture contains 6.0% (w/w) of each polymer including Dx-linked wheat germ agglutinin, 20 m*M* $H_3BO_3$–Tris, pH 7.8, diluted from stock solutions, total weight 10 g, and is prepared by mixing 50 mg of freeze-dried Dx–wheat germ agglutinin (200 $\mu$g ligand), 2.75 g of Dx T500 stock solution (20%), 1.50 g of PEG 3350 stock solution (40%), 1.00 ml of 0.20 *M* $H_3BO_3$–Tris, pH 7.8, and 4.70 ml water

The PEG-rich top phases used for reextraction of the bottom phases are siphoned off from systems prepared in the same manner as the homogenization and the affinity systems except that the squashed liver and the Dx-linked wheat germ agglutinin, respectively, are absent.

*Purification Procedure*

Adult male Sprague-Dawley rats are sacrificed by decapitation. Each liver is immediately chilled in ice-cold 0.25 *M* sucrose in 5 m*M* Tris-HCl, pH 8.0, squashed through a garlic press, and transferred to a glass–glass Dounce homogenizer or a glass–Teflon Potter–Elvehjem homogenizer containing the homogenization system. Up to 2 g of squashed liver can be included in a 15-g system. It should be noted that the polymer concentration in this system is critical.

After homogenization by 10 strokes with a loose-fitting Dounce pestle and 20 strokes with a tight-fitting one, or 3 strokes (1000 rpm) with a Teflon pestle in a Potter–Elvehjem homogenizer, the homogenization system is transferred to a centrifuge tube and centrifuged at low speed (150 *g*) for 5 min in a swing-out rotor to speed phase separation. The top phase is collected, and the bottom phase is reextracted with an equal volume of fresh top phase obtained by preequilibration with bottom phase (reextraction increases the final plasma membrane yield by ~30%). The system is thoroughly mixed as described under affinity partitioning, and the phases are separated. The top phases are combined and added to the 10-g affinity system. After thorough mixing and phase separation, speeded by centrifugation (see above), the bottom phase is reextracted twice with an equal

TABLE I
PURIFICATION OF PLASMA MEMBRANES BY TWO-PHASE PARTITIONING FOLLOWED BY AFFINITY PARTITIONING[a]

| | Homogenate | | Affinity partitioning | | Pellet | |
|---|---|---|---|---|---|---|
| Marker | Specific activity | Combined top phases (%) | Combined top phases (%) | Final bottom phase (%) | % | RSA |
| Protein | 114.6 ± 2.4 | 29.9 ± 1.1 | 24.2 ± 2.2 | 4.7 ± 0.8 | 1.7 ± 0.1 | — |
| Alkaline phosphodiesterase | 42.3 ± 3.6 | 73.8 ± 3.4 | 7.0 ± 0.8 | 72.0 ± 1.3 | 68.6 ± 3.3 | 40 |
| 5′-nucleotidase | 13.6 ± 1.8 | 70.1 ± 4.2 | 4.3 ± 2.4 | 69.4 ± 2.4 | 64.2 ± 1.1 | 38 |
| Asialoorosomucoid binding | 3.99 ± 0.82 | 61.0 ± 4.6 | 2.1 ± 0.8 | 59.9 ± 3.3 | 54.0 ± 2.9 | 32 |
| Arylesterase | 717 ± 35 | 5.2 ± 0.2 | 4.6 ± 0.8 | <2 | <2 | — |
| Galactosyltransferase | 0.44 ± 0.03 | 53.5 ± 4.9 | 40.2 ± 3.7 | 10.0 ± 1.6 | 9.8 ± 1.4 | 5.8 |
| Succinate–cytochrome-*c* reductase | 36.4 ± 3.1 | 2.5 ± 0.2 | <2 | <2 | <2 | — |
| *N*-Acetylglucosaminidase | 2.41 ± 0.12 | 43.7 ± 4.6 | 43.4 ± 9.3 | 4.0 ± 1.5 | 4.7 ± 1.0 | 2.8 |
| Lactate dehydrogenase | 3290 ± 308 | 16.4 ± 1.1 | 9.9 ± 0.6 | 5.0 ± 0.9 | <2 | |

[a] One gram of squashed liver was homogenized in a 15-g two-phase system and processed as described in the text. Membranes in the twice-extracted affinity bottom phase were pelleted and resuspended in 0.25 *M* sucrose in 5 m*M* Tris-HCl, pH 8.0. In addition to the markers measured in Fig. 1, the mitochondrial marker succinate–cytochrome-*c* reductase [G. L. Sottocasa, B. Kuylenstierna, L. Ernster, and A. Bergstrand, *J. Cell Biol.* **32,** 415 (1967)], the lysosomal marker *N*-acetylglucosaminidase [O. Touster, N. N. Aronson, J. T. Dulaney, and H. Hendrickson, *J. Cell Biol.* **47,** 604 (1970)], and the cytosolic marker lactate dehydrogenase [H. U. Bergmeyer and E. Bernt, *in* "Methoden der Enzymatischen Analyse" (H. U. Bergmeyer, ed.), p. 533. Verlag Chemie, Weinheim/Bergstrasse, 1970] were measured. Enzyme activities of the homogenate (sum of the top and bottom phases obtained after homogenization and phase separation) are expressed as nmol/min per mg protein, asialoorosumucoid binding as ng/mg protein, and protein as mg/g wet liver. Activities recovered in the different partitioning steps and after pelleting are expressed as percentage of the homogenate. Relative specific activities (RSA) were calculated relative to the homogenate. Values (±S.D.) are based on three independent preparations. From A. Persson and B. Jergil, *Anal. Biochem.* **204,** 131 (1992).

amount of preequilibrated top phase before pelleting membranes for 60 min at 100,000 *g*.

This protocol results in a preparation that is 30- to 40-fold enriched, compared to the homogenate, in plasma membrane markers for both the apical and basolateral domains and gives yields of 55–70% (Table I). The contamination by other membranes is low. The procedure can be completed within 90 min.

## Comments

The strategy for the presented affinity partitioning method is to attach the ligand to the large polymer (i.e., Dx) and to use conditions which cause membranes in general to partition into the PEG-rich top phase, while the desired membrane fraction is selectively extracted into the Dx-rich bottom phase by the ligand. Membranes will partition into the top phase in the presence of borate because borate has affinity for Dx, and the negatively charged membranes will be forced into the top phase by electrostatic repulsion.

The preparation procedure for material to be affinity extracted in the PEG/wheat germ agglutinin–Dx system is not critical. Thus, the ligand will bind plasma membranes from more or less crude preparations with similar avidity. The affinity extraction also works with crude homogenates as in the rapid purification procedure. Furthermore, so long as borate is present, other buffers are tolerated to some extent. The amount of plasma membrane protein pulled into the bottom phase is approximately 7 $\mu$g per microgram Dx-linked wheat germ agglutinin.

Affinity partitioning of rat liver plasma membranes has been described here. As the wheat germ agglutinin ligand binds *N*-acetylglucosamine and sialic acid, residues that are exposed on the surface of many types of cells, it should be possible, using the same ligand, to selectively partition plasma membranes from other sources as well.

# [48] Synaptic Membranes

*By* GÖTE JOHANSSON

## Introduction

It is of interest to isolate different parts of nerve endings to determine their composition (proteins and lipids) and organization. Fractionation of fragmented synaptic membranes also provides less heterogeneous material for studies on how receptors and ion channels function in their membrane-bound states. Traditional methods for fractionation of synaptic membranes are based on centrifugation. Partitioning of membranes between aqueous phases is another fractionation approach. Aqueous extraction also permits the use of polymer-bound affinity ligands (e.g., specific receptor-binding components) to obtain a selective extraction.[1] Synaptosomal membranes

[1] B. Olde and G. Johansson, *Neuroscience (Oxford)* **15,** 1247 (1985).

obtained by fragmentation of synaptosomes are contaminated by mitochondrial membranes because mitochondria are present inside synaptosomes.[2] Mitochondrial membranes can also be separated from synaptic membranes by aqueous two-phase extraction.[1]

A number of ligand–poly(ethylene glycol) derivatives useful for extraction of various kinds of synaptic membranes have been synthesized[3,4] and applied in affinity partitioning.[3,5–7] Dextran (Dx) has also been used as a carrier for ligands.[7–9] The two-phase systems used have been based on the polymer pairs Dx and poly(ethylene glycol) (PEG)[1,3,5] or Dx and Ficoll (the latter usually also containing a low concentration of PEG so as to increase the difference in density between the Dx and Ficoll phases).[6–9] Affinity partitioning of membranes rich in opiate-binding receptors is described here as an example of isolation of specific synaptic membranes. The isolation is achieved by use of a PEG-bound affinity ligand: naloxone. The approach outlined can be used for membrane extraction with other affinity ligands.

## Methods

### *Synthesis of Naloxone-Poly(ethylene Glycol)*

The pharmaceutical substance naloxone (Du Pont de Nemours, Wilmington, DE) has affinity for opiate receptors and can be used, bound to PEG, for affinity extraction of membranes containing such receptors.[1]

*Activation of Polymer.* PEG 8000 (BP Chemicals, Hythe, UK) is activated by reaction with tresyl chloride, $CF_3CH_2SO_2Cl$ (Synthelec AB, Lund, Sweden), giving tresyl-PEG, tresylate being an excellent leaving group.[10] Tresyl-PEG is prepared[1] by dissolving 100 g PEG 8000 in 250 ml dry dichloromethane containing 4.2 ml pyridine. The solution is cooled on ice, and tresyl chloride, 6.5 g, is added dropwise while mixing. After 2 hr on ice the solvent is evaporated under vacuum at room temperature.

[2] M. J. López-Pérez, this volume [38].

[3] B. Olde, Doctoral Thesis, University of Lund, Sweden (1988).

[4] G. Johansson, R. Gysin, and S. D. Flanagan, *J. Biol. Chem.* **256,** 9126 (1981).

[5] G. Johansson, H.-E. Åkerlund, and B. Olde, *J. Chromatogr.* **311,** 277 (1984).

[6] G. Johansson, M. Joelsson, B. Olde, and V. P. Shanbhag, *J. Chromatogr.* **331,** 11 (1985).

[7] M. T. Muiño Blanco, J. A. Cebrian Pérez, B. Olde, and G. Johansson, *J. Chromatogr.* **358,** 147 (1986).

[8] M. T. Muiño Blanco, J. A. Cebrian, B. Olde, and G. Johansson, *J. Chromatogr.* **547,** 79 (1991).

[9] J. A. Cebrian Pérez, M. T. Muiño Blanco, and G. Johansson, *Int. J. Biochem.* **23,** 1491 (1991).

[10] K. Nilsson and K. Mosbach, this series, Vol. 104, p. 56.

The polymer is recrystallized from 1 liter of absolute ethanol containing 50 m*M* dry HCl at 4°. It is collected by suction filtration and then recrystallized twice from 1.5 liters of dry methanol also at 4°.

*Synthesis.* Naloxone is converted to 6-aminonaloxone by reductive amination with sodium cyanoborohydride in methanol.[11] 6-Aminonaloxone, 0.8 g, is added to a solution of 10 g tresyl-PEG 8000 in 100 ml dry methanol. When the aminonaloxone is dissolved, 2 g of $Na_2CO_3$ (dried at 110°) is suspended in the solution, and the mixture is stirred 70 hr at room temperature. Solvent is removed under vacuum, and the polymer residue is recrystallized 5 times from 200-ml portions of absolute ethanol. Before use in affinity partitioning naloxone–PEG is dissolved in water (1–2% solution) and passed through a column of Dowex 50W-X4 to remove traces of free ligand.

### *Preparation of Synaptic Membranes*

Synaptosomes[1,2,12] are prepared from 100 g cortex of fresh calf brains. The synaptosomes are suspended in 60 ml ice-cold distilled water, and the suspension is kept on ice for 1 hr. The lysed synaptic membranes are collected by centrifugation at 45,000 $g_{max}$ for 1 hr at 3° using a Sorvall SS-34 rotor (Ivan Sorvall Inc., Norwalk, CN). The membrane pellet is suspended in 140 ml of distilled water.

### *Systems for Partitioning*

Two-phase systems used for extractions contain 8.8% (w/w) Dx T40 (molecular weight 40,000, Pharmacia, Uppsala, Sweden), 5.9% (w/w) PEG 3400 (BP Chemicals), 5 m*M* Tris–phosphoric acid (based on Tris), pH 7.4, and 0–10 m*M* KCl. The affinity partitioning is carried out in systems in which part of the PEG is replaced by naloxone–PEG 8000.

### *Adjustment of Partitioning with Salt*

The ground state partitioning of membranes is adjusted by addition of KCl. The membranes are increasingly excluded from the upper phase as the chloride concentration is increased. Mitochondrial membranes (followed via the marker enzyme succinate dehydrogenase, SDH) are excluded at a lower concentration of KCl than the synaptic membranes (followed via acetylcholinesterase, AChE; Fig. 1). By using a critical concentration, in this case 2 m*M* KCl, mitochondrial and synaptic membranes can be separated. Because of differences in membrane preparations

[11] J. B. Jiang, R. N. Hanson, and P. S. Portoghese, *J. Med. Chem.* **20,** 1100 (1977).

[12] F. Hajós, *Brain Res.* **93,** 485 (1975).

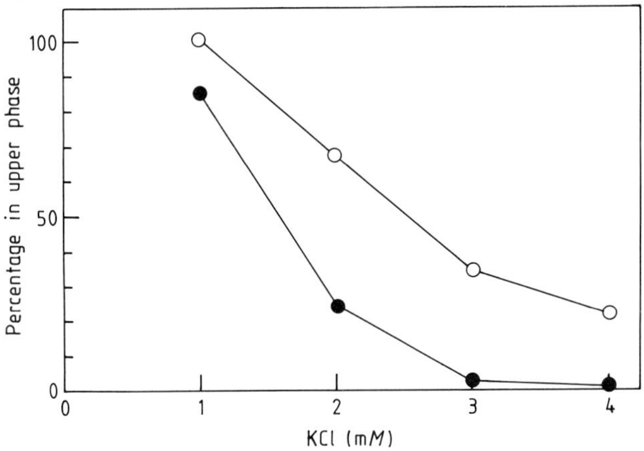


FIG. 1. Partitioning of synaptic membranes as a function of KCl concentration. ○, Acetylcholinesterase (marker for synaptic membranes); ●, succinate dehydrogenase (marker for mitochondrial membranes). The phase system contained 8.8% dextran T40, 5.9% PEG 3400, 5 m*M* Tris–$H_3PO_4$ (based on Tris), pH 7.4, and different concentrations of KCl. Temperature, 4°. [Adapted with permission from B. Olde and G. Johansson, *Neuroscience (Oxford)* **15,** 1247 (1985). Copyright 1985 Pergamon Press Ltd.]

and in traces of salt present in the polymers used, it is advisable to always determine the extraction curves in order to find an optimal salt concentration for the separation.

## *Partitioning Procedure*

For partitioning,[13] membranes containing 100–200 mg protein (assayed according to Bradford[14,15]) are suspended in 4 ml water and fragmented by two passages through a Yeda press (Scientific Instruments, Rehovot, Israel) using 100 atm nitrogen pressure. The membrane suspension (4 ml) is included in a 15-g two-phase system with the above composition (without KCl) at 3°, and the phases are carefully mixed. Phase settling is achieved by centrifugation at 1000 $g_{max}$ in a Sorvall SS-34 rotor for 6 min. The upper phase is collected (avoiding interfacial material) and combined with the same volume of fresh lower phase. Then 2 *M* KCl solution is added to give a final concentration of 1 m*M*. After equilibration and phase separation at 3° the upper phase, containing purified synaptic membranes, is again collected.

[13] H. Walter and C. Larsson, this volume [4].

[14] M. M. Bradford, *Anal. Biochem.* **72,** 248 (1976).

[15] B. Olde and G. Johansson, *Mol. Cell. Biochem.* **87,** 153 (1989).

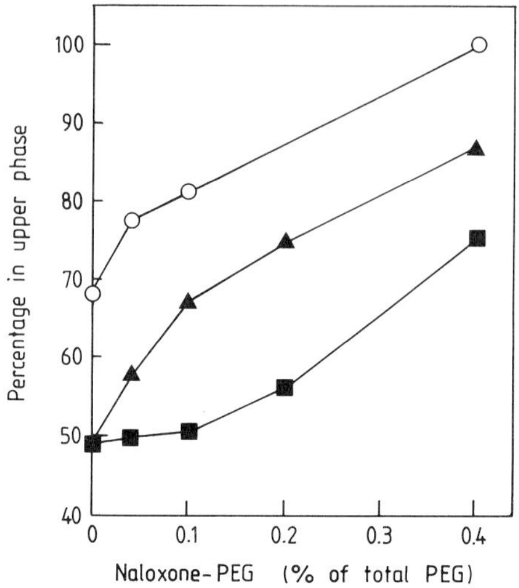


FIG. 2. Partitioning of synaptic membranes as function of naloxone–PEG concentration. ■, Membranes (determined by light scattering); ▲, stereospecific opiate-binding; and ○, acetylcholinesterase. The concentration of naloxone–PEG is expressed as a percentage of total PEG. The phase system contained 8.8% dextran T40, 5.9% PEG 3400, 5 m*M* Tris–$H_3PO_4$ (based on Tris), pH 7.4, and 1.7 m*M* KCl. Temperature, 4°. [Adapted with permission from B. Olde and G. Johansson, *Neuroscience (Oxford)* **15,** 1247 (1985). Copyright 1985 Pergamon Press Ltd.]

### *Affinity Extraction*

The upper phase from the system containing chloride is combined with a fresh lower phase, and the concentration of KCl is adjusted to 2–2.5 m*M*, so that less than 50% of the specific membranes (based on enzyme activity) remains in the upper phase. By adding ligand–PEG the membrane fragments containing the ligand-binding receptor can be extracted. The effect of using naloxone–PEG at different concentrations is shown in Fig. 2.

### *Countercurrent Distribution*

Countercurrent distribution (CCD) is a multiple extraction procedure[16] that permits resolution of membranes which differ only slightly in their partition ratios. Hence membrane heterogeneity and the effect of ligands on different parts of a membrane mixture can be studied. The influence of naloxone–PEG on the CCD of synaptic membranes is shown in Fig.

[16] H.-E. Åkerlund and P.-Å. Albertsson, this volume [8].

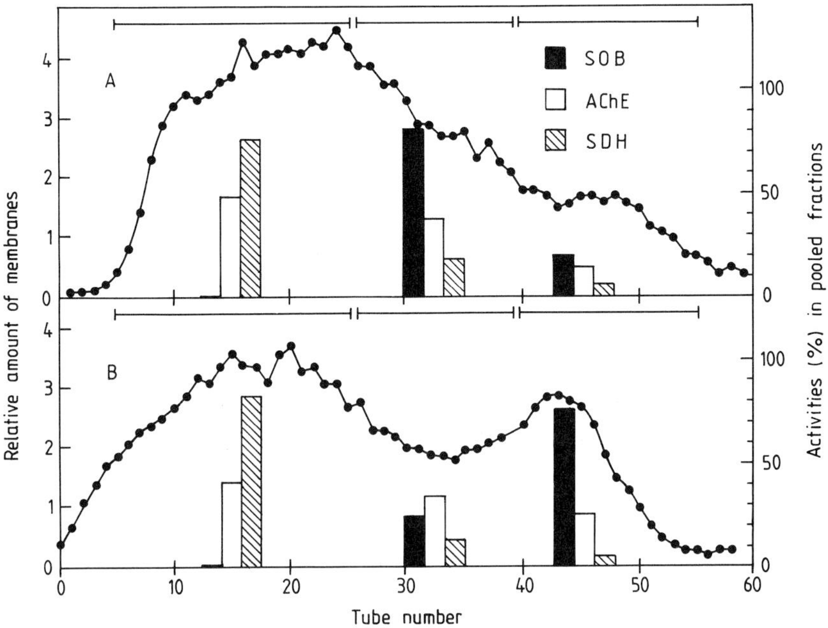


FIG. 3. Countercurrent distribution of synaptic membranes without (A) and with (B) naloxone–PEG. ●, Concentration of membranes measured as apparent absorbance due to light scattering. Pooled fractions, indicated by bars at top, were analyzed for stereospecific opiate binding (SOB), acetylcholinesterase (AChE), and succinate dehydrogenase (SDH), as indicated by bar graphs. The phase system was in Fig. 2 with 0.1% of the total PEG being naloxone–PEG 8000. [Adapted with permission from B. Olde and G. Johansson, *Neuroscience* (*Oxford*) **15,** 1247 (1985). Copyright 1985 Pergamon Press Ltd.]

3. The ligand–PEG causes membranes with (stereospecific) opiate-binding capability to shift from the middle of the diagram to the right, indicating increased affinity for the top phase. The position of mitochondrial membranes, identified by SDH, is not affected, whereas some of the AChE activity redistributes but to a lesser extent than that caused by membrane opiate binding.

*Instability of Membranes*

Fluid membranes, such as synaptic membranes, may alter their structure during the partitioning process which, in turn, may change their properties and cause aggregation.[5] This effect, especially apparent after CCD in which many transfers are carried out, can be minimized as follows.

TABLE I
RELATIVE MOBILITY, $R_f$, OF SYNAPTIC MEMBRANES (MAIN PEAK) IN COUNTERCURRENT DISTRIBUTION AS FUNCTION OF TIME[a]

| Ligand–PEG | Time (min) | Number of transfers | $R_f$ |
|---|---|---|---|
| None | 110 | 9 | 0.44 |
| | 110 | 55 | 0.42 |
| | 842 | 55 | 0.28 |
| Hexaethonium–PEG | 110 | 9 | 0.89 |
| | 110 | 55 | 0.81 |
| | 413 | 26 | 0.57 |
| | 842 | 55 | 0.41 |

[a] Adapted from G. Johansson, H.-E. Åkerlund, and B. Olde, *J. Chromatogr.* **311,** 277 (1984), by permission of Elsevier Science Publishers.

*Rapid Extraction.* The partitioning behavior of membranes may change with time during a CCD experiment. This change is reflected in the variation of the apparent relative mobility, $R_f$, that is, the relative position of the main peak after the CCD, as shown in Table I. The $R_f$ value decreases as a function of time but not with the number of transfers. This is due to an increasing affinity of membranes to the interface with time. To reduce the time for CCD the centrifugal apparatus described by Åkerlund can be used.[5,16]

*Choice of Polymers.* The polymers selected may be important. Membranes are more stable in systems with Dx of low molecular weight (e.g., Dx T40). Ficoll–Dx systems[6] are, in some cases, preferable to PEG–Dx systems.

*Addition of Proteins.* Addition of bovine serum albumin or ovalbumin to the systems improves the stability of membranes.

*Selection of Ligand Carrier.* Some ligand–PEG derivatives change the partitioning of membranes (increased accumulation at the interface) very rapidly even when no such effect is apparent with free ligand. Thus, polymers other than PEG might be more useful as ligand carriers.[17]

*Cryopartitioning.* An additional method to stabilize membranes is to carry out partitioning in glycerol-containing systems at $-10°$ to $-30°$. The low temperature may reduce membrane fluidity.[18]

[17] G. Johansson and M. Joelsson, *J. Chromatogr.* **411,** 161 (1987).
[18] G. Johansson and M. Joelsson, this volume [6].

## Discussion

The extraction of synaptic membranes containing opiate receptors shows that affinity partitioning, in combination with CCD, is a useful method for membrane fractionation. The directing effect of ligand–polymer is obvious despite the low density of receptors in the brain membranes. The choice of ligand carrier, phase-forming polymers and their molecular weights, system composition, temperature, and contact time may be crucial factors in achieving good separation.

## Acknowledgments

This work was supported by grants from NFR, the Swedish National Science Research Council.

# [49] Domain Structure of Biological Membranes Obtained by Fragmentation and Separation Analysis

*By* PER-ÅKE ALBERTSSON

## Introduction

Biomembranes are asymmetric structures. In addition to the transverse asymmetry they can also have lateral inhomogeneities. Many cell surfaces are highly polarized and consist of distinct macroscopic domains. For example, the basolateral and apical plasma membranes of polarized cells are separate domains with different compositions and functions. Membranes of cell organelles such as chloroplasts and mitochondria also consist of different domains (Fig. 1). Much information on this domain structure has been obtained by electron microscopy combined with immunogold labeling for localization of various membrane proteins. In this chapter, I describe a biochemical method in which fragmentation and separation analysis is used to test models for the domain structure of membranes.[1,2] As an example, the analysis of the chloroplast photosynthetic membrane, the thylakoid, is selected (see also the chapter by Gierow[3] in this volume).

[1] P.-Å. Albertsson, *Q. Rev. Biophys.* **21,** 61 (1988).

[2] P.-Å. Albertsson, E. Andreasson, P. Svensson, and S.-G. Yu, *Biochim. Biophys. Acta* **1098,** 90 (1991).

[3] J. P. Gierow, this volume [50].

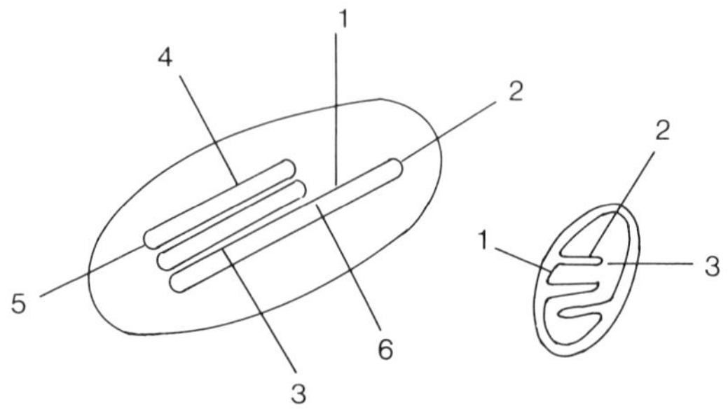


FIG. 1. From the appearance of organelles in the electron microscope one can predict that different domains will have different compositions and functions. (*Left*) Chloroplast thylakoid membrane: 1, stroma-exposed domain; 2, margin; 3, appressed domain of grana stack; 4, end membrane of grana stack; 5, margin of grana stack; 6, intermediate domain between nonappressed and appressed (fret). (*Right*) Inner mitochondrial membrane: 1, domain in contact with the outer membrane; 2, crista; 3, crista margin. [From P.-Å. Albertsson, *Q. Rev. Biophys.* **21,** 61 (1988).]

## Principle

When a biological membrane is fragmented mechanically (press treatment or sonication) it will yield vesicles of different sizes. If it is assumed that these vesicles originate from regions of the original membrane, and are not artificial products arising from fusion of fragments from different parts of the membrane, then separation and analysis of the vesicles gives information on the domain organization of the original membrane (Fig. 2). Biological membranes are closed structures and, for illustration, a model membrane containing two separate domains may be considered (Fig. 3a). Two components, A and B, are distributed unevenly between these domains. Random fragmentation will yield not only vesicles of different sizes but also of different compositions. Suppose that it is possible to separate these different vesicles and analyze them for their content of A and B. It can then be shown that, if the concentration of A is plotted against the concentration of B for all vesicles, the points will fit a straight line connecting the two points which represent the compositions, with respect to A and B, of the two pure domains (Fig. 3b).

Next, consider a membrane which consists of three domains (Fig. 4a). We assume again that our two components A and B are distributed differently among the three domains. Analysis of fragments obtained by random fragmentation shows that their compositions are represented by points within a triangle (Fig. 4b). The compositions of the pure domains are represented by the corners of the triangle.

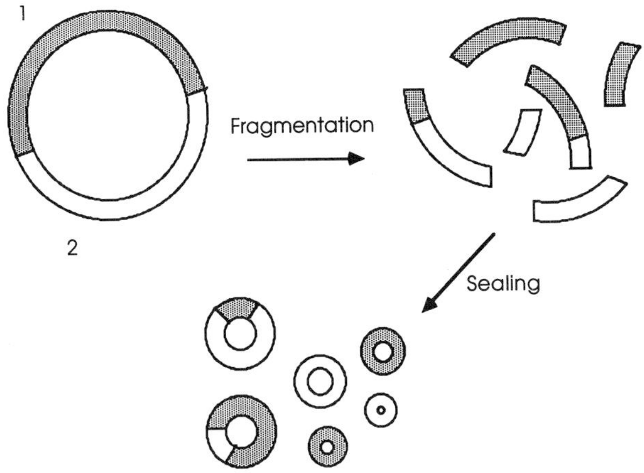


FIG. 2. Model of a membrane with two separate domains. Disintegration gives fragments which form vesicles. If the vesicles are formed only by intrafusion and not by fusion between different fragments, then separation and analyses of the separated vesicles with respect to domain components can give information on the original membrane structure.

Figure 5a shows another model with three different domains in which one, number 2, is located between the two others as a spacer. In this model, domain 1 is a neighbor of domain 2 which in turn is a neighbor of domain 3. However, domain 1 is not a neighbor of domain 3, in contrast

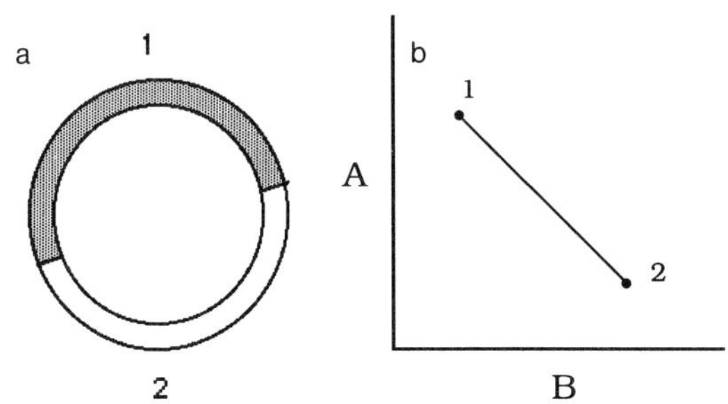


FIG. 3. (a) Model of a membrane containing two separate domains, 1 and 2. (b) Plot of the concentration of one component, A, versus another component, B, for all fragments obtained by random fragmentation of the membrane in (a). The points 1 and 2 represent compositions of pure domains 1 and 2, respectively. In this particular case A has a higher concentration in domain 1 while B has a higher concentration in domain 2; hence the negative slope of the line. [From P.-Å. Albertsson, E. Andreasson, P. Svensson, and S.-G. Yu, *Biochim. Biophys. Acta* **1098,** 90 (1991).]

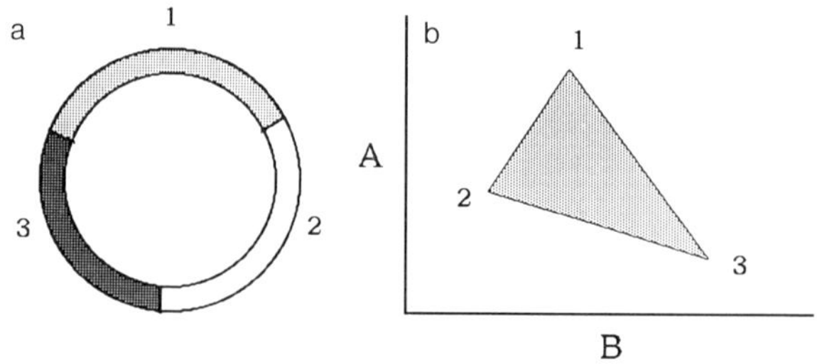


FIG. 4. (a) Model of a membrane containing three domains. (b) Plot of the concentration of one component, A, versus another component, B, for all fragments obtained by random fragmentation of the membrane in (a). The concentrations with respect to components A and B are represented by points within the shaded triangle. The corners 1, 2, and 3 represent the compositions of pure domains 1, 2, and 3, respectively. [From P.-Å. Albertsson, E. Andreasson, P. Svensson, and S.-G. Yu, *Biochim. Biophys. Acta* **1098,** 90 (1991).]

to the model in Fig. 4a in which all three domains are neighbors to one another. Analysis of the fragments obtained by random fragmentation of the membrane (Fig. 5) shows that their compositions are represented by points within the shaded area of Fig. 5b. Points representing the pure phases are the same as those in Fig. 4a, but there is an area representing compositions which are found in fragments from the model in Fig. 4a but not from the model in Fig. 5a. This area is bounded by lines between the points 1, $p$, and 3. For example, fragments consisting only of domains 1 and 3 will not be obtained from a membrane such as that in Fig. 5a.

By fragmentation of biological membranes followed by separation and analysis of the fragments we can gain information on the domain organization of the membrane when concentrations of different membrane compo-

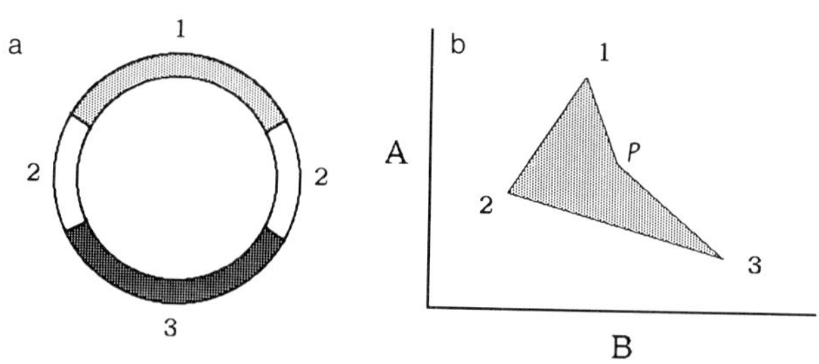


FIG. 5. (a) Model of a membrane containing three domains. Unlike the model in Fig. 4, the domains 1 and 3 are not neighbors. (b) Fragments of the model in (a) are represented by points in the shaded area with respect to the composition of the components A and B. [From P.-Å. Albertsson, E. Andreasson, P. Svensson, and S.-G. Yu, *Biochim. Biophys. Acta* **1098,** 90 (1991).]

nents are plotted against one another and the plots are compared with models.[1,2] By this method one can show whether the experimental results are consistent with a given model. One cannot prove a model to be correct, but one can prove that a model must be wrong.

## Materials and Methods

Isolation of spinach chloroplasts, sonication, and separation of thylakoid membrane vesicles are done as described by Andreasson *et al.*[4] Separation of grana-derived vesicles ($\alpha$ vesicles) from stroma lamellae vesicles ($\beta$ vesicles) is carried out either by countercurrent distribution or by a batch procedure.[3] Further sonication of the $\alpha$ vesicles, or inside-out vesicles obtained by press treatment, and separation into subfractions is carried out as described by Svensson and Albertsson.[5] Determining the concentration of P700 (reaction center of photosystem I) and measuring of the kinetics of P700 photooxidation are carried out as described by Albertsson *et al.*[2] Cytochrome *f* is determined immunologically as described by Andreasson *et al.*[4]

### *Fragmentation and Separation of Thylakoid Membrane*

A fragmentation and separation scheme for the thylakoid membrane is described elsewhere in this volume.[6] The procedure yields several samples with membrane vesicles originating from different parts of the thylakoid membrane. The first step is a sonication of whole, stacked thylakoids. Separation of the sonicate by partitioning in a dextran (Dx)–poly(ethylene glycol) (PEG) two-phase system results in two main fractions, B3 and T3, when a batch procedure is used or $\alpha$ vesicles and $\beta$ vesicles when countercurrent distribution is used.[4] B3 and $\alpha$ vesicles originate from the grana, whereas T3 and $\beta$ vesicles originate from the stroma lamellae. The grana vesicles (B3 or $\alpha$) can be fragmented further by repeated sonications, and the smaller vesicles so obtained can again be separated either by a batch procedure or by countercurrent distribution. With the batch procedure developed by Svensson and Albertsson,[5] fractions named for the length of sonication (120 sec, 180 sec, etc., for fractions sonicated for 120, 180 sec, etc.) are obtained together with BS, which is the fraction which resides in the bottom phase after several sonication–partitioning steps.

[4] E. Andreasson, P. Svensson, C. Weibull, and P.-Å. Albertsson, *Biochim. Biophys. Acta* **936,** 339 (1988).

[5] P. Svensson and P.-Å. Albertsson, *Photosynth. Res.* **20,** 249 (1989).

[6] P.-Å. Albertsson, E. Andreasson, H. Stefánsson, and L. Wollenberger, this volume [45].

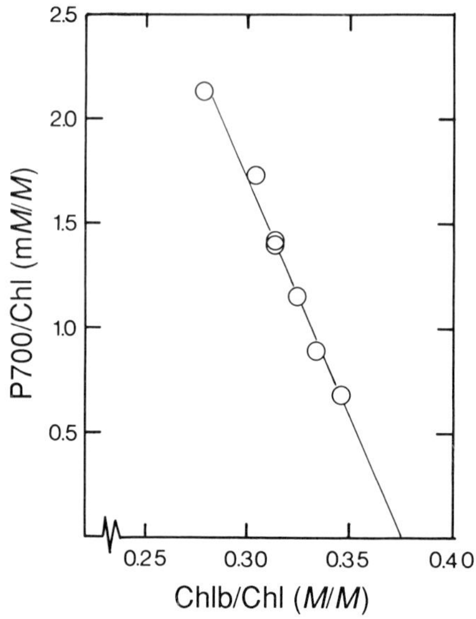


FIG. 6. Plot of P700 per chlorophyll ($a + b$) (Chl) against chlorophyll $b$ (Chl $b$) per chlorophyll ($a + b$). The line intersects the abscissa (P700 = 0) when Chl $b$/chlorophyll ($a + b$) equals 0.37, that is, when the chlorophyll $a/b$ ratio is 1.7. [From P. Svensson and P.-Å. Albertsson, *Photosynth. Res.* **20,** 249 (1989).]

## *Plotting of Data*

A plot of data for different vesicles originating from the grana region of the thylakoid is shown in Fig. 6. The concentration of two components, P700 and chlorophyll *b*, are plotted against each other (chlorophyll *b* is a component of the light-harvesting complex of photosystem II). The concentration is expressed as millimolar per molar total chlorophyll, the latter being a convenient, but approximate, way of measuring the amount of membrane material. One can also express the concentration per unit protein or any other component. The data in Fig. 6 fit a straight line very well and support a model in which the grana consist of at least two separate domains, one enriched in photosystem I and depleted in photosystem II and the other depleted in photosystem I and enriched in photosystem II. By extrapolation one can deduce that if there is a domain with zero concentration P700, that is, containing only photosystem II, then this domain would have a chlorophyll *a*/*b* ratio of 1.7 (chl *b*/chl = 0.37). This agrees well with the chlorophyll *a*/*b* ratio of isolated photosystem II with its light-harvesting complex.

A plot of P700 against cytochrome *f* (Cyt *f*) is shown in Fig. 7. In this case all fractions obtained by the fragmentation and separation procedure

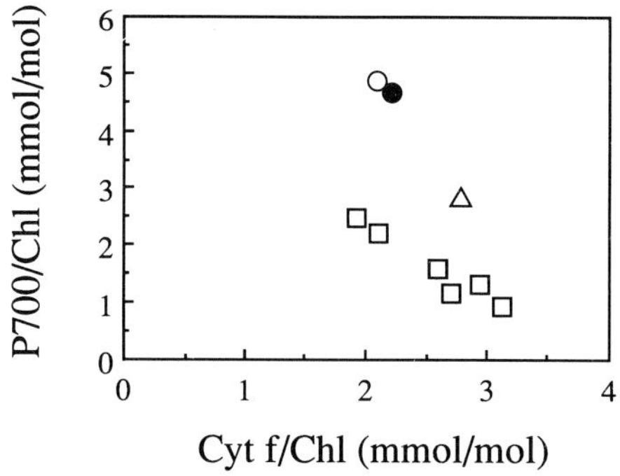


FIG. 7. P700/Chl as a function of Cyt *f*/Chl for thylakoid membrane fractions. □, Grana-derived membrane vesicles; △, thylakoids; and two stroma-derived membrane fractions: (●) β vesicles and (○) Y100 [From P.-Å. Albertsson, E. Andreasson, P. Svensson, and S.-G. Yu, *Biochim. Biophys. Acta* **1098,** 90 (1991).]

of whole thylakoids as starting material were used. The data can be fitted to a triangle rather than to a straight line and thus support a model of at least three separate domains for the whole thylakoid. One of these is the stroma lamellae (circles), and the other two originate in the grana. The grana fractions (squares) fit a straight line very well, as in the plot in Fig.

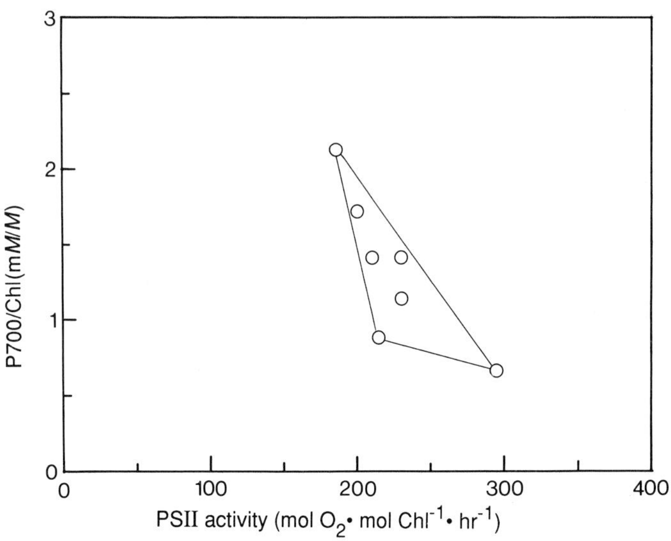


FIG. 8. Photosystem II (PSII) activity plotted against P700/Chl for different vesicles obtained by fragmentation and separation of grana vesicles (B3). [Data from procedure described by P. Svensson and P.-Å. Albertsson, *Photosynth. Res.* **20,** 249 (1989).]

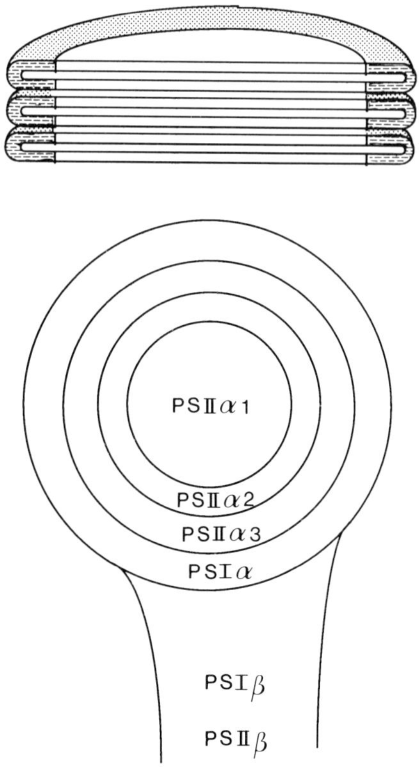


FIG. 9. Model of the thylakoid membrane from higher plants. (*Top*) Section through the interior of a granum. The gray, flat annulus is the PSI$\alpha$ domain which encloses the central (white) domain of PSII$\alpha$. The connections to the stroma lamellae have been omitted. (*Bottom*) Top view of a single thylakoid; PSII$\alpha_1$, PSII$\alpha_2$, and PSII$\alpha_3$ stand for domains with PSII$\alpha$ units having different antennae sizes ($\alpha_1 > \alpha_2 > \alpha_3$). PSI$\alpha$ stands for the gray annular zone at the top. [From P.-Å. Albertsson, E. Andreasson, and P. Svensson, *FEBS Lett.* **273,** 36 (1990).]

6, again supporting a grana model with at least two domains. One of these is the grana margins and the others are located inside the periphery of the grana (see below).

If the grana fractions are analyzed with respect to their photosystem II activity (oxygen evolution with phenyl-*p*-benzoquinone as acceptor) and P700 content, a plot as shown in Fig. 8 is obtained. These data do not fit a straight line but rather a triangle, and one has to conclude, therefore, that there are at least three domains in the grana. These results are supported by the finding that the average antenna size of photosystem II differs in the various vesicle populations originating from the grana.[7,8]

The model of the thylakoid membrane[9] shown in Fig. 9 is based on fractionation experiments of the type described above and on electron microscopy. The stroma lamellae harbor the two photosystems I$\beta$ and II$\beta$, whereas the photosystems I$\alpha$ and II$\alpha$ are localized in the grana. Photosystem I$\alpha$ is confined to an outer annulus of the grana, whereas photosystem II$\alpha$ is in the center of the grana disk. The latter, in turn, is suggested to be composed of concentric rings with photosystem II units having different antenna sizes, those with the largest antennae being localized in the center of the grana where the light intensity is lowest. The cytochrome *b*/*f* complex is found all over the thylakoid membrane but has different concentrations in the different domains as shown in Fig. 7. The ATP synthetase complex is localized exclusively in the stroma-exposed domains (the stroma lamellae and the periphery of the grana), that is, at the same location as the different photosystems I. This is supported by the fact that ATP synthetase always copurifies with photosystem I in the fractionation scheme.[5]

## Concluding Remarks

Fragmentation and separation analysis is a powerful tool for the study of lateral heterogeneity of biomembranes. It can be used to test models for the organization of biomembranes and thereby complement electron microscopy. An essential requirement is that the starting material (cells or cell organelles) which is to be analyzed be homogeneous. Also one should remember that the various plots of data cannot be used to prove that a certain model is correct. They can, however, be used to show that a certain model is probably wrong.

[7] P.-Å. Albertsson and S.-G. Yu, *Biochim. Biophys. Acta* **936,** 215 (1988).

[8] P.-Å. Albertsson and S.-G. Yu, *Biochim. Biophys. Acta* **1016,** 137 (1990).

[9] P.-Å. Albertsson, E. Andreasson, and P. Svensson, *FEBS Lett.* **273,** 36 (1990).

# [50] Relative Proximity of Domains in Plasma Membrane and Smooth Endoplasmic Reticulum from Rat Liver

*By* J. PETER GIEROW

## Introduction

A number of methods have been developed for the isolation of different membrane populations from rat liver. Most of these are based on a combination of differential and density gradient centrifugation and provide little information about the organization of the membrane populations obtained. The membranes can, in most cases, be further subfractionated by use of aqueous two-phase partitioning, which separates membranes on the basis of properties other than size or density. The fragmentation of a membrane which consists of discrete domains will yield fragments with differing compositions depending on the size of the fragment and the organization of the domains. A method for analyzing such a randomly fragmented membrane has been described by Albertsson,[1] in which the concentrations of membrane markers in different subfractions are plotted, one by one, against one another. This technique has been used to study the organization of the thylakoid membrane.[2,3] This chapter describes its application to plasma membranes[4] and smooth endoplasmic reticulum[5] from rat liver after subfractionation by countercurrent distribution in an aqueous two-phase polymer system.

## Methods

### *Preparation of Plasma Membranes*

*Solutions*

8.4% (w/w) Sucrose in 5 m*M* Tris-HCl, pH 8.0
34% (w/w) Sucrose in 5 m*M* Tris-HCl, pH 8.0

[1] P.-Å. Albertsson, *Q. Rev. Biophys.* **21,** 61 (1988).
[2] P.-Å. Albertsson, this volume [49].
[3] P.-Å. Albertsson, E. Andreasson, P. Svensson, and S.-G. Yu, *Biochim. Biophys. Acta* **1098,** 90 (1991).
[4] P. Gierow, B. Jergil, and P.-Å. Albertsson, *Biochem. J.* **249,** 369 (1988).
[5] P. Gierow and B. Jergil, *Biochem. J.* **262,** 55 (1989).

57% (w/w) Sucrose in 5 m*M* Tris-HCl, pH 8.0
15 m*M* Tris-$H_2SO_4$, pH 7.8

*Procedure.* The $P_2$ fraction of Aronson and Touster,[6] mainly containing plasma membranes from the blood sinusoidal and bile canalicular domains, is isolated by a modification of the method described earlier.[4] Typically two male Sprague-Dawley rats, weighing around 200 g, are starved overnight and sacrificed the following morning by decapitation. The livers are weighed and quickly transferred to ice-cold 8.4% sucrose in 5 m*M* Tris-HCl, pH 8.0. The tissue is cut into small pieces with a pair of scissors and passed through a tissue press. It is then homogenized at 4° in 3 ml of 8.4% sucrose buffer per gram wet tissue by six up-and-down strokes with a Dounce homogenizer (loose-fitting pestle). The homogenate is further processed by one up-and-down stroke in a glass–Teflon Potter–Elvehjem homogenizer at a pestle speed of 1000 rpm. The pellet obtained after centrifugation at 1000 *g* (SS-34 rotor, Sorvall) for 10 min is extracted twice by repeated Potter–Elvehjem homogenization and centrifugation as above.

The three supernatant solutions thus obtained are combined and centrifuged at 27,000 *g* for 11.5 min (timed from the start of acceleration to beginning of deceleration). The resulting supernatant solution and the loose upper part of the pellet are collected and pelleted by centrifugation at 105,000 *g* for 90 min at 4° in a Beckman 45 Ti-rotor. The microsomal pellet thus obtained is suspended by two up-and-down strokes with the Potter–Elvehjem homogenizer in 1.5 volumes of buffered 57% sucrose. The homogenizer is rinsed with a small volume of the same buffer. The two solutions are combined and further homogenized by 20 up-and-down strokes in a Dounce homogenizer with a tight-fitting pestle. The homogenizer is rinsed and the final volume adjusted to 2.5 volumes/g wet tissue with 57% sucrose.

The microsomal suspension is divided into centrifuge tubes, approximately 15 ml in each, and on each is layered 18 ml of 34% sucrose. The tubes are then filled to 0.5 cm from the rim by careful layering of 8.4% sucrose and centrifuged at 95,000 *g* in a SW-27 rotor (Beckman) for 16 hr. The layer at the 8.4/34% sucrose interface, containing predominantly plasma membranes, is harvested by aspiration, diluted 2- to 3-fold with 15 m*M* Tris-$H_2SO_4$, pH 7.8, and sedimented by centrifugation at 105,000 *g* for 90 min. The final pellet is brought to the desired volume by suspension in the latter buffer.

[6] N. N. Aronson and O. Touster, Jr., this series, Vol. 31, p. 90.

*Preparation of Smooth Endoplasmic Reticulum*

*Solutions*

0.35 *M* Sucrose in 10 m*M* $MgCl_2$, 25 m*M* KCl, 0.1 m*M* dithioerythritol, and 50 m*M* Tris-HCl, pH 7.4
0.9 *M* Sucrose (buffered as above)
1.3 *M* Sucrose (buffered as above)
15 m*M* LiOH and 1 m*M* Tris-$H_2SO_4$, pH 7.8

*Procedure.* Microsomes from smooth endoplasmic reticulum are prepared essentially as described by Lewis and Tata.[7] The liver from one rat, treated as above, is transferred to 2.5 volumes ice-cold, buffered 0.35 *M* sucrose, cut into small pieces with a pair of scissors, and homogenized by six up-and-down strokes in a Potter–Elvehjem homogenizer. The postnuclear supernatant solution, obtained after centrifugation at 640 *g* for 10 min, is centrifuged at 6700 *g* for 10 min at 4°. A microsomal fraction is obtained from the resulting supernatant solution by sedimentation at 105,000 *g* for 1 hr. The pellet is suspended by 20 up-and-down strokes in a Dounce homogenizer (tight-fitting pestle) in the buffer indicated above. The suspension (24 ml) is carefully layered on top of two gradients, each containing 10 ml of 0.9 *M* sucrose and 15 ml of 1.3 *M* sucrose in Spinco SW-27 tubes. The membranes at the 0.9–1.3 *M* sucrose interface are collected, after centrifugation at 95,000 *g* for 6.5 hr in a Beckman SW-27 swing-out rotor, diluted 3-fold with 15 m*M* LiOH and 1 m*M* Tris-$H_2SO_4$, pH 7.8, and sedimented at 105,000 *g* for 1 hr. The resulting pellet is suspended in minute amounts of the above-indicated buffer and used immediately in experiments outlined below.

*Preparation of Two-Phase Systems*

*Solutions*

20% (w/w) Dextran (Dx) T500 (Pharmacia-LKB, Piscataway, NJ)
40% (w/w) Poly(ethylene glycol) (PEG) 3350 (Carbowax 3350; Union Carbide, New York, NY)
0.5 *M* Tris-$H_2SO_4$, pH 7.8
0.1 *M* KCl
0.25 *M* LiOH and 25 m*M* Tris-$H_2SO_4$, pH 7.8
10 m*M* Tris-HCl, pH 7.4

*Procedure.* Stock solutions of Dx and PEG are prepared and the concentration of Dx is measured polarimetrically as described by López-

[7] J. A. Lewis and J. R. Tata, *J. Cell Sci.* **13,** 447 (1973).

Pérez *et al.*[8] and by Brooks and Norris-Jones.[9] The solutions can be stored below 0° until use. To determine the proper phase system composition needed to obtain a satisfactory separation of the membrane populations of interest, a series of miniature phase systems (total weight 2–4 g) which differ in polymer concentrations and ion composition are prepared.[10] It is usually desirable that approximately 50% of the membrane of interest be in the top phase. A discussion of the effects of polymer concentration and different ions on membrane partitioning is presented elsewhere in this volume.[10] A phase system containing 6.0% Dx T500, 6.0% PEG 3350, 15 m*M* Tris-$H_2SO_4$, pH 7.8, and 3 m*M* KCl has been found to separate efficiently plasma membrane domains.[4] A system containing 5.9% Dx T500, 5.9% PEG 3350, 10 m*M* LiOH, and 1 m*M* Tris-$H_2SO_4$, pH 7.8, was shown to be useful when studying smooth endoplasmic reticulum.[5] It is advantageous, if possible, to perform test runs comparing separations obtained on the thin-layer countercurrent distribution (CCD) apparatus with those obtained in the centrifugal version,[11] since there are differences in membrane partitioning behavior in the two models, probably owing to phase separation dynamics. In the case of plasma membranes the centrifugal CCD unit gave a better resolution, whereas a more useful profile of smooth endoplasmic reticulum was obtained using the standard thin-layer apparatus.

Two-phase systems in which the separation is to be carried out are obtained by weighing the phase system components into a separatory funnel. The total amount needed to load the centrifugal CCD unit is 175 g and for the 60-cavity thin-layer apparatus, 150 g. The solution is allowed to reach the temperature at which the separation is to be carried out, namely, 4°, and the contents are then thoroughly mixed. The phase system is allowed to settle overnight, and top and bottom phases are then collected separately. A sample system (see below) is prepared simultaneously. The membrane sample is added just prior to loading the CCD apparatus.

### *Centrifugal Countercurrent Distribution of Plasma Membranes*

One milliliter of freshly prepared plasma membranes (20 mg of protein) is sonicated for different periods of time (Branson Sonifer cell disruptor, Model B15, Branson Sonic Power Company, Danbury, CT; equipped with a microtip, intensity output 2, 50% pulsed duty cycle, total time 0, 1.5, or 6 min) to reduce the vesicle size and is then added to a sample system,

[8] M. J. López-Pérez, G. Paris, and C. Larsson, *Biochim. Biophys. Acta* **635,** 359 (1981).

[9] D. E. Brooks and R. Norris-Jones, this volume [2].

[10] H. Walter and C. Larsson, this volume [4].

[11] H.-E. Åkerlund and P.-Å. Albertsson, this volume [8].

with the composition described above and a total weight of 10 g. The sample system is designed so that the desired concentration is reached by the addition of the sample. Bottom phase (2.00 g) is added to obtain equal volumes of the two phases. The capacity of the bottom phase compartment of the partition block is 1.00 ml, but only 0.90 ml is used in order to keep both the bottom phase and the interface stationary.[11] Thus, the first five cavities (numbered 0–4) are each loaded with 1.80 ml of well-mixed sample system, and the remaining 55 cavities with 0.90 ml bottom phase and 0.90 ml top phase. Shaking is set to 30 sec, and centrifugation to 90 sec, resulting in a total time of less than 2 hr for 53 transfers. The phase system is broken by addition of 1.80 ml of 10 m*M* Tris-HCl, pH 7.4, to each cavity. The whole block is shaken for 1 min, and the fractions are collected.

*Thin-Layer Countercurrent Distribution of Smooth Endoplasmic Reticulum*

Smooth microsomes (30 mg) are added to a sample system with a total weight of 8.0 g and the composition described above. Bottom phase is added to obtain equal volumes of the two phases before the whole sample system is sonicated 5 times for 15 sec each, at 1-min intervals (intensity output 3, 50% pulsed duty cycle). The sample system and top and bottom phases are loaded onto a standard CCD apparatus equipped with a partition block with 60 cavities (numbered 0–59). The bottom phase capacity is 0.79 ml, but only 0.71 ml of each of the two phases is added to cavities 5–59 in order to keep the bottom phase and the interface stationary. The remaining five cavities (0–4) are loaded with 1.42 ml of well-mixed sample system. A mixing time of 20 sec and a settling time of 12 min is used, and 55 transfers are carried out. The phase system is broken by addition of 1 ml of 10 m*M* Tris-HCl, pH 7.4, and the whole block is shaken for 1 min before unloading. The fractions collected have to be analyzed immediately, owing to a limited stability of the enzymes, particularly glucose-6-phosphatase. Repeated freezing and thawing decrease the enzyme activity considerably.

*Assays*

The plasma membrane markers assayed for are 5-nucleotidase,[12] leucine aminopeptidase,[13] and $^{125}$I-labeled asialoorosomucoid receptors.[14,15]

[12] J. Avruch and D. F. H. Wallach, *Biochim. Biophys. Acta* **233,** 334 (1971).
[13] L. M. Roman and A. L. Hubbard, *J. Cell Biol.* **96,** 1548 (1983).
[14] A. L. Hubbard, D. A. Wall, and A. Ma, *J. Cell Biol.* **96,** 217 (1983).
[15] P. Gierow, M. Sommarin, C. Larsson, and B. Jergil, *Biochem. J.* **235,** 685 (1986).

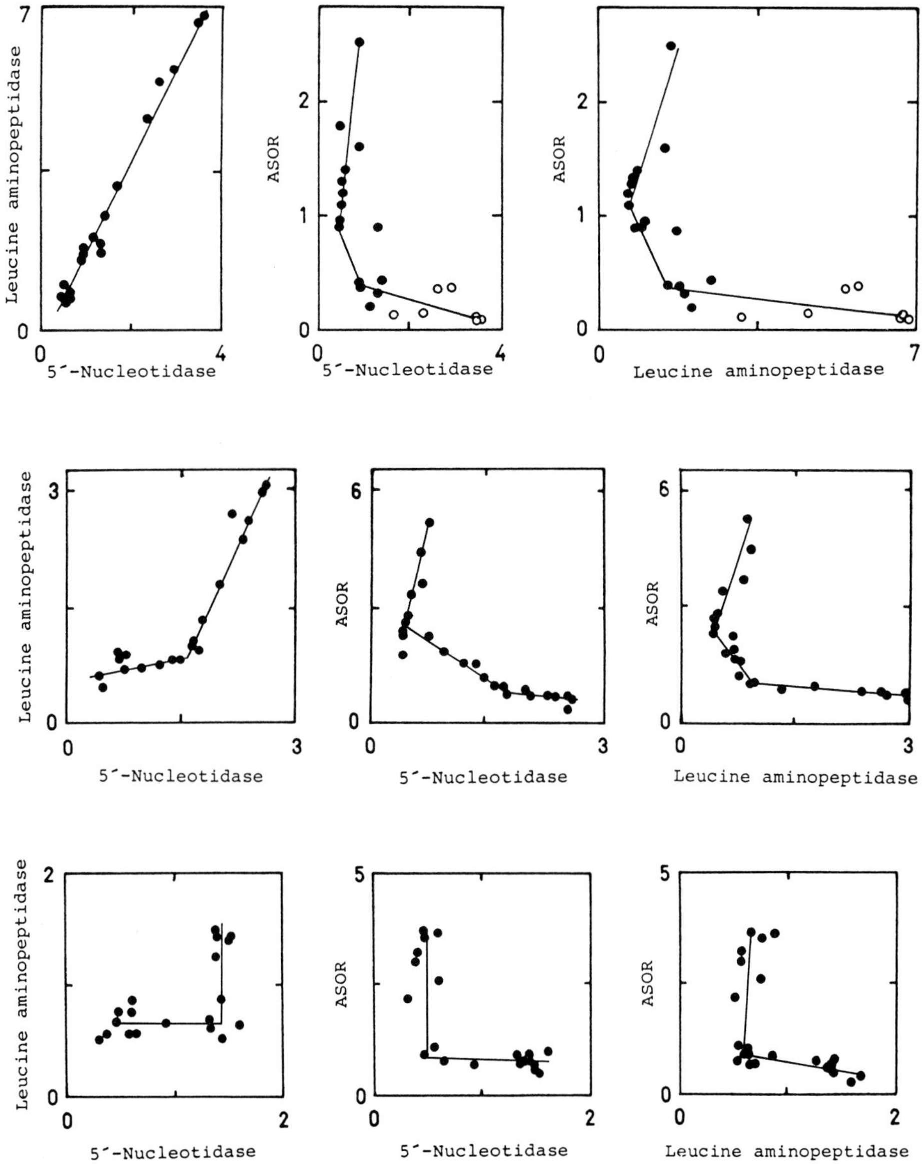


FIG. 1. Pairwise plots of relative specific enzyme activities. The relative specific activity of an enzyme in each countercurrent distribution fraction is calculated on the basis of the specific enzyme activity in the sample added and plotted against the corresponding activity of another enzyme. The top row depicts unsonicated membranes, the middle row membranes sonicated 1.5 min (total time), and the bottom row membranes sonicated for 6 min (total time). ASOR indicates asialoorosomucoid receptor binding. [From P. Gierow, B. Jergil, and P.-Å. Albertsson, *Biochem. J.* **249,** 369 (1988).]

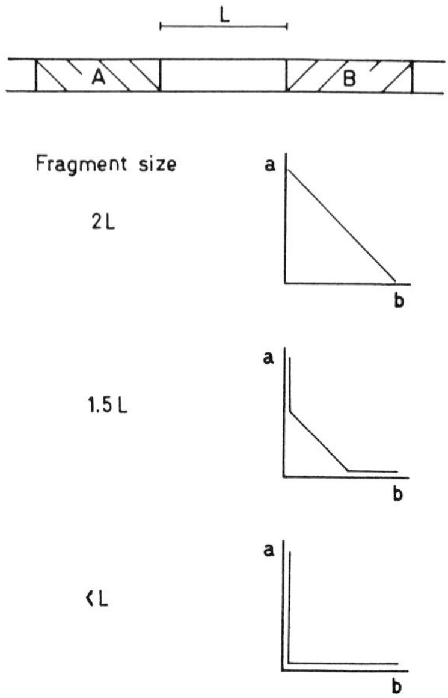


FIG. 2. Cross section of a hypothetical membrane consisting of three alternating domains (each of size $L$), one containing marker A, the second devoid of marker, and the third containing marker B. Depicted below are expected concentration plots of A versus B for fragments of different sizes. [From P. Gierow, B. Jergil, and P.-Å. Albertsson, *Biochem. J.* **249,** 369 (1988).]

In the study of the smooth endoplasmic reticulum the distribution of the following enzymes are analyzed: glucose-6-phosphatase,[16] NADPH–cytochrome-*c* reductase,[17] NADH–ferricyanide reductase,[17] and arylesterase,[18] all by previously described procedures. Protein can be determined either by a modification[15] of the Lowry method[19] or by the microassay of Bradford[20] as described elsewhere.[5]

### Analysis

Two-phase partitioning of plasma membranes results in a distribution profile with two dominating peaks, one to the far left and one to the far right. The left-hand peak, containing material partitioning into the bottom

[16] P. Gierow and B. Jergil, this series, Vol. 89, p. 44.

[17] E.-G. Hrycay and P. J. O'Brien, *Arch. Biophys. Biochem.* **160,** 230 (1974).

[18] E. H. Shephard and G. Hubscher, *Biochem. J.* **113,** 429 (1969).

[19] O. H. Lowry, N. J. Rosebrough, A. L. Farr, and R. J. Randall, *J. Biol. Chem.* **95,** 265 (1951).

[20] M. M. Bradford, *Anal. Biochem.* **72,** 248 (1976).

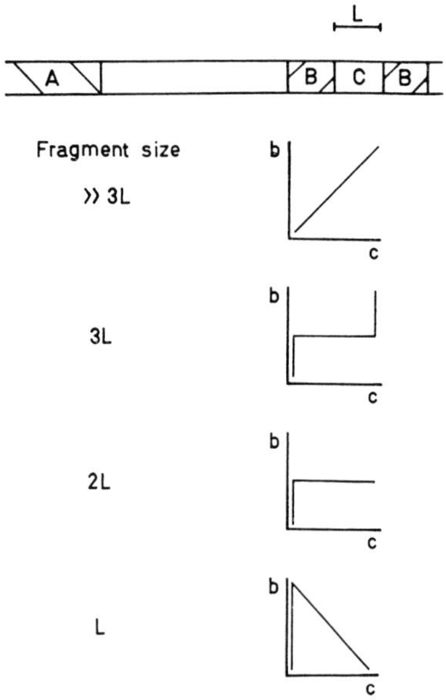


FIG. 3. Cross section of a hypothetical membrane consisting of domain B surrounding domain C, both having the size $L$, and being well separated from domain A by a fourth domain devoid of markers. Depicted below are concentration plots of B versus C for fragments of different sizes obtained by random fragmentation of the membrane. [From P. Gierow, B. Jergil, and P.-Å. Albertsson, *Biochem. J.* **249,** 369 (1988).]

phase and to the interface, is enriched in asialoorosomucoid binding sites, markers of the blood sinusoidal domain of the plasma membrane, whereas the right-hand peak is enriched in leucine aminopeptidase, a bile canalicular marker, and 5′-nucleotidase, which is regarded as a general plasma membrane marker. The domain organization of these markers can be further analyzed by plotting the specific activities of the three markers against one another, fraction by fraction, as shown in Fig. 1. These plots can be compared with plots of hypothetical models and fragment sizes. The results indicate that asialoorosomucoid binding sites are located in a domain separated from the domain containing 5′-nucleotidase and leucine aminopeptidase by an additional domain lacking these markers (Fig. 2). The configuration obtained by plotting leucine aminopeptidase against 5′-nucleotidase suggests that they are localized in close but separate domains, similar to the model in Fig. 3 with the axes reversed.

CCD of smooth endoplasmic reticulum results in a similar distribution profile with two dominating peaks. However, several populations can be detected by comparing the experimentally obtained curves with theoretical

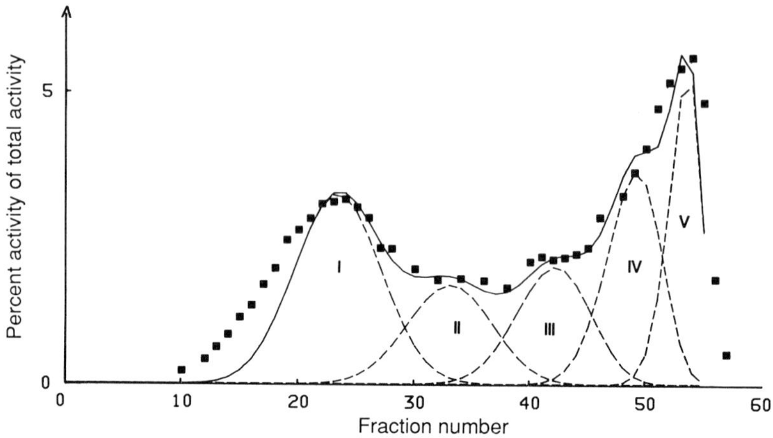


FIG. 4. Experimental data compared with theoretical curves. The experimentally obtained distribution of protein (■) is compared with the theoretically calculated populations (broken line, numbered I–V) and the sum of these populations (solid line). [From P. Gierow and B. Jergil, *Biochem. J.* **262,** 55 (1989).]

curves calculated by a computerized method based on a mathematical model of two-phase partitioning.[21,22] The comparison requires the determination of the apparent partition ratio, $G$, of the material to be analyzed. This can be calculated by using the experimentally determined quantity of the material in two consecutive fractions, $A_{n,i}$ and $A_{n,i+1}$, since

$$A_{n,i+1}/A_{n,i} = G(n - i)/(i + 1)$$

where $n$ is the number of transfers (here 55) and $i$ the fraction number. The $G$ value can then be used to calculate the theoretical amount, $T_{n,i}$, for each fraction, since

$$T_{n,i} = n![i!(n - i)!]^{-1} \times G^i(1 + G)^{-n}$$

Thus, several theoretical curves can be obtained until their summation curve matches the experimental curve.

When the analysis is applied to the CCD of smooth endoplasmic reticulum at least five membrane populations can be identified, as indicated by the protein profile in Fig. 4. Analysis of the marker enzymes also results in five populations. No significant differences are observed in the apparent

[21] L. C. Craig, *in* "Comprehensive Biochemistry" (M. Florkin and E. H. Stotz, eds.), Vol. 4, p. 1. Elsevier, Amsterdam, 1962.

[22] L. J. O. R. Morris and P. Morris, "Separation Methods in Biochemistry," 2nd Ed., p. 639. Pitman, Bath, 1976.

TABLE I

APPARENT PARTITION RATIO AND RELATIVE DISTRIBUTION OF COMPONENTS CALCULATED FROM THEORETICAL CURVES[a]

| Enzyme | Ratio or fraction | Peak | | | | |
|---|---|---|---|---|---|---|
| | | I | II | III | IV | V |
| Glucose-6-phosphatase | $G$ | 0.80 ± 0.04 | 1.45 ± 0.11 | 2.70 ± 0.23 | 6.13 ± 1.63 | 20.8 ± 9.6 |
| | $P$ | 0.39 ± 0.03 | 0.18 ± 0.03 | 0.14 ± 0.01 | 0.15 ± 0.04 | 0.16 ± 0.05 |
| NADPH–cytochrome-*c* reductase | $G$ | 0.75 ± 0.04 | 1.47 ± 0.06 | 2.76 ± 0.11 | 5.80 ± 0.56 | 17.8 ± 3.2 |
| | $P$ | 0.35 ± 0.05 | 0.15 ± 0.02 | 0.16 ± 0.02 | 0.16 ± 0.02 | 0.18 ± 0.04 |
| NADH–ferricyanide reductase | $G$ | 0.78 ± 0.05 | 1.47 ± 0.07 | 2.63 ± 0.09 | 5.88 ± 0.57 | 17.0 ± 3.2 |
| | $P$ | 0.32 ± 0.06 | 0.14 ± 0.01 | 0.17 ± 0.02 | 0.17 ± 0.03 | 0.20 ± 0.05 |
| Arylesterase | $G$ | 0.78 ± 0.02 | 1.53 ± 0.10 | 2.71 ± 0.21 | 5.90 ± 0.82 | 19.0 ± 4.3 |
| | $P$ | 0.43 ± 0.02 | 0.18 ± 0.01 | 0.14 ± 0.01 | 0.13 ± 0.02 | 0.12 ± 0.03 |

[a] Numbers refer to the peak numbers indicated in Fig. 4. $G$ is the apparent partition ratio, and $P$ the fraction of the enzyme in the different peaks (subpopulations), both expressed as means (±S.D.) from four experiments. From P. Gierow and B. Jergil, *Biochem. J.* **262,** 55 (1989).

partition ratios, *G* values, among the enzymes (Table I), but there is a tendency for glucose-6-phosphatase and arylesterase, both located on the lumenal side of the membrane, to be distributed to the left-hand part of the curve (peaks I and II). The two reductases, both exposed on the cytoplasmic surface, are distributed further to the right. Rather complex patterns are observed when the specific activities of the enzymes are plotted against one another, consistent with the occurrence of several populations. For a detailed discussion of these plots, see Ref. 5. It should be noted that the activity plots of glucose-6-phosphatase and arylesterase also suggested a codistribution of these enzymes. The two reductases seemed to be rather close in the distribution curve, but the more sensitive activity plots indicate differences in their distribution. Owing to complexity it is difficult to construct even a hypothetical membrane model corresponding to the activity plots and the theoretical curves. However, a multidomain membrane model has recently been proposed in a study of the thylakoid membrane,[2,3] in which the theoretical activity plots are very similar to some of the plots obtained with the smooth endoplasmic reticulum.

## Conclusion

This chapter has described the use of countercurrent distribution of membrane fragments in combination with plots of their specific enzyme activities to analyze the heterogeneity of both plasma membranes and the smooth endoplasmic reticulum. The technique should be applicable in studies on the domain organization of biological membranes in general as well as in constructing theoretical models of the membrane.

## Acknowledgments

This work was performed at the Department of Biochemistry, University of Lund, Sweden, and was supported by a grant from the Swedish Natural Science Research Council to Dr. Bengt Jergil. The computer program was kindly provided by Dr. Lars-Olof Persson.

# Section V

## Aqueous Phase Separations in Biological Systems

# [51] Lens Cytoplasmic Protein Solutions: Analysis of a Biologically Occurring Aqueous Phase Separation

*By* JOHN I. CLARK

## Introduction

The importance of phase equilibria and critical phenomena is widely recognized in physics, polymer chemistry, materials sciences, and biotechnology.[1-5] The 1991 Nobel Prize in Physics was awarded to Pierre-Gilles deGennes for contributions to the theory of phase equilibria in magnets, superconductors, and polymers. The applications of aqueous two-phase systems to the purification, characterization, and study of a wide variety of biomaterials have been documented,[5,6] and this volume summarizes some of the methodology used.

The physical properties of phase systems are best described by phase diagrams,[7-9] the construction of which is outlined in this volume[10] for two-polymer and polymer–salt aqueous two-phase systems. In contrast, the study of biological phase equilibria is still in its infancy. Aqueous highly concentrated protein solutions (as found in the cytoplasm of cells) may phase separate to yield aqueous protein-rich and protein-poor phases or distinct phases that are protein-rich in specific constituent proteins.[11-13]

[1] P. J. Flory, "Principles of Polymer Chemistry." Cornell Univ. Press, Ithaca, New York, 1953.

[2] H. E. Stanley, "Introduction to Phase Transitions and Critical Phenomena." Oxford Univ. Press, New York, 1971.

[3] P. Glansdorff and I. Prigogine, "Thermodynamic Theory of Structure, Stability and Fluctuations." Wiley(Interscience), New York, 1971.

[4] K. Kamide, "Thermodynamics of Polymer Solutions: Phase Equilibria and Critical Phenomena." Elsevier, New York, 1990.

[5] H. Walter, D. E. Brooks, and D. Fisher (eds.), "Partitioning in Aqueous Two-Phase Systems: Theory, Methods, Uses, and Applications to Biotechnology." Academic Press, Orlando, Florida, 1985.

[6] P.-Å. Albertsson, "Partition of Cell Particles and Macromolecules." Wiley(Interscience), New York, 1986.

[7] G. B. Benedek, *in* "Phase Transitions" *Proc. XIV Conf. Chem. (Univ. Brussels)* p. 149. Wiley(Interscience), New York, 1971.

[8] R. Van Der Haegen and B. W. Ready, *in* "Comprehensive Polymer Science" (G. Allen and J. C. Bevington, eds.), p. 121. Pergamon, New York, 1989.

[9] F. Tamas and I. Pal, "Phase Equilibria Spatial Diagrams." Iliffe Books, London, 1970.

[10] P.-Å. Albertsson and F. Tjerneld, this volume [1].

[11] V. B. Tolstoguzov, *Food Hydrocolloids* **2,** 339 (1988).

[12] J. I. Clark and G. B. Benedek, *Biochem. Biophys. Res. Commun.* **95,** 482 (1980).

Research on cell-free systems of lens cytoplasm demonstrates the value of using phase diagrams as a means for analyzing the biologial significance of the molecular interactions and thermodynamic properties of cytoplasmic proteins.[12–14] For cytoplasmic proteins, the phase diagram is divided into two regions by a coexistence curve.[12–14] In one region the cytoplasm exists as a single homogeneous phase. In the other region, separated protein-rich and protein-poor phases coexist. In lens cells the two separated phases are easily observed in electron micrographs (Fig. 1). In the phase diagram for lens cytoplasm a sharp boundary divides the conditions for single transparent phase from the conditions for an opaque two-phase system (Figs. 2 and 3).

The position of the phase boundary in a phase diagram depends on critical molecular interactions that determine the structure and function of cell cytoplasm.[15] The phase behavior of a protein solution is often determined by weak but very important molecular interactions having energies on the order of $kT$ (0.6 kcal/mol) where $k$ is the Boltzmann constant and $T$ is the absolute temperature.[16–18] Conformational changes in the protein molecules are unnecessary.[12,13,15] The collective influence of weak interactions on phase equilibria and cytoplasmic structure becomes apparent when minor modification in the proteins or in the cytoplasmic solvent that comprises the molecular environment of the proteins alters the position of the phase boundary in the phase diagram. The molecular interactions that control the location of the phase boundary in the phase diagram are essential for normal lens cell structure and function, especially in the transparent lens. The theoretical proposal that lens cytoplasm behaves as a binary polymer solution which undergoes a phase separation during opacification[15,19,20] was confirmed by the experimental determination of the coexistence curve for lens cytoplasm.[12–14] The experimental methods for mapping phase diagrams in protein solutions provide an important approach to understanding the specific molecular events that regu-

---

[13] M. Delaye, J. I. Clark, and G. B. Benedek, *Biochem. Biophys. Res. Commun.* **100,** 908 (1981).

[14] J. I. Clark, M. Delaye, and G. B. Benedek, *Proc. Tech. Program Electro–Opt./Laser Conf. Expo.* **80,** 365 (1980).

[15] T. Tanaka, C. Ishimoto, and L. T. Chylack, *Science* **197,** 1010 (1977).

[16] A. Gustafsson, J. Wennerström, and F. Tjerneld, *Polymer* **27,** 1768 (1986).

[17] V. G. Taratuta, A. Holschbach, G. M. Thurston, D. Blankschtein, and G. B. Benedek, *J. Phys. Chem.* **94,** 2140 (1990).

[18] J. Pande, C. Berland, M. Broide, O. Olutayo, J. Melhuish, and G. Benedek, *Proc. Natl. Acad. Sci. U.S.A.* **88,** 4916 (1991).

[19] T. Tanaka and G. B. Benedek, *Invest. Ophthalmol.* **14,** 449 (1975).

[20] C. Ishimoto, P. W. Goalwin, S. T. Sun, I. Nishio, and T. Tanaka, *Proc. Natl. Acad. Sci. U.S.A.* **76,** 4414 (1979).

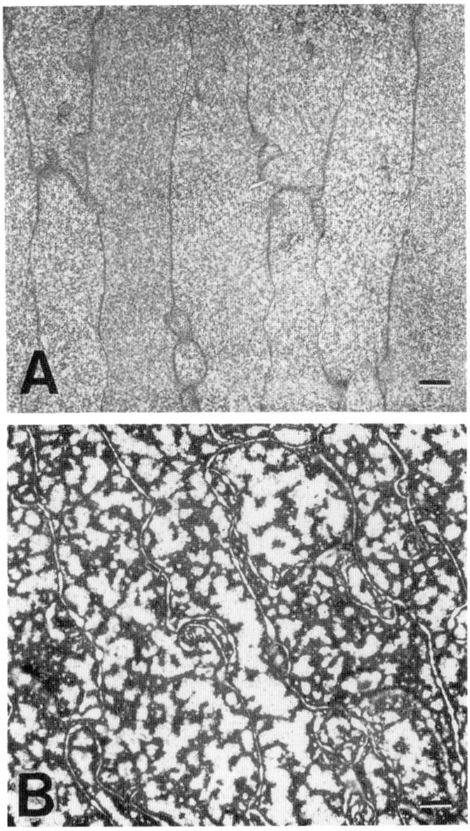


FIG. 1. In transparent lens cells (A) the cytoplasm exists as a single homogeneous phase. The spatial fluctuations within the cytoplasm are small relative to the wavelength of light. In opaque cells (B), a condensed cytoplasmic phase (darkly stained regions) is surrounded by a dilute phase (lightly stained regions). The two separated phases have spatial fluctuations that are large relative to the wavelength of light, which results in light scattering. Bar, 500 nm. Micrograph courtesy of J. M. Clark and S. Vaezy.

late the appearance of new phases and the compartmentalization of proteins in biological cells. The following sections outline methods used to construct phase diagrams for lens cytoplasm and cytoplasmic proteins.

## Methods

### *Experimental Determination of Coexistence Curve for Native Whole Lens Cytoplasm*

*Analysis of Separated Phases at Different Temperatures.* The experimental determination of the phase diagram for lens cytoplasm[12] requires

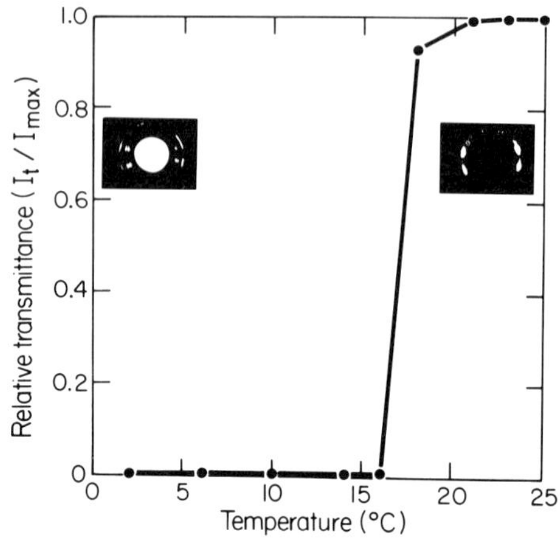


FIG. 2. Effect of temperature on the transparency of lens cytoplasm. A sharp transition from the transparent single-phase to the opaque two-phase state occurs in the cytoplasm at approximtely 17°. An opaque lens is shown in the left inset. The opacity in the lens nucleus is due to the presence of two separate cytoplasmic phases. A normal transparent lens is shown in the right inset. The transition from the single-phase to the two-phase cytoplasm occurs suddenly over a narrow temperature range. The transition temperature is defined as the phase separation temperature, $T_c$. $T_c$ increases when conditions favor phase separation and decreases when conditions favor a single transparent phase. [Reproduced with permission from J. I. Clark and G. B. Benedek, *Biochem. Biophys. Res. Commun.* **95,** 482 (1980).]

the isolation of the cytoplasm in its native state.[21] Twenty to forty calf lenses, from 2- to 3-week-old animals, are removed from fresh calf eyes and decapsulated. Whole lenses are chopped into small pieces approximately 5 mm on each side. A glass homogenizer with a Teflon pestle is warmed to approximately 37° in warm water, and the chopped lenses are placed in the homogenizer. Sodium azide is added to a concentration of approximately 0.02% (v/v) to inhibit bacterial contamination. Next the warm sample is homogenized gently using 3 to 6 strokes. The warm homogenate is poured into a 15-ml Corex tube and centrifuged at 30,000 *g* for 4 hr at 35° to remove cell fragments and membranous debris. This procedure results in 5 to 15 ml of undiluted native lens cytoplasm having a phase separation temperature, $T_c$, of approximately 15° and a protein concentration of approximately 40% (dry weight/total weight). The homogenate is transparent at temperatures above 17°, and the phase separation occurs over a narrow temperature range (Fig. 2).

The phase diagram for lens homogenate is constructed using the following procedures.[12,14] The temperature of the cytoplasmic homogenate is

[21] J. I. Clark, M. Delaye, P. Hammer, and L. Mengel, *Curr. Eye Res.* **1,** 695 (1981/1982).

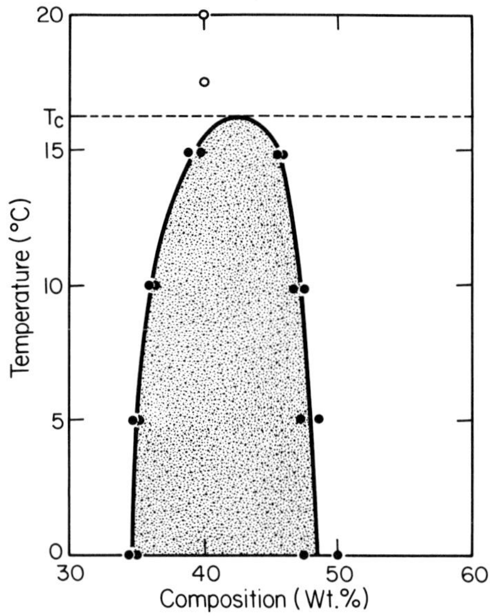


FIG. 3. Phase diagram of lens cytoplasm. The coexistence curve for the isolated cytoplasmic homogenate is similar to that of a phase separation in a two-polymer aqueous solution. The shaded area represents the conditions of temperature and composition where phase separation occurs and the cytoplasm is opaque. The area outside the curve represents the conditions for transparency of lens cell cytoplasm. The concentration of protein in the homogeneous transparent material was approximately 40% dry weight (open circles). At each temperature the concentrations measured in each of the separated phases of the opaque cytoplasm were plotted as filled circles. [Reproduced with permission from J. I. Clark and G. B. Benedek, *Biochem. Biophys. Res. Commun.* **95,** 482 (1980).]

decreased below $T_c$ to 15°, at which temperature the sample becomes opaque. The temperature is carefully maintained by keeping the sample in a Lauda water bath which controls the temperature to within 0.1°. The temperature of the sample is allowed to equilibrate. The physical separation of the phases in the opaque sample requires centrifugation at 6000 *g* for approximately 30 min in a temperature-controlled centrifuge. During centrifugation, the temperature is maintained at 15°. After centrifugation, the sample contains two physically distinct layers which correspond to the separated phases of the cytoplasm.

The protein concentration is determined for each phase (layer) using a simple spectrophotometric assay. A tiny aliquot from each layer is diluted by a known factor, and the absorption is measured at a wavelength of 280 nm. Using a standard curve, the value of the absorbance for each phase is converted to dry weight corrected by the dilution factor and plotted versus the temperature as shown in Fig. 3. After each measurement, the remaining sample is rewarmed to approximately 35° and mixed to form a single transparent phase. The remixed sample is cooled to 10°, the next experimental temperature, and permitted once again to phase separate. As before the two phases in the sample are layered using centrifugation. The concentration of each layer is measured and plotted in the phase diagram versus temperature. The procedure is repeated at successively lower temperatures until the coexistence curve for lens cytoplasm is determined (Fig. 3). The measurements are limited at low temperature because of the freezing of the cytoplasm at approximately −8°.

*Dilutions of Concentrated Nuclear Cytoplasm.* Nuclear cytoplasm refers to cytoplasm isolated from the central, nuclear cells of the lens. Historically, the lens was thought to be a single transparent cell. The central region of the lens was called the nucleus, and the periphery was called the cortex. It was not until the 1950s that electron microscopy revealed the existence of cell membranes throughout the lens and it was recognized that the lens is a cellular tissue. The term "nucleus," however, continues to be applied to the cells in the center of the lens.

A phase diagram is determined for nuclear cytoplasm using dilutions of concentrated cytoplasm isolated from cells of the lens center.[13] In this procedure fresh calf lenses obtained from animals 2 to 3 weeks of age are cooled to 4°. At this temperature, the cells of the nucleus are opaque, and the clear cortical cells are separated from the opaque cells. The opaque nuclei are chopped into small pieces and placed in a warm homogenizer. For nuclear cytoplasm, which is very viscous, the pestle is stainless steel and the tolerance is slightly greater than that of a Teflon homogenizer. The homogenization is performed at 35° to reduce the viscosity of the sample, and 0.01% sodium azide is added to prevent bacterial contamination. Following homogenization the warm sample is placed in transparent centrifuge tubes and centrifuged at 35° for 20 to 40 hr at approximately 180,000 *g* in a swinging-bucket rotor. Under these conditions the membranes float on top of the more dense cytoplasm, and the resulting homogenate is a transparent sample of highly concentrated lens cytoplasm. The protein concentration can exceed 60% by weight. The isolation procedure does not include dilution with buffers so there is no detectable denaturation of the proteins, which can occur when cytoplasmic homogenates are diluted with aqueous buffers. To dilute the sample and avoid denaturation, cytoplasmic fluid is extracted from cortical cells using the same centrifugation procedure described in the previous section, and the fluid is used for the dilution experiments. Dilution with cytoplasmic fluid is found to provide conditions that are similar to the physiological phases of lens cytoplasm, although 0.154 *M* phosphate buffer, pH 7.2, has been used effectively.

For determination of the coexistence curve, the phase separation temperature, $T_c$, of each successive dilution of the concentrated lens cytoplasm is measured using light scattering[13] and then plotted versus the protein concentration in the phase diagram. The protein concentration is determined by measuring the dry weight in each sample after lyophilization following measurement of the $T_c$. The concentration of each sample is defined as (dry weight)/(total wet weight). The $T_c$ is determined using laser spectroscopy.[12,13]

It is important to note that the coexistence curve determined using the cytoplasmic dilution method is similar to the coexistence curve determined using the isolated phases in the method described in the previous section. It appears that either method provides an accurate coexistence curve for cytoplasmic proteins. The methods in both this and the previous sections are reproducible, although they are subject to numerous subtle variations including the time and speed of centrifugation, buffer composition, age of calf lenses, freshness of calf lenses, and even the time of year that animals were brought to the abattoir. It is not surprsing that the subtle variations in the initial conditions of the preparation can alter the results. The $T_c$ and the coexistence curve measure the molecular interactions between protein and solvent constituents of the cells. It is expected that modifications of proteins alter their interactions and minor changes in the composition of the cytoplasmic fluid alter the molecular environment of the proteins. Such alterations are measured as changes in the $T_c$ and reflected in the precise form of the coexistence curve.

### *Experimental Determination of Coexistence Curve for Concentrated Lens Crystallins*

Additional understanding of coexistence curves for cytoplasmic proteins resulted from studies using isolated and purified lens proteins.[22] Lenses are obtained from 1- to 2-week-old calves. The $\gamma$-crystallins are isolated using Sephadex G-75 gel filtration. The $\gamma$-crystallin is fractionated into constituent subfractions, I, II, III, and IV, using sulfopropyl-Sephadex C-50. Purified fractions II, III, and IV are stored at $-20°$ in sodium acetate buffer (pH 5.0) with 1 m*M* 2-mercaptoethanol.

For determination of the coexistence curve, the crystallin fractions are thawed and dialyzed against 50 m*M* phosphate buffer, pH 7.0 and ionic

[22] R. J. Siezen, M. R. Fisch, C. Slingsby, and G. B. Benedek, *Proc. Natl. Acad. Sci. U.S.A.* **82,** 1701 (1985).

strength of 0.110, to remove the reducing agent. The resulting solution is filtered through a 0.22-$\mu$m Millipore (Bedford, MA) membrane. To achieve high protein concentrations, the samples are concentrated using an Amicon (Danvers, MA) B15 miniconcentrator. For light scattering measurements, the samples are centrifuged for 15 min at 10,000 *g* at a temperature well above the $T_c$ to remove particulates. Following the measurement of $T_c$, the protein concentrations are determined spectrophotometrically at 280 nm using an absorbance value of 21 for all $\gamma$-crystallin fractions.[22] The $T_c$ is determined using a laser spectrometer.[23] For any sample, the $T_c$ value could be reproduced with an accuracy of $\pm 0.5°$. The phase diagrams determined for solutions of purified $\gamma$-crystallin are very similar to those observed in nuclear cytoplasm. The results demonstrate the importance of $\gamma$-crystallin as a determinant of the $T_c$ in lens cytoplasm.

The coexistence curves for individual $\gamma$-crystallins are determined using the procedures described above. The results of these studies are used to estimate the contribution of each component protein to the phase separation temperature of the whole cytoplasm and demonstrate that the constituent $\gamma$-crystallins define the conditions for phase separation in the whole lens cell.[22]

### *Determination of Coexistence Curve and Spinodal Line for $\gamma_{II}$-Crystallin*

In an important demonstration of the effectiveness of protein solutions for the study of phase transitions and critical phenomena, the coexistence curve and the spinodal line have been determined for purified $\gamma_{II}$-crystallin.[24] The calf lens $\gamma_{II}$-crystallin is prepared by size-exclusion chromatography on Sephadex G-75 and by ion-exchange chromatography on sulfopropyl-Sephadex C-50. $\gamma_{II}$-crystallin fractions are pooled and concentrated by ultrafiltration at 20° using an Amicon PM10 concentrator and cleaned of contaminating peptides on a DEAE-cellulose column to 99% purity. The details of the biochemical preparation have been published in Thomson *et al.*[24] Protein concentrations are determined spectrophotometrically using the UV absorbance at 280 nm. The calibration of the 280 nm absorbance of known crystallin solutions is used to determine an extinction coefficient of $A_{280}$ at 0.1%/cm of 2.4.

The coexistence curve (Fig. 4, solid line) is determined using two methods: the $T_c$, the temperature of opacification, is measured at selected protein concentrations (first used in lens cytoplasm by Delaye *et al.*[13]), and

[23] M. Delaye, J. I. Clark, and G. B. Benedek, *Biophys. J.* **37,** 647 (1982).

[24] J. A. Thomson, P. Schurtenberger, G. M. Thurston, and G. B. Benedek, *Proc. Natl. Acad. Sci. U.S.A.* **84,** 7079 (1987).

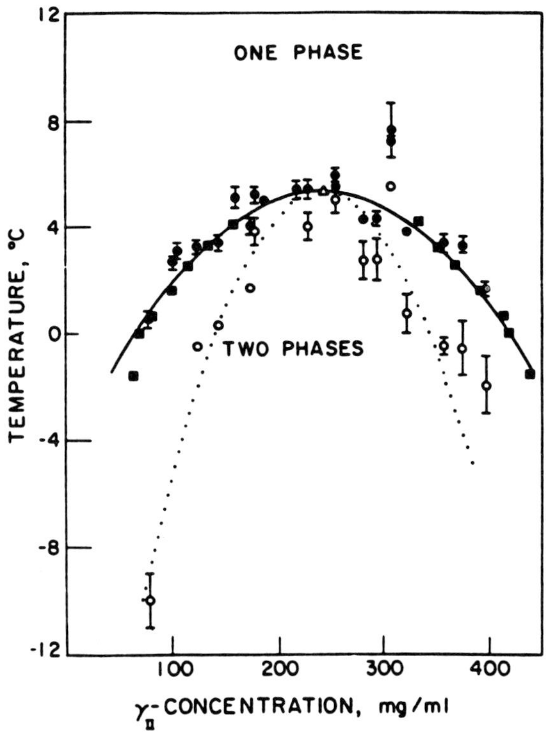


FIG. 4. Coexistence curve and spinodal line for the aqueous $\gamma_{II}$-crystallin system. The solid line is the fit to the coexistence curve. The dotted line is the fit to the spinodal line. ●, $T_c$ determinations; ■, concentration measurements of macroscopically separated phases; △, critical point; ○, spinodal temperatures. [Reproduced with permission from J. A. Thomson, P. Schurtenberger, G. M. Thurston and G. B. Benedek, *Proc. Natl. Acad. Sci. U.S.A.* **84,** 7079 (1987).] *Note:* For conditions within the spinodal line, the solution will spontaneously phase separate by a mechanism known as spinodal decomposition (P. Glansdorff and I. Prigogine, "Thermodynamic Theory of Structure, Stability and Fluctuations." Wiley, New York, 1971). The region between the spinodal line and the coexistence curve defines the metastable region, where the density fluctuations in the inhomogeneous solution are unstable and can appear as microscopic structures of various dimensions and morphologies (H. E. Stanley, "Introduction to Phase Transitions and Critical Phenomena." Oxford Univ. Press, New York, 1971).

the protein concentrations of separated phases are measured at selected temperatures below the critical temperature. The two sets of data are plotted in the same phase diagram, which shows that they are in very good agreement (Fig. 4, solid line). The spinodal line (Fig. 4, dotted line) is determined on the basis of light scattering measurements made as a function of temperature at different protein concentrations. As the temperature decreases to approach the coexistence curve the scattering intensity, $I$, increases dramatically and diverges at the spinodal line. The plot of $1/I$ versus temperature, $T$, is linear over a 20° temperature range until the existence curve is approached. Then it is not possible to take intensity measurements at lower temperatures near the spinodal line because of multiple scattering. Extrapolation of a plot of $1/I$ versus $T$ to $1/I = 0$ gives the $T_{sp}$, spinodal temperature, which is plotted in Fig. 4 for each protein concentration. Although the extrapolation of the data points provides some uncertainty in the shape of the spinodal line, there is good agreement with theory for both the coexistence curve and the spinodal line. There is no doubt about the similarity between the phase diagrams for purified $\gamma_{II}$-crystallins and for whole lens cytoplasm. Clearly the phase diagram for isolated crystallin proteins will provide new information on the fundamental molecular interactions between proteins in native cytoplasm.

### *Determination of Coexistence Curve for Highly Purified Crystallin Fractions*

$\gamma$-Crystallin is isolated from 1- to 6-week-old calf lenses using size-exclusion chromatography on Sephadex G-75 as described previously.[24] $\gamma$-Crystallin is further fractionated into $\gamma_{III}$- and $\gamma_{IV}$-crystallin using cation-exchange chromatography on sulfopropyl-Sephadex C-50. Anion-exchange chromatography on DEAE-Sephadex is used to subfractionate $\gamma_{III}$- into $\gamma_{IIIa}$- and $\gamma_{IIIb}$-crystallin. The purified fractions of $\gamma_{III}$- and $\gamma_{IV}$-crystallin are dialyzed extensively into 100 m*M* phosphate buffer, pH 7.0, which contains sodium azide to inhibit contamination. The purified proteins are concentrated by ultrafiltration using an Amicon Centricon and Centriprep-10 or by phase separation. When the phase separation method is used, the proteins are first concentrated to 100–250 mg/ml by ultrafiltration. Concentrated samples are placed in a temperature controlled bath at a temperature below $T_c$ so that the dense phase could be collected. Concentrations as high as 470 mg/ml are achieved using the phase separation method. Protein concentrations are determined by UV spectroscopy using extinction coefficients that are determined gravimetrically. The details of the biochemical procedures have been published.[25]

As in the previous section, two methods are used to determine the coexistence curves. The temperature of specific concentrations of the $\gamma$-crystallin fractions is decreased until opacity is observed. Then the observed $T_c$ is plotted against the protein concentration in a phase diagram. The resulting plot is the coexistence curve of the purified protein. In

[25] M. L. Broide, C. R. Berland, J. Pande, O. O. Ogun, and G. B. Benedek, *Proc. Natl. Acad. Sci. U.S.A.* **88,** 5660 (1991).

separate experiments, the protein concentrations of the two coexisting phases are determined at a fixed temperature below $T_c$. As expected, the two methods produce similar coexistence curves. An interesting point of these studies is that the purified calf $\gamma$-crystallins could be grouped into high $T_c$ proteins ($\gamma_{IIIa}$- and $\gamma_{IVa}$-crystallin with $T_c$ near 38°) and low $T_c$ proteins ($\gamma_{II}$- and $\gamma_{IIIb}$-crystallin with $T_c$ near 5°). Use of these methods with proteins produced by site-directed mutagenesis will allow a quantitative understanding of the key peptide sequences and amino acids that are responsible for the different molecular interactions and different $T_c$.

### *Experimental Methods for Determination of Coexistence Curves in Other Protein Solutions*

In addition to lens cytoplasm, phase diagrams have been determined for lysozyme (Fig. 5) and a few other aqueous protein solutions.[11,15,17] While discussion of these results is beyond the space limitations of this chapter, the similarity between results for other proteins and for lens cytoplasmic proteins is worth noting. The biophysical behavior of all protein solutions appears to be regulated by common thermodynamic factors. It will be valuable to extend the study to include phase diagrams for the major human lens cytoplasmic proteins at their physiological concentrations.

## Discussion and Summary

The major problem with any determination of phase diagrams for biological constituents has to do with how well they represent the physiological state. In the case of lens cytoplasmic proteins the results are persuasive that the solution behavior of the isolated cytoplasmic proteins is similar to the behavior of the proteins in cells. In these studies it is important to consider the effect of changes in pH, ionic strength, and general composition of the phase diagram. The analysis of phase diagrams in aqueous solutions of proteins has physiological and biomedical significance for problems in lens opacification,[26] aging,[27] and other disease processes.[28,29] The utility of these methods is not limited to pathology because the same molecular interactions that are responsible for defining the position of the coexistence curve in a phase diagram are responsible for partitioning,

[26] J. I. Clark and J. E. Steele, *Proc. Natl. Acad. Sci. U.S.A.* **89,** 1720 (1992).

[27] J. I. Clark, J. R. Neuringer, and G. B. Benedek, *J. Gerontol.* **38,** 287 (1983).

[28] C. T. Noguchi and A. N. Schechter, *Annu. Rev. Biophys. Chem.* **14,** 239 (1985).

[29] C. R. Middaugh, B. Gerber-Jensen, A. Harvitz, A. Paluszek, C. Scheffel, and G. W. Litman, *Proc. Natl. Acad. Sci. U.S.A.* **75,** 3440 (1978).

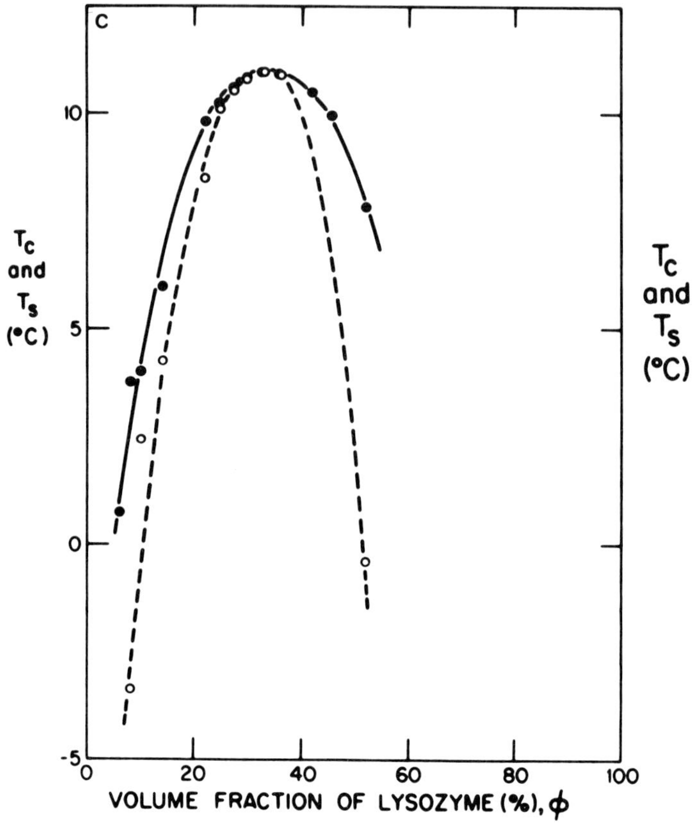


FIG. 5. Coexistence curve (solid line) for lysozyme–salt water mixtures determined using the phase separation temperature, $T_c$, at selected protein concentrations. The spinodal line (dashed line) was determined using the spinodal temperature, $T_s$, at selected protein concentrations. $T_c$ is the phase separation temperature at each concentration, and $T_s$ is the spinodal temperature. [Reproduced with permission from T. Tanaka, C. Ishimoto, and L. T. Chylack, *Science* **197,** 1010 (1977) Copyright 1977 by the AAAS.]

compartmentalization, aggregation, and condensation of cellular constituents.[30–33] Much more experimental research on phase equilibria of biological solutions is needed to understand the molecular interactions responsible for the physical and structural properties of cytoplasm. It is in support of these efforts that such methods are presented here.

[30] H. Walter, G. Johansson, and D. E. Brooks, *Anal. Biochem.* **197,** 1 (1991).

[31] B. R. Ganong and J. P. Delmore, *Anal. Biochem.* **193,** 35 (1991).

[32] J. D. Radolf, N. R. Chamberlain, A. Clausell, and M. V. Norgard, *Infect. Immun.* **56,** 490 (1988).

[33] C. Bordier, *J. Biol. Chem.* **256,** 1604 (1981).

## Acknowledgments

This work was supported by Grants EY 04542 and EY 01730 from the National Eye Institute and Research to Prevent Blindness, Inc. The author thanks J. M. Clark and S. Vaezy for contributions and J. Peetermans and H. Walter for reading the manuscript.

# Section VI

# Aqueous Two-Phase Systems in Large-Scale Processes: Biotechnology

A. Extractive Bioconversions
*Articles 52 through 54*

B. Proteins: Downstream Processing, etc.
*Articles 55 through 61*

C. Design of Proteins for Enhanced Extraction
*Article 62*

D. Other Applications of Aqueous Phases in Biotechnology
*Articles 63 through 66*

# [52] Cultivation of *Trichoderma reesei* and Production of Cellulolytic Enzymes

*By* FOLKE TJERNELD and INGRID PERSSON

## Introduction

Cells and microorganisms can be used in aqueous two-phase systems in bioconversions. The cells or microorganisms, in a phase system in which they partition to one of the phases, are supplied with a substrate. During mixing of the phases the bioconversion of substrate into product takes place. After phase separation the cell-free phase containing the product is removed. The cells remain in the reactor and are supplied with new substrate and complementary phase. Aqueous two-phase systems based on solutions of two polymers, for example, dextran (Dx) and poly(ethylene glycol) (PEG), have a high water content in both phases and are not detrimental to cells or microorganisms.

Cell recycling, in bioconversion processes, can be carried out by use of aqueous two-phase systems. By partitioning the cells to one phase and by recycling this phase the cells are effectively immobilized inside a reactor. One advantage, compared with immobilization of cells in a gel matrix, is that the cells can be supplied with macromolecular substrates in high concentration without steric hindrance.

A number of bioconversions using microorganisms in aqueous two-phase systems have been performed. Bioconversions involving low molecular weight substrates and products include ethanol production,[1,2] butanol and acetone production,[3] steroid conversion,[4,5] and fumarate conversion.[6] Production of enzymes in aqueous phase systems include $\alpha$-amylase,[7,8] cellulase,[9–11] and $\beta$-glucosidase.[12]

[1] I. Kühn, *Biotechnol. Bioeng.* **22,** 2393 (1980).
[2] B. Hahn-Hägerdal, B. Mattiasson, and P.-Å. Albertsson, *Biotechnol. Lett.* **3,** 53 (1981).
[3] B. Mattiasson, M. Souminen, E. Andersson, L. Häggström, P.-Å. Albertsson, and B. Hahn-Hägerdal, *Enzyme Eng.* **6,** 153 (1982).
[4] R. Kaul and B. Mattiasson, this volume [54].
[5] S. Flygare and P.-O. Larsson, *Enzyme Microb. Technol.* **11,** 752 (1989).
[6] Y. L. Yang, H. Hustedt, and M.-R. Kula, *Biotechnol. Appl. Biochem.* **10,** 173 (1988).
[7] E. Andersson, A.-C. Johansson, and B. Hahn-Hägerdal, *Enzyme Microb. Technol.* **7,** 333 (1985).
[8] E. Andersson and B. Hahn-Hägerdal, *Appl. Microbiol. Biotechnol.* **29,** 329 (1988).
[9] I. Persson, F. Tjerneld, and B. Hahn-Hägerdal, *Enzyme Microb. Technol.* **6,** 415 (1984).
[10] I. Persson, F. Tjerneld, and B. Hahn-Hägerdal, *Appl. Biochem. Biotechnol.* **27,** 9 (1991).

TABLE I
PARTITIONING OF CELLULOLYTIC ENZYME SYSTEM IN AQUEOUS TWO-PHASE SYSTEMS[a,b]

| Aqueous two-phase system | Molecular weight of bottom phase polymer | Partition coefficient |
|---|---|---|
| 7% PEG 8000[c]–17% Aquaphase PPT[d] | $3.5 \times 10^4$ | 0.83 |
| 5% PEG 8000–7% Dx T500[e] | $5.0 \times 10^5$ | 1.41 |
| 5% PEG 8000–7% hydrolyzed crude Dx[f] | $<5 \times 10^6$ | 1.60 |
| 5% PEG 3350[c]–6% crude Dx[f] | $5 \times 10^6$ | 3.56 |

[a] From I. Persson, F. Tjerneld, and B. Hahn-Hägerdal, *Appl. Biochem. Biotechnol.* **27,** 9 (1991), with permission.
[b] The partitioned enzyme system was from NOVO Celluclast 2.0L (76 FPU/g, NOVO A/S, Bagsvaerd, Denmark). The enzyme activity was measured as filter paper activity. The polymer concentration of the phase system is given as % (w/w) of the total weight of the phase system. The temperature was 20°. The phase systems included medium components as shown in Table III.
[c] Union Carbide, New York, NY.
[d] Hydroxypropyl starch (now Reppal PES 100, Reppe AB, Växjö, Sweden).
[e] Pharmacia Biotechnology AB, Uppsala, Sweden.
[f] Pharmacia AB, Staffanstorp, Sweden.

The most commonly used microorganism for cellulolytic enzyme production is the fungus *Trichoderma reesei.*[13] *Trichoderma reesei* Rutgers C30 has been used for enzyme production in phase systems in order to recycle the fungus and thus reduce enzyme production costs.[9] The influence of cultivation conditions on the production of cellulolytic enzymes has been studied,[10] and extractive cultivations with laboratory-scale fermentors have been made.[11]

## Partitioning of Cellulolytic Enzymes

When performing extractive fermentations with *Trichoderma reesei,* the fungus is partitioned to one of the phases and the produced cellulolytic enzymes are recovered from the other phase. *Trichoderma reesei* partitions completely to the bottom phase in several phase systems tested.[9,10] To obtain high enzyme yields it is advantageous to have high partition coefficients, $K$, for the produced enzymes. Table I shows the partitioning of the complete cellulolytic enzyme system from *T. reesei* (Celluclast

[11] I. Persson, H. Stålbrand, F. Tjerneld, and B. Hahn-Hägerdal, *Appl. Biochem. Biotechnol.* **27,** 27 (1991).
[12] I. Persson, F. Tjerneld, and B. Hahn-Hägerdal, *Biotechnol. Tech.* **3,** 265 (1989).
[13] I. Persson, F. Tjerneld, and B. Hahn-Hägerdal, *Process Biochem.* **26,** 65 (1991).

TABLE II
CELLULOLYTIC ENZYME PRODUCTION IN AQUEOUS TWO-PHASE SYSTEMS WITH *Trichoderma reesei* RUTGERS C30[a,b]

| Aqueous two-phase system | Productivity[c] (FPU/liter × hr) | Total yield (FPU/g cellulose) | Enzyme activity (FPU/ml) |
|---|---|---|---|
| 7% PEG 8000–17% Aquaphase PPT[d] | 9.3 | 115[e] | 0.72 |
| 5% PEG 8000–7% Dx T500 | 4.9 | 131[e] | 1.2 |
| 5% PEG 8000–7% hydrolyzed crude Dx | 6.5 | 88[e] | 0.79 |
| 5% PEG 3350–6% crude Dx | 4.5 | 66[e] | 0.56 |
| Regular medium | 2.0 | 97[f] | 0.86 |

[a] From I. Persson, F. Tjerneld, and B. Hahn-Hägerdal, *Appl. Biochem. Biotechnol.* **27,** 9 (1991), with permission.
[b] Substrate was 10 g/liter Solka Floc BW200; for cultivation medium, see Table III.
[c] Calculated from the enzyme activity in the withdrawn top phase.
[d] Hydroxypropyl starch (now Reppal PES 100, Reppe AB).
[e] Withdrawn enzyme activity plus residual activity in the bottom phase at the end of the cultivation, calculated from partition coefficient and phase volumes.
[f] Withdrawn enzyme activity plus residual activity at the end of the cultivation.

2.0L, NOVO A/S, Bagsvaerd, Denmark) in four different phase systems.[10] The enzyme activity is given as filter paper units (FPU), which is a measure of the combined activities of the different cellulases.[14] The highest $K$ was obtained in the phase system with a low molecular weight top phase polymer (PEG 3350) and a high molecular weight bottom phase polymer (crude Dx). This is in agreement with previous studies which show that high molecular weight bottom phase polymers cause proteins to partition to the top phase.[15]

## Influence of Phase Systems on Enzyme Production

### *Procedure*

The phase-forming polymers may have an influence on microbial enzyme production in aqueous two-phase systems.[8,10,12] Such influence can be studied as illustrated here for cellulolytic enzyme production with *Trichoderma reesei* Rutgers C30 in four aqueous two-phase systems (Table II).[10] The fungus is cultivated in the phase systems with cellulose (Solka

[14] T. M. Wood and K. M. Bhat, this series, Vol. 160, p. 87.
[15] P.-Å. Albertsson, A. Cajarville, D. E. Brooks, and F. Tjerneld, *Biochim. Biophys. Acta* **926,** 87 (1987).

TABLE III
CULTIVATION MEDIUM[a] USED IN AQUEOUS TWO-PHASE SYSTEMS FOR CELLULOLYTIC ENZYME PRODUCTION WITH *Trichoderma Reesei* RUTGERS C30[b,c]

| Component | Concentration |
|---|---|
| Sodium citrate dihydrate | 2.5 g/liter |
| $KH_2PO_4$ (anhydrous) | 5.0 g/liter |
| $NH_4NO_3$ (anhydrous) | 2.0 g/liter |
| $MgSO_4 \cdot 7H_2O$ | 0.2 g/liter |
| Biotin (0.01% aqueous solution) | 2.5 ml/liter |
| $CaCl_2 \cdot 2H_2O$ | 0.1 g/liter |
| Citric acid monohydrate | 5 mg/liter |
| $ZnSO_4 \cdot 7H_2O$ | 5 mg/liter |
| $Fe(NH_4)_2(SO_4)_2 \cdot 6H_2O$ | 1.0 mg/liter |
| $CuSO_4$ | 0.16 $\mu$g/liter |
| $MnSO_4 \cdot H_2O$ | 50 $\mu$g/liter |
| $H_3BO_3$ (anhydrous) | 50 $\mu$g/liter |
| $Na_2MoO_4 \cdot 2H_2O$ | 50 $\mu$g/liter |
| Proteose peptone | 1.0 g/liter |
| Tween 80 | 0.15 ml/liter |

[a] The medium is according to H. J. Vogel, *Am. Nat.* **98,** 435 (1964).
[b] I. Persson, F. Tjerneld, and B. Hahn-Hägerdal, *Appl. Biochem. Biotechnol.* **27,** 9 (1991).
[c] I. Persson, H. Stålbrand, F. Tjerneld, and B. Hahn-Hägerdal, *Appl. Biochem. Biotechnol.* **27,** 27 (1991).

Floc BW200, 10 g/liter, from James River Corp., Berlin, NH) as substrate in 1-liter flasks, with a cultivation volume of 100 ml, in a shaking water bath at 25°. The phase systems include the cultivation medium, which is Vogel's mineral salts medium N[16] (Table III) with the addition of proteose peptone (Difco Laboratories, Detroit, MI) at 1.0 g/liter and Tween 80 at 0.15 ml/liter. Shaking is stopped every 24 hr, and the cultures are transferred to measuring cylinders and allowed to settle for 30 min. The phase systems separate into a clear top phase containing produced enzyme and a bottom phase containing cellulose substrate and fungus. The volumes of top phases are measured. The whole top phase from each phase system is withdrawn and replaced with new top phase. Medium without phase-forming polymers is used as control. In the latter case the solids settle to the bottom of the cylinder, and the clear solution above the solids is

[16] H. J. Vogel, *Am. Nat.* **98,** 435 (1964).

withdrawn and replaced with new medium. Samples are taken and stored frozen for analysis of filter paper activity. The mixing is then resumed.[10]

*Results*

The highest enzyme concentration (Table II) was obtained in the PEG 8000–Dx T500 system, 1.2 FPU/ml, and the lowest in the PEG 3350–crude Dx system, 0.6 FPU/ml. The total yield of enzyme was also highest in the PEG 8000–Dx T500 system and lowest in the PEG 3350–crude Dx system. Both enzyme concentration and total yield were higher in the PEG 8000–Dx T500 system compared with the cultivation in a regular medium without polymers. The productivity was higher in cultivations in phase systems compared with regular medium. The highest productivity was obtained in phase system with Aquaphase PPT (hydroxypropyl starch, now Reppal PES 100, Reppe AB, Växjö, Sweden). The low enzyme activity and yield in the PEG 3350–crude Dx system can be explained by a reduced oxygen transfer rate in this system owing to the high viscosity of the crude Dx phase. Thus, the influence of the phase-forming polymers on the cultivation and enzyme production of the fungus must be considered when aqueous two-phase systems are used in enzyme production.[10] Based on these results we have selected the PEG 8000–Dx T500 phase system for extractive enzyme production in laboratory-scale fermentors with *Trichoderma reesei* Rutgers C30.

## Batch Fermentations in Aqueous Two-Phase Systems

*Procedure*

Batch fermentations in two-phase systems can be carried out in the following way. Fermentor cultivations with *Trichoderma reesei* are made in a 7-liter fermentor (Novaferm AB, Malmö, Sweden) in a phase system composed of 5% PEG 8000 and 7% Dx T500.[11] The working volume of the fermentor is 4 liters. The polymer concentrations in phase systems used in fermentor cultivations are calculated on a weight per volume basis. The cellulose substrate is Solka Floc BW200 at an initial concentration of 10 g/liter. Vogel's mineral salts medium N (Table III) is used in the cultivations with the addition of proteose peptone (1.0 g/liter) and Tween 80 (0.15 ml/liter). Temperature is set to 28°, and the pH is kept at 5.0 by addition of 2 *M* $NH_4OH$ or 0.1 *M* $H_2SO_4$. The air flow is chosen to be 10 liter/min and the stirring rate 350 rpm. A 20% silicone antifoam agent (BDH Chemicals Ltd., Poole, UK) is used when needed. The fermentor cultures are inoculated with 10% (400 ml) inocula cultivated for 65 hr in

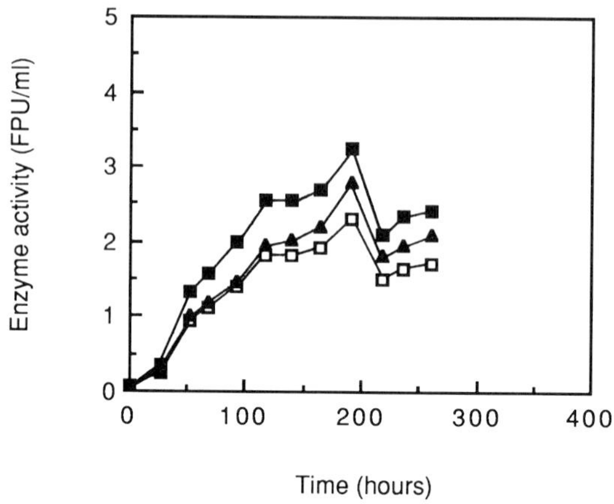


FIG. 1. Batch fermentation of *Trichoderma reesei* Rutgers C30 in an aqueous two-phase system composed of 5% PEG 8000–7% dextran T500. ■, Enzyme concentration in the top phase; □, enzyme concentration in the bottom phase; ▲, total enzyme concentration. The substrate was 10 g/liter Solka Floc BW200, the working volume 4 liters, temperature 28°, pH 5.0, air flow 10 liters/min, and agitation 350 rpm. [From I. Persson, H. Stålbrand, F. Tjerneld, and B. Hahn-Hägerdal, *Appl. Biochem. Biotechnol.* **27,** 27 (1991), with permission.]

shake flasks on 10 g/liter Solka Floc BW200 in Vogel's medium (Table III). The dilution of the phase system by titration, in order to keep a constant pH, is compensated by addition of PEG and Dx from sterile stock solutions of 40 and 20%, respectively. In the phase system used $K = 1.4$ for the cellulolytic enzymes as calculated from the filter paper activities (Table I). The phase volume ratio is 1 : 1 at the start of the cultivations. The composition of the entire phase system is Dx T500 70 g/liter, PEG 8000 50 g/liter, Solka Floc BW200 10 g/liter, and cultivation medium as shown in Table III.

### *Results*

Enzyme production in a batch cultivation with *Trichoderma reesei* Rutgers C30 in a PEG 8000–Dx T500 aqueous two-phase system is shown in Fig. 1. The enzyme concentration in the bottom phase and the total concentration in the phase system have been calculated from the partition coefficient and the actual volume ratios. The total enzyme concentration in the phase system reaches a plateau at approximately 2.0 FPU/ml after 200 hr of cultivation. An average of 2.5 FPU/ml is obtained in a batch cultivation without polymers using the same fermentation conditions.[11]

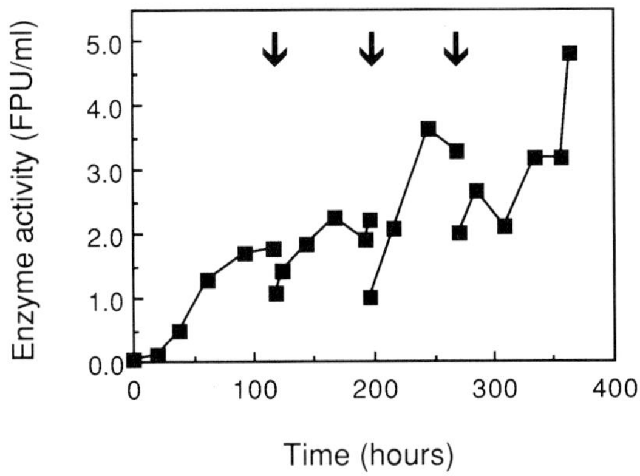


FIG. 2. Extractive fermentation in 5% PEG 8000–7% dextran T500 with *Trichoderma reesei* Rutgers C30. Arrows mark changes of top phase and addition of 40 g Solka Floc BW200. The culture conditions were the same as in Fig. 1. [From I. Persson, H. Stålbrand, F. Tjerneld, and B. Hahn-Hägerdal, *Appl. Biochem. Biotechnol.* **27,** 27 (1991), with permission.]

## Extractive Fermentations in Aqueous Two-Phase Systems

### *Procedure*

By cultivating microorganisms in aqueous two-phase systems it is possible to carry out extractive fermentations. The enzymes produced are recovered in the cell-free top phase. The effect of different parameters on cellulolytic enzyme production has been summarized elsewhere.[13] In fed-batch cultivations, where the cellulose substrate is added intermittently during fermentation, high cellulolytic enzyme concentrations are obtained.[17] By using aqueous two-phase systems it is possible to perform extractive fed-batch cultivations. After stopping the agitation in the fermentor and letting the phases separate, the top phase containing enzyme is removed. New substrate is added together with new top phase. This procedure is repeated intermittently during the fermentation.

For extractive cultivations the fermentor is fitted with a steam sterilizable valve (Novaferm AB) for exchange of top phase under sterile conditions. Figure 2 shows such an extractive fed-batch cultivation in a phase system during a total cultivation time of 360 hr.[11] The initial substrate concentration is 10 g/liter (Solka Floc BW200). The same phase system and cultivation conditions are used as in Fig. 1. The top phase containing enzyme is withdrawn when the enzyme concentration ceases to increase (120 hr). The stirring is stopped and the phases are left to separate for 1

[17] N. Hendy, C. Wilke, and H. Blanch, *Enzyme Microb. Technol.* **6,** 73 (1984).

hr. The enzyme extraction is repeated every 72 hr. A top phase volume of 1.0 liter is withdrawn on each occasion through the steam-sterilized valve. New top phase is added containing new cellulose substrate (40 g at each extraction) and medium components at original concentrations. The new top phase is prepared by using the phase diagram for the PEG 8000–Dx T500 system, which shows the polymer concentrations in the two phases.[18]

*Results*

The average enzyme concentration in the withdrawn top phase is around 3.0 FPU/ml.[11] The semicontinuous cultivation can be maintained for 360 hr with enzyme concentration increasing in the top phase (Fig. 2). The highest enzyme concentration observed was 4.8 FPU/ml, which is slightly higher than that obtained in an ordinary fed-batch cultivation in a medium without polymers (4.5 FPU/ml). The enzyme yield is approximately the same in the phase system and in the regular medium, 148 and 151 FPU/g cellulose, respectively.[11] When extractive cultivations are made in aqueous phase systems the enzymes are obtained in a cell-free top phase, whereas in an ordinary cultivation the enzymes need to be separated from the cells in the fermentation broth.

## Conclusions

Aqueous two-phase systems permit integration of enzyme production and enzyme purification. During fermentation the produced enzymes are continuously extracted into a cell-free PEG-rich top phase. The enzymes can be further purified by addition of a concentrated salt solution to this phase. In the resulting PEG–salt system the enzymes can be partitioned to the salt-rich phase and the PEG phase can be recycled.[19,20] By desalting in an ultrafiltration unit, a purified and concentrated enzyme solution is obtained.

## Acknowledgments

This research was supported by grants from The Swedish National Board for Industrial and Technical Development (NUTEK).

[18] P.-Å. Albertsson and F. Tjerneld, this volume [1].

[19] F. Tjerneld, G. Johansson, and M. Joelsson, *Biotechnol. Bioeng.* **30,** 809 (1987).

[20] G. Johansson, this volume [55].

# [53] Enzyme-Catalyzed Hydrolysis and Recycling in Cellulose Bioconversion

*By* FOLKE TJERNELD

## Introduction

Aqueous two-phase systems have been used for enzyme recycling in enzymatic cellulose hydrolysis processes so as to reduce process costs in cellulose bioconversion.[1-5] These systems offer several advantages in bioconversion processes.[6-8] The enzyme is partitioned to one of the phases together with the substrate. Because of the low interfacial tension[9] there is rapid equilibrium of the formed product across the interface. The low interfacial tension also leads to the formation of very small phase droplets during mixing of the phases. This minimizes migration distances and facilitates mass transfer of produced sugar. After phase separation the phase containing the product is removed while the enzyme is retained in the reactor in the second phase and can be resupplied with substrate. The advantages with bioconversions in aqueous phase systems are (1) the product is continuously removed, (2) the polymers stabilize enzymes, (3) the enzymes are recycled, and (4) macromolecular and particulate substrates can be used without steric hindrance being encountered as when enzymes are immobilized in a gel matrix.

In cellulose hydrolysis the cellulose particles are partitioned to the bottom phase of the system together with the degrading enzymes, and the sugar products are removed with the top phase. To design the hydrolysis process an optimal phase system has to be found, that is, one in which

[1] F. Tjerneld, I. Persson, P.-Å. Albertsson, and B. Hahn-Hägerdal, *Biotechnol. Bioeng.* **27,** 1036 (1985).

[2] F. Tjerneld, I. Persson, P.-Å. Albertsson, and B. Hahn-Hägerdal, *Biotechnol. Bioeng.* **27,** 1044 (1985).

[3] F. Tjerneld, I. Persson, P.-Å. Albertsson, and B. Hahn-Hägerdal, *Biotechnol. Bioeng. Symp.* **15,** 419 (1985).

[4] C. F. Mandenius, B. Nilsson, I. Persson, and F. Tjerneld, *Biotechnol. Bioeng.* **31,** 203 (1988).

[5] F. Tjerneld, I. Persson, and J. M. Lee, *Biotechnol. Bioeng.* **37,** 876 (1991).

[6] B. Hahn-Hägerdal, B. Mattiasson, and P.-Å. Albertsson, *Biotechnol. Lett.* **3,** 53 (1981).

[7] B. Mattiasson and B. Hahn-Hägerdal, *in* "Immobilized Cells and Organelles" Vol. 1 (B. Mattiasson, ed.), p. 121. CRC Press, Boca Raton, Florida, 1983.

[8] E. Andersson and B. Hahn-Hägerdal, *Enzyme Microb. Technol.* **12,** 242 (1990).

[9] P.-Å. Albertsson, "Partition of Cell Particles and Macromolecules," 3rd Ed., Wiley(Interscience), New York, 1986.

TABLE I
PHYSICAL DATA ON CELLULASES FROM *Trichoderma reesei*

| Enzyme | Molecular weight | Isoelectric point | Ref. |
|---|---|---|---|
| CBH I | 64,000 | 3.9 | *a* |
| CBH II | 53,000 | 5.9 | *a* |
| EG I | 55,000 | 4.5 | *a* |
| EG III | 48,000 | 5.5 | *a* |
| β-Glucosidase | 47,000 | 5.7 | *b* |

[a] R. Bhikhabhai, G. Johansson, and G. Pettersson, *J. Appl. Biochem.* **6,** 336 (1984).
[b] L. E. R. Berghem, and G. Pettersson, *Eur. J. Biochem.* **46,** 295 (1974).

the enzymes partition to the bottom phase. The affinity of the enzymes for the cellulose substrate enhances the retention of enzymes in the bottom phase. Selection of a low molecular weight bottom phase polymer, such as, dextran (Dx), and a high molecular weight top phase polymer, for example, poly(ethylene glycol) (PEG), also tends to give rise to low partition coefficients, $K$, for the enzymes. For large-scale processes inexpensive polymers have to be used.

The optimal process design is a combination of an aqueous two-phase system with an ultrafiltration membrane unit. The sugars produced are recovered by ultrafiltration of the top phase after phase separation. Enzymes and polymers are recycled, and only buffer and cellulose substrate have to be added to the reactor.

## Partitioning of Cellulose and Cellulases

Some physicochemical characteristics of the cellulases of *Trichoderma reesei* are given in Table I. Three enzyme activities involved in cellulose hydrolysis have been characterized: cellobiohydrolase (CBH), endo-β-glucanase (EG), and β-glucosidase. The optimal conditions for cellulose hydrolysis with *Trichoderma* enzymes occur at pH values of 4.5–5.0 in 50 m$M$ acetate or citrate buffers at 40–50°. To recycle the enzymes it is desirable to obtain a one-sided partitioning of enzymes and substrate to the bottom phase. The low molecular weight sugar products are partitioned evenly between the phases. The cellulose particles (Solka Floc BW200, James River Corp., Berlin, NH) are partitioned to the bottom phase in all phase systems studied.[1]

TABLE II
CELLULASE PARTITION COEFFICIENTS ($K$) AS FUNCTION OF MOLECULAR WEIGHT OF PHASE-FORMING POLYMERS[a]

| Molecular weight | | $K^b$ | |
|---|---|---|---|
| Dx[c] | PEG[c] | Endo-β-glucanase[d] | β-Glucosidase[d] |
| 10,000 | 8000 | 0.85 | 0.09 |
| 40,000 | 8000 | 0.91 | 0.18 |
| 110,000 | 8000 | 0.94 | 0.23 |
| 500,000 | 8000 | 1.11 | 0.30 |
| 2,000,000 | 8000 | 1.32 | 0.31 |
| >5,000,000[e] | 8000 | 1.36 | 0.55 |
| 40,000 | 20,000 | 0.80 | 0.10 |
| 40,000 | 40,000 | 0.71 | 0.07 |

[a] The phase systems used were composed of 8% (w/w) Dx, 6% (w/w) PEG, and 10 m$M$ sodium acetate, pH 4.8. [From F. Tjerneld, I. Persson, P.-Å. Albertsson, and B. Hahn-Hägerdal, *Biotechnol. Bioeng.* **27,** 1036 (1985), with permission.]

[b] Ratio between activity (IU/ml) in top phase and activity (IU/ml) in bottom phase.

[c] Fractionated dextrans were from Pharmacia (Uppsala, Sweden), PEG 8000 from Union Carbide (New York, NY), and PEG 20,000 and 40,000 from Serva (Heidelberg, Germany).

[d] From NOVO Celluclast 2.0L (NOVO A/S, Bagsvaerd, Denmark).

[e] Crude dextran (Pharmacia, Staffanstorp, Sweden).

By increasing the polymer concentrations in aqueous two-phase systems a more one-sided partitioning of proteins is obtained.[9] However, the inconvenience of increased viscosity at high polymer concentrations must be balanced against the lower $K$ values in this way. Thus, when the cellulases are used in a hydrolysis process the most important experimental parameters for the partitioning of the enzymes to the bottom phase are polymer molecular weight and enzyme adsorption to the cellulose particles.

### *Polymer Molecular Weight*

Polymer molecular weight influences partitioning of substances. Generally, if the bottom phase polymer molecular weight is increased the substance partitions more to the top phase; by increasing the top phase polymer molecular weight, the substance partitions more to the bottom phase.[10] Table II shows the partitioning of endo-β-glucanase and β-glucosi-

[10] P.-Å. Albertsson, A. Cajarville, D. E. Brooks, and F. Tjerneld, *Biochim. Biophys. Acta* **926,** 87 (1987).

dase activities (from NOVO Celluclast 2.0L, NOVO A/S, Bagsvaerd, Denmark) in a series of phase systems composed of Dx (Pharmacia Biotechnology AB, Uppsala, Sweden) of increasing molecular weights, together with PEG 8000 (Union Carbide, New York, NY). The phase system compositions are 8% (w/w) Dx and 6% (w/w) PEG. In these systems the two phases in equilibrium have the same concentrations of polymers irrespective of molecular weight.[9] Both enzymes partition more to the bottom phase when Dx of low molecular weight is used (Table II). When crude Dx (Pharmacia AB, Staffanstorp, Sweden) is used as phase-forming polymer, its high molecular weight leads to increased enzyme $K$ values.

The $K$ values of cellulases are also influenced by the molecular weight of PEG (Table II). Both endo-$\beta$-glucanase and $\beta$-glucosidase are distributed more to the bottom phase when a high-molecular-weight fraction of PEG is used. The optimal partitioning properties are obtained in a phase system with Dx of low molecular weight (Dx T40, Pharmacia Biotechnology AB) and PEG of high molecular weight (PEG 40,000, Serva, Heidelberg, Germany). However, in the hydrolysis process the less expensive crude dextran is used so as to reduce process costs.

### *Adsorption to Cellulose Particles*

Cellulose partitions to the bottom phase in Dx–PEG systems.[1] The cellulolytic enzymes have a strong affinity for cellulose fibers.[11] Figure 1 shows $K$ values of the cellulase enzymes at increasing ratios of particulate cellulose to enzyme (w/w) in Dx T40–PEG 8000 phase systems. The distribution of endo-$\beta$-glucanase activity to the bottom phase at higher cellulose/enzyme ratios is due to strong adsorption of the endo-$\beta$-glucanase components to the cellulose particles. $\beta$-Glucosidase does not specifically bind to cellulose,[12] however, and only slightly decreased $K$ values at high cellulose/enzyme ratios are observed (Fig. 1).

By combining the effects of polymer molecular weight and adsorption to cellulose, very low $K$ values can be obtained. In Fig. 2 the $K$ values for endo-$\beta$-glucanase and $\beta$-glucosidase are shown in phase systems with a Dx of low molecular weight (Dx T40) and a high cellulose/enzyme weight ratio (C/E = 40) as a function of PEG molecular weight.

### *Effect of Temperature*

The optimum temperature for cellulose hydrolysis using *Trichoderma* enzymes is between 40 and 50°.[13] The $K$ values of the enzymes are consid-

[11] J. Ståhlberg, G. Johansson, and G. Pettersson, *Bio/Technology* **9,** 286 (1991).

[12] T. K. Ghose and V. S. Bisaria, *Biotechnol. Bioeng.* **21,** 131 (1979).

[13] R. Eklund, M. Galbe, and G. Zacchi, *Enzyme Microb. Technol.* **12,** 225 (1990).

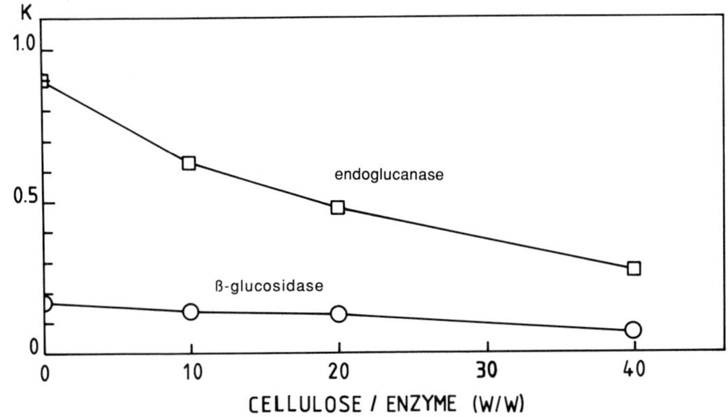


FIG. 1. The partition coefficient, $K$, for endo-$\beta$-glucanase and $\beta$-glucosidase as a function of the cellulose-to-enzyme ratio (w/w). The phase system contained 8% (w/w) dextran T40, 6% (w/w) PEG 8000, and 10 m$M$ sodium acetate buffer, pH 4.8, with 2% Avicel PH 105 cellulose. The cellulase enzyme preparation was NOVO Celluclast 2.0L (NOVO A/S). [From F. Tjerneld, I. Persson, P.-Å. Albertsson, and B. Hahn-Hägerdal, *Biotechnol. Bioeng.* **27,** 1036 (1985), with permission.]

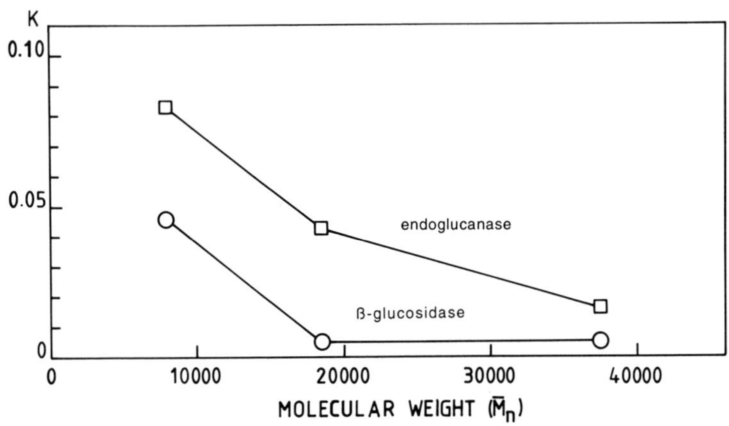


FIG. 2. The partition coefficient, $K$, for endo-$\beta$-glucanase and $\beta$-glucosidase as a function of the molecular weight of PEG with cellulose present. The phase system contained 8% (w/w) dextran T40, 6% (w/w) PEG, 10 m$M$ sodium acetate buffer, pH 4.8, and 5% Solka Floc BW 200 cellulose. The cellulase enzyme preparation was NOVO Celluclast 2.0L (NOVO A/S). The cellulose-to-enzyme weight ratio was 40. [From F. Tjerneld, I. Persson, P.-Å. Albertsson, and B. Hahn-Hägerdal, *Biotechnol. Bioeng.* **27,** 1036 (1985), with permission.]

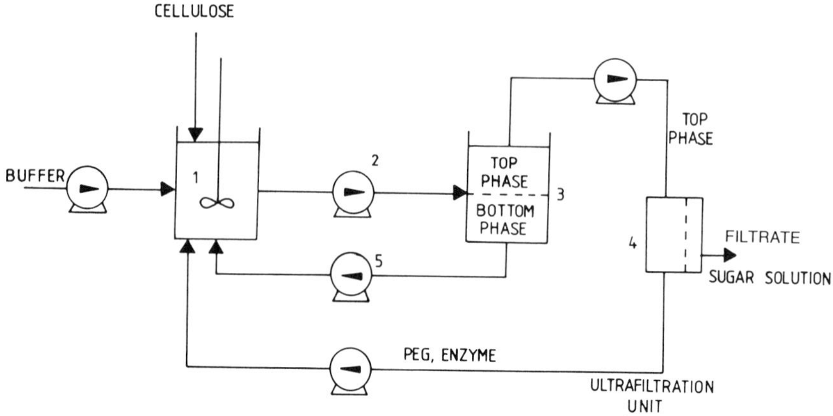


FIG. 3. Semicontinuous hydrolysis of cellulose. Cellulose and buffer enters the mixer (1), which contains a dextran–PEG aqueous two-phase system with cellulases in the bottom phase. The mixture is pumped (2) to the settler (3), where a cellulose-free top phase is obtained. The produced sugars are separated from the top-phase polymer (PEG) by ultrafiltration (4), and the PEG together with part of the cellulases is returned to the mixer. The bottom phase (containing enzyme and nondegraded cellulose) is pumped (5) back to the mixer. [From F. Tjerneld, I. Persson, P.-Å. Albertsson, and B. Hahn-Hägerdal, *Biotechnol. Bioeng. Symp.* **15,** 419 (1985), with permission.]

erably higher at these temperatures than at room temperature.[1] The loss of enzymes to the top phase in a hydrolysis process can be reduced by letting phase separation and the withdrawal of top phase occur at room temperature, while still carrying out the hydrolysis at 40–50°.[2]

## Semicontinuous Cellulose Hydrolysis in Aqueous Two-Phase Systems

### *Experimental Design*

The experimental setup is a mixer–settler (Fig. 3) with the hydrolysis taking place at 40–50° and phase separation at room temperature.[2,3] The sugar products are continuously withdrawn with the top phase. Because the cellulases do not completely partition to the bottom phase some enzymes remain in the top phase. To reduce enzyme loss and to recycle the top phase polymer an ultrafiltration unit is added to the mixer–settler.[3] After phase separation the top phase is passed over the ultrafiltration membranes (cutoff 10,000 MW), and the polymer (PEG 20,000) is recycled to the mixer together with any enzymes present in the top phase. The sugars produced are recovered in the filtrate. The ultrafiltration units used are Millipore Minitan (Millipore Corp., Bedford,

TABLE III
CONDITIONS FOR CELLULOSE HYDROLYSIS IN AQUEOUS TWO-PHASE SYSTEM COMBINED WITH ULTRAFILTRATION[a]

| Parameter (unit) | Value |
|---|---|
| Initial enzyme concentration (FPU/ml) | 2 |
| System volume (ml) | 1000 |
| Phase system | 5% (w/w) Crude dextran–3% (w/w) PEG 20000 |
| Initial cellulose concentration (g/liter) | 110 |
| Added cellulose (g) | 180 |
| Semicontinuous hydrolysis (hr) | 1200 |
| Dilution rate ($hr^{-1}$) | 0.002–0.004 |
| Average reducing sugar (g/liter) | 74 |
| Productivity (g/liter × hr) | 0.12–0.30 |
| Total filtrate volume (ml) | 3100 |
| Used enzyme/produced sugar (FPU/g RS) | 8.7 |

[a] From F. Tjerneld, I. Persson, P.-Å. Albertsson, and B. Hahn-Hägerdal, *Biotechnol. Bioeng. Symp.* **15,** 419 (1985), with permission.

MA) and Filtron Minisette (Filtron Technology Corp., Clinton, MA) systems.

*Cellulose Hydrolysis in Two-Phase Systems Combined with Ultrafiltration*

The experimental conditions for enzymatic cellulose hydrolysis in an aqueous two-phase system combined with ultrafiltration are described in Table III. The phase system is 5% (w/w) crude Dx, 3% (w/w) PEG 20,000, and 50 m*M* acetate buffer, pH 4.8. To prevent microbial growth, 0.03% sodium azide is added. The top and bottom phases are mixed in the mixer unit (Fig. 3) using a flat blade turbine, and the cellulose particles (Solka Floc BW 200, James River Corp., Berlin, NH) are added intermittently to the mixer. The mixer is immersed in a water bath at 50°. A peristaltic pump pumps the mixture to the settler unit where phase separation takes place. Before reaching the settler the mixture is water-cooled so as to improve partitioning conditions for the cellulases. The settler is at room temperature. From the settler the top phase is pumped to the ultrafiltration unit. The bottom phase is recirculated to the mixer by use of a peristaltic pump. The top phase is recirculated over the membrane while the filtrate containing the sugar products is collected. A volume, equal to that of the filtrate, of 50 m*M* acetate buffer is added to the mixer. A small loss of PEG through the membrane is sometimes detected. Solid PEG is added to the phase system to restore the volume of the top phase. Samples of the filtrate are analyzed for sugars. Total sugar concentration is determined as reducing sugars by the dinitrosalicylic acid method.[14,15] Glucose, cellobiose, and xylose are determined by high-performance liquid chromatography (HPLC).[16] The enzyme activities[15] in the system can be followed by analysis of samples withdrawn from the top phase.[3]

A cellulose concentration of approximately 80 g/liter is maintained in the reactor throughout the hydrolysis. Cellulose is then present in excess over enzymes, which is of advantage for enzyme adsorption and partitioning to the bottom phase containing cellulose. The cellulase enzymes (NOVO A/S)[17] are added initially at a concentration of 2 filter paper units (FPU)/ml.[15]

The ultrafiltration of the top phase is started after 50 hr when the concentration of reducing sugars is above 60 g/liter. The concentrations of glucose and cellobiose in the ultrafiltration filtrate are shown in Fig. 4A. In Fig. 4B the concentration of xylose in the ultrafiltration filtrate is shown. The semicontinuous hydrolysis can be continued for 1200 hr using only initially added enzymes.[3]

The activities of endo-$\beta$-glucanase and $\beta$-glucosidase have been measured during the hydrolysis both in the top phase and in the ultrafiltration filtrate. A small leakage of enzyme activity through the membrane was detected.[3] This can be explained by the presence of a low molecular mass component of endo-$\beta$-glucanase, EG IV, of 20–25 kD[18] which is not completely retained by the membrane. During the total period of 1200 hr semicontinuous hydrolysis 31% of the initial endo-$\beta$-glucanase activity and 0.5% of the initial $\beta$-glucosidase activity was lost in the filtrate.[3]

The sugar concentration is maintained between 70 and 80 g/liter by controlling the dilution rate of the reactor. Owing to the slow decline in enzyme activity the dilution rate has to be decreased during the hydrolysis, thereby reducing the productivity.[3] Constant productivity can be obtained by making small additions of enzymes during the reaction to compensate for the decrease in enzyme activity.[5]

## Discussion

When two-phase systems, without the ultrafiltration unit, were used for cellulase recycling the enzymes were gradually washed out with the

[14] G. L. Miller, R. Blum, W. E. Glennon, and A. L. Burton, *Anal. Biochem.* **2,** 127 (1960).

[15] T. M. Wood and K. M. Bhat, this series, Vol. 160, p. 87.

[16] J. K. Lin, B. J. Jacobson, A. N. Pereira, and M. R. Ladisch, this series, Vol. 160, p. 145.

[17] M. Schülein, this series, Vol. 160, p. 234.

[18] U. Håkansson, L. Fägerstam, G. Pettersson, and L. Andersson, *Biochim. Biophys. Acta* **524,** 385 (1978).

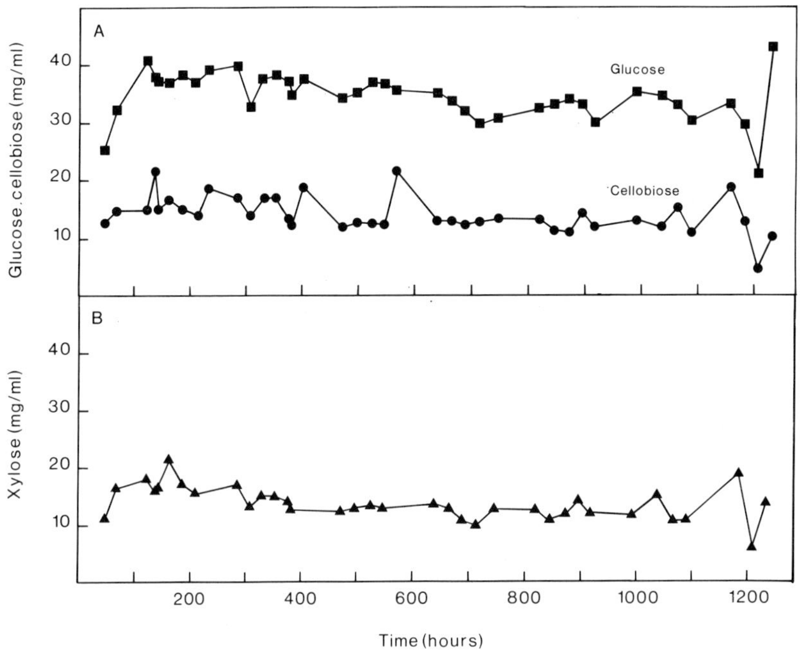


FIG. 4. Sugar concentrations in the ultrafiltration filtrate from semicontinuous hydrolysis of cellulose (Solka Floc BW 200) in an aqueous two-phase system combined with ultrafiltration. The experimental setup is shown in Fig. 3. The phase system was composed of 5% (w/w) crude dextran, 3% (w/w) PEG 20,000, 50 m*M* sodium acetate buffer, pH 4.8, and 0.03% sodium azide. Other conditions were listed in Table III. The temperature in the mixer was 50°. The sugars were analyzed by HPLC. (A) Concentrations of glucose and cellobiose. (B) Xylose concentration. [From F. Tjerneld, I. Persson, P.-Å. Albertsson, and B. Hahn-Hägerdal, *Biotechnol. Bioeng. Symp.* **15,** 419 (1985), with permission.]

top phase.[2] By using ultrafiltration membranes almost complete retention of the enzymes is achieved, allowing operation for 1200 hr without addition of enzyme. *Trichoderma* cellulases are very stable in solution,[19] and the use of polymeric phase systems in cellulose hydrolysis further increases cellulase stability.[9] Semicontinuous hydrolysis for 50 days using only initially added enzymes demonstrates excellent enzyme stability in the aqueous two-phase system.

The phase systems have the advantage that cellulose particles are completely partitioned to the bottom phase, which results in a clear top phase that can pass over ultrafiltration membranes without causing membrane fouling. Partitioning of enzymes to the bottom phase is also of advantage when the top phase is subject to ultrafiltration. The enzymes

[19] E. T. Reese and M. Mandels, *Biotechnol. Bioeng.* **22,** 323 (1980).

are not subjected to shear forces caused by pumping, and less protein is adsorbed on the membrane. Enzymatic starch hydrolysis in aqueous two-phase systems has also been described.[20]

The process design shown in Fig. 3 permits the complete recycling of phase forming polymers. Only buffer and cellulose have to be added to the reactor. The cutoff of the ultrafiltration membrane is chosen at a molecular weight of 10,000 to exclude the PEG 20,000 molecules. A small amount of PEG may pass the membrane. If that is observed solid PEG can be added to restore the top phase volume.

The degree of enzyme recovery is measured by the amount of enzyme used per gram of reducing sugars (RS) produced. In batch hydrolysis processes this is 30–35 FPU/g RS.[2,21] When a two-phase system is combined with ultrafiltration the enzyme consumption is reduced to 8.7 FPU/g RS, which is equivalent to an enzyme recovery of 70 to 75%.[3] By using ultrafiltration in a membrane reactor enzyme consumption is also low (9.6 FPU/g RS).[22] However, in the latter case severe problems with membrane fouling occur. It is then necessary to use high dilution rates which lead to low sugar concentrations (<10 g/liter).[22] These fouling problems are completely avoided by use of an aqueous two-phase system in combination with the ultrafiltration unit. A process design, such as described in this chapter, makes it possible to produce high sugar concentrations, 70–80 g/liter, in cellulose hydrolysis processes.[3]

## Acknowledgments

This research was supported by grants from the National Energy Administration (STEV), Sweden, and the Swedish National Board for Industrial and Technical Development (NUTEK).

[20] M. Larsson, V. Arasaratnam, and B. Mattiasson, *Biotechnol. Bioeng.* **33,** 758 (1989).
[21] D. Y. Ryu and M. Mandels, *Enzyme Microb. Technol.* **2,** 91 (1980).
[22] I. Ohlsson, G. Trägårdh, and B. Hahn-Hägerdal, *Biotechnol. Bioeng.* **26,** 647 (1984).

# [54] Biotransformation of Hydrocortisone into Prednisolone

*By* RAJNI KAUL and BO MATTIASSON

## Introduction

Biotransformations of organic compounds involve processes that are often characterized by dilute product streams owing to the low solubility of the substrate and/or product in the reaction medium. Several methods to improve these processes have been presented.[1-3] Aqueous two-phase systems offer a useful environment for biotransformations as evidenced by a model study on hydrocortisone transformation using *Arthrobacter simplex*.[4] Aqueous two-phase systems are biocompatible and even permit cells to grow.[5,6] These systems may also possess hydrophobic components as chains or side chains linked to phase-forming polymers and thereby enhance the solubility of the reactants. These properties make two-phase systems a favorable setting for bioconversions.

## Transformation in Aqueous Two-Phase Systems

### *Two-Phase Systems*

Most of the studies have been carried out with two-phase systems having fractionated dextran (Dx) as the bottom phase polymer. However, the use of Dx on a large scale is not economical. Hence, some relatively inexpensive polymers are being studied as substitutes for the Dx. Here, Reppal PES, a hydroxypropyl starch is used as such a replacement.

All the phase systems are prepared in 50 m*M* Tris-HCl solution, pH 7.0. The polymers forming the two-phase system are weighed and mixed in the buffer at the desired concentrations (w/w). Complete phase separation takes place within 30 min at room temperature. The volumes of the two phases are noted.

The phase systems used with Dx T40 and Dx T70 (Pharmacia, Uppsala, Sweden) as the bottom phase polymer and different top phase polymers, namely, poly(ethylene glycol) (PEG) 8000 (Union Carbide, New York,

[1] G. Carrea, *TIBTECH.* **2,** 102 (1984).
[2] E. Kondo and E. Masuo, *J. Gen. Appl. Microbiol.* **7,** 113 (1961).
[3] B. Mattiasson and P. Adlercreutz, *Trends Biotechnol.* **9,** 394 (1991).
[4] R. Kaul and B. Mattiasson, *Appl. Microbiol. Biotechnol.* **24,** 259 (1986).
[5] M. Larsson, V. Arasaratnam, and B. Mattiasson, *J. Microb. Biotechnol.* **3,** 22 (1988).
[6] M. Larsson, V. Arasaratnam, and B. Mattiasson, *Biotechnol. Bioeng.* **3,** 58 (1989).

TABLE I
EFFECTS OF VARIATIONS IN PHASE COMPOSITION ON PARTITIONING BEHAVIOR OF HYDROCORTISONE AND PREDNISOLONE[a]

| Phase system [% (w/w) : % (w/w)] | Volume ratio (top to bottom phase) | Partition coefficient ($K$) | |
|---|---|---|---|
| | | Hydrocortisone | Prednisolone |
| PEG 8000–dextran T40 | | | |
| 15 : 3 | 12.3 : 1 | 2.6 | 6.4 |
| 20 : 3 | 15.7 : 1 | 2.8 | 6.6 |
| 25 : 3 | 19.0 : 1 | 2.8 | 6.8 |
| 25 : 6 | 8.5 : 1 | 3.3 | 7.1 |
| Pluronic F68–dextran T40 | | | |
| 14 : 4 | 7.0 : 1 | 2.6 | 2.9 |
| 14 : 6 | 4.3 : 1 | 2.7 | 3.2 |
| PPG 2025–dextran T40 | | | |
| 25 : 3 | 0.5 : 1 | 11.5 | 15.6 |
| M-PEG 550–dextran T40 | | | |
| 25 : 3 | 24.0 : 1 | 2.0 | 2.2 |
| M-PEG 5000–dextran T70 | | | |
| 9.7 : 4 | 8.1 : 1 | 1.7 | 5.0 |
| 10.7 : 2 | 27.6 : 1 | 1.5 | 4.3 |

[a] From R. Kaul and B. Mattiasson, *Appl. Microbiol. Biotechnol.* **24,** 259 (1986), with permission.

NY), methoxypoly(ethylene glycol) (M-PEG) 550 and 5000 (Polysciences, Northampton, UK), poly(propylene glycol) (PPG) 2025 (BDH, Poole, UK), or Pluronic F-68 (Montoil AB, Stockholm, Sweden) are listed in Table I. Two-phase systems with Reppal PES 200 (Reppe Glykos, Växjö, Sweden) and PEG 8000 as phase polymers are also prepared in the same manner.

### *Estimation of Steroid Content*

A steroid sample (50 μl) is shaken with chloroform (0.2 ml).[7] The chloroform phase is subjected to analysis by high-performance liquid chromatography (HPLC) of hydrocortisone and prednisolone (Serva, Heidelberg, Germany) using an LDC chromatograph (Tillquist, Stockholm, Sweden). The steroid separation is carried out on a Spherisorb column (250 × 4.6 mm) with 5 μm particle size and eluting with a mobile phase of 98% (v/v) chloroform and 2% (v/v) methanol at a flow rate of 1.5 ml/min. Detection is by absorbance at 254 nm. Under these conditions the retention times for hydrocortisone and prednisolone are 6.1 and 7.7 min, respectively.

[7] R. Kaul, P. Adlercreutz, and B. Mattiasson, *Biotechnol. Bioeng.* **28,** 1432 (1986).

TABLE II
SOLUBILITY OF HYDROCORTISONE IN INDIVIDUAL PHASES OF TWO-PHASE SYSTEMS[a]

| System | Phase | Solubility (mg/ml)[b] |
|---|---|---|
| 25% PEG 8000–6% dextran T40 | PEG-rich | 1.0 |
| | Dextran-rich | 0.1 |
| 12% PEG 8000–3% Reppal PES 200 | PEG-rich | 0.69 |
| | Reppal-rich | 0.74 |

[a] From R. Kaul and B. Mattiasson, *Appl. Microbiol. Biotechnol.* **24,** 259 (1986), with permission.
[b] Solubility in pure water under these conditions was 0.2 mg/ml.

### *Determination of Partition Coefficients, K, of Steroids*

The $K$ values of hydrocortisone and prednisolone are determined by first stirring the respective steroids into a two-phase system for a few hours at 20°. The phase separation is carried out by allowing the systems to stand for 30 min or by centrifuging them at 1625 $g$ for 5 min at 20°. Aliquots are removed from both phases and their steroid concentration determined. $K$ is calculated as the ratio between the concentrations in top and bottom phases.

Table I shows the partitioning of both hydrocortisone and prednisolone in the two-phase systems using Dx and different top phase polymers. The steroids partition more toward the top phase. PPG was particularly effective in producing high $K$ values. The steroid crystals which did not go into solution settled onto the interface.

### *Determination of Hydrocortisone Solubility in Polymer Solutions*

The respective phases of the two-phase systems, 25% PEG 8000–Dx T40 and 12% PEG 8000–3% Reppal PES 200, are separated. Hydrocortisone (4 mg/ml) is added to the respective phases and stirred for about 24 hr at 20°. The undissolved steroid is sedimented by high-speed centrifugation (12,000 $g$ for 20 min) at 4°, and the amount dissolved is measured in the supernatant solution. The solubility in water is determined under identical conditions.

The solubility of hydrocortisone in the PEG solution was 10 times greater than in Dx (Table II). Reppal solubilized even a little more steroid than PEG. Reppal contains hydroxypropyl groups, which imparts to the polymer a weak hydrophobic character.

### *Organism and Culture Conditions*

*Arthrobacter simplex* ATCC 6946 (DSM, No. 20130, Braunschweig, Germany) is used for the transformation of hydrocortisone to prednisolone.[7] The bacteria are grown at 30° with shaking in a medium (pH 7.2–7.4) containing 1% (w/v) casein peptone tryptic digest (Difco, Detroit, MI), 0.5% yeast extract (Difco), 0.5% (w/v) NaCl, and 0.5% (w/v) glucose. After 24 hr of growth, the cells are induced for the enzyme, steroid $\Delta^1$-dehydrogenase, by addition of hydrocortisone (1 mg/ml) to the medium and continued incubation for another 6 hr. The bacteria are collected by centrifugation at 12,000 $g_{max}$ (4°) for 10 min, washed with 50 m*M* Tris-HCl solution, pH 7.0, and stored at −20° until use.

### *Transformation Reaction*

To 20 ml of the phase system, in a 100-ml beaker, is added hydrocortisone powder, and the mixture is stirred for a few minutes. The reaction is started by addition of bacterial cells. Samples are withdrawn periodically to follow the course of transformation. At the end of the reaction the phases are allowed to separate, and concentrations of the product in the phases are determined.

At a substrate concentration of 1 mg/ml and 22 mg (dry weight) bacterial cells, more than 99% of hydrocortisone could be transformed in the two-phase systems (Table III). Because phase volume ratios (top/bottom) of these systems are high (except for those of PPG–Dx) the major amount of product formed is in the top phase. The bottom phase contains a small percentage of product and some untransformed substrate (Table III).

The reaction efficiency in the PEG–Reppal system was found to be similar to that in the PEG–Dx system. However, with PEG–Reppal the bottom phase contained more product, namely, 12% in the 12% PEG–3% Reppal system and 20% in the 10% PEG–5% Reppal system, corresponding well to the steroid solubility.

Studies in a 25% PEG–6% Dx system with a constant substrate to cell ratio reveal that nearly complete transformation occurs only up to a substrate concentration of 2 mg/ml. At 5 mg/ml the conversion attained is 95% (Fig. 1). Incomplete hydrocortisone transformation at higher concentrations has been reported.[2,8] Many situations demand the complete conversion of a substrate, since ultimately the separation of product from unreacted substrate becomes critical.

[8] S. Ohlson, P.-O. Larsson, and K. Mosbach, *Eur. J. Appl. Microbiol. Biotechnol.* **7**, 103 (1979).

TABLE III
HYDROCORTISONE TRANSFORMATION IN AQUEOUS TWO-PHASE SYSTEMS[a]

| Phase system [% (w/w) : % (w/w)] | Reaction time (hr) | Hydrocortisone conversion (%) in | | | Amount of prednisolone in bottom phase (% of total) |
|---|---|---|---|---|---|
| | | Total system | Top phase[b] | Bottom phase[b] | |
| PEG 8000–dextran T40 | | | | | |
| 15 : 3 | 2.5 | 99.5 | 99.8 | 76 | 4.1 |
| 20 : 3 | 2.5 | 99.8 | 100 | 62 | 4.0 |
| 25 : 3 | 2.5 | 99.6 | 99.6 | 65 | 1.0 |
| 25 : 6 | 2.5 | 100 | 100 | 65 | 1.5 |
| Pluronic F68–dextran T40 | | | | | |
| 14 : 4 | 2.5 | 100 | 100 | 94 | 6.8 |
| 14 : 6 | 1.75 | 100 | 100 | 85 | 8.6 |
| PPG 2025–dextran T40 | | | | | |
| 25 : 3 | 3.0 | 99.5 | 99.3 | 93 | 20 |
| M-PEG 550–dextran T40 | | | | | |
| 25 : 3 | 6.5 | 99.4 | 99.4 | 99 | 2.1 |
| M-PEG 5000–dextran T70 | | | | | |
| 9.7 : 4 | 4.0 | 99.4 | 99.8 | 99.2 | 8.5 |
| 10.7 : 2 | 8.0 | 98.5 | 98.4 | 91 | 1.2 |

[a] From R. Kaul and B. Mattiasson, *Appl. Microbiol. Biotechnol.* **24,** 259 (1986), with permission.
[b] After completion of the reaction, the phases were separated, and the degree of conversion was determined in the respective phases.

## *Increasing Oxygen Supply*

One limiting factor during conversion of hydrocortisone to prednisolone is the lack of oxygen. By improving the oxygen supply to the reaction mixture it is possible to increase the rate of transformation, especially at high substrate concentrations. To increase the oxygen supply, the reaction can be carried out in the presence of perfluorochemicals, which are good solvents for gases.[9]

*Preparation of Emulsion of Perfluorochemical.* The emulsion (30% perfluorochemical, v/v) is prepared as described earlier.[10] The buffer solution (water phase), the perfluorochemical FC-72 (3M, St. Paul, MN) (organic phase), and Pluronic F-68 (emulsifier) are mixed in appropriate proportions. The amount of Pluronic F-68 is 0.21 g/ml FC-72. The mixture (50 ml) is placed in a 400-ml beaker and cooled in crushed ice. It is

[9] B. Mattiasson and P. Adlercreutz, *Trends Biotechnol.* **5,** 250 (1987).
[10] P. Adlercreutz and B. Mattiasson, *Eur. J. Appl. Microbiol. Biotechnol.* **16,** 165 (1982).

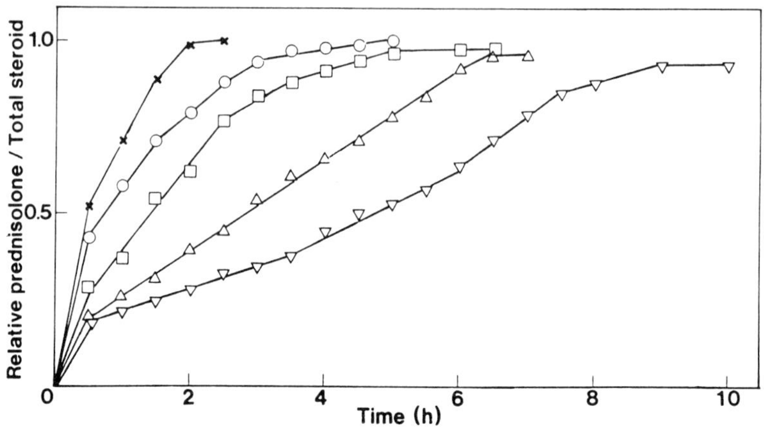


FIG. 1. Time course for hydrocortisone transformation by *Arthrobacter simplex* cells in a 25% (w/w) PEG 8000–6% (w/w) dextran T40 system at 20°. Hydrocortisone concentrations used were 1 (**x**), 2 (○), 3 (□), 4 (△), and 5 mg (▽) per milliliter of reaction medium. Experimental details are described in the text. [From R. Kaul and B. Mattiasson, *Appl. Microbiol. Biotechnol.* **24,** 259 (1986), with permission.]

then sonicated using MSE Soniprep 150 Ultrasonic disintegrator (MSE Scientific Instruments, Sussex, England) for 5 min until a white emulsion forms.

*Use of Perfluorochemical Emulsion During Bioconversion.* The emulsion (4 ml) is added to 20 ml of the two-phase system containing hydrocortisone (either 1 or 3 mg/ml), and the cells (22 or 66 mg dry weight, respectively) are then added. In the presence of the perfluorochemical emulsion, the transformation rate was found to increase, and the conversion of higher concentrations of hydrocortisone nearly reached 100% (Fig. 2). On completion of the reaction, as the phase system was allowed to stand, the emulsion formed a separate phase below that of Dx and could easily be recovered. Perfluorochemicals are very poor solvents for organic chemicals and the product was, hence, recovered in the polymer phases.

### *Product Extraction from Top Phase of Aqueous Two-Phase System*

Amberlite XAD-4 (BDH) is packed into a column (17 × 0.6 cm) and rinsed with 50 m*M* Tris-HCl solution, pH 7.0. The PEG-rich top phase from the 25% (w/w) PEG 8000–6% (w/w) Dx T40 system is separated after completion of hydrocortisone transformation, then passed through the Amberlite column at a flow rate of 0.5 ml/min. The column is washed with buffer and flushed with methanol for the desorption of prednisolone. The column is then regenerated by washing with buffer.

The PEG eluate is generally devoid of steroid product, and the latter

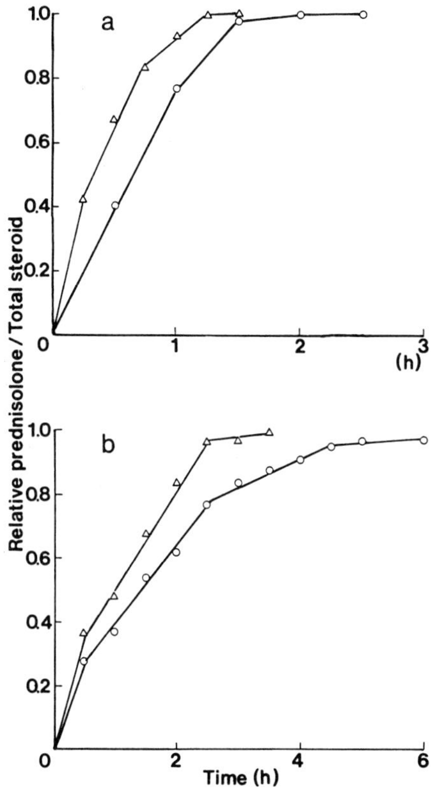


FIG. 2. Transformation of hydrocortisone to prednisolone by *Arthrobacter simplex* cells in 25% (w/w) PEG 8000–6% (w/w) dextran T40 system at 20° in the presence (△) and absence (○) of an emulsion of a perfluorochemical (F-72). The concentrations of hydrocortisone used were (a) 1 and (b) 3 mg/ml of the phase system. [From R. Kaul and B. Mattiasson, *Appl. Microbiol. Biotechnol.* **24,** 259 (1986), with permission.]

is totally desorbed by methanol. The solvent is evaporated so as to obtain prednisolone crystals. Sometimes prednisolone, up to 5% of the amount loaded, could be detected in the eluate. This is extracted by reloading the eluate onto the column. Adsorption to the Amberlite resin is a mild process and permits the phase polymer to be recycled. Amberlite could be used repeatedly after regeneration with buffer.

*Recycling of Cells and Phase Polymers*

The recovered PEG phase is supplemented with substrate and returned to the bioreactor, which contains the Dx-rich phase and the cells. PEG

TABLE IV

HYDROCORTISONE TRANSFORMATION BY *Arthrobacter simplex* CELLS IN AQUEOUS TWO-PHASE SYSTEM COMPARED TO OTHER METHODS

| System | Cell concentration for reaction | Hydrocortisone concentration (g/liter) | Organic solvent | Temperature (°C) | Reaction time (hr) | Conversion (%) | Ref. |
|---|---|---|---|---|---|---|---|
| Free growing cells | up to 3–4 g dry wt./liter | 2.0–20.0 | None | 30 | 18–24 | 89–91 | Chen *et al.*[a] |
| Cells adhered to glass slides | (~3.0 × $10^{11}$ ? cells/liter) | 0.36 | Methanol, 5% | 30 | 19 | 89 | Mozes and Rouxhet[b] |
| Cells immobilized in collagen; | | | | | | | |
| in packed bed reactor[c] | 5.75–35.5 g/liter | 0.67 | Ethanol, 15% | 30 | 0.5 (residence time)[d] | 79.6 | Venkatasubramanian *et al.*[e] |
| in batch[c] | 24 g/liter | 15.0 (+menadione) | None | 30 | 48 | 85 | |
| Cells entrapped in photocross-linkable resin prepolymers | 1.0 g/liter | 1.0 | Methanol, 1% | 30 | 1 | 90–100 | Sonomoto *et al.*[f] |
| Cells entrapped in polyacrylamide | (~1 g wet wt./liter)[c] | 0.36 | Methanol, 5% | 22 | 18[c] | 100 | Ohlson *et al.*[g] |

| | | | | | | | |
|---|---|---|---|---|---|---|---|
| Cells entrapped in alginate | 1–2 g wet wt./liter[c] | 0.36 | Methanol, 4% | 21 | 24 | Not stated | Ohlson *et al.*[h] |
| | 10–20 g wet wt./liter[c] | 9.0 | Methanol, 5% | 21 | 24[i] | 85 | |
| Cells coentrapped with substrate in alginate | 0.6–2.4 g dry wt./liter | 0.5–2.0 | None | 22 | 8–21 | 100 | Kaul *et al.*[j] |
| Living cells in aqueous two-phase system | 1.0–2.0 g dry wt./liter | 1.0–2.0 | None | 22 | 2.5–5.0[i] | 99.5–100 | Present study |

[a] K.-C. Chen, C.-C. Chang, C.-F. Chiu, and A. C. Ling, *Biotechnol. Bioeng.* **27,** 253 (1985).
[b] N. Mozes and P. G. Rouxhet, *Enzyme Microb. Technol.* **6,** 497 (1984).
[c] Calculated/estimated value from data given in the publication.
[d] The column (45 × 1 cm) was packed with 6 g collagen-immobilized cells with a cell content of 51%.
[e] K. Venkatasubramanian, A. Constantinides, and W. R. Vieth, *in* "Enzyme Engineering" (E. K. Pye and H. H. Weetal, eds.), Vol. 3, p. 29. Plenum, New York, 1978.
[f] K. Sonomoto, A. Tanaka, T. Omata, T. Yamane, and S. Fukui, *Eur. J. Appl. Microbiol. Biotechnol.* **6,** 325 (1979).
[g] S. Ohlson, P.-O. Larsson, and K. Mosbach, *Biotechnol. Bioeng.* **20,** 1267 (1978).
[h] S. Ohlson, P.-O. Larsson, and K. Mosbach, *Eur. J. Appl. Microbiol. Biotechnol.* **7,** 103 (1979).
[i] Time required for total hydrocortisone transformation is reduced after activation of cells.
[j] R. Kaul, P. Adlercreutz, and B. Mattiasson, *Biotechnol. Bioeng.* **28,** 1432 (1986).

lost during product extraction is replaced. Alternatively, a fresh PEG-rich phase can be used.

The transformation rate improved by 25–75% in a number of runs.[4] This was attributed to further induction of enzyme activity in the cells by the substrate. A similar observation had been reported earlier.[7] A decline in enzyme activity occurred, however, after about 7 runs, which resulted in slightly longer times for complete conversion.

Enhancement of $\Delta^1$-dehydrogenase activity of immobilized *A. simplex* is possible when the cells are incubated in nutrient media.[8] This activation is due to an increase in the number of cells that are simultaneously being induced for dehydrogenase activity by the substrate hydrocortisone. Application of a similar approach here, that is, adding the nutrients to the aqueous two-phase system, helped maintain the catalytic activity of the cells.[4] It is important, however, to maintain sterile conditions in long-term operations.

## Remarks

Table IV compares aqueous two-phase systems with some of the methods previously used for hydrocortisone transformation. It is evident that the reaction rate achieved in the two-phase system is among the highest reported.

Immobilized cells have earlier been employed to protect the biocatalyst from denaturation.[11] Conversion of hydrocortisone to prednisolone is also subject to product inhibition.[12] Aqueous two-phase systems offer suitable reaction conditions even when the viability of the microbial cells is to be maintained. The top phase serves as a steroid reservoir, whereas the cells are confined to the bottom phase. As the reaction proceeds the product is extracted simultaneously to the top phase and away from the biocatalyst. By manipulation of the phase volume ratios, it is possible to obtain most of the product in the top phase.

## Acknowledgments

The authors are grateful to The National Swedish Board for Technical and Industrial Development (NUTEK) for financial support.

[11] K. Sonomoto, A. Tanaka, T. Omata, T. Yamane, and S. Fukui, *Eur. J. Appl. Microbiol. Biotechnol.* **6,** 325 (1979).

[12] J. Kloosterman IV and M. D. Lilly, *Enzyme Microb. Technol.* **6,** 113 (1984).

# [55] Recovery of Proteins and Phase-Forming Chemicals

*By* GÖTE JOHANSSON

## Introduction

Extraction of biomaterials (e.g., proteins) using aqueous two-phase systems leads to a final product in one of the phases which usually contains 5–25% polymer components. Even if the extracted material can be stored in such an environment (the polymers often protect biosubstances such as enzymes), a separation of biomaterial and phase-forming polymers is often desirable. Particulate materials may be recovered, in most cases, by centrifugation, normally after a 2- to 4-fold dilution of phase to reduce its viscosity. For soluble biomaterial (e.g., proteins and nucleic acids), other means of separation are required. In biotechnical processes the reutilization of polymer, in addition to product recovery, is of essence for economic reasons. This is especially true when expensive polymers such as, for example, ligand–polymer derivatives are used for affinity partitioning.[1]

## Methods

The recovery procedures outlined below have been used with top phase, containing protein, from a system composed of 9.8% (w/w) dextran (Dx) T500 (Pharmacia-LKB, Uppsala, Sweden), 7% (w/w) poly(ethylene glycol) (PEG) 8000 (BP Chemicals, Hythe, UK), and 10 m*M* sodium phosphate buffer, pH 7.0, equilibrated at 23°. This top phase contains 12.0% (w/w) PEG and less than 0.1% Dx. Protein and PEG are separated as follows.

### *Separation of Protein and Poly(ethylene Glycol) by Formation of Polymer–Phosphate System*

From 0.2 to 1 liter of an aqueous solution containing 1.58 mol/kg $NaH_2PO_4$ and 1.58 mol/kg $Na_2HPO_4$ is added to 1-liter quantities of top phase (Table I). The mixtures are equilibrated at room temperature (23°) by careful mixing, and two phases form which settle in about 10 min when the height of the phase column is 8 cm. The upper, highly viscous phase contains the bulk of the PEG, whereas the lower, phosphate-rich phase

[1] G. Kopperschläger, this volume [11].

TABLE I
PHASE VOLUMES OF POLY(ETHYLENE GLYCOL)–PHOSPHATE SYSTEMS[a] AND RECOVERY OF POLYMER[b] IN TOP PHASE AND ALBUMIN[c] IN SALT PHASE

| Volume phosphate solution[d] added to 1 liter top phase (liter) | Phase volume | | Phase components recovered | |
|---|---|---|---|---|
| | Top phase (liter) | Bottom phase (liter) | PEG (% in top phase) | Albumin (% in bottom phase) |
| 0.20 | 0.43 | 0.75 | 68 | >99 |
| 0.40 | 0.31 | 1.07 | 97 | >99 |
| 0.60 | 0.26 | 1.30 | 99 | >99 |
| 0.80 | 0.23 | 1.52 | 100 | 83[e] |
| 1.00 | 0.22 | 1.73 | 100 | 84[e] |

[a] The system was made from 1 liter top phase containing 12.0% (w/w) PEG 8000, 0.05% dextran T500, and 10 m*M* sodium phosphate buffer, pH 7.0, also containing 10 g/liter of bovine serum albumin. Temperature, 23°.
[b] Determined by refractometry and corrected for phosphate.
[c] Determined photometrically at 280 nm. The phosphate can be removed from the albumin by dialysis.
[d] The sodium phosphate solution contained 2.9 mol/kg phosphate and contained equimolar quantities of $Na_2HPO_4$ and $NaH_2PO_4$.
[e] Remaining protein precipitated and was recovered at the interface.

usually contains most of the protein. Only at high concentrations of the phase-forming components may part of the protein precipitate. The volumes of the phases together with the quantity of PEG (in top phase) and of a model protein, bovine serum albumin (in bottom phase), are given in Table I. The separation will be more efficient when the equilibration temperature is raised, which reduces the solubility of PEG in phosphate solution. The same result can be effected by increasing the pH. Smaller amounts of phase (up to 5 liters) can be handled by use of a separatory funnel or other vessel from which lower phase can be drained. For larger volumes a centrifugal separator (e.g., LAPX 202, which has a capacity of 0.5–1 liter phase system per minute[2]; Alfa-Laval Separation, Bromma, Sweden) is useful.

The above process permits the PEG to be obtained in nearly quantitative fashion in the form of a concentrated (40–50%) aqueous solution. This solution can, in many cases, be used without further purification for

[2] F. Tjerneld, G. Johansson, and M. Joelsson, *Biotechnol. Bioeng.* **30,** 809 (1987).

a new system.[2,3] The method is especially suited for recovery of valuable PEG-bound ligands,[1] when their interaction with the extracted protein is due mainly to electrostatic forces. The protein in the salt phase can be concentrated by ultrafiltration (see below). The phosphate in the filtrate can, to a large extent, be recovered by crystallization at 6°.[3]

### *Separation of Protein and Poly(ethylene Glycol) by Using Ion-Exchangers*

Ten liters of the original PEG-rich top phase is diluted with 10–20 liters distilled water, and the solution is applied on a column (20 × 20 cm) of DEAE-cellulose (DE52, Whatman, Maidstone, UK), assuming that the proteins are negatively charged. The proteins bind to the ion exchanger but the uncharged PEG does not. The PEG is washed from the column with 12 liters of 5 m*M* sodium phosphate buffer, pH 7.0. The protein is then eluted by use of the same buffer which also contains KCl, 0.2–1 *M* depending on the protein. By eluting with a KCl gradient, protein mixtures can be additionally fractionated in this step. As an alternative to the ion exchanger, hydroxylapatite (Bio-Rad, Brussels, Belgium) can be used as a protein or nucleic acid adsorbent. The desorption, in this case, is done with increasing concentrations of phosphate buffer.

The method results in the recovery of protein in a concentrated form and with the possibility for immediate further purification. The PEG from the column is diluted (4–6% solution) and may be contaminated with proteins not retained by the column. Thus, this method is less applicable when the polymer is to be conserved.

### *Phase Formation with Highly Charged Polymers*

Ten liters of the original PEG-rich phase is supplied with 50 g of DEAE-dextran (DEAE-Dx) 500 (Pharmacia-LKB), which dissolves within a few minutes with gentle mixing. Phase settling is speeded by centrifuging the mixture (3000 *g* for 15 min) or by use of a centrifugal separator. The DEAE-Dx forms a small bottom phase, approximately 300 ml, in which 60–80% of the protein collects. To obtain a higher yield the amount of DEAE-Dx can be increased or the collected top phase can be reextracted by addition of a new 50-g portion of DEAE-Dx. This step is especially useful for the concentration of protein. The small lower phase can be applied on a DEAE-cellulose column, as above, which retains the nega-

[3] H. Hustedt, K. H. Kroner, and M.-R. Kula, *in* "Partitioning in Aqueous Two-Phase Systems: Theory, Methods, Uses, and Applications to Biotechnology" (H. Walter, D. E. Brooks, and D. Fisher, eds.), p. 529. Academic Press, Orlando, Florida, 1985.

tively charged protein but not the positively charged DEAE-Dx (L. Cheng and G. Johansson (1989), unpublished results).

### *Temperature-Induced Phase Separation*

By using a copolymer of ethylene oxide and propylene oxide, for example, Ucon 50-HB-5100 (Union Carbide, New York, NY), instead of PEG, the polymer can be recovered, after the extraction, by increasing the temperature of the upper phase. This is because the solubility of the copolymer decreases greatly with temperature. This effect becomes even more pronounced with an increased content of propylene oxide groups. The following process and recoveries have been reported.[4]

The primary extraction is made in a system containing 9.0% (w/w) Dx T40 (Pharmacia-LKB), 6.3% (w/w) Ucon 50-HB-5100, 0.2% (w/w) Ucon 50-HB-5100 carrying dye molecules (Procion Yellow HE-3G, I.C.I., Blackley, UK) as affinity ligands for glucose-6-phosphate dehydrogenase,[2] and 20 m*M* sodium phosphate buffer, pH 7.0, at 22°. Extract of baker's yeast (7% of the total system) is used as a source of enzyme. To recover both the extracted protein and the Ucon, two salts, $Na_2SO_4$ and NaCl, are dissolved in the top phase to a concentration of 0.2 *M* each. The phase is collected and warmed in a water bath to 40°. A new two-phase system forms with a water-rich top phase containing practically all the proteins (including the target enzyme) and a bottom phase rich in Ucon (approximately 40%, w/w) as well as the bulk of Ucon-bound ligand (~80% of total). This polymer mixture can be reused directly in a new extraction system.

### *Ultrafiltration*

When the partitioned biological macromolecules are large compared to the phase-forming polymer, the latter can be removed by ultrafiltration or dialysis using a membrane with suitable cutoff. Veide *et al.* have, in this way, separated β-galactosidase from PEG 3400.[5] The same method can be employed when the extracted material is of low molecular weight and the polymer is large, for example, PEG 40,000 (Serva, Heidelberg, Germany).[6]

[4] P. A. Alred, F. Tjerneld, A. Kozlowski, and J. M. Harris, *Bioseparation* **2,** 363 (1992).
[5] A. Veide, T. Lindbäck, and S.-O. Enfors, *Enzyme Microb. Technol.* **11,** 744 (1989).
[6] F. Tjerneld, I. Persson, and P.-Å. Albertsson, *Biotechnol. Bioeng. Symp.* **15,** 63 (1985).

*Electrophoresis: Method Useful in Small-Scale Preparations*

The phase-forming polymers are normally noncharged, and charged proteins and nucleic acids can be separated from them by electrophoresis.[7] The undiluted phase is placed at the lower part of a U tube and overlayed with buffer solution. The proteins migrate away from the polymers when an electric field is applied across the system. They can be collected in a section of the apparatus containing high salt concentration.[7]

## Conclusions

Proteins (and nucleic acids) can, in a number of ways, be freed from phase polymers, even when these are charged or carry affinity ligands. Such processes also permit the recovery of the polymers, allowing their reutilization.

## Acknowledgments

This work has been supported by a grant from the Swedish Council for Forestry and Agricultural Research.

[7] P.-Å. Albertsson, "Partition of Cell Particles and Macromolecules," 3rd Ed. Wiley(Interscience), New York, 1986.

# [56] Enzyme Recovery by Continuous Crosscurrent Extraction

*By* NEOPHYTOS PAPAMICHAEL and HELMUT HUSTEDT

## Introduction

We use the term "crosscurrent" to refer to extractions under conditions which vary between stages such that the product ends up alternately in one of the two different phases. This contrasts with cocurrent and countercurrent extractions, which are generally a series of operations carried out under identical conditions. The fundamental aspects of each of these methods are illustrated schematically for a two-stage process in Fig. 1.

From calculations of the yields, the consumption of phase-forming chemicals, and purification factors it is possible to compare the salient features of the three techniques (Table I). The countercurrent mode is

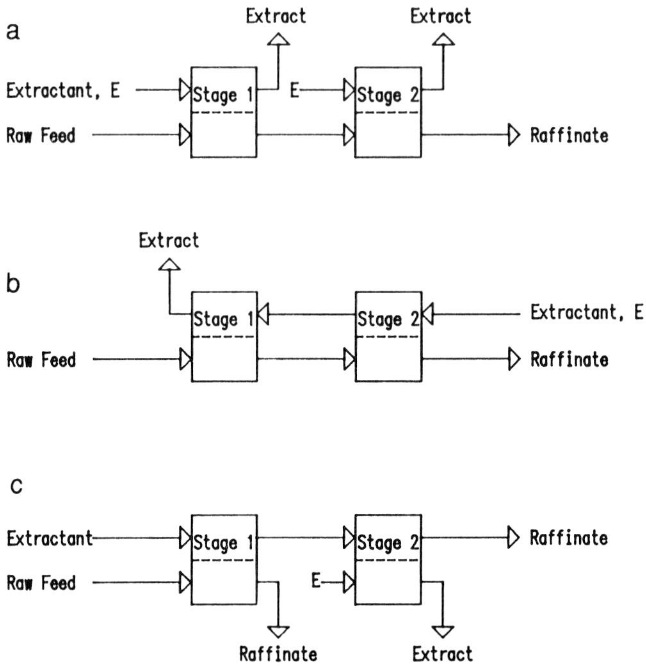


FIG. 1. Methods for continuous extraction. Boxes representing each stage consist of a mixer and settler. (a) Cocurrent extraction, each stage with fresh extractant; (b) countercurrent extraction, where the first stage is extracted using the upper effluent stream from the second stage; (c) crosscurrent extraction, where the second stage is extracted using fresh extractant and the product is shifted from the primary upper phase to the secondary lower phase.

more efficient at extraction than the cocurrent mode, but this occurs at the expense of the purification factor. It also requires less extractant to obtain the same yield of product, but the purification factor suffers here as well. Crosscurrent extraction has the same consumption requirements as cocurrent extraction but is the least efficient of the three methods in terms of yield. It should, therefore, be used only when the partition coefficient, $K$, is sufficiently high to overcome the inherently lower yield of the method. However, crosscurrent extraction easily outperforms the other two procedures with respect to purification factor. Cocurrent and countercurrent extractions would be the methods of choice when $K$ is relatively low.

We focus here on questions important for continuous operations, such as recycling and economics, by use of some examples from the literature in which processing was on a pilot or production scale. Aqueous phase

TABLE I
COMPARISON OF TWO-STAGE CONTINUOUS EXTRACTION METHODS FOR CONSUMPTION, YIELD, AND PURIFICATION FACTORS

| Mode | Consumption[a] | Yield (%) | Purification factor |
|---|---|---|---|
| Cocurrent | 2 | 93.8 | 1.25 |
| Countercurrent[b] | 2 | 97.7 | 1.14 |
| Countercurrent[c] | 1.135 | 93.8 | 1.02 |
| Crosscurrent | 2 | 56.3 | 2.25 |

[a] Consumption is in arbitrary units and the total calculated on the basis of 1 unit per stage of the cocurrent extraction method.
[b] Calculated for equal consumption compared to cocurrent extraction.
[c] Calculated for equal yield compared to cocurrent extraction ($c_{p,0}$ = 10%, $c_{i,0}$ = 20%, $K_p$ = 3 for extractions to the upper phase or $K_p$ = 0.333 for extractions to the lower phase, and $K_i$ = 1, where p designates product, i impurity, and 0 initial conditions).

systems themselves and the optimization of extraction conditions are discussed elsewhere in this volume.[1]

## Methods

Extraction as a unit operation requires two steps: mixing of the liquids and subsequent separation of the phases. This applies also to aqueous two-phase extraction.[2] In the laboratory this is simply done in a test tube followed by either centrifugation or gravity settling. On a larger scale, and when operating continuously, the mixing may be carried out in either a stirred vessel or in-line static mixers (which are basically tubes with internal baffles). Phase separation generally requires centrifugation if the viscosity is high or the density difference between the phases is low. The former occurs with disrupted cells at a high concentration, whereas low density differences are characteristic of polymer–polymer systems.[2] When product stability in the extraction medium during the period of separation is adequate, then separation in a settling tank is significantly more cost

[1] G. Johansson, this volume [3].
[2] H. Hustedt, K. H. Kroner, and M. R. Kula, *in* "Partitioning in Aqueous Two-Phase Systems" (H. Walter, D. E. Brooks, and D. Fisher, eds.), p. 529. Academic Press, Orlando, Florida, 1985.

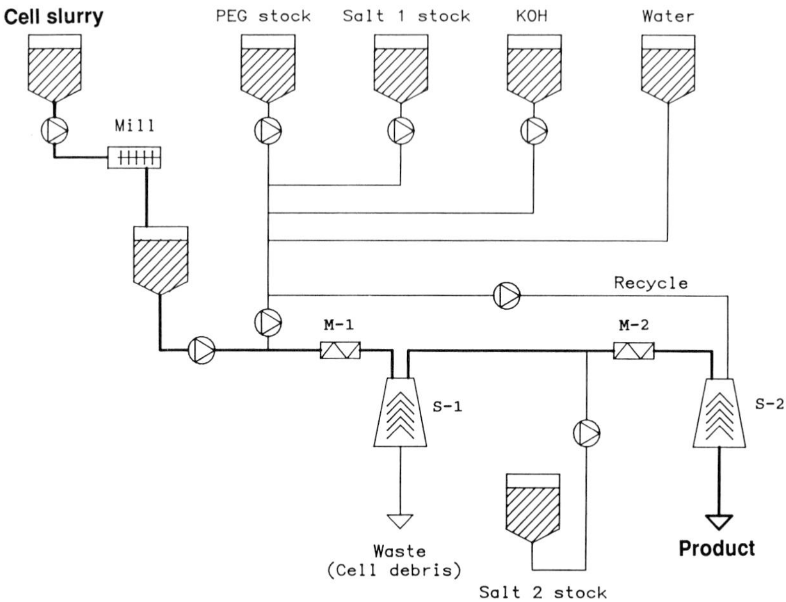


FIG. 2. General scheme for crosscurrent extraction with recycling of PEG. After cell disruption the cell slurry is mixed with stock solutions and the pH is adjusted by addition of alkali in a mixer (M-1); the mixture is separated by centrifugation (S-1), and the upper phase is mixed (M-2) with a second salt stock solution before being separated (S-2). The secondary upper effluent is partially recycled to the primary extraction. [Reproduced by permission from N. Papamichael, B. Börner, and H. Hustedt, *J. Chem. Technol. Biotechnol.* **54,** 47 (1992).]

effective. In most cases the increased residence time does not pose a problem.

Figure 2 shows a general scheme of crosscurrent extraction with recycling of the secondary upper phase as operated on a scale up to approximately 120 liter/hr throughput. The consumption of chemicals can become the determining factor in the economics of large-scale extractions. The reason for this is that, although equipment costs generally increase with a power of 0.6 with scale, those for chemical consumption in extraction increase linearly. Recycling is one method of reducing these costs, as discussed later.

## Examples

### *Fumarase from Saccharomyces cerevisiae*

Fumarase (fumarate hydratase, EC 4.2.1.2) catalyzes the conversion of fumarate to L-malate. The enzyme may be extracted from baker's yeast

TABLE II
STOCK SOLUTION CONCENTRATIONS AND EXTRACTION CONDITIONS FOR FUMARASE FROM *Saccharomyces cerevisiae*[a]

| Stream | Stock solution concentration (%, w/w) | Final concentration |
|---|---|---|
| Primary extraction (pH 8.5) | | |
| Cell homogenate | 50 | 20 (%, w/w) |
| PEG 1550 | 50 | 17 (%, w/w) |
| $K_2HPO_4$ | 40 | 8 (%, w/w)[b] |
| KOH (pH control) | 50 | As needed |
| Secondary extraction (pH 6.9–7.0) | | |
| TOP-1 | n.a.[e] | 70 (%, v/v)[c] |
| Potassium phosphate buffer, pH 6 | 22.3 | 30 (%, v/v)[c,d] |

[a] Reprinted from N. Papamichael, B. Börner, and H. Hustedt, *J. Chem. Technol. Biotechnol.* **54,** 47 (1992).
[b] Includes phosphate in cell homogenate.
[c] As % of total feed to secondary extraction.
[d] Corresponds to an additional 7.7% phosphate in the system.
[e] n.a., Not applicable.

in a two-stage process using a poly(ethylene glycol) (PEG)/potassium phosphate buffer phase system. The stock solution concentrations, system composition, and conditions of extraction are given in Table II.[3]

The extraction is carried out in a process plant shown schematically in Fig. 2. After wet milling (LME-4 bead mill, 12 eccentric rings, 85% lead-free glass bead loading, rotor speed 28.3 $sec^{-1}$; Netzsch Feinmahltechnik GmbH, Selb, Germany) at a biomass concentration of 50%, the cell homogenate is mixed with the stock solutions of PEG and buffer to a pH of 8.5 by in-line static mixers. Approximately 96% enzyme and 91% protein equilibration result at a flow rate of 60 liters/hr through a Kenics static mixer of 4.9 mm internal diameter with 21 mixing elements (Chemineer Ltd., Derby, UK).[4] The mixture is then separated by centrifugation in a disk stack separator [SA-1 with 24 disks, 1 mm spacing, equivalent settling area (Σ factor) 900 $m^2$; Westfalia Separator AG, Oelde, Germany]. The separator is required because the high biomass concentration results in a pastelike lower phase. The lighter effluent (TOP-1) is mixed with a second phosphate buffer to give a new phase system which is separated in an

[3] N. Papamichael, B. Börner, and H. Hustedt, *J. Chem. Technol. Biotechnol.* **54,** 47 (1992).
[4] H. Hustedt, K. H. Kroner, and N. Papamichael, *in* "Proceedings of the International Solvent Extraction Conference," Munich, September 1986, Vol. 3, p. 597. VCH, Weinheim, Germany, 1986.

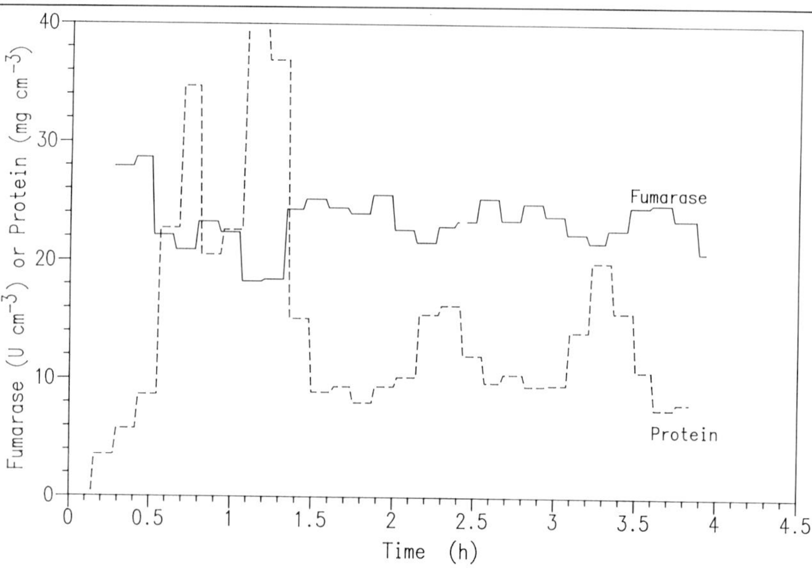


FIG. 3. Enzyme and total protein levels during extraction of fumarase from *Saccharomyces cerevisiae*. [N. Papamichael, B. Börner, and H. Hustedt, unpublished data, 1988.]

SA-1 separator. Figure 3 shows the enzyme and protein concentrations in a run conducted with recycling of TOP-2 to the primary extraction system. During operation the solids in the primary lower phase build up in the separator and must be intermittently removed. During such desludging procedures there is a significant change in the measured protein level in TOP-1, but the enzyme activity remains relatively stable. The enzyme yield is approximately 75% with a purification factor of 12. Recycling of this phase reduces process stability, an observation also made in the cases of fumarase extraction from *Brevibacterium ammoniagenes* and of penicillin acylase from recombinant *Escherichia coli*.[5]

*Superoxide Dismutase from Bovine Liver*

Boland *et al.*[6,7] have demonstrated that large-scale crosscurrent extractions of proteins are also possible from animal tissue. Superoxide dismutase (SOD) may be extracted from bovine liver using a PEG 1550 (Deutsche

[5] H. Hustedt, K.-H. Kroner, and N. Papamichael, *Process Biochem.* **23,** 129 (1988).
[6] M. J. Boland, P. G. M. Hesselink, N. Papamichael, and H. Hustedt, *J. Biotechnol.* **19,** 19 (1991).
[7] M. J. Boland, P. G. M. Hesselink, and H. Hustedt, *J. Biotechnol.* **11,** 337 (1989).

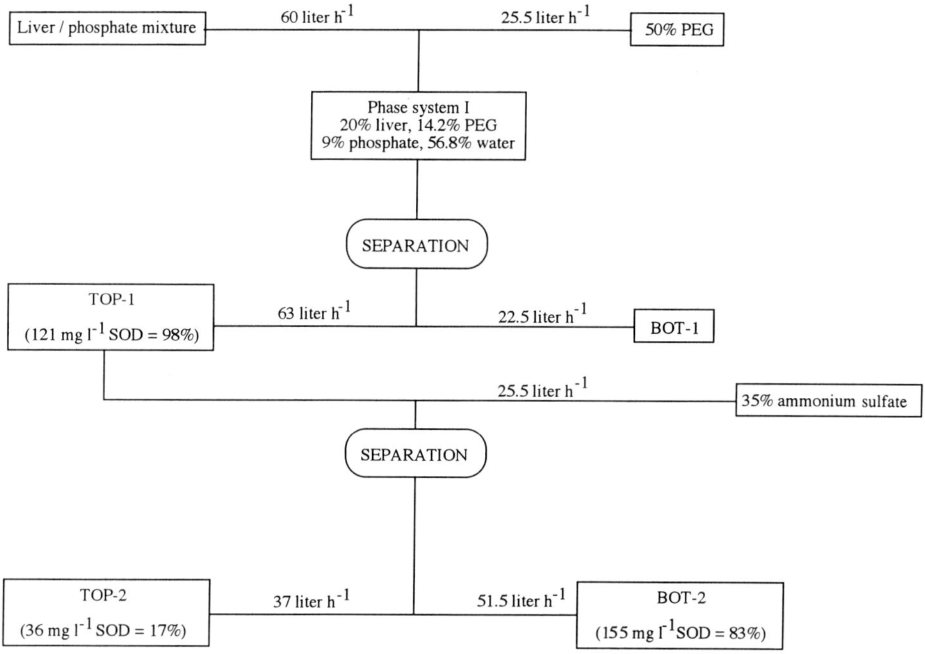


FIG. 4. Two-stage scheme for extraction of superoxide dismutase (SOD) from bovine liver. Primary extraction is with a PEG 1550–potassium phosphate phase system, with a SOD yield of 98%; secondary extraction is with a phase system comprising TOP-1 with ammonium sulfate, with a SOD yield of 83%. Phase component concentrations are given in % (w/w), and flow rates are quoted for a liver throughput of 18 kg/hr. [Adapted by permission from M. J. Boland, P. G. M. Hesselink, N. Papamichael, and H. Hustedt, *J. Biotechnol.* **19,** 19 (1991).]

BP, Düsseldorf, Germany) potassium phosphate system. The finely minced liver must be prefiltered to remove the larger fibrous strands and particles before making up the phase system. The extraction is carried out using the same equipment as above and conditions given in Fig. 4. A flow rate of 60 liters liver suspension per hour (18 kg liver/hr) results in a SOD concentration of 121 mg/liter (yield 98%) in the first extraction and 155 mg/liter (yield 83%) in the second extraction. The final enzyme yield is 80% with a purification factor of 4. Various other enzymes have been extracted on a small scale from animal tissues using this method, for example, $\beta$-galactosidase and lactate dehydrogenase from bovine liver and D-amino-acid oxidase and SOD from porcine kidney.[7] Scaleup should be possible in the manner described.

TABLE III
CONDITIONS OF EXTRACTION DURING LARGE-SCALE AFFINITY PARTITIONING OF LACTATE DEHYDROGENASE FROM PORCINE MUSCLE USING PROCION YELLOW HE-3G–POLY(ETHYLENE GLYCOL) AS AFFINITY LIGAND[a]

| Condition | Homogenized muscle | Muscle extract |
|---|---|---|
| Procion Yellow HE-3G–PEG concentration (as % of total PEG) | 2 | 1.5 |
| Muscle in system (g/liter) | 100 | 160 |
| Initial LDH concentration (kU/liter) | 66 | 70 |
| Recovery of LDH in top phase | | |
| kU/liter system | 32 | 58 |
| kU/kg muscle | 312 | 364 |
| kU/g ligand–PEG | 21 | 51 |
| Recovery of LDH in both phases (kU/liter) | 63 | 68 |
| Specific activity (U/mg) | 114 | 126 |
| Purification factor | 10 | 11 |

[a] Reprinted from F. Tjerneld, G. Johansson, and M. Joelsson, *Biotechnol. Bioeng.* **30,** 809 (1987). 

## *Affinity Extraction*

Tjerneld *et al.*[8] investigated the affinity extraction of lactate dehydrogenase (LDH) from porcine muscle using a phase system comprised of PEG 8000 (BP Chemicals, Hythe, UK) plus Procion Yellow HE-3G–PEG 8000, and a hydroxypropyl starch derivative (Aquaphase PPT, Perstorp Biolytica, Lund, Sweden). The extraction is carried out with a phase system containing 20% Aquaphase PPT, 7.6% PEG, 40 m*M* sodium phosphate buffer, pH 7.9, and either muscle homogenate (10%) or muscle extract (20%, corresponding to 16% meat). The components are mixed in a stirring vessel, and the phases are separated in a disk stack separator (LAPX 202, Alfa-Laval Separation, Tumba, Sweden) at a flow rate of 0.5 liter/min (Table III). Economic analysis of the process indicates that the major cost stems from the consumption of the starch derivative, which can only be partially recycled.

[8] F. Tjerneld, G. Johansson, and M. Joelsson, *Biotechnol. Bioeng.* **30,** 809 (1987).

TABLE IV

FLOW RATES AND CONCENTRATIONS DURING AFFINITY PARTITIONING OF LACTATE DEHYDROGENASE FROM RABBIT MUSCLE USING PROCION ORANGE H-ER–POLY(ETHYLENE GLYCOL) AS AFFINITY LIGAND[a]

| Stream | Flow rate (ml/hr) | PEG (%, w/w) | Procion Orange–PEG (%, w/w) | Reppal PES (%, w/w) | Sodium phosphate (m$M$) | Potassium phosphate (%, w/w) |
|---|---|---|---|---|---|---|
| Enzyme feed | 80 | — | — | — | 50 | — |
| Two-phase polymer feed | 238 | — | 0.05 | 20.1 | — | — |
| Reppal PES waste | 146 | 0.8 | 0.05 | 30.3 | 124 | — |
| TOP-1 (enzyme) | 334 | 11.4 | 0.51 | 0.5 | 124 | — |
| Phosphate eluent | 148 | 1.7 | — | — | — | 31.5 |
| BOT-2 product | 320 | 0.5 | — | — | — | 15.9 |
| TOP-2 (ligand-rich) | 162 | 23.3 | 0.95 | — | — | 5.0 |

[a] Reprinted by permission from S. P. Foaden, H. A. Chase, and C. R. Lowe, *in* "Separations Using Aqueous Phase Systems" (D. Fisher and I. A. Sutherland, eds.), p. 325. Plenum, New York, 1989.

Foaden *et al.*[9] described the purification of lactate dehydrogenase from rabbit muscle by affinity crosscurrent extraction using Procion Orange H-ER–PEG as ligand in a PEG 3350 (Sigma, Poole, UK)–Reppal PES (a hydroxypropyl starch derivative of molecular weight 114,000, Reppe, Växjö, Sweden) aqueous phase system. The extraction is carried under conditions given in Table IV. The process is in two stages, each consisting of a mixer (55 ml) and a settler (600 ml) separated by a weir. When a crude rabbit muscle extract (Sigma Chemicals, St. Louis, MO) with 38 U/mg lactate dehydrogenase activity and 1.1 mg/ml total protein concentration is applied, a yield of greater than 90% and a purification factor of over 2 for the enzyme is achieved. The ligand-rich top phase is completely recycled from the second stage to the primary stage. The consumption of the ligand is thus reduced from 1.79 to 0.12 g/hr.

### *Optimization of Large-Scale Crosscurrent Extraction*

Large-scale processing generally entails optimization of the extraction conditions as well as of the processing steps. Phase systems which offer cost advantages have been investigated, including the use of crude Dx[10]

[9] S. P. Foaden, H. A. Chase, and C. R. Lowe, *in* "Separations Using Aqueous Phase Systems" (D. Fisher and I. A. Sutherland, eds.), p. 325. Plenum, New York, 1989.

[10] K. H. Kroner, H. Hustedt, and M.-R. Kula, *Biotechnol. Bioeng.* **24,** 1015 (1982).

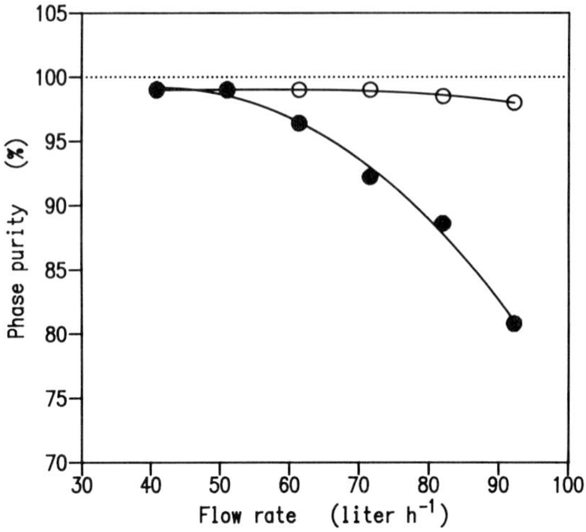


FIG. 5. Phase purity from an SA-1 disk stack separator as a function of throughput during the primary extraction of fumarase from *Saccharomyces cerevisiae*. ○, PEG-rich phase; ●, phosphate-rich phase. [N. Papamichael and H. Hustedt, unpublished data (1986).]

or other polymers[8,11] and replacing phosphate with citrate to reduce waste disposal costs.[12]

Use of disk stack centrifuges requires that the volumetric throughput be optimized to obtain satisfactory phase separation. To illustrate this point, Fig. 5 shows the phase purities obtained in the primary extraction of fumarase as a function of throughput in an SA-1 disk stack separator. In contrast to the upper phase, the lower phase purity drops sharply above 60 liters/hr owing to carryover of the upper phase with the viscous lower phase. Although the secondary extraction does not suffer in the same manner, this behavior restricts the flow rates which can be used in the complete, continuous process. Of course, the flow rates must be optimized for each individual process.

Recycling of the phase-forming components is another factor that is important to the optimization of the system. Fourfold batch recycling of the TOP-2 phase to the primary phase system in the fumarase/yeast extraction is possible without deleterious effects on product yield or purity.[13] Continuous recycling of this phase is also feasible, resulting in a reduction of 50% in PEG and 12% in potassium phosphate.[1] A cost analysis

[11] D. C. Szlag and K. A. Giuliano, *Biotechnol. Tech.* **2,** 277 (1988).

[12] J. Vernau and M.-R. Kula, *Biotechnol. Appl. Biochem.* **12,** 397 (1990).

[13] H. Hustedt, *Biotechnol. Lett.* **8,** 791 (1986).

TABLE V
ECONOMIC EVALUATION OF RECYCLING DURING FUMARASE PROCESS[a]

| Cost position | No recycling | | With recycling | | % Change[b] | |
|---|---|---|---|---|---|---|
| | Batch | Continuous | Batch | Continuous | Batch | Continuous |
| Materials | 2.48[c] | 2.48 | 1.41 | 1.89 | −43 | −24 |
| Labor | 1.34 | 1.0 | 1.66 | 0.91 | 24 | −8 |
| Capital | 0.14 | 0.18 | 0.14 | 0.23 | 0 | 28 |
| Utilities | 0.015 | 0.018 | 0.016 | 0.02 | 8 | 19 |
| Total | 6.5 | 5.69 | 6.19 | 4.88 | −5 | −14 |

[a] Values from N. Papamichael, B. Börner, and H. Hustedt, *J. Chem. Technol. Biotechnol.* **54,** 47 (1992).

[b] The values given are the % change in batch and continuous modes when recycling compared to nonrecycling operation.

[c] Values given are relative based on the multipliers for the individual cost points as given in Eq. (1). The biomass concentration in batch mode is 25% (w/v) and that in continuous mode 20% (w/v).

of this extraction has been carried out to compare batch and continuous operation with and without recycling. The estimated total cost is given by

$$T = 1.13M + 2.6L + 1.6I + 1.13U \quad (1)$$

where $T$ is the total, $M$ is materials, $L$ is labor, $I$ is the capital investment, and $U$ represents utilities. The multipliers take into account various factors such as interest payments on capital, plant overheads, and administration; however these factors can vary. Table V shows that the reduction in chemical consumption is high, 24% in the case of continuous and 43% with batch operation, but that the total savings are lower owing to the high labor component. Continuous processing with recycling requires increased capital expenditure, but this is offset by the additional savings which can be made in labor costs owing to computerization of the operation.

## Concluding Remarks

Continuous crosscurrent extraction has been discussed for a number of enzymes. The approach has several advantages when compared to batch processing. These include increased productivity and product homogeneity, higher yields of labile products, in-line recycling of the extraction media, and fully automated operation. In addition, the continuous process is easier to integrate with the cell disruption and ultrafiltration steps which usually precede and follow the extraction step. A potential disadvantage of the method is the increased capital outlay for machinery and process control, but the cost effectiveness of such an automated system justifies the additional expenditure.

# [57] Purification of Oxynitrilases from Plants

*By* RAINER WOKER, JÖRG VERNAU, and MARIA-REGINA KULA

## Introduction

Oxynitrilases are suitable catalysts for the synthesis of optically active (*R*)- and (*S*)-cyanohydrins from a broad range of structurally different aldehydes. These chiral cyanohydrins are precursors for numerous compounds such as pyrethroids, $\alpha$-hydroxy acids, and 2-amino alcohols.[1-5] The enzyme-catalyzed reaction is highly enantioselective (Fig. 1). Products with an enantiomeric excess exceeding 96% can be obtained if the chemical side reaction is sufficiently suppressed.[6-8] Here we describe the purification of (*R*)-oxynitrilase from bitter almonds (*Prunus amygdalus*) and (*S*)-oxynitrilase from *Sorghum bicolor*.

## Chemicals

Inorganic salts, sodium acetate, and Coomassie Brilliant blue are obtained from Merck (Darmstadt, Germany), and sodium citrate and citric acid are from Jungbunzlauer Ladenburg GmbH (Ladenburg, Germany). DL-*p*-Hydroxymandelonitrile is purchased from Aldrich (Steinheim, Germany), and bovine serum albumin is from Boehringer Mannheim (Mannheim, Germany). Poly(ethylene glycol) (PEG) with different molecular weights (400, 1550, 2000, and 20,000) is a gift from Chemische Werke Hüls (Marl, Germany). The Chelating-Sepharose Fast Flow and the Q-

[1] T. Matsuo, T. Nishiioka, M. Hirano, Y. Suzuki, K. Tsushima, N. Itaya, and H. Yoshioka, *Pestic. Sci.* **11,** 202 (1980).

[2] J. Brussee, W. T. Loos, C. G. Kruse, and A. Van der Gen, *Tetrahedron* **46,** 979 (1990).

[3] F. Effenberger, B. Gutterer, and T. Ziegler, *Liebigs Ann. Chem.*, p. 269 (1991).

[4] M. Elliott, N. F. Janes, B. P. S. Khambay, and D. A. Pulman, *Pestic. Sci.* **14,** 182 (1983).

[5] J. B. Jones, *Tetrahedron* **14,** 3351 (1986).

[6] F. Effenberger, T. Ziegler, and S. Förster, *Angew. Chem.* **99,** 491 (1987).

[7] U. Niedermeyer and M.-R. Kula, *Angew. Chem., Int. Ed. Eng.* **29,** 386 (1990).

[8] U. Kragl, U. Niedermeyer, M.-R. Kula, and C. Wandrey, *Ann. N.Y. Acad. Sci.* **613,** 167 (1990).

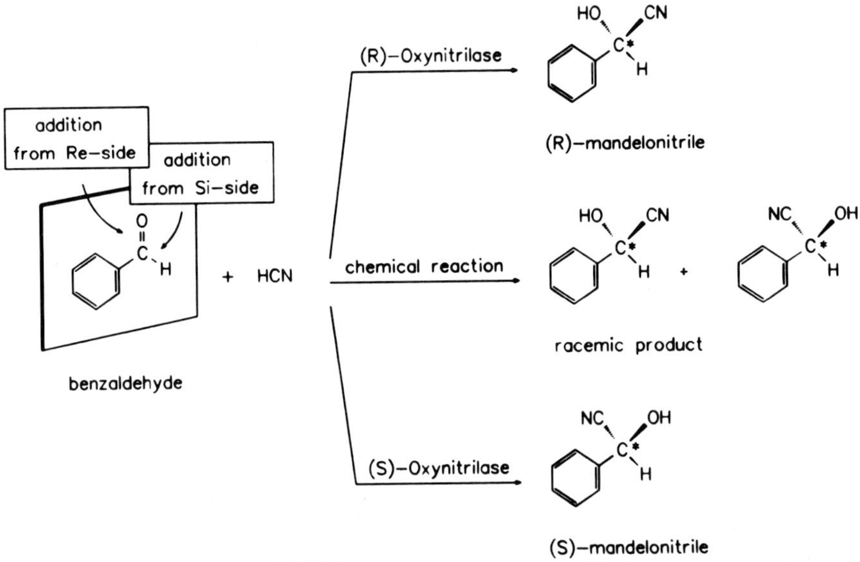


FIG. 1. Addition of hydrogen cyanide to benzaldehyde by enzyme-catalyzed processes and/or chemical reaction.

Sepharose are purchased from Pharmacia LKB (Uppsala, Sweden). Ion-exchange membranes (Q-Sartobind) are a gift from Sartorius AG (Göttingen, Germany).

## Analytical Procedures

### *Determination of (R)-Oxynitrilase Activity*

A solution of 2 m*M* DL-mandelonitrile free of benzaldehyde is freshly prepared in 50 m*M* sodium citrate buffer, pH 3.75. The substrate solution (990 $\mu$l) is mixed with 10 $\mu$l of protein solution to be tested, and the absorbance is followed at 250 nm in a spectrophotometer at 25°. An extinction coefficient of 12,350 $cm^{-1}$ $M^{-1}$ is used for calculation of enzyme activity. One unit is defined as the production of 1 $\mu$mol of benzaldehyde per minute under the conditions specified.

### *Determination of (S)-Oxynitrilase Activity*

For the *S*-oxynitrilase assay a 1 m*M* solution of *p*-hydroxymandelonitrile is freshly prepared in 50 m*M* sodium citrate buffer, pH 3.75. One milliliter of the substrate solution is mixed with 25 $\mu$l of the protein solution to be tested, and the absorbance at 285 nm is followed in a spectrophotometer at 25°. An extinction coefficient of 16,530 $cm^{-1}$ $M^{-1}$ is used for the calculation of enzyme activity. One unit is defined as the amount of enzyme that catalyzes the production of 1 mmol *p*-hydroxybenzaldehyde per minute under these conditions.

### *Protein Assay*

The protein concentration is determined according to Bradford[9] using bovine serum albumin (BSA) as standard.

## Experimental Methods

### *Preparation of Copper(II) Iminodiacetate–Poly(ethylene Glycol)*

The PEG derivative is synthesized from dibromo-PEG (molecular weight 20,000) and iminodiacetic acid (IDA) according to a published procedure.[10] The tridentate IDA–PEG is mixed with an excess Cu(II) sulfate and extensively dialyzed against distilled water before use.

### *Purification of (R)-Oxynitrilase Using Poly(ethylene Glycol)–Citrate Aqueous Two-Phase System*

The raw material for the enzyme extraction (bitter almond meal) is obtained from a commercial source (Caesar and Loretz, Hilden, Germany). To 50 g almond bran are added 105 g PEG 1550, 31.3 g sodium citrate/citric acid mixture (pH 7.0), and 313.7 g water, added to give a 500-g two-phase system containing 10% biomass (dry weight). The system is stirred for 30 min to ensure equilibrium of extraction. After centrifugation for 10 min at 1500 *g*, approximately 375 ml top phase 1 is obtained (Table I).

The top phase is separated and 67.5 g sodium citrate/citric acid mixture (pH 7.0), 14.4 g NaCl, and 293.1 g water are added to generate the second two-phase system. Under these conditions the enzyme is shifted from the PEG-rich top phase 1 to the salt-rich bottom phase 2. The mixture is stirred for 5 min. In the absence of solids the separation is accomplished in a separatory funel; 440 ml bottom phase 2 is drained after 60 min. The enzyme is stable in the salt solution and can be stored at 4° as a technical grade catalyst preparation. The procedure has been carried out success-

[9] M. M. Bradford, *Anal. Biochem.* **72,** 248 (1976).

[10] F. H. Arnold, M. E. van Dam, and G. E. Wuenschell, *Biotechnol. Appl. Biochem.* **11,** 492 (1989).

TABLE I
PURIFICATION OF (*R*)-OXYNITRILASE FROM ALMOND BRAN

| Purification step | (*R*)-Oxynitrilase activity (U) | Activity yield (%) | Specific activity (U/mg) |
|---|---|---|---|
| Extraction | 5700 | 100 | 0.82 |
| Two-phase system 1 | 5630 | 99 | 11 |
| Two-phase system 2 | 4840 | 85 | 16 |
| Ultrafiltration | 4550 | 80 | 24 |
| Anion-exchange chromatography (Q-Sepharose FF) | 3250 | 57 | 64 |

fully starting with 15 kg almond bran and using a continuously operating disk stack separator SA-1 (Westfalia AG, Oelde, Germany) with a flow rate of 50 liters/hr to separate the first phase system.

The enzyme is concentrated and desalted by ultrafiltration/diafiltration using a hydrophilic membrane with a cutoff of 20,000 Da. The bottom phase 2 is diluted 3-fold with water to minimize protein precipitation in the gel polarization layer on the membrane. More than 85% of the activity is recovered in the large-scale experiment. If desired, the enzyme can be further purified by anion-exchange chromatography using a Q-Sepharose Fast Flow column equilibrated with a 50 m*M* potassium phosphate buffer, pH 7.0. After a wash with starting buffer to remove unbound proteins, the (*R*)-oxynitrilase is eluted by a linear salt gradient from 0 to 500 m*M* sodium chloride in 50 m*M* potassium phosphate buffer (pH 7.0). The enzyme is desorbed around 300 m*M* salt. The procedure is summarized in Table I.[11,12] Owing to the rather high concentration of (*R*)-oxynitrilase in almond bran and the efficiency of the extraction system, application of an affinity step can be avoided.

## *Purification of (S)-Oxynitrilase*

For the purification of (*S*)-oxynitrilase from seedlings of *Sorghum bicolor* we make use of the affinity of (*S*)-oxynitrilase to immobilized divalent copper ions as the key step (Fig. 2).

*Enzyme Preparation.* Seeds of *Sorghum bicolor* are a gift from Semundo Saatzucht (Hamburg, Germany). In contrast to the (*R*)-oxynitrilase, which is found in the seeds of bitter almonds (*Prunus amygdalus*) and other members of the family Rosaceae, the corresponding (*S*)-oxynitri-

[11] I. Jansen, R. Woker, and M.-R. Kula, *Biotechnol. Appl. Biochem.* **15,** 90 (1992).
[12] J. Vernau and M.-R. Kula, *Biotechnol. Appl. Biochem.* **12,** 599 (1990).

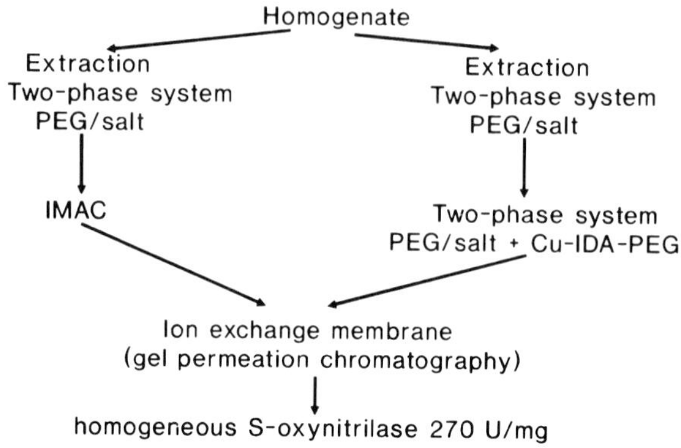


FIG. 2. Purification schemes for (*S*)-oxynitrilase from seedlings of *Sorghum bicolor*. IMAC, Immobilized metal affinity chromatography.

lase is present not in the seeds but in the developing shoot. Therefore seeds are soaked in water and allowed to germinate. Dark-grown shoots are harvested after 7 days, shock-frozen in liquid nitrogen, and ground to a powder.[11] About 400 g tissue are obtained from 100 g seeds. The frozen plant tissue serves as starting material.

*Purification of (S)-Oxynitrilase by Partitioning and Immobilized Metal Ion Affinity Chromatography*. Physical extraction of (*S*)-oxynitrilase[12] is possible, for example, in an aqueous two-phase system composed of 20% PEG 1000, 9% (w/w) potassium phosphate, at pH 8.0, and 8% powdered plant tissue. The enzyme is extracted into the PEG-rich top phase at room temperature. The phases are separated by centrifugation for 1 min at 1000 *g* at 20–22°. To generate the second phase system, 4 g PEG 20,000, 16 g potassium phosphate, pH 8.0, and 80 g distilled water are added per 100 g of separated top phase of the first extraction step. The phases are mixed for 5 min in an overhead shaker and centrifuged for 1 min at 1000 *g*. The salt-rich phase is diluted 3-fold with water to reduce the high viscosity of the solution before the sample is applied on a column of chelating Sepharose Fast Flow loaded with $Cu^{2+}$ ions.[5] The enzyme could be eluted with 50 m*M* EDTA, 150 m*M* NaCl, pH 7.0. Gel filtration serves to remove the Cu(II)–EDTA complex from the (*S*)-oxynitrilase. The procedure is illustrated in Fig. 2 and summarized in Table II.

## *Purification of S-Oxynitrilase by Affinity Extraction*

As a first step a physical extraction is carried out in an aqueous two-phase system composed of 17% PEG 2000, 7.2% sodium sulfate, 0.1% potassium phosphate, pH 8.0, and 8% powdered plant tissue. Under these

TABLE II
PURIFICATION OF (*S*)-OXYNITRILASE FROM *Sorghum bicolor*

| Purification step | (*S*)-Oxynitrilase activity (U) | Activity yield (%) | Specific activity (U/mg) |
|---|---|---|---|
| Raw material | 160 | 100 | 1.6 |
| Two-phase system of 20% PEG, 9% $Na_2SO_4$ | 128 | 80 | 6 |
| IMAC[a] column | 104 | 65 | 200 |
| Gel-permeation chromatography | 96 | 60 | 270 |

[a] IMAC, Immobilized metal ion affinity chromatography.

conditions the (*S*)-oxynitrilase partitions into the salt-rich bottom phase (Table III). This extraction step removes peptides, amino acids, and other small molecules which compete with the chelator or the enzyme for the binding of the metal ion.[13]

A second phase system is then generated by adding 8 g Cu–IDA–PEG, 34 g PEG 2000, and 58 g distilled water per 100 g of the separated bottom phase of the primary phase system. The partitioning of (*S*)-oxynitrilase readily changed in response to the affinity ligand, and all enzyme activity is collected with the PEG-rich top phase, which is separated by gravity for about 2 hr. Alternatively, a short centrifugation at low *g* values can be used to speed phase separation.

The third extraction step makes use of a shift in pH to 4.0, so as to protonate the histidines on the enzyme surface and thus abolish binding to the Cu–IDA–PEG. Per 100 g top phase of the second extraction step, 15 g sodium sulfate, 2 g sodium acetate, pH 4.0, and 73 g distilled water

[13] R. Woker, B. Champluvier, and M.-R. Kula, *J. Chromatogr.* **584,** 85 (1992).

TABLE III
PURIFICATION OF (*S*)-OXYNITRILASE FROM *Sorghum Bicolor*

| Purification step | *S*-Oxynitrilase activity (U) | Activity yield (%) | Specific activity (U/mg) |
|---|---|---|---|
| Raw material | 160 | 100 | 1.6 |
| Two-phase system 1 (0% Cu–IDA–PEG) | 148 | 92 | 4.1 |
| Two-phase system 2 (2.8% Cu–IDA–PEG) | 134 | 84 | 12.5 |
| Two-phase system 3 pH shift | 121 | 75 | 40 |
| Ion-exchange membrane | 97 | 60 | 269 |

are added. The ($S$)-oxynitrilase is recovered in the salt-rich phase, whereas the affinity ligand remains in the top phase and can be separated and reused. The enzyme can be stored at $-18°$ as a "technical grade" preparation.

After the affinity extraction by Cu–IDA–PEG, the lower phase of the third extraction step had to be desalted before it could be adsorbed on an ion-exchange membrane. Standard procedures (diafiltration, dialysis) are used for desalting. The ($S$)-oxynitrilase is bound on the membrane at pH 7. After rinsing to remove unbound proteins the ($S$)-oxynitrilase can be eluted as a homogeneous protein at pH 5.5 in 0.1 $M$ sodium acetate buffer, 0.15 $M$ NaCl. The purification scheme is presented in Fig. 2, and quantitative results are summarized in Table III.

If the first physical extraction of the enzyme is omitted, it is necessary to add 30% more of Cu–IDA–PEG to the affinity extraction system in order to enhance the partitioning and achieve similar yields as observed following the procedure described above. The purification factor by this approach is somewhat smaller than that reported in Table III. Because more Cu–IDA–PEG is needed for direct extraction of the ($S$)-oxynitrilase from biomass, we recommend the procedure described in detail above.

## Acknowledgments

We thank Monika Gehsing for skillful technical assistance, Sartorius AG for the generous gift of ion-exchange membranes, and Chemische Werke Hüls for poly(ethylene glycol). The financial support by the BMFT Schwerpunkt Stoffumwandlung mit Biokatalysatoren, University of Düsseldorf, and Fonds der Chemie is gratefully acknowledged.

# [58] Extractive Purification of Aspartase from *Escherichia coli* K12

*By* JENS PAULSEN and HELMUT HUSTEDT

## Introduction

Aspartase (L-aspartate ammonia-lyase, EC 4.3.1.1) catalyzes the reversible formation of L-aspartate from fumarate and $NH_4^+$. The isolated enzyme or whole cells are used for the industrial production of L-aspartate, which together with L-phenylalanine is the basis of the sweetener aspartame. Preparations of the enzyme purified to homogeneity were obtained from *Escherichia coli* B,[1] *Escherichia coli* W,[2] and *Pseudomonas fluorescens*.[3] All purification procedures comprise five to seven purification steps and are time consuming. Because of the industrial importance of aspartame production, we developed an improved purification scheme for processing large quantities of enzyme in a short time. This fast purification procedure consists of two liquid extraction and two chromatographic steps and can be completed within 12 hr with a yield of about 45%. The aspartase from *E. coli* K12 is purified from a high producing strain (JRG 1849) containing the *asp* gene on a multicopy plasmid (pGS94).[4]

## Methods

### *Materials*

Poly(ethylene glycol) (PEG) 1540 is purchased from Deutsche BP (Düsseldorf, Germany) and PEG 10,000 from Hoechst (Frankfurt, Germany). DEAE-Sepharose Fast Flow, Sephacryl S-200, Mono Q, Phenyl-Superose, and Phenyl-Sepharose are from Pharmacia/LKB (Freiburg, Germany). Bacto-tryptone and yeast extract are from Difco (Detroit, U.S.A.), and all other chemicals are from Merck (Darmstadt, Germany), analytical reagent grade.

### *Enzyme Assay*

*Principle*. The enzyme assay is based on the spectrophotometric determination at 240 nm of fumarate formation[5] at 30°.

*Reagents*

0.5 *M* Tris-HCl, pH 9.0
0.1 *M* $MgCl_2$
0.5 *M* sodium L-aspartate

The enzyme is diluted in 20 m*M* Tris-HCl, pH 9.0.

*Procedure*. The standard assay system, in a total volume of 1 ml, contains 0.2 ml of Tris-HCl buffer, 0.2 ml of sodium L-aspartate, and 30 μl of $MgCl_2$. The reaction is initiated by the addition of the enzyme, and the increase in absorption at 240 nm is determined. The molar extinction

[1] F. B. Rudolf and H. Fromm, *Arch. Biochem. Biophys.* **147,** 92 (1971).

[2] A. Suzuki, J. Yamaguchi, and M. Tokushige, *Biochim. Biophys. Acta* **321,** 369 (1973).

[3] J. S. Takagi, R. Fukunage, M. Tokushige, and H. Katsuki, *J. Biochem.* (*Tokyo*) **96,** 545 (1984).

[4] J. R. Guest, R. E. Roberts, and R. J. Wilde, *J. Gen. Microbiol.* **130,** 1271 (1984).

[5] M. Tokushige, K. Miyamoto, and H. Katsuki, *J. Biochem.* (*Tokyo*) **85,** 1415 (1979).

coefficient of fumarate at 240 nm is 2.530 $M^{-1}$ $cm^{-1}$ according to Emery.[6] One unit (U) of the enzyme is defined as the amount producing 1 $\mu$mol fumarate per minute under the assay conditions. Protein is determined according to Bradford[7] using bovine serum albumin as standard. Alternative assay methods are described by Tokushige.[8]

*Sodium Dodecyl Sulfate–Polyacrylamide Gel Electrophoresis*

Sodium dodecyl sulfate–polyacrylamide gel electrophoresis (SDS–PAGE) is carried out with the Pharmacia PHAST system using 8–25% gradient gels. Silver staining is carried out according to Butcher.[9]

*Partitioning*

Partition studies for the development of the extraction procedure are conducted on a 10-g scale at room temperature. Stock solutions of the components of the two-phase system are put into graduated centrifuge tubes together with either cell homogenate or top phase extract. The contents of the tubes are mixed on a Vortex agitator, then centrifuged for 5 min at 1850 g to speed phase separation. The volume of each phase is noted, and samples are taken from the top and bottom phases for determination of the enzyme concentrations using the activity assay described above. The partition coefficient, $K$, is defined as the ratio of the aspartase activities in the top and bottom phases.[10]

*Purification Procedure I*

All purification steps, except for the aqueous extraction, are carried out at 4°. The aspartase may be purified by two alternative methods. In both cases the cell debris and the bulk of contaminating proteins are removed by two liquid extraction steps. In one case the second liquid extraction step is followed by ultrafiltration, anion-exchange chromatography, and gel-permeation chromatography. In the other case hydrophobic interaction chromatography[11] is carried out directly after extraction followed by anion-exchange chromatography.

*Culture Growth: Escherichia coli* K12, strain JRG 1849, containing the plasmid pGS94, is grown aerobically at 30° in L-broth medium. The medium contains, per liter, 10 g Bacto-tryptone, 5 g NaCl, 5 g yeast extract,

[6] T. F. Emery, *Biochemistry* **2,** 1041 (1963).
[7] M. M. Bradford, *Anal. Biochem.* **72,** 248 (1976).
[8] M. Tokushige, this series, Vol. 113, p. 618.
[9] L. A. Butcher, *Anal. Biochem.* **148,** 384 (1986).
[10] H. Hustedt, K. H. Kroner, W. Stach, and M.-R. Kula, *Biotechnol. Bioeng.* **20,** 1989 (1978).
[11] P. Hubert, this volume [26].

TABLE I
COMPOSITION OF AQUEOUS PHASE SYSTEMS IN FIRST AND SECOND EXTRACTION

| Ingredient | Concentration (kg in 1 kg mixture)[a] | |
|---|---|---|
| Extraction I | | |
| Cell homogenate | 0.50 | |
| PEG 1540 | 0.22 | Prepared as solution |
| $NaH_2PO_4$ | 0.06 | |
| $K_2HPO_4$ | 0.02 | |
| Extraction II | | |
| Top phase from extraction I | 0.50 | |
| PEG 10,000 | 0.03 | Prepared as solution |
| NaCl | 0.02 | |
| $NaH_2PO_4$ | 0.04 | |
| $K_2HPO_4$ | 0.03 | |

[a] Remainder is water.

and 50 mg ampicillin. A 100-liter fermenter with 70 liters medium is inoculated with 2 liters of a culture grown overnight. At the end of the logarithmic phase (8 hr) the broth reaches an optical density of 5.0 at 546 nm, corresponding to 10 g wet cells per liter.

*Step 1: Preparation of Cell Homogenate.* The cells from the 100-liter fermenter are harvested by cross-flow microfiltration. The Pellicon system from Millipore (Eschborn, Germany) is used with 2 microfiltration membranes (each 0.46 $m^2$), pore size 0.45 $\mu$m. Approximately 70 liters is concentrated to 2 liters corresponding to a wet cell concentration of 30%. The average filtrate flux is 40 liters/hr $\times$ $m^2$. The cells are broken by two passages through a Lab 60 high-pressure homogenizer (APV-Schröder, Lübeck, Germany) at a pressure of 600 bar and a flow rate of 60 liters/hr.

*Step 2: First Aqueous Phase Extraction.* All liquid extraction steps are performed at room temperature. PEG is added in solid form and phosphate as a 40% solution (w/w) consisting of a mixture of dipotassium hydrogen phosphate and sodium dihydrogen phosphate. The proportion of the salts determine the pH. The cell homogenate is mixed with PEG 1540 and phosphate as described in Table I. The mixture is stirred for 20 min, then centrifuged for 10 min at 4000 *g*. The PEG phase (top phase) is decanted and used in the second extraction step. For larger scale operations phase settling can be conducted by a conventional centrifugal disk stack separator.[12]

[12] H. Hustedt, K. H. Kroner, and M.-R. Kula, *in* "Partitioning in Aqueous Two-Phase Systems: Theory, Methods, Uses, and Applications to Biotechnology" (H. Walter, D. E. Brooks, and D. Fisher, eds.), p. 529. Academic Press, Orlando, Florida, 1985.

*Step 3: Second Aqueous Phase Extraction.* The top phase is added to a solution of PEG 10,000, NaCl, and phosphate as described in Table I. The mixture is stirred for 20 min and then centrifuged for 10 min at 4000 *g*. Alternatively the two phases can be separated after settling at unit gravity for 1 hr. The salt-enriched phase (bottom phase) is obtained by decantation of the PEG phase or, when the separation is carried out using a separatory funnel, by letting this phase run out from the bottom of the funnel. For larger scale operations a disk stack separator can also be used for phase separation of this stage.

*Step 4: Anion-Exchange Chromatography.* The bottom phase from the second extraction is desalted by several concentration steps with an ultrafiltration module (cutoff 50,000) and subsequent dilution with 1 m*M* dithiothreitol (DTT) solution until the conductivity reaches 7 mS. The desalted bottom phase is applied to a DEAE-Sepharose Fast Flow column (5 × 15 cm) which has previously been equilibrated with 50 m*M* potassium phosphate, pH 7.2, 1 m*M* EDTA, and 1 m*M* DTT. The flow rate during the entire run is 1.2 liters/hr. After washing the column with 400 ml equilibration buffer, the aspartase is eluted with an increasing potassium phosphate concentration gradient from 50 to 500 m*M* in a total volume of 1200 ml. The aspartase appears as a symmetrical band at 0.27 *M* potassium phosphate (Fig. 1a). The fractions having a specific activity higher than 150 U/mg protein are pooled and concentrated by ultrafiltration (XM-50) to 10 ml, protein concentration 20 mg/ml.

*Step 5: Gel-Permeation Chromatography.* The concentrate (10 ml) is applied to a Sephacryl S-200 column (2.5 × 100 cm) equilibrated with 50 m*M* potassium phosphate, pH 6.8, 100 m*M* potassium chloride, 1 m*M* EDTA, and 1 m*M* DTT. Elution is with the same buffer. The flow rate for application and elution is 19.2 ml/hr. The aspartase appears after 48% of the column volume (236 ml). Fractions with a specific activity of 219 U/mg protein are pooled, giving a total volume of approximately 35 ml. Aspartase is stored at −20° in the elution buffer and is stable for several months.

### *Purification Procedure II*

As an alternative to steps 4 and 5 a hydrophobic interaction chromatography (instead of anion-exchange chromatography) can be used without prior desalting, followed by an anion-exchange chromatography (instead of gel-permeation chromatography).

*Hydrophobic Interaction Chromatography.* Four milliliters of the second bottom phase is applied to a Phenyl-Superose column (HR 5/5) which is equilibrated with 50 m*M* potassium phosphate, pH 6.1, 1.2 *M* ammo-

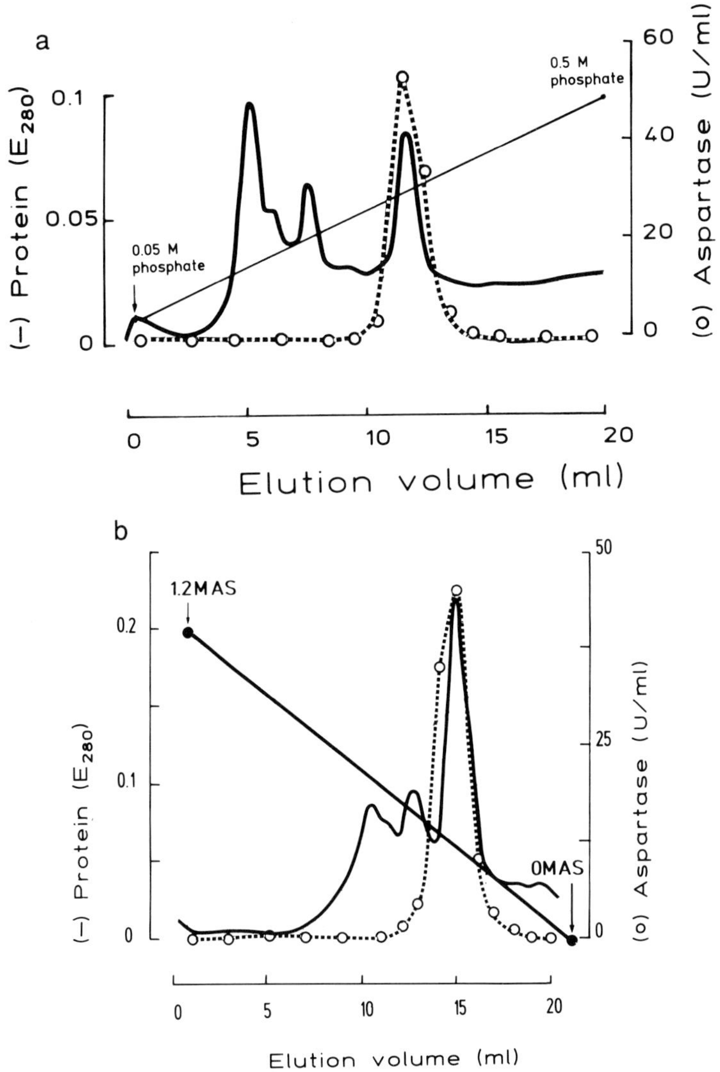


FIG. 1. Elution profile of aspartase activity and total protein after chromatography on Mono Q HR 5/5 (a) and Phenyl-Superose HR 5/5 (b). [From J. Paulsen, H. J. Hecht, D. Schomburg, and H. Hustedt, *in* "Enzyme in der Lebensmitteltechnologie" (K. H. Kroner, K. Lösche, and R. D. Schmidt, eds.), Symp. 2, p. 211. VCH, Weinheim, 1989, with permission.]

TABLE II
PREPARATION OF ASPARTASE FROM *Escherichia coli* K12 BY PURIFICATION METHOD I[a]

| Step | Activity (U) | Protein (mg) | Specific activity (U/mg protein) | Yield[b] (%) |
|---|---|---|---|---|
| Disrupted cells | 58,000 | 14,500 | 4 | 100 |
| Extraction I | 49,900 | 1560 | 32 | 86 |
| Extraction II | 42,340 | 814 | 52 | 73 |
| DEAE-Sepharose Fast Flow | 33,000 | 197 | 168 | 57 |
| Sephacryl S-200 | 26,100 | 120 | 219 | 45 |

[a] From J. Paulsen, H. J. Hecht, D. Schomburg, and H. Hustedt, *in* "Enzyme in der Lebensmitteltechnologie" (K. H. Kroner, K. Lösche, and R. D. Schmidt, eds.), Symp. 2, p. 211. VCH, Weinheim, 1989, with permission.
[b] The procedure started with 150 g wet cells.

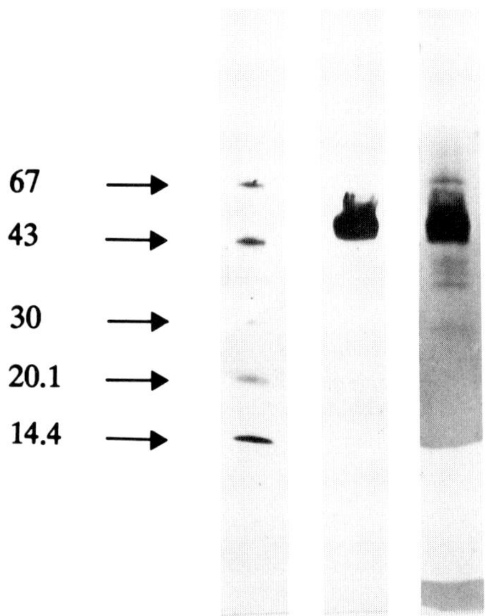


FIG. 2. SDS–polyacrylamide gel electrophoresis of *Escherichia coli* aspartase on 8–25% gradient gels from Pharmacia. Silver staining is carried out according to Butcher.[9] The first column shows the following standard proteins (kDa): bovine serum albumin (67), ovalbumin (43), carbonate dehydratase (30), soybean trypsin inhibitor (20.1), and α-lactalbumin (14.4). In the second column aspartase (0.11 g/liter) after the first purification method is shown; the third column shows aspartase (0.12 g/liter) after the second purification method.

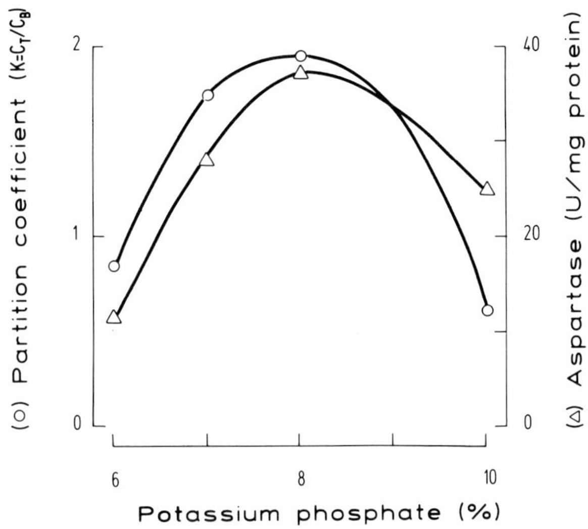


FIG. 3. Partition coefficient and specific activity of aspartase in the first extraction as a function of the phosphate concentration. The experiments were performed on a 10-g scale with a system containing 22% (w/w) PEG 1540, 50% (w/w) cell homogenate, and different concentrations of a 4 : 1 mixture of sodium and potassium phosphate ($NaH_2PO_4/K_2HPO_4$).

nium sulfate, 1 m*M* EDTA, and 1 m*M* DTT. After washing the column with 4 ml of the equilibration buffer, the aspartase is eluted with a decreasing concentration gradient of 1.2 to 0 *M* ammonium sulfate in a total volume of 20 ml. The enzyme appears as a symmetrical band at 0.3 *M* ammonium sulfate (Fig. 1b). The flow rate is 0.5 ml/min for all operations.

*Anion-Exchange Chromatography*. After 3-fold dilution with 50 m*M* potassium phosphate, pH 7.2, 1 m*M* EDTA, and 1 m*M* DTT, the active fractions are applied to a Mono Q column (HR 5/5) which has been equilibrated with the same buffer. After washing the column with 4 ml of equilibration buffer the aspartase is eluted with an increasing concentration gradient from 50 to 500 m*M* potassium phosphate in a total volume of 20 ml. The enzyme appears as a symmetrical band at 0.27 *M* potassium phosphate.

## Results

Table II presents the results of a representative purification procedure starting from 150 g wet cells corresponding to 14.5 g protein. A total of 120 mg of pure aspartase with a specific activity of 219 U/mg protein was obtained. The enzyme was purified 55-fold with an overall yield of 45%. When the alternative chromatographic steps are used (hydrophobic inter-

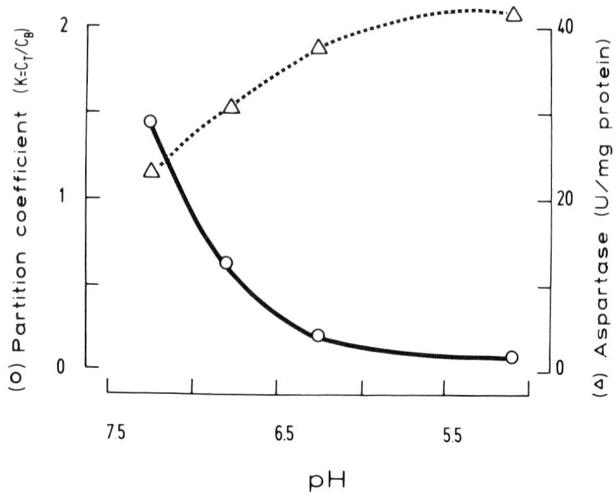


FIG. 4. Partition coefficient and specific activity of aspartase in the second extraction as a function of the pH. The experiments were performed on a 10-g scale containing 50% (w/w) top phase I, 3% (w/w) PEG 10,000, and 6% (w/w) sodium/potassium phosphate. The different pH values result from different ratios of the basic and acid salts. [From J. Paulsen, H. J. Hecht, D. Schomburg, and H. Hustedt, *in* "Enzyme in der Lebensmitteltechnologie" (K. H. Kroner, K. Lösche, and R. D. Schmidt, eds.), Symp. 2, p. 211. VCH, Weinheim, 1989, with permission.]

action followed by anion-exchange chromatography) similar results are obtained. The purity of the final preparation is indicated by SDS–PAGE (Fig. 2). The second and third columns show the enzyme purified by the first and second purification methods, respectively. SDS–PAGE reveals that the enzyme obtained by the first purification method is homogeneous, whereas the alternative method gives a product containing some impurities.

### Partitioning Studies

This section describes the partition experiments leading to the above purification protocol. Figure 3 shows the $K$ value and the specific activity of aspartase as a function of the phosphate concentration in the first aqueous extraction. In both cases the maximum is reached at a phosphate concentration of 8%. At this point the volume ratio of top to bottom phase is 3 : 1 and $K = 2$, resulting in a yield of 86% aspartase in the top PEG phase, whereas cell debris and 90% of the contaminating proteins partition into the bottom phase. The first extraction is carried out at pH 6.0 because the majority of total protein, at this pH, favors the bottom phase, whereas the aspartase is primarily in the top phase. To extract the aspartase into the bottom phase it is necessary to add small amounts of PEG 10,000. It is known that higher molecular weight PEG polymers tend to exclude most proteins from the PEG phase. Moreover, the *K* value and the specific activity of aspartase in the second extraction can be influenced to a large extent by pH (Fig. 4). By decreasing the pH from 7.25 to 5.15, *K* decreases from 1.5 to 0.14 and the specific activity in the bottom phase increases from 23 to 42 U/mg protein. A further decrease of *K*, which corresponds to a higher yield, is reached by the addition of 0.34 *M* sodium chloride. In this case the *K* value of the aspartase reaches the minimum value of 0.018. The second extraction results in a 1.6-fold purification and in a yield of over 90%. The overall yield after both extraction steps is 78% with a 13-fold purification.

## Conclusions

Improved procedures are described in this chapter for the purification of aspartase from *Escherichia coli* K12. In this process the conventional centrifugation steps for separation of cell debris, followed by heat treatment and/or ammonium sulfate fractionation, are replaced by aqueous two-phase extraction. This technique allows the cell homogenate to be used directly after cell disruption. The cell debris and the bulk of contaminating proteins are separated simultaneously. Compared with the other primary separation steps used for the purification of aspartase, aqueous extraction is more efficient and rapid. A large number of industrially important enzymes have been purified for use as technical biocatalysts using the liquid extraction technique (e.g., penicillin acylase and fumarase). The purification factor after liquid extraction is generally between 5 and 25.[13]

The second purification method was carried out only on a small scale with fast protein liquid chromatography (FPLC) columns, but similar results were obtained in initial scaleup experiments with Phenyl-Sepharose and DEAE-Sepharose Fast Flow columns. The second purification method has the advantage of allowing the salt-rich bottom phase to be applied directly to a hydrophobic column since high salt concentrations promote the binding of proteins to hydrophobic adsorbents. This latter purification scheme appears especially suited to automated semicontinuous processing which might be useful for the purification of large quantities of enzyme.[14]

[13] H. Hustedt, K. H. Kroner, and N. Papamichael, *Proc. Biochem.* **23,** 129 (1988).

[14] J. Paulsen, H. J. Hecht, D. Schomburg, and H. Hustedt, *in* "Enzyme in der Lebensmitteltechnologie" (K. H. Kroner, K. Lösche, and R. D. Schmidt, eds.), Symp. 2, p. 211. VCH-Verlag, Weinheim, 1989.

# [59] Large-Scale Purification of Formate Dehydrogenase

*By* ARNO CORDES and MARIA-REGINA KULA

## Introduction

The partitioning of proteins in favor of the poly(ethylene glycol) (PEG)-rich upper phase of an aqueous two-phase system can be increased by special ligands that are covalently bound to PEG.[1,2] Such ligands can be cofactors or inhibitors of the protein to be purified. As phase-forming polymers the combination of PEG and dextran (Dx) should be chosen so as to avoid high salt concentrations, which could lower the strength of the affinity binding between protein and ligand. Using this approach for the large-scale purification of dehydrogenases and kinases, high space–time yields as well as a partial purification of the desired enzyme can be achieved.[1-3]

A highly effective method based on affinity partitioning was developed for the purification of formate dehydrogenase (FDH) from *Candida boidinii*.[4] The enzyme could be purified 6-fold direct from a cell homogenate up to a scale of 45 kg of cells (Tables I and II). In this case the triazine dye Procion Red HE-3B coupled to PEG 6000 (yielding PEG–Red) was chosen as the ligand because measurements of the dissociation constants revealed that its affinity to FDH is very high.[3] To reduce the material costs crude Dx was used instead of purified Dx as one of the phase-forming polymers. The recycling of the upper phase polymers is possible and provides additional savings. The whole process is illustrated in Fig. 1.

## Methods

### *Production of Poly(ethylene Glycol)–Procion Red HE-3B on Kilogram Scale*

For the coupling of the triazine dye Procion Red HE-3B (Deutsche ICI, Frankfurt, Germany) to PEG 6000 (Serva, Heidelberg, FRG), an

[1] K. H. Kroner, A. Cordes, A. Schelper, M. Morr, A. F. Bückmann, and M.-R. Kula, *in* "Affinity Chromatography and Related Techniques" (T. J. C. Gribnau, J. Visser, and R. J. F. Nivard, eds.), p. 491. Elsevier, Amsterdam, 1982.

[2] G. Johansson, G. Kopperschläger, and P.-Å. Albertsson, *Eur. J. Biochem.* **131,** 584 (1983).

[3] A. Cordes, J. Flossdorf, and M.-R. Kula, *Biotechnol. Bioeng.* **30,** 514 (1987).

[4] A. Cordes and M.-R. Kula, *J. Chromatogr.* **376,** 375 (1986).

TABLE I
PROCESS FOR PURIFICATION OF FORMATE DEHYDROGENASE

| Method[a] | Yield (%) | Specific activity (U/mg) |
|---|---|---|
| Enzyme solubilization | 88 | 1.1 |
| Heat denaturation | 88 | 2.7 |
| Affinity partitioning | 70[b] | — |
| Enzyme/ligand separation | 80 | 6.0 |

[a] Experiments were carried out on a 5-g scale.
[b] Assay interfered with by high concentrations of PEG–Red.

unpublished method of Morr[5] is slightly modified. The synthesis is carried out as follows:

*Chlorination of Polymer*

1. Liquefy 1 kg of PEG 6000 in an incubator at 70° overnight and then use a vacuum pump (pressure < 100 Pa) to remove the water at a temperature of 90–95°.
2. Incubate the water-free PEG 6000 with 20 ml thionyl chloride ($SOCl_2$, Merck, Darmstadt, Germany) for 8 hr at 70° in a rotary evaporator. The reaction vessel should be flushed continuously with dry nitrogen to remove $SO_2$ and HCl and to avoid water condensation.
3. Remove remaining volatile products and excess thionyl chloride by extending the incubation time under vacuum (pressure < 6000 Pa) for 1 hr.

*Amination*

1. Dissolve the PEG 6000 chloride in a large excess of concentrated (25%) ammonia in a glass autoclave (Schott Glaswerke, Mainz, Germany) so that the vessel is filled to about 75%.
2. Incubate the vessel in an oil bath at 110° for approximately 30 hr.
3. Remove the excess ammonia under vacuum in a rotary evaporator at 80°.

*Coupling Procedure*

1. Dissolve the aminated PEG together with 380 g of Procion Red HE-3B (molar ratio 1 : 1.8, molecular weights 6750 and 1400, respectively) in 1 liter of water.

[5] M. Morr, personal communication (1982).

TABLE II
SCALEUP OF AFFINITY EXTRACTION OF FORMATE DEHYDROGENASE[a]

| Scale of phase system (kg) | Yield (%) | Specific activity[b] (U/mg) | Number of experiments | Recycling of top phase |
|---|---|---|---|---|
| 0.005 | 72 ± 3 | 5.9 ± 0.06 | 5 | 4 |
| 5.0 | 78 ± 2 | 5.0 ± 0.05 | 5 | 4 |
| 10.0 | 70 ± 4 | 4.7 ± 0.11 | 2 | 1 |
| 50.0 | 65 | 3.7 | 1 | — |
| 220.0 | 74 | 3.5 | 1 | — |

[a] From A. Cordes and M.-R. Kula, *J. Chromatogr.* **376,** 375 (1986), with permission.

[b] The specific activity of the final preparation depends to some extent on the enzyme concentration in the cells; for large-scale experiments a batch of *Candida boidinii* was employed with only 0.57 U/mg in the crude extract.

2. Adjust the pH to 11.0 by addition of 5 *M* NaOH and incubate the reaction mixture with stirring at 60° for 24 hr.

3. Remove the salts by dialysis or ultrafiltration against water.

The yield can be calculated by separating the small amounts of uncoupled dye (~1–2%) by gel filtration on Sephadex G-50 (Pharmacia, Uppsala, Sweden) and measuring the absorbance of the PEG–Red in water in a photometer at 542 nm [absorption coefficient ($\varepsilon$) of Procion Red HE-3B is 33.9 ml/($\mu$mol cm)].

### *Enzyme Assay*

1. Pipette the following into a test tube (all reagents are dissolved in distilled water): 2.00 ml of 0.075 *M* potassium phosphate buffer, pH 7.5, 0.50 ml of 1 *M* sodium formate, 0.50 ml of 10 m*M* $NAD^+$.

2. Add limiting amounts of enzyme (0.01–0.1 ml) and mix quickly on a Vortex mixer. If necessary the enzyme solution must be diluted in the potassium phosphate buffer to a level such that the increase of the absorbance at 340 nm does not exceed the value of 0.1 per minute.

3. Record the increase of the absorbance at 340 nm in a spectrophotometer for 1 min at 30° (using a 1-cm cuvette). The extinction coefficient of NADH at 340 nm is 6.22 ml/($\mu$mol cm). One unit of activity is defined as the amount of enzyme that reduces 1 $\mu$mol of $NAD^+$ to NADH per minute under reaction conditions.

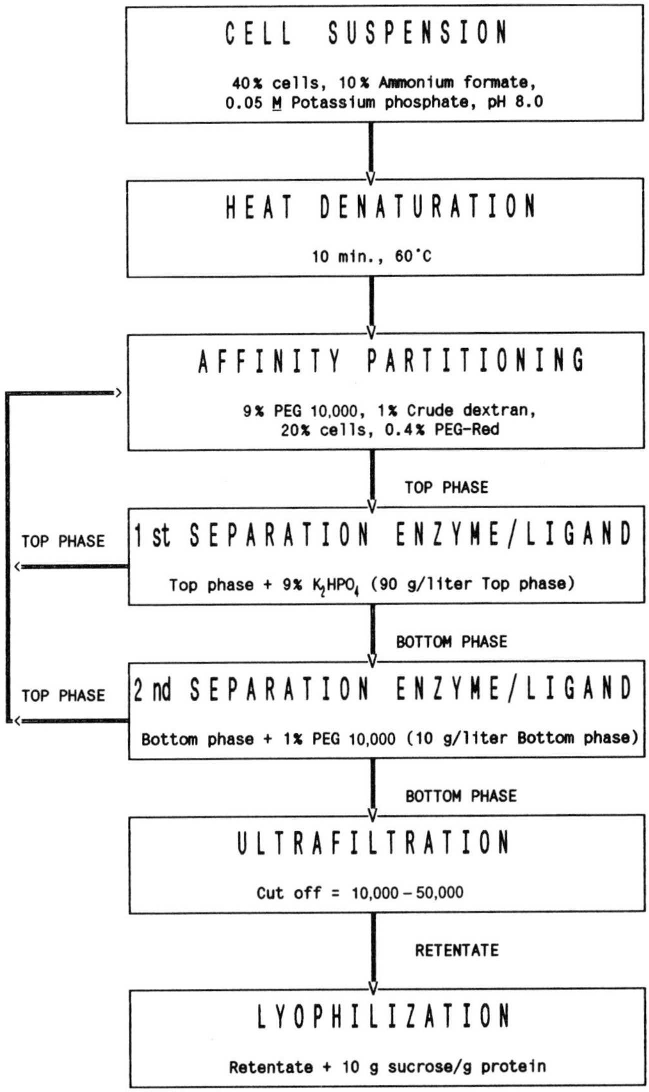


FIG. 1. Process for the purification of formate dehydrogenase with affinity partitioning as the key step. [From A. Cordes and M.-R. Kula, *J. Chromatogr.* **376,** 375 (1986), with permission.]

*Production of Cell Material*

The cultivation of *Candida boidinii* is carried out on a scaleup to 700 liters according to Sahm and Wagner.[6] Cells are stored frozen at −20° until further use.

*Purification of Formate Dehydrogenase*

*Step 1: Cell Suspension*

1. Suspend the frozen cells at a concentration of 40% (w/v) in a 50 m*M* potassium phosphate buffer, pH 8.0, which contains 10% ammonium formate (Riedel de Häen, Hannover, Germany).
2. Keep stirring until 80% of the enzyme can be detected in the supernatant solution (this usually occurs after 3–24 hr depending on the time cells are stored).
3. Adjust the pH to 8.0 with 6 *M* NaOH.

*Step 2: Heat Denaturation*

1. Heat the cell suspension in a stirring vessel in a water bath at 75° or in a temperature-controlled stirring tank, keeping the temperature at 60° for 10 min.
2. Cool the cell suspension quickly in an ice bath.

*Step 3: Affinity Partitioning (First Phase System).* The composition of the phase system is developed by varying the concentration of one component at a time, using constant buffer concentration 50 mmol/kg potassium phosphate, pH 7.5. The partition coefficients, $K$, of enzyme and protein ($K_E$ and $K_P$, respectively) are used to guide optimization (Figs. 2 and 3). They are calculated as the concentrations of enzyme or protein in the upper phase divided by the respective concentrations in the lower phase. The protein concentration is measured according to Bradford.[7] The sequence of the procedure and the results of optimization are given in Table III.

The phase system is prepared as follows.

1. Mix in a vessel the following components in the order listed:
   - 50% (w/w) Cell homogenate (prepared as described above)
   - 10% (w/w) Crude Dx solution [10% (w/w) crude Dx from Pfeiffer & Langen (Dormagen, Germany) in distilled water]
   - 18% (w/w) PEG 10,000 solution [50% (w/w) PEG 10,000 in distilled water]

[6] H. Sahm and F. Wagner, *Arch. Microbiol.* **84,** 29 (1972).

[7] M. Bradford, *Anal. Biochem.* **72,** 248 (1976).

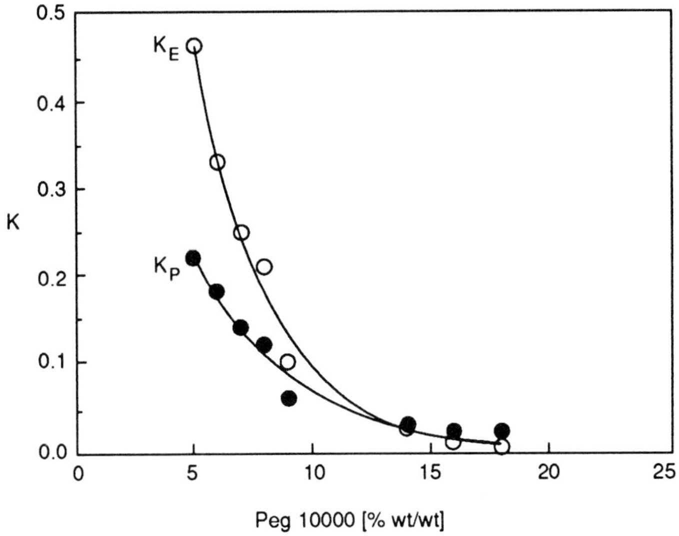


FIG. 2. Dependence of partitioning of formate dehydrogenase and protein on PEG concentration. $K_P$ is the partition coefficient of protein, and $K_E$ is the partition coefficient of formate dehydrogenase. The system contained 1% crude dextran, 50 m*M* potassium phosphate, pH 7.5, and 20% cells. [From A. Cordes and M.-R. Kula, *J. Chromatogr.* **376,** 375 (1986), with permission.]

21.6% (w/w) Distilled water
0.4% (w/w) PEG–Red

2. Keep stirring for about 1 hr at 20°.

3. Pump the homogeneous mixture into a settling tank and let the phase system separate. This normally takes about 12 hr. Alternatively, a centrifuge or continuous separator may be used for faster processing.

*Step 4: Separation of Enzyme and Ligand (Second and Third Phase Systems).* To separate the FDH from the ligand a PEG–salt system is generated by adding $K_2HPO_4$ to the upper phase (see below) of the first phase system, which contains the enzyme. Under these conditions the enzyme partitions into the salt-rich bottom phase while the PEG–Red remains in the top phase.

A tiny quantity of PEG–Red is also present in the lower phase, causing the lyophilized end product to have a light red color. If desired, residual PEG–Red can be extracted into the top phase of a third phase system by adding 1% (w/v) PEG 10,000 (see below) to the bottom phase of the second system. Recycling experiments on a 5-g and a 5-kg scale showed that the combined top phases of the second and third phase systems can be utilized directly for a new affinity partitioning step. The top phase can be recycled

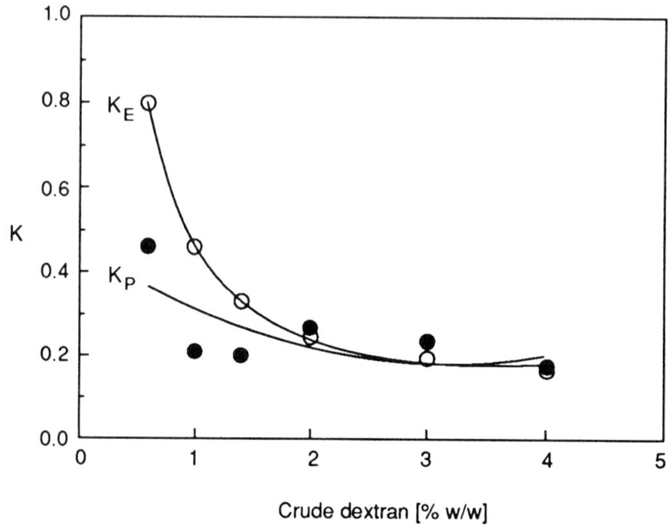


FIG. 3. Influence of crude dextran concentration on partitioning of formate dehydrogenase and protein. The system contained 9% PEG 10,000, 50 m*M* potassium phosphate, pH 7.5, and 20% cells.

three times without significant loss of enzyme (Table IV). Per affinity step the loss of 8% of the PEG 10,000 and 15% of the PEG–Red has to be compensated. This is done by adding 0.72% (w/w) PEG 10,000 and 0.06% (w/w) PEG–Red to the phase system of the new affinity partitioning step (see Table IV). The loss of PEG 10,000 has been calculated from the phase diagrams[8] of the PEG 10,000–crude Dx and the PEG 10,000–potassium phosphate phase systems. The decrease of PEG–Red can be followed by monitoring the absorbance of the top phases at 542 nm during the different recycling steps.

1. Pump the upper phase containing the enzyme into a vessel. Add 90 g $K_2HPO_4$ per liter upper phase and stir for 1 hr at 20°.
2. Let the new phase system settle in a settling tank for at least 2 hr.
3. Remove the bottom phase containing the enzyme and add to it 10 g PEG 10,000 per liter bottom phase to extract the remaining PEG–Red into the top phase. Separate as above.
4. Store the top phases with PEG–Red at −20° for further affinity extractions.

*Step 5: Ultrafiltration*. A membrane with a cutoff of 10,000 Da should be used.

[8] P.-Å. Albertsson and F. Tjerneld, this volume [1].

TABLE III
PHASE SYSTEM SELECTION FOR AFFINITY PARTITIONING

| Step | Optimized parameter | Constant conditions | Results |
|---|---|---|---|
| 1 | Molecular weight and concentration of PEG | 1% Crude dextran, 20% cells | 9% PEG 10,000 |
| 2 | Concentration of crude dextran | 9% PEG 10,000, 20% cells | 1% Crude dextran |
| 3 | Concentration of PEG–Red | 9% PEG 10,000, 1% crude dextran, 20% cells, 0.4% PEG–Red | 1 m*M* Procion Red HE-3B in the top phase |

1. Dilute the bottom phase containing the enzyme with distilled water to twice its volume.

2. Concentrate 3-fold and dilute with water to the initial volume. Repeat this step until a conductivity of not more than 10 mS is obtained.

3. Concentrate to one-tenth of the initial volume.

*Step 6: Lyophilization.* Before freeze-drying add 10 g sucrose per gram protein to the concentrate obtained on ultrafiltration.

## Economics

Costs for a production process for FDH based on affinity partitioning are increased by about 3% when PEG–Red is used, assuming a price of

TABLE IV
RECYCLING OF UPPER PHASE[a,b]

| Number of uses | $V_T$[c] (ml) | $V_B$[c] (ml) | Yield (%) FDH | Yield (%) PEG–Red | Specific activity (U/mg) |
|---|---|---|---|---|---|
| 1 | 3.7 | 1.1 | 75 | 100 | 6.0 |
| 2 | 3.7 | 1.1 | 80 | 84 | 6.1 |
| 3 | 3.5 | 1.0 | 78 | 80 | 5.8 |
| 4 | 3.7 | 1.1 | 88 | 85 | 5.9 |
| 5 | 3.6 | 1.2 | 67 | 77 | 3.5 |

[a] The phase systems contained 50% cell homogenate, 10% crude Dx solution (10%, w/w), 1.5% PEG 10,000 solution (50%, w/w), 0.06% PEG–Red, top phase from the PEG–salt system, and distilled water (to make up to total weight).

[b] From A. Cordes and M.-R. Kula, *J. Chromatogr.* **376,** 375 (1986), with permission.

[c] $V_T$ and $V_B$ are the volumes of the upper and lower phases of the affinity extraction, respectively.

U.S. $170/kg ligand.[4] If PEG–Red is recycled three times the increased cost is 1%. Thus, the utilization of an affinity ligand in a purification process does not automatically result in a dramatic increase in production costs. The slightly higher production costs should be weighed against the advantages of affinity partitioning. In the case of FDH these advantages lie in a high purification combined with a good yield.

From the literature two alternative processes for the purification of FDH from *Candida boidinii* are known.[9,10] One is based on a three-step extraction in aqueous two-phase systems.[9] Although yields comparable to the affinity partitioning process are achieved, the specific activity is only in the range of 2 U/mg. The second process makes use of traditional purification steps such as precipitations and ion-exchange chromatography.[10] This technique yields a very pure product but a low overall yield (23%). Furthermore, scaleup in this case may cause difficulties with regard to separation of small debris.

### Acknowledgments

We thank K. H. Kroner for helpful discussions, M. Morr for advice on the chemical synthesis of PEG–Red, and W. Stach for excellent technical assistance.

[9] K. H. Kroner, H. Schütte, W. Stach, and M.-R. Kula, *J. Chem. Technol. Biotechnol.* **32,** 130 (1982).

[10] H. Schütte, J. Flossdorf, H. Sahm, and M.-R. Kula, *Eur. J. Biochem.* **62,** 151 (1976).

# [60] Extraction and Purification of Leucine Dehydrogenase from *Bacillus cereus*

*By* HORST SCHÜTTE, DIETER SCHUSTOLLA, and HELMUT HUSTEDT

## Introduction

L-Leucine dehydrogenase (LeuDh) from *Bacillus cereus* was isolated by Sanwal and Zink[1] in 1961. Subsequently an improved production and purification procedure was developed, by Schütte *et al.*,[2] using a constitutive *Bacillus cereus* mutant. The enzyme catalyzes the oxidative deamination of L-leucine as well as the reverse reaction, namely, the reductive

[1] B. D. Sanwal and M. W. Zink, *Arch. Biochem. Biophys.* **94,** 430 (1961).

[2] H. Schütte, W. Hummel, H. Tsai, and M.-R. Kula, *Appl. Microb. Biotechnol.* **22,** 306 (1985).

amination of the corresponding keto acid with the reaction equilibrium being on the side of L-leucine. This fact, as well as the low substrate specificity of the enzyme,[2] make LeuDh useful as a technical biocatalyst. Here we describe aqueous two-phase partitioning, further extraction, and purification of the enzyme to virtual homogeneity by a number of chromatographic steps. The method derives mainly from a combination of work reported by Schütte *et al.*[2] and Schustolla *et al.*[3-5]

## Materials and Methods

### *Chemicals and Biomass*

Poly(ethylene glycol) (PEG) polymers with different molecular weights are obtained from Serva (Heidelberg, Germany). PEG 1540 used for large-scale extraction is from Deutsche BP (Düsseldorf, Germany); dextran (Dx) PL 500 is from Pfeiffer and Langen (Dormagen, Germany), and Dx T500 is from Pharmacia (Freiburg, Germany). Cibacron Blue F3G-A is delivered by Fluka (Neu-Ulm, Germany). Other chemicals used in partitioning experiments and in synthesis of PEG derivatives are from Sigma (Munich, Germany) and Merck (Darmstadt, Germany). The amino acids are from Sigma. $NAD^+$ is from Boehringer (Mannheim, Germany). All chromatographic materials used are obtained from Pharmacia. Glucose and yeast extract used for large-scale fermentation are from Maizena (Krefeld, Germany) and Ohly (Hamburg, Germany), respectively. All other fermentation materials are from Merck. *Bacillus cereus* biomass is produced according to Schütte *et al.*[2] using the mutant III/3 constitutive for LeuDh. Cell disintegration is mainly by bead milling.

### *Preparation of Cibacron Blue F3G-A–Poly(ethylene Glycol)*

*α-(2-Chloroethyl)-ω-chloro-PEG 8000.* The derivatization of the bifunctional PEG 8000 is carried out according to Bückmann *et al.*[6] One hundred grams (14.8 mmol) PEG 8000 is melted overnight in a 70° incubator and dried in a 500-ml flask of a laboratory rotating evaporator at 70° under high vacuum (<10 Pa). The chlorination is performed by adding 4.3 ml

[3] D. Schustolla, C. Ledoux, N. Papamichael, and H. Hustedt, *Ber. Bunsen-Ges. Phys. Chem.* **93,** 971 (1989).

[4] D. Schustolla and H. Hustedt, *in* "DECHEMA Biotechnology Conferences 3" (D. Behrens and A. Driesel, eds.), Part B, p. 1097. DECHEMA, Frankfurt, and VCH, Weinheim, 1989.

[5] D. Schustolla, Ph.D. Thesis, GBF, Braunschweig, and Technical University, Hannover (1992).

[6] A. F. Bückmann, M. Morr, and M.-R. Kula, *Biotechnol. Appl. Biochem.* **9,** 258 (1987).

(59.2 mmol) of freshly distilled thionyl chloride with continued slow rotation of the flask at 70° under a dry nitrogen atmosphere so as to avoid water contamination and decomposition of reagent. After 6 hr the reaction by-products, sulfur dioxide and hydrogen chloride, together with the remaining thionyl chloride, are drawn off under vacuum (final vacuum below 10 Pa). The product is cooled on a polyethylene tray.

*α-(2-Aminoethyl)-ω-amino-PEG 8000.* The chlorinated PEG is dissolved in 200 ml water by slight heating and mixing with a large excess of 25% aqueous ammonia (250 ml). The amination is carried out in a tightly closed 1-liter polyethylene vessel at 60° for 96 hr. Following the reaction the color of the solution changes from deep to light yellow. Excess ammonia is completely removed by use of a laboratory rotary evaporator and a reduced pressure of around 2500 Pa. The dry product is melted at 80° and poured onto a polyethylene tray to solidify. The aminated PEG is recrystallized twice from nonaqueous ethanol and then freeze-dried. This results in a material which can be stored for extended periods of time and used in most coupling reactions of ligands owing to the high reactivity of its terminal amino groups.

*Cibacron Blue F3G-A–PEG 8000.* In a nearly equimolar ratio, 10 g of aminated PEG 8000 and 2.5 g of pure Cibacron Blue F3G-A are dissolved in 100 ml water. The solution is stirred at 60° while maintaining the pH at 11 with 1 *M* potassium hydroxide. After 16 hr the mixture is neutralized with 2.5 *M* HCl. The free, unreacted dye is removed by gel filtration. The gel filtration column (5 × 52 cm) packed with Sephadex G-50 gel (fine) is loaded with 50 ml of reaction solution, diluted to a concentration of 5% (w/w) PEG, and eluted with 0.1 *M* NaCl solution (flow rate 100 ml/hr). The unreacted aminated PEG and free Cibacron Blue F3G-A passes through the column more slowly than the derivative owing to their lower molecular weights. The sodium chloride in the eluate is removed by dialysis against distilled water (Spectra/Por 3, cutoff 3500 Da, Spectrum Medical Industries Inc., Los Angeles, CA). The solution is concentrated by evaporation under vacuum at 60° and stored at 4°. The derivative obtained contains 0.22 mmol Cibacron Blue F3G-A per gram, estimated from the absorbance of the water solution at 612 nm (molar absorption coefficient $\varepsilon = 13{,}000\ M^{-1}\ \mathrm{cm}^{-1}$).

### *Partitioning Experiments*

Partitioning experiments are performed on a 5- or 10-g scale in calibrated centrifuge tubes at 20°. The two-phase systems are prepared from stock solutions of PEG [40 or 50% (w/w) depending on molecular weight], Dx [20% (w/w)], and potassium phosphate [40% (w/w)] at different pH values, by weight. After mixing the systems for 1 min on a Vortex mixer, they are centrifuged for 5–10 min at 2300 *g* at 20° in order to accelerate phase settling. Volumes of the phases are recorded, and the partitioning of enzyme and total protein is determined by taking out appropriate aliquots of each phase and assaying for enzyme activity and protein content. The partition coefficient, $K$, is defined as the ratio of the enzyme activity in the upper to that in the lower phase. The difference in the log $K$ values, $\Delta \log K$, obtained with and without the presence of PEG–ligand, is a measure of the affinity effect.

### *Analytical Methods*

The enzyme activity is assayed by following the increase in absorbance at 340 nm (LKB Ultrospec K 4053 spectrophotometer, Biochrom, Cambridge, UK) in the oxidative deamination reaction at 30°. The reaction mixture contains 10 m*M* L-leucine and 3.4 m*M* $NAD^+$ in a 100 m*M* glycine/NaCl/NaOH buffer, pH 10.7. The reaction is initiated by the addition of 0.1 ml enzyme solution to 3 ml test solution. One unit of L-leucine dehydrogenase is defined as the amount of enzyme that catalyzes the production of 1 μmol of NADH per minute under the standard assay conditions. Specific activity is defined as units per milligram of protein. Protein concentrations are determined by the method of Bradford[7] using bovine serum albumin (BSA) as standard. Disk gel electrophoresis is carried out according to Jovin *et al.*[8] at room temperature in a Tris/glycine buffer and using Coomassie Brilliant blue for staining gels.

## Experimental Data

### *Partitioning Behavior and Extraction of Leucine Dehydrogenase*

The major advantage of the aqueous two-phase extraction technology is its ability to process cell homogenates directly and to separate cell debris as well as the majority of contaminating proteins, nucleic acids, and microbial polysaccharides from the target protein in a single step. PEG–salt-based phase systems have proved to be especially useful for this purpose.[9] Two-polymer systems usually require the inclusion of a polymer-bound ligand to achieve a selective purification, and the affinity extraction might be hindered by cell debris and other cellular compounds.[5]

[7] M. M. Bradford, *Anal. Biochem.* **72,** 248 (1976).

[8] T. Jovin, A. Chrambach, and M. A. Maughton, *Anal. Biochem.* **9,** 351 (1964).

[9] H. Hustedt, K. H. Kroner, U. Menge, and M.-R. Kula, *Trends Biotechnol.* **3,** 139 (1985).

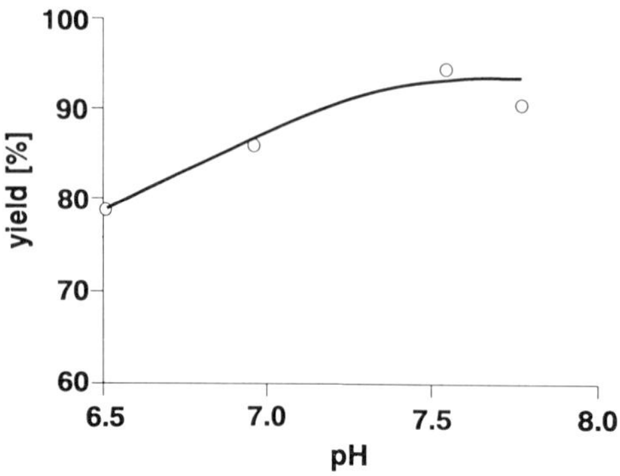


FIG. 1. Physical extraction of L-leucine dehydrogenase from *Bacillus cereus* cell homogenate, demonstrated by the activity yield in the top phase in relation to the pH of the phosphate stock solution used (18% PEG 1540, 7% potassium phosphate, and 50% cell homogenate containing 20% biomass) (H. Schütte, D. Schustolla, and H. Hustedt, unpublished results).

*Physical Extraction in PEG–Salt Systems*. Figure 1 shows as an example the extraction of LeuDh from a cell homogenate in a PEG 1540–potassium phosphate system as function of pH. High yields, more than 90%, are obtained in the top phase at a pH of at least 7.5, while the cell debris partitions into the bottom phase. The reextraction of the enzyme into the bottom phase, namely, the salt-rich phase, which permits the removal of the PEG and hydrophobic contaminants with the top phase, can be achieved by various means; addition of sodium chloride as in the purifica-

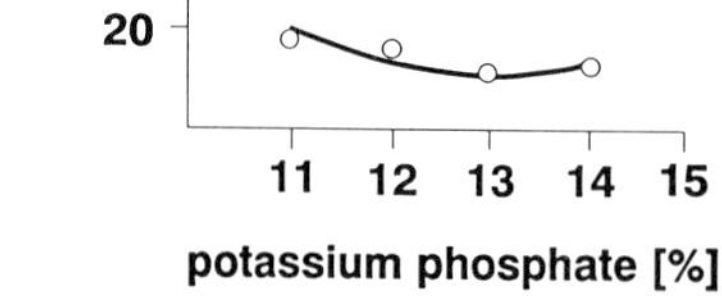


FIG. 2. Reextraction of L-leucine dehydrogenase into the salt-rich bottom phase; (●) activity yield in the bottom phase, (○) residual activity in the top phase (conditions: 50% top phase from the first extraction plus additional potassium phosphate as indicated and water added to 100% (H. Schütte, D. Schustolla, and H. Hustedt, unpublished results).

TABLE I
INFLUENCE OF DYE LIGANDS BOUND TO POLY(ETHYLENE GLYCOL) ON EXTRACTION[a] OF LEUCINE DEHYDROGENASE[b]

| PEG ligand | Δ log *K* |
|---|---|
| Procion Blue MX-3G | 0.19 |
| Procion Blue MX-7RX | 1.16 |
| Procion Blue MX-R | 1.37 |
| Cibabron Blue F3G-A | 1.05 |
| Procion Green HE4-BD | 1.11 |
| Procion Navy H-4R | 0.19 |
| Procion Yellow MX-GR | 1.18 |
| Procion Yellow MX-3R | 0.05 |
| Procion Yellow HE-3G | 1.14 |
| Procion Orange H-ER | 1.08 |
| Procion Red HE-3B | −0.17 |
| Procion Red MX-5B | −0.18 |
| Procion Turquoise MX-G | 1.09 |
| Procion Scarlet MX-G | 0.60 |

[a] Composition of systems: 6% (w/w) dextran T500, 5.4% (w/w) PEG 8000, 0.6% (w/w) PEG-dye, 20 mmol/kg potassium phosphate, pH 6.3, 40% (w/w) cell homogenate (40% (w/w) wet cell mass).

[b] Reproduced from D. Schustolla and H. Hustedt *in* "DECHEMA Biotechnology conferences 3" (D. Behrens and A. Driesel, eds.), Part B, p. 1097. DECHEMA, Frankfurt, and VCH, Weinheim, 1989, with permission.

tion procedure described below worked best[2] (another example is shown in Fig. 2).

*Affinity Partitioning and Extraction.* Affinity partitioning has been carried out with a large number of triazine dyes, $NAD^+$, NADH, and L-leucine, coupled to PEG 8000, as ligands.[3-5] Purified enzyme, cell extract, and cell homogenate have been used as starting materials. A few results are presented in Table I and in Figs. 3 and 4. From the 15 triazine dyes tested 6 had an affinity effect (expressed as Δ log *K*) greater than 1, with the highest value being 1.37. The largest purification effect, although very low with only about 1.5, was observed using Cibacron Blue F3G-A. Therefore this dye was studied in greater detail. Figure 3 shows the influence of PEG–Cibacron Blue concentration and reflects the limited affinity effect as well as the low purification. Figure 4 compares the effect of ligand concentration for purified LeuDh, cell extract, and cell homogenate.

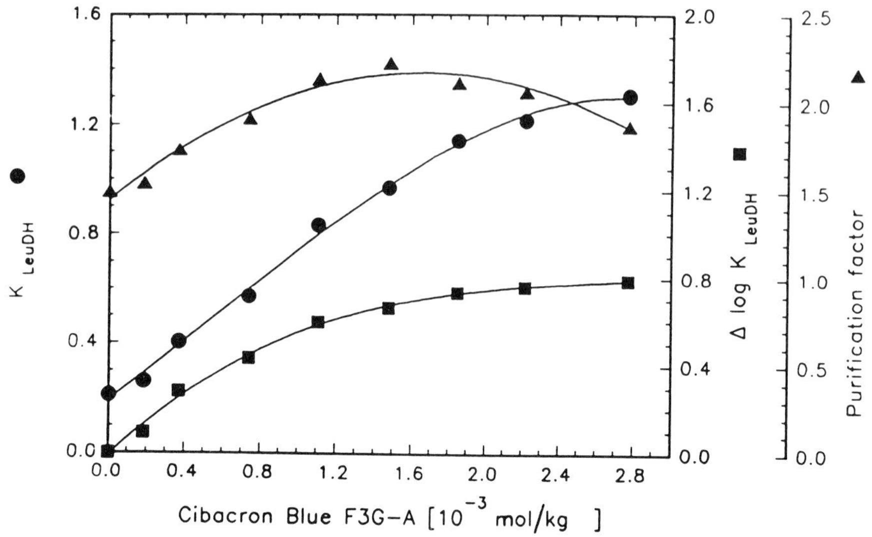


FIG. 3. Effect of PEG-bound Cibacron Blue F3G-A on the partition coefficient, $K$, and the purification of L-leucine dehydrogenase. The system contained 2% (w/w) dextran PL 500, 8% (w/w) PEG 8000 including dye–PEG, 40% (w/w) cell homogenate (containing 40%, w/w, wet cell mass), and 20 mmol/kg potassium phosphate buffer, pH 6.25. [From D. Schustolla, Ph.D. Thesis, GBF, Braunschweig, and Technical University, Hannover (1992), with permission.]

The purified enzyme solution and the cell extract give nearly the same result, whereas the $K$ value is much lower for the cell homogenate. This can be attributed partially to a much lower $K$ of the ligand in the presence of cell debris ($K < 2$) in contrast to "clear solutions" ($K \approx$ 12–14).[3-5] Obviously the dye adsorbs strongly to cell debris and is thus shifted into the bottom phase. The highest purification factors (3.4 for cell extract and 2.7 for cell homogenate) were achieved using PEG–NADH as ligand, but this is clearly too expensive to be used in large-scale extractions.[3,5] Thus, extraction in PEG–salt systems is more favorable for the large-scale downstream processing of LeuDH.

### *Purification of L-Leucine Dehydrogenase*

The following section is reproduced to a large extent from Schütte *et al.*,[2] with permission.

*Cell Disruption.* Thirty kilograms *Bacillus cereus* cells are disintegrated using a continuously operating industrial agitator mill (Netzsch LME 20,

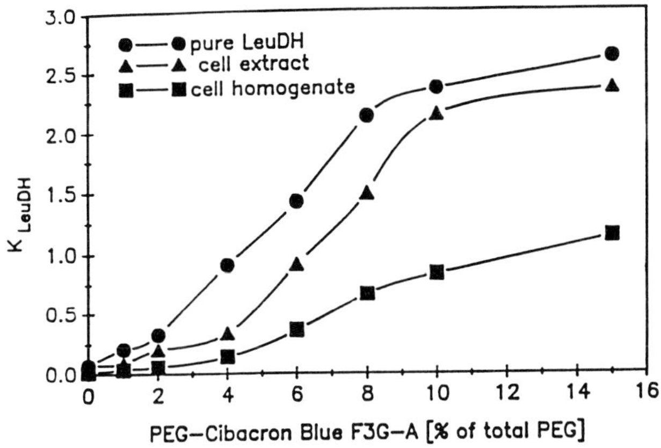


FIG. 4. Influence of the concentration of PEG-bound Cibacron Blue on the partition of L-leucine dehydrogenase using different starting materials. The system contained 5% (w/w) dextran T500, 8% (w/w) PEG 8000 including dye–PEG, 2 U/g pure enzyme or 40% (w/w) cell extract or 40% (w/w) cell homogenate containing 40% (w/w) wet cell mass, and 20 mmol/kg potassium phosphate buffer, pH 6.3. [From D. Schustolla, C. Ledoux, N. Papamichael, and H. Hustedt, *Ber. Bunsen–Ges. Phys. Chem.* **93,** 971 (1989), with permission.]

Selb, Germany) and then heat treated for 10 min at 63°.[10] The pH at which the heat denaturation procedure is carried out is important and should be in the range of pH 7.5 to 8.5 to obtain good recovery of the enzyme.[2]

*Removal of Cell Debris and Denatured Proteins.* For extraction of enzyme from *B. cereus* homogenate, PEG 1540 and potassium phosphate (pH 8.0) are added and stirred for 1 hr to form a PEG–salt system and to assure equilibrium of the partitioning (although a much shorter stirring time would be sufficient[11]). The whole extraction system consists of 18% (w/w) PEG 1540, 7% (w/w) potassium phosphate, 70 liters cell homogenate [biomass concentration ~40% (w/w) before disruption, wet weight], and deionized water to give a phase system of 140 kg (128 liters). Liquid–liquid separation is carried out employing a disk stack separator (Westfalia Separator, Model SA-1, Oelde, Germany) with a feed rate of 100 liters/hr. Approximately 98% of the LeuDh is found in the PEG-rich upper phase I (86 liters), which is essentially free of solid material. The specific enzyme activity is increased 11-fold when compared to the cell homogenate.

*Further Purification.* The addition of salt to top phase I led to the formation of a second phase system in which the enzyme partitioned into

[10] H. Schütte, K. H. Kroner, W. Hummel, and M.-R. Kula, *Ann. N.Y. Acad. Sci.* **413,** 270 (1983).

[11] P. F. Fauquex, H. Hustedt, and M.-R. Kula, *J. Chem. Technol. Biotechnol.* **35B,** 51 (1985).

the salt-rich lower phase. This second phase system has a final volume of 172 liters and contains 86 liters of upper phase I, 11% (w/v) additional potassium phosphate (pH 7.0), and sodium chloride to a final concentration of 0.3 *M*. After stirring for 1 hr phase separation is permitted to proceed at unit gravity in a settling tank. Complete phase settling occurs within 30 min. After separation the lower phase II is diafiltrated against 10 m*M* potassium phosphate buffer, pH 7.5, and concentrated using a Romicon Hollow Fiber Cartridge, Type HF 30–20-GM 80 (Romicon G.B., Oosterhout, Netherlands). Final concentration to 2.4 liters is accomplished using an Amicon Hollow Fiber Cartridge, type XM50, 2.5 $ft^2$ (Amicon, Witten, Germany). Then the enzyme solution is applied to a DEAE-Sephacel column (21 × 60 cm) equilibrated against dialysis buffer. After an extensive wash with starting buffer the protein is eluted by increasing the sodium chloride concentration gradually from 0 to 0.25 *M* in starting buffer. The LeuDh elutes at the end of the 100-liter gradient totally separate from the NADH oxidase. The enzyme-containing active fractions are concentrated by ultrafiltration and gel filtrated on a Sephacryl S-200 Superfine column (10 × 80 cm) at a flow rate of 88 ml/hr.

Final purification to virtual homogeneity is carried out using hydrophobic interaction chromatography or "interfacial salting out chromatography" on Sepharose 4B.[12] The enzyme solution, obtained by gel filtration, is dialyzed against 50 m*M* potassium phosphate buffer, pH 7.5, containing 35% saturated ammonium sulfate. The precipitate that forms overnight is separated by centrifugation and discarded. The supernatant solution is applied to a Sepharose 4B column (5 × 32 cm) equilibrated against the dialysis buffer. Unbound protein is washed off with starting buffer. LeuDh is eluted by applying a decreasing linear 6-liter gradient (40 to 20% ammonium sulfate). The enzyme appears in the effluent at an ammonium sulfate saturation of 35–30%. Homogeneous LeuDh fractions are collected, concentrated by ultrafiltration, and dialyzed against 50 m*M* potassium phosphate buffer, pH 7.5, containing 0.1% 2-mercaptoethanol. The enzyme is stored at 4° or after addition of 50% glycerol at −20°.

The purification of LeuDh from 30 kg (wet weight) *Bacillus cereus* cells is summarized in Table II. The final activity yield was 48%, and the overall purification factor 95.

## Concluding Remarks

We have described a process to obtain LeuDh from a *Bacillus cereus* mutant that includes partitioning, extraction, recovery, and purification.

[12] F. von der Haar, *Biochem. Biophys. Res. Commun.* **70,** 1009 (1976).

TABLE II
PURIFICATION OF L-LEUCINE DEHYDROGENASE[a]

| Step | Volume (liter) | Protein (g) | Total activity (U × $10^{-3}$) | Specific activity (U/mg) | Yield (%) | Purification factor |
|---|---|---|---|---|---|---|
| Homogenate | 70.0 | 723.0 | 434 | 0.6 | 100 | 1.0 |
| Heat treatment | 70.0 | 145.0 | 421 | 2.9 | 97 | 4.8 |
| Upper phase I | 86.0 | 61.0 | 412 | 6.8 | 95 | 11.3 |
| Lower phase II | 99.0 | 40.0 | 373 | 9.3 | 86 | 15.5 |
| Diafiltration | 2.4 | 22.0 | 360 | 16.0 | 83 | 26.7 |
| DEAE-Sephacel | 4.7 | 13.0 | 308 | 24.0 | 71 | 40.0 |
| Sephacryl S-200 | 0.48 | 6.3 | 260 | 41.0 | 60 | 68.3 |
| Sepharose 4B | 1.4 | 3.6 | 208 | 57.0 | 48 | 95.0 |

[a] Reproduced from H. Schütte, W. Hummel, H. Tsai, and M.-R. Kula, *Appl. Microb. Biotechnol.* **22,** 306 (1985), with permission.

The process delivers homogeneous enzyme and consists of eight steps. The overall enzyme activity yield of 48% is high. Affinity extraction of LeuDh was not too promising; extraction in a PEG–salt system appears to be much better suited for a large-scale downstream processing. The purification process described was developed in 1984 and can be carried out more rapidly today by use of modern chromatographic media.

## Acknowledgments

The authors thank Springer Verlag, VCH, and DECHEMA for kind permission to reproduce published material.

# [61] Purification of Proteins from Cell Culture Supernatant Solutions

*By* ULRICH MENGE

## Introduction

The main products of cultured animal cells are monoclonal antibodies and recombinant proteins (e.g., interferons and interleukins). Cytokines, the latter, are of great pharmaceutical interest and can be produced on a large scale only by recombinant cells. As most cytokines are glycosylated, only cloning in animal cells yields glycosylation of proteins similar to that found in the native protein.

Purification of the desired proteins requires their separation from the media components that are used to maintain viability and productivity of the animal cells.[1] The purification of cytokines from cell culture supernatant solutions presents special problems in that the quantity of product protein is small compared to the quantity of proteins that are part of the original cell culture medium. The product, often present at no more than 1 mg/liter, has to be concentrated (sometimes several thousandfold) from large volumes of cell culture medium. A maximum yield also has to be achieved in view of the high fermentation costs.

Precipitation techniques are generally unsuitable with respect to selectivity, practicability, and maintenance of product sterility. Ultrafiltration techniques and affinity adsorption procedures result, at high protein concentrations, in fouling support materials by nonspecific adsorption. These methods are also time-consuming. In contrast, protein extraction in aqueous two-phase systems is a rapid procedure that is not burdened by nonspecific adsorption to solid materials, mass transfer or fouling.[1,3,4] The purification of proteins from cell culture supernatant solutions is illustrated by the example of interferon-$\beta$ produced by cloned mouse L cells.

## Extraction in Aqueous Two-Phase Systems

### *Principles for Selection of Phase Compositions*

In Table I phase systems are listed that can be used for the extraction of cytokines, such as human interferon-$\beta$ and interleukin 2, from cell culture media.[1,3,4] In such systems very high partition coefficients, $K$ ($\geq 100$), can be realized both for interleukin 2 and interferons. Such $K$ values permit a quantitative extraction of cytokines into the top phases. An additional advantage of these phase systems is that albumin, a main constituent of many cell culture media, can be quantitatively extracted into the bottom phases. However, not only very high but also very low $K$ values can be realized for interferon-$\beta$ using phase systems with free or dextran-bound sulfate at high concentration (Table I). The manipulation of $K$ values for interferon-$\beta$ from greater than 100 to less than 1/100 provides excellent possibilities for control of the extraction of the protein.

Some other proteins (e.g., globulins, present in complete sera as used in cell cultivation), are also extracted into the top phase of poly(ethylene

[1] U. Menge, E. Fraune, J. Lehmann, and M.-R. Kula, *Dev. Biol. Stand.* **66,** 391 (1987).

[2] Deleted in proof.

[3] U. Menge, M. Morr, M.-R. Kula, and K. Anastassiadis, U.S. Patent 4,343,735 (1982).

[4] U. Menge, M. Morr, U. Mayr, and M.-R. Kula, *J. Appl. Biochem.* **5,** 75 (1983).

TABLE I
PHASE SYSTEMS USED IN PURIFICATION OF PROTEINS FROM CULTURE MEDIA AND FROM ANIMAL CELL CULTURE SUPERNATANT SOLUTIONS

| Product | Phase components | Concentrations (%, w/w) | pH | $K$ |
|---|---|---|---|---|
| Interleukin 2[a] | PEG 1550, sodium/potassium phosphate, NaCl | 5, 15, 5 | 7 | 99 |
| Interferon-β[b] | PEG 6000-$P_0$[c], sodium/potassium phosphate, NaCl | 1, 19.5, 2.9 | 6.9 | 350 |
| | PEG 6000, dextran sulfate | 15, 3, 10 | 7 | 0.002 |
| Interferon-α | PEG 1550, sodium/potassium phosphate, NaCl | 3, 19.5, 5 | 7 | ≥90 |
| | PEG 6000-$P_0$[c], sodium/potassium phosphate, NaCl | 1, 19.5, 2.9 | 6.7 | 30 |
| Media proteins | | | | |
| FCS, NCS[d] | PEG 1550, sodium/potassium phosphate, NaCl | 2.5, 19.5, 5 | 6–7 | 0.7 |
| Albumin | PEG 1550, sodium/potassium phosphate, NaCl | 14, 14, 0 | 6–7 | 0.01 |

[a] BHK cell-line, serum-free medium.
[b] Culture medium of induced human fibroblasts.[4]
[c] PEG phosphate ester.[3]
[d] FCS, Fetal calf serum; NCS, neonatal calf serum.

glycol) (PEG) 1550–6000 (Serva, Heidelberg, Germany)–orthophosphate systems. Their extraction into top phases diminishes the purification in the case of interferon-β significantly,[1] but such systems can be used to purify some proteins, for example, monoclonal antibodies from cell culture supernatant solutions.

If possible the $K$ of individual components of complex media should be determined separately, as these are usually present in excess when compared to the product. Their partitioning behavior is decisive for product isolation not only during the primary purification by phase extraction but also during subsequent chromatographic steps. Depending on the type and the concentration of the protein constituent in the cell culture medium, the purification factor for interferon-β, necessary to get a homogeneous product, can vary from 800 to 80,000.[1] Thus, for each product and for each cell culture medium, phase compositions must be optimized separately to yield an effective combination for purification. Some guidelines are presented below.

In the purification of products for pharmaceutical applications the partitioning behavior of endotoxins, immunogenic compounds, and nucleic acids must also be considered. Endotoxins have $K$ values of approximately 0.001 in PEG 1550–orthophosphate systems. Thus, efficient separation from such contamination is simple when the product partitions into the top phase.

A single extraction step is usually adequate for removal of noncritical contaminants such as albumin. Critical contaminants can be repeatedly extracted, e.g., from a top phase containing the product by using a counter phase of identical composition from systems prepared without sample. The residual mass of a contaminant in the top phase can be estimated from $Y = 100/[1 + (1/KQ_{vol})]$ where $Y$ corresponds to the yield (%) of the compound in the bottom phase and $Q_{vol}$ to the volume ratio of top to bottom phase. The following example illustrates the effect on purification of repeated extractions with a blank counter phase: At $K = 0.05$ and a volume ratio of $Q_{vol} = 0.1$ there are residual 0.49% of the mass of a contaminant in the top phase after one, 0.0024% after two, and 0.000012% after three extractions, etc. Such separations become more effective with larger bottom phase volumes. The loss of product in large bottom phase volumes is minimal when its $K$ is greater than 100.[1]

*Factorial Design of Experiments and Evaluation from Graphs*

Of great help in optimizing phase extractions of all kinds, and especially those of cytokines, is a factorial approach in combination with a graphic presentation of results (Fig. 1).[4,5] Interferon-$\beta$, as well as many other cell culture products, has to be analyzed by use of a biological test, and this causes complications. The results of such tests are subject to significant variations owing to the complexity of biological systems, the large dilutions (up to several hundred millionfold) of sample necessary for analyses, and the visual estimation of the biological effects.

Problems in the evaluation in part can be reduced by designing experiments factorially. Phase system composition, molecular weight of PEG, concentration of NaCl, and pH are simultaneously varied on distinct levels. The results are graphically presented in the form of a grid whose crossing planes each correspond to one level of one factor. This methodology permits a rapid analysis of the influence of phase parameters and of their interactions on different criteria (e.g., $K$ of product, yield, and specific activity), as depicted in Fig. 1.

In addition, the $K$ value of interferon-$\beta$ can often be given only as ">" value because the concentration of the protein in the bottom phases of these systems is below the detection limit. In other systems no protein is detected in the top phase (<5 $\mu$g/ml), and the specific activity has to be designated as ">." Thus, if one concentration is very small, the $K$ can be quite imprecise.

[5] U. Menge, *in* "Frontiers in Bioprocessing II" (P. Todd, S. Sikdar, and M. Bier, eds.), p. 331. ACS Books, Washington, D.C., 1992.

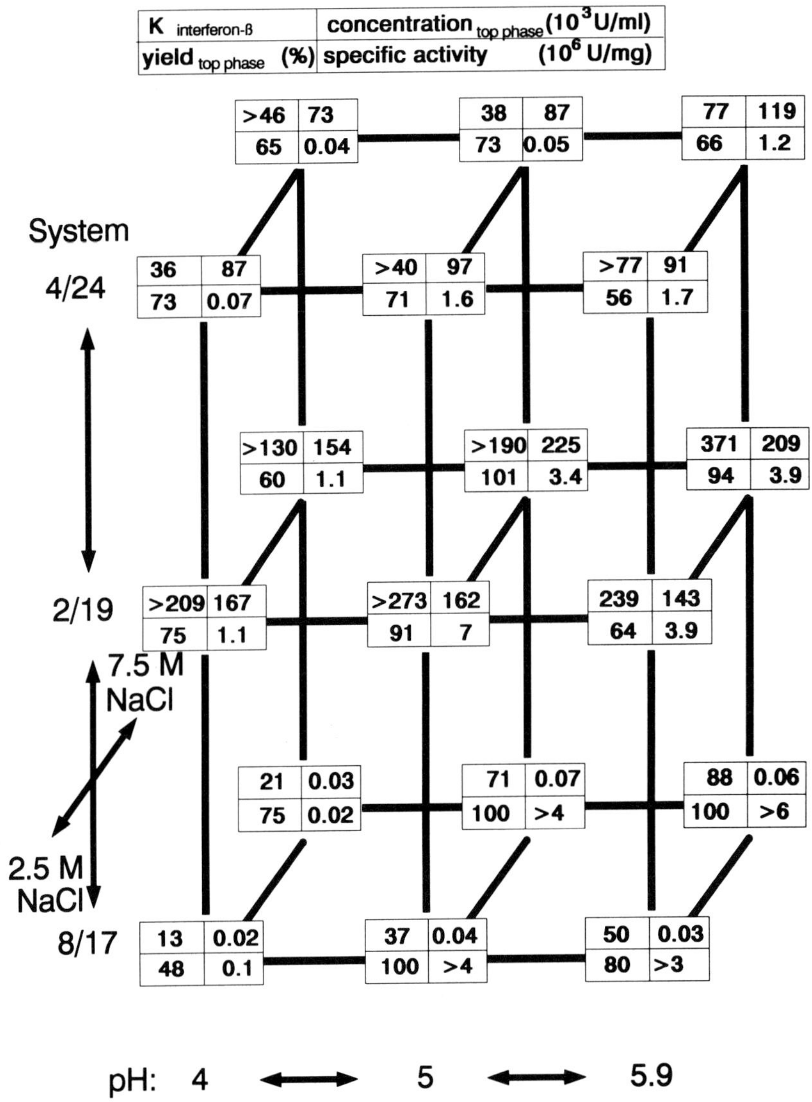


FIG. 1. Factorial design of partitioning experiments with interferon-$\beta$. The systems were as follows: 4/24, 4% (w/w) PEG 6000 phosphate ester, 24% (w/w) orthophosphate; 2/19, 2% (w/w) PEG 6000 phosphate ester, 19% (w/w) orthophosphate; 8/17, 8% (w/w) PEG 1550 phosphate ester, 17% (w/w) orthophosphate.

Despite the problems accompanying the partitioning of interferon-$\beta$ some conclusions can be drawn from Fig. 1. At pH 4, in most systems low yields and specific activities are observed. In systems containing 4/24% (w/w) PEG 6000 phosphate ester–orthophosphate the concentration of interferon-$\beta$ in the top phase is lower at 7.5 than at 2.5 *M* NaCl. However, in all other systems it is increased at 7.5 *M* NaCl. Systems showing the highest interferon-$\beta$ concentrations, yields, and specific activities in the top phases are in the center of the grid.

This evaluation has, to some extent, statistical support, and deviant results can be easily identified since each plane of the grid shows, in parallel, the influence of one experimental parameter in six or nine experiments. Additionally, four different experimental parameters, for example, in multiple phase experiments can be analyzed without resorting to different tables.

### *Synthesis of Poly(ethylene Glycol) Phosphate Ester*

PEG 6000 phosphate ester can be synthesized (M. Morr, personal communication, 1981) at 70° by addition of 40.2 g $P_2O_5$ to 170 g PEG 6000 with stirring over 1 hr. Then 50.2 g $H_3PO_4$ is added, and the mixture is shaken for 10 hr under nitrogen. Finally, the reaction is neutralized by addition of 118 ml of 50% (w/v) KOH. The content of free and bound phosphate can be analyzed by titration before and after dialysis (72 hr) of the product. Typical values are 10% free phosphate and 1.5 phosphate ester groups per PEG molecule.

### *Preparation of Phase Systems*

Phase extraction of crude cell culture supernatant solutions containing interferon-$\beta$ can be carried out as follows: 3.18 g PEG 6000-phosphate ester, 9.5 g NaCl, and 155 g of 40% (w/w) $K_2HPO_4$, adjusted to pH 6.8, are mixed at room temperature with 22.32 g demineralized water.[3] If necessary sterilized phase constituents are used. Finally cell culture supernatant solution is added to the mixture of phase components to give a total mass of 318 g. All this is stirred at room temperature for complete dissolution of phase components and is then poured into a centrifuge tube (Fig. 2). Phases are separated by centrifugation at less than 5000 *g* for a few minutes. The separated top phase is withdrawn by pipette.

Phase extraction of crude cell culture supernatant solutions containing interleukin 2 can be carried out as follows: 7.5 g PEG 1550, 7.5 g NaCl, 5.7 g $NaH_2PO_4 \cdot H_2O$, and 25.9 g $K_2HPO_4 \cdot 3H_2O$ are dissolved at about 60° in 43.4 g demineralized water. Ice-cooled cell culture supernatant solution (60 g) is added to the mixture of phase components, and residual

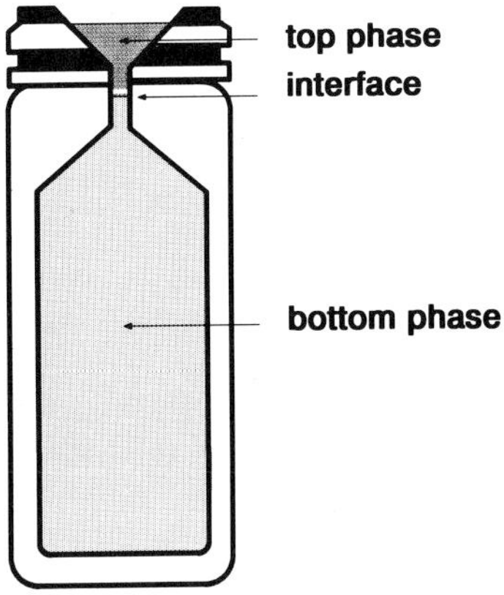


FIG. 2. Centrifuge tube made of acrylate for precise separation of phase systems (318 g total weight) having a small top and a large bottom phase. Actual size: 6.5 × 16 cm.

solids are dissolved by end-over-end rotation for 30 min at room temperature. Phases separate at 1 *g* within 1 hr or at low speed centrifugation in a few minutes.

### *Special Problems Relating to Phase Extraction of Cell Culture Products*

The purification of products from cell culture media engenders some specific problems when these are to be used for pharmaceutical purposes. (1) Sample sterility must be controlled, (2) the volume must be reduced (sometimes several thousandfold) to get highly active samples, and (3) pharmacologically critical contaminants must be removed so as to be below their detection limit. To meet these requirements sterile technique should be used with the phase systems, the volume of the top phase must be small, and the phase must be isolated without incurring contamination from the bottom phase.

The smallest volume that can be used depends on the yield required at a given $K$,[1] on the maximum sample concentration in the phase system, and on the minimum phase volume that can be handled. If the sample in the phase system amounts, for example, to 40% of the total mass, a 10-fold reduction of the process volume requires a volume ratio of top and bottom phase of 4/96. This ratio approaches the minimum in practice.

Because the separation of extremely small top phase volumes is difficult, special separatory devices are helpful. Top phases of a volume of less than or equal to 300 ml can be conveniently separated by use of special acrylic centrifuge tubes produced by any good workshop (Fig. 2).

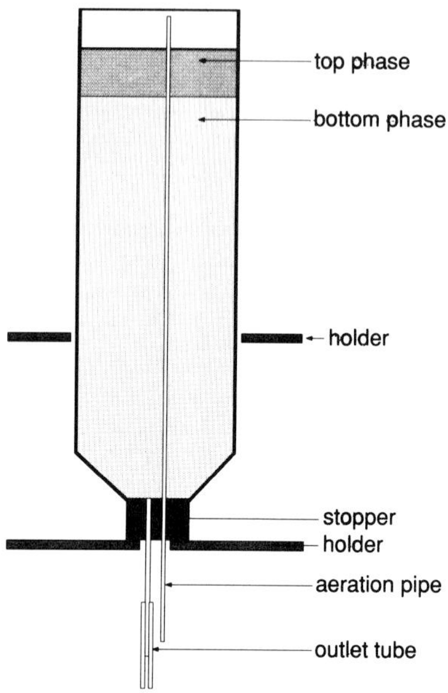


FIG. 3. Five-liter roller bottle for animal cell cultures used in the separation of phase systems having a small top phase. Actual size: 11.5 × 54 cm.

The top phase is concentrated in the upper conical part of such tubes by centrifugation and can be drawn off without contamination by bottom phase or by insolubilized materials that are adsorbed to the interface.

The interface can be positioned within the small central connecting tube by appropriate selection of tube geometry and phase volume ratios. For this purpose it is easiest to adjust the concentration of phase components along the tie line of the corresponding phase diagram[6] as the volume ratio can be varied without affecting phase composition and $K$. Small corrections in the position of the interface can also be made by removing some bottom phase or by filling 40% orthophosphate solution below it.

For small-scale preparations of pharmaceuticals 5-liter tissue culture roller bottles can be used (Fig. 3). Because all manipulations can be carried out in a laminar flow hood up to 5 kg of phase systems can be handled sterilely. Using such roller bottles phase separation can take place at unit gravity. After phase settling the bottom phase is drawn off via a lower outlet pipe; a second pipe serves to aerate the roller bottle. The conical

[6] P.-Å. Albertsson and F. Tjerneld, this volume [1].

narrowing of the bottle and the outlet tube permit precise separation of bottom and top phases.

Large-scale production can be carried out either by using settling tanks or, if the volume ratio is not too extreme, a separator. At a volume ratio of the phases of 1/10 and a total process volume of greater than or equal to 35 liters a separator type SA1 (Westfalia Separator AG, Oelde, Germany) separates top and bottom phases precisely.[1] At flow rates of greater than or equal to 1 liter/min large phase systems can be separated in a short time. Such an extraction plant can be made to comply with LP2 conditions without special efforts. The reliability of the separator technique has been demonstrated by the extraction of 30 mg interferon-$\beta$ from cloned mouse L cells in 10 runs from a total of 434 liters of cell culture supernatant solutions.[1]

### Subsequent Purification Steps

Product (e.g., interferon) purity is increased several hundredfold by phase partitioning.[1,3,4] Homogeneity will, however, rarely be attained in a single purification step. In addition it is necessary to separate phase constituents such as concentrated PEG, orthophosphate, and NaCl from the purified product. Suitable methods include ultrafiltration, hydrophobic interaction chromatography, or reversed-phase chromatography. Ultrafiltration is impractical, however, if the difference in molecular weight of product and PEG is too small.

Reversed-phase chromatography is a preferred method for the purification of interferons[2] and interleukins. If high amounts of albumin are present it will tail in a broad shoulder and coelute with the indicated products, as they display similar hydrophobicity.[1] Thus, for an adequate purification of these proteins an effective removal of albumin, as obtained by phase partitioning, is essential.

Top phases containing interferon-$\beta$ can be loaded directly on reversed-phase columns and the product can be eluted with an alcohol gradient. Volumes loaded on reversed-phase columns can be unconventionally large; for example, 18 liters top phase can be loaded on a 400-ml column. Thus 3 mg interferon-$\beta$ from a mouse cell line can be quantitatively separated from approximately 5 kg PEG phosphate ester in the top phase. By such a two-step procedure a degree of purity of about 50%, corresponding to a several thousandfold purification, and a yield of 70% can be realized.[1]

### Conclusion

During the development of purification processes for products from animal cell cultures, extraction steps offer great advantages, for example, they can be carried out on any scale. In the beginning a product will be available only in small quantities, and milliliter volumes are adequate to begin optimization of its phase extraction. Subsequently, extraction procedures developed on an analytical scale can be applied directly on a preparative scale without modification of phase compositions. A scaleup from 1 ml to 40-liter systems has been instituted with interferon-$\beta$ from mouse cell supernatant solutions.

The time necessary for product purification by phase extraction is less than 1 hr if a separator or a centrifuge is used for phase separation. Thus phase extraction is one of the most rapid biochemical separation methods. By selection of appropriate phase composition, proteases and other enzymes involved in posttranslational modifications of proteins can be separated from products within minutes, thus minimizing degradation processes. In addition, sensitive proteins may be significantly stabilized by the presence of PEG.

The production of pharmaceuticals from cloned cell lines requires the complete removal of endotoxins and nucleic acids, which can be realized by multiple extractions of the product phase with fresh counterphase. Such a procedure can be carried out without risk of recontamination (e.g., repeated use of chromatographic supports is avoided).

Phase extraction of interferon-$\beta$ is very effective, as indicated by $K$ values exceeding 100, a yield in the range of 100%, purification factors of several hundredfold, a 10-fold reduction of process volume, and reproducibility without problem.[1,3,4] These advantages of phase extractions for interferon-$\beta$ can be attributed to the fact that a high concentration of orthophosphate ($>16\%$, w/w) and NaCl ($\geq 2.9\%$, w/w) causes interferon-$\beta$ to become insoluble in the bottom phase ($\leq 0.5$ $\mu$g/liter), resulting in nonideal partitioning (i.e., the apparent $K$ increases at higher interferon-$\beta$ concentrations). However, the concentration of albumin, the main component of the medium, is very low in the top phase. The product is also stabilized by the phase components.

The optimization of extraction methods can be laborious, especially if the product has to be analyzed by use of biological tests. However, the advantages of phase extraction for the purification of animal cell culture products are determinant when this method is applied on a production scale.

# [62] Uses of Fusions of β-Galactosidase and Peptides to Proteins

*By* KRISTINA KÖHLER and ANDRES VEIDE

## Introduction

To combine successfully high recovery and concentration of a protein in a primary purification step with an aqueous two-phase system, a reasonably high target protein partition coefficient ($K$, concentration ratio top phase/bottom phase) is required. Parameters such as polymer molecular weight and concentration, salt type and concentration, and pH can be varied to affect $K$.[1] Selectivity can be further increased by covalent binding of an affinity ligand to one of the phase-forming polymers.[2] Another affinity purification concept makes use of recombinant DNA technology in which the target protein itself is furnished with a suitable ligand-binding function. This approach, which represents a general concept, has been applied to packed bed chromatography systems,[3] as well as to aqueous two-phase systems.[4-6]

*Escherichia coli* β-galactosidase can be partitioned very strongly to the poly(ethylene glycol) (PEG)-rich top phase in PEG 4000 (Merck-Schuchardt, Munich, Germany) or PEG 6000 (Merck-Schuchardt)/potassium phosphate aqueous two-phase systems. This led to the development of a continuous process for β-galactosidase production, where the majority of the other *E. coli* proteins, as well as nucleic acids and cell debris, partitioned into the potassium phosphate-rich lower phase.[7] The potential of β-galactosidase to direct the partitioning of β-galactosidase fusion proteins (or fusion partners) to the PEG-rich phase was investigated (Fig. 1).[8] A purification process for the recombinant protein AGβGal, composed of

[1] P.-Å. Albertsson, "Partition of Cell Particles and Macromolecules," 3rd Ed. Wiley(Interscience), New York, 1986.

[2] G. Kopperschläger and G. Birkenmeier, *Bioseparation* **1,** 235 (1990).

[3] M. Uhlén and T. Moks, this series, Vol. 185, p. 129.

[4] K. Köhler, C. Ljungquist, A. Kondo, A. Veide, and B. Nilsson, *Bio/Technology* **9,** 642 (1991).

[5] J. Luther and C. E. Glatz, *in* "7th International Conference on Partitioning in Aqueous Two-Phase Systems," Abstracts, p. 15. New Orleans, 1991.

[6] R. J. Todd, M. E. Van Dam, D. Casimiro, B. L. Haymore, and F. H. Arnold, *Proteins Struct. Funct. Genet.* **10,** 156 (1991).

[7] A. Veide, T. Lindbäck, and S.-O. Enfors, *Enzyme Microb. Technol.* **6,** 325 (1984).

[8] K. Köhler, A. Veide, and S.-O. Enfors, *Enzyme Microb. Technol.* **13,** 204 (1991).

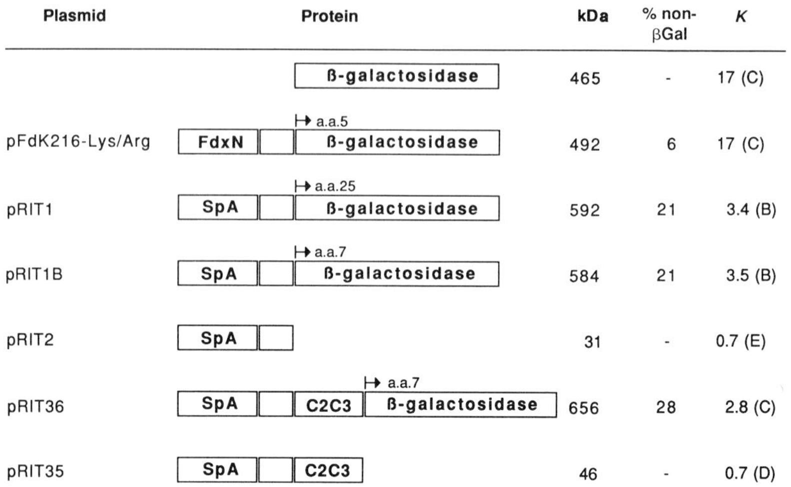


FIG. 1. Schematic description of the $\beta$-galactosidase fusion proteins and their partitioning in PEG 4000–potassium phosphate systems. The capital letters given with the partition coefficients ($K$ values) indicate the following phase system compositions, in % (w/w) PEG 4000/% (w/w) potassium phosphate: B, 6.8/10.5; C, 6.9/10.6; D, 7.0/10.8; E, 7.0/10.9. FdxN, *Rhizobium meliloti* ferredoxin-like protein; SpA, staphylococcal protein A; C2C3, IgG-binding regions from streptococcal protein G. [From K. Köhler, A. Veide, and S.-O. Enfors, *Enzyme Microb. Technol.* **13,** 204 (1991), Butterworth-Heinemann, with permission.]

five immunoglobulin G (IgG) binding domains of staphylococcal protein A, two IgG-binding domains from streptococcal protein G, and *E. coli* $\beta$-galactosidase, was subsequently developed (Fig. 2 and Table I).[9]

The work with $\beta$-galactosidase and $\beta$-galactosidase fusion proteins led to the development of a general peptide partitioning handle. A tryptophan (Trp)-based handle was designed (Figs. 3 and 4) based on the hypothesis of a specific interaction between PEG and Trp.[4,10,11] Trp and dipeptides containing Trp have been shown to partition strongly in favor of the PEG phase in PEG–salt aqueous two-phase systems.[12] Here we describe how recombinant DNA technology can be utilized to affect partitioning selectivity of intracellularly expressed target proteins in PEG–potassium phosphate systems. The $\beta$-galactosidase and the Trp-rich peptide partitioning handles are compared with regard to partitioning, solubility, stability, process recovery, and purification.

[9] L. Strandberg, K. Köhler, and S.-O. Enfors, *Process Biochem.* **26,** 225 (1991).

[10] M. Laskowski, Jr., *Fed. Proc.* **25,** 20 (1966).

[11] A. Radzicka and R. Wolfenden, *Biochemistry* **27,** 1664 (1988).

[12] A. D. Diamond, X. Lei, and J. T. Hsu, *Biotechnol. Tech.* **3,** 271 (1989).

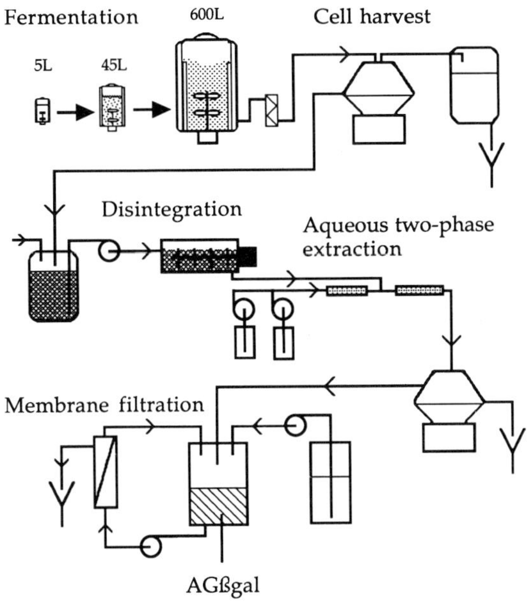


FIG. 2. Process scheme for AGβGal (pRIT36). [From L. Strandberg, K. Köhler, and S.-O. Enfors, *Process Biochem.* **26,** 225 (1991), Elsevier Science Publishers Ltd., with permission.]

TABLE I
RECOVERY AND SPECIFIC ACTIVITY OF AGβGal FUSION PROTEIN DURING ISOLATION AND PURIFICATION PROCESS[a]

| Process step | Volume (liters) | Enzyme activity (U/ml) | Specific activity (U/mg) | AGβGal recovery (%) |
|---|---|---|---|---|
| Cell suspension, French press | 16.2 | 6600 | 190 | 100 |
| Bead mill disintegration | 16.2 | 550 | 180 | 82 |
| Extraction (top phase) | 8.3 | 5800 | 450 | 66 |
| Diafiltration (retentate) | 35 | 1300 | 400 | 109[b] |
| Concentration (concentrate) | 2.5 | 11,300 | 430 | 64 |

[a] From L. Strandberg, K. Köhler, and S.-O. Enfors, *Process Biochem.* **26,** 225 (1991), Elsevier Science Publishers Ltd., with permission.

[b] The retentate volume was not measured during the diafiltration. The recovery given is calculated on the assumption that it was constant during the run. A decrease in volume was, however, noted without being quantified. The recovery is thus likely to be closer to 100%.

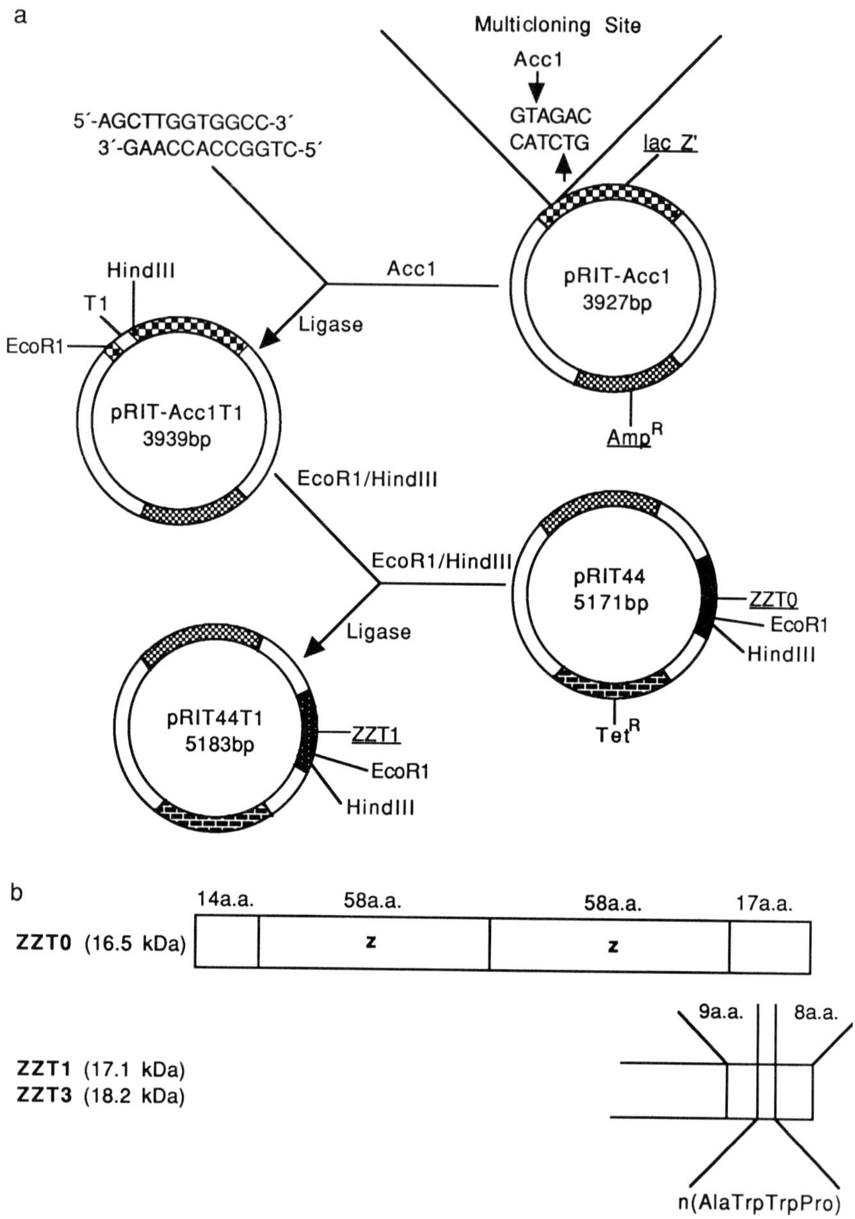


FIG. 3. (a) Schematic cloning scheme for pRIT44T1. The cloning of pRIT44T3 follows the same path. *lacZ'*, Gene fragment encoding the first 104 amino acids (aa) from the *lacZ* gene; *Amp*®, ampicillin resistance gene; *Tet*®, tetracycline resistance gene; T1, sequence encoding for partitioning peptide T; *ZZT0,* ZZT0 gene; *ZZT1,* ZZT1 gene; *Acc*I, *Eco*RI, and *Hin*dIII, restriction sites for the respective restriction enzyme. (b) A schematic illustration of the primary protein structures of ZZT0, ZZT1, and ZZT3. The amount of partitioning peptide T units can be varied as indicated.

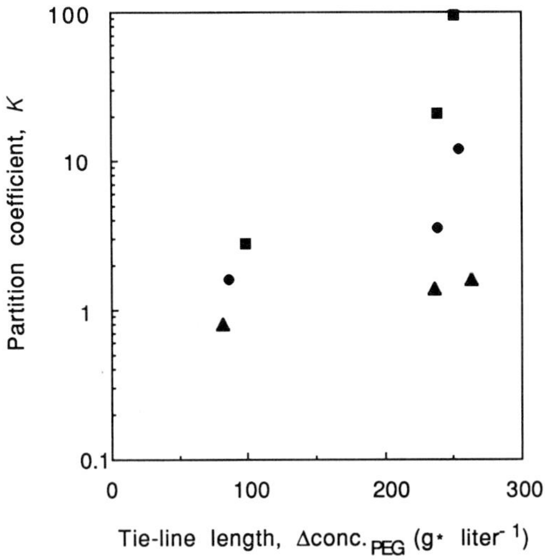


FIG. 4. Partitioning of IgG purified ZZT0 (▲), ZZT1 (●), and ZZT3 (■) in PEG 4000–potassium phosphate systems (pH 7) at three different tie-line lengths. [From K. Köhler, C. Ljungquist, A. Kondo, A. Veide, and B. Nilsson, *Bio/Technology* **9,** 642 (1991), Nature Publishing Co., with permission.]

## Methods

### *β-Galactosidase Fusion Proteins*

Fusions to the *E. coli lacZ* gene encoding β-galactosidase are relatively prevalent. Advantages in the use of β-galactosidase as a fusion partner are control of expression,[3] convenient enzyme assay, and the fact that the resulting fusion protein often exhibits increased resistance to proteolytic degradation.[13] Fusions can be made to both ends of the *lacZ* gene. Whether the fusion protein will be produced in soluble form depends on both the sequence of the fusion protein and the choice of host organism.[14] The β-galactosidase fusion proteins that we have studied are given in Fig. 1. All of these can be produced in soluble form in *E. coli,* and with the β-galactosidase enzyme activity intact. Thus, β-galactosidase activity can be a measure of fusion protein concentration. The different fusion proteins, native β-galactosidase, staphylococcal protein A (SpA), and staphylococcal protein A–streptococcal protein G (protein AG) were partitioned (see below).

[13] H. Hellebust, M. Murby, L. Abrahmsén, M. Uhlén, and S.-O. Enfors, *Bio/Technology* **7,** 165 (1989).

[14] H. M. Sassenfeld, *Trends in Biotechnol.* **8,** 88 (1990).

*Production System and Cultivation*

All proteins listed in Fig. 1 can be expressed from their respective plasmids in *E. coli* RR1ΔM15,[15] except for β-galactosidase which is produced constitutively by *E. coli* ATCC 15224.[7] In pRIT plasmids (Fig. 1), which encode a product and ampicillin resistance, expression is under the control of the λ PR promoter. In addition to the product plasmid, a plasmid, pRITcI857, encoding the temperature-sensitive repressor for the PR promoter and kanamycin resistance is included.[16] Induction is accomplished by raising the temperature from 30 to 42°, thereby destroying the repressor. The gene encoding the *Rhizobium meliloti* ferredoxin-like protein–β-galactosidase fusion (FdxNβGal) is under control of the *lac* promoter, which is induced by addition of 1 m*M* isopropyl-β-D-galactopyranoside [A. Pühler, personal communication, (1989)].

Cells transformed with the appropriate plasmids are cultivated in a stirring tank laboratory-scale bioreactor (5 liters), or in 1-liter shaking flasks containing 100 ml of medium.[8] A minimal salt medium supplemented with casamino acids, glycerol, trace elements, and antibiotics was used.[17] Cells are harvested by centrifugation at 3000 *g* for 20 min at 4°, resuspended in 50 m*M* potassium phosphate buffer (pH 7) containing 10 m*M* $MgCl_2$ to a 50% wet weight of cells, and stored at −80°.

*Partitioning Studies*

The partitioning behavior of target proteins from *E. coli* is examined in PEG 4000/potassium phosphate systems. Partitioning experiments on different proteins are carried out with purified protein samples and, more importantly, with cell homogenates or, alternatively, crude cell extracts. The latter is to simulate conditions that prevail when the extraction technique is applied as a primary purification step. Cell homogenates affect phase diagrams[18] and thereby the partitioning of proteins. Accurate determination of the phase diagram in such a system is also not straightforward.

Cell homogenates are prepared by high pressure in a French pressure cell press (FA 073, SLM Instruments, Inc., Urbana, IL) from frozen cells that have been thawed and subsequently suspended to 25 weight % in 50 m*M* potassium phosphate (pH 7) containing 10 m*M* $MgCl_2$. Cell debris is removed by centrifugation at 30,000 *g* for 20 min at 4°. Phase systems (10 g) are prepared in graduated centrifuge tubes, on weight basis from

[15] U. Rüther, *Nucleic Acids Res.* **10,** 5765 (1982).

[16] B. Nilsson and L. Abrahmsén, this series, Vol. 185, p. 144.

[17] L. Strandberg, S. Hober, M. Uhlén, and S.-O. Enfors, *J. Biotechnol.* **13,** 83 (1990).

[18] P.-Å. Albertsson and F. Tjerneld, this volume [1].

40 weight % stock solutions of PEG 4000 and potassium phosphate ($K_2HPO_4/KH_2PO_4$ = 1.42 mol/mol, pH 7). Phase components and distilled water are added to 5 g, followed by mixing, addition of 5 g cell disintegrate or crude extract free of cell debris (to 12.5 weight %), and finally careful mixing again. Typically this yields a total protein concentration of 0.6 weight % (7 g/liter). Phase systems are set to equilibrate at 20° for 10 min in a thermostatted bath. Phase separation is obtained by centrifugation at 500 *g* for 5 min. Phase volumes are read and samples from top and bottom phases are carefully withdrawn with Pasteur pipettes. β-Galactosidase activity is measured directly on the separated phases.

Protein solubilities at different concentrations of potassium phosphate buffer (pH 7) and PEG 4000 are examined. The solubilities are studied over a concentration of 0–250 g/liter potassium phosphate buffer and 0–200 g/liter PEG 4000. The protein concentrations used in the solubility experiments should be similar to the protein concentrations to be used in the partitioning experiments. The concentrations used for SpAβGal (pRIT1 and pRIT1B) and β-galactosidase were in the range 1.5–1.6 g/liter. The SpA concentration was 1.1 g/liter (Fig. 1).[8] Purified protein preparations in 50 m*M* potassium phosphate buffer (pH 7) containing 10 m*M* $MgCl_2$ are added to each respective sample. They are incubated at room temperature (21–22°) for 15–20 min. Centrifugation at 1000 *g* for 20 min at 21–22° yields supernatant solutions, which are analyzed for β-galactosidase activity. SpA is analyzed by absorbance at 280 and 310 nm with a protein-free sample as reference. The solubility is calculated as the amount of protein that can be recovered after precipitation compared to low potassium phosphate buffer- or PEG-free conditions [i.e., in 50 m*M* potassium phosphate buffer (pH 7) plus 10 m*M* $MgCl_2$].

Fusion protein concentration is measured by the β-galactosidase (β-D-galactoside galactohydrolase, EC 3.2.1.23) activity.[7] The activity is determined by following the hydrolysis of 2.3 m*M* *o*-nitrophenyl-β-D-galactopyranoside (ONPG) at 420 nm in 130 m*M* sodium phosphate buffer, pH 7.4, containing 1 m*M* $MgCl_2$ and 94 m*M* 2-mercaptoethanol. One unit of enzyme activity hydrolyzes 1.0 μmol/min of ONPG at 25°. SpA and protein AG are analyzed by enzyme-linked immunosorbent assay (ELISA) technique.[19] First, SpA and protein AG are allowed to bind to human IgG. SpAβGal is subsequently added and analyzed by β-galactosidase activity.

### *Process for AGβGal Fusion Protein*

A production process for a recombinant protein from *E. coli* on a pilot-scale is designed as follows (Fig. 2).[9] AGβGal, encoded by pRIT36[17]

[19] H. Hellebust, A. Veide, and S.-O. Enfors, *J. Biotechnol.* **7,** 185 (1988).

(Fig. 1), is produced in *E. coli* RR1ΔM15[15] on a 600-liter scale. Cells are harvested at a flow rate of 220 liters/hr in a continuous disk-stack centrifuge running at 12,000 *g* with the discharge interval set to 5 min (BTPX, Alfa-Laval, Tumba, Sweden). Prior to bead-mill disintegration (0.6 liter Dyno-Mill KDL, Willy A. Bachhofen AG Maschinenfabrik, Basel, Switzerland), the cells are suspended to 50 g dry cells per liter (25 weight % wet cell mass) in 50 m*M* potassium phosphate buffer (pH 7) containing 10 m*M* $MgCl_2$. The residence time in the mill is 3.6 min. The mill is filled to 85% with glass beads (0.25–0.5 mm in diameter).

Continuous extraction is performed at 20° in a 12.5 weight % PEG 4000–7.5 weight % potassium phosphate aqueous two-phase system. Stock solutions are 31 weight % PEG 4000 and 40 weight % potassium phosphate ($K_2HPO_4/KH_2PO_4$ = 1.42 mol/mol, pH 7). The final biomass concentration in the system is 12.5 weight %. Phase separation is obtained by running the phase system through a continuously working centrifuge (LAPX, Alfa-Laval). The total flow through the centrifuge is 16.8 liters/hr. AGβGal is recovered in the PEG-rich top phase, whereas particles, nucleic acids, and contaminating cell proteins are to a high degree removed with the salt phase.

Diafiltration, and subsequent concentration (Pellicon filter holder, Millipore Corporation, Bedford, MA), of the AGβGal-containing PEG top phase on a 100 kDa ultrafilter [1.84 $m^2$, 2 NOVA filters (Filtron Technology Corporation, Northborough, MA), 2 OMEGA filters (Millipore)] follow to remove the PEG. Prior to filtration, the PEG phase is diluted 5 times in 1 m*M* Tris (pH 7.2), 1 m*M* $MgCl_2$. The diluted top phase is washed with 7 volumes of buffer. The flux is 33 liters/hr/$m^2$, which corresponds to a filtrate flow of 1 liter/min. Then the retentate is reduced 14-fold, by running the filtration unit in a concentration mode. Finally, the concentrated AGβGal is freeze-dried after addition of 5% trehalose.

### *Tryptophan-Rich Fusion Proteins*

*Design.* The main requirement for the peptide handle is that it has surface-exposed Trp residues. The peptide handle design must be such that the exposure of Trp residues to the aqueous solution is optimized. To achieve conformational restriction proline (Pro) residues were used together with two bulky Trp residues at neighboring positions. Using alanine (Ala) as a spacer, the following peptide handle was selected, AlaTrpTrpPro, and designated the partitioning peptide, T.[4] Various methods, such as UV and fluorescence difference spectra and circular dichroism (CD) spectra, can be applied to assess the accessibility of the Trp residues, which depends on the folding of the peptide. Molecular dynamics simulations of the partitioning peptide can reveal its conformational properties.

*Cloning.* A cloning scheme is illustrated in Fig. 3a. Two oligonucleotides representing the partitioning peptide are synthesized (Gene Assembler Plus, Pharmacia, Uppsala, Sweden). By designing the oligonucleotides with sticky ends fitting a nonpalindromic site in the cloning plasmid (pRIT–*Acc*I),[20] a head-to-tail polymerization of oligonucleotides during the ligation step can be assured. In this way, genes encoding different repeats of the partitioning peptide sequence can be obtained. After transformation of the ligation mixture to *E. coli* HB101,[4] the nucleotide sequences of the partitioning peptides are tested by solid-phase dideoxy sequencing.[21]

The gene fusion vector pRIT44T0[4] is used not only for intracellular expression of the model target protein ZZT0 but also as a cloning partner into which multiples of the nucleotide sequence of the peptide can be ligated. ZZ is a two-domain derivative of SpA.[16] It is a small, hydrophilic, and IgG-binding protein, with a $K$ assumed to be equal to 1.[8] Two vectors, pRIT44T1 and pRIT44T3, into which one or three nucleotide sequences had been inserted, respectively, were selected, which coded for the fusion proteins ZZT1 and ZZT3.[4] The intracellular expression of the ZZT proteins (Fig. 3b) is under control of the *E. coli trp* promoter, and induction is accomplished by addition of indole-3-acrylic acid.

*Partitioning Studies.* For the partitioning experiments ZZT proteins are produced intracellularly in *E. coli* RV308.[4] After cell disintegration and centrifugation, the proteins are purified on IgG-Sepharose Fast Flow (Pharmacia). Elution is effected by 200 m$M$ acetic acid/ammonium acetate (pH 3.3). The eluted fractions, containing IgG-binding protein, are lyophilized and stored at −20° until use.

PEG 4000–potassium phosphate phase systems are prepared at room temperature (20–21°) as described above, except that the final weight is 5 g. Three systems having different tie-line lengths are used to estimate the dependence of $K$ on the difference between the top and bottom phases (which is related to the tie-line length)[18,22] (Fig. 4). Concentrated protein solutions are prepared by dissolving lyophilized protein in distilled water, followed by centrifugation at 13,000 $g$ for 5 min (20–21°). The absorbances at 280 and 310 nm are measured, and approximately 0.7 absorbance units ($\Delta A_{280/310} = A_{280} - A_{310}$) are added to each phase system. The resulting

[20] C. Ljungquist, A. Breitholtz, H. Brink-Nilsson, T. Moks, M. Uhlén, and B. Nilsson, *Eur. J. Biochem.* **186,** 563 (1989).

[21] T. Hultman, S. Ståhl, T. Moks, and M. Uhlén, *Nucleic Acids Res.* **7,** 629 (1988).

[22] D. E. Brooks and R. Norris-Jones, this volume [2].

protein concentrations are approximately 0.2 g/liter for ZZT3 and ZZT1, and 1 g/liter for ZZT0. Finally, the total weight is brought to 5 g with water. Temperature equilibration, phase separation, and sampling follow, as described above. Protein concentrations and $K$ values are calculated from $\Delta A_{280/310}$ and the extinction coefficients. Each top and bottom phase sample is read against an identical, but protein-free, blank. Sodium dodecyl sulfate–polyacrylamide gel electrophoresis (SDS–PAGE) is performed using standard procedures[23] with a 3.5% stacking gel and 15% separation gel. PEG is analyzed according to Skoog.[24]

## Discussion

### *Partition Coefficient*

The $K$ values of the peptides increased 4 to 5-fold on fusion to β-galactosidase (Fig. 1).[8] The fusion protein $K$ values were found to be adequately high for choosing aqueous two-phase extraction in the primary stages of a purification process. The impact of different multiples of the Trp peptide on fusion to the model target protein, ZZ, is shown in Fig. 4.[4] The $K$ of the fusion protein with a trimer (ZZT3) containing 6 Trp residues increased up to 100-fold compared to the target protein.[4] Preliminary experiments extracting ZZT3 from crude extracts of *E. coli* free of cell debris indicated, however, that the $K$ of ZZT3 was reduced to values more like those obtained for SpAβGal and AGβGal (K. Köhler and A. Veide, unpublished results). We have had similar results in earlier work with *E. coli* β-galactosidase partitioned in pure and in crude systems [K. Köhler and A. Veide (1988), unpublished results]. It is known that the phase diagram is affected by crude extracts, mainly by DNA,[25] which could have consequences also for protein partitioning. This emphasizes the importance of developing the partitioning procedure in an appropriate environment.

If a model for interacting molecules[26] is applied to fusion proteins, assuming that the fusion partners do not interact with each other, the $K$ for the fusion protein, $K_f$, can be expressed as

$$K_f = K_h K_t \tag{1}$$

[23] U. Laemmli, *Nature (London)* **227,** 680 (1970).

[24] B. Skoog, *Vox Sang.* **37,** 345 (1979).

[25] K. Köhler, L. von Bonsdorff-Lindeberg, and S.-O. Enfors, *in* "32nd IUPAC Congress," p. 1506. Stockholm, 1989.

[26] P.-Å. Albertsson, *Methods Biochem. Anal.* **29,** 1 (1983).

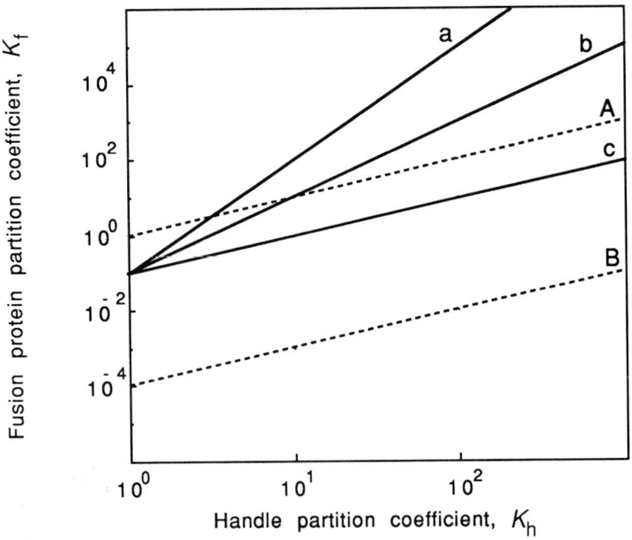


FIG. 5. A theoretical description of the fusion protein partition coefficient ($K_f$) as a function of the handle peptide/protein partition coefficient ($K_h$). The solid lines indicate the application of the Albertsson model to fusions where different amounts of peptide handles ($n$) (a, 3 peptide units; b, 2 peptide units; c, 1 peptide unit) are fused to a target protein with an assumed partition coefficient ($K_t$) of 0.1. The relationship can be described by $K_f = K_h{}^n K_t$. The dashed lines show the model when $\beta$-galactosidase is used as the handle for assumed target protein partition coefficients ($K_t$) of 1.0 (A) and 0.1 (B). In the latter case $K_f = K_h K_t^4$ describes the relationship between the partition coefficients.

where $K_h$ and $K_t$ are the $K$ values for the handle and the target protein, respectively (Fig. 5). If partitioning experiments in a given phase system are carried out with the fusion proteins and with their nonfused entities, calculated and measured $K_f$ values can be compared (Fig. 6) [K. Köhler and A. Veide (1991), unpublished results]. Measured values that are lower than those calculated may indicate interaction between the fused entities.

### *Solubility*

Solubility of the $\beta$-galactosidase fusion proteins decreased with increasing concentrations of phase components in the aqueous two-phase system (i.e., with increasing tie-line length).[8] Solubility experiments with SpA$\beta$Gal and its fusion partners were performed in pure PEG 4000 and in potassium phosphate aqueous solutions $\beta$-Galactosidase solubility was sensitive to potassium phosphate but compatible with high concentrations of PEG 4000. The reverse pattern was exhibited by SpA, which indicated the relationship between protein partitioning and solubility. The SpA$\beta$Gal

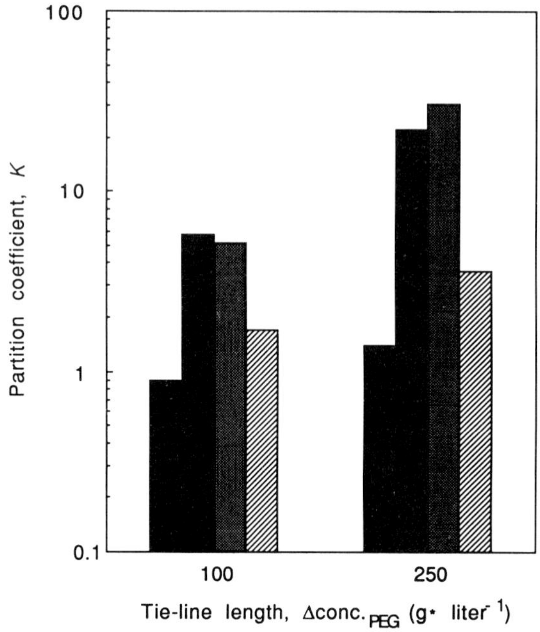


FIG. 6. Comparison of calculated ZZT1 fusion protein partition coefficients with their measured counterparts in two PEG 4000–potassium phosphate systems (pH 7) with different tie-line lengths. The calculated values are based on the partitioning of the fusion partner entities and the model for interacting molecules by Albertsson. The dipeptide TrpTrp was used as a model for the partitioning peptide T. The differently patterned bars indicate, from left to right, the measured partition coefficient of TrpTrp, measured partition coefficient of ZZT0, calculated partition coefficient of ZZT1, and measured partition coefficient of ZZT1.

fusions had lower solubility than $\beta$-galactosidase in PEG 4000 solutions as well as in potassium phosphate solutions. This may be explained in part by the larger size and higher degree of conformational asymmetry of the fusion proteins. A change in the linker region between the two units in SpA$\beta$Gal affected solubility, as indicated for two different versions of the protein (Fig. 1, pRIT1 and pRIT1B).[8] Thus, the design of the linker region is crucial for product solubility and, by extension, product recovery during aqueous two-phase system extraction,[8] as well as proteolytic stability[13] and *in vivo* solubility.[27] IgG-purified ZZT3[4] and ZZT3 in crude extracts [K. Köhler and A. Veide (1991), unpublished results] were completely soluble at the concentrations used (0.2 g/liter) for partitioning, as evidenced by protein recovery.

[27] L. Strandberg, and S.-O. Enfors, *Appl. Environ. Microbiol.* **57,** 1669 (1991).

*Stability*

Another important feature to consider when dealing with recombinant fusion proteins is their proteolytic stability. Both SpAβGal (the pRIT1B encoded form as mentioned above) and AGβGal can be produced to high soluble concentrations without significant degradation.[27] On the other hand, when the ZZT proteins were produced, *in vivo* degradation of the handle, and possibly the target protein, was a reality. Degradation increased with increasing quantities of fused handles (as revealed by SDS–PAGE, gel not shown). At harvest of a ZZT3-producing culture approximately 80% of IgG-binding protein is full-length ZZT3 protein.[4] The rate of ZZT3 degradation *in vivo* is under investigation.

*Recovery and Purification*

Product solubility and stability in the aqueous two-phase system have important implications for product recovery. *K* values for the β-galactosidase fusion proteins increased with system tie-line length.[8] This was predominantly due to decreasing solubility of the product in the salt phase. Initially, the product could be located in the PEG phase, which resulted in a slight increase in product *K*, recovery, and purification. Eventually, PEG phase solubility maximum was reached and product recovery declined. For these reasons, we utilized an aqueous two-phase system relatively close to the the critical point in the large-scale applications to optimize recovery and purification.[9] Such systems are more sensitive to changes in process parameters (e.g., temperature, flow, and biomass composition). Thus, increasing the biomass [in the AGβGal process from 12.5% (wet weight) to 25–30% (wet weight)], and thereby the fusion protein concentration, in the aqueous two-phase system must be controlled carefully with respect to fusion protein recovery. On the other hand, the addition of biomass reduces the amount needed of phase-forming components, which favors fusion protein solubility.

TABLE II
COMPARISON OF β-GALACTOSIDASE WITH PARTITIONING PEPTIDE T AS HANDLES[a]

| Parameter | β-Galactosidase | Partitioning peptide T |
|---|---|---|
| Design flexibility | − | ++ |
| Solubility | + | + |
| *In vivo* stability | + | − |

[a] (−) Unsatisfactory; (+) acceptable; (++) good.

When IgG-purified ZZT3 was partitioned the degradation products primarily distributed to the salt phase while the full-length fusion protein was recovered predominantly in the PEG phase. This indicated that degradation took place at the C terminus and cleaved off the partitioning peptide. When partitioning the crude extract ZZT3 was recovered in the PEG phase with a yield of 60% and a purity of 60–80% as judged by SDS–PAGE [K. Köhler and A. Veide (1991), unpublished results].

## Concluding Remarks

It is feasible to fuse short peptides or native proteins to a target protein to effect the partitioning of the latter in a desirable manner. Although the use of peptide handles is far from fully explored, a first assessment and comparison with $\beta$-galactosidase as a handle has been made (Table II). With peptides the number of peptide handle T units can be selected in a relatively free manner. This makes them more flexible than use of the $\beta$-galactosidase handle, which has a fixed handle to target protein molecular ratio. The much smaller size of the peptide compared to $\beta$-galactosidase makes the synthesis of the former energetically more ideal for the cell. However, one has to pay attention to the possible increase of proteolytic susceptibility of the target protein when it is fused to a short nonnative peptide.

# [63] Concentration of Biomaterials: Virus Concentration and Viral Protein Isolation

*By* LENA HAMMAR

## Introduction

Viruses, like other particles, can be steered to partition unevenly between the phases of aqueous two-polymer phase systems. If one phase is strongly favored, for example, the bottom phase, systems with a small bottom to top phase volume ratio can be used to concentrate and purify the virus.

Purification of viruses and phages was among the first applications of aqueous two-phase extraction. The results gave important insights on the concentration and purification of viruses as well as on the principles involved. These early reports have been reviewed and discussed by several authors.[1–6] Table I lists animal viruses studied; plant viruses and bacteriophages are covered in earlier reviews.[5,7] Extraction in aqueous polymer systems remains an attractive option when virus purification from large volumes is involved[8–14] and in dealing with labile viruses.[12,15,16]

It is remarkable that phase extraction in two-polymer aqueous systems has not found wider use in virology. The technique is easy to set up and, in its basic form, does not require expensive equipment. The availability of a large number of water-soluble polymers permits the selection of two-phase systems with different properties. Therefore it should be possible to find conditions for concentration of most viruses and to obtain purified virions with well-preserved morphology and biological properties. Although the variables involved are complex, it is comforting that some established systems work well with several viruses. Usually the system can be adjusted to fit a particular application by simple means, such as a change in salt concentration or of buffer ions.[17]

The general principles of two-phase extraction are discussed in other chapters of this volume.[18] Here I focus on protocols and procedures that are useful for concentration and purification of viruses. The purification of retroviruses is discussed in some detail. These relatively large enveloped viruses carry a loosely attached glycoprotein, the external *env* protein,

[1] P.-Å. Albertsson and G. Frick, *Biochim. Biophys. Acta* **37,** 230 (1960).
[2] P.-Å. Albertsson, *Methods Virol.* **2,** 303 (1967).
[3] L. Philipson, *in* "Fundamental Techniques in Virology" (K. Habel and P. Salzman, eds.), p. 109. Academic Press, New York, 1969.
[4] P.-Å. Albertsson, *Water Resour. Symp.* **7,** 16 (1974).
[5] K.-E. Magnusson and O. Stendahl, *in* "Partitioning in Aqueous Two-Phase Systems: Theory, Methods, Uses, and Applications to Biotechnology." (H. Walter, D. E. Brooks, and D. Fisher, eds.), p. 415. Academic Press, Orlando, Florida, 1985.
[6] H. Walter and G. Johansson, *Anal. Biochem.* **155,** 215 (1986).
[7] P.-Å. Albertsson, "Partition of Cell Particles and Macromolecules." Wiley(Interscience), New York, 1986.
[8] I. V. Pavlenko, A. I. Krasheniuk, T. A. Salnikova, A. A. Brezgun, and I. P. Smorodinova, *Vopr. Virusol.* **30,** 732 (1985).
[9] V. G. Pomelova, S. Gajdamovich, and V. A. Demenev, *Vopr. Virusol.* **31,** 584 (1986).
[10] T. Pöyry, M. Stenvik, and T. Hovi, *Appl. Environ. Microbiol.* **54,** 371 (1988).
[11] L. Hammar, G. Gilljam, and B. Wahren, *Report to EC Concerted Research Program, European Vaccine against AIDS, Rome* (1992).
[12] L. Hammar and G. Gilljam, *Aids Res. Hum. Retroviruses* **6,** 1379 (1990).
[13] G. Gilljam, S. Eriksson, J. Hinkula, and L. Hammar, in preparation (1993).
[14] J. Grindrod and D. O. Chiver, *Arch. Gesamte Virusforsch.* **31,** 365 (1970).
[15] L. Hammar, S. Eriksson, K. Malm, and B. Morein, *J. Virol. Methods* **24,** 91 (1989).
[16] L. Hammar, M. Merza, K. Malm, S. Eriksson, and B. Morein, *Biotechnol. Appl. Biochem.* **11,** 296 (1989).
[17] H. Walter and C. Larsson, this volume [4].
[18] D. E. Brooks and R. Norris-Jones, this volume [2].

TABLE I
ANIMAL VIRUS SUBJECTED TO PARTITIONING IN AQUEOUS TWO-PHASE SYSTEMS

| Type | Genome | Size of virion (nm) | Classification | Virus | Refs. |
|---|---|---|---|---|---|
| NAKED | | | | | |
| Site of capsid assembly | | | | | |
| Cytoplasm | RNA, single stranded | 17–30 | Picorna | Enteroviruses | |
| | | | | Poliovirus | *a–e* |
| | | | | Coxsackie virus | *d* |
| | | | | ECHO virus | *d, f, g* |
| | | | | Foot-and-mouth disease virus | *h* |
| Nucleus | DNA, double stranded | 70–90 | Adeno | Adenovirus | *f, g* |
| ENVELOPED | | | | | |
| Membrane type | | | | | |
| Intracytoplasmic vacuoles | RNA, single stranded | 40–50 | Flavi | Japanese encephalitis virus | *i* |
| Endoplasmic reticulum | RNA, single stranded | 50–70 | Toga | Venezuelan equine encephalomyelitis virus | *j* |
| Cell membrane | RNA, single stranded | 90–120 | Orthomyxo | Influenza virus | *d, f, g* |
| | | 80–150 | Retro | Feline leukemia virus | *k* |
| | | | | Bovine leukemia virus | *l* |
| | | | | Human immunodeficiency virus (HIV-1, HIV-2) | *m, n* |
| | | | | Simian immunodeficiency virus ($SIV_{mac}$) | *o* |

| | | | | | |
|---|---|---|---|---|---|
| | | 120–300 | Paramyxo | Parotitis virus | *f* |
| | | | | Newcastle disease virus | *f* |
| | | | | Measles virus | *p* |
| Intracytoplasmic inclusion bodies | DNA, double stranded | 230–300 | Pox | Vaccinia virus | *f, q* |

[a] E. C. J. Norrby and P.-Å. Albertsson, *Nature (London)* **188,** 1047 (1960).
[b] S. Bengtsson, L. Philipson, and P.-Å. Albertsson, *Biochem. Biophys. Res. Commun.* **9,** 318 (1962).
[c] S. Bengtsson and L. Philipson, *Virology* **20,** 176 (1963).
[d] J. Grindrod and D. O. Chiver, *Arch. Gesamte Virusforsch.* **31,** 365 (1970).
[e] T. Pöyry, M. Stenvik, and T. Hovi, *Appl. Environ. Microbiol.* **54,** 371 (1988).
[f] T. Wesslén, P.-Å. Albertsson, and L. Philipson, *Arch. Gesamte Virusforsch.* **9,** 510 (1959).
[g] L. Philipson, P.-Å. Albertsson, and G. Frick, *Virology* **11,** 553 (1960).
[h] The Wellcome Foundation Limited, Belgian Patent No. 636,492 (1964).
[i] H. Nakai, *Acta Virol.* **9,** 89 (1965).
[j] V. G. Pomelova, S. Gajdamovich, and V. A. Demenev, *Vopr. Virusol.* **31,** 584 (1986).
[k] L. Hammar, S. Eriksson, K. Malm, and B. Morein, *J. Virol. Methods* **24,** 91 (1989).
[l] L. Hammar, M. Merza, K. Malm, S. Eriksson, and B. Morein, *Biotechnol. Appl. Biochem.* **11,** 296 (1989).
[m] L. Hammar and G. Gilljam, *AIDS Res. Hum. Retroviruses* **6,** 1379 (1990).
[n] G. Gilljam, S. Eriksson, J. Hinkula, and L. Hammar, in preparation (1993).
[o] L. Hammar, G. Gilljam, and B. Wahren, *Report to EC Concerted Research Program, European Vaccine against AIDS, Rome* (1992); G. Gilljam and L. Hammar, *J. Chromatogr.* in press (1993).
[p] I. V. Pavlenko, A. I. Krasheniuk, T. A. Salnikova, A. A. Brezgun, and I. P. Smorodinova, *Vopr. Virusol.* **30,** 732 (1985).
[q] P.-Å. Albertsson and G. Frick, *Biochim. Biophys. Acta* **37,** 230 (1960).

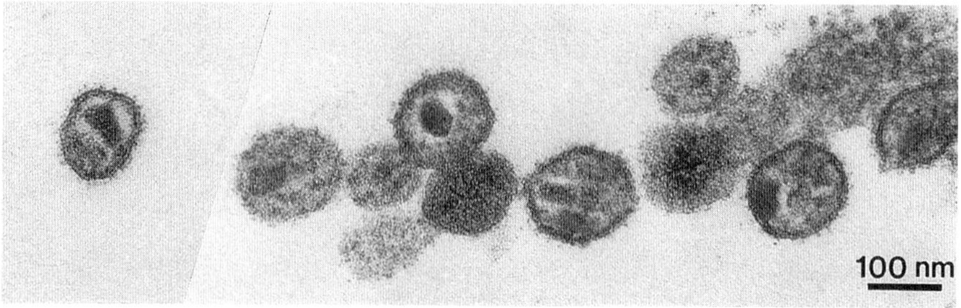


FIG. 1. HIV-2 at the interface after extraction in the system 0.2% dextran T500–7.0% PEG 6000. Material from the interface was fixed by the addition of glutaraldehyde to 1.25% (v/v) and formaldehyde to 1% (v/v), final concentrations, postfixed in 2% (w/v) osmium tetroxide, treated with 1% (w/v) tannic acid, and embedded in Spurr epoxy resin. Thin sections were cut and further contrasted by uranyl acetate and lead citrate. (Electron micrograph by Dr. Sigrid Eriksson, Department of Veterinary Virology, Uppsala, Sweden.)

which forms the protrusions from the virion seen in carefully prepared specimens (Fig. 1). This protein is important for binding virus to target cells and harbors the major determinants for neutralizing antibodies. The external *env* protein is usually recovered in poor yield with the virion particle when purification is done by pelleting and banding in an ultracentrifuge. Thus, less than 25% of gp120 is recovered with pelleted human immunodeficiency virus type 1 (HIV-1) particles. However, when the virus is concentrated by extraction in a two-phase system composed of dextran (Dx) and poly(ethylene glycol) (PEG) the yield of gp120 increases 2.5-fold.[12] The recovery of HIV-1 gp120 and the capsid protein, *gag* p24, in the top and (interface plus) bottom phase of such a system is shown in Fig. 2A.

A schematic diagram, for the practical handling of two-phase extraction of infectious material, is depicted in Fig. 3. It is used in the protocol, given below, for purification of a simian immunodeficiency virus ($SIV_{mac}$). The effects of some commonly used polymers on virus assay systems are also presented and discussed. Polymer concentrations are given as %, which refers to % (w/w).

## Polymer Systems

Two-phase systems composed of dextran sulfate and either PEG[18a] or poly(vinyl alcohol) (PVA) usually permit efficient concentration of virion particles in the bottom phase and/or at the interface. The systems

[18a] The conventions used by different companies for molecular weight notations of poly(ethylene glycol) may differ.

can be manipulated by varying the salt concentration and have been successfully used with small naked viruses.[19–22] Philipson *et al.*[20] showed that viruses can be steered to the bottom phase or to the top phase by selection of the NaCl concentration of the system and utilized this method for the concentration of adenoviruses and Echo viruses. A similar strategy was adopted by Pomelova *et al.*[9] for purification of the small enveloped Venezuelan equine encephalomyelitis (VEE) virus (see below).

Among several polymer systems tested for concentration of feline leukemia virus (FeLV), the optimal recovery obtained was with a dextran sulfate–PVA (Polyviol 48/20, BDH Chemicals Ltd., Poole, UK) system. The external glycoprotein and the *gag* protein cofractionated to the dextran sulfate-rich bottom phase.[15]

Dextran sulfate can usually be removed by precipitation with potassium ions.[20] This seems to do no harm to small or naked viruses like the adeno- or picornaviruses nor to the rather stable enveloped influenza virus, but the larger retroviruses may not be recovered in an intact form since their *env* proteins have a tendency to coprecipitate with the polymer. In the conventional setup of a dextran sulfate–PEG system (which uses a small bottom phase) the HIV-1 gp120 is recovered in the bottom phase.[12] It is difficult to remove the dextran sulfate from the glycoprotein by precipitation or by ion-exchange chromatography. However, when Triton X-100 (Union Carbide, Danbury, CT) is introduced and the system changed to give a reversed volume ratio (i.e., top phase is small) the glycoprotein is recovered in the PEG-rich top phase while the *gag* protein p24 remains in the bottom phase (Fig. 2B). With PVA as the polymer predominant in the top phase both proteins remain in the bottom phase (Fig. 2C).

Dx and PEG of various molecular weights are usually harmless to biological systems. Two-phase systems based on these polymers have been used with many viruses (see below). Virus infectivity is usually retained. Large viruses have a greater tendency to partition to the bottom phase or interface than do smaller viruses, but they also have a greater tendency to be precipitated by PEG. The phase diagram should be consulted to obtain a desired phase volume ratio without having an unnecessarily high PEG concentration.[23]

[19] T. Wesslén, P.-Å. Albertsson, and L. Philipson, *Arch. Gesamte Virusforsch.* **9,** 510 (1959).
[20] L. Philipson, P.-Å. Albertsson, and G. Frick, *Virology* **11,** 553 (1960).
[21] E. C. J. Norrby and P.-Å. Albertsson, *Nature (London)* **188,** 1047 (1960).
[22] The Wellcome Foundation Limited, Belgian Patent No. 636,492 (1964).
[23] P.-Å. Albertsson and F. Tjerneld, this volume [1].

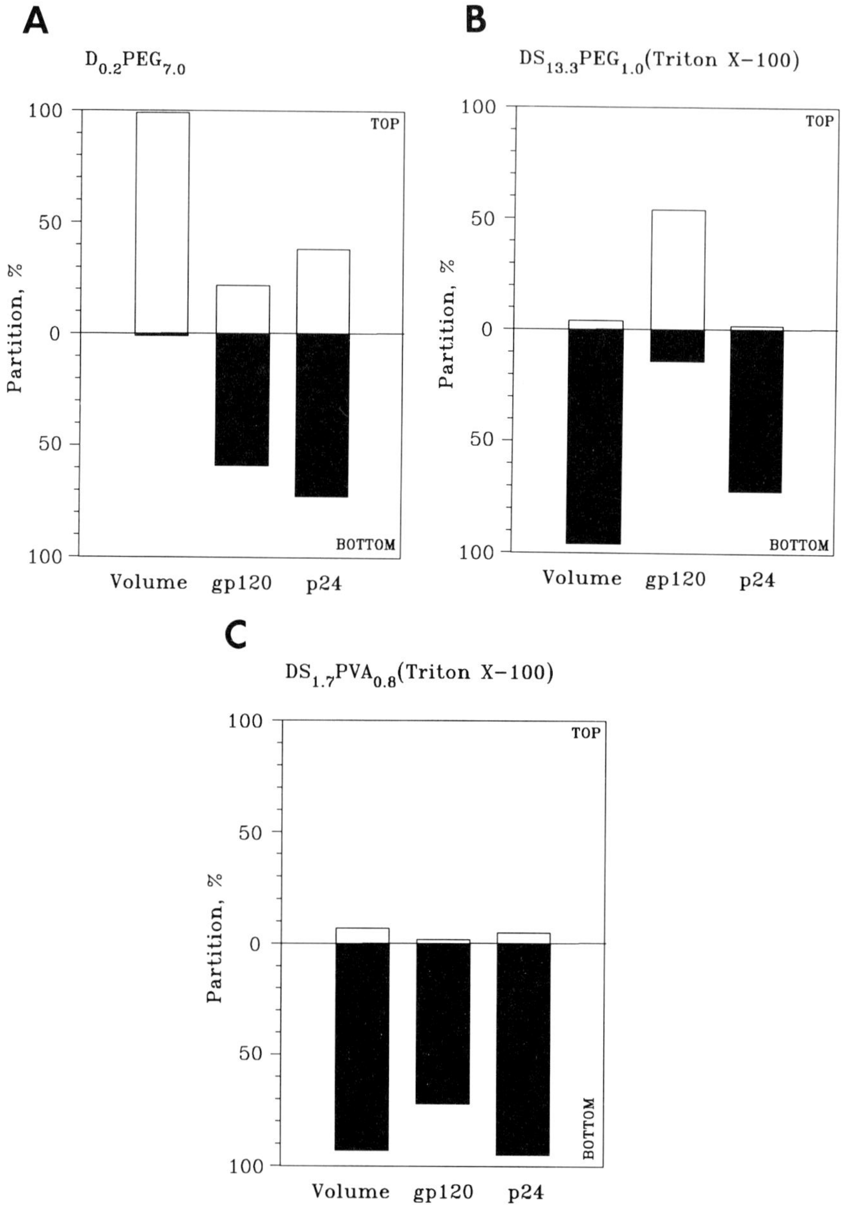


FIG. 2. (A) Two-phase extraction of HIV-$1_{IIIB}$ in the system 0.2% dextran T500–7.0% PEG 8000. The relative volumes of and the recovery of gp120 and p24 in the top phase and at the interface are indicated. [Data from L. Hammar and G. Gilljam, *AIDS Res. Hum. Retroviruses* **6,** 1379 (1990).] (B, C) Concentrated HIV-$1_{IIIB}$ was lysed by the addition of

Hydroxypropyl starch (Reppal PES, Reppe Glykos, Växjö, Sweden) seems to be a promising alternative to Dx when used in combination with PEG. The use of such a system is illustrated with SIV in Fig. 4.

Enveloped viruses obtain their coat as they bud from one of the membranes in the host cell, or during formation of inclusion bodies (Table I). Some of the original membrane characteristics are retained, but the virus-coded glycoproteins, inserted in or attached to the surface of the virus membrane, are specific. These glycoproteins are potential targets for binding by lectins. A preliminary study[24] with $SIV_{mac}$ indicates that the recovery of its *env* protein, gp148, is increased about 3-fold (from 25 to 80%) when snowdrop (*Galanthus nivalis*) lectin (GNA Boehringer Mannheim GmbH, Mannheim, Germany)-substituted Dx is incorporated in the system routinely used for SIV purification (see below). If the detergent Empigen BB (Allbright & Wilson, Whitehaven, Cumbria, UK) is also included in the system, less than 10% of the glycoprotein remains in the top phase.

## Procedures

### *Purification of Venezuelan Equine Encephalomyelitis Virus*

The three-step purification procedure, designed by Pomelova *et al.*,[9] is followed by electron microscopy. A 300-fold total concentration is reported based on infectious activity.[9]

*System.* The system contains 0.7% dextran sulfate (molecular weight 500,000) (Serva, Heidelberg, Germany), 8.6% PEG 4000 (Loba Fein Chemie GmbH, Fischamend, Austria), and 0.5 *M* NaCl.

*Procedure.* At each step the phase mixture is equilibrated for 30 min by end-over-end mixing, then left for 1 hr in the cold (6–8°), and finally centrifuged at 1200 *g* for 15–30 min.

1. PEG and NaCl are added as dry powders and dissolved in the culture medium containing virus. Dextran sulfate is then added using a 10% stock solution. In this first system the virus partitions to the bottom phase, which is collected.

[24] A. Persson and L. Hammar, in preparation (1993).

Triton X-100 and subjected to two-phase extraction in dextran sulfate in combination with poly(vinyl alcohol) or PEG. The relative volume of the top phase and interface and the recovery of the two viral proteins are indicated. In (B) the system was 13.3% dextran sulfate–1.0% PEG 8000 containing 0.01% (v/v) Triton X-100. In (C) the system was 1.7% dextran sulfate–0.8% poly(vinyl alcohol) containing 0.01% (v/v) Triton X-100.

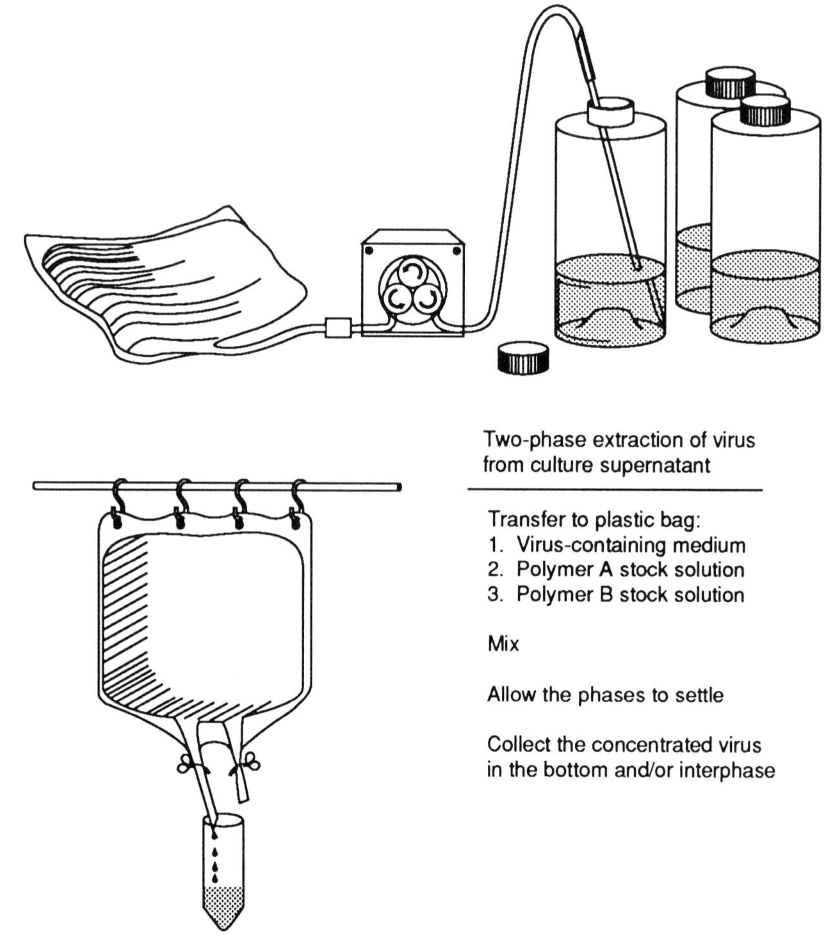


FIG. 3. Arrangement for the two-phase extraction of virus from culture supernatant solutions. A peristaltic pump is used to transfer culture medium containing virus and polymer stock solutions to a plastic bag with attached tubing. The bag is agitated while filling and left hanging to permit the phases to settle. The virus in the bottom phase and/or at the interface is then collected.

2. To three parts of bottom phase, two parts of a solution containing 5 *M* NaCl in 8% PEG are added. The pH is adjusted to 9.0 with Tris. The final NaCl concentration is 3 *M*. In this phase system the virus partitions to the top phase.

3. The top phase is collected and diluted with a 10% solution of PEG to decrease the salt concentration to 0.5 *M*. Then a 10% dextran sulfate

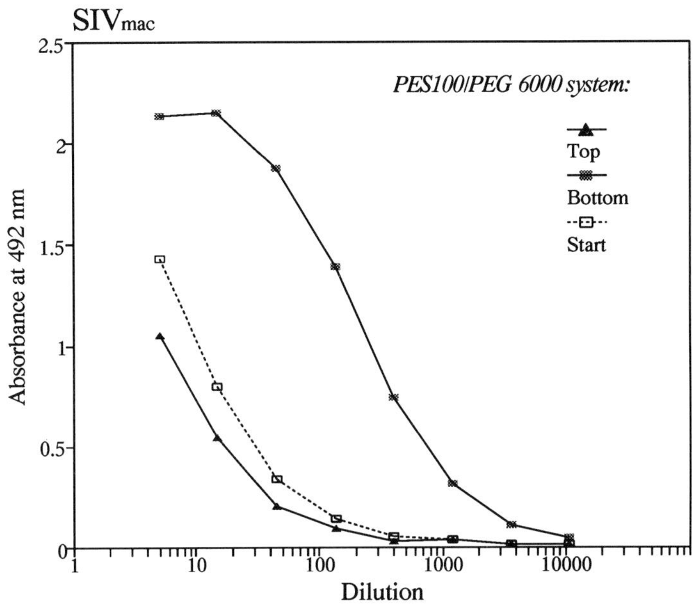


FIG. 4. Enzyme-linked immunosorbent assay data from a two-phase partitioning experiment with $SIV_{mac}$. The system contained 1.0% Reppal PES 100 (molecular weight ~100,000) and 7.0% PEG 6000. The phase system was obtained by mixing 20 ml supernatant solution (Start), from cultures of $SIV_{mac}$-infected C8166 cells, with 1 ml of 25% (w/w) PES 100, 3.5 ml of 50% (w/w) PEG 6000, and 0.5 ml of 100 m*M* Tris-HCl, 500 m*M* NaCl, pH 8.0. After centrifugation at 1000 *g* for 20 min at room temperature samples were taken from the top phase (Top) and the bottom phase (Bottom). The bottom phase volume was about 0.5 ml. For the gp148 ELISA, polystyrene microtiter plates coated with GNA (*Galanthus nivalis* agglutinin) were used to bind SIV gp148. The bound glycoprotein was quantified using a monoclonal antibody in a peroxidase-linked immunoassay. [Data from Gilljam, *AIDS Res. Hum. Retroviruses* **9,** 431 (1993).]

solution is added so as to obtain a final concentration of 0.7% dextran sulfate. The virus now again partitions to the bottom phase containing the dextran sulfate. The bottom phase is collected and the dextran sulfate precipitated by the addition of 3 *M* KCl (to about 2 mmol per milliliter bottom phase).

## *Concentration of Sewage Water Specimen for Virus Isolation*

Pöyry *et al.*[10] have devised the following procedure for screening sewage for cytopathic viruses. A 100-fold concentration is obtained.

*System.* The system contains 1% Dx T40 (Pharmacia, Uppsala, Sweden), 10% PEG 6000 (Fluka, Buchs, Switzerland), and 0.2 *M* NaCl.

*Procedure.* Mix 30 g of 20% Dx T40 (in water), 200.4 g of 30% PEG 6000, and 24 ml 5 *M* NaCl with the sewage specimen to make 600 ml total volume. Adjust the pH to 7–8 and agitate vigorously for 1 hr at room temperature. Transfer to a separatory funnel and leave overnight at 4°. Collect the bottom phase (2–3 ml) and a sample from the interface, extract with chloroform, and inoculate cell cultures.

### *Purification of Measles Virus*

Pavlenko *et al.*[8] obtained a 1500-fold purification with 32% recovery of measles virus as based on RNA transcriptase measurements. The virus is obtained in 1% of the original volume.

*System.* The system contains 0.2% Dx T500 (Pharmacia), 6.5% PEG 6000 (Serva), 10 m*M* sodium phosphate, and 150 m*M* NaCl, pH 7.5.

*Procedure.* Polymers are added to the culture medium containing virus to obtain the first phase system. The interface and bottom phase are collected and diluted to the original volume. More of the polymers are added in a quantity needed to obtain a second phase system. The virus is then recovered in the bottom phase.

### *Purification of Retroviruses*

HIV-1, HIV-2, and $SIV_{mac}$ have all been purified from 3 to 25 liters of culture media using the following protocol.[11–13] It is designed to give about a 300-fold concentration of the virus at the interface. This is obtained with Dx T500 at 0.20–0.24% (w/w) and PEG 6000 (E. Merck-Schuchardt, Darmstadt, Germany) or PEG 8000 (Union Carbide) at 7.0–7.22% (w/w), final concentrations.

### *Protocol for Two-Phase Extraction of Simian Immunodeficiency Virus from Culture Supernatant Solutions*

*Principle.* Mix culture medium containing virus with polymer stock solutions, preferably in a plastic bag as outlined in Fig. 3, and allow the phases to settle. Withdraw the bottom phase and the interface. Wash the virus from the interface with fresh top phase. Remove the polymers by centrifugation through a sucrose cushion[11] or by chromatography on Sephacryl S-500 or S-1000.[13] Results with the latter are depicted, using HIV-2, in Fig. 5.

*System.* The system contains 0.24% Dx T500 and 7.2% PEG 6000 (E. Merck-Schuchardt).

*Virus Material.* Seeding ampoules of simian immunodeficiency virus, $SIV_{mac}$ 251 32H isolate, were from the Center for Applied Microbiology and Research (Porton Down, UK). The virus is produced in roller bottle cultures (500 ml/flask) of the human T-cell line C8166. The growth medium is RPMI 1640 supplemented with 5% fetal bovine serum and 2 m*M* L-glutamine.

Cell-free supernatant solution containing virus is used. This is obtained by leaving the flasks containing cultures standing upright in the incubator overnight. The cells sediment, and 400 ml of the 500 ml culture fluid can easily be removed, with the aid of a peristaltic pump, as a cell-free solution. The remaining culture fluid is used after centrifugation at 800 *g* for 10 min at room temperature to remove the cells.

*Stock Solutions*

PBS (10×), Phosphate-buffered saline, 10-fold stock solution: 100 m*M* sodium phosphate, pH 7.4, 1.55 *M* NaCl

10% Dx: For 100 g, mix 10 ml PBS (10×) and 80 ml water and heat to 100° in a water bath; add 10 g Dx T500 while stirring until it is completely dissolved

50% PEG: For 1000 g, mix 100 ml PBS (10×) and 400 ml water and heat to 100° as above, then add 500 g PEG 6000 while stirring until completely dissolved; store at room temperature and use within 1 week

8% PEG: Prepare fresh from the 50% PEG stock and PBS and add dithiothreitol (DTT) to 1 m*M*

30% and 60% Sucrose: Make up in 10 m*M* Tris-HCl, pH 8.0, 155 m*M* NaCl, 1 m*M* DTT

Phase system mix for each bag (4000 g total weight): Mix 3330 ml medium containing virus, 96 g of 10% Dx, and 576 g of 50% PEG

*Equipment*

Plastic bags (5-liter) with attached tubing (Transfer bags, Baxter Medical AB, Bromma, Sweden)

Peristaltic pump, allowing a flow of about 500 ml/min, with sterile tubing of appropriate strength and dimensions and secured fittings

Balance, 4000 g capacity, for weighing the bags while filling them

Support for the bags in the lamellar flow hood (most hoods have a T profile around the ceiling; suspend a stainless steel rod, about 1 cm in cross section, over the T profile as support for S hooks to hold the bags)

*Procedure*. Work with SIV is performed under Biosafety level 3 conditions.[24a] Prepare the solutions and weigh out the amounts of stock solutions needed for each bag in polypropylene flasks (under sterile conditions). Make sure all solutions, including the virus, are equilibrated to the same

[24a] U.S. Department of Health and Human Services, Biosafety in Microbiological and Biomedical Laboratories, 1st ed., HHS Publication no. (CDC) 84-8395, U.S. Government Printing Office, Washington, D.C. March (1984).

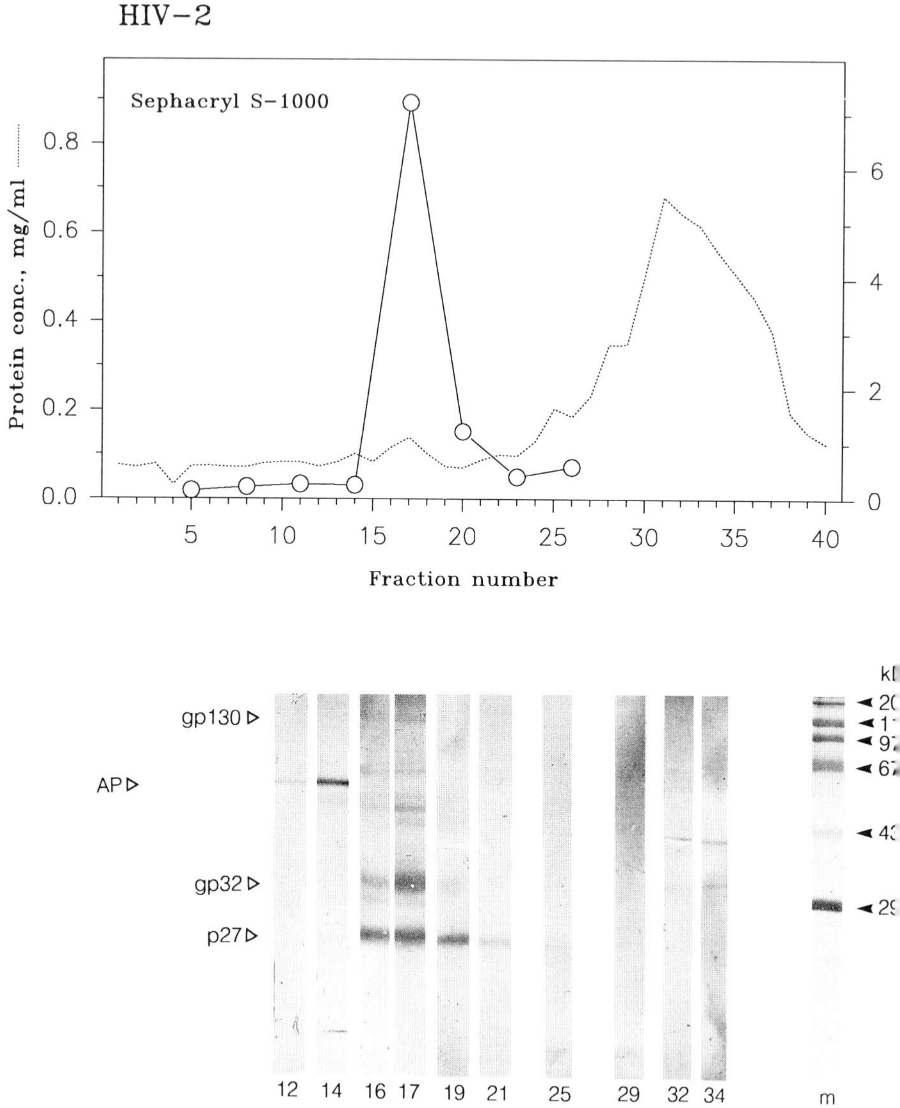


FIG. 5. Removal of polymers by chromatography on Sephacryl S-1000. The sample, HIV-2, was concentrated by two-phase extraction in the system 0.2% dextran T500–7.0% PEG 6000 and recovered at the interface. It was diluted 3-fold with 10 m*M* sodium phosphate, 155 m*M* NaCl, pH 7.5, and applied to a Sephacryl S-1000 column (1.4 cm diameter, total volume 110 ml; sample volume was 3 ml) in the same buffer. The flow was adjusted by hydrostatic pressure to 10–12 ml/hr. Fractions of 3 ml were collected and assayed for reverse transcriptase (RT) activity and protein content. The presence of other viral proteins under

temperature (room temperature). Arrange bags, balance, and pump in a lamellar flow hood. Attach the pump inlet via a silicone tubing to a 10 ml sterile pipette (with cotton plug removed) and the outlet to the plastic bag with secured fittings.

Transfer about half the amount of the medium containing virus to the bag. Add Dx stock solution, some more of the virus medium, so that the two polymers do not mix in the tubing, and then the PEG stock solution. Rinse the tubing with the remaining amount of virus medium (and/or PBS). Assure that all PEG solution has been completely introduced into the bag and that the total weight of the system is 4000 g. Agitate the mixture while filling the bag by tapping. Finally secure the tubing of the bag with a clamp. Disconnect it from the pump and cover the fittings securely. Turn the bag upside down a couple of times and leave it hanging for 4 hr, or overnight, to allow the phases to settle. Material from a 25-liter culture will make 8 bags and is a suitable quantity to work up in a half-day period.

Collect the bottom phase and the interface into separate Falcon tubes. Wash the interface containing virus twice with equal volumes of 8% PEG and centrifuge at 800 *g* for 10 min at room temperature to sharpen the band at the interface. The interface represents about 0.3% of the initial culture fluid volume.

For removal of polymers, suspend the final interface material, from a 25-liter preparation, in PBS with 1 m*M* DTT to a final volume of 120 ml. The material may be viscous if the bottom phase is not fully separated from the interface. In such a case suspend the material in more PBS–1 m*M* DTT before proceeding. The virus is freed from polymers and soluble proteins by centrifugation through a 30% sucrose cushion. Layer 3 ml of 60% sucrose under 15 ml of 30% sucrose and on top of that add 12 ml of the interface suspension in a centrifuge tube (30 ml). Centrifuge at 12,000 *g* for 16 hr at 4°. Collect the virus on top of the 60% sucrose cushion.

*Results.* The SIV accumulates in the interface, as demonstrated by reverse transcriptase (RT) assay (Fig. 6A), and enzyme-linked immunosorbent assay (ELISA) for *env* gp148 and for *gag* proteins. The RT recovery in the interface fraction is about 80% and that of gp148 about 25%. This recovery is maintained after repeated extractions by additions of 8% PEG solution to the interface and centrifugation through a 20% sucrose layer.

---

the RT peak was confirmed in Western blots probed with an HIV-2-positive human serum (bottom; lanes are numbered according to the fractions). AP is a peptide with phosphatase activity and indicates a vesicle in fractions 12 and 14 derived from cells. [From G. Gilljam, S. Eriksson, J. Hinkula, and L. Hammar, in preparation (1993).]

A

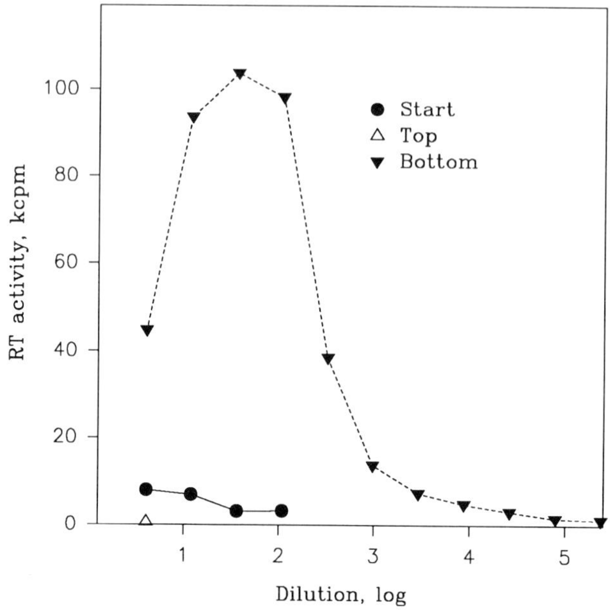


FIG. 6. (A) Reverse transcriptase assayed on samples from a two-phase extraction experiment with $SIV_{mac}$. The system contained 0.22% dextran T500 and 7.0% PEG 6000. The two-phase extraction was done according to the procedure described in the text for the concentration and purification of SIV. The reverse transcriptase (RT) assay was by the method of L. Vrang, H. Bazín, G. Remaud, and J. Chattopadhyaya, *Antiviral Res.* **7,** 139 (1987), as modified in L. Hammar and G. Gilljam, *AIDS Res. Hum. Retroviruses* **6,** 1379 (1990). (B) Effect of some polymers on reverse transcriptase activity. A constant amount of a recombinant HIV-1 reverse transcriptase was mixed with PEG 6000, dextran T500, or dextran sulfate to obtain the concentrations indicated. Aliquots were assayed as described under (A).

## Effect of Polymers on Virus Detection and Analyses

### *Effect on Virus–Cell Interaction*

Tests based on the effect of viruses on cells in culture are basic in virology. Dextran sulfate has been reported to severely inhibit plaque formation and cytopathic effects of several enteroviruses on primary rhesus monkey kidney (PMK) cells.[14,25] In contrast, Dx increased, or had no effect on, plaque formation and was moderately inhibitory, depending on type and strain of virus, in tests for cytopathic effects.[14] On testing

[25] J. Grindrod and D. O. Cliver, *Arch. Gesamte Virusforsch.* **28,** 337 (1969).

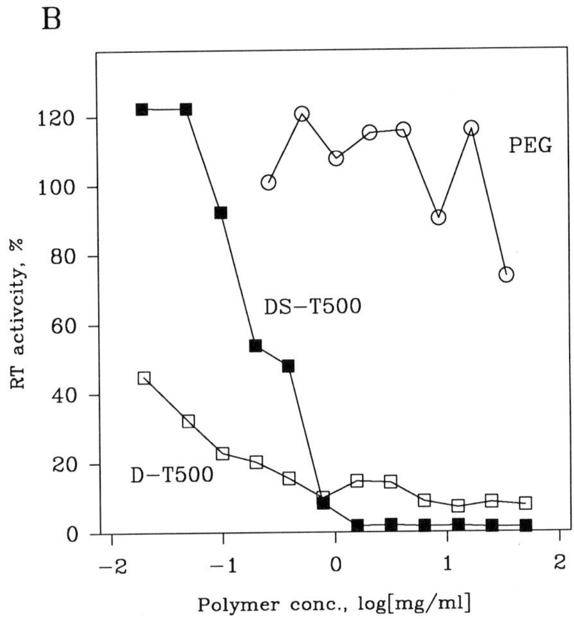


FIG. 6. *(continued)*

infectivity of HIV using the human lymphocyte line H9, we found dextran sulfate (molecular weight 500,000) to be toxic to the cells down to 400 $\mu$g/ml.[12] The polymer blocked HIV-1 infection at concentrations down to or below 1 $\mu$g/ml in accord with the literature. Mitsuya et al.[26] demonstrated a suppressive effect on HIV of dextran sulfate (molecular weight 8000) at 10 $\mu$g/ml, which concentration did not affect the viability of the cells. This phenomenon applies to some other sulfated polymers as well. It seems to involve regions in the HIV gp120 molecule interacting with a cell receptor other than CD4.[27] Dx and PEG were not harmful to the H9 cells at 1 mg/ml or below as evidenced by microscopy. The enhancement of viral infectivity by DEAE-Dx is well documented[28–30] and has been utilized to increase the sensitivity in plaque-forming assays.[31,32]

[26] H. Mitsuya, D. J. Looney, S. Kuno, R. Ueno, F. Wong-Staal, and S. Broder, *Science* **240,** 646 (1988).

[27] L. N. Callahan, M. Phelan, M. Mallinson, and M. A. Norcross, *J. Virol.* **65,** 1543 (1991).

[28] J. S. Pagano and J. S. McCutchan, *Prog. Immunobiol. Stand.* **3,** 152 (1969).

[29] K. Sasaki, T. Furukawa, and S. A. Potkin, *Proc. Soc. Exp. Biol. Med.* **166,** 281 (1981).

[30] C. R. Rossi and G. K. Kiesel, *Arch. Virol.* **56,** 227 (1978).

[31] C. Wallis and J. L. Melnick, *J. Virol.* **2,** 267 (1968).

[32] J. C. Booth, *Arch. Virol.* **55,** 251 (1977).

### *Assay for Reverse Transcriptase Activity*

When samples from two-phase extractions are assayed for RT activity, the dilution curve often has the appearance shown in Fig. 6A. A blocking effect disappears at high dilution. The generally harmless and inert Dx T500 was found to inhibit the RT to an extent exceeding the effect of the negatively charged dextran sulfate (Fig. 6B). PEG showed only a modest effect on the test system.

### *Dot Blot Analysis*

The dot blot technique is simple and useful for fast analysis of viral proteins. It is usually more sensitive than a direct ELISA. Polymers like dextran sulfate, PVA, or methyl cellulose at high concentrations block the binding of the virus to the membrane. This effect may be overcome by dilution.[12,15]

### *Enzyme-Linked Immunosorbent Assay*

Several polymers compete with the virus for binding to the polystyrene surface. Therefore a capture ELISA is more reliable. The plates are then coated with a capturing molecule like an antibody or a suitable lectin. The lectin from *Galanthus nivalis* (GNA) binds terminal mannose residues.[33] These are present on several retroviruses but not, usually, on antibodies of the IgG type, thus providing ELISAs for the quantitation of HIV and SIV and their external glycoproteins[11,13,34,35] (see Fig. 4).

### *Electron Microscopy*

Most polymers block the binding of the virus to the grids. The solutions may also be too viscous to permit easy penetration of negative contrast media. For embedding and sectioning the presence of Dx and PEG in the sample causes no problem when conventional fixing and contrast media are used (Fig. 1).

### *Polyacrylamide Electrophoreses*

Charged polymers in the samples should be removed in order not to distort fractionation. Dextran sulfate can be removed from the sample by

[33] N. Shibuya, I. J. Goldstein, and E. J. M. Van Damme, *J. Biol. Chem.* **263,** 729 (1988).
[34] B. E. Weiler, H. Schacke, M. Bachmann, L. Brigido, M. Gilbert, J. Mills, E. Matthes, J. M. Forrest, and W. E. Müller, *J. Virol. Methods* **32,** 287 (1991).
[35] M. Mahmood and A. J. Hay, *J. Immunol. Methods* **151,** 9 (1992).

precipitation with potassium or barium ions.[20] An excess of these ions will precipitate sodium dodecyl sulfate.

## Comments

The suppressive or stimulating effects of the polymers on cell systems have discouraged microbiologists from using two-phase extraction for purification of viral material. This holds in particular for material to be used *in vivo*. However, polymer effects have also been explored. PEG has been coupled to potential immunogens to obtain tolerance during therapy.[36,37] In contrast, some charged Dx derivatives have been used as potential adjuvants for immunization. These include dextran sulfate[38] and DEAE-Dx, the latter used with foot-and-mouth disease virus[39,40] and VEE virus.[41]

Affinity partitioning with biospecific ligands immobilized on phase polymers have been used for membrane vesicle purification. The different membrane systems in the cell differ in their glycosylation. With the lectin wheat germ agglutinin covalently attached to Dx and introduced in a Dx–PEG two-phase system, Persson and Jergil[42] purified rat liver plasma membranes. A similar approach can be used for the purification of enveloped viruses and viral glycoproteins. The use of lectins with selective affinity for particular carbohydrate structures should open a new field in the purification of enveloped viruses and viral glycoproteins by affinity partitioning.

Despite the theoretical complexity of aqueous two-phase systems, a few well-characterized systems can be used for the concentration of most viruses without or with only minor modifications. Two-phase extraction methods are competitive with other techniques for virus concentration and purification. Ultracentrifugation of large volumes can be replaced by two-phase extraction, a great advantage when working under the constraint of sterility or with hazardous infectious material. It is possible to work on a scale from less than a milliliter up to tens of liters in a normally equipped virus laboratory.

[36] A. H. Sehon, *Adv. Exp. Med. Biol.* **251,** 341 (1989).
[37] S. Dreborg and E. B. Åkerblom, *Crit. Rev. Ther. Drug Carrier Syst.* **6,** 315 (1990).
[38] J. W. B. Bradfield, R. L. Souhami, and J. L. Addison, *Immunology* **26,** 383 (1974).
[39] E. C. Anderson, R. C. Master, and G. N. Mowat, *Res. Vet. Sci.* **12,** 351 (1971).
[40] G. Wittmann, B. Dietzschold, and K. Bauer, *Arch. Virol.* **47,** 225 (1975).
[41] W. E. Houston, C. L. Crabbs, and R. J. Kremer, *Infect. Immunol.* **13,** 1559 (1976).
[42] A. Persson and B. Jergil, this volume [47].

## Acknowledgments

This study was supported by grants from the Swedish Medical Research council (B93-16H-09932 and K92-16H-10236), the Swedish Cancer Society (3205-B91), and the EC concerted research programme, European vaccine against AIDS, Programme EVA.

# [64] Extraction of Chromatophores

*By* ANNA-LISA SMEDS and SVEN-OLAF ENFORS

## Introduction

In this chapter we describe partitioning of chromatophores, which are membrane vesicles from photosynthetic bacteria. Separation of a variety of membrane vesicles in dextran (Dx)–poly(ethylene glycol) (PEG) aqueous two-phase systems has been described earlier.[1] Transfer of vesicles to either top or bottom phase can be achieved by inclusion of salts and small amounts of polymer derivatives in the phase system. This methodology can be utilized for the isolation of chromatophores from a crude extract of disintegrated cells.[2] A semicontinuous ATP-regeneration reactor has also been suggested, in which photophosphorylation by the chromatophores occurs in one of the phases and the reactants distribute through the whole emulsion.[3]

## Methods

### *Cultivation, Harvest, and Disintegration*

*Rhodospirillum rubrum* is cultivated in a medium according to Ormerod *et al.*[4] in a completely sealed 3-liter bottle or, on a larger scale, in a 50-liter light fermenter as described earlier.[5] The light intensity is increased during the course of the cultivation to give a high photomembrane productivity. The cells are harvested by centrifugation and then prepared according to the following alternatives. (A) The cells are washed in 0.2 *M* glycylgly-

[1] C. Larsson, B. Andersson, and H.-E. Åkerlund, *in* "Partitioning in Aqueous Two-Phase Systems" (H. Walter, D. E. Brooks, and D. Fisher, eds.), p. 497. Academic Press, Orlando, Florida, 1985.

[2] A.-L. Smeds and S.-O. Enfors, *Enzyme Microb. Technol.* **7,** 601 (1985).

[3] A.-L. Smeds, A. Veide, and S.-O. Enfors, *Enzyme Microb. Technol.* **5,** 33 (1983).

[4] J. G. Ormerod, K. S. Ormerod, and H. Gest, *Arch. Biochem. Biophys.* **94,** 449 (1961).

[5] A.-L. Smeds and S.-O. Enfors, *Appl. Microbiol. Biotechnol.* **22,** 83 (1985).

cine buffer, pH 7.5, and then suspended in the same buffer also containing a small amount of DNase and RNase. The suspension is sonicated 2 times for 30 sec each at 60 W (Sonifier B-12, Branson Sonic Power Company, Danbury, CT). In alternative (B), the cells are suspended in sodium phosphate, 0.1 mol/kg, pH 7.5, to a concentration of 50% (w/w), subsequently frozen as droplets in liquid nitrogen, and stored at −70°. After thawing, the cells are washed in the same buffer, then homogenized in disintegration buffer (Ultraturrax, Ika-werk, Staufen, Germany). Disintegration is performed in a glass bead mill (Dyno Mill, W. A. Bachofen AG, Basel, Switzerland) or in a high-pressure homogenizer (French Press FA 073, Aminco, Urbana, IL).

### *Isolation of Chromatophores by Centrifugation*

Centrifugation of the disintegrated cells is done at 0–5° in one of two ways: (A) first for 30 min at 18,000 *g* for the removal of cell debris, then for 60 min at 66,000 *g* for the recovery of the chromatophores; or (B) first for 30 min at 20,000 *g* for the removal of the debris and finally for 60 min at 70,000 *g* for the sedimentation of chromatophores.

### *Estimation of Bacteriochlorophyll*

Estimation of the amount of bacteriochlorophyll (Bchl) is made according to Pace *et al.*[6] by extraction in acetone–methanol (7 : 2, v/v) followed by absorbance measurements at 772 nm using a Perkin-Elmer (Norwalk, CT) 124 spectrophotometer (an extinction coefficient of 75 $mM^{-1}$ $cm^{-1}$ is used) or, in alternative (B) by direct absorbance measurement at 880 and 820 nm, using a Perkin-Elmer Lambda 3 spectrophotometer. The Bchl molarity is calculated from the difference of these absorbances and the *in vivo* $\varepsilon_{Bchl}$ value of 140 $mM^{-1}$ $cm^{-1}$ according to Clayton.[7] The reaction medium for photophosphorylation contains sodium phosphate, 50 mmol/kg; fumaric acid, 1 mmol/kg; $MgCl_2$, 10 mmol/kg; ascorbic acid, 17 mmol/kg; and glucose, 17 mmol/kg; pH 7.8.

### *Aqueous Two-Phase Extraction*

Phase diagrams and phase system binodals are determined according to Albertsson.[8] PEG palmitate (P-PEG) is synthesized according to the

[6] G. W. Pace, Ho. Seung Yang, S. R. Tannenbaum, and M. C. Archer, *Biotechnol. Bioeng.* **18,** 1413 (1976).

[7] R. K. Clayton, in "Bacterial Photosynthesis" (H. Gest, A. San Pietro, and L. P. Vernon, eds.), p. 495. Antioch Press, Yellow Springs, Ohio, 1963.

[8] P.-Å. Albertsson, "Partition of Cell Particles and Macromolecules," 3rd Ed. Wiley (Interscience), New York, 1985.

method of Shanbhag and Johansson[9] (esterification of palmitoyl chloride with triethylamine as proton acceptor). PEG is obtained from Merck (Darmstadt, Germany) and Dx and DEAE-Dx from Pharmacia-LKB (Uppsala, Sweden). The interfacial tension is measured at 25° in a spinning drop apparatus (SITE, Krüss, Germany). The Dx phase is used as the continuous phase.

Aqueous two-phase extraction is performed by mixing the crude extract or chromatophores prepared by centrifugation, as indicated, with stock solutions of phase-forming components, salts, and water in glass tubes by end-over-end rotation for 15 min at 20 rpm at the temperature indicated. After centrifugation at 1400 *g* for 15 min, samples are withdrawn from the bottom phase using a Pasteur pipette.

On a larger scale, the components are mixed in a stirring tank for 15 min at 0–2°. Phase separation is performed using a tubular bowl centrifuge (CEPA Labor CLE, Carl Padberg GmbH, Lahr, Germany). The emulsion is pumped at a flow rate of 9 liters/hr into a previously chilled 0.25-liter purification rotor rotating at 10,000 rpm (2400 *g*). The proportions of top and bottom flow rates are varied by choosing the diameter of a flow restriction hole in an exchangeable end piece of the rotor. The chromatophore-containing (top) phase is dispensed into small tubes and stored at −15 to −20°.

## Results

In a PEG–Dx system the chromatophores can be partitioned to the bottom or top phase by inclusion of polymer derivatives. Table I shows that in a sodium phosphate-buffered system (top phase positive relative to bottom phase) the partition of the chromatophores was changed from the top to the bottom phase when DEAE-Dx (3% N content) was included in the phase system at a final concentration of 0.1%. No such shift was observed in a glycylgycine-buffered system. It is also shown that at a higher concentration of phase-forming polymers (i.e., at a higher interfacial tension), the chromatophores were partly adsorbed to the interface. At lower polymer concentrations the phase system is close to the binodal, providing the temperature is kept at 0–2°, whereas the higher polymer concentrations correspond to a position in the phase diagram more distant from the binodal.

In reaction medium the partitioning of the chromatophores was slightly different (see Table II). At a PEG–Dx composition of 6.25–6.25% in reaction medium and a temperature of 20–22° the chromatophores partitioned to a large extent to the interface despite the fact that this polymer

[9] V. P. Shanbhag and G. Johansson, *Biochem. Biophys. Res. Commun.* **61,** 1141 (1974).

TABLE I
DISTRIBUTION OF CHROMATOPHORES IN POLY(ETHYLENE GLYCOL) 4000–DEXTRAN T500 PHASE SYSTEMS[a]

| Phase system[b] | | | | | Distribution of chromatophores: Phase[c] | | |
|---|---|---|---|---|---|---|---|
| PEG (%, w/w) | Dextran (%, w/w) | Buffer | Buffer concentration (mmol/kg) | DEAE-dextran (%, w/w) | T | I | B |
| 5.4 | 5.4 | Sodium phosphate | 5 | 0 | +[d] | | |
| 5.4 | 5.4 | Sodium phosphate | 5 | 0.1 | | | + |
| 5.4 | 5.4 | Sodium phosphate | 25 | 0 | + | | |
| 5.4 | 5.4 | Sodium phosphate | 25 | 0.1 | | | + |
| 5.4 | 5.4 | Glycylglycine | 5 | 0 | + | | |
| 5.4 | 5.4 | Glycylglycine | 5 | 0.1 | + | | |
| 5.4 | 5.4 | Glycylglycine | 25 | 0 | + | | |
| 5.4 | 5.4 | Glycylglycine | 25 | 0.1 | + | | |
| 6.4 | 6.4 | Sodium phosphate | 5 | 0 | + | (+)[e] | |
| 6.4 | 6.4 | Sodium phosphate | 5 | 0.1 | | + | (+) |
| 6.4 | 6.4 | Sodium phosphate | 25 | 0 | + | + | |
| 6.4 | 6.4 | Sodium phosphate | 25 | 0.1 | | + | (+) |
| 6.4 | 6.4 | Glycylglycine | 5 | 0 | (+) | + | (+) |
| 6.4 | 6.4 | Glycylglycine | 5 | 0.1 | (+) | + | |
| 6.4 | 6.4 | Glycylglycine | 25 | 0 | (+) | + | (+) |
| 6.4 | 6.4 | Glycylglycine | 25 | 0.1 | (+) | + | |

[a] Chromatophores prepared according to method (A) described in the text (disintegration and centrifugation). From A.-L. Smeds, A. Veide, and S.-O. Enfors, *Enzyme Microb. Technol.* **5**, 33 (1983).
[b] All systems also contained sucrose, 0.2 mol/kg. The temperature was 0–2° and the pH 7.5.
[c] T, Top, PEG-rich phase; I, interface; B, bottom, dextran-rich phase.
[d] + indicates the main part of the chromatophores.
[e] (+) indicates small amounts of chromatophores.

concentration is close to the binodal at that temperature. However, if the hydrophobic PEG derivative PEG palmitate was included at a concentration of 0.2%, all the chromatophores were recovered in the top phase. The simultaneous inclusion of DEAE-Dx and PEG palmitate in the phases resulted in partitioning among top and bottom phase and the interface.

### *Isolation of Chromatophores from a Crude Extract of Disintegrated Cells*

A crude extract of *Rhodospirillum rubrum* cells disintegrated in a glass bead mill was mixed with phase system constituents. Table III shows that by inclusion of sodium chloride to a final concentration of 50 m*M*, the

TABLE II
DISTRIBUTION OF CHROMATOPHORES IN REACTION MEDIUM[a]

| Polymer composition | Phase volume ratio (T/B) | Recovery (% of totally added) T | I | B |
|---|---|---|---|---|
| PEG 6.25%, dextran 6.25% | 1 | 36 | 64 | — |
| PEG 6.25% including 0.2% PEG palmitate, dextran 6.25% | 1 | 120 | — | — |
| PEG 6.25%, dextran 6.25% including 0.8% DEAE-dextran | 1 | 10 | 34 | 56 |

[a] Conditions: chromatophore concentration 10 $\mu$mol Bchl/kg, pH 7.8, 20–22°. Symbols as in Table I. Chromatophores were prepared and Bchl measured according to method (A) described in the text (disintegration, centrifugation, and absorbance measurement). From A.-L. Smeds, A. Veide, and S.-O. Enfors, *Enzyme Microb. Technol.* **5,** 33 (1983).

partition coefficient of materials absorbing at 260 and 280 nm decreased. Also Bchl-containing material was transferred to some extent to the bottom phase. However, after inclusion of PEG palmitate at a final concentration of 0.2%, no Bchl partitioned to the bottom phase. The Bchl recovery was 60% in the top phase, indicating that some material adhered to the

TABLE III
PARTITION OF COMPONENTS OF A DISINTEGRATED CELL SUSPENSION OF *Rhodospirillum rubrum* IN AQUEOUS TWO-PHASE SYSTEMS

| Composition[a] Sodium chloride mmol/kg | P-PEG (%) | Recovery of Bchl[b] in the top phase (%) | Partition coefficient 280 nm | 260 nm | Purity of chromatophores ($\mu M$ Bchl/$A_{280}$) | Purification ratio[c] |
|---|---|---|---|---|---|---|
| 0 | 0 | 35 | 2.6 | 1.9 | 1.7 | 0.9 |
| 0 | 0.2 | 75 | 3.1 | 2.0 | 3.4 | 1.8 |
| 50 | 0 | 13 | 0.8 | 0.6 | 2.0 | 1.1 |
| 50 | 0.2 | 60 | 1.8 | 1.2 | 4.3 | 2.3 |

[a] All phase systems contained PEG 4000, 5.1%; dextran T40, 9.5%; sodium phosphate, 5 mmol/kg, pH 7.5; and components indicated. The temperature was 0–2°. For disintegration, frozen stored cells were disintegrated in a glass bead mill for 2 min at 2000 rpm with 0.1–0.2 mm beads in a 5 mmol/kg sodium phosphate phase system. From A.-L. Smeds and S.-O. Enfors, *Enzyme Microb. Technol.* **7,** 601 (1985).

[b] Bchl was estimated according to method (B) described in the text.

[c] The purification ratio was expressed as $\mu M$ Bchl/$A_{280}$ in the top phase over $\mu M$ bchl/$A_{280}$ in the crude extract.

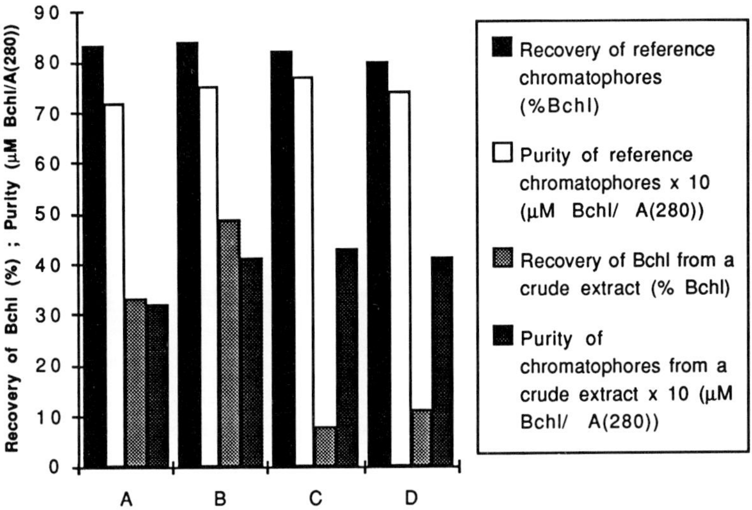


FIG. 1. Recovery and purity of chromatophores in the top phase after partitioning of disintegrated cells or reference chromatophores obtained by differential centrifugation. All systems included sodium chloride, 50 mmol/kg; PEG palmitate, 0.2%; sodium phosphate, 5 mmol/kg, pH 7.5; and biological material corresponding to 20 g cells (wet weight)/kg phase system. Temperature, 0–2°. Polymers were included to give final concentrations of (A) 4.5% PEG 4000 and 7.0% dextran T500, (B) 5.1% PEG 4000 and 9.5% dextran T40, (C) 2.5% PEG 10,000 and 5.5% dextran T500, and (D) 3.4% PEG 10,000 and 7.5% dextran T40. The suspension of disintegrated cells was prepared as described in Table III. [From A.-L. Smeds and S.-O. Enfors, *Enzyme Microb. Technol.* **7,** 601 (1985).]

interface. UV-absorbing compounds from the crude extract, like nucleic acids, and proteins, partitioned to the bottom phase, the chromatophores to the top phase, and cell debris to the interface.

The distributions of Bchl-containing material from a crude extract were compared with chromatophores prepared by differential centrifugation ("reference chromatophores") in four different phase systems, in all of which the polymer concentrations were adjusted to correspond to the same positions relative to the binodals. Figure 1 shows that the recovery of chromatophores from a crude extract was very different in different two-phase systems, whereas the reference chromatophores were in the top phase in all systems. In the systems containing PEG 10,000 the chromatophore recovery from crude extract was very low compared to the systems containing PEG 4000. The purity of the chromatophores, expressed as micromolar Bchl per $A_{280}$ unit, was considerably lower after extraction of crude extract compared to reference chromatophores. However, it did not change appreciably with phase system used.

TABLE IV
INFLUENCE OF PEG PALMITATE ON INTERFACIAL TENSION IN TWO-PHASE SYSTEMS OF PEG AND INCREASING DEXTRAN CONCENTRATIONS

| Composition[a] | | | |
|---|---|---|---|
| PEG 4000 (%, w/w) | Dextran T500 (%, w/w) | P-PEG in top phase (%, w/w) | Interfacial tension (mN/m) |
| 5.5 | 7 | 0 | $9.9 \times 10^{-3}$ |
| 5.5 | 7 | 0.4 | $11.0 \times 10^{-3}$ |
| 5.5 | 8.5 | 0 | $17.0 \times 10^{-3}$ |
| 5.5 | 8.5 | 0.4 | $20.0 \times 10^{-3}$ |
| 5.5 | 10 | 0 | $36.0 \times 10^{-3}$ |
| 5.5 | 10 | 0.4 | $40.0 \times 10^{-3}$ |

[a] All systems also contained 5 mmol/kg sodium phosphate, pH 7.5, and 0.2 mol/kg sucrose at 20°. Using these polymer concentrations a top/bottom volume ratio of approximately 1 was obtained. From A.-L. Smeds and S.-O. Enfors, *Enzyme Microb. Technol.* **7**, 601 (1985).

Furthermore, the partitioning of the debris (20,000 *g*) fraction and the particle-free 70,000 *g* supernatant solution were studied separately (not shown in Fig. 1). The debris was confined to the interface, whereas the partition coefficient for 260 and 280 nm absorbing material was 0.15–0.3 in all four phase systems.

For isolation of chromatophores on a larger scale the crude extract of disintegrated cells was first centrifuged at 20,000 *g* for 30 min at 0–5°. The debris-free suspension was mixed with constituents of a phase system designed to give a small top phase containing chromatophores at high concentration and a large bottom phase: PEG 4000, 2.2% (w/w); Dx T500, 14% (w/w); sodium chloride, 50 mmol/kg; PEG palmitate, 0.2% (w/w); sucrose, 0.2 mol/kg; and sodium phosphate, 5 mmol/kg, pH 7.5. Temperature was 0–2°. The concentration of crude extract corresponded to 20 g bacteria (wet weight)/kg phase system. The emulsion was loaded into a CEPA centrifuge at a flow rate of 9 liters/hr. The resulting phase flow rates were 8% for the top and 92% for the bottom phase. The contamination of the top phase with bottom phase was 9%.

The influence of PEG palmitate on the interfacial tension was studied in three PEG/Dx systems at different positions relative to the binodal. Addition of PEG palmitate up to a final concentration of 0.4% (w/w) did not change the interfacial tension appreciably (Table IV).

## Discussion

In an aqueous two-phase extraction the main separation criteria for particles are their surface properties. The partition results in Table I are typical for negatively charged particles. Table II shows that the addition of a small amount of the hydrophobic polymer derivative PEG palmitate had a substantial influence on the partitioning. Because different surface domains can be used as targets for different polymer derivatives, aqueous two-phase system extraction provides a versatile tool for the separation of particles exposing complicated structures, like chromatophores.

For large-scale isolation of chromatophores aqueous two-phase extraction is advantageous because only low gravitational forces are required to effect separation. Extraction of a crude extract of disintegrated cells is shown in Table III and Fig. 1. When the separate fractions, obtained by differential centrifugation, were extracted, the debris fraction (20,000 *g*) was partitioned to the interfaces and the chromatophore fraction (70,000 *g*) to the top phases of all sytems tested. However, extraction of whole crude extract resulted in different recoveries in different phase systems. One reason for that might be aggregation of the biological material. Larger particles often partition to the interface, and their partitioning is also more sensitive to small changes in interfacial tension.[8] In addition, a large quantity of biological material in a phase system contributes to phase formation.[10] Therefore, an isolation procedure including extraction of disintegrated cells may become sensitive both to small changes in polymer composition and to concentration and status of the biological material. However, when a debris-free fraction (20,000 *g* supernatant solution) was extracted in a two-phase system the process was more robust.

[10] K. Köhler, L. von Bonsdorff-Lindeberg, and S.-O. Enfors, *Enzyme Microb. Technol.* **11,** 730 (1989).

# [65] Extraction of Proteins from Sources Containing Tannins and Anionic Mucilages

*By* HANS VILTER

## Introduction

### *Extraction of Plant Enzymes*

The isolation of plant enzymes is attended by several problems. The disruption of plant tissue is difficult because of the rigidity of the cell walls. The rigidity is due to cellulose fibers which are embedded into a matrix of phycocolloids like pectins, carrageenans, agar, and alginates. Phycocolloids are used as gel-forming or thickening agents and emulsifiers (e.g., in food and pharmaceutical products). The viscosity of extracts from plant materials rich in phycocolloids is high. Solutions of high viscosity hinder chromatographic processes, adding to the difficulty of using liquid–solid separations. The presence of anionic polysaccharides may convert an anion exchanger into a mixed-bed ion exchanger, thus causing still more problems during the separation procedure. Moreover, cell wall vacuoles contribute to the bulk of most plant tissues, and plant vacuoles contain a large number of secondary plant products which can affect the stability of enzymes during extraction. Most of the latter problems are caused by the presence of a large variety of phenolic compounds.[1,2] Phenolic compounds can modify and even denature a number of proteins. Additional pigments are frequently located in the plastids (e.g., lipophilic compounds such as chlorophylls). Such compounds attach, through hydrophobic interaction, to proteins during the extraction process. These problems can be avoided as far as possible by choosing material with a low content of slime, pigments, and phenols.

No general method for the isolation of enzymes from plant material can be given. However, there is a general approach. The tissues are disrupted in the extraction medium, which is usually an aqueous buffer. Additives are used to stabilize or to solubilize the enzymes of interest.[3,4] Several phenol scavengers are known; primarily insoluble poly(vinylpyrrolidone) (trade name, POLYCLAR AT, Serva, Heidelberg, Germany) is used for this purpose. Several liquid–solid separation steps are then used. The presence of insoluble poly(vinylpyrrolidone) makes the homogenization medium containing phycocolloids reminiscent of soft cheese. The extract is obtained by pressing the homogenate through muslin. Scaleup of the process is, therefore, limited.

### *Use of Aqueous Two-Phase Systems*

The advantage of applying aqueous two-phase systems[5] to the purification of enzymes from plants has been described. The extraction of plant polyphenol oxidases using temperature-induced phase separation of Triton

[1] J. B. Harborne (ed.), "Methods in Plant Biochemistry" (Plant Phenolics), Vol. 1. Academic Press, London, 1989.

[2] M. A. Ragan and K.-W. Glombitza, *in* "Progress in Phycological Research" (F. E. Round and D. J. Chapman, eds.), Vol. 4, p. 129. Biopress, Bristol, England, 1986.

[3] W. D. Loomis, this series, Vol. 31, p. 528.

[4] P. Gegenheimer, this series, Vol. 182, p. 174.

[5] P.-Å. Albertsson and F. Tjerneld, this volume [1].

X-114 (TX-114) (Fluka, Buchs, Switzerland) solutions is feasible.[6,7] Bromoperoxidases were isolated from brown algae using two-phase systems generated by poly(ethylene glycol) (PEG) (Hüls, Marl, Germany) and salt solutions.[8,9] Cytochromes were obtained as a byproduct. The bromoperoxidases belong to a novel class of nonheme peroxidases with vanadium(V) as the prosthetic group and were first discovered in the brown alga *Ascophyllum nodosum*.[10] At first[11–14] peroxidases from brown algae were isolated in line with the general approach for plant enzyme purification outlined above. However, this procedure did not yield enzymes in sufficient quantity or quality for applications and for structure elucidation.

## Process Development

Partitioning and stability of the enzymes were studied in several PEG–dextran (Dx) (Pfeiffer & Langen, Dormagen, Germany) as well as PEG–salt systems using volumes from 1 to 5 ml. Reducing compounds in the algal material and required additives interfere with the peroxidase assay. Therefore pilot experiments were performed using purified peroxidases isolated by conventional means.[15] In the absence of interfering compounds the peroxidase activity could be assayed directly in the phases. Based on the results from these pilot experiments the most promising PEG–salt systems were selected.

Potassium phosphate, ammonium sulfate, potassium citrate, and potassium carbonate were efficient in causing peroxidases to partition to the top phase. These salts provide a broad pH range. Depending on the pH and on the salt used the peroxidases were extracted as either active holoenzymes or inactive apoenzymes. However, the inactivation of peroxidases resulting from treatment with potassium phosphate or potassium carbonate, up to pH 12, was reversible. Transfer of the apoenzymes into a Tris-buffer containing vanadate reconstituted the holoenzymes. No artifacts

[6] A. Sánchez-Ferrer, R. Bru, and F. García-Carmona, *Plant Physiol.* **91,** 1481 (1989).
[7] A. Sánchez-Ferrer, R. Bru, and F. García-Carmona, *Anal. Biochem.* **184,** 279 (1990).
[8] H. Vilter, *Bioseparation* **1,** 283 (1990).
[9] P. Jordan and H. Vilter, *Biochim. Biophys. Acta* **1073,** 98 (1991).
[10] H. Vilter, *Phytochemistry* **23,** 1387 (1984).
[11] M. J. Murphy and C. Ó Hecha, *Phytochemistry* **12,** 2645 (1973).
[12] H. Vilter, K.-W. Glombitza, and A. Grawe, *Bot. Mar.* **26,** 331 (1983).
[13] E. De Boer, Y. Van Kooyk, M. P. M. Tromp, H. Plat, and R. Wever, *Biochim. Biophys. Acta* **869,** 48 (1986).
[14] E. De Boer, M. P. M. Tromp, H. Plat, B. E. Krenn, and R. Wever, *Biochim. Biophys. Acta* **872,** 104 (1986).
[15] H. Vilter, *Bot. Mar.* **26,** 451 (1983).

arising from the treatment were observed on polyacrylamide gel electrophoresis (PAGE).

To obtain a high yield the magnitude of the partition coefficient, $K$, is important, but this represents only one aspect. First, the enzymes had to be solubilized from the plant material. The efficiency of the extraction depends on the disintegration of tissues, the amount of biomass in the extraction medium, the pH, and the additives. Because large amounts of peroxidases in brown algae are associated with alginates in the cell wall, alkaline solutions which solubilize alginates, as well as peroxidases, are necessary. Systems containing ammonium sulfate were not suitable for the initial extraction procedures. Homogenization of fresh algae is difficult. Therefore, in the past, the extraction was repeated up to seven times when Tris buffers were used.[14] Using a centrifugal mill (Retsch, Haan, Germany), freeze-dried algae were easily and efficiently converted into a powder in large quantities. Using an algal powder during process development has advantages. Because the same powder was used in all experiments, variations due to different enzyme and water content or different size of particles were avoided.

The process development was directed to obtaining a high yield per liter of extraction medium. Systems with high $K$ values are very useful for this purpose. With $K$ values above 10 an almost quantitative extraction into top phase is possible. As algae are easily available in ton quantities, complete extraction was unnecessary. To increase the input of biomass, systems were formed with top phase volumes of 25–30% with respect to the total volume of the extraction medium. Compared to conventional extractions, using Tris buffers, two-phase systems permit 5 to 6 times more biomass to be extracted in the same volume. This means that to obtain the same amount of enzyme from 50 kg fresh *A. nodosum,* a two-phase system with a total volume of 300 liters partitioned the enzymes into about 80 liters of top phase, whereas the conventional extraction method using Tris buffer requires 1600 liters. Furthermore, to clarify the extract by centrifugation at high $g$ values only a few liters per day could be processed with the conventional method. Two-phase systems thus offer a potential for scaleup.

## Analytical Procedures

### *Peroxidase Assay*

In the absence of compounds which react with iodine, the peroxidases can be detected by the oxidation of 6 m$M$ iodide in the presence of 0.8 m$M$ hydrogen peroxide at pH 6.2 (citrate/phosphate buffer).[12] The assay is rapid. Hydrogen peroxide stabilizes the enzyme, which permits the use of phosphate. In the absence of hydrogen peroxide there is a reversible inactivation of peroxidases owing to the (competitive) binding of phosphate [rather than vanadium(V)], which acts as the prosthetic group.

### *Reconstitution of Holoenzyme from Apoenzyme*

The reconstitution of the holoenzyme from the apoenzyme by vanadate is a slow process. On a preparative scale the reconstitution should be performed by dialysis against 20 $\mu M$ $Na_3VO_4$ in 5 m$M$ Tris/HCl, pH 9.0, and the protein should be diluted to be below 1 mg/ml. The extent of reconstitution is determined by using about 1 $\mu$g enzyme per milliliter with 2 m$M$ $Na_3VO_4$ in 50 m$M$ Tris-HCl, pH 9.0, at 20°; and the activity is measured after 24 hr of incubation with the vanadate. Using immobilized chelators is a convenient way to remove nonspecifically bound vanadate.

## Extraction Processes

### *Method A: Extraction of Peroxidases from Brown Algae Belonging to Fucaceae*

The following method has been developed to extract peroxidases from the brown algae *Ascophyllum nodosum*. It is also suitable for the extraction of peroxidases from other brown algae belonging to the *Fucaceae*.

An aqueous solution is prepared containing 17.5% $K_2HPO_4$, 0.2% KOH, and 0.3% thioglycerol. At room temperature 3% algal powder (dry weight to water) is added and mixed for 30 min. The first phase system is prepared by addition of 17.5% PEG 1550. The mixing is continued for 30 min. The extraction medium is prepared using technical grade chemicals.

Alginates stabilize emulsions; therefore, the phase system is separated by low-speed centrifugation or in a centrifugal separator. The alginates and the sulfated fucans are enriched in the bottom phase (see Table I), and their viscosity slows the settling of cell debris in the bottom phase. The settling of cell debris is a problem during conventional extraction using buffers without two-phase systems. However, when two-phase systems are separated in a centrifugal separator the slow settling is of advantage because the particles leave the separator suspended in bottom phase. If the particles had formed a sediment in the separator they would, in a short time, have blocked the flow through the equipment.

Peroxidases partition to the top phase together with phenols and chlorophylls. It is possible to recover the peroxidases from the top phase by addition of acetone; however, the precipitates also contain phenols in

TABLE I
EXTRACT FROM *Ascophyllum nodosum* IN AQUEOUS TWO-PHASE SYSTEM[a]

| Parameter | Value |
|---|---|
| System | |
| PEG 1550 | 10% |
| $K_2HPO_4$ | 17.5% |
| pH | 9.3 |
| Partition coefficient | |
| Peroxidases | 52 |
| Sulfated fucans | 0.24 |
| Alginates | 0.0035 |

[a] From H. Vilter, *Bioseparation* **1,** 283 (1990), with permission.

large amounts. Therefore a second phase system is used to remove the phenols, chlorophylls, and other compounds. For this purpose magnesium sulfate is found to be suitable. Per 100 liters top phase a solution containing 80 liters water and 40 kg $MgSO_4 \cdot 7H_2O$ is added, thereby generating the second phase system. To remove the remaining phenols and chlorophylls from the second bottom phase, a solution containing 5 kg PEG 1550 in 5 liters water is added per 100 liters bottom phase. The peroxidases remain in the bottom phase.

Magnesium sulfate is not the salt of choice for preparation of the second phase system. However, it gives rise to the most efficient system for extracting the peroxidases from the PEG phase containing phenols. The disadvantage of magnesium sulfate is its chaotropic effect. The peroxidases could not be adsorbed from the bottom phase which contained the magnesium sulfate onto a hydrophobic matrix. Desalting is a time-consuming process, and therefore a transfer into a third phase system was deemed to be more convenient.

The third system is generated by the addition of a solution containing 20 kg PEG 1550, 0.48 kg tris-2-amino-2(hydroxymethyl)propane-1,3-diol, and 60 kg ammonium sulfate in 100 liters water per 100 liters bottom phase of the second system. By addition of PEG and ammonium sulfate the peroxidases are transferred to the top phase of the third system.

In all the steps, mixing time is 30 min. The separation of the second and third phase systems took place overnight in 100-liter plastic bottles. Adding acetone to the top phase of the third system precipitated the enzymes (300 liters acetone per 100 liters top phase). The precipitate is extracted several times using Tris-HCl, pH 9.0, and the solution is then dialyzed. From the initial volume of 300 liters the enzymes are concentrated to 3 liters by use of these procedures. Using hydrophobic interaction and anion-exchange chromatography the peroxidases are separated. A pure isoenzyme is obtained in quantity and quality sufficient for crystallization.[16]

*Method B: Extraction of Peroxidases from Brown Algae Belonging to Laminariales*

The following method has been developed to extract peroxidases from the brown algae *Laminaria digitata* and *Laminaria saccharina,* because results with Method A were, in these cases, unsatisfactory. Method B is also suitable for isolation of peroxidases from other brown algae belonging to the Laminariales.

The method is simple. The extraction medium is prepared using an aqueous solution containing 15 kg $K_2CO_3$ per 100 liters water, at room temperature. Three kilograms of algal powder is added and, after mixing 15 min, so is 20 kg of PEG 1550. The phases are separated using a low-speed centrifuge or a centrifugal separator. Some Laminariales have a low phenol content, and, in this case, the addition of thioglycerol is unnecessary. Per 100 liters top phase a solution containing 6 kg ammonium sulfate in 10 liters water is added. By the addition of 300 liters acetone to 100 liters top phase the peroxidases are recovered by precipitation.

A further purification using a second phase system is desirable. Addition of an aqueous solution containing 5 kg $K_2HPO_4$ and 20 kg $KH_2PO_4$ (100 liters water per 100 liters top phase) transfers some of the peroxidase isoenzymes into the bottom phase of the second system. The peroxidases are easily recovered from the phosphate-rich bottom phase by use of hydrophobic interaction chromatography with Phenyl-Sepharose.[9]

## Conclusions

Separations in aqueous two-phase systems are a powerful technique for plant enzyme purification. The technique is particularly useful when the extraction of proteins entails separation from large amounts of anionic carbohydrates and/or phenolic compounds, both of which hamper conventional protein extraction methodology. Extractions using aqueous two-phase systems are not only very efficient, but they can easily be adapted to large-scale preparations. The extractions can be carried out with inexpensive chemicals and equipment using simple manipulations. With an

[16] A. Müller-Fahrnow, W. Hinrichs, W. Saenger, and H. Vilter, *FEBS Lett.* **239,** 292 (1988).

efficient strategy, the development of an extraction scheme is as easy as that of a chromatographic purification procedure.

Acknowledgments

This work was supported by the Deutsche Forschungsgemeinschaft (DFG).

# [66] Purification of Water-Based Cutting Fluids

*By* GUNNAR B. STRÖM, BJÖRN MÄLARSTIG, and GÖRAN BLOMQUIST

## Introduction

Water-soluble cutting fluids represent a common type of industrial lubricants that are widely used in the tool and die trades for cutting, turning, drilling, grinding, and similar machining of materials. Their primary function is to increase the lifespan of tools by acting as cooling and lubricating agents. Cutting fluids consist of so-called base oils, which may be based on mineral oils or synthetic polymers.

Owing to the high water content (90–97%) microbial growth easily occurs in water-based cutting fluids, causing biodegradation. To prevent this, bactericidal and fungicidal agents are continuously added. Because of the occupational hazards that such additives pose, the concentration of biocides is kept low. Operation of machinery generates aerosols, which contain not only the fluid components but also bacteria, predominantly gram-negative *Pseudomonas* species. These organisms can, when inhaled by the operators, cause work-related respiratory allergies.[1]

To avoid the hazards of microorganisms and to prolong the lifetime of cutting fluids, techniques other than biocide treatment, for example, gamma ray or ultraviolet irradiation treatment, pasteurization, and ozone injections have also been tested.[2] None of these methods has, because of their complexity, gained interest from the tool and die industry. Methods that could selectively separate and concentrate contaminants including microorganisms and inorganic particles in order to prolong the life of the fluids would therefore be useful.

[1] S. A. Travers, B. Crook, and P. Griffin, *in* "4th International Conference on Aerobiology," p. 80. Stockholm, Sweden, 1990.

[2] C. Evans, *Tribol. Int.* **10,** 33 (1977).

Aqueous two-phase systems have been widely used in the separation of macromolecules, cell particles, and whole cells.[3] In microbiology, the technique has been successfully applied to the separation of microorganisms and organic, nonmicrobiological particles[4] and to that of fungal spores within genera as well as between genera.[5,6] Closely related fungi have been classified by cross-partitioning in aqueous two phase systems,[7] combined with multivariate data analysis.[8,9] Here we describe a method for the large-scale removal of microorganisms and inorganic particles from industrial cutting fluids by use of aqueous polymer two-phase systems.

## Materials and Methods

### *Phase Polymers*

Breox TB 120 (BP Chemicals, Hythe, UK): polyoxyalkylene glycol ether with a viscosity of 50,000 centistokes (cSt) and with an ethylene oxide/propylene oxide (EO/PO) ratio of 3 : 1

Breox 380-EP (BP Chemicals): Acrylic acid grafted polyoxyalkylene glycol ether with a viscosity of 1250–3000 cSt and with an EO/PO ratio of 1 : 1

Crude dextran (Dx) (Sorigona, Arlöv, Sweden) with a molecular weight of $5 \times 10^6$ to $40 \times 10^6$

### *Separation of Oil-in-Water Emulsions*

Stock solutions of Breox TB 120 (40%, w/w) and crude Dx (20%, w/w) are prepared. The cutting fluid (5%, w/w, mineral oil emulsion 829B 1 : 4; Castrol, Pangbourne, UK) is inoculated with *Pseudomonas aeruginosa* cells to a final concentration of $3.2 \times 10^8$ cells/ml. Polymer solutions are added to the cutting fluid to a final concentration of 2.5% (w/w) Breox TB 120 and 0.1% (w/w) crude Dx. The two-phase system is thoroughly mixed in a 50-liter plastic vessel. After phase separation the Dx-rich lower phase is discarded. Then 0.25 kg of 20% (w/w) crude Dx is added, and the phase system is thoroughly mixed and allowed to settle. This procedure

[3] P.-Å. Albertsson, "Partition of Cell Particles and Macromolecules," 3rd Ed. Wiley(Interscience), New York, 1986.

[4] G. Ström, U. Palmgren, and G. Blomquist, *Appl. Environ. Microbiol.* **53,** 860 (1987).

[5] G. Blomquist, G. Ström, and B. Söderström, *Appl. Environ. Microbiol.* **47,** 1316 (1984).

[6] G. Ström and G. Blomquist, *Appl. Environ. Microbiol.* **52,** 723 (1986).

[7] H. Walter and D. Forciniti, this volume [20].

[8] G. Ström, G. Blomquist, and B. Söderström, *J. Appl. Bacteriol.* **66,** 461 (1989).

[9] G. Blomquist and G. Ström, this volume [32].

is repeated three times. At each step 0.1-ml samples are removed from the upper phase for quantitative analysis of bacteria.

### *Separation of Synthetic Cutting Fluids*

A 1-$m^3$ plastic (PVC) vessel with a V-shaped bottom is filled with 925 kg of a 3% (w/w) synthetic cutting fluid in water (Aral Multrol 400; Unik Nordiska, Gothenburg, Sweden). To the vessel 30 kg of 20% (w/w) Breox TB-120 and 10 kg of 20% (w/w) Breox 380-EP are added. The solution of cutting fluid and polymers is mixed using a circulation pump. The concentration of the polymers in the final system is 2% (w/w) (1.5% Breox TB-120 plus 0.5% Breox 380-EP).

The vessel containing the Breox polymers and cutting fluid is connected with a fully automated grinding machine (Corocam, Sandvik, Sweden). On each separation occasion 5 kg of 20% (w/w) crude Dx is added. The phase system thus formed is thoroughly mixed and allowed to settle over the weekend while no work is carried out. The height of the tank is 125 cm and the settling rate 4 cm/hr. After phase separation the Dx-rich lower phase is collected through a valve in the V-shaped bottom of the vessel. The lower phase volume is 10 liters and is discarded. In addition, 0.1-ml samples for determination of microorganisms and 2.0-ml samples for particle analysis are collected prior to and after phase separation. The same cutting fluid is used over a 10-week period.

### *Quantitative Analysis of Particles*

Two milliliters of the cutting fluid is collected prior to and after separation, and the particle density is determined using a HACH turbidity meter (Svenska Merkanto, Uppsala, Sweden).

### *Quantitative Analysis of Microorganisms*

For analysis, 0.1 ml of the cutting fluid prior to and after separation is stained with 100 $\mu$l of 0.01% acridine orange (Bio Mérieux, Carbonnières les Bains, France) in a glass microanalytic funnel containing a black-stained Nucleopore filter (pore size 0.4 $\mu$m; Nucleopore Corp., Pleasanton, CA). After 3 minutes the liquid is drawn off and mounted on a slide for microscopic epifluorescence analysis. At least 40 fields and 400 microorganisms are counted at $\times 1250$ magnification.[10]

[10] U. Palmgren, G. Ström, G. Blomquist, and P. Malmberg, *J. Appl. Bacteriol.* **61,** 401 (1986).

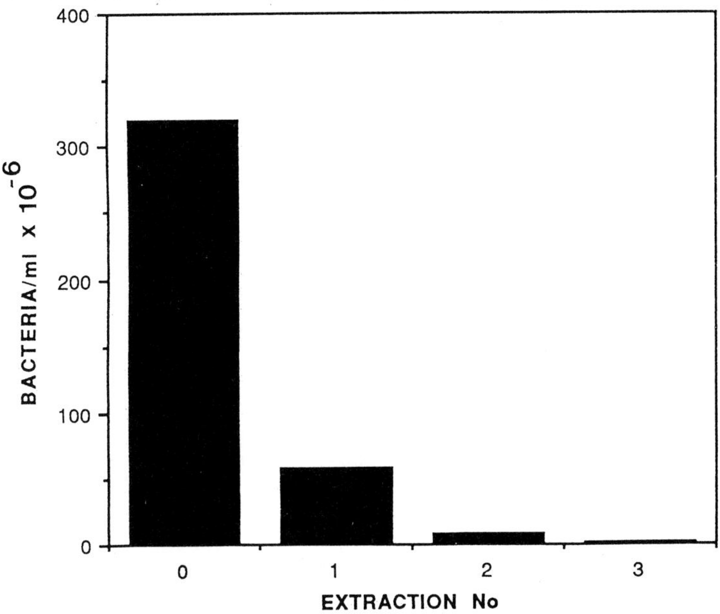


FIG. 1. Stepwise extractions of *Pseudomonas aeruginosa* from an oil-in-water cutting fluid emulsion (5%, w/w, 829 B, 1:4) using a two-phase system composed of 2.5% (w/w) Breox TB 120 and 0.1% (w/w) crude dextran.

## Results and Discussion

To prevent biodeterioration, a batch extraction of a microbially contaminated oil-in-water emulsion, incorporated in an aqueous polymer two-phase system, was carried out (Fig. 1). The two-phase system was composed of Breox TB 120, crude Dx, and water. The bacterial concentration prior to extraction was $3 \times 10^8$ cells/ml and consisted entirely of *Pseudomonas aeruginosa* cells. After one extraction the bacterial concentration was reduced by 90%, and after three extractions only $1.6 \times 10^6$ cells/ml remained in the upper phase emulsion.

Water-based mineral oil emulsions contain, in general, nonionic and anionic emulsifiers. In the two-phase system the emulsifiers are concentrated in the upper phase, resulting in an upper phase emulsion. Introduction of cationic lower phase polymers, for example, DEAE-dextran T500 (Pharmacia, Uppsala, Sweden), to increase the separation efficiency causes the anionic detergents to partition to the lower phase, thereby splitting the emulsion. On the other hand, introduction of negatively charged upper phase polymers, such as Breox 380-EP, results in a stable upper phase emulsion. By increasing the negative charge of the upper

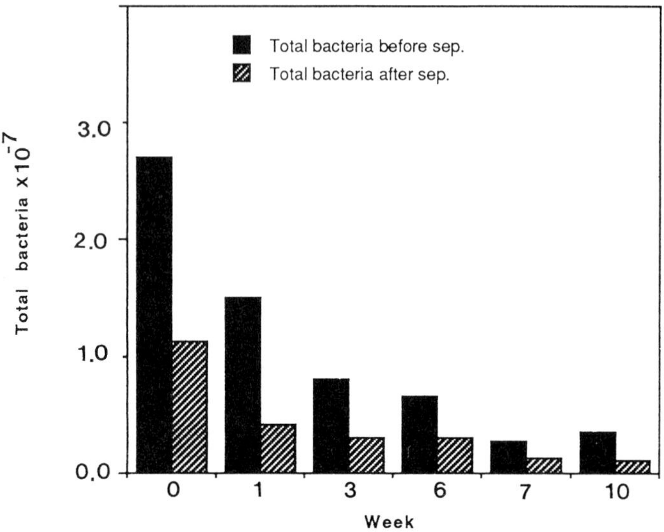


FIG. 2. Total number of bacterial cells prior to and after separation in a vessel containing 1 $m^3$ synthetic cutting fluid, during a 10-week period. The cutting fluid was composed 3% (w/w) Aral multrol, 2% (w/w) Breox TB 120, 0.5% (w/w) Breox 380-EP, and 0.1% (w/w) crude dextran.

phase, replacing 0.5% of the nonionic Breox TB 120 with 0.5% Breox 380-EP, no decrease of bacterial cells in the upper emulsion phase could be observed. This indicates that the primary basis for the partitioning behavior of cells between these phases are hydrophobic/hydrophilic interactions rather than repulsive forces between the negatively charged upper phase polymer and negative charges on the bacterial cell surface.

The possibility of applying two-phase separation on an industrial scale was tested by use of a 1-month-old, continuously used, synthetic fluid. The experiment was set up at a local industry using the grinding-fluid used in daily operations. Separations were carried out on six occasions during a 10-week period. As shown in Fig. 2, there is a gradual decrease in bacterial concentrations from $1.4 \times 10^7$ to $6 \times 10^5$ cells/ml. The bacterial concentrations monitored in a 20-$m^3$ central tank containing the same grinding fluid during the 10-week period showed an increase in bacterial counts from $1.4 \times 10^7$ to $6.8 \times 10^7$ cells/ml. The microflora consisted primarily of bacterial cells. Yeast cells and molds were below the detection limit ($<10^3$ cells/ml).

The delimiting factor in the lifetime of grinding fluids is often considered to be the accumulation of inorganic particles originating in the grinding operations. Measurements of the particle density carried out prior to and

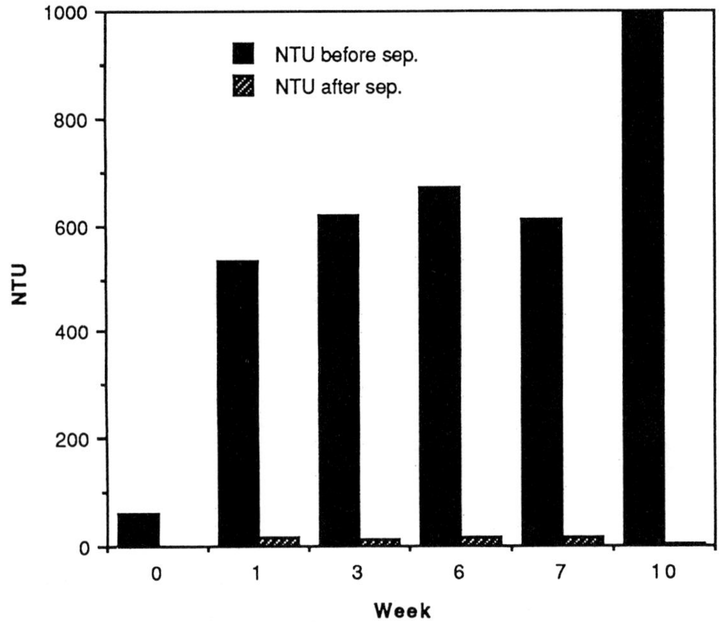


FIG. 3. Concentration of particles expressed as NTU (national turbidity units) prior to and after separation in a vessel containing 1 $m^3$ synthetic cutting fluid, during a 10-week period. The cutting fluid was composed 3% (w/w) Aral multrol, 2% (w/w) Breox TB 120, 0.5% (w/w) Breox 380-EP, and 0.1% (w/w) crude dextran.

after two-phase separation on six occasions during the 10-week period are presented in Fig. 3. After the first separation the number of particles was reduced from 60 to 2 NTU (national turbidity units). After 10 weeks, prior to addition of the lower phase polymer, the grinding fluid contained 965 NTU/ml, which was reduced to 6 NTU/ml after phase separation, resulting in a completely transparent fluid. The extreme partition coefficient of the particles also applied to submicron particles as determined by light scattering. Partitioning studies carried out using 1-$\mu$m ferromagnetic particles has shown that these particles in general prefer the lower phase when partitioned in aqueous two-phase systems composed of a variety of different polymer combinations.[11]

## Conclusions

The polymers used in the two-phase purification may well serve as lubricating components of grinding or cutting fluids. An increased separa-

[11] S. Flygare, G. Wikström, G. Johansson, and P.-O. Larsson, *Enzyme Microb. Technol.* **12,** 95 (1990).

tion efficiency can be achieved when the composition of the polymers as well as other components of the cutting fluid are optimized for phase separation. The cost for the polymers used for the purification during the 10-week period should be measured against an increased lifetime of the product (3-fold) and the reduced number of breaks in production. The costs for exchanging and disposing of the grinding fluid should be considered, and the positive health effect of using a cutting fluid containing a minimum of microbial and particulate contaminants should not be neglected.

# Author Index

Numbers in parentheses are footnote reference numbers and indicate that an author's work is referred to although the name is not cited in the text.

## A

## B

## C

## D

## G

## H

## I



## J

## K

## L

## N

## O



## P

## Q



## R

## S

## W

## Y



## Z

# Subject Index

## A

## D

## E

## F

## G



## H

## I

## M

## T

## U



## V



## W



## Y



## Z

ISBN 0-12-182129-3